A Companion
to Paleopathology

The *Blackwell Companions to Anthropology* offers a series of comprehensive syntheses of the traditional subdisciplines, primary subjects, and geographic areas of inquiry for the field. Taken together, the series represents both a contemporary survey of anthropology and a cutting edge guide to the emerging research and intellectual trends in the field as a whole.

Forthcoming

A Companion
to Paleopathology

Edited by
Anne L. Grauer

WILEY Blackwell

Library of Congress Cataloging-in-Publication Data

A companion to paleopathology / edited by Anne L. Grauer.
     p.   cm. – (Blackwell companions to anthropology)
   Includes bibliographical references and index.
   ISBN 978-1-4443-3425-8 (hardback : alk. paper) – ISBN 978-1-1191-1163-4 (paperback)
 1. Paleopathology.   I. Grauer, Anne L., 1958–
   R134.8.C65 2012
   614'.17–dc23
                                                     2011018221

A catalogue record for this book is available from the British Library.

Cover image: (1) X-ray image showing a fifth metacarpal fracture © Scott Camazine / Alamy;
(2) Scanning Electron Micrograph of cancellous or spongy bone tissue © Susumu Nishinaga / Science
Photo Library; (3) Human skull showing the roots of the upper & lower teeth © John Watney / Science
Photo Library

Set in 10/12.5pt Galliard by SPi Global, Pondicherry, India

1   2016

To Peter, Evelyn, and Thomas
for their love (and patience)

Contents

List of Illustrations

(C) Abnormal size apparent in the cranium in the left of the figure compared with a normal cranium in the right portion of the figure. The abnormal size is caused by a defect in the pituitary gland which secreted excessive growth hormone during development. (Photo courtesy of Dr. Takao Suzuki, Shinjyuku-ku, Tokyo, Japan.) (D) Severe deformity in a case of rickets resulting from a deficiency in vitamin D in a male about 18 years of age at the time of death. Note the inward collapse of the pelvis caused by the weight of the body on the femoral heads. (Department of Pathology, University of Strasbourg, France catalog number 7664d.) (E) Left lateral view of a cranium with premature fusion of the sagittal suture resulting in an elongated cranium. Adult male from a pre-Columbian site in Cinco Cerros region of Peru (NMNH 293841).

List of Tables

Notes on Contributors

Adauto Araújo is Senior Researcher at Fundação Oswaldo Cruz, Rio de Janeiro, Brazil. His research focus is on paleoparasitology and the origin and evolution of infectious diseases. He is the author of numerous publications, the most recent being Araujo et al. (2009) "Paleoparasitology of Chagas disease: a review," *Memórias do Instituto Oswaldo Cruz*. He is a member of the Academia de Medicina do Rio de Janeiro, Sociedade Brasileira de Parasitologia, and the Paleopathology Association.

George J. Armelagos is Goodrich C. White Professor of Anthropology at Emory University. His research has focused on diet and disease in human adaptation. He has co-authored *Demographic Anthropology* with Alan Swedlund, *Consuming Passions* with Peter Farb, *Paleopathology at the Origins of Agriculture* with Mark Cohen. He has been president of American Association of Physical Anthropologists and Chair of the Anthropology Section of the American Association for the Advancement of Science. He received the Viking Fund Medal (2005), anthropology's highest honor, the Franz Boas Award for Exemplary Service (2008) from the American Anthropological Association, and the Charles Darwin Award for Lifetime Achievement from the American Association of Physical Anthropologists (2009). In addition to all of this, he is considered a relatively "good guy."

Gila Kahila Bar Gal is Senior Lecturer at the Koret School of Veterinary Medicine, The Hebrew University of Jerusalem. Her research interests include molecular evolution, co-evolution of host–pathogen, ancient DNA, domestication and conservation genetics. She has published articles in *Nature*, and *PLoS ONE*, and most recently has co-authored a paper (Polani et al. 2010) entitled, "Evolutionary dynamics of endogenous feline leukemia virus proliferation among species of the domestic cat lineage," *Virology*.

Ethne Barnes is a physical anthropologist and paleopathologist, consultant, and independent researcher in Tucson, Arizona. She is recognized for establishing the morphogenetic approach to analyzing developmental defects of the skeleton in paleopathology. Her research and consultation work includes archaeological projects in Greece, Turkey, China, South America and North America. Major publications include *Developmental Defects of the Axial Skeleton in Paleopathology* (1994), *Diseases and Human Evolution* (2005), and "Congenital Anomalies" in *Advances in Human Paleopathlogy* (Pinhasi and Mays 2008).

Jesper Boldsen is Professor of Anthropology and head of ADBOU at the Institute of Forensic Medicine, University of Southern Denmark. His research interests revolve around medieval population biology, epidemiology, demography, evolution and leprosy. Recent publications include: "Early childhood stress and adult age mortality—a study of dental enamel hypoplasia in the medieval Danish village of Tirup," *American Journal of Physical Anthropology* (2007); "Leprosy in Medieval Denmark— Osteological and epidemiological analyses," *Anthropologischer Anzeiger* (2010).

Don Brothwell is Emeritus Professor of Human Palaeoecology in the Department of Archaeology at the University of York, U.K. His research interests are broad-based, incorporating the archaeological and geological context of human remains, along with study of disease across history, space, and a variety of species. His publications include 17 books and over 190 chapters and articles, including *Food in Antiquity: A Survey of the Diet of Early Peoples* (1998), and *Handbook of Archaeological Sciences* (2005).

Michele R. Buzon is Associate Professor of Anthropology at Purdue University. Her research interests include the interplay of health and identity with sociopolitical changes in Nile Valley during the New Kingdom and Napatan periods using paleopathological, archaeological, and isotopic methods. Recent publications include Buzon and Bowen (2010) "Oxygen isotope analysis of migration in the Nile Valley," *Archaeometry*, and Buzon and Bombak (2010) "Dental Disease in the Nile Valley during the New Kingdom," *International Journal of Osteoarchaeology*.

Francisca Alves Cardoso is Lecturer in Biological Anthropology in the Institute of Philosophy and Human Sciences in the Federal University of Pará, Brazil, and a Research Fellow of CRIA—Centre for Research in Anthropology, Portugal. Her most recent research focuses on the importance of socioeconomic and cultural variables in the interpretation of human remains, having previously discussed gender and activity-related issues as her PhD topic: "A Portrait of Gender in Two 19th/20th Portuguese Populations: A Paleopathological Perspective."

Della Collins Cook is Professor of Anthropology at Indiana University. Her interests include paleopathology, mortuary practices and history of physical anthropology. While she is an Americanist, she has worked with remains from South Africa, Egypt, Greece, and Portugal. She and her colleagues have published on dental modification, isotopic evidence for nutrition, and biological distance. Her recent publications include *The Myth of Syphilis: A Natural History of North American Treponematosis* (2005, co-authored with Mary Lucas Powell), and "The Evolution of American Paleopathology," in *Bioarchaeology: The Contextual Study of Human Remains*, Buikstra and Beck, eds. (2006).

Katharina Dittmar is Assistant Professor of Evolutionary Biology at SUNY—Buffalo. Her research focus includes the evolution of parasitism and evolutionary biology. Her recent publications include "Rapid evolution of protein kinase alters the sensitivity to viral inhibitors," *Nature Structural and Molecular Biology* (2009) (co-authored with S. Rothenburg et al.), and Dittmar et al. (2006) "Molecular phylogenetic analysis of nycteribiid and streblid batflies (Diptera: Brachycera, Calyptratae): Implications for host association and phylogeographic origins," in *Molecular Phylogenetics and Evolution*.

Keith Dobney is Sixth Century Chair of Human Palaeoecology in the newly established Archaeology Department at the University of Aberdeen, UK. The main material focus of his work is the study of animal (including human) remains, his principal research being: the origins and spread of agriculture, migration and dispersal, palaeoeconomy, palaeopathology and palaeoepidemiology. He is currently one of two project leaders of a CNRS-funded Projet de Groupement De Recherche Européan (GDRE): "*BIOARCH- Bioarchaeological Investigations of the Interactions between Holocene Human Societies and their Environments*," and the Director of a similar (Co-Reach funded) Chinese-European research grouping (*EUCH-BIOARCH*).

James H. Gosman is Adjunct Assistant Professor in Anthropology at The Ohio State University, and an orthopedic surgeon by training. His research interest encompasses skeletal biology and bioarchaeology, with particular focus on human trabecular bone ontogeny and locomotor development. Recent publications include, Gosman and Ketcham (2009), "Patterns in ontogeny of human trabecular bone from SunWatch village in the prehistoric Ohio Valley: General features of microarchitectural change," *American Journal of Physical Anthropology*, and Gosman and Stout (2010) "Current concepts in skeletal biology," in *A Companion to Biological Anthropology*, ed. Larsen.

Anne L. Grauer is Professor of Anthropology at Loyola University Chicago. Her research interests include issues of gender in human skeletal analyses, particularly in North American historic and British medieval populations. She has served as an Associate Editor for the *AJPA* and on the Executive Board of the AAPA. She is currently the Past-President of the Paleopathology Association. Her publications include *Bodies of Evidence: Reconstructing History Through Skeletal Analysis* (editor) (1995), *Sex and Gender in Paleopathological Perspective* (edited with Stuart-Macadam) (1998), and most recently, Fitch, Grauer, and Augustine (2010), "Lead Isotope Ratios: Tracking the Migration of European Americans to Grafton, Illinois in the 19th Century," *International Journal of Osteoarchaeology*.

Charlotte Henderson is an Honorary Research Associate in the Department of Archaeology, Durham University (U.K.). She is a member of the International Working Group on methods for recording entheseal changes (EC) (http://www.uc.pt/en/cia/msm/msm_after). Her research focuses on quantitative methods for studying EC along with their etiology, particularly pathological. Her Ph.D. dissertation (2009) was titled "Musculo-skeletal stress markers in bioarchaeology: indicators of activity levels or human variation? A re-analysis and interpretation."

Margaret A. Judd is Associate Professor in the University of Pittsburgh's Department of Anthropology. She is currently excavating a Byzantine crypt at Mount Nebo in Jordan, following several years of excavation in Sudan and Jordan. Research interests include trauma and health consequences of social, ideological and technological change. Most recent publications include *Growing up Gabati* (in press), and "Pubic symphyseal face eburnation: An Egyptian sport story?" *International Journal of Osteoarchaeology* (2010).

Robert Jurmain is Professor Emeritus of Anthropology at San Jose State University, California. His research focus concerns paleopathology of humans and nonhuman primates, most specifically, degenerative joint disease and trauma. He has authored articles in numerous peer-reviewed journals and contributions to edited volumes. He is author of *Stories from the Skeleton: Behavioral Reconstruction in Human Osteology* (1999) as well as coauthor of three textbooks in physical anthropology and archaeology, now in a total of 30 editions.

M. Anne Katzenberg is Professor of Physical Anthropology, Department of Archaeology, University of Calgary, Canada. Her research focuses on reconstructing past diet and health using stable isotopes, and refining interpretations of stable isotope data. She is a Fellow of the Royal Society of Canada, and co-editor of the book, *Biological Anthropology of the Human Skeleton* (2000), with Shelley R. Saunders. She has published numerous stable isotope studies of past peoples from North and South America, Europe and Asia.

Tomasz Kozłowski is Assistant Professor, Department of Anthropology, at Nicolaus Copernicus University in Toruń (Poland). His interests include bioarchaeology and paleopathology, focusing on the early mediaeval settlement complex in Kałdus (Poland) and studying the relics of Neolithic settlement in Catalhoyuk in Turkey. He is a member of the Global History of Health Project. His recent publications include "Human bone remains," in W. Chudziak (ed.), *Early Mediaeval Skeleton Cemetery in Kaldus, Mons Sancti Laurentii*, Vol. 5 (2010), and co-editor with M. Grupa, *Kwidzyn Cathedral—The Mystery of the Crypts* (2009).

Patricia M. Lambert is Professor of Biological Anthropology and Associate Dean of Research in the College of Humanities and Social Sciences at Utah State University. She served as Associate Editor of the *AJPA* from 2002 to 2008 and Executive Committee Member for the American Association of Physical Anthropologists from 2006 to 2009. Her research focuses on prehistoric health and violence in the Americas. Recent publications include "Health versus fitness: competing themes in the origins and spread of agriculture," *Current Anthropology* (2009).

Larisa M. Lehmer is Research Associate in the Dermatopathology division of Central Coast Pathology Consultants in San Luis Obispo, CA, and investigates topics in dermatopathology and osteopathology. Her publications include "MEC of the parotid gland presenting as periauricular cystic nodules," *Journal of Cutaneous Pathology*, "Expectorated rhabdomyosarcoma: case report and review of the literature," *Human Pathology*, and "Cutaneous metastasis of osteosarcoma in the scalp" *American Journal of Dermatopathology*.

John R. Lukacs is Professor Emeritus of Anthropology at the University of Oregon in Eugene, OR. His research focuses on the paleopathology and dental anthropology of prehistoric and living South Asians. Recent publications appear in: *Current Anthropology, Comparative Dental Morphology,* and *Clinical Oral Investigations.* A monograph on the bioarchaeology of early Holocene foragers of India is nearing completion for publication in *British Archaeological Reports.*

Niels Lynnerup is Professor of Forensic Anthropology, and head of the Unit of Forensic Anthropology at the University of Copenhagen. His research comprises both the living (photogrammetry, gait analyses) as well as the dead (paleodemography, stable isotopes and CT-scanning and 3-D visualisation techniques). Key publications include "Computed tomography scanning and three-dimensional visualization of mummies and bog bodies," *Advances in Human Paleopathology* (2008), and "Mummies," *Yearbook of Physical Anthropology* (2007). He has recently served as the Vice-president of the Paleopathology Association and is co-editor of the Journal of Paleopathology.

Simon Mays is Human Skeletal Biologist for English Heritage. His research covers all areas of archaeological human remains. He is currently an Associate Editor for the American Journal of Physical Anthropology. Key publications include *Advances in Human Palaeopathology* (edited with R. Pinhasi 2008), and *The Archaeology of Human Remains,* second edition (2010).

George R. Milner is Professor of Anthropology at The Pennsylvania State University. His osteological research focuses on paleopathology (including trauma) and paleodemography, and includes the development of new means of skeletal age estimation for both archaeological and forensic purposes. Recent publications include (with Buikstra and Wiant) "Archaic burial sites in the midcontinent," in *Archaic Societies: Diversity and Complexity Across the Midcontinent* (Emerson, McElrath, and Fortier, eds., 2009), and (with Wood and Boldsen) "Advances in paleodemography," in *Biological Anthropology of the Human Skeleton,* 2nd edition (Katzenberg and Saunders, eds., 2008).

Piers D. Mitchell is Affiliated Lecturer at the University of Cambridge. He has a doctorate in medical history and is also a medical practitioner specializing in children's orthopedic surgery. His recent publications include *Medicine in the Crusades: Warfare, Wounds and the Medieval Surgeon* (2004) and *Anatomical Dissection in Enlightenment Britain and Beyond: Autopsy, Pathology and Display* (2011).

Donald J. Ortner is Biological Anthropologist in the Department of Anthropology, Smithsonian Institution. His research emphasis is on disease in archeological human skeletal remains. He has done fieldwork in Jordan and has conducted research projects in the United States, Europe, and Australia. From 1999 to 2001, he was president of the Paleopathology Association. Recent major publications include: *Identification of Pathological Conditions in Human Skeletal Remains* (2003), *EB I Tombs and Burials of Bâb edh-Dhrâ, Jordan* (2008) with Frohlich, and "Ecology, culture and disease in past human populations," (with Schutkowski), in H. Schutkowski, ed, *Between Biology and Culture* (2008).

Christina Papageorgopoulou is Alexander von Humboldt Postdoctoral Fellow at the Workgroup of Palaeogenetics, Institute of Anthropology, Johannes Gutenberg-University, Mainz, Germany and a Research Fellow at the Centre for Evolutionary Medicine, University of Zurich, Switzerland. Her main interests center on new methods in palaeopathology, variability in human growth and development in past populations, and the Mesolithic-Neolithic transition in southeastern Europe from a genetic perspective. She is currently the editor of the *Bulletin der Schweizerischen Gesellschaft für Anthropologie,* and the *Newsletter of the American Dermatoglyphics Association.*

Mary Lucas Powell is Former Director/Curator of the W. S. Webb Museum of Anthropology at the University of Kentucky, and Editor Emerita of the *Paleopathology Newsletter.* Her research has focused on the relationship between diet and dental health, and the natural history and paleoepidemiology of treponemal disease and tuberculosis in prehistoric Native American populations in the Southeastern United States and Torre de Palma, a late Classical/medieval site in eastern Portugal. She recently published *The Myth of Syphilis: The Natural History of Treponematosis in North America* (2005) with co-author, D.C. Cook.

Bruce D. Ragsdale is Smithsonian Research Associate and Adjunct Professor of Anthropology at Arizona State University, and a practicing pathologist. He apprenticed with Walter Putschar at Massachusetts General Hospital, worked a decade with Lent Johnson at AFIP, has been on the faculty of nine medical schools. He has co-chaired 20 workshops at Paleopathology Association meetings, established a bone collection at ASU, and continues to contribute academically at local, state and international levels while directing Western Dermatopathology in California.

Rebecca Redfern is Curator of Human Osteology at the Museum of London. Her research interests include trauma, lifecourse and gender studies, focusing on Iron Age and Roman communities in Britain. Publications include: *Spitalfields: A Bioarchaeological Study of Health and Disease From a Medieval London Cemetery* (MoLA Monograph), and "A re-appraisal of the evidence for violence in the late Iron Age human remains from Maiden Castle hillfort, Dorset, England" in *Proceedings of the Prehistoric Society.*

Karl J. Reinhard is Professor of Environmental Archaeology, School of Natural Resources, University of Nebraska. His research interests include archaeoparasitology, palynology, and paleonutrition. He is the author and co-author of numerous publications, including most recently: Vinton, Perry, Reinhard, Santoro, and Teixeira-Santos (2009) "Impact of empire expansion on household diet: the Inka in northern Chile's Atacama Desert," in *PLoS ONE 4: e8069,* and Araújo, Reinhard, Ferreira, and Gardner (2008) "Parasites: probes for evidence of prehistoric human migrations," in *Trends in Parasitology.*

Charlotte Roberts is Full Professor of Bioarchaeology at Durham University, U.K., Her research focuses on the interaction of people with their environments in the past, the use of pathogen aDNA analysis to explore the origin and evolution of infections, the impact of air quality and mobility on health, and the Global History of Health (Ohio State University). She is the author of numerous books and articles, including most recently, *Human Remains in Archaeology: A Handbook* (2009), and *The Bioarchaeology of Tuberculosis. A Global View on a Reemerging Disease* (2008), co-authored with J. E. Buikstra.

Frank Rühli is Head of the Centre for Evolutionary Medicine at the Institute of Anatomy, University of Zurich, Co-Heads the "Swiss MDiagnostic Imaging of Mummy Project", and is a research fellow of the Institute of History of Medicine, University of Zurich. His research includes the development and use of imaging techniques, the microevolution of anatomical variations in normal and pathological tissues, and assessing the biological standard of living and state of health of conscripts of the Swiss armed forces. He is the President of the German Society of Anthropology, and Vice-President of the Swiss Society of Anthropology. His many publications appear in both anthropological and medical journals.

Dong Hoon Shin is Associate Professor at Seoul National University. He has led the Anthropology and Paleopathology Laboratory, Department of Anatomy/Institute of Forensic Medicine. During the past decade, he has performed aDNA, paleoparasitological, anthropometric and paleoradiological work on the samples from archaeological fields of Korea. He was a scientific committee member of the Korean Association of Physical Anthropologists and aDNA, 2010, in Munich, and is the author of numerous articles.

Mark Spigelman is Visiting Professor, Department of International Health, Royal Free and University College London Medical School, and Hebrew University Medical School, Jerusalem, Department of Microbiology, and Associate Professor in the Department of Anatomy and Anthropology, Sackler Medical School, Tel Aviv University, Israel. His research centers on paleomicrobiology, survival of biomolecules, the relationship between microbial diseases of the past and today, and developing techniques for minimally destructive sampling of human remains. He is the author of numerous articles, recently co-authoring with Hershkovitz et al. (2008), "Detection and molecular characterization of 9000-Year-Old *Mycobacterium tuberculosis* from a neolithic settlement in the eastern Mediterranean," *PlosOne*.

Ann L.W. Stodder is an Archaeologist / Osteologist with the Office of Archaeological Studies at the Museum of New Mexico, Adjunct Associate Professor of Anthropology at the University of New Mexico, and a Research Associate of the Field Museum of Natural History. Her research, in the U.S. Southwest and various parts of Oceania, addresses paleoepidemiology, mortuary ritual, and human taphonomy. She is the editor of *Reanalysis and Reinterpretation in Southwestern Bioarchaeology* (2008) and *The Bioarchaeology of Individuals* (2012).

Bethany L. Turner is Assistant Professor in Anthropology at Georgia State University. Her research focuses on multi-isotopic and osteological analyses of Pre-Columbian Peruvian populations to reconstruct diet, residential mobility, and overall health related to cultural transitions in ancient Andean imperial states. Recent publications include "Partnerships, pitfalls, and ethical concerns in international bioarchaeology," in Agarwal and Glencross (eds.) *Social Bioarchaeology* (2010), and "Insights into immigration and social class at Machu Picchu, Peru based on oxygen, strontium and lead isotopic analysis," *Journal of Archaeological Science* (2009).

Beth Upex is Research Fellow in the newly established Archaeology Department at the University of Aberdeen. She has a background in human and animal osteoarchaeology and palaeopathology. Her PhD research explored the use of enamel hypoplasia in caprines as a means of interpreting past climatic change and changing animal husbandry practices. Her current research is focused on understanding various aspects of animal domestication and other human/animal interactions through the study of skeletal pathology.

Sébastien Villotte is Post-Doctoral Researcher at the University of Exeter, Department of Archaeology. His research focuses on human behavior during European prehistory, including a significant component on methodology. In the last two years, he co-organized four workshops on entheseal changes. His recent publications include "Enthesopathies and activity patterns in the early medieval Great Moravian population: Evidence of division of labour," *International Journal of Osteoarchaeology* (2010), and (with Castex, Couallier, Dutour, Knusel, and Henry-Gambier) "Enthesopathies as occupational stress markers: evidence from the upper limb," in the *American Journal of Physical Anthropology* (2010).

Tony Waldron is Honorary Professor at the Institute of Archaeology, University College London. His research focuses on both methodological theoretical issues in bio-archaeology and paleopathology. His books include *Counting the Dead* (1994), *Paleoepidemiology: The Measure of Disease in the Past* (2007), and *Palaeopathology* (2009).

Johann Wanek is Medical Physicist Assistant at the Anatomical Institute of the University of Zurich. He is particularly interested in X-ray imaging and its impact on ancient DNA using a Monte Carlo based simulation, and the physicochemical properties of mummified tissues. He is currently a reviewer at the AUTOMED 2010 (Automation in Medicine), Swiss Federal Institute of Technology Zurich.

Darlene A. Weston is Assistant Professor in the Department of Anthropology at the University of British Columbia and Associated Scientist at the Max Planck Institute for Evolutionary Anthropology. Her research focuses on the biocultural interpretation of infectious disease and stress indicators and the interactions between health and paleodemography in European and Caribbean populations. Key publications include papers in the *American Journal of Physical Anthropology*, the *Journal of Human Evolution* and the *Journal of Forensic Sciences*.

Henryk W. Witas is Professor and Head of Department of Molecular Biology, Faculty of Biomedical Sciences and Postgraduate Education at the Medical University of Lodz, Poland. His research concentrates on allelic determinants associated or responsible for pathologic phenotype of monogene and polygene diseases, and those responsible for susceptibility to infectious diseases in historic and prehistoric gene pools. His recent publications include Witas et al. (2007) "Extremely high frequency of autoimmune-predisposing alleles in medieval specimens" *J. Zhejiang*

Univ Sci B; and Witas et al. (2010) "Changes in frequency of IDDM-associated HLA DQB, CTLA4 and INS alleles," *International Journal of Immunogenetics.*

Michael R. Zimmerman is Adjunct Professor of Anthropology at the University of Pennsylvania and Adjunct Professor of Biology at Villanova University. He was recently a Visiting Professor at the University of Manchester's KNH Centre for Biomedical Egyptology. He is also a retired pathologist. His research interest is in mummy paleopathology. His most recent publication is "Cancer: A new disease, an old disease, or something in between?" with R.A. David, in *Nature Reviews Cancer* (2010).

Molly K. Zuckerman is Assistant Professor at Mississippi State University. Her research focuses on the evolution and social and ecological history of acquired syphilis in early modern England, the evolution of infectious disease and cancer, epidemiological transitions, and the bioarchaeology of gender, inequality, and identity. Recent publications include Harper, Zuckerman, Harper M, Kingston, and Armelagos (in press), "The origin and antiquity of syphilis revisited: an appraisal of Old World Pre-Columbian evidence for treponemal infection," *Yearbook of Physical Anthropology*, and Harper, Zuckerman, and Armelagos (in press), "Correspondence: a possible (but not probable?) case of treponemal disease (Response to Mays et al. 2010)," *International Journal of Osteoarchaeology.*

Acknowledgements

I extend my heart-felt thanks to the contributors who so quickly and enthusiastically agreed to participate in this project. Their commitment to the field of paleopathology and respect for scientific discourse has made it an honor to be their colleague. Deep gratitude also goes to the chapter reviewers who helped strengthen the volume and shared each author's desire to represent the promise of our field. I thank Rosalie Robertson, Senior Editor at Wiley-Blackwell, for the humbling invitation to edit this volume, and Julia Kirk, Project Editor, for cheerfully answering each of my gazillion questions. And last, but not least, many thanks are owed to Alec McAulay, Project Manager, whose keen eye and sense of humor made the final editing and production details enjoyable to complete.

CHAPTER **1**

Introduction: The Scope of Paleopathology

Anne L. Grauer

The field of paleopathology, quite simply, entails the study of ancient disease. However, nothing is "quite simple" within this scientific discipline. As noted by Buikstra (2010), defining the term "ancient" can be as complex as defining the term "disease." In part, this is due to our preconceived notions about these terms. "Ancient," for instance, conjures thoughts of prehistoric or early historic life, dating hundreds to thousands of years ago. The term "disease" is often used to imply harmful changes caused by invading pathogens. Within paleopathology, however, the terms "ancient" and "disease" hold more nuanced, and even contested meaning. For instance, determining what material will be paleopathologically examined often relies upon the origin of the sample and/or the question being posed, rather than the date the individual(s) died. In many states throughout the U.S., human remains are considered "old" and are recovered archaeologically, rather than forensically (or under the auspices of funeral directors), when they are deemed to be over 100 years old. Hence, large numbers of human remains from as recently as the late 19th and early 20th centuries have been studied by paleopathologists. These are hardly "ancient" by most definitions. Yet, the information gained from these skeletal remains about human disease in the past is enormous.

The term "disease" is similarly complex. The colloquial use of the word, which often alludes to "infection," ignores the complex processes and body changes that a wider definition of "disease" would encompass. If discrete pathogens (such as viruses or bacteria) are viewed as the sole cause of disease, then paleopathological investigation would be limited to exploring remnants of the body's immunological response. Indeed, understanding what triggered an immune response in the past, how the

A Companion to Paleopathology, First Edition. Edited by Anne L. Grauer.
© 2012 John Wiley & Sons, Ltd. Published 2016 by John Wiley & Sons, Ltd.

response was triggered, and/or variation in the responses between individuals and populations are important goals which provide great insight into the past. But they are also extremely limiting. The term "disease" within paleopathology (and the medical community) is broader and more encompassing. Stedman's Medical Dictionary (26th edition, 1995) defines disease, in part, as "an interruption, cessation, or disorder of body functions, systems, or organs ... A morbid entity characterized usually by at least two of these criteria: recognized etiological agent(s), identifiable group of signs and symptoms, or consistent anatomical alterations..." (p. 492). Adopting this definition leads researchers to evaluate many types of conditions affecting the human body, including those with potential sociocultural etiologies, such as malnutrition, interpersonal violence, and deliberate body modification. Thus, paleopathological focus might begin with recognition of bony changes that are quantified and qualified: that is, recognition of "consistent anatomical alterations"; but continues with the exploration of singular or multifactoral causes of these alterations and the ramifications of the conditions on our understanding of human life. It is with these broad perspectives and goals that paleopathologists view the past.

A Brief History of Paleopathology

As Dorothy Salisbury Davis, the 20th-century crime fiction writer quipped, "History's like a story in a way: it depends on who's telling it." Chronicling the development of paleopathology is no exception. Individual perspectives shape the scope, direction, and intent of the discourse, resulting in many crafted explanations and interpretations. Within this volume a number of researchers look back to the roots, and follow the development of particular aspects of our field. Alongside their visions of the past, are histories told by a number of other researchers, including Moodie (1923), Williams (1929), Wells (1964), Janssens (1970), Jarcho (1966a), Brothwell and Sandison (1967), Buikstra and Cook (1980), Angel (1981), Ubelaker (1982), Buikstra and Beck (2006), Grauer (2008), and Buikstra (2010). The multiple interpretations reflect the complex and multidisciplinary nature of paleopathology.

To many, the inception of paleopathology can be found in the 19th century, with physicians and anatomists such as Rudolf Virchow (1821–1902), Frederic Wood Jones (1879–1954), and Grafton Elliot Smith (1871–1937) developing interests in recognizing disease in ancient human bone and mummies. Their training and expertise in human anatomy, identification of disease, and diagnosis, allowed them to make associations between the presence of lesions on bone (and in mummified tissues) and potential causes. In particular, Sir Marc Armand Ruffer (1859–1917), a professor of bacteriology and the first director of the British Institute of Preventive Medicine (argued to be the true "pioneer of palaeopathology" by Sandison (1967)), published numerous articles on disease in Egyptian mummies. His careful approach, which relied on comparisons of clinical manifestations of disease to lesions found in ancient remains, paved the way for histological and radiographic investigation into ancient disease within paleopathology today.

As the early 20th century took hold, a new approach towards understanding disease in the past developed, much to the credit of Earnest Hooton (1887–1954). While most of the work conducted on human skeletal remains in North America was intertwined with agendas within physical anthropology (that is, heavily focused on metric analyses and the recording of human physical variation), Hooton set off in a new direction. Trained in the classics, he examined the presence of disease in wider context. His unit of analysis moved from the individual to the population. Along with publishing his detailed metric and morphological data on Native American remains from Pecos Pueblo (Hooton 1930), he incorporated a broader body of data into his work: archaeological data. The presence of pathological conditions such as trauma, dental disease, and osteoarthritis, for instance, were inventively viewed across space and time, and interpreted alongside the effects of a changing environment, food preparation, diet, and the presence of infectious disease. As promising as Hooton's synthesis appeared, however, it was not until decades later that his approach was rigorously adopted by others.

This is not to imply that work within paleopathology remained stagnant in the mid 20th century. On the contrary, Møller-Christensen, working with Danish material and particularly interested in leprosy, brought to paleopathology rigorous criteria for diagnosis coupled with a population approach (e.g. Møller-Christensen 1961). Similarly, Calvin Wells, publishing prolifically for two decades starting from the 1950s, provided paleopathologists with a large body of work that systematically evaluated the presence of a wide variety of skeletal conditions, and provided a benchmark from which other researchers could measure their rigor. The work of Ackerknecht (1953, 1965) offered a synthesis of many paleopathologists' work, as he sought to highlight the field's contributions to our understanding of the past.

During these same decades a division within paleopathology, as practice in North America and in Europe, took root (Roberts 2006; Mays 2010). In Europe, researchers tended to have medical backgrounds and focused on the presence of disease in particular individuals, leading to the publication of descriptive case studies. In North America, researchers were commonly trained in physical anthropology, and employed an anthropological approach to their inquiry. They had access to large samples of skeletons derived from archaeological excavations. Both trajectories had their flaws. Both came under fire in the 1960s. Brothwell and Sandison (1967), for instance, speaking to the "European approach" posited that "the time has come for some form of palaeopathological stock-taking and pooling of recently collected data" (Brothwell and Sandison 1967:xiv). Their volume sought to provide the field with a compendium of skeletally recognized conditions and diseases. Here, insights into the antiquity and geographical scope of skeletal lesions, as well as radiographic, histological, and photographic documentation were provided. On the other side of the Atlantic, Jarcho (1966b) was calling for a "revival of paleopathology" in the U.S. He criticized the direction that field had taken, which contributed to weak methodological work, the unsystematic collection of data, and the increasing marginalization of paleopathology from both the medical and archaeological communities.

In the decades that followed, substantial strides were taken within paleopathology. Great efforts were made to strengthen diagnostic criteria of diseases. This included providing careful descriptions of lesions in dry bone and firm clinical foundations for differential diagnosis (see Steinbock 1976; Ortner and Putschar 1981; Aufderheide and Rodriguez-Martin 1998; Ortner 2003). Methods to standardize data recording were offered by Ragsdale (1992), who argued for the careful use of description and classification of disease, and by Buikstra and Ubelaker (1994) and Brickley and

McKinley (2004) who sought to devise comprehensive and systematic means of recording osteological data. Theoretical strides were also made. Slowly, paleopathological attention moved away from questions such as "what disease is this?" and "when was this disease first found in humans?" to questions of "how" and "why". Huss-Ashmore et al. (1981), Armelagos et al. (1982), and Cohen and Armelagos (1984), for instance, offered ways in which paleopathological investigation could shed light on human nutrition and subsistence strategies, and how subsistence strategies could impact the presence of disease. Brothwell (1982), while exploring the promise of the field of environmental archaeology, offered provocative ways in which understanding details of complex urban environments might lead to a new understanding of host–pathogen relationships.

Throughout these decades new questions regarding the history of human disease were posed. Why, for instance, do some populations suffer from conditions that others do not? Why has the presence of some diseases seemingly changed over time? How might a wide range of variables affect the presence of diseases and disease processes? How might differences in human social interaction affect the host–pathogen relationship? Perhaps, it can be argued, the most profound changes within paleopathology have not come from focusing on populations rather than individuals, but rather have arisen with the influence and inclusion of different theoretical paradigms. These have recently been adopted in efforts to answer questions, and were borrowed from and expanded upon from other fields.

NEW DIRECTIONS IN PALEOPATHOLOGY

The multidisciplinary and multifocal approaches which characterized the history of paleopathology remain today and serve, in part, as sources of potential and tension within the field. As in the past, researchers today are not uniformly trained to be paleopathologists. They can be trained in medicine, dentistry, archaeology or physical anthropology, and choose to focus (sometimes only occasionally) on paleopathological questions and the examination of skeletal or mummified remains. Increasingly, however, making a productive contribution today to paleopathology not only requires superb familiarity with bone and/or preserved tissue, it requires a firm knowledge of microbiology, physiology, biochemistry, medicine, archaeology, history, and human culture, to name a few. Hence, the formulation of new questions within paleopathology has required even more of researchers, as they seek new means of obtaining answers. Recent efforts, for instance, to diagnose disease more accurately and to place it into an evolutionary context have led researchers to adopt technologically sophisticated techniques, such as tissue imaging and genetic testing. Similarly, efforts to understand environmental variables contributing to health and disease in the past are using sophisticated biochemical tests to explore stable isotopes and trace elements in bone. The technical knowledge needed to complete the requisite techniques, let alone to allow reasonable interpretation of the results (or manage the financial expense), has encouraged researchers to become more specialized and/or to seek to collaborate with specialists who may or may not be familiar with paleopathology at all. The promise of new information fuels a growing tension between the specialist and the generalist. Hence, it can be argued that the singular term "paleopathology" truly masks the variation and complexity within the field.

Productively contributing to paleopathology does not solely rely on technology. It can also be achieved through the development and implementation of new theories and approaches. Interestingly, for all the past published reflections on the field of paleopathology, few have delved into the theoretical premises upon which the discipline has been built. Taking it a step further, it could be argued that the topic is staunchly ignored. There are a number of ways that this oversight can be interpreted. First, perhaps one could argue that paleopathological research is an a-theoretical endeavor—meaning that there are no underlying assumptions being made or that multiple interpretations of data are not possible. This seems a bit preposterous. One of the key tenets upon which paleopathologists build their arguments is that disease processes occurring today are similar to those in the past, i.e. that pathogenesis does not change over time. We clinically identify the presence of a disease in modern humans, and assert that finding the same indicators on tissue samples from the past indicates that the identical pathogen or disease process was responsible. This is, in fact, a key reason why clinical case studies have been so important to paleopathological research. But recent work on the genera *Mycobacterium* (e.g., Monot et al. 2009) and *Treponema* (e.g., Hardham et al. 1997), calls this assumption into question, as the evolutionary paths of these pathogens are complex. If we argue that pathogens are capable of genetically changing over time, which might alter how they affect their hosts, then we must grapple with the assumption made in paleopathology that they do not.

An alternative interpretation is that exploring theoretical underpinnings is relatively new to paleopathology. Although paleopathology today is considered a scientific endeavor, its early practitioners were men of medicine, with their strengths resting in the application of biological knowledge to healing, not scientific research. But a huge component of paleopathology has incorporated scientific analyses of archaeo-logical data for at least forty years. Shouldn't it have been inevitable that we also adopt archaeological theory? We haven't. Paleopathologists have been selective in their use of archaeological theory, opting consistently to apply the "processual approach" in spite of archaeology's (and other fields such as anthropology, evolutionary biology, and genetics) ensuing labile and provocative paradigms. The processual approach, stemming from a labeled "New Archaeology", germinated in the works of Willey and Phillips (1958), and more specifically tackled issues of mortuary analysis with publications by Binford and Binford (1968), Saxe (1970), and Binford (1971). The premise of the processual approach was that through rigorous and quantitative scientific investigation of mortuary remains, researchers could begin to piece together social dimensions of the past. Social dimensions, such as status, were viewed as predictable and recognizable responses to human environmental adaptation. The approach was deemed "positivist" and "deterministic" by opponents, who argued, in part, that developing scientific and statistical correlations between material artifacts within space and time did not equal "objective truth" about the past (e.g. Hodder 1982; Shanks and Tilley 1982; Earle and Pruecel 1987; and Hodder and Hutson 2003).

In spite of the new archaeological paradigms developed since the 1960s and 1970s, and the rise of the bioarchaeological approach in human skeletal analysis (see Armelagos and Van Gerven 2003, and Buikstra and Beck 2006 for comprehensive reviews), paleopathologists tend to be content with the processual approach. High frequencies of skeletal lesions associated with nutritional deficiency, growth disruption, trauma, or degenerative joint disease, for instance, within a population, are posited to reflect a "less healthy" group. Individuals of lower social status are expected to have greater frequencies of pathological lesions, and the presence of pathological lesions predictably suggest lower status. The argument is circular and leaves little room for multifactorial cause, complex social interactions, individual choice, and/or change over time due to human action. Even as early as 1970, Gabriel Lasker, quoting his mentor Eliot Chappell, feared that this approach would render physical anthropology (and hence paleopathology) "the handmaiden to human history" (Lasker 1970:2).

While changes in archaeological theory didn't sway most paleopathologists to entertain alternative interpretations of their data, the publication of *The Osteological Paradox* in 1992 by Wood et al., and the ensuing dialogue between researchers (see Goodman 1993; Storey 1997; Wright and Chew 1998) certainly did. Wood et al. brought to light the ramifications of demographic nonstationarity (that population size does not remain constant over time and is especially sensitive to changes in fertility), selective mortality (that populations under investigation represent those who died, not those who were living), and hidden heterogeneity (that individuals vary in their susceptibility to disease). The authors show how datasets can be reinterpreted. For instance, they argue that individuals displaying pathological lesions do not necessarily represent the "sickest" individuals, but rather might indicate immunological strength, as the manifestation of lesions within the skeletal system requires the individual to live through the disease episode, rather than to die immediately from it. These concepts, alongside a growing emphasis on exploring the cultural context of disease (see, for instance, Huss-Ashmore et al. 1981; Angel et al. 1987; Larsen 1997; Goodman and Leatherman 1999) crystallized the need within paleopathology to explore disease processes, and to develop more complex interpretations of disease in the past (Wright and Yoder 2003).

Whether we have adequately met the challenge of developing new directions within paleopathology is debatable. On one hand, overly simplistic interpretations of the presence of skeletal lesions are still amplified through the microphones at professional meetings and appear in print. On the other, as seen in a wide variety of publications and within the chapters of this volume, the field of paleopathology has matured. Researchers such as Sofaer (2006), Geller (2008), Knudson and Stojanowski (2008), and Agarwal and Glencross (2011) offer provocative new ways to integrate social theory with skeletal analyses. Likewise, recently published work exploring the association between epistasis (the interaction between two or more genes) and pathological conditions (Carlborg and Haley 2004; Vieira 2008), epigenetics (which explores the external influences on gene behavior) and disease (Jones and Baylin 2002), and the evolution of disease (Frank 2002; Crespi 2010), to name a few, directly and indirectly offer paleopathologists tantalizing new prospects for skeletal analysis and interpretation.

What Is a "Companion"?

There is a wide range of books published on the topic of paleopathology designed to guide and inform the novice and advanced reader. Superb compendia such as *Human Paleopathology* (Ortner 2003) and *The Cambridge Encyclopedia of Human*

Paleopathology (Aufderheide and Rodriguez-Martin 1998) provide us with resources to undertake differential diagnosis and a glimpse into the antiquity of particular conditions. Other books, such as *Biological Anthropology of the Human Skeleton* (Katzenberg and Saunders 2000), and *Advances in Human Palaeopathology* (Pinhasi and Mays 2008), offer the reader foundations in specific topics germane to the field. While books such as *The Archaeology of Disease* (Roberts and Manchester 2005), *Bioarchaeology: Interpreting Behavior from the Human Skeleton* (Larsen 1997), *The Backbone of History: Health and Nutrition in the Western Hemisphere* (Steckel and Rose 2002), and *Health and Disease in Britain* (Roberts and Cox 2003) offer syntheses of a large volume of data generated over decades by many researchers. Equally important are contributions made by authors delving into particular diseases, such as in *The Myth of Syphilis* (Powell and Cook 2005), *The Bioarchaeology of Tuberculosis* (Roberts and Buikstra 2003), *Tuberculosis Past and Present* (Palfi et al. 1999), *Developmental Defects of the Axial Skeleton in Paleopathology* (Barnes 1994), and *The Bioarchaeology of Metabolic Bone Disease* (Brickley and Ives 2008). Publications focusing on specific themes, such as *The Bioarchaeology of Children* (Lewis 2007), *Sex and Gender in Paleopathological Perspective* (Grauer and Stuart-Macadam 1998), *In the Wake of Contact: Biological Responses to Conquest* (Larsen and Milner 1994), and *Disease and Demography of the Americas* (Verano and Ubelaker 1992), also provide important, and often thematic, insight into the past. And lastly, but certainly not least, there are monographic publications which explore health and disease within singular individuals or within particular populations, such as *Bones in the Basement: Postmortem Racism in Nineteenth-Century Medical Training* (Blakely and Harrington 1997), *The Archaeology of Individuals* (Stodder and Palkovich 2010), and *The Skeletal Biology of the New York African Burial Ground* (Blakey and Rankin-Hill 2009).

This book is offered as a complement to those above: a volume to read alongside the vast periodical literature and growing collection of books which provide great breadth and depth to our field. Each contributor to this volume has been asked to provide a "snapshot" of a topic, and to expose issues and controversies together with their vision of a particular aspect of paleopathology. Their voices are varied, but their own: ranging from dense and detailed, to more casual and introspective. They reflect the discourse in our field. The result, we hope, is a compendium highlighting the breadth and promise of paleopathology.

Parsing paleopathology into discrete topics is not simple. Is it possible to discuss a particular disease without talking about the biological and physiological processes taking place within the host? Is it possible to avoid overlapping content between chapters when skeletal responses to change and the mechanical properties of bone are so limited and rarely unique to a single disease? Can full discussion of theories or approaches take place without incorporating examples that might be discussed elsewhere in the volume in different contexts? And all this begs the question of whether we agree on the classification of diseases in the first place, since many metabolic and endocrine diseases can be classified as growth disorders; and erosive joint disorders, suspected to be triggered by infectious pathogens, can be arguably classified as infectious disease. Distinctions, which might appear at first to be absolute, begin to blur and overlap within the pages of this volume. Inevitably, however, celebrating the overlap allows each contribution to stand independently, while enticing conversation and discourse when information is supplementary or contradictory to that provided in other chapters.

This volume begins with broad brushstrokes. The use of human skeletal and mummified remains within paleopathology cannot be undertaken without introspection into the social and political ramifications of material, as Lambert points out in Chapter 2, "Ethics and Issues in the use of Human Skeletal Remains in Paleopathology." But the types of questions that paleopathologists seek to answer also need our attention. They are neither identical between researchers, nor uniform through time. As discussed above, paleopathology can be characterized by its ability to unite many threads into a common cloth. Researchers from a variety of fields, with a variety of backgrounds come together to ask questions. All questions center on disease in the past.

Hence, Part I of this book seeks to highlight these varied views. Zuckerman, Turner, and Armelagos in Chapter 3, "Evolutionary Thought in Paleopathology and the Rise of the Biocultural Approach," provide a detailed account of evolutionary thought within paleopathology, emphasizing the complex interactions between humans and their environment that inform the biocultural approach. They argue that in spite of the promise of this approach, it remains underutilized, if not neglected. Buzon in Chapter 4, "The Bioarchaeological Approach to Paleopathology," highlights the role and concomitant issues that bringing archaeological techniques and data pose for paleopathology. From issues regarding sampling and recovery of specimens, to the interpretation of social identity via archaeological analysis, paleopathology has much to gain by fully integrating archaeology into the study of disease. In Chapter 5: "The Molecular Approach in Paleopathology," Gosman goes beyond paleopathologists' usual discussion of basic bone biology and into new ways molecular biology might impact our interpretation of pathological lesions on skeletal remains. His message is sobering, as he argues that skeletal interpretations made by paleopathologists are often far too simplistic. Katzenberg in Chapter 6, "The Ecological Approach: Understanding Past Diet and the Relationship Between Diet and Disease," entices the reader to look beyond the usual anthropocentric interpretations of disease in the past, which have humans as the both the instigator and victim of change, towards an ecological approach—an approach where numerous environmental variables impact one another, many outside humans' direct control. Boldsen and Milner in Chapter 7, "An Epidemiological Approach to Paleopathology," guide the reader through the perils and requisite premises of interpreting disease beyond its presence in individuals and on to a population level. Their straightforward approach, coupled with many examples, effectively highlights the key role that paleoepidemiology plays in our understanding of disease in the past. Cast in a promising light, Chapter 8, "The Promise, the Problems, and the Future of DNA Analysis in Paleopathology Studies," by Spigelman, Shin, and Bar-Gal, captures the ways in which DNA analyses have begun to answer questions germane to paleopathology, while simultaneously outlining imperative protocols that must be met in order for the work to provide genuine results. Zimmerman in Chapter 9, "The Analysis and Interpretation of Mummified Remains," summarizes a wide body of data to draw attention to the contributions that mummified remains have made to our understanding of disease. In Chapter 10, "The Study of Parasites Through Time: Archaeoparasitology and Paleoparasitology," Dittmar, Araujo, and Reinhard chronicle the work of parasitologists from around the world and make evident the theoretical interdependency between many branches of science, such as archaeology, evolutionary biology, microbiology, ecology, and the social sciences. Chapter 11, "More Than Just Mad Cows: Exploring Human–Animal Relationships Through Animal Paleopathology," by Upex and Dobney, effectively argue for the much-warranted, but often overlooked need to include the analysis of faunal remains in paleopathological investigations. Through these analyses, complex interactions between human and animal hosts and pathogens are better understood, and the dynamic relationship between humans and animals becomes more readily appreciated. Lastly, it is fitting to end the section dedicated to approaches and perspectives with introspection into how paleopathology's past might inform its future. Hence, in Chapter 12, "How Does the History of Paleopathology Predict Its Future?" Powell and Cook provide the reader with a perspicacious look into controversies and paths that serve as this field's greatest obstacles, but may render its greatest promise.

Equally germane to the field of paleopathology is discussion centered on methods and techniques by which we amass data and develop interpretations. Part II of this book seeks to address these issues. At the cornerstone of these discussions rests a set of issues involving the recognition and classification of disease. Offering one means of classification and a discussion on cellular processes that precede our ability to diagnose the presence of disease, Ragsdale and Lehmer, in Chapter 13 "A Knowledge of Bone at the Cellular (Histological) Level is Essential to Paleopathology," argue that studying mechanisms of disease may be more critical to paleopathology than attempts at diagnosis. Likewise, Ortner in Chapter 14, "Differential Diagnosis and Issues of Disease Classification," conveys a similar message but hinges his arguments on the role that both cause and pathogenesis play in our ability to classify disease, which ultimately, he argues, influences paleopathologists' interpretations of the significance of finding a disease in the past.

Other key variables have become important to wrestle within paleopathology; in particular, the determination of age-at-death and sex in skeletal remains. Milner and Boldsen, in Chapter 15 "Estimating Age and Sex From the Skeleton: A Paleopathological Perspective," review the techniques used to make these determinations, grapple with the biases, and present the need for careful paleodemographic analysis within paleopathological studies. In Chapter 16, "The Relationship Between Paleopathology and the Clinical Sciences," Mays argues that case studies and clinical research play an important role in paleopathology and that tension between qualitative and quantitative analysis is unlikely to end soon; perhaps to the benefit of paleopathology. Another key resource in paleopathology is historical documents. Mitchell in Chapter 17, "Integrating Historical Sources with Paleopathology," critically evaluates potential biases and brings the social dimensions of human disease to the forefront. While advances in imaging techniques have greatly assisted paleopathological investigation, Wanek, Papageorgopoulou, and Rühli in Chapter 18, "Fundamentals of Paleoimaging Techniques: Bridging the Gap Between Physicists and Paleopathologists," seek to disclose both the positive and negative aspects of using particular techniques. In Chapter 19, "Data and Data Analysis: Issues in Paleopathology," Stodder draws our attention to complexities of data collection and the ramifications that small decisions, either made or ignored, can have on paleopathological research.

Part III seeks to tackle particular conditions and diseases that have garnered much attention in paleopathological research. The section begins with a discussion of the presence and interpretation of trauma, offered by Judd and Redfern in Chapter 20. The authors make particular note of the controversies centering on the classification, recording, and interpretation of the deceptively complex suite of skeletal changes referred to as "trauma." In Chapter 21, "Developmental Disorders in the Skeleton," by Barnes, the reader is informed that variability exists in every part of the human skeleton, with the concept of "disorder" being created at times artificially in an effort to derive meaning from human variation. Kozlowski and Witas in Chapter 22, "Metabolic and Endocrine Diseases," offer the reader insight into these widely encompassing disease classifications which, within paleopathological research, often focus on the association between nutritional deficiencies and disease. In Chapter 23, "Tumors: Problems of Differential Diagnosis in Paleopathology," Brothwell, through the use of a case-study approach, addresses how bone preservation and lesion interpretation affect the diagnosis and eventual understanding of tumors in the past. Roberts in Chapter 24, "Re-Emerging Infections: Developments in Bioarchaeological Contributions to Understanding Tuberculosis Today," takes a holistic approach to the study of tuberculosis by placing it in historical, social, political, and economic contexts. This demonstrates the paleopathological examination of the disease to be a critical contributor to our understanding of its future trajectory in human populations. Providing a clinical and epidemiological approach, alongside a paleopathological assessment, Lynnerup and Boldsen in Chapter 25, "Leprosy (Hansen's disease)," discuss multiple facets of this disease, which still afflicts huge numbers of people today. In Chapter 26, "Treponematosis: Past, Present, and Future," Cook and Powell review the suite of conditions classified as treponemal disease and the debates that have developed within paleopathology concerning the evolution of syphilis. Venturing past a clinical discussion, the authors draw attention to the social dimensions which have influenced our understanding of the disease in the past and might color our views today. The growing discourse on a pathological lesion of complex etiological origin, referred to loosely as "periosteal reactions" is the primary focus of Weston in Chapter 27, "Nonspecific Infection in Paleopathology: Interpreting Periosteal Reactions." Here, the author casts a critical eye on paleopathological interpretations of these lesions by exploring the physiological and biochemical bases of bone production. Waldron, in Chapter 28, "Joint Disease," assists the reader in understanding the bases of classification of the disease and evaluates efforts to interpret their presence in human skeletal remains. Chapter 29, "Bioarchaeology's Holy Grail: The Reconstruction of Activity," by Jurmain, Cardoso, Henderson, and Villotte tackles a "hot" topic in paleopathology: can quantifiable and/or qualifiable changes to bone serve as markers for occupation, activity, or mechanical stress? Clearly, many researchers have argued "Yes". However, these authors question both the diagnosis and interpretation of lesions used to reconstruct activity in the past, providing the reader a glimpse into the heat of the controversy. To end, the analysis of dental remains in human skeletal remains is addressed by Lukacs in Chapter 30, "Oral Health in Past Populations: Context, Concepts and Controversies in the Study of Dental Disease." Like bone, teeth provide insight into health and disease in the past, but only, as argued by the author, if we recognize the intricacies of diagnosis and the perils of interpretation.

REFERENCES

Ackerknecht, E. H., 1953 Paleopathology. *In* Anthropology Today. An Encyclopedic Inventory. A. L., Kroeber, ed. pp. 120–126. Chicago: University of Chicago Press.

Ackerknecht, E. H., 1965   History and Geography of the Most Important Diseases. New York: Hafner.

Agarwal, S. C., and Glencross, B. A., eds., 2011   Social Bioarchaeology. Chichester: Wiley-Blackwell.

Angel, L. J., 1981   History and Development of Paleopathology. American Journal of Physical Anthropology 56:509–515.

Angel, L. J., Kelley, J., Parrington, M., and Pinter, S., 1987   Life Stresses of the Free Black Community as Represented by the First African Baptist Church, Philadelphia, 1823–1841. American Journal of Physical Anthropology 74:213–229.

Armelagos, G. J., Carlson, D. S., and Van Gerven, D. P., 1982   The Theoretical Foundation and Development of Skeletal Biology. *In* A History of American Physical Anthropology, 1930–1980. F. Spencer, ed. pp. 305–328. New York: Academic Press.

Armelagos, G. J., and Van Gerven, D. P., 2003   A Century of Skeletal Biology and Paleopathology: Contrasts, Contradictions, and Conflicts. American Anthropologist 105(1):51–62.

Aufderheide, A. C., and Rodriguez-Martin, C., 1998   The Cambridge Encyclopedia of Human Paleopathology. Cambridge: Cambridge University Press.

Barnes, E., 1994   Developmental Defects of the Axial Skeleton in Paleopathology. Niwot, CO: University Press of Colorado.

Binford, L., 1971   Mortuary Practices: Their Study and Their Potential. *In* Approaches to the Social Dimensions of Mortuary Practices. J. A. Brown, ed. pp. 6–29. Washington, DC: Society for American Archaeology.

Binford, S. R., and Binford, L., 1968   New Perspectives in Archaeology. Chicago: Aldine Press.

Blakely, R. L., and Harrington, J. M., 1997   Bones in the Basement: Postmortem Racism in Nineteenth-Century Medical Training. Washington, DC: Smithsonian Institution Press.

Blakey, M. L., and Rankin-Hill, L. M., 2009   The Skeletal Biology of the New York African Burial Ground: Part I. The New York African Burial Ground: Unearthing the African Presence in Colonial New York, Volume 1. Washington, DC: Howard University Press.

Brickley, M., and Ives, R., 2008   The Bioarchaeology of Metabolic Bone Disease. Oxford: Academic Press.

Brickley, M., and McKinley, J. I., 2004   Guidelines to the Standards for Recording Human Remains. IFA Paper Number 7. Southampton: BABAO.

Brothwell, D., 1982   Linking Urban Man with His Urban Environment. *In* Environmental Archeology in the Urban Context. R. A. Hall, and H. K. Kenward, eds. pp. 126–129. Oxford: Council of British Archaeology Report No. 43.

Brothwell, D., and Sandison, A. T., 1967   Editorial Prolegomenon: The Present and Future. *In* Disease in Antiquity: A Survey of the Diseases, Injuries and Surgery of Early Populations. D. Brothwell, and A. T. Sandison, eds. pp. xi–xiv. Springfield, IL: Charles C. Thomas.

Buikstra, J. E., 2010   Paleopathology: A Contemporary Perspective. *In* Companion to Biological Anthropology. C. S. Larsen, ed. pp. 395–411. Chichester: Wiley–Blackwell.

Buikstra, J. E., and Beck, L., 2006   Bioarchaeology: The Contextual Analysis of Human Remains. San Diego, CA: Elsevier Inc.

Buikstra, J. E., and Cook, D. C., 1980   Palaeopathology: An American Account. Annual Reviews in Anthropology 9:433–470.

Buikstra, J. E., and Ubelaker, D. H., 1994   Standards for Data Collection from Human Skeletal Remains: Proceedings of a Seminar at The Field Museum of Natural History Organized by Jonathan Haas. Research Series No. 44. Fayetteville: Arkansas Archaeological Survey.

Carlborg, O., and Haley, C. S., 2004   Epistasis: Too Often Neglected in Complex Trait Studies? Nature Reviews Genetics 5:618–625. doi:10.1038/nrg1407.

Crespi, B. J., 2010   The Origins and Evolution of Genetic Disease Risk in Modern Humans. Annals of the New York Academy of Sciences 1206(1):80–109.

Earle, T. K., and Preucel, R. W., 1987   Processual Archaeology and the Radical Critique. Current Anthropology Vol. 28(4):501–538.

Frank, S. A., 2002   Immunology and Evolution of Infectious Disease. Princeton, NJ: Princeton University Press.

Geller, P. L., 2008   Conceiving Sex: Fomenting a Feminist Bioarchaeology. Journal of Social Archaeology 8:113–138.

Goodman, A., 1993   On the Interpretation of Health from Skeletal Remains. Current Anthropology 34(3):281–288.

Goodman, A., and Leatherman, T., eds. 1999   Building a New Biocultural Synthesis: Political-Economic Perspectives on Human Biology. Ann Arbor, MI: The University of Michigan Press.

Grauer, A. L., 2008   Macroscopic Analysis and Data Collection in Palaeopathology. In Advances in Human Palaeopathology. R. Pinhasi, and S. Mays, eds. pp. 57–76. Chichester: John Wiley and Sons.

Grauer, A. L., and Stuart-Macadam, P., eds., 1998   Sex and Gender in Paleopathological Perspective. Cambridge: Cambridge University Press.

Hardham, J. M., Frye, J. G., Young, N. R., and Stamm, L. V., 1997   Identification and Sequences of the Treponema pallidum flhA, flhF, and orf304 Genes. DNA Sequence 7:107–116.

Hodder, I., ed., 1982   Symbolic and Structural Archaeology. Cambridge: Cambridge University Press.

Hodder, I., and Hutson, S., 2003   Reading the Past: Current Approaches to Interpretation in Archaeology. 3rd edition. Cambridge: Cambridge University Press.

Hooton, E. A., 1930   The Indians of Pecos Pueblo: A Study of their Skeletal Remains. New Haven, CT: Yale University Press.

Huss-Ashmore, R., Goodman, A., and Armelagos, G. J., 1981   Nutritional Inference From Paleopathology. In Advances in Archaeological Method and Theory Volume 5. M. Schiffer, ed. pp. 395–474. New York: Academic Press.

Janssens, P. A., 1970   Paleopathology: Diseases and Injuries of Prehistoric Man. Ida Dequeecker, trans. New York: Humanities Press.

Jarcho, S., ed., 1966a   Human Paleopathology. New Haven: Yale University Press.

Jarcho, S., 1966b   The Development and Present Condition of Human Palaeopathology in the United States. In Human Paleopathology. S. Jarcho, ed. pp. 3–30.New Haven, CT: Yale University Press.

Jones, P. A., and Baylin, S. B., 2002   The fundamental Role of Epigenetic Events in Cancer. Nature Reviews Genetics 3, 415–428. doi:10.1038/nrg816.

Katzenberg, M. A., and Saunders, S. R., 2000   Biological Anthropology of the Human Skeleton. New York: Wiley-Liss.

Knudson, K. J., and Stojanowski, C. M., 2008   New Directions in Bioarchaeology: Recent Contributions to the Study of Human Social Identities. Journal of Archaeological Research 16:397–432.

Larsen, C. S., 1997   Bioarchaeology: Interpreting Behavior from the Human Skeleton. Cambridge Studies in Biological Anthropology 21. Cambridge: Cambridge University Press.

Larsen, C. S., and Milner, G. R., 1994   In the Wake of Contact: Biological Responses to Conquest. New York: Wiley-Liss.

Lasker, G., 1970   Physical Anthropology: Search for General Processes and Principles. American Anthropologist 72:1–8.

Lewis, M. E., 2007   The Bioarchaeology of Children: Perspectives from Biological and Forensic Anthropology. Cambridge: Cambridge University Press.

Mays, S. A., 2010   Human Osteoarchaeology in the UK 2001–2007: A Bibliometric Perspective. International Journal of Osteoarchaeology 20:192–204.

Møller-Christensen, V., 1961   Bone Changes in Leprosy. Copenhagen: Munksgaard.

Monot, M., Honoré, N., Garnier, T., Zidane, N., Sherafi, D., Paniz-Mondolfi, A., Matsuoka, M., Taylor, G. M., Donoghue, H. D., Bouwman, A., et al., 2009   Comparative Genomic and Phylogeographic Analysis of *Mycobacterium leprae*. Nature Genetics 41:1282–1289.

Moodie, R. L., 1923   Paleopathology: An Introduction to the Study of Ancient Evidences of Disease. Urbana: University of Illinois Press.

Ortner, D. J., 2003   Identification of Pathological Conditions in Human Skeletal Remains. Second Edition. San Diego, CA: Academic Press.

Ortner, D. J., and Putschar, W. G. J., 1981   Identification of Pathological Conditions in Human Skeletal Remains. Smithsonian Contributions to Anthropology, No. 28. Washington, DC: Smithsonian Institution Press.

Palfi, G., Dutour, O., Deak, J., and Hutas, I., 1999   Tuberculosis Past and Present. Szeged, Hungary: Golden Book Publisher, and Tuberculosis Foundation.

Pinhasi, R., and Mays, S., 2008   Advances in Human Palaeopathology. Chichester: John Wiley and Sons.

Powell, M. L., and Cook, D. C., 2005   The Myth of Syphilis: The Natural History of Treponematosis in North America. Gainesville: University of Florida Press.

Ragsdale, B. L.   1992   Task Force on Terminology: Provisional Word List (March 1992). Paleopathology Newsletter 78:7–8.

Roberts, C. A., 2006   A View From Afar: Bioarchaeology in Britain. *In* Bioarchaeology: The Contextual Analysis of Human Remains. J. E. Buikstra, and L. Beck, eds. pp. 417–439. San Diego, CA: Elsevier Inc.

Roberts, C. A., and Buikstra, J. E., 2003   The Bioarchaeology of Tuberculosis: A Global View On a Reemerging Disease. Gainesville: University of Florida Press.

Roberts, C. A., and Cox, M., 2003   Health and Disease in Britain: Prehistory to the Present Day. Gloucester: Sutton Publishing.

Roberts, C. A., and Manchester, K., 2005   The Archaeology of Disease. 3rd edition. Stroud: Sutton Publishing.

Sandison, A. T., 1967   Sir Marc Armand Ruffer (1859–1917) Pioneer of Palaeopathology. Medical History 11(2):150–156.

Saxe, A. A., 1970   Social Dimensions of Mortuary Practices. Ph.D. dissertation, University of Michigan.

Shanks, M., and Tilley, C., 1982   Ideology, Symbolic Power and Ritual communication: A Reinterpretation of Neolithic Mortuary Practices. *In* Symbolic and Structural Archaeology. I. Hodder, ed. pp. 129–154. Cambridge: Cambridge University Press.

Sofaer, J., 2006   The Body as Material Culture. A Theoretical Osteoarchaeology. Cambridge: Cambridge University Press.

Steckel, R. H., and Rose, J. C., 2002   The Backbone of History: Heath and Nutrition in the Western Hemisphere. Cambridge: Cambridge University Press.

Stedman's Medical Dictionary. 26th edition, 1995   Baltimore MD: Williams and Wilkins.

Steinbock, R. T., 1976   Paleopathological Diagnosis and Interpretation. Springfield, IL: Charles C. Thomas.

Stodder, A. L. W., and Palkovich, A. M., 2010   The Bioarchaeology of Individuals. Gainesville: University Press of Florida.

Storey, R., 1997   Individual Frailty, Children of Privilege, and Stress in Late Classic Cop'an. *In* Bones of the Maya: Studies of Ancient Skeletons. S. Whittington, and D. Reed, eds. pp. 116–126. Washington, DC: Smithsonian Institution Press.

Ubelaker, D. H., 1982   The Development of American Paleopathology. *In* A History of American Physical Anthropology, 1930–1980. F. Spencer, ed. pp. 337–356. New York: Academic Press.

Verano, J. W., and Ubelaker, D. H., 1992 Disease and Demography in the Americas. Washington, DC: Smithsonian Institution Press.

Vieira, A. R., 2008 Unraveling human Cleft Lip and Palate Research. Journal of Dental Research 87:119–125.

Wells, C., 1964 Bones, Bodies and Diseases: Evidence of Disease and Abnormality in Early Man. London: Thames and Hudson.

Willey, G. R., and Phillips, P., 1958 Method and Theory in American Archaeology. Chicago: University of Chicago Press.

Williams, H. U., 1929 Human Paleopathology, With Some Original Observations on Symmetrical Osteoporosis of the Skull. Archives of Pathology 7:839–902.

Wood, J., Milner, G., Harpending, H., and Weiss, K., 1992 The Osteological Paradox: Problems of Inferring Prehistoric Health from Skeletal Samples. Current Anthropology 33:343–370.

Wright, L. E., and Chew, F., 1998 Porotic Hyperostosis and Paleoepidemiology: A Forensic Perspective on Anemia Among the Ancient Maya. American Anthropologist 100:924–939.

Wright, L. E., and Yoder, C. J., 2003 Recent Progress in Bioarchaeology: Approaches to the Osteological Paradox. Journal of Archaeological Research 11(1):43–70.

PART I

Approaches, Perspectives and Issues

CHAPTER **2**

Ethics and Issues in the Use of Human Skeletal Remains in Paleopathology

Patricia M. Lambert

INTRODUCTION

Scholars have been using human remains for scientific purposes for hundreds of years, and it is likely that the practice has raised significant ethical questions from the time of the earliest documented studies by Greek medical practitioners in the third century B.C. (Von Staden 1992). This is because human remains, whether whole cadavers or skeletons, are the physical embodiment of once-living people situated within a social context of kinship relations, unique culture history, religious beliefs, and value systems. People's sentiments about death and toward the dead therefore profoundly influence their attitudes towards the collection and use of human remains in scientific research, and these attitudes in turn impact the social climate in which researchers work and publish (Walker 2000). Ethical issues surrounding the use of human remains in science usually emerge when people with an interest in or claim to these remains differ in their attitudes about appropriate handling and use of human bodies or body parts. They often overlap with those encountered in the field of medical research (Kunstadter 1980; Pellegrino 1999) and recognition of the importance of ethical principles to guide human skeletal research has increased in tandem with developments in bioethics (Domen 2002). In both cases, beliefs that conflict with the secular view of science can present significant challenges to the conduct of the research, particularly where they are widely held and based in religious or cultural traditions. The purpose of this chapter is to provide a historical context for understanding modern ethical issues in the scientific use of human skeletal remains,

A Companion to Paleopathology, First Edition. Edited by Anne L. Grauer.

to review the ethical standards and laws that guide this research in the United States and elsewhere, and to touch on some of the other areas of osteological research where questions of ethics have emerged.

HISTORICAL BACKGROUND

According to the archaeological record, mortuary rituals are a mechanism by which humans have coped with the uncertainties of death for thousands of years (Morris 1992; Davies 1999; Pearson 2000). The belief in an afterlife appears to be quite ancient, as indicated by grave accompaniments such as food items and utilitarian wares in both Old and New World contexts suggesting preparation for a life beyond death. It is this very practice that makes the study of the dead so important in modern archaeology. In mortuary contexts, the remains of people identified by age, sex, stature, health, and genetic affiliation can preserve a record of human relationships to others in the burial community and with the accoutrements of both daily life and ritual practice. These relationships enable scientists to reconstruct subtle and often intimate aspects of the lives and deaths of ancient people, and collectively to reconstruct the history of populations and of our species (Larsen and Walker 2004). It is this personalization of the dead through the mortuary ritual, however, that can strike a strong chord with the living. Depending on how the death experience is conceptualized, disturbance of the dead can be viewed as an act of discovery—or one of desecration. Broadly speaking, these perspectives can be seen to characterize the view of science, on the one hand, and that of religion, on the other, with adherents at either extreme firmly and passionately convinced of the moral and ethical correctness of their stance.

In Western Judeo-Christian traditions, belief in the afterlife has been framed around the concept of a "soul," or "life force," which exists in life and preserves the essence of a person after death (Spellman 1994). In medieval times, this belief was coupled with the concept of "bodily resurrection," the belief that the body needed to be maintained intact after death for salvation to occur (Bynum 1998:590). Oddly enough, this did not prevent the development of a growing interest and trade throughout the medieval period in the body-parts of saints and martyrs (Bynum 1997). After the Reformation of the 16th century, the belief in bodily resurrection and the importance of the corpse persisted most fully in Protestant faiths (Spellman 1994; Laderman 1996). Although the strength of these beliefs has been tempered over time (Bynum 1998), attitudes towards the dead and thus towards the study of human remains continue to be influenced by these religious beliefs.

Use of human remains in Western science can be traced back to the early dissections of Herophilus of Chalcedon and Erasistratus of Ceos in the third century B.C., Greek physicians who performed systematic dissection to advance the art of medicine (Von Staden 1992). At this time, skeletal studies were probably secondary and would have occurred primarily in the context of soft-tissue dissection. For hundreds of years thereafter, the study of human remains was primarily the purview of physicians who sought to educate themselves about the human body in order to advance medicine. This work was often clandestine, due to predominant religious views about the consequences of desecrating the dead, and it was the bodies of criminals that most often came "under the knife" for this purpose (Frank 1976; Uchino 1983; Von Staden 1992). This is illustrated by the Murder Act of 1751, which limited dissection in England and Wales to the corpses of those executed for murder. Later, when this source proved insufficient for the needs of a growing medical establishment, physicians in the U.S. and Britain turned to body-snatchers, who obtained cadavers from graves of the recently interred (Sawyer 1964; Frank 1976; McDonald 1995; Schultz 2005). Their primary source was the graves of the poor (e.g., Richardson 2001) and in the U.S., the nonwhite, who were powerless to stop the desecration (Walker 2000). It is only much more recently that people in the U.S. and elsewhere seem to have overcome the aversion to dissection, and now it is the middle and upper classes that most commonly donate their bodies to science (Uchino 1983; Garment et al. 2007). The long association of the scientific study of the dead with the criminal element, and probably more importantly with infringement of the rights of the impoverished, continues to cast a negative light on scientific study of the dead as an act of denigration—particularly when the remains being studied relate to a sector of society that has been the target of discrimination.

Although human remains have been used to inform medical studies for hundreds of years, the use of skeletal elements in anthropologically oriented research traces back to the late 18th century, when scientists began to turn their attention to the classification of human biological diversity. These early efforts by scholars such as Johann Friedrich Blumenbach (1752–1840), a German professor of anatomy, focused almost exclusively on the skull as a basis for racial classification. While Blumenbach's studies were qualitative in nature (Augstein 1996), the American physician Samuel Morton (1799–1851) introduced a new level of rigor to the study of human diversity in early 19th-century America through application of craniometric techniques. Paul Broca, a French professor of clinical surgery and founder of the Anthropological Society of Paris (1859), was an admirer of Morton's work and continued to refine the method in his own research. These and other 19th-century researchers made laudable steps towards greater empiricism in anthropometric research on human diversity (Brace 2005), but their research was not immune from Eurocentric biases of the day. While the impact of these biases on their published research has been the subject of considerable debate (e.g., Gould 1981; Brace 2005), it is clear, none the less, that their work was embraced by contemporaries who sought to foster racist views and thus provided a scientific basis for discrimination (Gould 1981; Augstein 1996; Brace 2005; Little and Sussman 2010). It is a legacy that has been difficult to shed entirely, despite paradigmatic shifts over the last century in science generally and in biological anthropology more specifically.

From these inauspicious origins, the anthropological study of human remains has matured into a multidisciplinary, strongly empirical science involving the entire skeleton at macroscopic, microscopic, and biochemical scales of analysis. These empirical studies have often challenged the racist nature of many early classification schemes (e.g., Rankin-Hill and Blakey 1994; and see various authors in a special issue of the *American Journal of Physical Anthropology* (2009) 139:1:1–107) and created awareness within the discipline of the need for professional standards and ethics to guide the conduct of research using human skeletal remains.

Today, all disciplines in the social and biological sciences are guided by codes of ethics designed to ensure that practitioners conduct professional activities according to shared principles of behavior. This is particularly true of disciplines engaged in research on human subjects. In anthropology, such codes can be found on the websites of the American Association of Physical Anthropologists (AAPA 2003), the Society for American Archaeology (SAA 1996), and the American Anthropological Association (AAA 2009), the three largest professional anthropological societies in the United States. The principles ranked most highly by all three organizations include respect and accountability to those involved in or affected by the research, and the long-term conservation of the archaeological / paleoanthropological record. The code exists, in the words of the AAPA Code of Ethics, "to provide ... guidelines for making ethical choices in the conduct of their physical anthropological work." However, as the AAPA code notes: "because physical anthropologists can find themselves in complex situations and subject to more than one code of ethics, the AAPA Code of Ethics provides a framework, not an ironclad formula, for making decisions" (AAPA 2003).

Therein lies the challenge facing researchers committed to the conduct of ethical research on human remains: when the codes conflict, which principle has priority? For example, archaeological human remains constitute a significant and irreplaceable part of the archaeological record, so the first principle of the SAA's code (and one of the top principles of both the AAA and AAPA codes) requiring long-term conservation and protection of this resource calls upon researchers to document the remains and work to preserve them for future study. However, human remains derive from ancestral humans who may be linked genetically or culturally to living descendants. Depending on how descendent groups view their relationship to, and scientific study of, the dead, the study and preservation of ancestral human skeletal remains may be seen as desirable, irrelevant, undesirable, or even harmful. If scientific study of the dead is seen unfavorably by descendent groups, then scholars whose research involves the study of these human skeletal remains are faced with the difficult task of interpreting the meaning of the first principle of the AAA (2009:2) code—"to avoid harm or wrong"—and of the AAPA (2003) code—"to respect the well-being of [living] humans"—in terms of how these apply to archaeological human skeletal remains and descendent groups. As discussed in subsequent sections, some governmental bodies have attempted to legislate on the treatment of archaeological human remains, but these laws can also conflict and tend to include ambiguities that continue to complicate human skeletal research in the United States and other countries.

BIOARCHAEOLOGY, ETHICS, AND THE LAW IN THE UNITED STATES

Some of the most significant legal and ethical issues faced by modern scholars who study human skeletal remains relate to the question of who controls the dead and their possessions in archaeological contexts. Nowhere has this question received more attention than in the United States, where a series of increasingly specific laws enacted in the 20th century attempted to legislate on the treatment of archaeological human remains. These laws reveal significant changes in attitudes concerning the scientific value of archaeological human remains over the course of the 20th century, and illustrate the complexities that can arise when secular and religious ethics collide in the conduct of scientific research.

National concern with archaeological remains first found a voice in the United States in legislation aimed at protecting archaeological resources and sites from the destruction wrought by treasure hunters, looters, and vandals. The Antiquities Act of 1906 provided the first federally mandated protection of archaeological resources on federal lands by imposing requirements for permits and penalties for violations. This act laid the foundation for ethics in American archaeology by mandating the protection of archaeological resources and validating their importance for scientific study and education.

The Antiquities Act was followed by the Archaeological Resources Protection Act of 1979 (ARPA, Public Law 96-95, October 31, 1979, 16 U.S. Code Sec. 470aa et seq.), whose purpose was to toughen existing laws protecting archaeological resources for the purpose of securing "for the present and future benefit of the American people, the protection of archaeological resources and sites" on public and Indian lands. Unlike the Antiquities Act, ARPA specifically mentions "graves" and "human skeletal remains" in the list of archaeological resources warranting protection (Section 3.1), thereby clearly identifying these remains as archaeological resources in need of and finding protection under ARPA. Both the Antiquities Act and ARPA enforce ethics central to the conduct of professional archaeology, with a particular emphasis on protection, preservation and public benefit.

Laws that followed focused more specifically on human skeletal remains and had a very different purpose and ethic. On November 28, 1989, the National Museum of the American Indian Act (NMAIA) was signed into law. This statute provided a provision for the repatriation of human remains and associated funerary objects in the Smithsonian Institution collections "that could be culturally linked to extant Native American groups, going well beyond descendants of named individuals" (Ortner 1994:12). The focus of this act is on empowering indigenous Americans to lay claim to the remains of their ancestors for purposes of reburial, negotiating terms of curation, or other dispositions. In this sense, it marks a shift in archaeology law from the protection of archaeological resources to the legal acknowledgement of private ownership (or rights of stewardship) by a sector of the American population of a component of the archaeological record.

The Native American Graves Protection and Repatriation Act (NAGPRA, Public Law 101-601, November 16, 1990, 25 U.S. Code 3001 et seq.) was enacted the following year, expanding the federal mandate for repatriation claims to include human remains, associated grave goods, sacred objects, and objects of cultural patrimony from federal and Indian lands, as well as those housed in federal agencies and federally funded institutions. Its passage was an acknowledgment on the part of the Federal Government that there were items in museums that had been obtained in a manner inconsistent with modern American ethics, even if their acquisition had been legal at the time. NAGPRA provides a legal mechanism by which federally recognized "Indian tribes" and "Native Hawaiian organizations" can gain control over the disposition of Native American human skeletal remains on federal and tribal lands, and obtain a detailed accounting of such remains and other specified items in federally funded curation facilities for purposes of submitting claims for repatriation. It is also legal impetus for the initiation of conversations about these remains and cultural materials between federally recognized Native American / Native Hawaiian peoples and the scientific and museum communities (Larsen and Walker 2004; Walker 2004).

In light of the history of disenfranchisement of indigenous peoples in the North America, a history well known to those trained in North American anthropology, NAGPRA's passage was an ethical victory that could be shared by all communities impacted by the law (e.g., Dongoske 2000; Dongoske et al. 2000b).

This is not to say that the proposed statute was universally popular among scientists (e.g., Meighan 1992; see also discussion in Walker 2004:13–14), some of whom feared the loss of important archaeological collections and their rights to study them—nor in all probability among all Native Americans, especially individuals and groups without federal standing. However, NAGPRA was ultimately supported by the major scientific anthropology organizations, including the Society for American Archaeology and the American Association of Physical Anthropologists, whose representatives had lobbied hard on behalf of their preservation ethic even as they supported compromise on behalf of indigenous rights.

In its 1990 configuration, NAGPRA was a relatively balanced piece of legislation that represented a compromise between advocates for indigenous rights and advocates for science. Senator McCain, a member of the U.S. Senate Select Committee on Indian Affairs, committed this perspective on legislative intent to record at the time the law was passed:

> The passage of this legislation marks the end of a long process for many Indian tribes and museums. The subject of repatriation is charged with high emotions in both the Native American community and the museum community. I believe this bill represents a true compromise ... In the end, each party had to give a little in order to strike a true balance and to resolve these very difficult and emotional issues [*Congressional Record*, October 26, 1990, 17173].

The balance was effected in this way: the statute directed control over Native American/Native Hawaiian human remains found after its passage on November 16, 1990 to descendent and culturally affiliated groups on whose tribal or federally recognized aboriginal lands the remains were found, and facilitated claims for the repatriation of human remains in federally funded repositories in cases where lineal descent or cultural affiliation could demonstrated. The limits defined for repatriation claims thus provided the mechanism by which remains and cultural items for which lineal descent or cultural affiliation could not be demonstrated would be preserved for future scientific study and preservation.

The regulation of this mechanism for balance was also legislated through the formulation of an advisory committee comprising equal numbers of representatives from the Native American/Native Hawaiian and Science/Museum communities. In theory, this was a way to insure that all interested parties were represented in the decision-making process and to maintain the balance of interests that the statute was meant to implement. In reality, although nominees were solicited from these different communities, members were actually appointed by a governmental body (Department of the Interior), thus rendering the delicate balance vulnerable to political pressure and third-party intervention (Larsen and Walker 2004). Another unintended consequence of the committee's formulation and perhaps of the statute as well was that it codified two opposing camps—one spiritual, one secularist—and thus fostered the incorrect views that indigenous Americans did not support scientific study of indigenous human remains, and that scientists did not support indigenous interests and spiritual views. This was an unfortunate outcome for all concerned.

The linchpin of the balance that led to the passage of NAGPRA was the concept of cultural affiliation, which was to serve as a basis for decisions about which human remains in federal repositories were open to claims of repatriation. Cultural affiliation is defined in NAGPRA as "a relationship of shared group identity which can be reasonably traced historically or prehistorically between a present day Indian tribe or Native Hawaiian organization and an identifiable earlier group" (25 U.S.C. 3001, Sec. 2[2]). In the absence of clear evidence for such a relationship, NAGPRA specifies the basis for establishing cultural affiliation as "a preponderance of the evidence based upon geographical, kinship, biological, archaeological, anthropological, linguistic, folkloric, oral traditional, historical, or other relevant information or expert opinion" (25 U.S.C. 3005, Sec. 7[4]). However, there was from the start a lack of clarity concerning this fundamental term that has never been resolved and lies at the heart of legal and ethical debates concerning the meaning and intent of the statute.

An opportunity to refine the legal meaning of cultural affiliation and thus to clarify how aspects of the statute should be implemented in relation to scientific inquiry arose in the case of Kennewick Man, a legal battle over an ancient skeleton that pitted scientists and a group of citizens (Friends of America's Past) against five Native American tribes and bands of the Columbia River Basin. The Early Holocene (>8,000 years B.P.) male skeleton was found in 1996 eroding out of a riverbed along a stretch of the Columbia River in the state of Washington managed by the Army Corps of Engineers. Both scientific and Native American interest in the remains was heightened after James Chatters, a local archaeologist and forensic anthropologist, reported the remains as those of a "Caucasian" male (Preston 1997). Several Columbia River Basin tribes asserted claims over the remains under NAGPRA, and the U.S. Army Corps of Engineers concurred and hastened to repatriate the skeleton (Bruning 2006). This action led to a legal challenge by a group of scientist-plaintiffs comprising both archaeologists and biological anthropologists, whose members disputed the Corp's determination and sought permission to conduct further studies on the skeleton (Bonnichsen et al. v. United States et al. 969 F. Supp. 614, 618 [Dist. OR] [1997]:618). The Oregon District Court decided in favor of the plaintiffs, a decision upheld in the U.S. Court of Appeals for the Ninth Circuit, but not on the basis of a determination on cultural affiliation—which would have helped to clarify how these decisions are made for ancient remains under NAGPRA. Instead, the Court's decision rested on the argument that the Federal Government (U.S. Army Corps) had failed to prove that the skeleton was Native American and thus it was not subject to NAGPRA. The court authorized the study of Kennewick Man under the provisions of ARPA, the law that governs archaeological resources that fall outside NAGPRA's purview. In other words, the case failed to clarify how NAGPRA might be expected to govern the disposition and scientific study of ancient Native American remains (Bruning 2006). It was an important decision for science in the sense that it recognized the legal significance of the statute's language in terms of the intent of the law, but the case further polarized people with divergent religious and ethical views on the scientific study of human skeletal remains and on what it means to be Native American. It also set the stage for subsequent governmental intervention in the interpretation and implementation of the statute.

The first effort to do so emerged in 2004 with the proposal of Senate Bill 2843, Section 14, the Native Americans Technical Corrections Act of 2004. This bill proposed an amendment to NAGPRA that would add the words "or was" to the statutory definition of Native American, thus changing the statute to read "of or relating to a tribe, people or culture that is *or was* indigenous to the United States." The purpose of the amendment was both philosophical and practical: it would serve to rectify statutory language that upon interpretation in a court of law many in the Native American community felt was disenfranchising; it would include under the umbrella of NAGPRA all ancient human remains from the continental United States and thus mitigate if not nullify the legal impact of the Bonnichsen vs. the U.S. decision. The amendment was not enacted, however, and for a time human skeletal researchers in the U.S. continued to work in accordance with NAGPRA guidelines for remains associated with tribal lands and culturally affiliated human remains, and to conduct research on culturally unidentifiable remains under the guidelines of individual curation facilities.

In 2007, a new aspect to the ethical debate that has long surrounded NAGPRA emerged when the U.S. Department of the Interior (DOI) turned its attention to the question of "culturally unidentifiable human remains (CUHR)," the largest category of archaeological human remains in the United States. Culturally unidentifiable human remains are those remains that museums and federal repositories have not been able to affiliate with a lineal descendent or any modern Native American group based on the statute's definition of "cultural affiliation" and the lines of evidence outlined to make such determinations. These remains tend to be of greater antiquity than culturally affiliated remains and have from the time NAGPRA was drafted been at the center of scientific concern regarding the impact of repatriation law on the study of American prehistory. The ultimate disposition of this category of human remains was left open to future clarification in the 1990 statute and up to this time had received limited attention from the bureaucratic units set up to implement the statue. In 2007, the DOI promulgated the proposed CUHR rule (43CFR Part 10, Federal Register Vol. 72, No. 54, 13184–13189) to direct a course of action for repatriating these remains. A period of comment followed, during which a number of scientific and museum organizations, as well as many Native American groups, expressed concern over aspects of the rule, both practical and ethical. From a scientific perspective, many considered the proposed rule to be legally and ethically dubious because it violated the legal intent of NAGPRA—the balance that preserved American interest in the scientific study of archaeological human remains (e.g., see AAPA 2007).

The final rule for culturally unidentifiable human remains was enacted in May, 2010. A close reading of the comments section of this rule (Federal Register 75:49:12378–12405, March 15, 2010; see also AAPA 2010), which clarifies how the DOI responded to comments on the 2007 proposed rule, reveals a clear and consistent bias against scientific interests and concerns. In consequence, the rule appears to have set the stage for future legal challenges. Perhaps one of the most contentious elements of this rule is the narrow definition "disposition," a crucial term in the implementation of NAGPRA, to mean "transfer of control" instead of the more common meaning, "final arrangement." In consequence, the rule focuses on how the transfer of control is to be effected rather than on different possible final arrangements, which might include the retention of human remains in federal repositories. The rule includes several extralegal recommendations for how a museum or federal agency might dispose of culturally unidentifiable human remains, including repatriating them to unrelated "Indian tribes or Native Hawaiian organizations," or reinterring them without any Native American involvement if none of the former is willing to accept control over particular remains. Although NAGPRA does not mandate the nature of disposition for successful repatriation claims and by intention avoids the term "reburial," the CUHR Final Rule essentially calls for the reburial of all Native American human remains currently in museums and federal repositories, even in cases where evidence for a relationship of shared group identity is lacking. While this view is supported by some Native American and Native Hawaiian leaders and communities—and some anthropologists as well—it is contrary to the legal intent of the statute. Many American scientists and museum personnel who have been involved in the implementation of NAGPRA for many years are now in the difficult position of having to legally challenge a law that they supported in principle, for if they do not they will have abdicated their professional and ethical responsibility to protect the archaeological resources of America on behalf of all Americans, past, present, and future.

The purpose of discussing these laws is that they underlie the complex ethical dilemmas that many scientists who study Native American remains in the United States and elsewhere find themselves in: to protect archaeological resources for current and future research, and to respect and "do no harm" to people impacted by their research. This issue is far from unique to the United States, however (Larsen and Walker 2004). In recent years, repatriation issues related to human remains have also emerged in a number of other countries, including Argentina (Endere 2004), Australia (Aird 2002; Turnbull 2002; Wallis et al. 2008), Canada (McAleese 1998), Great Britain (Brothwell 2004; Payne 2004; Sayer 2009), Israel (Watzman 1996; Nagar 2004), Norway (Schanche 2004; Sellevold 2004), and South Africa (Fish 2004; Nemaheni 2004). In Australia, Canada, and the United States, the scientific collections at issue derive from minority indigenous populations where treatment of the dead figures prominently in larger issues of indigenous rights. Repatriation efforts in these nations have occurred on a broader scale and reflect national efforts to redress past wrongs associated with Western European colonialism. Similar historical situations also exist in countries such as Argentina, Norway, and South Africa, but to date repatriation efforts have been much more limited, often involving the return of specifically named, historically significant individuals (e.g., Endere 2004; Fish 2004; Sellevold 2004). The argument for the preservation ethic in repatriation disputes in some of these nations is complicated by the fact that whereas most of the human remains in scientific collections are indigenous—because these are the people who populated the prehistoric landscape—most of the scientists conducting the research are not. Until the ethnic composition of the scientific community shifts, this disparity will likely continue to influence perceptions of human skeletal research within these countries as biased towards study of the "other."

The repatriation experience is more diverse elsewhere in the world. In Israel, opposition to the excavation of human remains has come from the orthodox Jewish community, whose members presumably have no greater relationship to the archaeological dead than that of other Israelis. This community believes that the graves of the dead should remain undisturbed and has lobbied successfully for severe constraints on research involving archaeological human skeletal remains (Nagar 2004).

In Great Britain, considerable dialogue about repatriation/reburial has been inspired by historical connections to the three countries at the forefront of repatriation movements (Brothwell 2004; Sayer 2009; Smith 2004), but to date activity on this front has primarily related to Christian remains (Payne 2004; Sayer 2009). In most other parts of the world, human remains from archaeological contexts do not appear to have been a major source of contention to date, even in countries such as Mexico and Peru with oppressive colonial histories and ongoing human rights struggles. Perhaps it is the predominance of religious views that imbue the physical remains of the dead with a spiritual identity that predisposes people in some places to enjoin the dead in battles among the living—even when these religious views emanate from the majority population. It will be interesting to see where the repatriation movement is twenty years from now, whether it has spread and impacted the study of archaeological human remains globally, or seen a reversal due to changing political climates, religious views, or status of descendent populations. From a scientific perspective it is alarming to consider the possibility that we may look back on the late 20th century as the period in which all evidence of the biocultural history of some world populations was forever erased.

ETHICAL ISSUES IN THE USE OF HUMAN BONE TISSUE

The legal and ethical debates concerning repatriation and reburial have largely overshadowed other important ethical issues in the scientific use of human skeletal remains. One area in which these increasingly occur pertains to the sampling and use of human bone tissue. Over the past three decades, the study of human skeletal remains has entered a new dimension of analysis involving histology, molecular biology, and biochemistry, and in so doing has encountered ethical dilemmas more common in medical (Domen 2002) and genome research (Bathe and McGuire 2009). Foremost among these are issues surrounding bone sampling and informed consent. For archaeological specimens, the sampling process may simply represent an agreement between a curation facility and a researcher regarding which skeletal element(s) will be sampled, in what quantities, and for what purpose. The primary concern in these agreements is likely to relate to the preservation ethic; that is, does the research justify destruction of skeletal material and are the selected specimens most appropriate for the analysis to be undertaken. However, in countries such as the United States, where NAGPRA or similar legislation governs the disposition of many human skeletal remains, such agreements increasingly involve descendent groups who have a vested interest both in the information obtained and in what happens to samples after the research is completed (Kaestle and Horsburgh 2002). The need for informed consent may be more obvious and straightforward in studies where living subjects contribute DNA samples as part of the research. However, informed consent can be more complicated when the study exclusively involves ancient human skeletal remains. For example, a tribal group may have a particular belief or oral history about their place of origins, and ancient DNA (aDNA) research may be sought to validate this history. The researcher conducting the study may be aware that alternative outcomes that do not support traditional views are possible or even likely. In this case the principle of informed consent has more to do with potential outcomes than with how the sample will be used, and the researcher has an obligation to make the descendent group aware of the possibilities.

In their review article, "Ancient DNA in Anthropology," Fredrika Kaestle and K. Ann Horsburgh grapple with some of the ethical issues that arise in the study of ancient DNA. Based on their experience and that of others working in this fast-growing area of biological anthropology, they offer a series of questions that can serve to guide the formulation of an ethically sound aDNA research project:

1. Does the application of the method address an anthropological question?
2. Are there nondestructive methods that can be used to achieve the result?
3. Do the conditions of the remains or other material suggest aDNA is more likely to be present than not?
4. How will different stakeholders view the destruction of the remains in question?
5. What are the ELSI [Ethical, Legal, and Social Issues] implications of possible study results, if any, for living groups?
6. Has a reasonable attempt been made to define and receive informed consent from different stakeholders? [Kaestle and Horsburgh 2002:109]

These questions actually have much broader application in human skeletal research, and provide a good example of the kind of reasoning that should predicate the formulation of any research plan that calls for the destructive analysis of human skeletal tissue.

One additional question is what the final disposition of unused tissue samples will be. Often, a consent agreement will include specifications for how the sample can be used and what is to be done with unused portions. However, if additional research is planned or envisioned, and consent is a necessary component of accessing the samples, then this long-term use should be included in the agreement; otherwise, permission will need to be sought for new analyses on existing samples.

THE USE OF HUMAN SKELETAL REMAINS IN THE STUDY OF SENSITIVE TOPICS

An area that is seldom discussed in the context of ethics in the use of human skeletal remains and which is particularly relevant in paleopathology pertains to the study and reporting of "sensitive" topics, for example, certain diseases or specific types of violence. These can become an issue when the affected human remains have a relationship of shared group identity with living people, because the behavior or experiences of ancestral peoples can be seen to reflect on the descendants. If the behavior is judged by modern standards to be undesirable, inappropriate, or worse, then there may be concern on the part of scientists or descendants (even broadly construed) that findings will negatively impact the living, particularly in cases where negative stereotyping has historically played a role in discrimination.

A clear case in point is the study of treponematosis, one of the disease syndromes that paleopathologists regularly investigate, particularly in the Americas (see chapter 26). This is a disease with several distinct forms, three of which affect the skeleton with recognizable lesions. Two of the syndromes are endemic and, while these forms may cause disfiguring lesions, they do not generally have any significant social stigma associated with them. However, the third is venereal syphilis, a disease associated with sexual activity and in some social contexts with promiscuity and shame (Powell and Cook 2005). For this reason, there tends to be a cloak of silence surrounding this infection in the living, even in the current, Post-Antibiotic Era of this easily treated disease. It is this association of shame and expectation of privacy that makes the revelation of this disease in a skeleton potentially subject to ethical concerns, particularly in cases where a familial or cultural relationship exists between the living and dead. While the privacy accorded the dead is limited in forensic contexts, where information is needed to solve a crime, this justification cannot plausibly be cited in all research contexts. Biomedical studies among the living, whether of this disease or any other, are subject to ethical standards enforced by institutional review boards—but the dead from archaeological contexts generally fall outside of this regulatory process. A powerful justification for this distinction is that the dead provide a unique opportunity to study a disease such as syphilis unimpeded by modern social mores and the necessary protective constraints that come with working among the living. In this sense, human skeletal research skirts some of more serious ethical issues encountered in medical studies among the living and yet has the justification of contributing in practical ways to medical knowledge about disease progression and pathogen evolution.

The reporting of osteological evidence of human violence also has been seen as controversial in some contexts, although there appears to be no universal sentiment concerning what violent behaviors are appropriate, which are socially unacceptable, or what constitutes unethical reporting. Probably no subject in this regard has received as much attention, nor been questioned to the same degree as reports of body mutilation and cannibalism in the American Southwest. This is illustrated in dialogue surrounding Christy and Janine Turner's compendium volume *Man Corn*, which evaluates in detail the evidence for violence or cannibalism, or both, at 76 archaeological sites in the Anasazi culture region of the northern Southwest. In *Man Corn*, the Turners argue that the evidence for cannibalism is compelling at 38 of the 76 sites and conclude that "Southwestern cannibalism seems to be explainable, at least for the moment, by a hypothesis that combines social control, social pathology, and ritual purpose within the Chacoan sphere of influence" (Turner and Turner 1999:484). While reviewers have both lauded and criticized aspects of their research, Andrew Darling (1999:442) has taken the Turners to task on ethical grounds. Based on the Turner's (selectively quoted) conclusion that, "the majority of Chaco Anasazi cannibal episodes resulted from acts of violent terrorism... incited by a few zealous cultists from Mexico and their descendent followers," he chastises the authors for presenting a "speculative" interpretation that in his mind does "a disservice to the history of the descendants of the *Mexica* and the Anasazi...[and] betrays a lack of consideration of the anthropological literature that demonstrates that cannibalistic practice, whether real or imagined, can exist as part of coherent, and healthy, cultural systems." Darling's own view of the "cannibalism" assemblages is that some at least are remnants of socially sanctioned witch-killings, a perspective gleaned from the ethnographic record (Darling 1998) and one that he presumably considers more palatable to descendant groups in the social climate of the late 20th century. Both views contain elements of moral judgment, but also attempts by the researchers to deflect "charges" of cannibalism away from mainstream Puebloan culture.

The author gained personal experience with this debate during a collaborative research project in southwestern Colorado in the late 1990s, which provided ample opportunity to consider the ethics of reporting controversial findings (Billman et al. 2000; Lambert et al. 2000). At Cowboy Wash, a 12th-century site on the southern piedmont of Sleeping Ute Mountain, the field team discovered the disarticulated and heavily processed remains of seven people on the floors of two pit structures. The osteological evidence led the team to conclude that the villagers had met a violent death, during or after which their bodies were processed for consumption. When the results were published (Billman et al. 2000) a critique appeared alongside the primary article that challenged this interpretation on several levels (Dongoske et al. 2000a). As articulated by the authors, the effort was driven by a concern that: "Any characterization of Indians as 'cannibals' dredges up a long history of oppression and racism, and we have to realize that this may generate ill-will and negative emotions about archaeology in many descendant communities" (Dongoske et al. 2000a). Curiously, though, the authors invoked genocidal behavior as a legitimate and preferable alternative to cannibalism, much as Darling argued in favor of witch-killing.

The response of critics in both cases emphasized the Western European view of violent cannibalism as a socially unacceptable behavior. The problem with this stance was that if and when evidence came to light that showed that cannibalism had occurred in the Southwest, as it did (Marlar et al. 2000), then their well-intended efforts at deflection actually cast more negative aspersions on the behavior than any of the original research. For the author, at least, the lessons were these: do the best science possible in all contexts; report observations accurately and to the best of one's ability, without judgment. If anthropology teaches us anything it is that systems of morality are neither universal, nor unchanging. If we want to understand the full range of human behaviors, then we have to be willing to acknowledge and document what we see and to do our best to refrain from judging those behaviors according to modern value systems.

FINAL COMMENTS

The ethical issues covered in this chapter are not unique to paleopathology. However, the cross-disciplinary nature of modern human skeletal research requires a broader consideration of how ethics influence our research—both in terms of access to and handling of skeletal materials, and in the interpretation and reporting of results. One area that has been touched on but not explored in great detail here is the more subtle ways that ethics shape our studies. For example, in the U.S. wealthy and powerful people have traditionally been buried separately and in a distinct manner from the poor and working classes. In consequence their graves are seldom disturbed or excavated because, in addition to being better marked and sheltered, they are protected by an ethic of preservation and reverence for the dead that does not extend to the unmarked cemeteries and unidentified remains of the poor. Research on various aspects of health, disease, violence, work-load and life-expectancy during the historic period is therefore biased toward the experiences of the poor and disenfranchised in ways that may significantly impact our ability to address important research questions.

For example, the author encountered this problem when attempting to address questions concerning the biological impacts of slavery on 19th-century African Americans in northern North Carolina (Lambert 2006). The study focused on remains recovered from an unmarked cemetery during salvage archaeological work along the Roanoke River. The skeletal remains of both children and adults showed signs of extreme stress during the growth years—an expected finding given the social circumstances. However, no comparative data exist on the skeletal health of wealthy slave-owners from this area to provide a broader context for interpreting the results. There may have been mitigating factors that positively influenced the health of the enslaved relative to slave-owners. For example, foods such as collard greens and field peas that slaves grew in their small gardens were quite nutritious, whereas the diets of wealthy slave-owners may have emphasized sugar, starch, salt, and fats. In addition, given the absence of antibiotics and lack of knowledge about the causes of disease and disease transmission, the children of slave owners may not have fared much better in some respects than those of the enslaved (e.g., Thomas et al. 1977). However, the remains of the former are protected in churchyards and private family cemeteries and thus not available for comparative analysis. This situation is not limited to the United States, though in some cases it is the wealthy whose remains are studied because those of the poor are deemed unworthy of protection and study. Either bias influences our understanding of health and disease in the past in potentially significant ways.

One final area of ethics that warrants some discussion here relates to the rights of the dead, and this is a fitting question on which to conclude this treatment of ethics in the use of human skeletal remains for research. As Kaestle and Horburgh (2002) point out in their discussion of the consent process in aDNA research, the permission process presumes that those giving it are the most likely to know what the dead would want and that they do so in the best interest of the dead. This is not necessarily true, and it raises an unanswerable but important philosophical question about the rights of the dead to have their story told. This is an ethical dilemma that the author has struggled with many times in attempting to negotiate the complex path of NAGPRA, because her studies focus on warfare and, in particular, on the victims of war. In modern forensic contexts (e.g., Baraybar and Gasior 2006), the war dead are often accorded such rights and their bodies sought out to reveal the history that others have tried to hide. The forensic cases of war victims in places like Bosnia and Guatemala remind us that the stories of the dead are powerful and important, and that they belong to the people who lived them. Should the same not be true of people who lived in the more distant past? It is this awareness, coupled with the knowledge that human populations continue to face many of the same kinds of environmental challenges as ancestral humans and that we can therefore glean useful insights from studying human biological history that keeps so many scholars in the field despite the sometimes challenging political, legal, and ethical issues that can make doing research on human skeletal remains difficult or even impossible in some contexts. In the end, most of us would agree that ethical standards are important to the conduct of our research. If as individuals and groups we disagree on exactly what those standards should be, then as anthropologists we need to remind ourselves that it was a fascination with human diversity, past and present, that drew most of us into the field in the first place.

REFERENCES

AAA, 2009   AAA Code of Ethics. Electronic document. http://www.aaanet.org/issues/policy-advocacy/upload/AAA-Ethics-Code-2009.pdf.

AAPA, 2003   AAPA Code of Ethics. Electronic document. http://physanth.org/association/position-statements/code-of-ethics.

AAPA, 2007   American Association of Physical Anthropologists Position Statement on the Department of the Interior's Proposed Rule for the Disposition of Culturally Unidentifiable Human Remains. Electronic document. http://physanth.org/association/position-statements/NAGPRA%20AAPA%20CHUR%20Position%20Statement.pdf.

AAPA, 2010   AAPA Comments on 43CFR Part 10: Native American Graves Protection and Repatriation Act Regulations—Disposition of Culturally Unidentifiable Human Remains; Final Rule, Federal Register 75:49:12378 (March 15, 2010). Electronic document. http://physanth.org/association/position-statements/AAPA%20Comment%20on%20CUHR%20Rule%205-10-10-1.pdf.

Aird, M., 2002   Developments in the Repatriation of Human Remains and Other Cultural Items in Queensland, Australia. *In* The Dead and Their Possessions: Repatriation in Principle, Policy and Practice. C. Fforde, J. Hubert, and P. Turnbull, eds. pp. 303–311. One World Archaeology 43. New York: Routledge.

Augstein, H. F., 1996   Race: The Origins of an Idea, 1760–1850. Bristol: Thoemmes Press.

Baraybar, J. P., and Marek G., 2006   Forensic Anthropology and the Most Probable Cause of Death in Cases of Violations Against International Humanitarian Law: An Example from Bosnia and Herzegovina. Journal of Forensic Sciences 51:103–108.

Bathe, O. F., and McGuire, A. I., 2009   The Ethical Use of Existing Samples for Genome Research. Genetics in Medicine 11:712–715.

Billman, B. R., Lambert, P. M., and Banks L., 2000   Warfare, Cannibalism, and Drought on the Colorado Plateau in the Twelfth Century A.D. American Antiquity 65:145–178.

Brace, C. L., 2005   "RACE" is a Four-Letter Word: the Genesis of the Concept. Oxford: Oxford University Press.

Brothwell, D., 2004   Bring Out Your Dead: People, Pots and Politics. Antiquity 78:414–418.

Bruning, S. B., 2006   Complex Legal Legacies: The Native American Graves Protection and Repatriation Act, Scientific Study, and Kennewick Man. American Antiquity 71:501–521.

Bynum, C. W., 1997   Body-Part Reliquaries and Body Parts in the Middle Ages. Gesta 36(1):3–7.

Bynum, C. W., 1998   Death and Resurrection in the Middle Ages: Some Modern Implications. Proceedings of the American Philosophical Society 142:4:589–596.

Darling, J. A., 1998   Mass Inhumation and the Execution of Witches in the American Southwest. American Anthropologist 100:732–752.

Darling, J. A., 1999   *Review of* Man Corn: Cannibalism and Violence in the Prehistoric American Southwest. Latin American Antiquity 10:441–442.

Davies, J., 1999   Death, Burial and Rebirth in the Religions of Antiquity. New York: Routledge.

Domen, R. E., 2002   Ethical and Professional Issues in Pathology: A Survey of Current Issues and Educational Efforts. Human Pathology 33:779–782.

Dongoske, K. E., 2000   NAGPRA: A New Beginning, Not an End for Osteological Analysis—A Hopi Perspective. *In* Repatriation Reader: Who Owns American Indian Remains? D. A. Mihesuah, ed. pp. 282–291. Lincoln: University of Nebraska Press.

Dongoske, K. E., Martin, D. L., and Ferguson, T. J., 2000a   Critique of the Claim of Cannibalism at Cowboy Wash. American Antiquity 65:179–190.

Dongoske, K. E., Aldenderfer, M., and Doehmer, K., eds., 2000b   Working Together: Native Americans and Archaeologists. Washington, DC: Society for American Archaeology Press.

Endere, M. L., 2004   The Reburial Issue in Argentina: A Growing Conflict. *In* The Dead and Their Possessions: Repatriation in Principle, Policy and Practice. C. Fforde, J. Hubert, and P. Turnbull, eds. pp. 266–283. One World Archaeology 43. New York: Routledge.

Fish, W. S., 2004   "Ndi nnyi ane a do dzhia marambo?" – "Who Will Take The Bones?": Excavations at Matoks, Northern Province, South Africa. *In* The Dead and Their Possessions: Repatriation in Principle, Policy and Practice. C. Fforde, J. Hubert, and P. Turnbull, eds. pp. 261–265. One World Archaeology 43. New York: Routledge.

Frank, J. B., 1976   Body Snatching: A Grave Medical Problem. The Yale Journal of Biology and Medicine 49:399–410.

Garment, A., Lederer, S., and Roger, N., 2007   Let the Dead Teach the Living: The Rise of Body Bequethal in 20th Century America. Academic Medicine 82:1000–1003.

Gould, S. J., 1981   The Mismeasure of Man. New York: W. W. Norton and Co.

Kaestle, F. A., and Horsburgh, K. A., 2002   Ancient DNA in Anthropology: Methods, Applications, and Ethics. Yearbook of Physical Anthropology 45:92–130.

Kunstadter, P., 1980   Medical Ethics in Cross-Cultural and Multi-Cultural Perspectives. Social Science and Medicine 14B:289–296.

Laderman, G., 1996   The Sacred Remains: American Attitudes Toward Death, 1799–1883. London: Yale University Press.

Lambert, P. M., 2006   Infectious Diseases Among Enslaved African Americans at Eaton's Estate, Warren County, North Carolina, 1830–1850. Memorias do Instituto Oswaldo Cruz 101:(Sup.II):107–117.

Lambert, P. M., Banks L. L., Billman, B. R., Marlar, R. A., Newman, M. E., and Reinhard, K. J., 2000   Response to Critique of the Claim of Cannibalism at Cowboy Wash. American Antiquity 65:397–406.

Larsen, C.S., and Walker, P. L., 2004   The Ethics of Bioarchaeology. *In* Ethical Issues in Biological Anthropology. T. Turner, ed. pp. 111–119. Albany: State University of New York Press.

Little, M. A., and Sussman, W. W., 2010   History of Biological Anthropology. *In* A Companion to Biological Anthropology. C. S. Larsen, ed. pp. 13–38. Chichester: Wiley-Blackwell.

Marlar, R. A., Banks L. L., Billman, B. R., Lambert, P. M., and Marlar, J. E., 2000   Biochemical Evidence of Cannibalism at a Prehistoric Puebloan Site in Southwestern Colorado. Nature 407:74–78.

McAleese, K., 1998   The Reinterment of Thule Inuit Burials and Associated Artifacts: IdCr-14 Rose Island, Saglek Bay, Labrador. Etudes Inuit Studies 22:2:41–52.

McDonald, S. W., 1995   The Life and Times of James Jeffray, Regius Professor of Anatomy, University of Glasgow 1790–1848. Scottish Medical Journal 40:121–124.

Meighan, C. W., 1992   Some Scholars' Views on Reburial. American Antiquity 57:704–710.

Morris, I., 1992   Death-Ritual and Social Structure in Classical Antiquity. Cambridge: Cambridge University Press.

Nagar, Y., 2004   Bone Reburial in Israel: Legal Restrictions and Methodological Implications. *In* The Dead and Their Possessions: Repatriation in Principle, Policy and Practice. C. Fforde, J. Hubert, and P. Turnbull, eds. pp. 87–90. One World Archaeology 43. New York: Routledge.

Nemaheni, T. I., 2004   The Reburial of Human Remains at Thulamela, Kruger National Park, South Africa. *In* The Dead and Their Possessions: Repatriation in Principle, Policy and Practice. C. Fforde, J. Hubert, and P. Turnbull, eds. pp. 256–260. One World Archaeology 43. New York: Routledge.

Ortner, D. J., 1994   Scientific Policy and Public Interest: Perspectives on the Larsen Bay Repatriation Case. *In* Reckoning with the Dead: The Larsen Bay Repatriation and the Smithsonian Institution. T. L. Bray and T. W. Killion, eds. pp. 10–14. Washington, DC: Smithsonian Institution.

Payne, S., 2004   Handle with Care: Thoughts on the Return of Human Bone Collections. Antiquity 78:419–420.

Pearson, M. P., 2000   The Archaeology of Death and Burial. College Station: Texas A & M University Press.

Pellegrino, E. D., 1999   The Origins and Evolution of Bioethics: Some Personal Reflections. Kennedy Institute of Ethics Journal 9.1:73–88.

Powell, M. L., and Cook D. C., eds., 2005   The Myth of Syphilis: The Natural History of Treponematosis in North America. Gainesville: University of Florida Press.

Preston, D., 1997   A Reporter at Large: The Lost Man. The New Yorker, June 16, 1997:70–81.

Rankin-Hill, L. M., and Blakey, M. L., 1994   W. Montague Cobb. American Anthropologist 96:74–96.

Richardson, R., 2001   Death, Dissection and the Destitute. Chicago: University of Chicago Press.

SAA, 1996   Principles of Archaeological Ethics. Electronic document. http://www.saa.org/ AbouttheSociety/PrinciplesofArchaeologicalEthics/tabid/203/Default.aspx

Sawyer, R., 1964   Body Snatching in Britain. Marquette Medical Review 30:1:38–9.

Sayer, D., 2009   Is There a Crisis Facing British Burial Archaeology? Antiquity 83:199–205.

Schanche, A., 2004   Saami Skulls, Anthropological Race Research and the Repatriation Question in Norway. In The Dead and Their Possessions: Repatriation in Principle, Policy and Practice. C. Fforde, J. Hubert, and P, Turnbull, eds. pp. 47–58. One World Archaeology 43. New York: Routledge.

Schultz, S., 2005   Body Snatching: The Robbing of Graves for the Education of Physicians in early Nineteenth Century America. Jefferson, NC: McFarland and Company.

Sellevold, B. J., 2004   Skeletal Remains of the Norwegian Saami. In The Dead and Their Possessions: Repatriation in Principle, Policy and Practice. C. Fforde, J. Hubert, and P. Turnbull, eds. pp. 59–62. One World Archaeology 43. New York: Routledge.

Smith, L., 2004   The Repatriation of Human Remains – Problem or Opportunity? Antiquity 78:404–413.

Spellman, W. M., 1994   Between Death and Judgment: Conflicting Images of the Afterlife in Late Seventeenth-century English Eulogies. Harvard Theological Review 87:49–65.

Thomas, D. H., South, S., and Larsen, C. S., 1977   Rich Man, Poor Men: Observations on Three Antebellum Burials from the Georgia Coast. Anthropological Papers of the American Museum of Natural History 54, part 3.

Turnbull, P., 2002   Indigenous Australian people, Their Defense of the Dead and Native Title. In The Dead and Their Possessions: Repatriation in Principle, Policy and Practice. C. Fforde, J. Hubert, and P. Turnbull, eds. pp. 63–86. One World Archaeology 43. New York: Routledge.

Turner, C. G., and Turner, J. A., 1999   Man Corn: Cannibalism and Violence in the Prehistoric American Southwest. Salt Lake City: University of Utah Press.

Uchino, S., 1983   The History of Body Donation and Changes in the Mental Attitude of Donors. Annals of the Japanese Association for Philosophical and Ethical Researches in Medicine 15:157–166.

Various, 2009   Special Issue: Race Reconciled, American Journal of Physical Anthropology 139:1:1–107.

Von Staden, H., 1992   The Discovery of the Body: Human Dissection and its Cultural Contexts in Ancient Greece. The Yale Journal of Biology and Medicine 65:223–241.

Walker, P. L., 2000   Bioarchaeological Ethics: A Historical Perspective on the Value of Human Remains. In Skeletal Biology of Past Peoples: Research Methods. 2nd Edition. M. A. Katzenberg, and S. R. Saunders, eds. pp. 3–39. New York: John Wiley and Sons.

Walker, P. L., 2004   Caring for the Dead: Finding a Common Ground in Disputes over Museum Collections of Human Remains. In Documenta Archaeobiologiae 2: Conservation Policy and Current Research, Yearbook of the State Collection of Anthropology and Palaeoanatomy. G. Grupe and J. Peters, eds. pp. 13–27. Rahden, North Rhine-Westphalia: Verlag M. Leidorf.

Wallis, L. A., Moffat, I., Trevorrow, G., and Massey, T., 2008   Locating Places for Repatriated Burial: A Case Study from Ngarrindjeri ruwe, South Australia. Antiquity 82:750–860.

Watzman, H., 1996   Israel's Reburial Debate. Archaeology 49:5.

CHAPTER 3

Evolutionary Thought in Paleopathology and the Rise of the Biocultural Approach

Molly K. Zuckerman,
Bethany L. Turner,
and George J. Armelagos

INTRODUCTION

The definition of paleopathology has evolved over time through the field's many incarnations. Over the past 200 years it has developed from a pastime of Victorian physicians to an individualistic enterprise focused on clinical diagnoses and case studies, and, by the late 20th century, into its modern incarnation. Contemporary paleopathology investigates the evolution of human diseases, the dynamic interactions between human societies and infectious and noninfectious disease, and ways in which humans have adapted to changes in their environments. It provides primary evidence for the state of health of humans and their ancestors, is empirical and multidisciplinary, and, by combining biological, cultural, ecological, and epidemiological data, it has to a certain extent become wide-ranging and holistic (Roberts and Manchester 1999). Therefore, it seems obvious that evolutionary thought should be central to this enterprise. However, it is not; despite multiple urgings during the lifetime of the field, paleopathology continues to make little use of evolutionary perspectives.

In this chapter, we consider an evolutionary perspective as one in which evidence of disease and variation in its expression, as well as population-level patterns in disease,

A Companion to Paleopathology, First Edition. Edited by Anne L. Grauer.

trauma, and skeletal evidence of physiological stress, are interpreted in an adaptive context. This may include the principles of plasticity, natural selection, adaptation, gene flow and/or genetic drift. This approach also views evidence of stress and disease as products of environmental inadequacies, whether socially, economically, politically, or ecologically generated. As such, an evolutionary perspective grounds itself in consideration of the many social, cultural, environmental, and economic factors within and between populations that can significantly influence host–pathogen–environment relationships (Brown 1998). Differences in patterns of disease or stress between skeletal populations can thus be interpreted as indicators of the adaptiveness or maladaptiveness of cultural strategies in the past; the ways in which different cultural systems adaptively buffered against inadequacies, compromised between multiple threats, exposed their constituents to new ones, or exacerbated existing deficiencies.

Over the past century, paleopathology has become vastly more sophisticated in its methodologies and analytical techniques. It has drawn substantially on advances in clinical research to better understand the etiology and interaction of the range of pathological conditions documented throughout the archaeological record. This range—and the geographical areas, cultural contexts, and time periods in which they have been analyzed—has also expanded at an exponential rate. However, equivalent advances in evolutionary thought are nowhere to be found. This relative absence has been noted in several reviews and content analyses of published literature in paleopathology, and is discussed in depth throughout the following chapter.

Here, we trace the development of evolutionary thought within the scope of the emergence and maturation of paleopathology and survey its usage in contemporary research. As the biocultural approach has played a significant role in the development of paleopathology and evolutionary thought, we pay additional attention to its contributions. While we acknowledge and discuss the limitations and constraints involved in incorporating evolutionary thought and the biocultural approach into paleopathological research, we argue that doing so confers disproportionate advantages to both paleopathology as a whole and to contemporary societies.

HISTORICAL FRAMING: THE DEVELOPMENT OF EVOLUTIONARY THOUGHT AND THE BIOCULTURAL APPROACH IN PALEOPATHOLOGY

Several scholars have divided the development of paleopathology into four periods, each defined by their dominant research interests (see Ubelaker 1982). Briefly, the first (1774–1870) was characterized by a focus on pathologies in Quaternary fauna; the second (1870–1900) on documentation of traumatic lesions and syphilis in humans; and the third (1900–1930) on infectious disease and evidence for medical treatment in prehistory. The fourth phase (1930–present) has been characterized by a gradual development and adoption of ecological, evolutionary, and epidemiological approaches to the study of past disease. Angel (1981) overlays these with more conceptual divisions. According to this schema, the 19th century was characterized by purely descriptive research, the early 20th by analytic, and the period between 1930 and 1970 by increasingly synthetic and specialized foci. The decades between the 1970s and the present are characterized by an increasingly interdisciplinary and investigative orientation. While these phases remain relevant, to clarify the often confusing, meandering development of paleopathology and the slow emergence of evolutionary thought within, the following discussion is divided into the relevant decades, with some degree of overlap.

The 18th and 19th centuries

Many researchers regard the 18th and 19th centuries (the first and second periods) as the "dark ages" of paleopathology. Practitioners were primarily physicians and focused narrowly on craniology and osteometry. However, among these early studies, Wyman's (1868) work on cranial pathologies shows the beginnings of a comparative population-level approach. Among other contributions, Jones's (1876) work on the antiquity of acquired syphilis also marked the beginning of investigations into the origins and antiquity of the disease. Questions concerning the origin and antiquity of diseases remained a driving force in paleopathology, but later would serve as stimuli for the development of evolutionary thought (Cook and Powell 2006).

1900 to the 1960s

The third period exhibited a similar dearth of evolutionary thought, but with a few shining exceptions. Moodie (1923) and Williams (1929) contributed the first major syntheses of contemporary pathological evidence. Hooton (1930) also published his landmark population-level analysis of regional evidence of disease among the Pueblo. Hooton's study foreshadowed the tenets of the epidemiological approach in modern paleopathology, even though three decades would pass before similar methodologies became widely adopted. As indicated by this developmental lag, the theoretical advances of the fourth period did not achieve fruition until the late 1960s and early 1970s.

In the first formal evaluations of the field, Jarcho (1966: 28) and Brothwell and Sandison (1967) called the 1930s through the mid 1960s a "doldrums." All three characterized research during this period as bereft of theoretical and methodological innovations and overwhelmed by descriptive and typological approaches. Case studies dominated the literature, presenting only particularistic diagnoses of singular lesions and disease episodes in solitary skeletons. The small remainder of publications consti-tuted socially, spatially, and ecologically decontextualized collections of raw data (Larsen 1997). Overall, according to Armelagos and colleagues (1982), the cumula-tive intellectual goal of the field was to use these studies to chart the prevalence of a limited list of conditions and thus demonstrate the evolution of the modern disease burden, but few researchers gave any consideration to the role of ecological or social factors in this trajectory. Instead, the overwhelming majority of publications isolated their biological data from the contextual archaeological data. They also gave little consideration to the role of culture in mediating disease occurrences (Cook and Powell 2006). Mays (1997) and others attribute this to the medical or anatomical backgrounds of most contemporary researchers, leaving them without the knowledge—or desire—to explore the wider implications of their work.

While most paleopathologists persisted in this static state into the 1960s, significant paradigmatic shifts were occurring throughout the rest of the field. These formed the foundations of evolutionary thought and the biocultural approach in biological anthropology and, derivatively, in paleopathology. This shift was largely due to a declining interest in pure typological description and racial classification in the social and natural sciences and movement towards population-level analyses. The reverberating effects of Nazism and institutionalized, genocidal racism in Germany in the 1940s contributed; the horrifying applications of a race-based, pseudo-evolutionary eugenics paradigm encouraged many post-war scientists to reorient the ways that in which they conceptualized and applied their methods to societal issues (Blakey 1987). More significantly, the "Origins and Evolution of Man" symposium, at the Cold Springs Harbor Biological Laboratory, and its corresponding proceedings (Warren 1951), effectively introduced the populational approach to biological anthropology and human disease ecology, which was a recent adoption from population and evolutionary biology (Armelagos et al. 1982; Smocovits 1996). Using the population as the unit of study allowed researchers to move beyond the clinical perspective in order to consider diseases in their broader biological and social context (Armelagos et al. 1978). Shortly afterwards, Washburn (1951; 1953) introduced the "new physical anthropology." He proposed a strategic redirection from existing modes of research in the field, characterized as anthropometry and "speculation," towards synthetic, theory-driven research motivated by hypothesis testing, and premised on models of evolution and adaptation.

Spurred by Washburn and the novelty of the populational approach, some anthropologists eagerly—and successfully—adopted an evolutionary paradigm in the 1950s (Haraway 1990). Others were less successful, like Boyd (1950), who attempted to apply it to the new method of blood-group analysis, but ultimately only propagated a generation's-worth of typological reconstructions of cultural and genetic relationships between human populations (Armelagos et al. 1982). For the most part however, the discipline—and paleopathologists in particular—neglected these paradigms for decades (Armelagos 2003). Some carried on the legacy of Hooton, Moodie, and Williams, but overall the field remained mired in a particularistic concern with single lesions and the limited range of conditions these could be assigned to using contemporary medical technology. The result was that only a very limited range of inferences were made about past peoples and their societies for several decades (Buikstra and Cook 1980).

The 1960s

The 1960s marked the slow emergence of modern paleopathology and in turn, the first strong signs of both biocultural and evolutionary thought. In three nearly synchronous re-evaluations, Angel (1981), Ubelaker (1982), and Buikstra and Cook (1980) stated that this decade marked a renaissance in the field. Several significant paradigmatic shifts very gradually transformed paleopathology from a descriptive enterprise to an interpretive, interrogative, and independent one that is increasingly focused on ecological, epidemiological, and, finally, evolutionary considerations (Ortner 1991).

Stimuli came from a variety of directions, primarily from developments in related fields. They include the development of processual archaeology, the emergence of ecological and biocultural approaches, and most importantly, recognition of skeletal stress indicators. In archaeology, the development of processual archaeology, or "new archaeology," in the 1960s had a progressively transformative effect on paleopathology. Unlike previous paradigms in archaeology that were rooted in description and studies of cultural diffusion, processual archaeology adopted an empirical and ecological approach for investigating the adaptive relationship between cultural systems and their environments (Binford and Binford 1968). Processual archaeologists emphasized hypothesis-testing and creation of general principles of adaptation that could be applied to both archaeological and contemporary cultures. In paleopathology, this translated to an interest in ecological perspectives, process, hypothesis-testing, and regional-level analysis (Armelagos and Van Gerven 2003). As in bioarchaeology, this linkage between archaeological and skeletal analyses has made it possible for paleopathologists to address some significant questions on adaptation in past populations. These originally included investigations on a regional level, but have progressed to analyses of the effects of trade, agriculture—which will be discussed below—and culture contact on past health, as well as informative comparisons of multiple populations (Armelagos 2003).

The emergence of the ecological approach in the 1960s within anthropology also granted paleopathology a nascent emphasis on environmental context and on interpreting evidence of health and disease as indicators of the biological outcomes of adaptive processes. The ecological approach conceptualizes the cultural, biological, and physical components of humans' environments as an integrated whole which influences human biology and behavior (Moran 1982). It was conceived of in Livingstone's (1958) seminal paper, which is also regarded as a landmark in the development of the biocultural approach (Armelagos 2008). This paper unraveled the linkages between population growth, subsistence strategy, and distribution of the sickle cell trait, which grants a heterozygotic advantage against malaria, in West Africa. Livingstone demonstrated that selection for the trait—and thus gene frequencies—was greatest in areas with high densities of mosquito malarial vectors. Mosquito densities had been exacerbated by swidden agriculture in these areas, making them inhospitable to humans, but, eventually, selection for the heterozygotic form allowed groups to reutilize these areas. Importantly, this was one of the first anthropological studies to conceptualize the "environment" as more than external physical conditions (Dufour 2006). Inspired, researchers produced a slew of studies on "human adaptability": genetic adaptation, nongenetic acclimatization responses, and phenotypic plasticity to various environmental and socially influenced challenges in human populations. In doing so, a materialist, holistic approach was generated that bridged the subdisciplines. This lent biological anthropology an unprecedented four-field orientation and generated a commitment to an adaptive and evolutionary perspective, often in a cross-cultural setting (Goodman 1998; Goodman and Leatherman 1998). Skeletal biology and paleopathology provided longitudinal depth to understanding the adaptive process (Armelagos and Van Gerven 2003).

However, neither approach had a profound, immediate effect on contemporary research in paleopathology nor the development of evolutionary thought in the field; many researchers simply rejected them. While contemporary observers may have spoken of a "processual biological anthropology," Lasker (1970) and Armelagos and Van Gerven (2003) argue that paleopathology remained mired in typology and descriptive historicism throughout the 1950s and early 1960s.

Instead, perhaps the two greatest influences on the mid-century renaissance and rise of evolutionary thought were development of the biocultural approach and, more importantly, recognition of skeletal stress indicators. The biocultural approach arose within anthropology between 1960 and the 1980s, from the ecological approach and the "new" physical anthropology and archaeology (Goodman and Leatherman 1998). Definitions of the approach have changed over time, but overall it explicitly emphasizes the dynamic interaction between humans and their larger social, cultural, and physical environments. Human variability is viewed as a function of responsiveness to environment factors that both mediate and produce each other. Effectively, biology and culture are held as dialectically intertwined (Levins and Lewontin 1985). Perhaps the greatest distinction between this paradigm and others in biological anthropology is that it explicitly considers social and cultural components of the environment in regards to human adaptation. For health and disease, the biocultural approach embraces the idea that a society's technology, social system, and even ideologies could generate biological outcomes, such as patterns of disease (Zuckerman and Armelagos 2010a).

This new approach found fertile, though limited, ground in paleopathology, especially among those researchers promoting evolutionary thought (Armelagos 1997). Many studies focused on functional (adaptive) anatomy and morphology and biomechanics, such as the effects of undernutrition on skeletal morphology, arthritis, and trauma. Behavioral interpretations of these pathologies were also often explicitly generated from culturally specific ethnohistoric accounts (Cook and Powell 2006). They were united by a central concern with evolutionary processes; bioculturally oriented paleopathological research focuses on examining patterns of pathologies in order to elucidate the effects of social, ecological, and political processes on health within and between populations. In contrast, more traditional, descriptive paleopathological research places the emphasis on questions about the presence, absence, or degree of a given pathology in a given context (see Goodman 1998).

Nonetheless, despite the obvious conceptual and theoretical benefits conveyed by the biocultural perspective, as the following sections demonstrate, it has been little utilized in the field. There are diverse reasons for this under-utilization, but the consequences have been clear. Perhaps the most important is that the version of the biocultural approach employed in paleopathology is generally far less sophisticated than that in bioarchaeology and biological anthropology (Goodman and Leatherman 1998; Zuckerman and Armelagos 2010a). For example, from the 1970s to the 1990s, critiques of biocultural and ecological approaches in anthropology arose from new perspectives of political-economy and processual ecology. These new perspectives posited that biocultural and ecological research reflected an "adaptationist program" and tendency to naturalize social processes, rather than viewing social processes as products of larger social and political processes—a weakness common to functionalist, evolutionary perspectives (Gould and Lewontin 1979). In biological anthropology and, soon afterwards, bioarchaeology, these concerns contributed to a biocultural approach that was engaged, action-oriented, and concerned with issues of power, social relations, and adaptive constraints (Goodman and Leatherman 1998). While this trend had a transformative effect in bioarchaeology, producing a continuing emphasis on framing hypotheses within political, social, and economic contexts and issues, it had little effect on the development of biocultural or evolutionary thought in paleopathology. Even in the 1970s and 1980s, Armelagos et al. (1982) continued to make dire prognostics about the state of the field. Despite the increasing sophistication—and applicability—of the biocultural approach and other available theoretical and methodological developments, much of contemporary paleopathological research continued on a blinkered, descriptive path rather than engaging with biocultural, evolutionary or processual models.

The most powerful influence on the mid-century renaissance and rise of evolutionary thought was the recognition of physiological stress indicators on the skeleton (Buikstra 1991). This topic has also received the most bioculturally oriented attention. After first being recognized in the 1970s (Steinbock 1976), skeletal stress indicators were integrated into a systemic "stress concept" gleaned from human biology (Selye 1950). From there, Goodman and colleagues have incorporated them into an increasingly sophisticated set of models for estimating interactions between host resistance, cultural systems, and environmental and sociopolitical stressors (see Goodman and Martin 2002). Stress indicators can be used to gage the success of a society's cultural and biological adaptation. Overall, Cook and Powell (2006) and others have argued that this development revolutionized paleopathology. Research began to move away from an individualistic attention to single disease episodes and specific diagnoses towards an anthropologically oriented, longitudinal, and evolutionary attention to population processes.

The 1970s and 1980s

Spurred by the critiques of Jarcho and Brothwell and Sandison, by the 1970s and 1980s a few researchers had begun to generate large-scale, epidemiological, and culturally and ecologically relevant interpretations of diachronic trends in health (Larsen 1997). These studies integrated archaeological, ethnographic, historical, and skeletal data to interpret patterns of stress, nutrition, trauma, disease, and activity-related pathologies within an adaptive framework. This development was also faciliated by dramatic improvements in dating technologies, the recovery of large skeletal samples with high-quality temporal-spatial documentation, and a growing recognition of the futility of attempting to diagnose specific diseases in the skeleton (Ubelaker 1982). Much of this work focused on samples from the New World and Sudanese Nubia; studies of these regions continue to have a great influence on the development of evolutionary thought (Cook and Powell 2006). One of the period's dominant foci was the use of stress indicators, and evidence of trauma, nutritional deficiencies, and infectious disease to examine the effects of agricultural intensification in prehistory on human health. This enterprise began with Cohen and Armelagos's (1984) collaborative work. It continues as a major stimulant of evolutionary analyses in paleopathology as well as a source of continued debate (see Gage and DeWitte 2009). The standardization of data and use of large samples in Cohen and Armelagos's (1984) volume allowed for unprecedented cross-cultural comparisons of diachronic trends in morbidity and mortality. Large-scale, diachronic epidemiological analyses consequently surged in the late 1980s, only to fall by the wayside in the 1990s. They have made a comeback in the early 2000s, revived in part by the Global History of Health Project for the western and eastern hemispheres and the first related volume, Steckel and Rose's (2002) *The Backbone of History*.

Paleopathology in the Modern Era: The Current State of Evolutionary and Biocultural Thought

Content analyses of paleopathological research over the past two decades present an ambiguous picture of the state of bioculturally oriented research and evolutionary thought. For example, Hens and Godde's (2008) analysis of papers published in biological anthropology's flagship journal, the *American Journal of Physical Anthropology* (*AJPA*), between 1980 and 2004, showed that many authors use processual or evolutionary frameworks in their work. They directly attributed this to Armelagos et al.'s (1982) critique of existing descriptive paradigms in the field and call for an increase in analytical biocultural, processual, or evolutionary studies. However, Park et al.'s (2010) analysis of paleopathological papers published in the *AJPA*, the *Journal of Archaeological Sciences* (*JAS*), and the *International Journal of Osteoarchaeology* (*IJO*) by British researchers between 1997 and 2006, and Mays's (1997; 2010) of osteoarchaeological (including paleopathological) papers published by British, American (USA), German, and Japanese researchers in the *AJPA*, *IJO*, *JAS*, and several German and Japanese journals between 1991 to 1995 and 2001 to 2007 do not report such optimistic findings. Park and colleagues and Mays found no increase in the use of either biocultural or evolutionary approaches in the surveyed papers. Instead, the proportion of studies employing these strategies seems to have remained static, maintaining equivalent proportions with methodological and case study-based papers. Within this limited segment however, evolutionary and biocultural research has advanced considerably in sophistication and scope.

Research continues on a variety of topics of long-standing interest. For example, new evidence and interpretations have been added to the productive debate over the evolution of cancer. Researchers continue to push towards the overall goal of relating frequencies of neoplasms and other environmentally related health conditions, such as growth arrest and metabolic disease, to environmental quality and the rise of modern environments. Following in the steps of Cockburn et al. (1980), investigations also continue on mummies, soft tissue pathologies, and the evolution of skeletally-invisible conditions such as cardiovascular disease. Likewise, work continues on the evolutionary trajectories of several conditions of long-standing interest to paleopathologists, such as leprosy, brucellosis, osteoporosis, rheumatoid arthritis, tuberculosis, and the treponematoses. Uniquely, many of these studies employ an integration of skeletal and biomolecular evidence. This has enabled researchers to make novel inferences about the phylogeny and evolutionary trajectory of several of these conditions, though sometimes with varying levels of success (see de Melo et al. 2010; contra Harper and Zuckerman 2010). Many previously less-studied conditions are also receiving increased attention. These include smallpox, Lyme disease, and following in Ruffer's (1914) tradition, a variety of parasitological infections, like Chagas disease and schistosomiasis.

Methodological advances

Many of these new investigations of the evolution, phylogeny, and ecology of both long-studied and relatively unexamined conditions have been made possible by innovative uses of biomolecular and biochemical techniques over the past two decades. Innovations in stable isotopic reconstructions of both individual and population-level dietary profiles have been used to augment pathological data, contributing to multifocal studies of noninfectious diseases. Isotopic analyses have been used to document past diets and mobility patterns throughout the course of human history, including that linked to cultural or behavioral evolution. Recently, they have also been used to broadly assess causal etiologies in conditions such as porotic hyperostosis, osteoporosis, and linear enamel hypoplasia. Biochemical and histological techniques have been used to detect evidence of biochemical prevention and treatment for various conditions, and, arguably, to improve diagnoses for various diseases, including anemia and syphilis. The recent development of polymerase chain reaction (PCR)-related techniques for identifying ancient DNA (aDNA) has allowed the mapping of the complete genome of several ancient pathogens, such as leprosy, the treponematoses, and tuberculosis, with attendant insights into their evolutionary trajectories (see Chapter 8 by Spigelman et al., this volume). While current research indicates that the treponematoses cannot be studied using aDNA—though not for lack of trying—characterization of aDNA from mummified tissues has also been critical to elucidating the evolution of *Mycobacterium tuberculosis* and *M. leprae*. Advances in aDNA analysis have also enabled characterization of the causal organisms for several of the world's greatest epidemics, such as the Black Death, the Plague of Athens, and the 1918 Spanish Flu. However, while studies of influenza virus RNA have given insights into its virulence and the potential for future pandemics, the accuracy of results for the Black Death and the Plague of Athens remain subjects of debate.

Paleoepidemiological and paleoecological studies of the evolution of parasitism are particularly notable for their frequent and sophisticated usage of the biocultural approach and evolutionary thought. These studies have placed paleoparasitology and archaeoparasitology at the interface of a wide array of studies, including climate change, human evolution and disease ecology, and evolutionary processes (Dittmar 2009, and Chapter 10 by Dittmar et al., this volume). Researchers have adopted methods, like aDNA analysis and enzyme-linked immunosorbent assay (ELISA), from other fields to trace the evolutionary trajectory of various parasites and infer the subsistence strategies, behaviors, and levels of mobility that have selected for and against these colonists at different times in history (Kloos and David 2002; Gonçalves et al. 2003). Studies have examined the origin and dispersion of Old World parasites between Africa and Europe; the effect of social, ecological, and economic factors on the history of malaria in Rome; the paleobiogeography of parasites in relation to ancient human migrations; the geographic and temporal origin of visceral leishmaniasis and head lice; and the ecological evolutionary origins of humans and nonhuman primates (see Dittmar 2009).

The concentration of evolutionary thought in methodologically advanced studies reflects a larger phenomenon of increased numbers of method-focused work within paleopathology. Several content analyses, like Park et al. (2010) and Mays (1997, 2010, and Chapter 16, this volume) have noted this trend. More specifically, Mays found a striking, statistically significant increase in the number of methodological papers on bone chemistry, primarily isotopic reconstructions and aDNA, published in the 2000s in comparison to those from the 1990s. Both attribute this to the increasing sophistication, applicability, and popularity of both techniques, the latter of which is still in its infancy (Roberts and Ingham 2008). It may also be a response to Ortner (1991) and Ubelaker's (2003) calls for greater technological attention to differential diagnosis in the field (Park et al. 2010). Park and colleagues also question the role of funding opportunities in this trend; according to Roberts and Cox (2003: 266), funding for skeletal studies is often directed towards "high tech, media attracting research, which is exemplified by the often media-friendly findings of aDNA and isotopic" studies. However, rather than being used to generate *novel* research questions, both techniques—like many other methodological advances in the field—are too often used to address *existing* questions. They are primarily employed to circumnavigate the limitations of skeletal evidence, such as the relative insensitivity of bone to most diseases (Ortner 1992). For example, aDNA has been used to resolve the long-standing question of the historical decline of leprosy, a series of mid-20th-century debates over the New World Pre-Columbian presence of tuberculosis, and address the centuries-old question of what caused the Black Death. While Park and colleagues believe that this trend will undoubtedly contribute to the development of new research projects, generating more nuanced epidemiological data, we are not so certain. As the history of skeletal biology shows (Armelagos and Van Gerven 2003), novel methodologies do not necessarily produce novel analytical research questions.

Theoretical advances

Research continues on both novel and established questions as well as the application of skeletal material to established queries in other fields. For example, work continues on the concept of epidemiological transitions. Citing the "Old Friends" hypothesis (Rook 2007; 2009), Armelagos and Harper (2010) have emphasized the importance of heirloom pathogens such as lice and malaria, and souvenir zoonoses, such as trypanosomiasis, in shaping the evolution of human immune systems and habitats. Under the persistent influence of processual archaeology, many researchers are also doggedly pursuing an improved understanding of the etiology and interpretive significance (plasticity vs. adaptation) of stress indicators, particularly linear enamel hypoplasias (LEH), porotic hyperostosis (PH), and cribra orbitalia (CO).

The formation and significance of LEH have been reviewed extensively elsewhere (Goodman and Rose 1990), but are generally understood as disruptions in enamel striae that occur during tooth crown formation in response to systemic physiological stress. As this can be induced by several factors, they serve as nonspecific stress indicators. Several studies have demonstrated a clear link between frequencies of LEH and markedly reduced longevity, both those formed during uterine development and in the post-natal period. This suggests that, rather than benign indicators, LEH represent stress-induced growth disruptions with long-term health effects (Duray 1996). There are several possible explanations for this association, none of which are unambiguously supported. Individuals with LEH and increased frailty (age standardized relative risk of death) may have a pre-existing inherent susceptibility to biological insults (Rothman and Greenlander 1998), or this may indicate differential, life-long patterns of social, cultural, and behavioral exposure to stressors; individuals with defects experienced high levels of both early and later-life stress. Lastly, their higher frailty may be due to "biological damage" to the immune system in early life, resulting in greater susceptibility to later life insults (see Armelagos et al. 2009; Duray 1996). Armelagos and colleagues have argued that this explanation provides support for the Barker Hypothesis (i.e., the Developmental Origins of Health and Disease Hypothesis), which holds that fetal and early-life environments have an impact on adult morbidity and mortality. Since its inception in the early 20th century this hypothesis has gained increasing support and attention in studies of mortality and chronic disease among contemporary humans; Armelagos and colleagues suggest that it can be further supported and explored in studies of those in the past as well. This novel approach to a well-established hypothesis demonstrates the potential of using paleopathology to measure differential survival within a framework of developmental phenotypic plasticity in the past as well as to contribute productively to discussions of the phenomenon in modern populations.

PH and CO have also been subject to decades-long debate over their etiology and implications for understanding human plasticity. Both CO, porosity on the orbital roofs, and PH, porosity on the cranial vault, have long been viewed as evidence of bone marrow hypertrophy in response to chronic anemia during childhood. However, much like LEH, they are not specific and have thus been attributed to various types of anemia and non-anemia-related etiologies, including, most recently, waterborne parasite-induced hemoblastic anemia (see Walker et al. 2009). Understanding the etiology of these lesions is critical as they are nearly ubiquitous in skeletal samples from around the globe.

In the 1990s, a provocative application of evolutionary thought to the etiology and implications of these lesions sparked prolonged debate in paleopathology. Stuart-Macadam (1992; 1998) and others posited that the causal hypertrophy was an adaptive response to parasitic infection. This was premised on the idea of iron withholding by hosts to reduce its bioavailability to pathogenic microbes. The resulting anemia was a trade-off, with the ultimate benefit being a reduction in host pathogen burden. This embodies many of the pitfalls of a functionalist, ecological approach: an adaptationist naturalization of social processes. Holland and O'Brien (1997) disputed this, arguing that the lesions signified the presence of a chronic condition with substantial biocultural consequences for long-term function, immunocompetence, and cognition. The crux of the debate therefore, was not only lesion etiology but also the interpretive weight given to evidence of plasticity: short-term survival or long-term survival and reproductive success. Similar critiques have been raised by other scholars; Goodman (1994) posited that the concept of adaptation (*vs.* maladaptation) should not apply merely to survival, but to larger issues of suffering and overall health. In essence, the authors argued that analyses of survival have to incorporate a consideration of physiological well-being and quality of life; that living is about more than simply not dying.

These debates, though contentious and even rancorous at times, stand as excellent examples of where evolutionarily oriented research in paleopathology—and arguably, the field itself—should move towards. They draw upon diverse skeletal samples from multiple contexts and interrogate population-level data against an in-depth understanding of modern, clinical research. They also engage directly with issues of theoretical interpretation in regards to physiological adaptation, such as selection for immunocompetence, differential survival, and function in short-term and long-term perspectives. While there are many virulent debates in paleopathology, few of them engage so thoroughly with these broader microevolutionary threads. Nor do they have so much relevance to understanding the evolution of human health or health and plasticity in modern human populations.

CONSTRAINTS, PITFALLS, AND LIMITATIONS TO EVOLUTIONARY THOUGHT AND THE BIOCULTURAL APPROACH

One of the explanations for this shortfall is that applications of evolutionary thought can be fraught with potential confounders and theoretical, methodological, and material concerns. The same applies, though to an arguably much lesser extent, to the biocultural approach. Some of these are unique to the field but others are more general in scope. The following discussion highlights and expands upon several of these limitations and challenges in paleopathology.

General limitations

Perhaps the most fundamental limitation to any analysis of evolution in human populations, living or ancient, is that imposed by human life-history patterns. The prolonged period between birth and age at reproduction, the length of interbirth intervals, and the relatively low number of offspring produced over individual lifespans means that tracking changes in allele frequencies in populations of humans is much more difficult than among fast-reproducing species (Mace 2000). Recent advances in demonstrating evolutionary change in human populations within the past 10,000 years underscore the importance of evolutionary processes in analyzing human physiology throughout history. However, they also highlight the difficulty in observing processes such as natural selection or genetic drift in this context through examples of exceptionally strong positive selection on identified mutations or theorized selection on highly polymorphic systems such as the human leukocyte antigen (HLA) system. Studies of other species with faster reproductive rates and smaller genomes (Nevo 1997; Gorbunova et al. 2008) often show subtle patterns of allelic frequencies; it is therefore unreasonable to expect that all evolved traits in human populations would exhibit such highly visible and quantitatively measurable patterns.

Theoretical and material limitations

Some of the most fundamental limitations are those imposed by the unique characteristics of skeletal samples derived from the archaeological record; they impede evolutionary and biocultural analyses in any field that employs this type of material. These include sampling biases, or biases generated by differential preservation (taphonomic) of various skeletal elements; biological mortality biases; persistent difficulties in accurately aging skeletons; and the variable representativeness of skeletal samples to their original living populations. These fundamental concerns remain sources of active and continuous investigation in paleopathology as well as bioarchaeology, paleodemography, and paleoepidemiology (see Wright and Yoder 2003).

One of the most fundamental limitations in paleopathology is that imposed by the fact that very few diseases manifest on the skeleton, which limits the number of diseases that can be studied in an evolutionary context (Dustugue 1980). Many infectious diseases act too acutely to affect the skeleton. This means that most traditionally studied conditions are chronic, though as discussed above, rapid advances in biomolecular techniques are gradually granting researchers access to macroscopically invisible conditions.

Evolutionary analyses of specific conditions that do manifest macroscopically are also often confounded by diagnostic uncertainty. In part, this is due to the skeleton's limited physiological responses to insult; bone morphology is affected only by resorption and proliferation of bone, inhibition of bone formation, and by abnormalities in its growth and formation. This means that different diseases can affect the skeleton in very similar ways. Many closely related conditions, such as the treponematoses, generate very similar, overlapping sets of skeletal pathologies, effectively confounding differential diagnosis. Differential diagnosis can also be impeded by a host of other factors, such as whether the lesions were active or healing at the time of death, by the individual's immune response to the causal pathogen, the severity and duration of the condition, and the presence of coexisting pathologies. This issue has greatly impeded evolutionary analyses of specific conditions. For example, despite great effort by many researchers, it still remains impossible to differentiate between the treponematoses in dry bone (Ortner 2003). Until recent advances in phylogenetic analysis, this has obfuscated analyses of the evolutionary history of the treponematoses and resolution of one of the greatest historico-scientific questions of all time: the origin and antiquity of syphilis. While the emphasis placed on improving differential diagnosis by several researchers (Ortner 1991; Ubelaker 2003) has likely facilitated the current methodological bloom in the field, whether this has also facilitated the use of evolutionary thought remains yet to be seen.

Methodological and analytical issues

Evolutionary analyses have also been impeded by methodological and analytical considerations. First and foremost among these are issues surrounding the collection, presentation, and interpretation of data. For example, in their discussion of factors that circumscribe the use of evolutionary thought in paleopathology, Ortner (1991) and Ubelaker (2003) included the need for increased standardization in data collection and of criteria used in differential diagnosis, more objective descriptive terminology, and improvements in both the capabilities of differential diagnosis and in the ability to assess the individual and population-level consequences of disease. The following discussion highlights and expands upon these concerns as well considering several other related challenges to use of evolutionary thought and the biocultural approach.

Paleopathology has a long—though hopefully now receding—history of employing nonstandardized forms of data collection and variable and ambiguous diagnostic criteria. The constraints generated by this fatal flaw came to a head in the 1990s: in a series of brief but influential evaluations of the field, Ortner (1991; 1992; 1994) argued that its lack of methodological and theoretical rigor—particularly the absence of standardized diagnostic criteria and data collection—had effectively halted any potential theoretical advances, particularly in evolutionary thought. Guides for standardized data collection began emerging in the 1980s and 1990s, particularly in the U.K., but were forced mostly by circumstances rather than by endogenous attempts at increased empiricism: the Native American Graves Protection and Repatriation Act (NAGPRA) and the looming repatriation of entire North American skeletal collections in the 1990s (Rose et al. 1996). Published standardized diagnostic criteria also lagged behind the field's development; while a steady stream has emerged since the 1970s, standardized criteria are not uniformly employed (Miller et al. 1996) and even now many often risk being self-referential and biomedically uninformed (Cook and Powell 2006).

At least until the mid 1990s, this meant that researchers largely could not make meaningful comparisons between data collected from different contexts—or even by different researchers (Roberts 1994). For Ortner (1991), this effectively precluded them from asking—and answering—fundamental questions about human microevolution and adaptation, the biocultural implications of disease, and the evolutionary trajectory and ecology of infectious diseases. Armelagos (1994) called these pronouncements fatalistic and uninformed and responded with a host of historical and contemporary evolutionarily themed publications in paleopathology. However, nonstandardization has been a major impediment to the field's use of an evolutionary perspective. As in any other scientific field, methodologies constrain utility, and they continue to do so in paleopathology, though hopefully to a progressively lesser degree (Armelagos 1994; Powell and Cook 2005b; Rothschild 2005). Only in the past three decades, with the analysis of large skeletal samples and authors' and editors' explicit use of standardized data and criteria, have several of the most important and productive questions in biological anthropology, such as the effects of intensive agriculture on health (Cohen and Armelagos 1984) and the origins and antiquity of syphilis (Powell and Cook 2005a; Harper et al. in press), been amenable to analysis.

Trends in the presentation and publication of data have also impeded evolutionary analyses. This is particularly true for the historical emphasis placed on case studies, the only very recent rise of large databases of paleopathological data, and the low publication rate for negative evidence. Content analyses demonstrate that case studies persist as one of the primary publication types in leading journals (Hens and Godde 2008; Park et al. 2010). While they are not unproductive or detrimental contributions *de facto* (Lovejoy et al. 1982; Stojanowksi and Buikstra 2005) as they provide the basis for synthetic work (Stodder et al. 2006), they can only contribute to syntheses and evolutionary interpretations after being aggregated into population-level analyses and embedded in a comparative framework. The recent emergence of large databases, like The Wellcome Osteological Research Database, at the Museum of London, the Global History of Health Project for the Western and Eastern Hemispheres at The Ohio State University, and the Standard Osteological Database at the University of Arkansas, is also critical to evolutionary research, though they remain few and underutilized. Though Mays (2010) found that fewer than 1 percent of studies published in leading journals between 2001 and 2007 made use of them, they provide exceptional opportunities for comparative evaluations of the health effects of various adaptive strategies in the past. Importantly, these also contain negative evidence, which is a crucial component of population-level interpretations. Only by knowing where and when a condition was present or absent is it possible to document the adaptiveness or maladaptivness of a cultural strategy or, in relation to host–pathogen interactions, infer the temporal, phylogenetic, and geographical trajectory of a disease. For example, investigations of the Pre-Columbian treponemal disease in the Old World have been greatly impaired by the relative scarcity of publications documenting the absence of the disease in regional skeletal samples (Zuckerman and Armelagos 2010b; Harper et al. in press), though notable exceptions exist (e.g., Møller-Christensen 1969; Crane-Kramer 2002).

The sheer difficulty of employing an approach can also be a discouraging factor. The biocultural approach in particular can be challenging to implement. For example, in a biocultural analysis, researchers are typically out to assess the effects of a culturally defined, independent variable on some aspect of human biology. For Dufour (2006), these are often very difficult to operationalize, especially when they are composed of multiple, intersecting social, ecological, and economic variables. Successfully operationalizing them in ways that are ethnographically or historically valid and replicable requires location-specific ethnographic knowledge (or archaeological and historical) (Dressler 1995). For many skeletal samples this degree of contextual information may simply be nonexistent (e.g., Djurić-Srejić and Roberts 2001). When it is available, researchers must also exercise significant caution in this process, given the varying but omnipresent degrees of incompleteness and biases of historical, archaeological, and skeletal data (Cox 1995). Researchers also have to grapple with understanding the complex mechanics and effects of major constructs such as health or poverty (Dufour 2006). Poverty is a multidimensional social, economic, material, and even psychological phenomenon and specifics of the context—ethnographic, epidemiological, and nutritional, as well as characteristics of the human-built and physical environments—can lead to a multiplicity of research questions and approaches (see Narayan 2000). Lastly, understanding complex interactions between biology and culture necessitates defining and measuring multiple causal pathways, which can be extremely challenging (Dufour 2006).

Perhaps the greatest impediment is that many researchers simply do not ask evolutionarily or bioculturally themed research questions (Goodman 1998; Armelagos and Van Gerven 2003; Park et al. 2010). This is due to several ideological factors. For example, paleopathological analyses are premised on the underlying assumption that biological characteristics of known disease organisms and parasite cycles follow uniformitarian principles. This is key to evolutionary studies—it enables diagnosis of different conditions in various contexts through time with an inherent but acceptable degree of uncertainty. However, these principles tend to minimize the fact that pathogens can evolve (Cohen and Crane-Kramer 2003) and that diseases cannot be expected to manifest in the same way in every environment or human population (Buikstra and Cook 1980; Ortner et al. 1992; Heathcote et al. 1998). While some argue that this is a negligible concern (Cohen and Crane-Kramer 2003), and others translate it into diagnostic caution (Weaver et al. 2005), this could also be turned on its head to enable paleopathologists to explore evidence for host–pathogen coevolution in the past. Skeletal and biomolecular evidence of disease in the past is eminently well suited to investigations of this phenomenon (Baum and Kahila Bar-Gal 2003). However, such investigations are infrequently conducted and have primarily focused on nonhuman hosts (Dittmar 2009). Other ideological constraints are legacies of research foci in the past. For example, Goodman (1998) argues that the majority of studies are steered away from either pursuing or fully realizing either approach by a persistent view of societies as integrated and functional wholes and by a focus on evolutionary questions within a narrow ecological framework. These, he states, are legacies of the ecological bias from processual archaeology and human adaptability studies. In general, these limitations prematurely prevent researchers from achieving a fully fleshed-out consideration of the relationships between sociopolitical factors, such as access to ideology, and patterns of disease in past populations. In the U.K.,

Roberts (2006) writes that researchers acknowledge that a variety of interacting social, environmental, economic, and temporal factors can predispose populations and their members to different diseases, but that these issues have not always been considered in their analyses.

Since the 1950s, American and British paleopathology, as well as skeletal biology and bioarchaeology, have been characterized by a distinct directional schism. One side tends towards anthropological, processual, biocultural, and evolutionary analytical (hypothesis-driven and theoretical) approaches. The other employs a descriptive approach (theoretically uncontextualized and data- or method-driven) and uses biomedical paradigms to determine the implications of disease and stimulate research (Armelagos et al. 2003; Larsen 2002). Each reflects a discrete take on what Ortner (1991; 1992; 1994) defines as one of the two fundamental questions in paleopathology: "What do lesions mean for individuals and their populations?" (The other—"What is it?"—is usually too difficult to answer.)

Both paradigms have been recommended with equal ardor, but their actual impact on research in the field is best determined through content analyses (bibliometric analyses) (Stojanowksi and Buikstra 2005). For example, Lovejoy et al.'s (1982) survey of skeletal biology papers published in the *AJPA* between 1930 and 1980 documented an overall increase in "analytical" papers but an overall excess of "descriptive" and insufficiently theoretical ones. Proportionately, paleopathological papers were likewise deemed excessively descriptive. Armelagos and Van Gerven's (2003) revisitation reached a similar conclusion. In a survey of later *AJPA* papers on human osteology (1980–1984; 1996–2000) they found that descriptive approaches remained dominant. In paleopathology, interest had reverted from evolutionary and biocultural questions to a focus on methodological advances, case studies, and differential diagnosis. In a rejoinder, Stojanowski and Buikstra's (2005) evaluation of human osteology papers in *AJPA* during the same period instead found little change in the visibility of analytical research and that what exists has a significantly greater impact than descriptive research. Hens and Godde (2008), who replicated Lovejoy et al.'s approach for *AJPA* and *IJO* publications (1980–2004), found a significant increase in analytical papers, including those which employed broader evolutionary or ecological contexts, though this trend was not found within paleopathological papers specifically. Several other scholars have also found little evidence of change in the field. Park and colleagues (2010), for example, found that British researchers tended to produce equal portions of case studies, methodological papers, and population-level analyses. While the proportion of case studies has decreased substantially, population-level analyses have only increased marginally. Mays (1997; 2010) found that proportions of case studies have not significantly altered and that national-level differences in their proportions are due to national academic legacies; British researchers tend to produce more case studies because of a continuing biomedical tradition. American researchers produce more population-level analyses because of their field's more anthropological, analytical bent (Robb 2000). Despite this, several British researchers have argued for a turn towards this more population-level, analytical approach (Roberts and Manchester 2005). Overall, Mays, Park and colleagues, Buikstra and Stojanowski, and Hens and Godde conclude that while proportions vary, human osteology and paleopathology reside in a healthy balance between descriptive and analytical, evolutionary approaches.

CONCLUSION: WHY EVOLUTIONARY THOUGHT SHOULD BE INCORPORATED INTO PALEOPATHOLOGY

Despite the limitations and constraints that too often impede use of evolutionary thought or the biocultural approach in paleopathology, employing either confers disproportionate benefit. Proximately, evolutionary approaches, like biocultural approaches, provide opportunities to explore hypothesis-driven research questions, which confer direct advantages to the field. In a wider political climate of anti-evolutionary critiques and rejection of cultural adaptation theories and ecological interpretations (McAnany and Yoffee 2009), and lingering divisive and (anti-ecological) sentiments within anthropology (Segal and Yanagisako 2005), the ability to objectively test hypotheses offers the field a harbor from these critiques (Armelagos 2003). Additionally, given the established ethical climate surrounding the study of human remains, especially in regards to issues of repatriation and reburial (Walker 2008) and growing public interest in the findings of skeletal research, many researchers have stated that justifications for doing so have to be clear to avoid potential misunderstandings (Parker Pearson 1999). According to Park and colleagues (2010) and others, the clear research objectives provided by evolutionary and biocultural approaches can be used to satisfy many of these concerns by demonstrating the clear value of these analyses.

Ultimately, using either approach can also make paleopathology more relevant to contemporary societies. The biocultural approach in particular can be used to deconstruct and consequently, denaturalize human suffering. With an incorporation of political economic perspectives, careful analysis of the social relations that structure access to material resources can reveal that unequal distributions of disease, undernutrition, trauma, and biological adaptation are the products of human action and interest rather than natural and inevitable stressors equally born by all members of a society (Leatherman and Goodman 1997). Revealing these underlying social processes can also indicate solutions for coping with and combating them in the modern era (Goodman 1998). Bioculturally oriented paleopathological analyses can also act dialectically. In other words, investigations of the linkage between sociopolitical processes and their biological effects on past populations can be used to reveal the causes of health disparities in contemporary societies and vice versa. For example, collaborative and multidisciplinary research projects between contemporary North American indigenous groups, paleopathologists, and researchers from other scientific fields have been used to investigate the historical origins of health problems currently endemic in these groups. Many of them, like diabetes, are archaeologically invisible, but can be inferentially tied to historical changes in dietary quality and activity levels and have led to pragmatic solutions for preventing health disparities (Martin 1998).

By better allowing researchers to understand challenges to human health in the past, the use of evolutionary thought may also enable them to predict and interpret them more effectively in the present and the future (Armelagos et al. 2005). As many scholars have emphasized, we currently exist in an era wherein the optimism of the antibiotic age is gradually fading and being replaced with a newfound respect for our long-term adversaries: pathogens. In the face of fast-paced environmental change and emerging and re-emerging infectious diseases—and with many of them displaying new virulence, antibiotic resistance, and a capacity for rapid evolution—an understanding of the interplay of pathogens, humans, and environments in the past is growing increasingly critical (Fischer and Klose 1996; Morens et al. 2004, 2008). Reconstructing evolutionary patterns of infectious diseases such as tuberculosis, syphilis, malaria, and others holds enormous potential for understanding the nature of our present encounters with these conditions as well as guidance to rational public health policy decisions for their control.

Many scholars (Tylor 1881; White 1965; Goodman 1998; Armelagos and Van Gerven 2003; Walker 2008) have argued that to constitute a socially, scientifically, and ethically valid endeavor, anthropological research must be relevant and useful to contemporary societies. According to Walker, this imperative is even more critical for research dealing with human skeletal remains. For Armelagos and Van Gerven and others, this means that paleopathology must be oriented towards understanding adaptation and evolution from an evolutionary and biocultural perspective. This requires a synthetic approach that recognizes disease as a response to environmental inadequacy and a given cultural system's failure to protectively buffer its denizens from harm. In turn, this requires inter- and intradiscliplinary attention to all of the factors involved in human plasticity and adaptation in the human environmental milieu across time and place. Paleopathology can help us to understand the factors in evolution that have generated contemporary patterns in the global disease burden and as such, researchers should be attentive to on-the-ground applications for their research and the potential of their results to prevent or accommodate disease and lessen human suffering. Only by illuminating the social, cultural, and environmental origins of disease can paleopathology be at the forefront of a relevant, practical, and synthetic anthropology.

REFERENCES

Angel, J. L., 1981   History and Development of Paleopathology. American Journal of Physical Anthropology 56(4):509–515.
Armelagos, G., 1994   Review: Human Paleopathology: Current Syntheses and Future Options. Journal of Field Archaeology 21(2):239–243.
Armelagos, G., 1997   Paleopathology. In History of Physical Anthropology: An Encyclopedia, vol. 2. F. Spencer, ed. pp. 790–796. New York: Garland.
Armelagos, G., 2003   Bioarcheology as Anthropology. In Archaeology as Anthropology. S. Gillespie, and D. Nichols, eds. pp. 27–40. Arlington, VA: Archaeological Papers of the American Anthropological Association Series.
Armelagos, G., 2008   Biocultural Anthropology At Its Origins: Transformation of the New Physical Anthropology in the 1950s. In The Tao of Anthropology. A. Kelso, ed. pp. 269–282. Gainesville: University of Florida Press.
Armelagos, G., Brown, P., and Turner, B., 2005   Evolutionary, Historical and Political Economic Perspectives on Health and Disease. Social Science and Medicine 61:755–765.
Armelagos, G., Carlson, D., and Van Gerven, D., 1982   The Theoretical Foundations and Development of Skeletal Biology. In A History of American Physical Anthropology 1930–1980, vol. 2. F. Spencer, ed. pp. 305–329. New York: Academic Press.
Armelagos, G., Goodman, A., Harper, K., and Blakey, M., 2009   Enamel Hypoplasia and Early Mortality: Bioarcheological Support for the Barker Hypothesis. Evolutionary Anthropology 18:261–271.

Armelagos, G., Goodman, A., and Jacobs, K., 1978   The Ecological Perspective in Disease. *In* Health and the Human Condition. M. Logan, and E. Hunt, eds. pp. 71–83. North Scituate, MA: Duxbury Press.

Armelagos, G., and Harper, K., 2010   The Changing Disease-Scape in the Third Epidemiological Transition. International Journal of Environmental Research and Public Health 7(2):675–697.

Armelagos, G., and Van Gerven, D., 2003   A Century of Skeletal Biology and Paleopathology: Contrasts, Contradictions, and Conflicts. American Anthropologist 105(1):51–62.

Baum, J., and Bar-Gal, G. K., 2003   The Emergence and Co-Evolution of Human Pathogens *In* Emerging Pathogens: Archaeology, Ecology and Evolution of Infectious Diseases. C. Greenblatt, and M. Spigelman, eds. pp. 67–79. Oxford: Oxford University Press.

Binford, S., and Binford, L., 1968   New Perspectives in Archaeology. Chicago: Aldine Publishing Company.

Blakey, M., 1987   Intrinsic Social and Political Bias in the History of American Physical Anthropology, with Special Reference to the Work of Aleš Hrdlička. Critique of Anthropology 7(2):7–35.

Boyd, W., 1950   Genetics and the Races of Man. Boston: Heath.

Brothwell, D., and Sandison, A., 1967   Editorial Prologomenon. *In* Diseases in Antiquity: A Survey of the Diseases, Injuries, and Surgery of Early Populations. D. Brothwell, and A. Sandison, eds. pp. xi–xiv. Springfield, IL: C. C. Thomas.

Brown, P., 1998   Understanding and Applying Medical Anthropology. Mountain View, CA: Mayfield Publishing Co.

Buikstra, J. E., 1991   Out of the Appendix and Into the Dirt: Comments on Thirteen Years of Bioarchaeological Research. *In* What Mean These Bones? Studies in Southeastern Bioarchaeology. M. Powell, P. Bridges, and A. Mires, eds. pp. 172–189. Tuscaloosa, AL: The University of Alabama Press.

Buikstra, J. E., and Cook, D., 1980   Paleopathology: An American Account. Annual Review of Anthropology 9:433–470.

Cockburn, A., Cockburn, E., and Reyman, T., eds., 1980   Mummies, Disease, and Ancient Cultures. New York: Cambridge University Press.

Cohen, M., and Armelagos, G., eds., 1984   Paleopathology at the Origins of Agriculture. Orlando, FL: Academic Press.

Cohen, M., and Crane-Kramer, G., 2003   The State and Future of Paleoepidemiology. *In* Emerging Pathogens: Archaeology, Ecology and Evolution of Infectious Diseases. C. Greenblatt, and M. Spigelman, eds. pp. 79–93. Oxford: Oxford University Press.

Cook, D., and Powell, M., 2006   The Evolution of American Paleopathology. *In* Bioarchaeology: The Contextual Analysis of Human Remains. J. Buikstra, and L. Beck, eds. pp. 281–323. Amsterdam: Academic Press.

Cox, M., 1995   A Dangerous Assumption: Anyone Can Be a Historian! The Lessons from Christ Church Spitalfields. *In* Grave Reflections: Portraying the Past through Skeletal Studies. S. Saunders, and A. Herring, eds. Toronto: Canadian Scholars' Press.

Crane-Kramer, G., 2002   Was there a Medieval Diagnostic Confusion between Leprosy and Syphilis? An Examination of the Skeletal Evidence. *In* The Past and Present of Leprosy: Archaeological, Historical, Palaeopathological and Clinical Approaches. C. A. Roberts, M. Lewis, and K. Manchester, eds. Oxford: BAR Archaeopress.

de Melo, F., de Mello, J., Fraga, A. M., Nunes, K., and Eggers, S., 2010   Syphilis at the Crossroad of Phylogenetics and Paleopathology. PLoS Neglected Tropical Diseases 4(1):e575.

Dittmar, K., 2009   Old Parasites for a New World: The Future of Paleoparasitological Research. A Review. Journal of Parasitology 95(2):365–371.

Djurić-Srejić, M., and Roberts, C. A., 2001   Palaeopathological Evidence of Infectious Disease in Skeletal Populations from Later Medieval Serbia. International Journal of Historical Archaeology 11(5):311–320.

Dressler, W., 1995 Modeling Biocultural Interactions: Examples from Studies of Stress and Cardiovascular Disease. Yearbook of Physical Anthropology 38:27–56.

Dufour, D., 2006 Biocultural Approaches in Human Biology. American Journal of Human Biology 18(1):1–9.

Duray, S., 1996 Dental Indicators of Stress and Reduced Age at Death in Prehistoric Native Americans. American Journal of Physical Anthropology 99:275–286.

Dustugue, J., 1980 Possibiities, Limits, and Prospects in Paleopathology of the Human Skeleton. Journal of Human Evolution 9:3–8.

Fischer, E., and Klose, S., 1996 Infectious Diseases. Munchen: Piper.

Gage, T, and DeWitte, S., 2009 What Do We Know About the Agricultural Demographic Transition? Current Anthropology 50(5):649–655.

Gonçalves, M. L. C., Araújo, A., and Fernando Ferreira, L., 2003 Human Intestinal Parasites in the Past: New Findings and a Review. Memorias de Instituto Oswaldo Cruz 98(1):103–118.

Goodman, A., 1994 Cartesian Reductionism and Vulgar Adaptationism: Issues in the Interpretation of Nutritional Status in Prehistory. *In* Paleonutrition: The Diet and Health of Prehistoric Americans. Occasional Paper No. 2. K. Sobolik, ed. pp. 163–177. Carbondale, IL: Center for Archaeological Investgations.

Goodman, A. 1998 The Biological Consequences of Inequality in Antiquity. *In* Building a New Biocultural Synthesis: Political–Economic Perspectives on Human Biology. A. Goodman, and T. Leatherman, eds. pp. 141–169. Ann Arbor, MI: The University of Michigan Press.

Goodman, A., and Leatherman T., eds., 1998 Building a New Biocultural Synthesis: Political–Economic Perspectives on Human Biology. Ann Arbor, MI: University of Michigan Press.

Goodman, A., and Martin, D., 2002 Reconstructing Health Profiles from Skeletal Remains. *In* The Backbone of History: Health and Nutrition in the Western Hemisphere. R. Steckel, and J. Rose, eds. pp. 11–61. Cambridge: Cambridge University Press.

Goodman, A., and Rose, J., 1990 Assessment of Systemic Physiological Perturbations from Dental Enamel Hypoplasias and Associated Histological Structures. Yearbook of Physical Anthropology 33:59–110.

Gorbunova, V., Bozzella, M. J., and Seluanov, A., 2008 Rodents for Comparative Aging Studies: From Mice to Beavers. Age 30(2–3):111–119.

Gould, S. J., and Lewontin, R., 1979 The Spandrels of San Marco and the Panglossian Paradigm: A Critique of the Adaptationist Programme. Proceedings of the Royal Society of London B 205:581–598.

Haraway, D., 1990 Remodelling the Human Way of Life: Sherwood Washburn and the New Physical Anthropology, 1950–1980. *In* Bones, Bodies, Behavior: Essays on Biological Anthropology. G. J. Stocking, ed., pp. 206–260. Madison: University of Wisconsin Press.

Harper, K., and Zuckerman, M., 2010 Comments From Group that Supplied Sequence Data (Comment on "Syphilis at the Crossroad of Phylogenetics and Paleopathology"). PLoS Neglected Tropical Diseases 4(1):e575.

Harper, K., Zuckerman, M., Harper, M., Kingston, J., and Armelagos, G., in press. The Origin and Antiquity of Syphilis Revisited: An Appraisal of Old World Pre–Columbian Evidence for Treponemal Infection. Yearbook of Physical Anthropology.

Heathcote, G., Stodder, A., Buckley, H., Hanson, D., Douglas, M., Underwood, J., Taisipic, T., and Diego, V., 1998 On Treponemal Disease in the Western Pacific: Corrections and Critique. Current Anthropology 39(3):359–368.

Hens, S., and Godde, K., 2008 Brief Communication: Skeletal Biology Past and Present: Are We Moving in the Right Direction? American Journal of Physical Anthropology 137:234–239.

Holland, T., and O'Brien, M., 1997 Parasites, Porotic Hyperostosis, and the Implications of Changing Perspectives. American Antiquity 62(2):183–193.

Hooton, E., 1930   The Indians of Pecos Pueblo: A Study of Their Skeletal Remains. New Haven, CT: Yale University Press.

Jarcho, S., 1966   The Development and Present Condition of Human Paleopathology in the United States. *In* Human Paleopathology. S. Jarcho, ed. pp. 3–30. New Haven, CT: Yale University Press.

Jones, J., 1876   Exploration of the Aboriginal Remains of Tennessee. Smithsonian Contributions to Knowledge 259:1–17.

Kloos, H., and David, R., 2002   The Paleoepidemiology of Schistosomiasis in Ancient Egypt. Human Ecology Review 9(1):14–26.

Larsen, C., 1997   Bioarchaeology: Interpreting Behavior from the Human Skeleton. New York: Cambridge University Press.

Larsen, C., 2002   Bioarchaeology: The Lives and Lifestyles of Past People. Journal of Archaeological Research 10(2):119–166.

Lasker, G., 1970   Physical Anthropology: Search for General Processes and Principles. American Anthropologist 72:1–8.

Leatherman, T., and Goodman, A., 1997   Expanding the Biocultural Synthesis: Toward a Biology of Poverty. American Journal of Physical Anthropology 102:1–3.

Levins, R., and Lewontin, R., 1985   The Dialectical Biologist. Cambridge, MA: Havard University Press.

Livingstone, F., 1958   Anthropological Implications of Sickle Cell Gene Distribution in West Africa. American Anthropologist 60:533–562.

Lovejoy, C., Mensforth R., and Armelagos G., 1982   Five Decades of Skeletal Biology as Reflected in the American Journal of Physical Anthropology. *In* A History of American Physical Anthropology, 1930–1980, vol. 2. F. Spencer, ed., pp. 329–336. New York: Academic Press.

Mace, R., 2000   Evolutionary Ecology of Human Life History. Animal Behaviour 59(1):1–10.

Martin, D., 1998   Owning the Sins of the Past: Historical Trends, Misssed Opportunities and New Directions in the Study of Human Remains. *In* Building a New Biocultural Synthesis: Political–Economic Perspectives on Human Biology. A. Goodman, and T. Leatherman, eds., pp. 171–190. Ann Arbor, MI: The University of Michigan Press.

Mays, S., 1997   A Perspective on Human Osteoarchaeology in Britain. International Journal of Osteoarchaeology 7:600–604.

Mays, S., 2010   Human Osteoarchaeology in the U.K. 2001–2007: A Bibliometric Perspective. International Journal of Osteoarchaeology 20:192–204.

McAnany, P., and Yoffee, N., eds., 2009   Questioning Collapse: Human Resilience, Ecological Vulnerability, and the Aftermath of Empire. Cambridge: Cambridge University Press.

Miller, E., Ragsdale, B., and Ortner, D.,1996   Accuracy in Dry Bone Diagnosis: A Comment on Palaeopathological Methods. International Journal of Osteoarchaeology 6(3):221–229.

Møller-Christensen, V., 1969   The History of Syphilis and Leprosy – An Osteoarchaeological Approach. Abbotempo 1:20–25.

Moodie, R., 1923   Paleopathology: An Introduction to the Study of Ancient Evidences of Disease. Chicago: University of Chicago Press.

Moran, E., 1982   Human Adaptability: An Introduction to Ecological Anthropology. Boulder, CO: Westview Press.

Morens, D., Folkers, G., and Fauci, A., 2004   The Challenge of Emerging and Re–Emerging Infectious Diseases. Nature 430:242–249.

Morens, D., Folkers, G., and Fauci, A., 2008   Emerging Infections: A Perpetual Challenge. Lancet Infectious Diseases 8:710–719.

Narayan, D., 2000   Voices of the Poor: Can Anyone Hear Us? New York: Oxford University Press.

Nevo, E., 1997   Evolution in Action Across Life at "Evolution Canyons," Israel. Theoretical Population Biology 52(3):231–243.

Ortner, D., 1991   Theoretical and Methodological Issues in Paleopathology. *In* Human Paleopathology: Current Syntheses and Future Options. D. Ortner, and A. Aufderheide, eds. pp. 5–11. Washington, DC: Smithsonian Institution Press.

Ortner, D., 1992   Skeletal Pathology: Probabilities, Possibilities, and Impossibilities. *In* Disease and Demography in the Americas. J. Verano, and D. Ubelaker, eds. pp. 5–13. Washington, DC: Smithsonian Institution Press.

Ortner D., 1994   Descriptive Methodology in Paleopathology. *In* Skeletal Biology in the Great Plains: Migration, Warfare, Health, and Subsistence. D. Owsley, and R. Jantz, eds. pp. 73–80. Washington, DC: Smithsonian Institution Press.

Ortner, D. J., 2003   Identification of Pathological Conditions in Human Skeletal Remains, 2nd. ed. Academic Press.

Ortner, D., Tuross N, and Stix A., 1992   New Approaches to the Study of Disease in Archeological New World Populations. Human Biology 64:337–360.

Park, V., Roberts, C. A., and Jakob, T., 2010   Palaeopathology in Britain: A Critical Analysis of Publications With the Aim of Exploring Recent Trends (1997–2006). International Journal of Osteoarchaeology 20:497–507.

Parker Pearson, M., 1999   The Archaeology of Death and Burial. Gloucester: Sutton Publishing.

Powell, M., and Cook, D., eds., 2005a   The Myth of Syphilis: The Natural History of Treponematosis in North America. Gainesville: University Press of Florida and the Florida Museum of Natural History.

Powell, M., and Cook, D., 2005b   Treponematosis: Inquiries into the Nature of a Protean Disease. *In* The Myth of Syphilis: The Natural History of Treponematosis in North America. M. Powell, and D. Cook, eds. pp. 9–63. Gainesville: University Press of Florida and the Florida Museum of Natural History.

Robb, J., 2000   Analysing Human Skeletal Data. *In* Human Osteology in Archaeology and Forensic Science. M. Cox, and S. Mays, eds. pp. 475–490. London: Greenwich Medical Media.

Roberts, C. A., 1994   Treponematosis in Gloucester, England: A Theoretical and Practical Approach to the Pre–Columbian Theory. *In* The Origin of Syphilis in Europe: Before or After 1493? O. Dutour, G. Palfi, J. Berato, and J. Brun, eds. Toulon: Centre Archeologique du Var.

Roberts, C. A., 2006   A View from Afar: Bioarchaeology in Britain. *In* Bioarchaeology: The Contextual Analysis of Human Remains. J. Buikstra, and L. Beck, eds. pp. 417–439. Amsterdam: Academic Press.

Roberts, C. A., and Cox, M., 2003   Health and Disease in Britain: From Prehistory to the Present Day. Thrupp, U.K.: Sutton Publishing.

Roberts, C. A., and Ingham, S., 2008   Using Ancient DNA Analysis in Palaeopathology: A Critical Analysis of Published Papers, With Recommendations for Future Work. International Journal of Osteoarchaeology 18(6):600–613.

Roberts, C. A., and Manchester, K., 1999   The Archaeology of Disease. Ithaca: Cornell University Press.

Roberts, C. A., and Manchester, K., 2005   The Archaeology of Disease, 3rd edition. Ithaca: Cornell University Press.

Rook, G., 2007   The Hygiene Hypothesis and the Increasing Prevalence of Chronic Inflammatory Disorders. Transactions of the Royal Society of Tropical Medicine and Hygiene 101(11):1072–1074.

Rook, G., 2009   Review Series on Helminths, Immune Modulation and the Hygiene Hypothesis: The Broader Implications of the Hygiene Hypothesis. Immunology 126(1):3–11.

Rose, J., Green, T., and Green, V., 1996   NAGPRA is Forever: Osteology and the Repatriation of Skeletons. Annual Review of Anthropology 25:81–103.

Rothman, K, and Greenlander, S., 1998   Modern Epidemiology, 2nd edition. Philadelphia, PA: Lippincott–Raven.

Rothschild, B., 2005   Scientific Integrity, Conflict of Interest and Mythology Related to Analysis of Treponematoses: The Importance of Being Ernest. Chungará (Arica) 37(2):265–267.

Ruffer, M., 1914   Studies in Paleopathology, Note on the Diseases of the Sudan and Nubia in Ancient Times. Mitteilungen der Berliner Gesellschaft fur Anthropologie, Ethnologie und Urgeschichte 13:453–460.

Segal, D., and Yanagisako, S., 2005   Unwrapping the Sacred Bundle: Reflections on the Disciplining of Anthropology. Durham, NC: Duke University Press.

Selye, H., 1950   Stress. Montreal: Medical Publishers.

Smocovits, V., 1996   Unifying Biology: The Evolutionary Synthesis and Evolutionary Biology. Princeton, NJ: Princeton University Press.

Steckel, R., and Rose, J., eds., 2002   The Backbone of History: Health and Nutrition in the Western Hemisphere. Cambridge: Cambridge University Press.

Steinbock, R., 1976   Paleopathological Diagnosis and Interpretation. Springfield, IL: C. C. Thomas.

Stodder, A., Johnson, K., Chan, A., Handwerk, E., and Rudolph, K., 2006   Publishing Patterns in Palaeo–Pathology: Findings of the Publications Explorations Committee. Palaeopathology Newsletter 134:6–13.

Stojanowksi, C., and Buikstra, J. E., 2005   Research Trends in Human Osteology: A Content Analysis of Papers Published in the American Journal of Physical Anthropology. American Journal of Physical Anthropology 128:98–109.

Stuart-Macadam, P., 1992   Porotic Hyperostosis – A New Perspective. American Journal of Physical Anthropology 87(1):39–47.

Stuart-Macadam, P., 1998   Iron Deficiency Anemia: Exploring the Difference. In Sex and Gender in Paleopathological Perspective. A. L. Grauer, and P. Stuart–Macadam, eds. pp. 45–63. Cambridge: Cambridge University Press.

Tylor, E., 1881   Anthropology: An Introduction to the Study of Man and Civilization, vols. 1 and 2. London: Watts Publishing Group Ltd.

Ubelaker, D., 1982   The Development of Human Paleopathology. In A History of American Physical Anthropology, 1930–1980, vol. 2. F. Spencer, ed., pp. 337–356. New York: Academic Press.

Ubelaker, D., 2003   Anthropological Perspectives on the Study of Ancient Disease. In Emerging Pathogens: Archaeology, Ecology and Evolution of Infectious Diseases. C. Greenblatt, and M. Spigelman, eds. pp. 93–102. Oxford: Oxford University Press.

Walker, P., 2008   Bioarchaeological Ethics: A Historical Perspective on the Value of Human Remains. In Biological Anthropology of the Human Skeleton. M. Katzenberg, and S. Saunders, eds. pp. 3–41. New York: Wiley–Liss.

Walker P., Bathurst R., Richman R., Gjerdrum T., and Andrushko V., 2009   The Causes of Porotic Hyperostosis and Cribra Orbitalia: A Reappraisal of the Iron-Deficiency–Anemia Hypothesis. American Journal of Physical Anthropology 139(2):109–125.

Warren, K., ed., 1951   Origin and Evolution of Man. New York: Long Island Biological Association.

Washburn, S., 1951   The New Physical Anthropology. Transactions of the New York Academy of Science 13(7):298–304.

Washburn, S., 1953   The Strategy of Physical Anthropology. In Anthropology Today: An Encyclopedic Inventory. A. Kroeber, ed., pp. 714–727. Chicago: University of Chicago Press.

Weaver, D., Sandford, M., Bogdan, G., Kissling, G., and Powell, M., 2005   Prehistoric Treponematosis on the North Carolina Coast In The Myth of Syphilis: The Natural History of Treponematosis in North America. M. Powell, and D. Cook, eds., pp. 77–91. Gainesville: University Press of Florida and the Florida Museum of Natural History.

White, L., 1965   Anthropology 1964: Retrospect and Prospect. American Anthropologist 67:629–637.

Williams, H., 1929 Human Paleopathology, With Some Original Observations on Symmetrical Osteoporosis of the Skull. Chicago: American Medical Association.

Wright, L., and Yoder, C., 2003 Recent Progress in Bioarchaeology: Approaches to the Osteological Paradox. Journal of Archaeological Research 11(1):44–70.

Wyman, J., 1868 Observations on Crania. Boston: A.A. Kingman.

Zuckerman, M., and Armelagos, G., 2010a The Origins of Biosocial Dimensions in Bioarchaeology. *In* Social Bioarchaeology. S. Agarwal, and B. Glencross, eds. pp. 15–44. New York: Wiley-Blackwell Publishers.

Zuckerman, M., and Armelagos, G., 2010b Treponemal Infection in Skeletal Samples From Ancient Nubia. Paleopathology Association Annual Meeting. Albuquerque, NM April 13.

CHAPTER 4

The Bioarchaeological Approach to Paleopathology

Michele R. Buzon

INTRODUCTION

As a dynamic and rapidly developing anthropological specialization, bioarchaeology (synonymous with human osteoarchaeology) integrates the study of human skeletal remains in an interdisciplinary approach with theory, methods and data from archaeology, biological anthropology, history, cultural anthropology, medical science, geography and other related disciplines (Buzon et al. 2005). Paleopathology, the investigation and reconstruction of past human health and disease, is one of the primary foci of bioarchaeology. However, as a field of study, paleopathology has not always been grounded in archaeological research. Concomitantly, archaeological investigations have often ignored or dismissed paleopathological contributions. The goal of this chapter is to illustrate the advantages of taking a bioarchaeological approach in all stages of paleopathological research for developing a more complete and integrative interpretation of past human health and disease. It begins with a discussion of theoretical approaches in paleopathology and the contribution of research to larger anthropological and archaeological questions. Next, it details the necessity of following proper archaeological techniques when excavating human remains. Finally, it demonstrates the benefits of integrating archaeological data with paleopathological analyses for more comprehensive interpretations.

A Companion to Paleopathology, First Edition. Edited by Anne L. Grauer.
© 2012 John Wiley & Sons, Ltd. Published 2016 by John Wiley & Sons, Ltd.

APPROACHES TO PALEOPATHOLOGICAL RESEARCH

History

The early phases of paleopathological research were primarily case studies conducted by physicians. Occurrences of conditions were reported but these cases provided little information on the broader scope of disease. As discussed in a number of recent publications, this work generally focused on individual, rather than populational, experiences (Aufderheide and Rodríguez-Martín 1998; Roberts and Manchester 2005; Cook and Powell 2006). As collections from archaeological contexts accumulated in institutions such as the Smithsonian and the Peabody Museum at Harvard University, researchers who played active roles in amassing and curating these samples began to take a more contextual approach to the study of past human disease. Aleš Hrdlička promoted the complete collection of the skeleton and finer chronological control of specimens including photographic documentation and stratigraphic analyses (e.g., Hrdlička 1904; discussed in Roberts and Manchester 2005; Buikstra 2006a; Cook and Powell 2006). Earnest Hooten, too, is known for the integration of archaeological excavation and regional research questions in his work. In his study of Pecos Pueblo he thoroughly recorded skeletal lesions and contextually interpreted them using cultural, behavioral and temporal information (e.g., Hooten 1930; discussed in Beck 2006; Buikstra 2006a; Cook and Powell 2006). J. Lawrence Angel is also remembered for his theoretically oriented research that explicitly combined archaeology with paleopathological study incorporating environmental, ecological and historical data in eastern Mediterranean and African-American populations (e.g., Angel 1946; discussed in Buikstra 2006b; Cook and Powell 2006). Calvin Wells and Don Brothwell advocated the inclusion of relevant archaeological evidence relating to the environment and behavior in order to interpret British skeletal remains in context (Brothwell 1963; Wells 1964; discussed in Roberts 2006).

Larsen (2006) and Goldstein (2006) argue that despite these examples of contextual approaches in the 20th century, many researchers continued to study skeletal material from archaeological sites without having seen the context of recovery or incorporating it into their work. Often ending up in the appendix of site reports, paleopathological research was viewed more as a technique rather than a theoretical approach (Buikstra 1991; Sofaer 2006). Seen by some as service providers, osteologists often did not incorporate their work into interpretive archaeological perspectives. While the utility of the body through skeletal analysis was appreciated in the "new archaeology" perspective, there was rarely a holistic integration of skeletal analyses into archaeological procedures (Sofaer 2006). This lack of shared research interests between some archaeologists and osteologists continues today, sometimes resulting in the exclusion of both types of data and alternative explanations (Goldstein 2006; Knüsel 2010). Even in multifaceted investigations that integrate paleopathology with archaeological research, the efforts are often unidirectional, with paleopathological publications rarely being cited by archaeologists (Cook and Powell 2006). Sofaer (2006) contends that osteoarchaeologists have been less than proactive and have been unsuccessful in demonstrating the potential of these data outside of their own subdiscipline.

Bioarchaeology

As first articulated in the 1970s, Buikstra has promoted an approach to paleopathology and the broader field of bioarchaeology as an interdisciplinary undertaking that not only incorporates archaeological data in reconstructing the lives of past populations but begins from an equal partnership between archaeology and osteology during the formulation of research questions and planning. This "biocultural" paradigm emphasizes social theory in paleopathological research (Buikstra 1991; 2006b). The study of the skeleton from archaeological sites has been seen to lack theoretical insight, being more technical and atheoretical, while interpretative archaeology is viewed as theoretically oriented (Sofaer 2006). However, Sofaer (2006) makes a strong case for the type of research espoused by Buikstra by highlighting the tacit theory in the study of human remains. The plasticity of the skeleton allows for functional adaptation to environmental changes, whether natural or cultural. The skeleton is thus a reflection of a lifetime of interaction with the world, displaying physical responses to damage, mechanical stress and disease. This plasticity signifies that the skeleton is a record of the history of social relationships (Sofaer 2006).

Treating the human skeleton as a form of material culture on par with the analyses of ceramics, lithics and any other finds re-establishes it in the archaeological project (Sofaer 2006). By viewing skeletons and objects as material and social, the integration of osteological research in an equal partnership with archaeology as promoted by Buikstra is both possible and worthwhile. Larsen and Walker (2010) advise that the most successful bioarchaeological research incorporates methods and insights from other disciplines that address similar interests. Placing paleopathological studies within a bioarchaeological approach allows for a more fruitful interpretation and a better promotion of the importance of paleopathological research (Grauer 2008).

THE IMPORTANCE OF ARCHAEOLOGICAL TECHNIQUE IN PALEOPATHOLOGICAL RESEARCH

The proper excavation of human skeletal remains from an archaeological site requires cautious procedures in order to maximize the amount of information obtainable for paleopathological investigations. Called *l'anthropologie de terrain* (field anthropology), Duday (2006; discussed in Knudson and Stojanowski 2008; Knüsel 2010) has developed suggestions for archaeological excavation specific to funerary contexts. In line with Buikstra's holistic view of bioarchaeology, Duday advocates careful recording of the skeleton, its elements, and its relationship to burial features, grave goods, and taphonomic changes.

In many cases, the lack of recording makes it impossible to establish the context of a sample, without any means to resolve the issue (Roberts and Manchester 2005; Pinhasi and Bourbou 2008). Integrated participation of osteologists in archaeological excavation can help to avoid these difficulties.

Although the demographic dynamics of the past certainly cannot be controlled, a better understanding of the biases in the samples is possible. Careful excavation and recording can provide important information on, for instance, the differential disposal of particular groups, such as children. Infants may be buried away from the main cemetery in places such as house floors, entryways or separate cemeteries. Burial practices may also vary by social status, sex, disease status or other aspects. Diverse burial conditions may result in differential preservation. The archaeological context in these circumstances is important when making inferences about the meaning of various practices, such as infanticide (Smith and Bar-Gal 1992; Saunders 2008).

Because it was previously thought that immature remains could yield little information, many skeletal collections in museums contain mostly adults (Ubelaker 1989). The dearth of juvenile skeletons also may be due to incomplete recovery as a result of biased and/or inadequate excavation techniques. When present, incompleteness of juvenile remains may be attributed to the degree of skill of the excavators and the need to be able to recognize juvenile bones. The lack of training may contribute to inadequate recovery of skeletal remains (Saunders 2008). Age-related variation in skeletal resistance to decay can distort mortality profiles (Walker et al. 1988; Nawrocki 1995; Walker 1995); some skeletal material does not preserve or may be affected by taphonomic factors. Roots of plants can secrete acids that decalcify bone or leave a network of root grooves on the bone surface that may be confused with other conditions (pseudopathology). For instance, indentations left by plant roots may be misinterpreted as antemortem damage such as tumor metastasis, pathway growth of a fistula, trepanation or injury caused by an arrow. In addition, the actions of small roots can mimic blood vessel tracks associated with an epidural hematoma or inflamed meninges (Schultz 1997). Activities of bacteria, fungi, worms, and insects also can cause bone to degrade. Gnawing by rodents and carnivores can produce pits and grooves in bone that can be mistaken for cultural processing. While taphonomic factors may differentially affect juvenile skeletal materials, observations by a trained osteologist can help to understand the sample. Assessing the preservation of specimens from various archaeologically defined areas or strata can assist in determining whether differential preservation may bias samples. The collection of soil samples and subsequent chemical analysis of composition and acidity may provide information regarding preservation, erosion and other factors affecting the bones (Ubelaker 1989; Buzon et al. 2005).

Thus, from a practical standpoint, having an osteologist as archaeologist is tremendously useful; otherwise, data may be lost (Roberts 2009). Exposure of skeletal material requires knowledge of human osteology, as this knowledge helps to guide the strategy of exposure. Various types of burials may be encountered, for instance complete skeletons, fragmented or disturbed remains, cremation and isolated elements (Glassow 2005). In addition, the context of the burial is important. Special procedures are needed to document the mortuary practices associated with burial and associated objects: for instance, the position of grave goods may offer information about mortuary ritual. The location of individuals with respect to other burials can be useful in understanding cemetery use, relationship between individuals and later intrusive burials (Glassow 2005). The skeleton should be treated like any other archaeological feature through documentation. Synthesizing biological and cultural aspects of the funerary record with skeletal remains can aid in placing research within the social and political context of the archaeological site (Knüsel 2010).

The complete and accurate recovery of all skeletal parts is an important first step in maximizing the data available. All material should be saved, as an experienced osteologist can often estimate age at death, sex, as well as take measurements from even poorly preserved remains and small fragments that may possess diagnostic features (Ubelaker 1989). Observations of fragile remains before excavation can prevent the loss of these vital data in cases of poor preservation. Sieving the soil during excavation of skeletal material and of the soil in the grave after the bones have been removed can help recover small elements, including teeth and ancient calcified tissues (Hillson 2008, see Steinbock 1990 for a discussion of archaeological gallstones). Subsequent careful handling can also help maintain the sample, as fragmentary and poorly preserved skeletal materials commonly suffer post-mortem damage from excavation tools and rough handling that can be confused with pathological conditions (Roberts and Manchester 2005). The collection of soil samples from the burial area as well as sieving can recover calcified cysts caused by tapeworms and other parasites (Mays 1998; Roberts and Manchester 2005; and see Chapter 10 by Dittmar et al., this volume for discussion of archaeo- and paleoparasitology). Soil samples taken from the abdominal and sacral areas of the burial can also provide palynological specimens for dietary and medicinal research (Berg 2002).

Larger and more representative samples yield more reliable interpretations. Thus, an effort should be made to acquire as much data as possible (Ubelaker 1989). While the ideal condition would be a skeletal sample that represents the total deaths in the population during the time period under study, this is very rarely the case, for a variety of reasons. Oftentimes, a whole cemetery is not excavated due to budgetary, temporal, or logistical constraints, or due to analytical design (Pinhasi and Bourbou 2008). There is some preference for testing part of a site systematically and leaving the rest undisturbed for future research. However, this approach is not suitable for unprotected sites that might attract attention and increase the risk of destruction by looting. Regardless of the excavation details, all relevant data on mortuary practices and grave locations based on earlier observers, previous investigators or through examination of the ground surface during current excavation should be assembled to understand the circumstances of the sample. Photographs and descriptions of burial location and condition are vital. Efforts should be made to describe perishable materials before removal (Ubelaker 1989). In addition, a clear understanding of the timescale involved can provide critical information regarding the representativeness of the sample (Pinhasi and Turner 2008). Poor preservation will limit the ability to observe pathological conditions (Waldron 1994; Stodder 2008, and Chapter 19, this volume).

The sample may also be biased due to specific collection practices with regard to pathological conditions. In earlier investigations, it was common for normal individuals to be saved less often than those with pathological conditions. Analysis of such a sample by a researcher unaware of these selection criteria may erroneously lead to a conclusion of a remarkably high frequency of disease (Ubelaker 1989).

The Enhancement of Paleopathological Interpretation Through a Bioarchaeological Approach

The last few decades have brought about intensive discussion of the interpretive issues in paleopathological research (e.g., Kent 1986; Stuart-Macadam 1992; Wood et al. 1992; Goodman 1993, 1994; Waldron 1994; Wright and Yoder 2003; Cook and Powell 2006; Walker et al. 2009). Researchers have challenged conventional wisdom about connections between skeletal lesions and poor health, the etiology of various lesions, heterogeneity in susceptibility to disease (frailty), and our understanding of the samples we use. An integrative bioarchaeological approach can improve our ability to observe the various factors that play a role in the presence, absence and patterns of skeletal lesions, bone size and shape, and demography, as well as help address some the above-mentioned limitations. This section provides an overview of the contextual bioarchaeological approach to paleopathological analyses, with selected examples of research that explicitly use archaeological data.

Social roles

Sofaer (2006) has highlighted the relationship between the body and its social as well as its physical environment: the body is the product of human action. The skeleton is a record of the various roles held by an individual during life. For instance, age, gender, status and ethnic or cultural identity distinguish individuals in society. This variation can result in differential exposure to disease and access to resources that can be reflected in the frequency and severity of skeletal lesions. Because of biological immaturity, children are particularly susceptible to morbidity and mortality. Furthermore, they are dependent on the energy input from other adults and children to provide social arrangements for their care (Halcrow and Tayles 2008). The study of children is useful when investigating patterns of health and disease in prehistory because of the demographic variation and sensitivity to environmental and cultural changes (Van Gerven and Armelagos 1983; Goodman and Armelagos 1989; Buikstra and Ubelaker 1994). Additionally, as we age, our exposure to disease and trauma increases. Thus, older individuals will show higher frequencies of conditions (with all else being equal) due to their greater length of exposure (Glencross and Sawchuk 2003).

In addition to the effects that physical age has on health, social meanings of age-categories vary across cultures. Concepts of infancy and childhood are closely related to social, cultural, and historical factors in the society (Halcrow and Tayles 2008). The construction of these definitions differs between medical practitioners, osteologists and anthropologists. Often, categories are chosen for ease of use with the particular sample, but may not be necessarily relevant to biological development, social identity or roles in life (Halcrow and Tayles 2008). Because age-categories and roles are culturally defined, it is best not to assume that they correspond to Western models (Kamp 2001:4). Social age can be explored through archaeological analysis, though this endeavor can be difficult. Burial remains can be analyzed in order to elucidate the social roles of individuals at various ages. Dietary analysis, isotopic investigation of weaning, historical records and ethnographic data can also provide important information to help define social age. Perry (2006) highlighted the importance of examining social age in her study of health in the Byzantine Near East. The cultural transition to adulthood, which occurred in this culture with marriage, at around 13–15 years, marked an important difference in health. Higher mortality rates in individuals just becoming adult in society may be attributable to increased labor and self-sufficiency. Applying conventional age-categories, these individuals would have been classified as subadults and this information would have been obscured (Halcrow and Tayles 2008).

Sex and gender are also important factors in paleopathological analyses (see for instance, Grauer and Stuart-Macadam 1998). Biological differences between males and females impact the frequency of diseases in these groups; similarly, the social roles occupied by each gender create differential exposure to risks and resources. Research has demonstrated that the immune response of females to infectious disease is greater and more successful than that of males due to the selective pressures linked with pregnancy and childbirth in women and the role of sex hormones. The function and value of each gender in society also affects health. One's culture and society may be helpful in providing beneficial ways in dealing with environmental stressors; or, individuals may be denied access to resources and treatment or have an increased exposure to disease due to their social role (Ortner 1998).

Researchers are continually refining methods and techniques for sex determination in the skeleton, especially for individuals who died before puberty. The identification of gender in archaeological contexts is further fraught with difficulty. Generally, archaeologists have treated "gender" (a social and cultural construction) synonymously with "sex" (a biologically-based determination limited to two categories: male or female), and thus traditionally divide burial artifacts into two mutually exclusive categories (Arnold and Wicker 2001; Sofaer 2006). This approach is problematic, as it conflates sex with gender and assumes a bimodal distribution. Discussions regarding archaeology and the study of gender have been useful and provided new and integrative ways of approaching the topic for past populations (Conkey and Spector 1984; Conkey and Gero 1991; Arnold and Wicker 2001; Sofaer 2006). Gowland (2006) explored both age and gender in her study of Anglo-Saxon cemetery patterns finding that gendered grave good assemblages begin with individuals 13–17 years old, marking a shift in social status at this age threshold. O'Gorman (2001) presented an examination of gender differences using a comparative analysis of domestic economics and mortuary patterns in Oneonta burials to examine social inequalities. Sofaer (2006) suggests reading the body as material culture, apart from its sex determination, as a reflection of human actions. She provides examples of foot-binding, food-preparation, labor and other gendered activities that leave their mark on the skeleton as ways of investigating patterns of gender rather than allocating people in a categorical fashion.

In addition to investigating identity through biological (metric and nonmetric traits, DNA) and biochemical means (elemental and isotopic analyses), the archaeological context of burials can provide crucial data related to cultural, ethnic and group affinities. By examining the mortuary context, skeletal samples can be broken into social or kin groups. Through studying these groups in combination with diet and lesion abundance, the implications of social differentials can be investigated. For example, foreign migrants may represent another subgroup that differs in frailty (Wright and Yoder 2003). Information regarding ethnic or cultural identity can be gleaned from the analysis of burial architecture, pottery and other grave inclusions as well as burial position and orientation. Although equating archaeological cultures and ethnic groups is less than straightforward, identities of ancient peoples can be examined archaeologically through a careful contextual approach (e.g., Hodder 1982; Santley et al. 1987; Meskell 1994; Lightfoot and Martinez 1995; Emberling 1997; Jones 1997). Funerals, monuments and burial practices are key areas of the expression of ethnic identities because they provide a materialization of primordial ties that form such an important dimension of the construction of ethnic identity (Santley et al. 1987; Emberling 1997; Hall 1997; DeCorse 1999). Tombs can provide a highly visible statement of place of origin (Blake 1999). It is important to remember, however, that burial ritual does not necessarily directly replicate a person's identity during life. Burials may allow for the renegotiation of identity, providing an opportunity to reinterpret one's identity (Hodder 1982).

Individuals from various social groups may have had different life experiences, with unequal access to resources. The inclusion of immigrants in a population may indicate increased exposure and susceptibility to disease as people adapt to the region and other groups (Roberts and Manchester 2005). Torres-Rouff (2008) investigated levels of traumatic injuries with body modifications and data from grave and mortuary assemblages to understand the influence of the Tiwanaku on life in northern Chile. By combining information on burial contexts with traumatic injuries and biodistance analysis, Sutter and Verano (2007) suggest that Moche sacrificial victims were nonlocal warriors captured in territorial combat. Lewis (2010) found that individuals buried in pagan and Christian-style graves in Roman Britain showed similarities and differences their disease frequencies. Fay (2006) examined the "disease culture" in late Medieval and Tudor Britain by studying the precise burial context of individuals afflicted with leprosy, providing information regarding social and economic roles during life. Buzon (2006a) investigated the effects of Nubian and Egyptian interaction and sociopolitical changes at Tombos, finding that the ethnic groups did not differ significantly in health. These examples underscore how individual experiences can be examined in the context of larger social issues to offer nuanced interpretations of past health and disease (Knudson and Stojanowski 2008). Synthesizing this kind of archaeological analysis with biological data can provide the basis for rich interpretations not possible with either data set alone (e.g., Smith 2003; Buzon 2006b; Knudson and Stojanowski 2009).

Social status

Levels of social status and the effect of inequality on disease interpretations may also be gleaned from these archaeological analyses. Compared with high-status individuals, lower-status individuals may suffer additional biological stresses with regard to undernutrition, disease-load, unhealthy living environments, lack of medical care and physically demanding lives (Robb et al. 2001). Storey (1997) combined development defects of dental enamel with mortuary characteristics to understand how status affected individual frailty at Copán, concluding that even privileged children were not buffered against stress. Buzon and Judd (2008) found that "sacrificial" status at Kerma did not differentially affect health. Similarly, Powell (1991) found that status differentiation at Moundville brought no substantial health costs or benefits. Using a political-economic perspective, Martin (1998) found that the mortuary contexts of injured females at La Plata differed from the rest of the generally high-status population. Again, a simple correspondence between biological status and social status is not anticipated. But, a more nuanced reading of the archaeological and skeletal data can contribute a more contextual picture of past communities (Robb et al. 2001; Buzon and Judd 2008).

Living environments

Data regarding the daily lives of the individuals in the sample may also help identify subgroups and offer ideas about how the living environment may have predisposed individuals to poor health (Roberts 2000). Domestic structures may provide information about the living environment experienced by the sample. For instance, housing structures in urban communities that are crowded together may offer close contact, poor ventilation and lack of hygiene. Villages and towns in early stages of development may be dealing with waste disposal issues and contaminated water (Keene 1983). Characteristics of residential architecture may indicate levels of exposure to disease (Wright and Yoder 2003). These factors greatly affect the range of disease experienced by the inhabitants, such as exposure to infectious pathogens (Walker 1986). The location of habitation sites with regard to coastal versus inland, elevation, temperature, humidity, and access to water can also be a critical factor in the development of disease. Environmental changes such as deforestation, land cultivation and urbanization can also offer clues to understanding these experiences (Roberts and Manchester 2005). Redfern and Roberts (2005) effectively reveal the usefulness of archaeological data in their study of Romano-British urban health highlighting the connection between high rates of stress indicators with evidence for squalid settlements, possibly contaminated water sources, sanitation issues and parasitic remains.

Activity patterns

Information obtained from archaeological indications of occupation can provide important details of the health hazards due to the type of work conducted by individuals in the past. Industries such as pottery, textiles and mining, as well as exposure to smoky fire, can create particles that can induce inflammation and infection in the respiratory tract (Lancaster 1990; Larsen and Koenig 1994; Dietz et al. 1995). Occupations that involve contact with animals such as hunting wild game, tanning, butchery and farming increase exposure to zoonoses (Waldron 1989; Reber 1999) and dangers of trauma. Subsistence strategies and changes in economy and diet all differentially affect potential health hazards in terms of disease vectors and risk of injury (Roberts and Manchester 2005).

Data regarding "occupations" can aid in the interpretation of osteoarthritis and markers of activity on the skeletal material (see Chapter 28 by Waldron, and Chapter 29 by Jurmain et al., in this volume for further discussion on this topic). Evidence from various types of subsistence strategies such as hunting and gathering, fishing, and agriculture can help us to understand the effects of these daily activities on the skeleton. Artifacts found with burials or in domestic sites offer clues regarding the types of activities in which the population participated. For instance, the excavation of the Mary Rose ship revealed numerous longbows and arrows indicative of their archer occupation (Stirland 2000). These artifacts have been linked with the presence of *os acromiale*, as well as increased robustness and size of the left humerus. The large *latte* stones used at pillars in Chamorro, Guam houses may have caused spondylolysis in the prehistoric population due to the necessary lifting, pulling and moving of these very heavy objects (Arriaza 1997). Molnar (2006) linked musculoskeletal marker patterns with artifactual evidence for archery and harpooning in a population from

Gotland. Modifications of the joints and ankle bones found in individuals from Abu Hureyra are associated with the use of quernstones and/or mortars to prepare food (Molleson 1989).

Evidence for activity patterns also includes indications and interpretations of trauma on the skeleton. Interpreting the context of injury can be done through the integration of archaeological data such as settlement patterns, weaponry and iconography. The location and character of settlements may indicate defensive behaviors aimed at safety such as relocating to a steep slope or a more compact settlement (Haas and Creamer 1993; Lambert 2002). Settlements may also include defensive features such as walls, forts or moats. For instance, Wari iconography from the Peruvian Andes depicting military themes and the distribution of Wari administrative sites in conjunction with cranial trauma are used to support the interpretation of high levels of violence in the society (Tung 2007). Identifying the behavior behind the injury is difficult, though compelling contextual evidence, such as the presence of artifacts, can help. The presence of artifacts that appear to be designed and used for human-directed violence can offer useful collaborative evidence for interpersonal violence. Projectile points embedded in bone or recovered from the abdominal cavity as well as sinew around the neck provide strong evidence of injuries due to interpersonal violence (Roberts and Manchester 2005). Walker and Lambert (1989) found arrow points associated with and imbedded in the bone of individuals from a prehistoric Ventura site in California. Archaeological evidence for local craft specialization, intensification of intervillage competition for resources and ranked social systems in combination with skeletal analysis revealed the poor health of this group. Iconography can provide details of battle scenes or other activities, though these depictions must be interpreted carefully by examining the intended purpose (Lambert 2002). Data regarding subsistence activities, technology and occupation can also help us understand traumatic injury patterns (Lovell 2008). The correlation of injury types with contextual social information allows for more accurate interpretation. For example, the wooden staff represented in tomb paintings of ancient Nubian and Egyptian artistic representations of sport and battle and the throwstick found in burials at the Kerma site can be linked with specific skeletal injuries (Judd 2004, 2006). Similarly, depictions of sports and exercise by Egyptians, such as wrestling, may help explain the presence of pubic symphyseal face eburnation at Hierakonpolis (Judd 2010).

Infectious disease and diet

Our understanding of infectious disease and diets in the past can be furthered by the inclusion of archaeological data. Parasites and their eggs can be found in burial soil and mummified remains, as well as in the excavation of settlement sites features such as latrines and trash pits, where parasites are passed in feces of humans and animals (Reinhard 1990, 1992; Bouchet et al. 2003). Through microscopic examination and DNA analysis specific diagnoses and confirmation of infection can be made. Preserved paleoparisitological remains can aid in our knowledge of past disease and the evolution of human biology and social behaviors (Bouchet et al. 2003). Innovative methods for identifying parasites are expanding our ability to construct the etiology of osteological and dental indicators of physical stress (Reinhard and Bryant 2008).

Palynological research has identified three categories of pollen that can be found: background pollen rain from plant types that have usually travelled long distances, pollen from plants incorporated into drinking water, and dietary plants consumed as food sources. Digestive tract analyses in conjunction with what is found in the midden are important for investigating diet, mortuary practices, seasonality and taphonomy (Berg 2002). When proper methods for the pollen study of archaeological materials are followed, such as the inclusion of control samples, extraction in a contamination-free laboratory, comparison with modern reference samples, and critical interpretation, this type of analysis can provide meaningful and useful reconstructions of past diets (Reinhard et al. 2007). Through the microscopic examination of food remains the contamination of fruit with many species of *Streptomyces* at the Roman site of Herculaneum was revealed. This contamination may have produced natural antibiotics and may have been responsible for natural fluorescence found in the human skeletal material and low level of non-specific infection in the sample (Capasso 2007). At Canyon de Chelly, elevated levels of porotic hyperostosis were correlated with coprolite analysis results revealing the dietary impact of environmental deterioration, increased reliance on wild plant foods and greater parasite diversity (Reinhard and Bryant 2008). Similarly, Bathurst (2005) linked skeletal signs of nutritional deficiency with evidence of intestinal parasites along the Pacific coast of Canada.

The types of tools used to grind grain can have an impact on rates of tooth-wear. Through the analysis of microwear, Teaford and Lytle (1996) have demonstrated that grain ground by large-grained rocks such as sandstone results in higher wear rates than fine-grained igneous rocks. Wear rates of various populations can be interpreted in conjunction with different food-processing techniques as well as microscopic analysis of plant dietary abrasives and processing abrasives in order to gain insight on past diets (Reinhard and Bryant 2008; Teaford and Lytle 1996). The combination of archaeoparisitology, botanical and faunal analyses, settlement types and patterns, and tool use with nutritional conditions reflected in skeletal and dental tissues as well as stable isotope data, provides us with a more complete reconstruction of diet and understanding of health conditions experienced by past peoples (Hillson 2008; Katzenberg 2008; Knüsel 2010). Buzon and Grauer (2002) integrated paleopathological analysis with archaeological indications of subsistence strategies at the SU Site in New Mexico to reveal a population transitioning to agricultural intensification. Working with prehistoric Japanese samples in conjunction with information regarding tools for food procurement, botanical and faunal data, Temple (2010) found few differences in systemic stress between groups practicing foraging and agriculture. Klaus et al. (2010) combined information regarding settlement patterns, diet, ancient DNA and lesions associated with tuberculosis on the north coast of Peru, to suggest that immunity to the disease was closely linked with diet.

Injuries and treatment

Injuries and other physical ailments have affected all societies in the past; skeletal remains and artifacts provide evidence for ancient treatments. For example, finds from ancient Egypt substantiate suggestions of medical therapy. Nerlich and colleagues (2000) describe a New Kingdom mummy from Thebes-West with an amputated right big toe that had been replaced by a prosthetic wood toe attached by leather strings. Smith (1908) reported bark splints tied with linen bandages on unhealed limbs of Egyptian mummies. It has been suggested that trepanation, surgery of the scalp and removal of the underlying bone, was used to treat aliments due to its association with increased intracranial pressure and head wounds (Roberts and Manchester 2005). The many examples of trepanned skulls worldwide can sometimes be associated with the tools used in the procedure. Rifkinson-Mann (1988) discusses the instruments found in the burial caves of Paracas finding that early tools were triangular knifelike obsidian instruments in a wooden handle and later tools were T-shaped or anchor-shaped. Evidence for the use of medicines is difficult to establish and reconstruct for past populations. The analysis of burial soils samples and coprolites can provide sources of data. For example, plant species consumed near the time of death may have been used for medicinal purposes. The investigation of medicinal use can shed light on social, economic and health factors in the population (Berg 2002). Several medicinal plants were found in Danish burial soil samples, including *Hypericum*, now used as an antidepressant. In Arizona burial soil samples, *Larrea*, used to treat diarrhea and bladder ailments, was the most frequently found medicinal plant species (Berg 2002). The inclusion of these material remains helps us to understand the context of these ancient treatments. The analysis of treatments also opens a window into our understanding of disability in the past—how individuals were affected by their impairments. The archaeological context of human remains is important in assessing the social reaction to physical impairment (Knüsel 1999). Although the study of disability in the past in its early stages, this research path can contribute to our view of the social effects of health conditions (Buikstra 2010).

CONCLUSION

The incorporation of the bioarchaeological perspective in paleopathological studies provides for deeper and more meaningful research questions and interpretations. Additionally, this approach allows for a broader contribution of past health and disease to larger anthropological topics. The practice of contextual bioarchaeological research in paleopathology will serve to strengthen the discipline and promote its integration into related fields. All aspects of paleopathological research can benefit from this approach. The development of research questions investigating health and disease in the past is improved with a broad contextual approach with archaeology. The full participation of the paleopathologist in archaeological excavation and analysis from the planning stages through the completion of the project is also extremely beneficial, as an osteologist can offer essential skills and theoretical perspective. The inclusion of archaeological data in the interpretation of paleopathological data offers a means to deal with interpretative challenges. It makes it possible to build up a better, more complete understanding of sample representativeness and factors that result in individual heterogeneity and subgroups, such as age, sex and gender, ethnic background, living conditions, diet, activities and medical treatment. Collaborative endeavors present paleopathology with a bright future full of valuable and thought-provoking contributions to reconstruct the history and lifeways of humans.

ACKNOWLEDGMENTS

I am thankful to Anne Grauer for the invitation to contribute to this volume and for her helpful advice both on this chapter and the development of my bioarchaeological research over the years. I dedicate this chapter to the late Phillip L. Walker (1947–2009), my graduate advisor and friend, who actively practiced this approach and inspired others in bioarchaeology.

REFERENCES

Angel, J. L., 1946   Social Biology of Greek Culture Growth. American Anthropologist 48:493–553.
Arnold, B., and Wicker, N. L., 2001   Introduction. *In* Gender and the Archaeology of Death. B. Arnold, and N. L. Wicker, eds. pp. vii–xxi. Lanham, MD: AltaMira Press.
Arriaza, B. T., 1997   Spondylolysis in Prehistoric Human Remains from Guam and its Possible Etiology. American Journal of Physical Anthropology 104:393–397.
Aufderheide, A. C., and Rodríguez–Martín, C., 1998   The Cambridge Encyclopedia of Human Paleopathology. Cambridge: Cambridge University Press.
Bathurst, R. R., 2005   Archaeological Evidence of Intestinal Parasites from Coastal Shell Middens. Journal of Archaeological Science 32:115–123.
Beck, L. A., 2006   Kidder, Hooton, Pecos, and the Birth of Bioarchaeology. *In* Bioarchaeology: The Contextual Analysis of Human Remains. J. E. Buikstra, and L. A. Beck, eds. pp. 83–94. New York: Academic Press.
Berg, G. E., 2002   Last Meals: Recovering Abdominal Contents from Skeletonized Remains. Journal of Archaeological Science 29:1349–1365.
Blake, E., 1999   Identity-Mapping in the Sardinian Bronze Age. European Journal of Archaeology 2:35–55.
Bouchet, F., Guidon, N., Dittmar, K., Harter, S., Fernando Ferreira, L., Chaves, S. M., Reinhard, K., and Araújo, A., 2003   Parasite Remains in Archaeological Sites. Memorias do Instituto do Oswaldo Cruz 98:47–52.
Brothwell, D., 1963   Digging Up Bones. London: British Museum (Natural History).
Buikstra, J. E., 1991   Out of the Appendix and into the Dirt: Comments on Thirteen Years of Bioarchaeological Research. *In* What Mean These Bones? Studies in Southeastern Bioarchaeology. M. L. Powell, P. S. Bridges, and A. M. Wagner, eds. pp. 172–188. Tuscaloosa: University of Alabama Press.
Buikstra, J. E., 2006a   A Historical Introduction. *In* Bioarchaeology: The Contextual Analysis of Human Remains. J. E. Buikstra, and L. A. Beck, eds. pp. 7–26. New York: Academic Press.
Buikstra, J. E., 2006b   On to the 21st century: Introduction. *In* Bioarchaeology: The Contextual Analysis of Human Remains. J. E. Buikstra and L. A. Beck, eds. pp. 347–358. New York: Academic Press.
Buikstra, J. E., 2010   Paleopathology: A Contemporary Perspective. *In* A Companion to Biological Anthropology. C. S. Larsen, ed. pp. 395–411. Oxford: Blackwell.
Buikstra, J. E., and Ubelaker, D. H., 1994   Standards for Data Collection from Human Skeletal Remains: Proceedings of a Seminar the Field Museum of Natural History. Arkansas Archaeological Survey Research Series No. 44. Fayetteville: Arkansas Archaeological Survey.
Buzon, M. R., 2006a   Health of Non-Elites at Tombos: Nutritional and Disease Stress in New Kingdom Nubia. American Journal of Physical Anthropology 130:26–37.
Buzon, M. R., 2006b   Biological and Ethnic Identity in New Kingdom Nubia: A Case Study from Tombos. Current Anthropology 47:683–695.

Buzon, M. R., and Grauer, A. L., 2002 A Bioarchaeological Analysis of Subsistence Strategies at the SU site, New Mexico. Kiva 68:5–24.

Buzon, M. R., and Judd, M. A., 2008 Investigating Health at Kerma: Sacrificial versus Non-sacrificial Burials. American Journal of Physical Anthropology 136:93–99.

Buzon, M. R., Eng, J. T., Lambert, P., and Walker, P. L., 2005 Bioarchaeological Methods. *In* Handbook of Archaeology Methods, vol. II. H. Maschner, and. C. Chippindale, eds. pp. 871–918. Lanham, MD: AltaMira Press.

Capasso L., 2007 Infectious Disease and Eating Habits at Herculaneum (1st Century AD, southern Italy). International Journal of Osteoarchaeology 17:350–357.

Conkey, M. W., and Gero, J. M., 1991 Tensions, Pluralities and Engendering Archaeology: An Introduction to Women and Prehistory. *In* Engendering Archaeology: Women and Prehistory. J. M. Gero, and M. W. Conkey, eds. pp. 3–30. Oxford: Blackwell Publishers.

Conkey, M. W., and Spector, J. D., 1984 Archaeology and the Study of Gender. Advances in Archaeological Method and Theory 7:1–38.

Cook, D. C., and Powell, M. L., 2006 The Evolution of American Paleopathology. *In* Bioarchaeology: The Contextual Analysis of Human Remains. J. E. Buikstra, and. L. A. Beck, eds. pp. 281–322. New York: Academic Press.

DeCorse, C. R., 1999 Oceans Apart: Africanist Perspectives on Diaspora Archaeology. *In* "I, Too, Am America": Archaeological Studies of African-American Life. T. A. Singleton, ed. pp. 142–55. Charlottesville: University Press of Virginia.

Dietz, A., Senneweld, E., and Maier, H., 1995 Indoor Air Pollution by Emissions of Fossil Fuel Single Stoves: Possibly a Hitherto Underrated Risk Factor in the Development of Carcinomas in the Head and Neck. Otalaryngology – Head and Neck Surgery 112:308–351.

Duday, H., 2006 L'archéothanatologie ou L'archéologie de la Mort (Archaeothanatology or the Archaeology of Death). Translated by C. J. Knüsel. *In* Social Archaeology of Funerary Remains. R. L. Gowland, and C. J. Knüsel, eds. pp. 30–56. Oxford: Oxbow.

Emberling, G., 1997 Ethnicity in Complex Societies: Archaeological Perspectives. Journal of Archaeological Research 5: 295–344.

Fay, I., 2006 Text, Space and the Evidence of Human Remains in English Late Medieval and Tudor Disease Culture: Some Problems and Possibilities. *In* Social Archaeology of Funerary Remains. R. L. Gowland, and C. J. Knüsel, eds. pp. 190–208. Oxford: Oxbow.

Glassow, M. A., 2005 Excavation. *In* Handbook of Archaeological Methods, vol. 1. H. D. G. Maschner, and. C. Chippindale, eds. pp. 133–175. Lanham: Altamira Press.

Glencross, B., and Sawchuk, L., 2003 The Person-Years Construct: Ageing and the Prevalence of Health-Related Phenomena from Skeletal Samples. International Journal of Osteoarchaeology 13:369–374.

Goldstein, L., 2006 Mortuary Analysis and Bioarchaeology. *In* Bioarchaeology: The Contextual Analysis of Human Remains. J. E. Buikstra, and L. A. Beck, eds. pp. 375–388. New York: Academic Press.

Goodman, A. H., 1993 On the Interpretation of Health from Skeletal Remains. Current Anthropology 34:281–288.

Goodman, A. H., 1994 Cartesian Reductionism and Vulgar Adaptationism: Issues in the Interpretation of Nutritional Status in Prehistory. *In* Paleonutrition: The Diet and Health of Prehistoric Americans. K. D. Sobolik, ed. pp. 163–177. Carbondale: Southern Illinois University.

Goodman, A. H., and Armelagos, G. J., 1989 Infant and Childhood Morbidity and Mortality Risks in Archaeological Populations. World Archaeology 21:225–243.

Gowland, R., 2006 Ageing the Past: Examining Age Identity from Funerary Evidence. *In* Social Archaeology of Funerary Remains. R. L. Gowland, and C. J. Knüsel, eds. pp. 143–154. Oxford: Oxbow.

Grauer, A. L., 2008 Macroscopic Analysis and Data Collection in Paleopathology. *In* Advances in Human Paleopathology. R. Pinhasi, and S. Mays, eds. pp. 57–76. Chichester: John Wiley.

Grauer, A. L., and Stuart-Macadam, P., eds., 1998   Sex and Gender in Paleopathological Perspective. Cambridge: Cambridge University Press.

Haas, J., and Creamer, W., 1996   The Role of Warfare in the Pueblo III Period. *In* Pueblo Cultures in Transition. M. A. Adler, ed. pp. 205–213. Tucson: University of Arizona Press.

Halcrow, S. E., and Tayles, N., 2008   The Bioarchaeological Investigation of Childhood and Social Age: Problems and Prospects. Journal of Archaeological Method and Theory 15: 190–215.

Hall, J. M., 1997   Ethnic Identity in Greek Antiquity. Cambridge: Cambridge University Press.

Hillson, S., 2008   Dental Pathology. *In* Biological Anthropology of the Human Skeleton. 2nd edition. M. A. Katzenberg, and S. R. Saunders, eds. pp. 301–340. Hoboken NJ: John Wiley.

Hodder, I., 1982   The Present Past: An Introduction to Anthropology for Archaeologists. London: Batsford.

Hooten, E. A., 1930   The Indians of Pecos Pueblo: A Study of their Skeletal Remains. Papers of the Southwestern Expedition, vol. 4. New Haven: Yale University Press.

Hrdlička, A., 1904   Directions for Collecting Information and Specimens for Physical Anthropology. Bulletin of the United States National Museum. No. 39. Washington DC: Government Printing Office.

Jones, S., 1997   The Archaeology of Ethnicity: Constructing Identities in the Past and Present. London: Routledge.

Judd, M. A., 2004   Trauma in the City of Kerma: Ancient versus Modern Injury Patterns. International Journal of Osteoarchaeology 14:34–51.

Judd, M. A., 2006   Continuity of Interpersonal Violence between Nubian Communities. American Journal of Physical Anthropology 131: 324–333.

Judd, M. A., 2010   Pubic Symphyseal Face Eburnation: An Egyptian Sport Story? International Journal of Osteoarchaeology 20:280–290.

Kamp, K. A., 2001   Where Have All the Children Gone? The Archaeology of Childhood. Journal of Archaeological Method and Theory 8:1–34.

Katzenberg, M. A., 2008   Stable Isotope Analysis: A Tool for Studying Past Diet, Demography, and Life History. *In* Biological Anthropology of the Human Skeleton. 2nd edition. M. Anne Katzenberg, and Shelley R. Saunders, eds. pp. 413–441. Hoboken: John Wiley & Sons.

Keene, D., 1983   Medieval Urban Environment in Documentary Records. Archives 16:137–144.

Kent, S., 1986   The Influence of Sedentism and Aggregation on Porotic Hyperstosis and Anaemia: A Case Study. Man 21:605–636.

Klaus, H. D., Wilbur, A. K., Temple, D. H., Buikstra, J. E., Stone, A. C., Fernandez, M., Wester, C., and Tam., M. E., 2010   Tuberculosis on the North Coast of Peru: Skeletal and Molecular Paleopathology of Late Pre-Hispanic and Postcontact Mycobacterial Disease. Journal of Archaeological Science 37:2587–2597 doi:10.1016/j.jas.2010.05.019.

Knudson, K. J., and Stojanowski, C. M., 2008   New Directions in Bioarchaeology: Recent Contributions to the Study of Human Social Identities. Journal of Archaeological Research 16:397–432.

Knudson, K. J., and Stojanowski, C. M., 2009   Bioarchaeology and Identity in the Americas. Gainesville: University Press of Florida.

Knüsel, C. J., 1999   Orthopaedic Disability: Some Hard Evidence. Cambridge Archaeological Review 15:31–53.

Knüsel, C. J., 2010   Bioarchaeology: A Synthetic Approach. Bulletin et Memoires de la Societe d'Anthropologie de Paris 22:62–73.

Lambert, P. M., 2002   The Archaeology of War: A North American Perspective. Journal of Archaeological Research 10:207–241.

Lancaster, H. O., 1990   Expectations of Life: A Study on the Demography, Statistics, and History of World Mortality. London: Springer-Verlag.

Larsen, C. S., 2006   The Changing Face of Bioarchaeology: An Interdisciplinary Science. *In* Bioarchaeology: The Contextual Analysis of Human Remains. J. E. Buikstra, and L. A. Beck, eds. pp. 359–374. New York: Academic Press.

Larsen, C. S., and Walker, P. L., 2010   Bioarchaeology: Health, Lifestyle and Society. *In* A Companion to Biological Anthropology. C. S. Larsen, ed. pp. 379–394.Oxford: Blackwell.

Larson, T. V., and Koenig, J. Q., 1994   Wood Smoke: Emissions and Noncancer Respiratory Effects. Annual Review of Public Health 15:133–156.

Lewis, M. E., 2010   Life and Death in a *Civitas* Capital: Metabolic Disease and Trauma in the Children from Late Roman Dorchester, Dorset. American Journal of Physical Anthropology 142:405–416.

Lightfoot, K. G., and Martinez, A., 1995   Frontiers and Boundaries in Archaeological Perspective. Annual Review of Anthropology 24:471–92.

Lovell, N. C., 2008   Analysis and Interpretation of Skeletal Trauma. *In* Biological Anthropology of the Human Skeleton. 2nd edition. M. A. Katzenberg, and S. R. Saunders, eds. pp. 341–386. Hoboken, NJ: John Wiley & Sons.

Martin, D. L., 1998   Owning the Sins of the Past: Historical Trends, Missed Opportunities, and New Directions in the Study of Human Remains. *In* Building a New Biocultural Synthesis: Political-Economic Perspectives on Human Biology. A. H. Goodman, and T. L. Leatherman, eds. pp. 171–190. Ann Arbor: The University of Michigan Press.

Mays, S., 1998   The Archaeology of Human Bones. London: Routledge.

Meskell, L. M., 1994   Dying Young: The Experience of Death at Deir el Medineh. Archaeological Review from Cambridge 13:35–45.

Molleson, T., 1989   Seed Preparation in the Mesolithic: The Osteological Evidence. Antiquity 63:356–62.

Molnar, P., 2006   Tracing Prehistoric Activities: Musculoskeletal Stress Marker Analysis of a Stone-Age Population on the Island of Gotland in the Baltic Sea. American Journal of Physical Anthropology 136:423–431.

Nawrocki, S. P., 1995   Taphonomic Processes in Historic Cemeteries. *In* Bodies of Evidence: Reconstructing History through Skeletal Analysis. A. L. Grauer, ed. pp. 49–66. New York: Wiley–Liss.

Nerlich, A. G., Zinc, A., Szeimies, U., and Hagedorn, H. G., 2000   Ancient Egyptian Prosthesis of the Big Toe. Lancet 356:2176–2179.

O'Gorman, J. A., 2001   Life, Death, and the Longhouse: A Gendered View of Oneonta Social Organization. *In* Gender and the Archaeology of Death. B. Arnold, and N. L. Wicker, eds. pp. 23–50. Lanham, MD: AltaMira Press.

Ortner, D. J., 1998   Infectious Disease, Sex, and Gender: The Complexity of It All. *In* Sex and Gender in Paleopathological Perspective. A. L. Grauer, and P. Stuart-Macadam, eds. pp. 79–92. Cambridge: Cambridge University Press.

Perry, M., 2006   Refining Childhood Through Bioarchaeology: Toward an Archaeological and Biological Understanding of Children in Antiquity. Archaeological Papers of the American Anthropological Association 15:89–111.

Pinhasi, R., and Bourbou, C., 2008   How Representative Are Human Skeletal Assemblages for Population Analysis? *In* Advances in Human Paleopathology. R. Pinhasi, and S. Mays, eds. pp. 31–44. Chichester: John Wiley & Sons.

Pinhasi, R., and Turner, K., 2008   Epidemiological Approaches in Paleopathology. *In* Advances in Human Paleopathology. R. Pinhasi, and S. Mays, eds. pp. 45–56.Chichester: John Wiley & Sons.

Powell, M. L., 1991   Ranked Status and Health in the Mississippian Chiefdom at Moundville. *In* What Mean These Bones? Studies in Southeastern Bioarchaeology. M. L. Powell, P. S. Bridges, and A. M. Wagner, eds. pp. 22–51. Tuscaloosa: The University of Alabama Press.

Reber, V. B., 1999   Blood, Coughs and Fever: Tuberculosis and the Working Classes of Buenos Aires, Argentina 1885–1915. Social History of Medicine 12:73–100.

Redfern, R., and Roberts, C., 2005   Health in Romano-British Urban Communities: Reflections from Cemeteries. *In* Fertile Ground: Papers in Honour of Susan Limbrey. D. N. Smith, M. B. Brickley, and W. Smith, eds. pp. 115–129. Oxford: Oxbow Books.

Reinhard, K. J., 1992   Parasitology as an Interpretive Tool in Archaeology. American Antiquity 57:231–245.

Reinhard, K. J., and Bryant, V. M. Jr., 2008   Pathoecology and the Future of Coprolite Studies in Bioarchaeology. *In* Reanalysis and Reinterpretation in Southwestern Bioarchaeology. A. W. M. Stodder, ed. pp. 205–224. Tempe: Arizona University Press.

Reinhard, K. J., 1990   Archaeoparasitology in North America. American Journal of Physical Anthropology 82:145–162.

Reinhard, K. J., Bryant, V. M., and Vinton, S. D., 2007   Comment on Reinterpreting the Pollen Data from Dos Cabezas. International Journal of Osteoarchaeology 17:531–541.

Rifkinson-Mann, S., 1988   Cranial Surgery in Ancient Peru. Neurosurgery 23:411–416.

Robb, J., Bigazzi, R., Lazzarini, L., Scarsini, C., and Sonego, F., 2001   Social Status and Biological Status: A Comparison of Grave Goods and Skeletal Indicators from Pontecagnano. American Journal of Physical Anthropology 115:213–222.

Roberts, C. A., 2000   Infectious Disease in Biocultural Perspective: Past, Present and Future Work in Britain. *In* Human Osteology in Archaeology and Forensic Science. M. Cox, and S. Mays, eds. pp. 145–162. London: Greenwich Medical Media Ltd.

Roberts, C. A., 2006   A View from Afar: Bioarchaeology in Britain. *In* Bioarchaeology: The Contextual Analysis of Human Remains. J. E. Buikstra, and L. A. Beck, eds. pp. 417–439. New York: Academic Press.

Roberts, C. A., 2009   Human Remains in Archaeology: A Handbook. York: Council for British Archaeology.

Roberts, C. A., and Manchester, K., 2005   The Archaeology of Disease. 3rd edition. Ithaca, NY: Cornell University Press.

Santley, R. S., Yarborough, C. M., and Hall, B. A., 1987   Enclaves, Ethnicity, and the Archaeological Record at Matacapan. *In* Ethnicity and Culture. R. Auger, M. F. Glass, S. MacEachern, and P. H. MacCartney, eds. pp. 85–100. Calgary: Archaeological Association of the University of Calgary.

Saunders, S. R., 2008   Juvenile Skeletons and Growth-Related Studies. *In* Biological Anthropology of the Human Skeleton. 2nd edition. M. A. Katzenberg, and S. R. Saunders, eds. pp. 117–148. Hoboken: John Wiley & Sons.

Schultz, M., 1997   Microscopic Investigation of Excavated Skeletal Remains: A Contribution to Paleopathology and Forensic Medicine. *In* Forensic Taphonomy: The Postmortem Fate of Human Remains. W. D. Haglund, and M. H. Sorg, eds. pp. 201–222. Boca Raton. LA: CRC Press.

Smith, G. E., 1908   The Most Ancient Splints. The British Medical Journal 1:732–734.

Smith, P., and Bar-Gal, K. G., 1992   Identification of Infanticide in Archaeological Sites: A Case Study from the Late Roman–Early Byzantine Periods at Ashkelon, Israel. Journal of Archaeological Science 19:667–675.

Smith, S. T., 2003   Wretched Kush: Ethnic Identities and Boundaries in Egypt's Nubian Empire. London: Routledge.

Sofaer, J. R., 2006   The Body as Material Culture: A Theoretical Osteoarchaeology. Cambridge: Cambridge University Press.

Steinbock, R. T., 1990   Studies in Ancient Calcified Tissues and Organic Concretions III: Gallstones (Cholelithiasis). Journal of Paleopathology 3:95–106.

Stirland, A. J., 2000   Raising the Dead: The Skeleton Crew of Henry VIII's Great Ship, the Mary Rose. Chichester: John Wiley & Sons.

Stodder, A. L. W., 2008   Taphonomy and the Nature of Archaeological Assemblages. *In* Biological Anthropology of the Human Skeleton. 2nd edition. M. A. Katzenberg, and S. R. Saunders, eds. pp. 71–114. Hoboken. NJ: John Wiley & Sons.

Storey, R., 1997   Individual Frailty, Children of Privilege, and Stress in Late Classic Copán. *In* Bones of the Maya: Studies of Ancient Skeletons. S. L. Whittington, and D. M. Reed, eds. pp. 116–126. Washington: Smithsonian Institution Press.

Stuart-Macadam, P., 1992   Porotic Hyperostosis: a New Perspective. American Journal of Physical Anthropology 87:39–47.

Sutter, R., and Verano, J., 2007   Biodistance Analysis of the Moche Sacrificial Victims from Huaca de la Luna Plaza 3C: Matrix Method Test of their Origins. American Journal of Physical Anthropology 132(2): 193–206.

Teaford, M. F., and Lytle, J. D., 1996   Brief Communication: Diet-Induced Changes in Rates of Human Tooth Microwear: A Case Study Involving Stone-Ground Maize. American Journal of Physical Anthropology 100:143–147.

Temple, D. H., 2010   Patterns of Systemic Stress during the Agricultural Transition in Prehistoric Japan. American Journal of Physical Anthropology 142:112–124.

Torres-Rouff, C., 2008   The Influence of Tiwanaku on Life in the Chilean Atacama: Mortuary and Bodily Perspectives. American Anthropologist 110:325–337.

Tung, T. A., 2007   Trauma and Violence in the Wari Empire of the Peruvian Andes: Warfare, Raids, and Ritual Fights. American Journal of Physical Anthropology 133:941–956.

Ubelaker, D. H., 1989   Human Skeletal Remains: Excavation, Analysis, and Interpretation. 2nd edition. Washington, DC: Taraxacum.

Van Gerven, D. P., and Armelagos, G. J., 1983   "Farewell to Paleodemography?" Rumors of Its Death Have Been Greatly Exaggerated. Journal of Human Evolution 12:353–360.

Waldron, T., 1989   The Effects of Urbanisation on Human Health. *In* Diet and Crafts in Towns: The Evidence of Animal Remains from the Roman to the Post-Medieval Periods. D. Serjeantson, and T. Waldron, eds. pp. 55–73. British Archaeological Reports British Series 199. Oxford: British Archaeological Reports.

Waldron, T., 1994   Counting the Dead: The Epidemiology of Skeletal Populations. Chichester: John Wiley & Sons.

Walker, P. L., 1986   Porotic Hyperostosis in a Marine-Dependent California Indian Population. American Journal of Physical Anthropology 69:345–354.

Walker, P. L., 1995   Problems of Preservation and Sexism in Sexing: Some Lessons from Historical Collections for Paleodemographers. *In* Grave Reflections: Portraying the Past through Skeletal Studies. A. Herring, and S. R. Saunders, eds. pp. 31–47. Toronto: Canadian Scholars' Press.

Walker, P. L., and Lambert, P., 1989   Skeletal Evidence for Stress During a Period of Cultural Change in Prehistoric California. *In* Advances in Paleopathology. L. Capasso, ed. pp. 207–212. Chieti: Marino Solfanelli.

Walker, P. L., Bathurst, R. R., Richman, R., Gjerdrum, T., and Andrushko, V. A., 2009   The Causes of Porotic Hyperostosis and Cribra Orbitalia: A Reappraisal of the Iron-Deficiency Anemia Hypothesis. American Journal of Physical Anthropology 139:109–125.

Walker, P. L., Johnson, J. R., and Lambert, P. M., 1988   Age and Sex Biases in the Preservation of Human Skeletal Remains. American Journal of Physical Anthropology 76:183–188.

Wells, C., 1964   Bones, Bodies and Disease: Evidence of Disease and Abnormality in Ancient Man. London: Thames and Hudson.

Wood, J. W., Milner, G. R., Harpending, H. C., and Weiss, K. M., 1992   The Osteological Paradox: Problems of Inferring Prehistoric Health from Skeletal Samples. Current Anthropology 33(4):343–370.

Wright, L. E., and Yoder, C. J., 2003   Recent Progress in Bioarchaeology: Approaches to the Osteological Paradox. Journal of Archaeological Research 11:43–70.

CHAPTER 5

The Molecular Biological Approach in Paleopathology

James H. Gosman

INTRODUCTION

The interests of paleopathologists are broad-ranging, many of which ultimately converge upon bone. This chapter has been conceived to present current research and emerging fundamental concepts of bone science, intended to demonstrate the underlying molecular and biological mechanisms for many of the skeletal changes observed. These include evolving perspectives on the function of the various bone cells, molecular biology and regulatory/signaling systems; mechanobiological processes including remodeling, microdamage, and functional adaptation; and recent research into immune system and bone biology interactions. What follows is not meant to be a primer nor an exhaustive review of the biology of bone, but rather a consideration of selected aspects of this vast literature from a systems biology perspective, which focuses on complex interactions at a scale of analysis relatively unfamiliar to paleopathology. The reader is referred to basic texts for a refresher (if necessary) on basic aspects of bone structure and composition (e.g. Ortner and Putschar 1985; White and Folkens 2000; Agarwal and Stout 2003; and http://depts.washington.edu/bonebio/ASBMRed/ASBMRed.html). In addition, the content of this chapter may be best digested in small segments, possibly with a "Wikipedia" at hand.

This chapter is organized as a three-part sonata form: exposition of essential players and processes, development of complexities, and theme and variations by example. The theme of the chapter is to highlight the importance of an integrative approach linking bone science, bone morphology, and biocultural interpretation. The objective of this chapter is to emphasize the recent and developing understandings of the

A Companion to Paleopathology, First Edition. Edited by Anne L. Grauer.
© 2012 John Wiley & Sons, Ltd. Published 2016 by John Wiley & Sons, Ltd.

biological systems in play at the crossroads of paleopathology. Recent reappraisals of well-established paleopathological interpretations in regards to such conditions as porotic hyperostosis (Walker et al. 2009), periosteal reaction (Weston 2008), musculo-skeletal stress markers (Villotte et al. 2010), and degenerative joint disease (Weiss and Jurmain 2007, and see Chapter 29 by Jurmain et al., this volume) have suggested that an understanding of the fundamental biological processes at work may shift the significance of these conditions onto a new trajectory.

CELLULAR BIOLOGY AND REGULATORY SYSTEMS

Clarification and expansion of the roles of the essential bone cells and their precursors have been, and remain, topics of intense interest to researchers investigating bone morphogenesis, growth, modeling, remodeling, repair, adaptation, and response to disease. The fundamental triad of differentiated cells comprises osteoblasts, osteoclasts, and osteocytes. The activity of these cells is influenced and regulated by genetic frameworks, mechanical forces, system hormones (e.g. growth hormone (GH), parathyroid hormone (PTH)), and various cytokines (Datta et al. 2008).

Osteoblasts

The osteoblast lineage consists of mesenchymally derived cells including osteoblasts, osteocytes, and bone-lining cells. In general terms, osteoblasts are mononuclear cells responsible for bone formation. These cells are now shown to be multifunctional, forming bone by synthesizing collagen matrix and secreting calcium phosphate mineral, regulating the activity of osteoclasts, supporting the hematopoietic stem cell niche, and playing an important role in various bone diseases. Osteocytes and bone-lining cells, derived from osteoblasts, are abundantly networked within the bone structure. These cells are thought to regulate metabolic and mechanobiology-related functions such as sensation and transduction of mechanically-derived signals (Pearson and Lieberman 2004; Bonewald 2007). Each cell type is phenotypically unique, with particular identifying biochemical markers (Massaro and Rogers 2004). The osteoblastic differentiation process is triggered by a genetic regulatory cascade (St-Arnaud 2003). This journey can be divided in sequential stages: proliferation, extracellular matrix deposition, matrix maturation and mineralization (Stein and Lian 1993).

Osteoblasts are genetically sophisticated fibroblasts (Ducy et al. 2000) originating in a variety of tissues including the periosteum and endosteum. The only morphological feature specific to osteoblasts versus fibroblasts is the presence of a specific extracellular matrix. Differentiated osteoblasts produce a mucoprotein matrix, called osteoid, in which collagen fibrils are enmeshed. This is followed with mineralization of the osteoid by deposition of inorganic crystals of calcium phosphate on the collagen fibers. Bone formation by differentiated osteoblasts is controlled by complex, hierarchical, homeostatic, biological systems. Modulators of bone mass include a polygenic regulatory system, calcium availability, sex steroids, nutrition, and mechanical usage. The central regulation of osteoblast function features the endocrine system comprising parathyroid hormone (PTH); 1, 25(OH)$_2$ vitamin D; calcitonin; and the sex steroids. These hormones have varied roles in different contexts. Estrogen is particularly important in bone formation and resorption, up-regulating osteoblast activity. Estrogen receptors (ERα and ERβ) in osteocytes and osteoblasts have critical roles in mechanotransduction and the osteogenic response to strain (Pearson and Lieberman 2004). Mechanotransduction is the coupling of extracellular mechanical signals (i.e. fluid flow) with cell signaling molecules to produce changes in gene expression influencing bone metabolism. PTH and vitamin D are mineral-sensitive hormones stimulating osteoclasts and inhibiting osteoblasts. Calcitonin is a mineral-building hormone secreted by the thyroid which up-regulates osteoblasts and down-regulates osteoclasts. Recent research has demonstrated a possible central "master-regulator" of bone formation in the hormone Leptin (Ducy et al. 2000), produced in adipose tissue, inhibiting osteoblast function through hypothalamic and sympathetic nervous system signaling pathways (Harada and Rodan 2003).

Osteoblastic differentiation and function is controlled with temporal and spatial specificity by systems of autocrine and paracrine cell signaling. The search for osteoblastic differentiation factors has lead to the identification of transcription factor Runx2 (named for encoding a nuclear protein homologous to *Drosophila* Runt) as the master switch bridging the gap between osteoblast differentiation and osteoblast function (Karsenty 2001; Harada and Rodan 2003). Runx2, in turn, is regulated by several growth factors (Krause et al. 2008). By interacting with many activators, repressors, and co-regulatory proteins, Runx2 influences the differentiation of pluripotent mesenchymal stem cells toward the osteoblast lineage and determines the efficiency of bone formation (Schroeder et al. 2005).

Another important signaling pathway involved in osteoblast activity is Wnt (named for the *wingless* gene originally identified in *Drosophila*) signaling by way of the cell surface Wnt co-receptor LRP5 (low-density lipoprotein receptor-related protein 5) (Hu et al. 2005; Williams and Insogna 2009). The intracellular Wnt/β-catenin pathway is frequently referred to as the canonical pathway, which activates gene expression in the cell nucleus stimulating osteoblast differentiation and bone formation. This system is negatively modulated (antagonized) by a number of molecules including Dickkopf family members (Dkk1, Dkk2) and sclerostin (an osteocyte-derived SOST gene product) (see Datta et al. 2008 for a review). A simple summary indicates that the important signaling player for up-regulating (increasing) bone formation is the Wnt/ Lrp5/β-catenin pathway. It is the balance between these various components of a complex molecular signaling system that determines, in part, relative bone formation by regulation of osteoblasts and bone metabolism in response to the normal physiological environment, mechanical loading, and/or pathological conditions (Robling et al. 2006; Sawakami et al. 2006).

Research on the effects of aging on osteoblast and osteoprogenitor cells has demonstrated the possible effects of senescence and oxidative damage on these cells in respect to diminished synthetic capacity, a decline in the number of progenitor cells which can be recruited to differentiate into osteoblasts (Nashida and Endo 1999; Almeida et al. 2007b), and a reduced sensitivity to mechanical signals (Donahue et al. 2003). These are important factors (in addition to changes in hormone levels) in the age-related imbalance of bone deposition and resorption. This will be discussed in more detail later in this chapter.

Research now indicates that osteoblasts have functions in addition to their bone-formation duties. A number of concerns in paleopathology revolve around hematopoiesis and the corresponding bone responses to anemia of various etiologies. The bone marrow blood cells all arise from the hematopoietic stem cell (HSC), which is dependent on the maintenance of a supportive microenvironment or niche (Schofield 1978). This niche is characterized by the intersection of hematopoietic, skeletal, and vascular cells and biology (Wu et al. 2009). In essence, all of these cells are living together and continuously "talking" to one another. Recent research has demonstrated a critical role for osteoblasts for regulating and supporting the HSC niche (Adams and Scadden 2006). In addition, within the niche, bone-resorbing osteoclasts are recruited, thus further linking bone remodeling with hematopoiesis (Kollet et al. 2006). A great deal of research is under way in order to understand the complex cross-talk among a diverse population of participant cells. One fascinating aspect of this concept may lie in its application as an explanatory mechanistic model for paleopathology interpretations of bone changes and anemia in ancient (and extant) humans.

Osteocytes

The standard understanding of osteocytes is that they are relatively inactive, differentiated osteoblasts that have become embedded in the mineralized matrix they have created as osteoblasts. However, the concept emerging from recent research is quite the opposite. These cells appear to be "control freaks" as "major regulators of bone mass by integrating hormonal and mechanical loading signals" (Dallas and Bonewald 2008:443), linking PTH and Wnt signaling pathways, and orchestrating modeling/remodeling. In addition, osteocytes have an important role in mineral homeostasis (Bonewald 2007). These abundant cells (Mullender et al. 1996) reside within cavities termed lacunae, making physical contact with each other, osteoblasts, bone lining cells, and possibly osteoclasts by way of cellular processes passing through tunnels called canaliculi (Martin et al. 1998; Donahue et al. 2003). This structure is termed the lacunocanalucular system. Gap-junctions, which are specialized connections between cells, allow osteocytes to form a functional syncytium, linking all cells within bone, forming the basis of the emerging consensus of their role as mechanosensory cells.

Mechanosensitivity of osteocytes has been demonstrated by studies *in vitro* and *in vivo* of pulsating, steady, and oscillating fluid flow, as well as changing substrate strain levels, resulting in increased mRNA expression, gene up-regulation, and meta-bolic synthesis of various active compounds (Donahue et al. 2003). Research data have established the functionality of the coupled gap-junctions between osteocytes themselves and osteoblasts (Yellowley et al. 2000; Alford et al. 2003), supporting the hypothesis that osteocytes appraise mechanical signals and regulate bone adaptation targeting osteoblastic function (Mullender and Huiskes 1997; Bonewald 2007). Kim-Weroha and co-researchers (2008) have demonstrated that *in vivo* osteocytes respond rapidly to mechanical load with an increase in β-catenin signaling within one hour. This is associated with the down-regulation of the two inhibitors of Lrp5-mediated Wnt signaling—sclerostin (Sost) and Dkk1 (Kim-Weroha et al. 2008; Bonewald and Johnson 2008). The end result of this regulatory progression is bone formation in response to mechanical strain, thus providing a molecular framework for the observed morphological changes seen in bone functional adaptation.

Osteocytes have demonstrated the capacity to modify their lacunar surroundings by both removing and/or replacing bone matrix. The implications of these relatively new insights are not fully understood, but changes in lacunar size, conformation, and mineralization are likely to be important in regard to phosphate and possible calcium homeostasis, sensitivity to mechanotransduction, and the mechanical properties of bone. Current research in reference to mineral balance is directed towards the hypothesis that the osteocyte network can function as a mineral-regulating endocrine gland. Researchers have demonstrated that lacunae enlarge under conditions of calcium demand such as pregnancy and lactation, and quickly refill thereafter (Dallas and Bonewald 2008; Qing et al. 2008).

Aging effects on osteocytes include the loss of their ability to remodel their lacunae, a reduction in the density of lacunae and number of osteocytes, and decreased ability of the surviving cells to respond to biochemical and mechanical stimuli (Pearson and Lieberman 2004; Nicolella et al. 2008). Concurrent with these aging changes and apoptosis (programmed cell death) is the accumulation of microcracks/damage in bone, rendering the bone prone to fracture. The potential role of osteocytic cell death in bone diseases, such as osteoporosis, osteoarthritis, and osteonecrosis is thought to involve the degradation of the connectivity and structure of the osteocyte lacuocanalicular system combined with perturbations in molecular interactions (Knothe-Tate et al. 2004).

Osteoclasts

Osteoclasts are the cells responsible for bone removal. They are derived from a fusion of multiple (10–20) hematopoietic precursor cells (mononuclear phagocytes) originating in the bone marrow (Boyle et al. 2003). As part of the macrophage lineage, these cells differentiate at or near the bone surface and function to resorb bone. The mononuclear osteoclast precursors have been shown to reside in a number of tissues including the endosteum and periosteum (Pettit et al. 2008). Osteoclasts, as the dedicated bone resorptive cell, have a crucial role in a wide range of pathological conditions of bone, such as osteoporosis, bone response to cancer, inflammatory arthritis, and Paget's disease of bone.

The mechanism of osteoclastic bone degradation depends upon close physical osteoclast-bone interaction provided by integrins. Integrins are cell/matrix attachment molecules which mediate bone/osteoclast recognition and are responsible for intimate physical attachment (Ross 2008). Osteoclasts adhere to bone by a ruffled surface which creates a seal (Teitelbaum 2000), allowing bone resorption to occur from the effects of decreasing the local pH (secreted hydrochloric acid) and from the secretion of various anti-collagen proteolytic enzymes. The degraded matrix and protein fragments are transported by intracellular vesicles to the nonattached region of the osteoclast, where they are released into the surrounding intracellular fluid (Stenbeck and Horton 2004).

Osteoclast differentiation, activation, and function (osteoclastogenesis) are regulated by at least 24 known genes. Two hematopoietic factors which are necessary and sufficient for osteoclastogenesis and maturation are the TNF (tumor necrosis factor)-related cytokine RANKL (receptor activator for nuclear factor κ β ligand) and the polypeptide growth factor M-CSF-1 (macrophage colony-stimulating factor-1)

(Boyle et al. 2003). These proteins are produced by bone-residing cells including, but not limited to, marrow stromal cells, osteoblasts, and immune cells. RANKL is the critical osteoclastogenic cytokine, required for osteoclast formation and activation. M-CSF-1 has been demonstrated to be important to proliferation, survival, and differentiation of the osteoclast progenitors (Datta et al. 2008).

Regulation of osteoclast function includes important molecules synthesized by osteoblasts (and other cells). Two important molecules are RANK/RANK-L and osteoprotegerin (OPG). Osteoclastogenesis and bone resorption are coordinated by the RANKL/RANK/OPG regulatory axis and RANKL:OPG ratio (Boyle et al. 2003; Robling et al. 2006). RANK-L (cytokine) and RANK (transmembrane signaling receptor) are required for osteoclast differentiation, activation, and bone resorption. OPG (soluble protein) blocks osteoclast formation, bone resorption, and induces apoptosis by acting as a decoy receptor and blocking RANKL binding to its cellular receptor RANK. OPG over-expression blocks osteoclast production resulting in osteopetrosis. OPG deletion results in enhanced bone resorption and osteoporosis. "Expression of RANKL and OPG is therefore coordinated to regulate bone resorption and density positively and negatively by controlling the activation state of RANK on osteoclasts" (Boyle et al. 2003:338). Additional stimulators of osteoclast function are PTH; 1, 25 $(OH)_2$ vitamin D; thyroid hormone; glucocorticoids; IGF-1, and BMP-2 and -4. Additional inhibitors of osteoclast function are calcitonin, nitric oxide, gonadal steroids, and interleukin-1 and -6 (Boyle et al. 2003). There are many other important molecules as well as intracellular signaling pathways (known and yet to be known) contributing to the complex interactions in the bone microenvironment (see Tanaka 2008 for recent review).

The effect of aging on osteoclasts is usually framed in terms of adult skeletal disease (osteoporosis, periodontal disease, rheumatoid arthritis, multiple myeloma, and metastatic cancers) in which there is evidence for excess osteoclastic activity, leading to an imbalance in bone remodeling which favors resorption (Duong and Rodan 2001). Age-related metabolic changes in these cells have recently been identified as related to the cellular effects of oxidative stress (Almeida et al. 2007b). Oxidative stress is the imbalance between the production of damaging reactive molecules (reactive oxygen species) and the cells' ability to mitigate that damage. Bone homeostasis is characterized by the coupling of the bone forming actions of osteoblasts and the bone resorbing actions of osteoclasts. Bone disease frequently is a result of a derangement of this interaction. Research is progressing rapidly in the direction of reconstructing the osteocyte/osteoclast/osteoblast signaling network and its role in bone remodeling and disease.

Implications for paleopathology

The cellular actors and regulatory processes just discussed include emerging and often difficult concepts for bone scientists in many different disciplines. This chapter argues that the *introduction* to the molecular aspects of skeletal biology is an essential component of the *process* of understanding and eventual application of these perspectives to the discipline of paleopathology. These biological systems do have a defining characteristic essential to the development of new research agendas and insights in regards to paleopathological interpretations, namely an integrated network of bone structure

82   JAMES H. GOSMAN

and process. The examination of the multiple and complex biocultural inputs influencing the regulatory control of bone formation and/or resorption can be linked to the morphological changes in skeletal tissue. These connections can stimulate critical thinking for paleopathology, characterized by consideration of an integrated, complex, and multifactorial causation as opposed to the more traditional monocausal reductionist approach.

MODELING/REMODELING

Bone is composed of four surfaces upon which modeling and remodeling occur: the trabecular, endosteal, periosteal, and Haversian envelopes (Martin et al. 1998). Although the effects of the modeling and remodeling processes differ at each surface, a general description of modeling, remodeling, and the organized cellular instrument of remodeling, the basic multicellular unit (BMU) is useful. In addition, this chapter will review recent data describing bi-directional signals between bone cells, which are proving to be important to the control of the bone-remodeling cycle.

Modeling is "the principal mode of bone cell coordination in the growing skeleton" (Parfitt 2003:4). During ontogeny, bone is formed in one location and resorbed in another to accommodate the changes of bone size and shape. This is orchestrated by the combination of genetic, systemic/local biological, and local mechanical factors. As bones grow in length and width, the result of endochondral ossification and periosteal intramembranous bone apposition, they are sculpted by periosteal and endosteal resorption and formation at different locations (Martin et al., 1998). The modeling process involves independent actions of osteoblasts and osteoclasts. The periosteal envelope is especially active in the modeling process during ontogeny and to a much lesser extent during adult life. It is responsible for increasing bone diameter. This redistribution of bone is a relatively continuous process resulting in a net gain in bone mass (Parfitt 2003) corresponding to the increase in body mass during growth and development. The rate and extent of modeling is greatly reduced after skeletal maturity. The end-result of the modeling process is skeletal shape, architecture, and mass appropriate to biological and mechanical requirements. Examples include tibial metaphyseal "cut-back," diaphyseal enlargement or drift, and cranial reshaping to accommodate increase in brain size. The mechanical control of this growth and adaptation process is thought to be explained by the osteocyte/osteoclast/osteoblast signal transduction syncytium responding to increased strain in the local matrix, triggering bone formation or resorption until the strains are normalized (Huiskes et al. 2000; Sommerfield and Rubin 2001).

Bone remodeling (bone maintenance) is the "mechanism of bone replacement in the vertebrate skeleton" (Parfitt 2002:5). It has three apparent purposes: (1) its metabolic function provides a mechanism to maintain calcium homeostasis by promoting the exchange of calcium ions at the bone surface (Martin et al. 1998); (2) its structural function provides a mechanism for skeletal adaptation to the mechanical environment; and (3) its maintenance function provides a mechanism to repair fatigue damage created in bone by repetitive cycles of mechanical loading (Burr 2002). It is argued that the "main purpose of remodeling is to prevent degradation of function (microdamage) as bone becomes older" (Parfitt 2002). The load bearing function of bone is threatened by the accumulation of fatigue microdamage which is targeted for remodeling (Burr 2002). Remodeling continues throughout life with periosteal, Haversian, trabecular, and endosteal envelope specificity. It generally does not affect the size and shape of the bone, but does affect the amount of bone present.

The remodeling process is characterized by the sequentially synchronized "coupled" actions of osteoclasts and osteoblasts occurring on the same surface. This was described by Harold Frost and co-workers (Takahashi et al. 1964). Both systemic and local regulatory systems keep the two processes (bone resorption and bone formation) in balance: perturbations contribute to bone disease. Systemic regulators of remodeling (e.g. PTH, growth hormone, sex hormones) are relatively well understood and are not further discussed in this chapter, other than to comment that they have been shown to modulate the relative expression of RANKL and OPG (among other effects) (Ryser et al. 2009). In addition, PTH is thought to have a regulatory role throughout the remodeling cycle (Henriksen et al. 2009). The discussion to follow focuses on local bone cell structures and cellular interactions.

Basic Multicellular Unit (BMU)

The sequential presentation of remodeling is referred to as the Activation–Resorption–Formation (ARF) cycle (see Henriksen et al. 2009 for recent review). The temporary, cyclic anatomic structure responsible for this cycle has been named the basic multicellular unit (BMU) by Frost (1969) (see also Parfitt 2002). These anatomical units are most readily identified in cortical bone (osteonal). They have also been described in cancellous bone (hemi-osteonal) (Parfitt 1994). Parfitt argues that the BMU in cancellous bone "travels across the surface digging a trench rather than a tunnel, but maintaining its size, shape and individual identity by the continuous recruitment of new cells, just as in cortical bone" (Parfitt 1994:273). A fully developed BMU consists of a team of osteoclasts forming the cutting cone, a team of osteoblasts behind forming the closing cone, some form of vascular supply, and associated connective tissue (Parfitt 1994). The capillary in cortical bone or specialized sinusoid in cancellous bone (Melsen et al. 1995) is at the heart of the BMU, ideally situated to coordinate the coupled functions activation, resorption, and formation (Parfitt 2000). Preosteoclasts, originating in the bone marrow, arrive at the resorption site via the circulation. The individual differentiated osteoclasts are short-lived (12 days), turning over at a rate of 8 percent per day. Osteoblasts from precursors in the local connective tissue, refill the resorbed bone at each successive cross-sectional location, maintaining three-dimensional organization. Some of the osteoblastic team become buried as osteocytes, some die (average life span is measured in weeks), and some become relatively quiescent bone lining cells. The BMU exists and travels in three dimensions, excavating and refilling a trench in cancellous bone of 200μm at about 10μm/day for 100 days, while maintaining the proper spatial and temporal relationships among its cellular elements (Parfitt 1994; Ryser et al. 2009).

The lifespan of the BMU has a beginning (origination), middle (progression), and end (termination); its duration is approximately three months. Activation of remodeling leading to the origination of a BMU is thought to be mediated by the osteocytic production of RANKL in the vicinity of a microfracture. It is described as beginning on a small area of quiescent bone surface and involves digestion of the endosteal membrane by enzymes released from lining cells. These changes in lining cell morphology expose the mineralized bone surface. The process of neoangiogenesis (new blood vessel growth) provides the capability for the egress of circulating mono-nuclear osteoclast precursors at precisely the correct location. Their attraction to the region of exposed mineral and subsequent fusion to form osteoclasts allows the assembly of a sufficient number of osteoclasts to form the cutting cone (Parfitt 2002). Progression is travel in a particular direction for a particular time. Constant re-supply of osteoclast precursors is necessary.

Steering of BMUs is guided in two ways, in the principal stress direction (Burger et al. 2003) and towards microdamage (Martin 2007). Based upon results of finite element analysis, van Oers (2008) propose a model in which strain induced osteocyte signals inhibit osteoclast activity and stimulate osteoblast activity. The model can explain several characteristics of BMUs: osteonal and trabecular load alignment, resorption of dead osteocytes, smaller osteons' diameters in high-strain regions, and double-ended and drifting osteons (Van Oers 2008). After the osteoclasts have completed their tasks, the BMU stops moving forward and switches from bone resorption to bone formation. This is the reversal phase, thought to be guided, in part, by the specific temporal and spatial pattern of the RANKL:OPG ratio and bi-directional Ephrin-Eph signaling (cell-surface molecules implicated in bone homeostasis (Zhao et al. 2006)). The supply of osteoclast precursors is turned off and the existing osteoclasts die by apoptosis. Osteoblast precursors are recruited; the formation phase begins. Bone formation is regulated by osteoblast-derived OPG as well as osteoclast-derived signaling molecules such as IGF-1 and TGF-β. When bone formation is complete, the cycle is terminated. Recent research suggests that the molecule sclerostin, expressed by the newly formed osteocytes, is important to shutting down the BMU cycle (Winkler et al. 2003).

BMU remodeling can function in two modes, namely the conservation mode, in which the completed BMU has resorbed and formed equal amounts of bone, and the disuse mode, in which the completed BMU forms less bone than has been resorbed, resulting in net bone loss (Frost 2003). The BMU remodeling process is thought to have targeted and nontargeted components (Burr 2002). Parfitt (2002) and Burr (2002) have recently summarized concepts: (1) some remodeling is targeted for the replacement of fatigue-microdamaged bone; and (2) a substantial amount (70 percent) of total remodeling is not targeted for this specific purpose. Nontargeted remodeling is surplus to load-bearing issues, provides a margin of safety, and may have several purposes or mechanisms including removal of hypermineralized bone, related to initially targeted BMUs which may overshoot their target, and stochastic BMU origination.

A conceptual model describing the temporal and spatial characteristics of BMU dynamics indicates significant roles for the RANKL/RANK/OPG pathway and autocrine/paracrine bi-directional interactions (e.g. TGF-β, IGF-1, Wnt ligands, sclerostin, and EphrinB2-EphB2 signaling) between osteoclasts and osteoblasts as important actors (Henriksen et al. 2009). Many unsolved issues exist, especially as related to the molecular mechanisms for regulating bone remodeling in homeostasis and disease. Recent studies have described novel local molecular interactions important to intracellular communication influencing remodeling cycle transitions and coupling (Karsdal and Henriksen 2007). The importance of understanding the details and mechanisms of the remodeling cycle for paleopathology rests on the emerging consensus that many of the pathological skeletal conditions of interest to paleopathologists have disordered remodeling as a significant factor.

Reactive Oxygen Species (ROS)

A recent focus on the importance of the age-associated increase in reactive oxygen species (ROS) (free radical, reactive molecules containing the oxygen atom; see Crews 2003 for an extended discussion on ROS) on skeletal homeostasis is contributing to the shift from relatively oversimplified monocausal, estrogen-based models of bone loss pathological changes to the multifactorial/microenvironmental perspective (Manolagas 2009). Data are accumulating which indicate that oxidative stress plays a significant role in bone disease (e.g. osteoporosis). Recent research has demonstrated that changes in ROS generated by physiological aging are associated with a decreased remodeling rate characterized by decreased osteoblast number, increased osteoblast and osteocyte apoptosis, and decreased bone formation (Almeida et al. 2007b). In addition, osteoclast differentiation and activation have been shown to be mediated by ROS, enhancing bone resorption (Garrett et al. 1990; Lee et al. 2005).

Estrogen and/or androgen deficiency augment this effect by influencing the bone cellular defense against oxidative stress (Manolagas 2009). The mechanism for this is thought to involve the ROS-induced activation of the Forkhead box O (FoxO) transcription factors (important in oxidative stress resistance); the downstream effect of this is an attenuation of Wnt signaling and decreased osteoblastogenesis (Almeida et al. 2007a). The combined effect is loss of bone mass and microarchitectural deterioration. This evidence points to increased ROS as being part of a significant mechanism leading to the acceleration of aging and sex steroid deficiency effects on bone pathology (Almeida et al. 2007b). The significance of this brief discussion on ROS and bone for paleopathologists is that bone research now indicates a broad molecular and mechanistic framework for age-associated bone loss, from which new multidisciplinary research questions may be developed. Bone loss identified in ancient human groups may be interpreted within this framework. This could include specific consideration of lifestyle conditions that can contribute to an increased level of oxidative stress (e.g. nutritional factors, disease, and psychosocial disruption).

Osteoimmunology

Perturbations in the active processes of skeletal maintenance, repair, and homeostasis may be revealed in skeletal samples as pathological excessive bone loss (osteoporosis, periodontal disease, inflammatory arthritides, and tumor-induced bone disease), inappropriate new bone formation (spondyloarthritis), or a combination of both (Paget's disease of bone). These processes are not strictly limited to the skeletal system, but involve complex interactions between bone and other organ systems (Mensah et al. 2009).

Osteoimmunology is a rapidly developing subdiscipline studying the interactions between bone and immune cells and considering these interactions as part of a single integrated system. The term "osteoimmunology" was suggested by Arron and Choi (2000) to describe the interface between these biological disciplines, based on the observations that activated T cells (lymphocytes fundamental to cell-mediated immunity) express the protein RANKL, which promotes osteoclastogenesis and bone resorption as well as IFN-γ (a cytokine providing adaptive immunity against infectious agents; also associated with various autoinflammatory and autoimmune conditions)) which inhibits osteoclastogenesis (Takayanagi et al. 2000). T cells, in addition, have been found to express other immunological responsive factors which influence osteoclastogenesis (see Lorenzo et al. 2008 for a review). The discussion which follows serves as an introduction to the synergies between bone and immune cell function and how these mechanisms may contribute to the fundamental understanding of many common bone-related disorders using selected examples.

The foundation upon which osteoimmunology rests is the rapidly expanding research data set indicating that bone provides the microenvironment for many of the complex interactions involving immune cells, hematopoietic stem cells (multipotent stem cells which can differentiate into any of the blood cell types), and numerous regulatory cytokines. These players and biological systems have a profound effect on the function and fate of bone cells, structure, and disease (Walsh et al. 2006). The integrated nature of the immune system with bone biology has been exemplified by several recent observations: (1) that osteoblast-related cells form and regulate the hematopoietic stem cell (HSC) niche which is the source and support for blood and immune cells (Jung et al. 2008); (2) osteoclasts originate from myeloid precursor cells, which also differentiate into macrophages and dendritic cells (antigen-presenting cells necessary for T-cell-dependent immune responses) (Alnaeeli et al. 2006); and (3) many of the same biochemical mediators of immune cell function (inflammatory cytokines and growth factors) also regulate bone cell activity (osteoblasts and osteoclasts) (Lorenzo et al. 2008).

Bone microenvironment

The scope of osteoimmunology interactions is only beginning to be revealed. Many functions are likely to be served including bone's role in adaptive immunity through maintenance of the bone marrow space by hematopoietic cell regulation of bone turnover. Most relevant from the perspective of bone biology and paleopathology is how bone homeostasis is influenced by immune responses, particularly the augmentation of bone responsiveness to inflammatory states resulting from an activated or disease-affected immune system (Alnaeeli et al. 2007). Recent research has demonstrated that immune cells and inflammatory cytokines are important factors in the skeletal changes involving bone turnover and bone mass in such conditions as postmenopausal osteoporosis, rheumatoid arthritis, periodontal disease, osteomyelitis, and tumor-related bone disease. These metabolic, inflammatory, and neoplastic pathological conditions are characterized by variations in local bone microenvironmental conditions featuring infiltrating lymphocytes and other immunologically active cells which release an array chemical factors affecting bone metabolism by disturbing the equilibrium between bone formation (osteoblasts) and bone resorption (osteoclasts) (Walsh et al. 2006).

The study of the RANKL/RANK/OPG remodeling axis has revealed complex regulatory interactions between immune cells and bone-remodeling cells. RANKL is expressed on osteoblasts, activated T cells, and various tumor cells (e.g. breast and prostate). It binds to its receptor RANK on osteoclasts and through several intracellular signaling pathways induces these cells to mature and activate; bone loss is the expected result under most inflammatory conditions involving an activated T cell response (Mensah et al. 2009). Two counter-regulatory mechanisms to preserve bone mass have been identified. The first involves the decoy receptor OPG produced by osteoblast-lineage cells, which binds RANKL and prevents that molecule from activating the osteoclasts. This system is influenced by numerous upstream hormones and cytokines contributing to the local variability of the bone marrow microenvironment (Hofbauer et al. 2000). The second mechanism discovered and reported on by Takayanagi and co-authors (2000) involves the activated T cells which, in addition to expressing RANKL, also synthesize the protein interferon-γ, which blocks RANKL-induced osteoclast activation, thus acting to prevent uncontrolled bone loss initiated through inflammatory T-cell reactions (Arron and Choi 2000).

What is currently known about the nature of these converging biological systems is perhaps best further explained by a series of summary examples focused on the molecular and cellular bone biology which is relevant to the discipline of paleopathology: postmenopausal osteoporosis, periodontal disease/infection, joint disease, and tumor-related bone changes. These common conditions of interest represent to varying degrees the intersection of modern bone science and bone disease. These conditions are also presented in a broader, more traditional framework elsewhere in this volume.

Bone Science and Disease

Infectious disease: periodontitis, osteomyelitis, and immunity

Dental disease and antemortem tooth loss are important considerations toward understanding the health and life style of ancient human groups (Oztunc et al. 2006). Periodontitis is a bacterial-derived plaque-related condition characterized by both soft tissue and alveolar bone loss (Kirkwood 2008). This may result in the loss of integrity of the supporting structures of the teeth and tooth loss. Our interest here is the mechanism by which the susceptibility to oral infection combined with the immune and inflammatory responses generated by the byproducts of the microbial tissue invasion produce bone loss. Periodontal bone loss can be considered a model for infection-related osteoimmunological interactions.

The presence of periodontal microbial pathogen surface proteins and molecules elicit an immune response which includes T cells, macrophages, and polymorphonuclear lymphocytes. These cells release various immune and inflammatory molecules (interleukin [IL]-1β, IL-6, and tumor necrosis factor [TNF]-α) as well as matrix metalloproteases (MMPs) which degrade the connective tissue matrix (Lappin et al. 2001). Under the influence of these (and other) inflammatory signals, the activated T cells within the periodontal tissues demonstrate an increased expression of RANKL. This triggers osteoclastogenesis through the RANKL/RANK/OPG signaling system, resulting in continuing bone resorption and eventual periodontal tissue destruction with tooth loss.

This emerging model for the pathogenesis of periodontal disease has the potential of opening novel research opportunities for understanding important regulatory systems (Teng 2003). The biomolecular perspective may enhance (or change) paleopathological interpretations of periodontal disease by bringing immune-related influences to the table when analyzing this condition characterized by bone loss and/ or possible bone infection. For example, recent research suggests that periodontal disease may serve as a broader proxy for the general health and mortality in archaeological populations (DeWitte and Bekvalac 2010).

Osteomyelitis serves as an additional example. Osteomyelitis is a bacterial infection frequently distinguished by progressive inflammatory destruction of bone (Marriott 2004). Common causative organisms are *Staphylococcus aureus* and *Salmonella* species. These types of infections damage bone by several mechanisms: (1) stimulation of RANKL and osteoclastogenesis, (2) bacterial elaboration of virulence factors such as matrix-degrading proteases, and (3) bacterial induction of osteoblast apoptosis (Alexander et al. 2003). These pathogens employ measures to evade humoral immunity responses by becoming internalized by host osteoblasts, thus promoting ongoing viability (Ellington et al. 2003).

Intracellular pathogens typically activate a cell-mediated immune response by the osteoblasts secreting chemokines which mobilize and localize T cells and macrophages (Bost et al. 2001). It is important to recall that the T lymphocytes produce the osteo-clastogenesis factors RANKL and IL-17 as well as the intracellular infection fighting cytokine IFN-γ. Furthermore, the bacterially challenged osteoblasts directly produce pro-inflammatory cytokines and various growth factors that are known to augment osteoclast activity (Bost et al. 1999). Taken together, these data support the concept that osteoblasts play a critical role in contributing to the host immune response and aberrant bone remodeling in bone bacterial infections (Marriott 2004). This, in turn, has implications for the abnormal bone loss (increased) and bone formation (decreased) exhibited at sites of bacterial infection. The research data that indicate that bacterial bone infection is characterized predominantly by bone resorption and secondarily by bone formation should be taken into consideration, for example, as part of the interpretive reappraisal of periosteal bone formation as an indicator of nonspecific infection.

Peri/postmenopausal osteoporosis and T cells

Perimenopausal bone loss is thought to be related to increases in bone turnover resulting in net bone resorption (Heaney 2003). These changes in bone biology have been in part linked to both the direct and indirect effects of estrogen deficiency on bone cells. The direct effects of estrogen are mediated through the two receptors (ERs), ERα and ERβ: ERα is most active in bone. The receptors are present on osteoblasts, osteocytes, osteoclasts, and bone marrow (BM) stromal cells. Bone mar-row stromal cells are osteoblast precursors which supply niche support for most of the cells in human bone marrow including pre-osteoclasts, T cells, and B cells (Pacifici 2008). The bone loss related to the direct action of estrogen deficiency is caused primarily by increased osteoclast formation and action secondary to an increase in the osteoblast/stromal cell production of RANKL and inflammatory cytokines, a decrease in the expression of OPG, and a reduction in osteoclast apoptosis.

Sex steroids have long been recognized as influencing the immune system generally and T cells specifically (Weitzmann and Pacifici 2006). It is only recently that these connections have been studied in relation to estrogen-mediated bone loss. Pertinent to this discussion of osteoimmunology is the indirect mechanism of estrogen deficiency-related bone loss through estrogen's up-regulation of T-cell activation; attenuation of antioxidant pathways, resulting in an increase of reactive oxygen species; and production of osteoclastogenic/inflammatory factors, such as RANKL, IL-1, IL-6, and TNF-α (see Weitzmann and Pacifici 2006 for review). These concepts have been tested in experimental animal models, with emergent validation in humans (Abrahamsen et al. 1997; Eghbali-Fatourechi et al. 2003). The importance of this osteoimmunological framework lies in the hypothesis that postmenopausal osteoporosis may be in part a result of "an autoimmune inflammatory disorder triggered by E [estrogen] deficiency" (Pacifici 2008:76). It may possible as a result of this type of approach that consideration and interpretation of bone loss in archaeological collections can come be viewed within a more complex environmental context. Additional data points are likely to enhance the outcome of this type of research.

Joint disease

Joint disease with varying degrees of joint architectural destruction is best viewed within the larger framework of the mechanisms of remodeling and repair. Lying within this framework is the continuum of joint disease phenotypes, ranging from primarily bone destructive (rheumatoid arthritis [RA]) to predominantly bone anabolic (osteoarthritis [OA]), characterized by osteophyte formation. The role of cytokines and various bone-specific regulatory signaling pathways involved in bone formation and resorption is under intensive research scrutiny for both the locally-inflammatory condition of osteoarthritis (Bonnet and Walsh 2005) as well as the systemically-inflammatory joint disorders such as rheumatoid arthritis and associated spondyloarthropathies. Rheumatoid arthritis serves as a disease model for bone resorption linked to RANKL-mediated osteoclastogenesis induced by the production of proinflammatory cytokines by immune cells in the synovium (joint lining) and other periarticular tissues (Lorenzo et al. 2008). This condition is characterized by several different manifestations of disordered remodeling: (1) focal marginal articular bone erosions and periarticular osteopenia, (2) subchondral bone loss, and (3) systemic osteoporosis (Goldring 2008). The osteoclast-driven bone resorption is attributed to pathological changes in the local bone microenvironment which includes a wide spectrum of cytokines, chemokines, and growth factors. RANKL/OPG and TNF-α have received the most attention to date (Walsh et al. 2005).

While rheumatoid arthritis has been observed to have essentially no bone repair activity; osteoarthritis has been observed to have excessive bone repair activity. As previously discussed, the wingless (Wnt) signaling pathway serves as a master-regulator for bone formation. Recent research has indicated the importance of modulation of the Wnt signaling by Dickkopf-1 (DKK-1), a Wnt inhibitor protein (Diarra et al. 2007). These authors have demonstrated that the expression level of DKK-1 plays an important role in "determining whether a diseased joint undergoes destruction or reacts by forming new bone" (Diarra et al. 2007:157). DKK-1 is up-regulated in inflamed RA synovial tissues and down-regulated in joint diseases characterized by new bone formation (OA) (Goldring and Goldring 2007).

Experimental models of osteoarthritis as well as human clinical material provide data suggesting that an increase in Wnt signaling activity may adversely affect the cartilage-bone biomechanical unit (Dell'Accio et al. 2006; Yuasa et al. 2008). The increased bone formation results in amplified bone stiffness and cartilage strain (Luyten et al. 2009). The importance of these tissue- and spatially-dependent modulations in the bone microenvironment remains central to ongoing research efforts. These findings provide a glimpse into the still largely unknown molecular mechanisms and determinants of joint disease, which are increasingly viewed within the context of physiological and pathological disturbances in joint maintenance, repair, and remodeling regulatory systems, as opposed to the more simplistic "wear and tear" concept.

Cancer-related bone disease

The disruption of normal bone physiology and structure by cancer cells spreading to the skeleton provides an opportunity to examine the effects of unique multifactorial aspects of specific bone microenvironments. This type of perspective provides linkages of molecular biological systems to observable bone changes of interest to those studying ancient and extant humans. Interestingly, the 19th-century surgeon Stephen Paget (1889) provided an explanation of the phenomenon of metastatic bone disease that remains operative as a guiding principle of the present day's research agenda. The "seed and soil" hypothesis describes how the microenvironment of the organ (soil) serves as fertile and "congenial" soil for the cancer cells (seeds) (Paget 1889:571; Clines and Guise 2008). Cancer-related bone disease lies within the spectrum of osteolytic and osteoblastic limits. Not infrequently the bone-cancer mechanisms have a positive feedback reciprocity, which can result in inexorable disease progression.

Breast cancer cells spreading to bone provide a model for an osteolytic interaction between cancer cells and the bone microenvironment based on the actions of parathyroid hormone-related protein (PTHrP) and transforming growth factor β (TGF-β). PTHrP and other cytokines and growth factors are produced by breast cancer cells in bone (Clines and Guise 2008). The increased expression of these molecules stimulates osteoblasts to up-regulate RANKL and down-regulate OPG, activating osteoclastogenesis (Thomas et al. 1999). Osteoclastic bone resorption allows TGF-β and other tumor-related factors to be released from the bone matrix; these factors, in turn, directly stimulate the breast cancer cells—the vicious cycle begins and continues.

Contrary to breast cancer, prostate cancer cells spreading to bone provide a model for an osteoblastic interaction by stimulation of the osteoblast and production of pathological new bone. The Wnt osteoblast regulatory pathway comes into play as the Wnt antagonist DKK-1 (Clines et al. 2007) is down-regulated by the prostate-tumor-produced protein endothelin (ET-1), resulting in unopposed Wnt signaling and the osteoblastic response of pathological new bone. ET-1 is currently thought to be a principal factor in this response (Yin et al. 2003). Other molecules under investigation include bone morphogenetic protein (BMP-7), known for its positive effect on bone formation (Buijs et al. 2007), and prostate-specific antigen (PSA), which may indirectly influence osteoblasts by activating various inactive factors (Williams et al. 2007). The importance of bone microenvironment is an essential message, providing paleopathologists with biological mechanisms supporting their observations.

CONCLUSION

This chapter has provided a portal into the translation of current perspectives in bone science to an improved pathophysiological understanding of the biological mechanisms of many of the skeletal conditions observed and analyzed by paleopathologists. Selected key signaling pathways have been identified and discussed: RANKL/OPG, Wnt/β-catenin, Lrp5 inhibitors Sost/Dkk-1, and immune interactions. The significance of variations in the bone microenvironment has been emphasized. The insights derived from this type of approach and scale of analysis can serve as catalysts for anthropologically-related researchers to move towards these complexities, stimulating the development of novel research questions and agendas. The understanding of modern bone science concepts is likely to be crucial to advancements in paleopathological observations and critical interpretations.

REFERENCES

Abrahamsen, B., Bendtzen, K., and Beck-Nielsen, H., 1997   Cytokines and T-lymphocyte Subsets in Healthy Post-menopausal Women: Estrogen Retards Bone Loss without Affecting the Release of IL-1 or IL-1ra. Bone 20:251–258.

Adams, G., and Scadden, D., 2006   The Hematopoietic Stem Cell in its Place. Nature Immunology 7:333–337.

Agarwal, S., and Stout, S., eds., 2003   Bone Loss and Osteoporosis: an Anthropological Perspective. New York: Kluwer Academic.

Alexander, E., Rivera, F., Marriot, I., Anguita, J., Bost, K., and Hudson, M., 2003   *Staphylococcus aureus*-induced Tumor Necrosis Factor-related Apoptosis-inducing Ligand Expression Mediates Apoptosis and Caspase-8 Activation in Infected Osteoblasts. BMC Microbiology 3:5–16.

Alford, A., Jacobs, C., and Donahue, H., 2003   Oscillating Fluid Flow Regulates Gap Junction Communication in Osteocytic MLO-Y4 Cells by an ERK1/2 MAP Kinase-Dependent Mechanism. Bone 33:64–70.

Almeida, M., Han, L., Martin-Millan, M., O'Brien, C. A., and Manolagas, S. C., 2007a   Oxidative Stress Antagonizes Wnt Signaling in Osteoblast Precursors by Diverting β-Catenin from T Cell Factor to Forkhead Box O-mediated Transcription. Journal of Biological Chemistry 282:27298–27305.

Almeida, M., Han, L., Martin-Millan, M., Plotkin, L. I., Stewart, S. A., et al., 2007b   Skeletal Involution by Age-associated Oxidative Stress and Its Acceleration by Loss of Sex Steroids. Journal of Biological Chemistry 282:27285–27297.

Alnaeeli, M., Park, J., Mahamed, D., Penninger, J., and Teng, Y., 2007   Dendritic Cells at the Osteo-immune Interface: Implications for Inflammation-induced Bone Loss. Journal of Bone and Mineral Research 22:775–780.

Alnaeeli, M., Penninger, J., and Teng, Y., 2006   Immune Interactions with CD4+ T Cells Promote the Development of Functional Osteoclasts from Murine CD11c+ Dendritic Cells. Journal of Immunology 177:3314.

Arron, J., and Choi, Y., 2000   Bone Versus Immune System. Nature 408:535–536.

Bonewald, L., 2007   Osteocytes as Dynamic Multifunctional Cells. Annals of the New York Academy of Science 1116:281.

Bonewald, L., and Johnson, M., 2008   Osteocytes, Mechanosensing and Wnt Signaling. Bone 42:606–615.

Bonnet, C., and Walsh, D., 2005   Osteoarthritis, Angiogenesis and Inflammation. Rheumatology 44:7–16.

Bost, K., Bento, J., Petty, C., Schrum, L., Hudson, M., and Marriott, I., 2001   Monocyte Chemoattractant Protein-1 Expression by Osteoblasts Following Infection with Staphylococcus Aureus or Salmonella. Journal of Interferon and Cytokine Research 21:297–304.

Bost, K., Ramp, W., Nicholson, N., Bento, J., Marriott, I., and Hudson, M., 1999   *Staphylococcus aureus* Infection of Mouse or Human Osteoblasts Induces High Levels of Interleukin-6 and Interleukin-12 Production. Journal of Infectious Diseases 180:1912–1920.

Boyle, W., Simonet, W., and Lacet, D., 2003   Osteoclast Differentiation and Activation. Nature 423:337–342.

Buijs, J., Rentsch, C., van der Horst, G., van Overveld, P., Wetterwald, A., et al., 2007   BMP7, a Putative Regulator of Epithelial Homeostasis in the Human Prostate, is a Potent Inhibitor of Prostate Cancer Bone Metastasis *In Vivo*. American Journal of Pathology 171:1047.

Burger, E. H., Klein-Nulend, J., and Smit, T. H., 2003   Strain-Derived Canalicular Fluid Flow Regulates Osteoclast Activity in a Remodeling Osteon: A Proposal. Journal of Biomechanics 36:1453–1459.

Burr, D. B., 2002   Targeted and Nontargeted Remodeling. Bone 30:2–4.

Clines, G., and Guise, T., 2008   Molecular Mechanisms and Treatment of Bone Metastasis. Expert Reviews in Molecular Medicine 10:e7.

Clines, G., Mohammad, K., Bao, Y., Stephens, O., Suva, L., et al., 2007   Dickkopf Homolog 1 Mediates Endothelin-1-stimulated New Bone Formation. Molecular Endocrinology 21:486–498.

Crews, D., 2003   Human Senescence: Evolutionary and Biocultural Perspectives. Cambridge: Cambridge University Press.

Dallas, S., and Bonewald, L., 2008   Osteocytes Play to Standing Room Only: Meeting Report from the 30th Annual Meeting of the American Society for Bone and Mineral Research. IBMS BoneKey 5:441–5.

Datta, H., NG, W., Walker, J., Tuck, S., and Varanasi, S., 2008   The Cell Biology of Bone Metabolism. Journal of Clinical Pathology 61:577–587.

Dell'Accio, F., De Bari, C., El Tawil, N., Barone, F., Mitsiadis, T., et al., 2006   Activation of WNT and BMP Signaling in Adult Human Articular Cartilage Following Mechanical Injury. Arthritis Research and Therapy 8:R139.

Dewitte, S., and Bekvalac, J., 2010   Oral Health and Frailty in the Medieval English Cemetery of St Mary Graces. American Journal of Physical Anthropology 142:341–354.

Diarra, D., Stolina, M., Polzer, K., Zwerina, J., Ominsky, M., et al., 2007   Dickkopf-1 is a Master Regulator of Joint Remodeling. Nature Medicine 13:156–163.

Donahue, H., Chen, Q., Saunders, C., and Yellowley, C., 2003   Bone Cells and Mechanotransduction. *In* Molecular Biology in Orthopaedics. R. Rosier, and C. Evans, eds. pp. 179–190. Rosemont, IL: American Academy of Orthopedic Surgeons.

Ducy, P., Schinke, T., and Karsenty, G., 2000   The Osteoblast: A Sophisticated Fibroblast Under Central Surveillance. Science 289:1501–1504.

Duong L., and Rodan, A., 2001   Regulation of Osteoclast Formation and Function. Reviews in Endocrine and Metabolic Disorders 2:95–104.

Eghbali-Fatourechi, G., Khosla, S., Sanyal, A., Boyle, W., Lacey, D., and Riggs, B., 2003   Role of RANK Ligand in Mediating Increased Bone Resorption in Early Postmenopausal Women. Journal of Clinical Investigation 111:1221–1230.

Ellington, J., Harris, M., Webb, L., Smith, B., Smith, T., et al., 2003   Intracellular *Staphylococcus aureus*. Journal of Bone and Joint Surgery (British volume) 85:918–921.

Frost, H., 1969   Tetracycline-Based Histological Analysis of Bone Remodeling. Calcified Tissue Research 3:211–237.

Frost, H., 2003   On Changing Views about Age-Related Bone Loss. *In* Bone Loss and Osteoporosis: An Anthropological Perspective. S. Agarwal, and S. Stout, eds. pp. 19–32. New York: Kluwer Academic.

Garrett, I., Boyce, B., Oreffo, R., Bonewald, L., Poser, J., and Mundy, G., 1990   Oxygen-derived Free Radicals Stimulate Osteoclastic Bone Resorption in Rodent Bone *In Vitro* and *In Vivo*. Journal of Clinical Investigation 85:632.

Goldring, M. B., and Goldring, S. R., 2007   Osteoarthritis. Journal of Cellular Physiology 213:626–634.

Goldring, S., 2008   Inflammation-induced Bone Loss in the Rheumatic Diseases. *In* Primer on the Metabolic Bone Diseases and Disorders of Mineral Metabolism. C. Rosen, ed. pp. 272–276. Washington, DC: American Society for Bone and Mineral Research.

Harada, S., and Rodan, G., 2003   Control of Osteoblast Function and Regulation of Bone Mass. Nature 423:349–355.

Heaney, R., 2003   Is the Paradigm Shifting? Bone 33:457–465.

Henriksen, K., Neutzsky-Wulff, A., Bonewald, L., and Karsdal, M., 2009   Local Communication on and within Bone Controls Bone Remodeling. Bone 44:1026–1033.

Hofbauer, L., Khosla, S., Dunstan, C., Lacey, D., Boyle, W., and Riggs, B., 2000   The Roles of Osteoprotegerin and Osteoprotegerin Ligand in the Paracrine Regulation of Bone Resorption. Journal of Bone and Mineral Research 15:2–12.

Hu, H., Hilton, M., Tu, X., Ornitz, D., and Long, F., 2005   Sequential Roles for Hedgehog and Wnt Signaling in Osteoblast Development. Development 132:49–60.

Huiskes, R., 2000   If Bone is the Answer, Then What is the Question? Journal of Anatomy 197:145–156.

Huiskes, R., Ruimerman, R., Van Lenthe, G., and Janssen, J., 2000   Effects of Mechanical Forces on Maintenance and Adaptation of Form in Trabecular Bone. Nature 405:704–706.

Jung, Y., Song, J., Shiozawa, Y., Wang, J., Wang, Z., et al., 2008   Hematopoietic Stem Cells Regulate Mesenchymal Stromal Cell Induction into Osteoblasts Thereby Participating in the Formation of the Stem Cell Niche. Stem Cells 26:2042–2051.

Karsdal, M., and Henriksen, K., 2007   Osteoclasts Control Osteoblast Activity. BoneKEy-Osteovision 4:19–24.

Karsenty, G., 2001   Minireview: Transcriptional Control of Osteoblast Differentiation. Endocrinology 142:2731–2733.

Kim-Weroha, N., Ferris, A., Holladay, B., Kotha, S., Kamel, M., and Johnson, M., 2008   In-Vivo Load-Activated Propagation of B-catenin Signaling in Osteocytes Through Coordinated Down-regulators of Inhibitors of Lrp5. Journal of Bone and Mineral Research 23(suppl):S13.

Kirkwood, K., 2008   Periodontal Diseases and Oral Bone Loss. *In* Primer on the Metabolic Bone Diseases and Disorders of Mineral Metabolism. 7th edition. C. Rosen, ed. pp. 510–513. Washington, DC: American Society for Bone and Mineral Research.

Knothe-Tate, M., Adamson, J., Tami, A., and Bauer, T., 2004   The Osteocyte. International Journal of Biochemistry and Cell Biology 36:1–8.

Kollet, O., Dar, A., Shivticl, S., Kalinkovich, A., Lapid, K., et al., 2006   Osteoclasts Degrade Endosteal Components and Promote Mobilization of Hematopoietic Progenitor Cells. Nature Medicine 12:657–664.

Krause, C., de Gorter, D., Karperien, M., and Dijke, P., 2008   Signal Transduction Cascades Controlling Osteoblast Differentiation. *In* Primer on the Metabolic Bone Diseases and Disorders of Mineral Metabolism, C. Rosen, ed. pp. 10–16. Washington, DC: American Society for Bone and Mineral Research.

Lappin, D., MacLeod, C., Kerr, A., Mitchell, T., and Kinane, D., 2001   Anti-inflammatory Cytokine IL-10 and T Cell Cytokine Profile in Periodontitis Granulation Tissue. Clinical and Experimental Immunology 123:294.

Lee, N. K., Choi, Y. G., Baik, J. Y., Han, S. Y., Jeong, D., et al., 2005   A Crucial Role for Reactive Oxygen Species in RANKL-induced Osteoclast Differentiation. Blood 106:852–859.

Lorenzo, J., Horowitz, M., and Choi, Y., 2008   Osteoimmunology: Interactions of the Bone and Immune System. Endocrine Reviews 29:403–40.

Luyten, F., Tylzanowski, P., and Lories, R., 2009   Wnt Signaling and Osteoarthritis. Bone 44:522–7.

Manolagas, S., 2009   Oxidative Stress, Cell Apoptosis, Glucocorticoids and Osteoporosis. Bone 45:120.

Marriott, I., 2004   Osteoblast Responses to Bacterial Pathogens: A Previously Unappreciated Role for Bone-Forming Cells in Host Defense And Disease Progression. Immunologic Research 30:291–308.

Martin, R., 2003   Functional Adaptation and Fragility of the Skeleton. *In* Bone Loss and Osteoporosis: An Anthropological Perspective. S. Agarwal, and S. Stout, eds. pp. 121–138. New York: Kluwer Academic.

Martin, R., 2007   Targeted Bone Remodeling Involves BMU Steering As Well As Activation. Bone 40(6):1574–1580.

Martin, R., Burr, D., and Sharkey, N., 1998   Skeletal Tissue Mechanics. New York: Springer.

Massaro, E., and Rogers, J., 2004   The Skeleton: Biochemical, Genetic, and Molecular Interactions in Development and Homeostasis. Totowa, NJ: Humana Press.

Melsen, F., Mosekilde, L., and Eriksen, E., 1995   Spatial Distribution of Sinusoids in Relation to Remodeling Sites: Evidence for Specialized Sinusoidal Structures Associated with Formative Sites. Journal of Bone and Mineral Research 10:S209.

Mensah, K., Li, J., and Schwarz, E., 2009   The Emerging Field of Osteoimmunology. Immunologic Research 45:100–113.

Mullender, M., and Huiskes, R., 1997   Osteocytes and Bone Lining Cells: Which are the Best Candidates for Mechano-Sensors in Cancellous Bone? Bone 6:527–532.

Mullender, M., Meer, D., Huiskes, R., and Lips, P., 1996   Osteocyte Density Changes in Aging and Osteoporosis. Bone 18:109–113.

Nashida, S., and Endo, N., 1999   Number of Osteoprogenitor Cells in Human Bone Marrow Markedly Decreases after Skeletal Maturation. Journal of Bone and Mineral Metabolism 17:171–177.

Nicolella, D., Yao, W., and Lane, N., 2008   Estrogen Deficiency Alters the Localized Material Properties of the Peri-lacunar Bone Matrix in Old Rats. Journal of Bone and Mineral Research 23(suppl):S400.

Ortner, D., and Putschar, W., 1985   Identification of Pathological Conditions in Human Skeletal Remains. Washington, DC: Smithsonian Institution Press.

Oztunc, H., Yoldas, O., and Nalbantoglu, E., 2006   The Periodontal Disease Status of the Historical Population of Assos. International Journal of Osteoarchaeology 16:76.

Pacifici, R., 2008   Estrogen Deficiency, T Cells and Bone Loss. Cellular Immunology 252:68–80.

Paget, S., 1889   The Distribution of Secondary Growths in Cancer of the Breast. Lancet 133:571–2.

Parfitt, A. M., 1994   Osteonal and Hemi-Osteonal Remodeling: The Spatial and Temporal Framework for Signal Traffic in Adult Human Bone. Journal of Cellular Biochemistry 55: 273–286.

Parfitt, A. M., 2000   The Mechanism of Coupling: A Role for the Vasculature. Bone 26: 319–323.

Parfitt, A. M., 2002   Targeted and Nontargeted Bone Remodeling: Relationship to Basic Multicellular Unit Origination and Progression. Bone 30:5–7.

Parfitt, A. M., 2003   New Concepts of Bone Remodeling: A Unified Spatial and Temporal Model with Physiologic and Pathophysiologic Implications. *In* Bone Loss and Osteoporosis: An Anthropological Perspective. S. Agarwal, and S. Stout, eds. pp. 3–17. New York: Kluwer Academic.

Pearson, O., and Lieberman, D., 2004   The Aging of Wolff's "Law": Ontogeny and Responses to Mechanical Loading in Cortical Bone. Yearbook of Physical Anthropology 47:63–99.

Pettit, A., Chang, M., Hume, D., and Raggatt, L., 2008   Osteal Macrophages: A New Twist on Coupling During Bone Dynamics. Bone 43:976–982.

Qing, H., Ardeshirpour, L., Dusevich, V., Dallas, M., Wysolmerski, J., and Bonewald, L., 2008   Osteocytic Perilacunar Remodeling as a Significant Source of Calcium During Lactation. Journal of Bone and Mineral Research 23(suppl):S401.

Robling, A. G., Castillo, A. B., and Turner, C. H., 2006   Biomechanical and Molecular Regulation of Bone Remodeling. Annual Review of Biomedical Engineering 8:455–498.

Ross, F., 2008   Osteoclast Biology and Bone Resorption. In Primer on the Metabolic Bone Diseases and Disorders of Mineral Metabolism. C. Rosen, ed. pp. 16–22. Washington, DC: American Society for Bone and Mineral Research.

Ryser, M. D., Nigam, N., and Komarova, S. V., 2009   Mathematical Modeling of Spatio-Temporal Dynamics of a Single Bone Multicellular Unit. Journal of Bone and Mineral Research 24:860–870.

St-Arnaud, R., 2003   Transcriptional Control of the Osteoblast Phenotype. In Molecular Biology in Orthopaedics. R. Rosier, and C. Evans, eds. pp. 191–209. Rosemont, IL: American Academy of Orthopedic Surgeons.

Sawakami, K., Robling, A., Ai, M., Pitner, N., Liu, D., et al., 2006   The Wnt Co-receptor LRP5 is Essential for Skeletal Mechanotransduction but not for the Anabolic Bone Response to Parathyroid Hormone Treatment. Journal of Biological Chemistry 281:23698.

Schofield, R., 1978   The Relationship Between the Spleen Colony-forming Cell and the Haematopoietic Stem Cell. Blood Cells 4:7–25.

Schroeder, T., Jensen, E., and Westendorf, J., 2005   Runx2: A Master Organizer of Gene Transcription in Developing and Maturing Osteoblasts. Birth Defects Research C 75:213–225.

Sommerfeldt, D., and Rubin, C., 2001   Biology of Bone and How It Orchestrates the Form and Function of the Skeleton. European Spine Journal 10: S86–S95.

Stein, G., and Lian, J., 1993   Molecular Mechanisms Mediating Proliferation/Differentiation Interrelationships during Progressive Development of the Osteoblast Phenotype. Endocrine Reviews 14:424–442.

Stenbeck, G., and Horton, M., 2004   Endocytic Trafficking in Actively Resorbing Osteoclasts. Journal of Cell Science 117:827.

Takahashi, H., Epker, B., and Frost, H., 1964   Resorption Preceded Formative Activity. Surgical Forum 15:437–438.

Takayanagi, H., Ogasawara, K., Hida, S., Chiba, T., Murata, S., et al., 2000   T-cell-mediated Regulation of Osteoclastogenesis by Signaling Cross-talk Between RANKL and IFN-γ. Nature 408:600–604.

Tanaka, S., 2008   Osteoclasts: Meeting Report from the 30th Annual Meeting of the American Society of Bone and Mineral Research. BoneKEy-Osteovision 5:454–457.

Teitelbaum, S. L., 2000   Bone Resorption by Osteoclasts. Science 289(5484):1504–1508.

Teitelbaum, S. L., and Ross, F. P., 2003   Genetic Regulation of Osteoclast Development and Function. Nature Reviews Genetics 4:638–649.

Teng, Y., 2003   The Role of Acquired Immunity and Periodontal Disease Progression. Critical Reviews in Oral Biology and Medicine 14:237–252.

Thomas, R., Guise, T., Yin, J., Elliott, J., Horwood, N., et al., 1999   Breast Cancer Cells Interact with Osteoblasts to Support Osteoclast Formation. Endocrinology 140:4451–4458.

van Oers, R., 2008   A Unified Theory for Osteonal and Hemi-Osteonal Remodeling. Bone 42(2): 250–259.

Villotte, S., Castex, D., Couallier, V., Dutour, O., Knüsel, C., and Henry-Gambier, D., 2010   Enthesopathies as Occupational Stress Markers: Evidence from the Upper Limb. American Journal of Physical Anthropology 142:224–234.

Walker, P. L., Bathurst, R. R., Richman, R., Gjerdrum, T., and Andrushko, V. A., 2009   The Causes of Porotic Hyperostosis and Cribra Orbitalia: A Reappraisal of the Iron-Deficiency-Anemia Hypothesis. American Journal of Physical Anthropology 139(2):109–125.

Walsh, M. C., Kim, N., Kadono, Y., Rho, J., Lee, S. Y., et al., 2006   Osteoimmunology: Interplay Between the Immune System and Bone Metabolism. Annual Reviews in Immunology 24:33–63.

Walsh, N., Crotti, T., Goldring, S., and Gravallese, E., 2005   Rheumatic Diseases: The Effects of Inflammation on Bone. Immunological Reviews 208:228–251.

Weiss, E., and Jurmain, R., 2007   Osteoarthritis Revisited: A Contemporary Review of Aetiology. International Journal of Osteoarchaeology 17(5):437.

Weitzmann, M., and Pacifici, R., 2006   Estrogen Deficiency and Bone Loss: An Inflammatory Tale. Journal of Clinical Investigation 116:1186–1194.

Weston D.A., 2008   Investigating the Specificity of Periosteal Reactions in Pathology Museum Specimens. American Journal of Physical Anthropology 137(1):48–59.

White, T., and Folkens, P., 2000   Human Osteology. 2nd Edition. San Diego: Academic Press.

Williams, B., and Insogna, K., 2009   Where Wnts Went: The Exploding Field of Lrp5 and Lrp6 Signaling in Bone. Journal of Bone and Mineral Research 24:171–178.

Williams, S., Singh, P., Isaacs, J., and Denmeade, S., 2007   Does PSA Play a Role as a Promoting Agent During the Initiation and/or Progression of Prostate Cancer? Prostate 67:312–329.

Winkler, D., Sutherland, M., Geoghegan, J., Yu, C., Hayes, T., et al., 2003   Osteocyte Control of Bone Formation via Sclerostin, a Novel BMP Antagonist. EMBO 22:6267–6276.

Wu, J. Y., Scadden, D. T., and Kronenberg, H. M., 2009   Role of the Osteoblast Lineage in the Bone Marrow Hematopoietic Niches. Journal of Bone and Mineral Research 24:759–764.

Yellowley, C. E., Li, Z., Zhou, Z., Jacobs, C. R., and Donahue, H. J., 2000   Functional Gap Junctions between Osteocytic and Osteoblastic Cells. Journal of Bone and Mineral Research 15:209–217.

Yin, J., Mohammad, K., Käkönen, S., Harris, S., Wu-Wong, J., et al., 2003   A Causal Role for Endothelin-1 in the Pathogenesis of Osteoblastic Bone Metastases. Proceedings of the National Academy of Sciences of the USA 100:10954–10959.

Yuasa, T., Otani, T., Koike, T., Iwamoto, M., and Enomoto-Iwamoto, M., 2008   Wnt//[beta]-Catenin Signaling Stimulates Matrix Catabolic Genes and Activity in Articular Chondrocytes: Its Possible Role in Joint Degeneration. Laboratory Investigation 88:264–274.

Zhao, C., Irie, N., Takada, Y., Shimoda, K., Miyamoto, T., et al., 2006   Bidirectional EphrinB2-EphB4 Signaling Controls Bone Homeostasis. Cell Metabolism 4:111–121.

CHAPTER 6

The Ecological Approach: Understanding Past Diet and the Relationship Between Diet and Disease

M. Anne Katzenberg

INTRODUCTION

What is the "ecological approach"?

Ecology is the study of how living organisms interact with one another and with their environments. The intricacies of living systems and their resilience are truly amazing, and to understand the complexity of the interrelations among organisms is to appreciate the wonder of natural selection. In this chapter, I focus on the ecological approach as it relates to human diet and subsistence, the interactions of humans with the plants and animals that make up their food, and with the environments in which those plants and animals are found.

Within the realm of ecology is the field of stable isotope ecology. This field, through the use of stable isotopes, follows the flow of nutrients from the base of a food web up through the trophic levels. Stable isotopes have also been used to reconstruct the diets of past humans. By understanding diet, we gain knowledge of environmental interactions between humans and the plants and animals that are their foods. To truly understand stable isotope ecology, we must explore as much of the food web as possible. Thus, this chapter will consider the ecological approach at two levels: human

A Companion to Paleopathology, First Edition. Edited by Anne L. Grauer.
© 2012 John Wiley & Sons, Ltd. Published 2016 by John Wiley & Sons, Ltd.

interactions with other organisms and the resulting exposure to disease-causing organisms, and stable isotope analysis of human tissues for the purpose of reconstructing past diet.

Exposure to disease-causing organisms

As a student of paleopathology with the anthropocentric bias of most undergraduate students, I remember being struck with the concept that disease-causing organisms seek to find food, to reproduce and to "be successful" in the evolutionary game, just as humans do. This is elegantly presented by Burnet and White in their book *The Natural History of Infectious Disease* (1972) and is one example of an ecological approach in which the competing as well as sometimes mutual needs of microorganisms and humans are considered together. It is useful to view interactions among different species from all perspectives. For example, we might view the domestication of plants from several different perspectives. From the perspective of humans, domestication of plants allows them to have better control over their food supply, which increases the carrying capacity of the land, resulting in increased population density and an increase in overall population size. From the perspective of rats, domestication of plants by humans provides a more stable food supply through the human practice of storing grains in storage bins or silos. Another perspective is that of the flea that lives on the blood of the rat who eats the grain. The flea transmits bubonic plague, carried by the rat, to humans, who live in close proximity to rats by providing a constant food supply through their practice of domesticating plants. The plague bacillus has the ability to reside in the rat, in the flea and in humans and to pass among these species to better enhance its own survival and reproduction. Thus, the ability to detect the timing of the domestication of plants in prehistory leads to a wealth of information relevant to understanding past interactions between humans and disease organisms through an understanding of their ecology. Diet and disease are intimately linked through such relationships.

Human subsistence practices

Clearly, understanding the mode of subsistence is important to understanding human disease experience. Much has been written about the differential exposure to disease that results from human subsistence practices and changes in such practices (e.g., Cohen and Armelagos 1984). In a classic paper, Aidan Cockburn, founder of the Paleopathology Association, provided both an evolutionary and an ecological approach to human disease experience (1971). He discussed diseases that humans share with nonhuman primates, suggesting that these are the diseases most likely to be of the greatest antiquity for humans and their ancestors. He then considered diseases likely to be contracted from hunting and from improper storage of meat, then diseases contracted from exposure to the soil during agricultural pursuits, and from proximity to domesticated animals and finally, to crowd infections, found with increasing population size and density. Thus, Cockburn provides a comprehensive view of human disease experience from our earliest ancestors to the development of complex civilizations. Many other researchers have picked up on this theme and have focused on particular modes of subsistence and also on disease experience during shifts from one mode of subsistence to another (e.g., Dunn 1968; Armelagos and McArdle 1975; Lallo et al. 1977; Cohen and Armelagos 1984 and individual chapters therein; Goodman et al. 1984; and reviewed by Buikstra and Cook, 1980).

While subsistence and disease exposure are surely linked, the impact of particular subsistence shifts to changes in settlement patterns and to population size and density are important factors in understanding patterns of disease, especially patterns of infectious diseases. Consider, from the disease organism's perspective, the difference in opportunities for spread and survival provided by small-scale, highly mobile hunter-gatherer groups versus large, sedentary and dense villages of agriculturalists. With the development of complex civilizations, we see status differences and differential access to foods, thus diet and health become important within segments of populations.

Another important aspect to human disease experience involves a thorough understanding of environments and their pathogens. For example, in the history of Western medicine, studies of "tropical diseases" became necessary as people native to temperate climates ventured into tropical regions, frequently becoming ill with conditions not previously encountered by their health practitioners. In the same way, students of paleopathology must become familiar with diseases specific to particular regions (i.e., organisms adapted to specific ranges of temperature, humidity, or carried by local fauna). Migrations of past humans and the development and expansion of trade networks can result in encounters with new diseases, and with introductions and exchanges of diseases among newly encountered people (Dobyns 1992).

While there are many sources of evidence for detecting modes of subsistence and subsistence shifts from the archaeological record, it is important to actually establish dietary change through more direct methods of analysis on human remains. Direct methods provide information on diet variation within skeletal samples which is not evident through analysis of associated plant and animal remains and artifacts used in food procurement and preparation.

ANALYSIS OF HUMAN DIET USING STABLE ISOTOPES

As the ecological and biocultural approaches became better established within the field of paleopathology during the 1970s, new methods for revealing past diet were emerging. Analyses of trace elements and stable isotopes from preserved bones and teeth showed great promise for differentiating past diet directly from the individuals of interest (Price et al. 1985). That is, rather than obtaining information about diet and subsistence only from archaeological remains of tools, animal bones and plant remains, diet could also be determined directly from human remains. This makes it possible to investigate age and sex differences as well as status differences within skeletal samples. Later, stable isotope methods were used to study mobility. Stable isotopes of carbon and nitrogen reveal past dietary information that can be linked to life history processes such as nursing, weaning, growth and health status. Stable isotope studies from mummified remains allow researchers to view diet at different times by analyzing tissues with different turnover rates. Stable isotopes of oxygen and strontium can be used to detect mobility and migration. This chapter emphasizes what we can learn from stable isotope data but it is equally important to understand the limitations of this method of analysis. As is true of many pursuits, a little knowledge

$\delta^H X = [(R_{sample} - R_{standard} - 1)] \times 1000$

Where R = the ratio of the heavier to lighter isotopes, so that for carbon isotopes, the equation is:

$$\delta^{13}C\text{‰ PDB} = \left[\frac{^{13}C/^{12}C \text{ sample}}{^{13}C/^{12}C \text{ standard}} - 1 \right] \times 1000$$

Figure 6.1   Delta notation for stable isotope ratios.

can be dangerous. A lack of understanding of the many physiological and taphonomic factors that result in variation in isotope values can easily be misinterpreted as dietary variation. Similarly, interpretations of oxygen and strontium isotope data must be grounded in understanding the geological processes that result in the variation observed in different regions of the earth.

What we can learn from stable isotopes

There are many thorough reviews of the use of stable isotopes for paleodiet reconstruction and evidence of past mobility that are written for undergraduate and graduate students (e.g., Schwarcz and Schoeninger 1991; Sealy 2001; Katzenberg 2008). These provide the basics of history, terminology, scientific principles and applications of stable isotope studies. The ratio of stable isotopes of carbon (^{13}C and ^{12}C, expressed as $\delta^{13}C$; see Figure 6.1) differentiate plants that use different pathways for photosynthesis (C_3, C_4, and CAM plants). The $\delta^{13}C$ values of C_3 and C_4 plants are distinctive and do not overlap. The range of $\delta^{13}C$ values of C_3 plants is –20‰ to –35‰ (‰ = permil, meaning per thousand) and the range of $\delta^{13}C$ values of C_4 plants is –9‰ to –14‰ (Deines, 1980). CAM plants (Crassulacean Acid Metabolism) over-lap C_3 and C_4 plants in their $\delta^{13}C$ values. The difference in the stable isotope ratios of plants is passed on to the consumers of those plants with an offset which varies, depending on the tissue (Vogel 1978; Ambrose and Norr 1993:6). Thus the introduction of maize, a C_4 plant, into North and South America from Meso-America, has been documented by analyzing stable carbon isotope ratios in preserved bone collagen of dated human remains from all over the Americas (see Staller et al. 2006, "Histories of Maize" and individual chapters in the section on stable isotopes). In regions where no other C_4 plants were used prior to the domestication of maize, stable carbon isotope analysis coupled with radiocarbon dating provides an accurate determination of the timing of the introduction, and intensification of maize use (e.g., van der Merwe and Vogel 1978; Schwarcz et al. 1985). In the Old World, stable carbon isotopes can reveal the spread of millet, a C_4 plant that extends across Europe and Asia (Hunt et al. 2008).

Stable carbon isotopes are also useful when trying to distinguish the relative importance of marine foods in the diet when people were also consuming terrestrial foods. Here the difference originates in the offset between dissolved inorganic carbon in the ocean ($\delta^{13}C = 0$‰) versus atmospheric carbon dioxide ($\delta^{13}C = -7.8$‰). Due to the widespread burning of fossil fuels, the $\delta^{13}C$ value of atmospheric carbon dioxide has decreased by approximately 1.2‰ over the last 130 years (Boutton 1991). It is important to consider this difference when comparing modern and prehistoric species, and when using modern plants and animals as proxies for prehistoric species.

Organisms living in the ocean reflect the $\delta^{13}C$ value of source carbon and any offset between diet and source. Human consumers of marine and terrestrial foods can be distinguished using stable carbon isotopes (e.g., Tauber 1981; Chisholm et al. 1982).

In some of the first stable carbon isotope studies of prehistoric humans, assumptions were made about the stable isotope ratios of freshwater fish, which were assumed to be similar to the stable carbon isotope ratios of C_3 plants (i.e., not enriched in the heavier isotope of carbon) (e.g., van der Merwe 1982). This is often true, but it is not always the case. In large and deep lakes with varying habitats, there may be a wide range of $\delta^{13}C$ values among fish and this variation may occur within species as well as between species. This may be traced to the source carbon, which varies due to the range of inputs (atmospheric CO_2, dissolved organic and inorganic carbon from plants and animals and surrounding rocks, and fractionation due to variations in temperature and pH in different areas of lakes) (Hecky and Hesslein 1995). Paleodiet studies of fauna and people living in and near Lake Baikal in Siberia illustrate the wide range of variation in the $\delta^{13}C$ values of fish species from shallow and open waters of the lake ($\delta^{13}C$ values from –25‰ to –15‰) and the resulting variation in human consumers of these fish (Katzenberg and Weber 1999; Katzenberg et al. 2009). This phenomenon does not occur in rivers, since the multiple sources of carbon are well mixed.

Stable isotopes of nitrogen provide a second source of information with respect to diet reconstruction and are particularly useful in situations where there may be more than one source of food that is enriched in the heavier isotope of carbon (C_4 plants and marine foods, for example). Nitrogen isotopes (^{15}N and ^{14}N, also expressed using delta notation, $\delta^{15}N$) vary within a food web by trophic level, with an increase of approximately 3‰ with each step (Minigawa and Wada 1984; Schoeninger and DeNiro 1984). Since marine food webs are more complex (have more steps) than terrestrial food webs, $\delta^{15}N$ values among high trophic level marine animals such as seals and whales, are considerably higher than $\delta^{15}N$ values of high trophic level terrestrial animals. Freshwater ecosystems also show trophic level differences, with $\delta^{15}N$ at top levels in large lakes higher than in terrestrial ecosystems but not so high as marine ecosystems where there are various marine mammals (Katzenberg 1989; Katzenberg and Weber 1999).

Nitrogen comes from dietary protein which may come from both plants and animals. It is a common misconception that protein comes only from foods of animal origin and obviously this is not possible. Herbivores, such as moose, elk and deer, all survive on plant foods and have large bodies with muscles, collagen, and other proteins, manufactured from the amino acids ingested in the plant protein in their diets. For human diets, animal sources of protein contain the complete complement of essential amino acids (those which cannot be manufactured in the body and must be obtained from the diet) while plants may be deficient in one or more of the amino acids essential for humans. Most traditional diets in which staple crops such as rice, wheat, or maize, are important combine plant foods, for instance corn with beans in the Americas, or lentils with rice in the Middle East, and these combinations provide all essential amino acids when consumed together. Similarly, cuisine that largely consists of plant-based foods, is often complemented with small amounts of animal foods such as cheese, egg, meat or fish, again, providing all essential amino acids in one meal. In this case, some of the amino acids will be obtained from plant sources and some will be from animal sources. Another common misconception with respect to stable isotope studies is that the $\delta^{15}N$ provides information on the amount of protein in the diet. $\delta^{15}N$ provides information on the source of protein but not the amount of protein, because $\delta^{15}N$ is a ratio of ratios (i.e., the relative amounts of the heavier and lighter isotope in the substance of interest, in this case a food item, relative to a standard substance—atmospheric nitrogen). $\delta^{15}N$ is not a measure of the total amount of nitrogen ingested. $\delta^{15}N$ may provide information on the relative amounts of plant and animal protein when several individuals or samples are compared, since animals tend to be enriched in the heavier isotope (^{15}N) relative to plants from within the same environment.

In high latitudes, people obtain almost all of their protein from animal sources (Cordain et al. 2000). The contrast between marine and terrestrial isotope signatures in high latitude hunters is nicely illustrated in a study by Coltrain and colleagues (2004) on Thule individuals from the high Arctic. Stable nitrogen isotope ratios are as high as 20‰ among consumers of whales and seals in contrast to individuals consuming terrestrial mammals, who would be expected to have $\delta^{15}N$ values no greater than 9‰. Northern hunter-fisher-gatherers around Lake Baikal, a high-latitude, freshwater lake, exploited a mix of terrestrial and aquatic resources including freshwater seals. There, $\delta^{15}N$ values exhibit a wide range due to the mixed diet, but they do not exceed 16‰ (Katzenberg et al. 2009).

It is important to be aware of some caveats with respect to interpretations of nitrogen isotopes. The trophic level shift of 3‰ is not constant. Vanderklift and Ponsard (2003) report a range of variation in the differences between $\delta^{15}N$ of diet and tissues for various taxa, cautioning that comparisons from lab studies to natural studies of other species, and comparisons of different tissues may result in incorrect interpretations. Some of the sources of variation in the spacing between diet and tissue are not yet well understood, but this is a subject of active research. Researchers have also found a wide range of variation in the $\delta^{15}N$ of plant tissues, which may be influenced by such factors as rainfall, sea spray and saline soils (Virginia and Delwiche 1982; Shearer et al. 1983; Heaton 1987) as well as disturbed soils (Schwarcz et al. 1999; Comisso and Nelson 2006, 2008). Thus plant $\delta^{15}N$ values may be as high, or even higher than, expected $\delta^{15}N$ values in some animal species (Figure 6.2). This is another illustration of the importance of understanding the isotope ecology of a region, which is best accomplished by sampling the available food sources.

In summary, to understand stable isotope variation, one must understand the source(s) of the element at the base of food webs, and the phenomena that result in partitioning or discrimination through the food web. The entire terrestrial food web traces its carbon uptake back to atmospheric carbon dioxide (CO_2). This CO_2 is taken up by plants and is converted to sugar. But the way in which it is taken up varies depending on the photosynthetic pathway used by the plant. The C_4 pathway results in water conservation for the plant and is found in certain plant species that are adapted to hot, dry or saline conditions. Animals consuming these plants incorporate carbon into the collagen in their bones with a reasonably predictable offset. In the marine food web, the source carbon is primarily dissolved inorganic carbon which differs from atmospheric CO_2 by about 7.8 ‰. In freshwater lakes, the source carbon comes from multiple sources, as explained above. Understanding the flow of carbon and

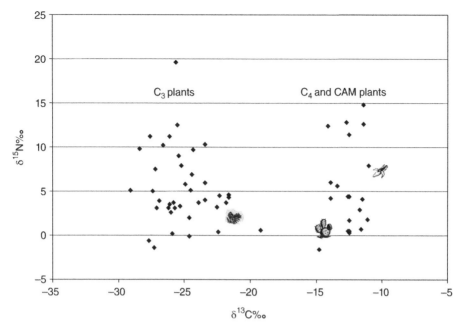

Figure 6.2   Stable isotope analysis of modern plants collected from West Central Chihuahua, Mexico. There is a wide range of variation in both $\delta^{13}C$ and $\delta^{15}N$ values. C3 plants include juniper, pinon pine, onion and walnut. C4 plants include maize, sorghum and dropseed grass. CAM plants include agave, stool and prickly pear cactus. (Webster and Katzenberg 2009).

nitrogen through food webs from the bottom up, the approach taken by ecologists (Fry 2006), is also very useful in understanding human diets. A knowledge of diets leads to a better understanding of exposure to pathogens, potential trauma, repetitive activities that might leave marks on bone and deficiency diseases.

Paleodiet and paleonutrition

Long before people worried about specific nutrients in their diets and what balance of protein, fat and carbohydrates to consume, they were quite successful in living, reproducing and populating the earth. It is clear from ethnographic studies and from studies of primate biology, that we are adapted to a generalized diet and that we can thrive on diets ranging from almost entirely animal-based (Inuit and other high-latitude peoples) to plant-based (tropical inhabitants and modern vegans). Nutritionists have determined the nutrients essential for adequate nutrition (though these requirements are occasionally revised), but these nutrients may be obtained from a wide range of foods. While there is ample evidence of broad-scale dietary adaptations on the part of past human groups, there is also evidence of seasonal scarcity, and starvation undoubtedly occurred in the past. Some pathological conditions are caused by interruptions in growth, which may be due to nutritional stress as well as other causes. For example, enamel hypoplasia, an interruption in the deposition of enamel during tooth formation, may occur repeatedly and since the time of formation of hypoplastic bands can be determined, it may suggest annual periods of stress (Reid and Dean 2006; see also Chapter 30 by Lukacs, this volume). Skeletal and dental indicators that suggest food scarcity, periods of malnutrition or lack of specific essential nutrients have been investigated by paleopathologists for over 100 years, but more intensively since the 1970s (Goodman and Armelagos 1988; Goodman and Rose 1990). Temporary cessation of growth of long bones may be exhibited as radio-opaque Harris lines, and temporary cessation of growth of teeth may be exhibited as enamel hypoplasia, but it is important to be familiar with the various problems that may arise in using and interpreting these indicators (see discussion by Lewis 2007). Short stature suggests chronic food shortages but may be caused by other factors (Lewis 2007). By adding regional study of the plants and animals upon which human diets depended one can reveal the potential for such conditions. In turn, the presence of the conditions supports hypotheses about seasonal stress. For example, Speth and Spielman (1983) describe the seasonal variation in fatness of game animals and argue for dietary stress in spring when terrestrial mammals hunted by humans are extremely lean. If one then observes a high prevalence of annual hypoplastic bands in the enamel of hunter-gatherers who rely on terrestrial mammals in a cold region, then it is reasonable to suggest a link between nutritional stress and the pathological lesions of enamel hypoplasia in that population.

ARE STABLE ISOTOPES USEFUL IN DIAGNOSING PAST DISEASE?

Stable isotopes are clearly useful in revealing past diets, but are they useful in revealing past diseases? The most promising isotopic indicator of disease is stable nitrogen isotope ratios, because $\delta^{15}N$ provides information about the source of protein in the diet, and one source of protein is catabolism of one's own tissues. When insufficient dietary protein is ingested, body protein, such as muscle tissue, is utilized. This body protein is already enriched in the heavier isotope of nitrogen (^{15}N) relative to dietary sources and the result of metabolizing it is the preferential excretion of the lighter isotope (^{14}N) in urine, and therefore even further enrichment in the heavier isotope of nitrogen in the pool of body protein. As a result, $\delta^{15}N$ of bone collagen from an individual suffering from protein stress may be elevated. However, bone collagen has a long residence time and slow turnover, so dietary protein insufficiency would have to occur for a long period of time or be very severe for it to be reflected in bone collagen. Hobson and colleagues (1993) and; Hobson and Clark (1992) found elevated $\delta^{15}N$ in various tissues of protein-stressed geese and quail. In humans, analysis of $\delta^{15}N$ in tissues which record a relatively short period of time, such as hair and fingernails, might be expected to show periods of protein stress.

White and Armelagos (1977) found differences in $\delta^{15}N$ between individuals suffering from osteopenia (low bone mineral density) in a Nubian sample and those with normal bone mass. Both males and females had elevated $\delta^{15}N$ but the difference was greater among females in both the third and fifth decades of life. A wide range of factors are considered as potential causes, including physiological mechanisms related to pregnancy and lactation as well as calcium intake, utilization and kidney function. Individuals with normal bone mass did not show age or sex differences in $\delta^{15}N$ values. Thus, in this study, it was possible to identify stable nitrogen isotope ratios that were different by linking them to a recognizable alteration in bone mass.

Katzenberg and Lovell (1999) studied bones taken at autopsy from individuals with pathological lesions and compared diseased and normal segments of bones for $\delta^{13}C$ and $\delta^{15}N$ from bone collagen. One individual who suffered from a wasting disease exhibited elevated $\delta^{15}N$ in a bone affected with osteomyelitis. The $\delta^{15}N$ of collagen taken from the lesion was almost 2‰ higher than the two unaffected segments from the same skeletal element. The authors interpret this finding as reflecting formation of the collagen in the lesion from catabolized body protein, thus elevating $\delta^{15}N$ in the collagen of the segment most recently formed. Another individual sampled for this study had paralytic atrophy with presumably low bone turnover due to an absence of forces that stimulate remodeling. The three bone segments sampled varied by up to 0.6‰ and exhibited greater variation than the control samples taken from individuals lacking any bony lesions (maximum variation 0.4‰).

Based on these differences between normal and diseased segments within one skeletal element, it is unlikely that an unaffected bone from an individual with a chronic disease will show an obvious difference in stable carbon and nitrogen isotope ratios compared to healthy individuals from the same population. So the important question is, "How sick does someone have to be to exhibit altered stable isotope values, and how altered would they be if they are not to be confused with dietary variation within a population?" The answer to the first question has been addressed with respect to protein stress in animal studies and on tissues with faster turnover times than bone collagen (e.g., Hobson et al. 1993; Fuller et al. 2004). The answer to the second question will vary with each different population. Generally, stable isotope variation in a sample of human skeletal remains is greater when the dietary sources have highly variable stable isotope signatures. Thus, human groups consuming both C_3 and C_4 plants will exhibit more variation in the range of $\delta^{13}C$ values than groups consuming only C_3 plants and animals who consumed C_3 plants. Similarly, people consuming foods from terrestrial and marine sources, and from both C_3 and C_4 plants, will have more variable $\delta^{15}N$ and $\delta^{13}C$ values than people consuming inland, C_3-based diets with no marine foods. Further complications to establishing the "normal" range of variation include the length of time a cemetery was in use and potential dietary changes over time, climatic variation during the use of a cemetery, and the degree of homogeneity of diets among people who were interred together. These are the very questions that are often interesting to archaeologists but they add uncertainty to interpretations of pathology.

Weaning, infant mortality

An important question that has been addressed using stable nitrogen isotopes, and to a lesser degree, stable carbon isotopes, is the determination of age at weaning in the past. The process of weaning (reduction, then cessation of nursing) has long been associated with a high risk of mortality due to the loss of passive immunity and the introduction of other foods, which may introduce infectious agents to the infant or young child (Katzenberg et al. 1996; Herring et al. 1998). Fogel and colleagues (1989) introduced the concept of using stable nitrogen isotope ratios for determining the age of weaning and demonstrated the validity of the method using fingernails from living mothers and their infants and documenting the time of weaning. They provided a clear demonstration of a pattern of elevated $\delta^{15}N$ among infants with a reduction toward maternal values at the approximate time that nursing was stopped following the principle that infants are consuming maternal tissue in the form of milk, and therefore exhibit a trophic level effect as long as they are consuming breast milk. Fingernails, largely composed of the protein keratin, record such dietary changes more precisely than does bone collagen, since nails do not turn over and the growth rate is predictable. Fuller and colleagues (2006) carried out a larger study on living individuals, sampling both hair and fingernails of mothers and infants. Their results supported the earlier work of Fogel and colleagues (1989), and also pointed out variation in the spacing between the δ^{15}N of mothers and infants. One possible cause of this variation is nitrogen balance during growth and development, a topic explored by Waters-Rist and Katzenberg (2010) in a paper that also reviews the various studies using nitrogen isotopes to explore nursing and weaning practices on prehistoric and early historic samples. In these samples, bone collagen is the material available and a large number of studies have illustrated a pattern of increasing δ^{15}N in infants and young children that decreases somewhere between one and three years of age. The significance of this research to paleopathology, is in tying weaning to morbidity and mortality in past populations. For example, Schurr and Powell (2005) tested the hypothesis that the duration of nursing decreased with the appearance of intensive food production. They found no differences in the timing of weaning, based on δ^{15}N values, for two sites predating and two sites postdating the introduction of agriculture into the Ohio Valley.

In contrast to stable nitrogen isotopes, there is no reason to expect that the ratio of stable carbon isotopes will vary directly due to disease. There are, however, useful indirect links, such as the link between the domestication and consumption of maize, a sticky carbohydrate, and the concomitant increase in dental caries observed in many skeletal samples (Larsen 1983; Powell 1985). So while dental caries do not result in altered δ^{13}C values, there is a link between stable carbon isotope ratios and carious lesions. For example, Figure 6.3 illustrates the relationship between the incidence of dental caries and δ^{13}C for four sites from southern Ontario, dating from A.D. 200–300 to A.D. 1275 and thus spanning the introduction of maize agriculture into the region (caries data from Cybulski 1968 and stable isotope data from Katzenberg et al. 1995).

MUMMIES

Studies of mummified remains provide an opportunity to analyze stable isotope ratios in tissues with varying turnover times (for other types of analyses using mummified remains see Chapter 9 by Zimmerman, in this volume). Hair and fingernails, which can be analyzed for stable isotopes of carbon and nitrogen as well as sulfur, do not turn over and represent distinctive and measureable time periods prior to death. For example, a 1-centimeter segment of hair is estimated to represent one month of life, with hair growing between 0.44 and .035 mm per day (Saitoh et al. 1967; Willams 2005). Muscle and tendon provide information from a longer period of time prior to death with some turnover, while bone provides the longest time depth with turnover of 10–25 years (Stenhouse and Baxter 1979) or more (Hedges et al. 2007). These differences in timing allow researchers to address questions related to short-term dietary change, such as season of death and migration between regions of different diet.

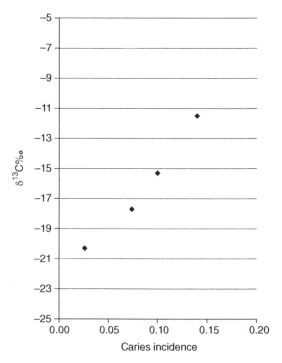

Figure 6.3 Caries incidence increases as more maize is consumed, as reflected in increasing $\delta^{13}C$ values from four sites in southern Ontario: (from left to right) Serpent Mounds, Surma, Serpent Pits and Bennett.

White (1993) studied tissues of desiccated mummies from prehistoric Nubia from two time periods (Christian and X Group) in an attempt to determine the season of death. Later, White and Schwarcz (1994) demonstrated seasonal variation in diet among these individuals based on segmental analyses of hair samples. Similar work has been done on Peruvian mummified remains (Williams 2005; Finucane 2007). An advantage of working with hair is that the relationship between diet and hair stable isotope ratios can be studied on living individuals. Such studies (O'Connell and Hedges 1999; O'Connell et al. 2001) demonstrate differences in $\delta^{15}N$ between humans who consume animal products and those who do not (vegans), but not between ovo-lacto vegetarians and omnivores. Physiological variables such as activity and exercise have also been considered with respect to stable isotope variation in hair of humans (Huelsemann et al. 2009) and nonhuman primates (Schoeninger et al. 1997).

MOBILITY

The ecological approach must also consider changes in residence over the course of the lifespan, and stable isotopes of several elements have been used to address questions of residence and mobility. Strontium isotopes, usually expressed as $^{87}Sr/^{86}Sr$, vary in the environment based on the age of the underlying bedrock. ^{87}Sr is a product of the radioactive decay of ^{87}Rb (rubidium) and therefore increases over time. Older rocks will accumulate more ^{87}Sr and therefore have higher $^{87}Sr/^{86}Sr$ than younger rocks, and rocks with a higher Rb content will also have more [87]Sr than rocks with lower Rb contents. Unlike stable isotopes of lighter elements such as carbon and nitrogen, which are fractionated during the physiological processes of photosynthesis, digestion and utilization by the body, Sr isotopes are not fractionated. Bentley (2006) provides a thorough review of the use of strontium isotopes in archaeological mobility studies. For studies of human remains, it is necessary to determine the Sr isotope ratio of bioavailable strontium within a region and this is best evaluated by analyzing bones of fauna with limited ranges rather than the rocks themselves (Price et al. 2002). Strontium enters the body by way of food and water so variation in diet may result in variation in strontium isotopes even within the same region if, for example the proportion of marine foods varies relative to terrestrial foods. The application to human remains and to paleopathology is indirect but important. It is possible to tell if individuals moved into a region from another region if there is variation in the Sr isotope ratios between the two regions. Most often, the method evaluates the range of variation for a sample of individuals and anyone that is more than two standard deviations outside the range is considered to be from somewhere else (Bentley et al. 2003; Buzon et al. 2007). Thus, an increase in population interactions, migration and potential spread of disease may be discernable with information about the homogeneity or heterogeneity of any particular group of people. It is also possible that there is insufficient variation in Sr isotope ratios among potential locations where people have traveled, and other methods may be more successful in these cases (e.g., Knudson et al. 2004).

Oxygen isotopes, expressed using the delta notation as $\delta^{18}O$, vary based on precipitation and surface waters. The highest values are found at low latitudes along the coast. $\delta^{18}O$ values decrease with increasing latitude, increasing distance from the coast and increasing altitude, since the heavier isotope is preferentially lost in precipitation. As clouds move inland and to higher altitudes, the remaining water becomes depleted in the heavier isotope of oxygen. Evaporation results in preferential loss of the lighter isotope, so water sources must be considered as well as geographical location. There are two different standards for reporting $\delta^{18}O$. The one used for water is Standard Mean Ocean Water (SMOW). Stable oxygen isotope measurements from carbonates, including the carbonate in bone mineral, are reported relative to the VPDB standard, the same one that is used for carbon (VPDB refers to a formation of Cretaceous belemnite at Peedee in South Carolina, USA). Oxygen isotopes may be measured from either carbonate or phosphate. (Basic information on the analysis of oxygen isotopes from bone is presented by Schwarcz and colleagues, 2010 along with a review of archaeological applications). Like strontium isotopes, a local range can be established from analysis of individuals buried together. In order to be able to identify someone from outside the local range, they would have to come from a place with different $\delta^{18}O$ from their source of water.

The combination of strontium and oxygen isotope analysis may provide discrimination among locations when only one of these tracers will not (Evans et al. 2006; White et al. 2007). Mobility studies provide contextual information about individuals in past populations that can be correlated with data on trauma and disease, and therefore enhance the suite of chemical tracers used to better understand past people. They may also provide information about past trade and exchange, which may be linked to opportunities for the spread of diseases.

SUMMARY

The ecological approach to paleopathology involves gaining an understanding of the interactions of people and their environments. The use of stable isotopes, which are effective environmental tracers, allows the researcher to learn about past diet, including nursing and weaning. They also provide information about the exploitation of plants and animals that were available as potential food sources. As with other approaches, there are limitations to stable isotope studies. They do not provide information about specific diseases or specific nutritional deficiencies, with the possible exception of insufficient dietary protein. Ecological studies of animals may be done using fast-turnover soft tissues whereas archaeological studies are most often carried out on teeth, formed early in life, and bone collagen, which records many years of life. Thus, bioarchaeologists may not see some of the more subtle variation that is found in ecological studies. With knowledge of basic principles, stable isotopes provide extensive information about past ecology and that knowledge provides a richer context for the interpretation of paleopathological data.

REFERENCES

Ambrose, S. H., and Norr, L., 1993   Experimental Evidence for the Relationship of the Carbon Isotope Ratios of Whole Diet and Dietary Protein to Those of Bone Collagen and Carbonate. Berlin: Springer-Verlag.

Armelagos, G. J., and McArdle, A., 1975   Population, Disease and Evolution. American Antiquity 40:1–10.

Bentley, R. A., 2006   Strontium Isotopes from the Earth to the Archaeological Skeleton: A Review. Journal of Archaeological Method and Theory 13(3):135–187.

Bentley, R. A., Krause, R., Price, T. D., and Kaufmann, B., 2003   Human Mobility at the Early Neolithic Settlement of Vaihingen, Germany: Evidence From Strontium Isotope Analysis. Archaeometry 45(3):471–486.

Boutton, T. W., 1991   Stable Carbon Isotope Ratios of Natural Materials: II. Atmospheric, Terrestrial, Marine, and Freshwater Environments. *In* Carbon Isotope Techniques. D. C. Coleman, and B. Fry, eds. pp.173–185. San Diego: Academic Press.

Buikstra, J. E., and Cook, D. C., 1980   Paleopathology: An American Account. Annual Review of Anthropology 9:433–470.

Burnet, M., and White, D. O., 1972   Natural History of Infectious Disease. Cambridge: Cambridge University Press.

Buzon, M. R., Simonetti, A., and Creaser, R. A., 2007   Migration in the Nile Valley During the New Kingdom Period: A Preliminary Strontium Isotope Study. Journal of Archaeological Science 34(9):1391–1401.

Chisholm, B. S., Erle, N. D., and Schwarcz, H. P., 1982   Stable Carbon Isotope Ratios as a Measure of Marine Versus Terrestrial Protein in Ancient Diets. Science 216:1131–1132.

Cockburn, T. A., 1971   Infectious Diseases in Ancient Populations. Current Anthropology 12:45–62.

Cohen, M., and Armelagos, G. J., 1984   Paleopathology and the Origins of Agriculture. New York: Academic Press.

Coltrain, J. B., Hayes, M. G., and O'Rourke, D. H., 2004   Sealing, Whaling and Caribou: The Skeletal Isotope Chemistry of Eastern Arctic Foragers. Journal of Archaeological Science 31(1):39–57.

Commisso, R. G., and Nelson, D. E., 2006   Modern Plant δ^{15}N Values Reflect Ancient Human Activity. Journal of Archaeological Science, 33(8):1167–1176.

Commisso, R. G., and Nelson, D. E., 2008   Correlation Between Modern Plant δ^{15}N Values and Activity Areas of Medieval Norse Farms. Journal of Archaeological Science 35(2):492–504.

Cordain, L., Miller, J. B., Boyd Eaton, S., Mann, N., Holt, S. H. A., and Speth, J. D., 2000   Plant–Animal Subsistence Ratios and Macronutrient Energy Estimations in Worldwide Hunter-Gatherer Diets. American Journal of Clinical Nutrition 71(3):682–692.

Cybulski, J. S., 1968   Analysis of the Skeletal Remains from the Surma Site, Fort Erie, Ontario. Ontario Archaeology 11:8–26.

Deines, P., 1980   The Isotopic Composition of Reduced Organic Carbon. In Handbook of Environmental Isotope Geochemistry. P. Fritz and J. C. Fontes, eds. pp. 329–406. Amsterdam: Elsevier.

Dobyns, H. F., 1992   Native American Trade Centers as Contagious Disease Foci, In Disease and Demography in the Americas. J. W. Verano, and D. H. Ubelaker, eds. pp. 215–222 Washington, DC: Smithsonian Institution Press.

Dunn, F. L., 1968   Epidemiological Factors: Health and Disease in Hunter-Gatherers. In Man the Hunter. R. Lee, and I. DeVore, eds. pp. 221–228. Chicago: Aldine.

Evans, J., Stoodley, N., and Chenery, C., 2006   A Strontium and Oxygen Isotope Assessment of a Possible Fourth Century Immigrant Population in a Hampshire Cemetery, Southern England. Journal of Archaeological Science, 33(2): 265–272.

Finucane, B. C., 2007   Mummies, Maize, and Manure: Multi-Tissue Stable Isotope Analysis of Late Prehistoric Human Remains from the Ayacucho Valley, Peru. Journal of Archaeological Science, 34(12):2115–2124.

Fogel, M. L., Tuross, N., and Owsley, D. W., 1989   Nitrogen Isotope Tracers of Human Lactation in Modern and Archaeological Populations. Carnegie Institution, Annual Report of the Director; Geophysical Laboratory. Washington, DC: Carnegie Institution.

Fry, B., 2006   Stable Isotope Ecology. New York: Springer.

Fuller, B. T., Fuller, J. L., Harris, D. A., and Hedges, R. E. M., 2006   Detection of Breastfeeding and Weaning in Modern Human Infants with Carbon and Nitrogen Stable Isotope Ratios. American Journal of Physical Anthropology 129:279–293.

Fuller, B. T., Fuller, J. L., Sage, N. E., Harris, D. A., O'Connell, T. C., and Hedges, R. E. M., 2004   Nitrogen Balance and δ^{15}N: Why You're Not What You Eat During Pregnancy. Rapid Communications in Mass Spectrometry, 18(23):2889–2896.

Goodman, A. H., and Armelagos, G. J., 1988   Childhood Stress and Decreased Longevity in a Prehistoric Population. American Anthropologist 90(4):936–943.

Goodman, A. H., Lallo, J., Armelagos, G. J., and Rose, J. C., 1984   Health Changes at Dickson Mounds, Illinois (A.D. 950–1300). In Paleopathology at the Origins of Agriculture. M. Cohen, and G. J. Armelagos, eds. pp. 271–306. New York: Academic Press.

Goodman, A. H. and Rose, J. C., 1990   Assessment of Systemic Physiological Perturbations from Dental Enamel Hypoplasias and Associated Histological Structures. Yearbook of Physical Anthropology 33:59–110.

Heaton, T. H. E., 1987   The ^{15}N/^{14}N Ratios of Plants in South Africa and Namibia: Relationship to Climate and Coastal/Saline Environments. Oecologia 74:236–246.

Hecky, R. E., and Hesslein, R. H., 1995   Contributions of Benthic Algae to Lake Food Webs as Revealed by Stable Isotope Analysis. Journal of the North American Benthological Society 14(4):631–653.

Hedges, R. E. M., Clement, J. G., Thomas, D. L., and O'Connell, T. C., 2007   Collagen Turnover in the Adult Femoral Mid-Shaft: Modeled from Anthropogenic Radiocarbon Tracer Measurements. American Journal of Physical Anthropology 133(2):808–816.

Herring, D. A., Saunders, S. R., and Katzenberg, M. A., 1998   Investigating the Weaning Process in Past Populations. American Journal of Physical Anthropology 105:425–439.

Hillson, S., 1986   Teeth. Cambridge: Cambridge University Press.

Hobson, K. A., Alisauskas, R. T., and Clark, R. G., 1993   Stable Nitrogen Isotope Enrichment in Avian Tissues Due to Fasting and Nutritional Stress: Implications for Isotopic Analyses of Diet. The Condor 95:388–394.

Hobson, K. A., and Clark, R. G., 1992   Assessing Avian Diets Using Stable Isotopes II: Factors Influencing Diet–Tissue Fractionation. The Condor 94:187–195.

Huelsemann, F., Flenker, U., Koehler, K., and Schaenzer, W., 2009   Effect of a Controlled Dietary Change on Carbon and Nitrogen Stable Isotope Ratios of Human Hair. Rapid Communications in Mass Spectrometry 23(16):2448–2454.

Hunt, H., Van der Linden, M., Liu, X., Motuzaite–Matuzeviciute, G., Colledge, S., and Jones, M. K., 2008   Millets Across Eurasia: Chronology and Context of Early Records of the Genera *Panicum* and *Setaria* From Archaeological Sites in the Old World. Vegetation History and Archaeobotany 17(Suppl. 1):5–18.

Katzenberg, M. A., 1989   Stable Isotope Analysis of Archaeological Faunal Remains from Southern Ontario. Journal of Archaeological Science 16:319–329.

Katzenberg, M. A., 2008   Stable Isotope Analysis: A Tool for Studying Past Diet, Demography, and Life History. *In* Biological Anthropology of the Human Skeleton. M. A. Katzenberg, and S. R. Saunders, eds. pp. 413–442. Hoboken, NJ: John Wiley & Sons.

Katzenberg, M. A., Goriunova, O. I., and Weber, A. W., 2009   Paleodiet Reconstruction of Bronze Age Siberians from the Mortuary Site of Khuzhir-Nuge XIV, Lake Baikal. Journal of Archaeological Science 36(3):663–674.

Katzenberg, M. A., Herring, D. A., and Saunders, S. R., 1996   Weaning and Infant Mortality: Evaluating the Skeletal Evidence. Yearbook of Physical Anthropology 39:177–199.

Katzenberg, M. A., and Lovell, N. C., 1999   Stable Isotope Variation in Pathological Bone. International Journal of Osteoarchaeology 9:316–324.

Katzenberg, M. A., Schwarcz, H. P., Knyf, M., and Melbye, F. J., 1995   Stable Isotope Evidence for Maize Horticulture and Paleodiet in Southern Ontario, Canada. American Antiquity 60:335–350.

Katzenberg, M. A., and Weber, A. W., 1999   Stable Isotope Ecology and Paleodiet in the Lake Baikal Region of Siberia. Journal of Archaeological Science 26(6):651–665.

Knudson, K. J., Price, T. D., Buikstra, J. E., and Blom, D. E., 2004   The Use of Strontium Isotope Analysis to Investigate Tiwanaku Migration and Mortuary Ritual in Bolivia and Peru. Archaeometry 46(1):5–18.

Lallo, J. W., Armelagos, G. J., and Mensforth, R. P., 1977   The Role of Diet, Disease, and Physiology in the Origin of Porotic Hyperostosis. Human Biology 49:471–483.

Larsen, C. S., 1983   Behavioural Implications of Temporal Change in Cariogenesis. Journal of Archaeological Science 10(1):1–8.

Lewis, M. E., 2007   The Bioarchaeology of Children: Perspectives from Biological and Forensic Anthropology. Cambridge: Cambridge University Press.

Minagawa, M., and Wada, E., 1984   Stepwise Enrichment of ^{15}N Along Food Chains: Further Evidence and the Relation Between δ^{15}N and Animal Age. Geochimica Et Cosmochimica Acta, 48:1135–1140.

O'Connell, T. C., and Hedges, R. E. M., 1999   Investigations into the Effect of Diet on Modern Human Hair Isotopic Values. American Journal of Physical Anthropology 108(4):409–425.

O'Connell, T. C., Hedges, R. E. M., Healey, M. A., and Simpson, A. H. R. W., 2001   Isotopic Comparison of Hair, Nail and Bone: Modern Analyses. Journal of Archaeological Science, 28(11):1247–1255.

Powell, M. L., 1985   The Analysis of Dental Wear and Caries for Dietary Reconstruction. *In* The Analysis of Prehistoric Diets. R. I. Gilbert, and J. H. Mielke, eds. Orlando: Academic Press.

Price, T. D., Burton, J. H., and Bentley, R. A., 2002   The Characterization of Biologically Available Strontium Isotope Ratios for the Study of Prehistoric Migration. Archaeometry 44(1):117–135.

Price, T. D., Schoeninger, M. J., and Armelagos, G. J., 1985   Bone Chemistry and Past Behaviour: An Overview. Journal of Human Evolution 14:419–447.

Reid, D. J., and Dean, M. C., 2006   Variation in Modern Human Enamel Formation Times. Journal of Human Evolution 50(3):329–346.

Saitoh, M., Makoto, U., and Sakamoto, M., 1967   Rate of Hair Growth. *In* Advances in Biology of Skin, Vol. IX. Hair Growth. W. Montagna, and R. L. Dobson, eds. pp. 183–201. Oxford: Pergamon Press.

Schoeninger, M. J., and DeNiro, M. J., 1984   Nitrogen and Carbon Isotopic Composition of Bone Collagen From Marine and Terrestrial Animals. Geochimica Et Cosmochimica Acta 48:625–639.

Schoeninger, M. J., Iwaniec, U. T., and Glander, K. E., 1997   Stable Isotope Ratios Indicate Diet and Habitat Use in New World Monkeys. American Journal of Physical Anthropology, 1031:69–83.

Schurr, M. R., and Powell, M. L., 2005   The Role of Changing Childhood Diets in the Prehistoric Evolution of Food Production: An Isotopic Assessment. American Journal of Physical Anthropology 126(3):278–294.

Schwarcz, H. P., Dupras, T. L., and Fairgrieve, S. I., 1999   [15]N Enrichment in the Sahara: In Search of a Global Relationship. Journal of Archaeological Science 26(6):629–636.

Schwarcz, H. P., Melbye, F. J., Katzenberg, M. A., and Knyf, M., 1985   Stable Isotopes in Human Skeletons of Southern Ontario: Reconstructing Paleodiet. Journal of Archaeological Science 12:187–206.

Schwarcz, H. P., and Schoeninger, M. J., 1991   Stable Isotope Analyses in Human Nutritional Ecology. Yearbook of Physical Anthropology 34:283–322.

Schwarcz, H. P., White, C. D., and Longstaffe, F. J., 2010   Stable and Radiogenic Isotopes in Biological Archaeology: Some Applications. *In* Isoscapes: Understanding Movement, Pattern, and Process on Earth Through Isotope Mapping. J. B. West, G. J. Bowen, T. E. Dawson, and K. P. Tu, eds. pp. 335–356. Dordrecht: Springer.

Sealy, J., 2001   Body Tissue Chemistry and Palaeodiet, *In* Handbook of Archaeological Sciences. D. R. Brothwell and A. M. Pollard, eds. pp. 269–279. Chichester: John Wiley & Sons.

Shearer, G., Kohl, D. H., Virginia, R. A., Bryan, B. A., Skeeters, J. L., Nilsen, E., Sharifi, M. R., and Rundel, P. W., 1983   Estimates of N_2–Fixation From Variation in the Natural Abundance of ^{15}N in Sonoran Desert Ecosystems. Oecologia 56:365–373.

Speth, J. D., and Spielmann, K. A., 1983   Energy Source, Protein Metabolism, and Hunter–Gatherer Subsistence Strategies. Journal of Anthropological Archaeology 2:1–31.

Staller, J. E., Tykot, R. H., and Benz, B. F., 2006   Histories of Maize: Multidisciplinary Approaches to the Prehistory, Linguistics, Biogeography, Domestication and Evolution of Maize. Amsterdam: Elsevier.

Stenhouse, M. J., and Baxter, M. S., 1979   The Uptake of Bomb ^{14}C in Humans. *In* Radiocarbon Dating. Proceedings of the 9th International Conference on Radiocarbon Dating. R. Berger, and H. E. Suess, eds. Los Angeles: University of California Press.

Tauber, H., 1981   [13]C Evidence for Dietary Habits of Prehistoric Man in Denmark. Nature 292:332–333.

van der Merwe, N. J., 1982   Carbon Isotopes, Photosynthesis, and Archaeology. American Scientist 70:596–606.

van der Merwe, N. J., and Vogel, J. C., 1978   [13]C Content of Human Collagen as a Measure of Prehistoric Diet in Woodland North America. Nature 276:815–816.

Vanderklift, M. A., and Ponsard, S., 2003   Sources of Variation in Consumer–Diet $\delta^{15}N$ Enrichment: A Meta–Analysis. Oecologia 136(2):169–182.

Virginia, R. A., and Delwiche, C. C., 1982   Natural ^{15}N Abundance of Presumed N_2-Fixing and Non-N_2-Fixing Plants From Selected Ecosystems. Oecologia 54:317–325.

Vogel, J. C., 1978   Isotopic Assessment of the Dietary Habits of Ungulates. South African Journal of Science 74:298–301.

Waters-Rist, A. L., and Katzenberg, M. A., 2010   The Effect of Growth on Stable Nitrogen Isotope Ratios in Subadult Bone Collagen. International Journal of Osteoarchaeology 20(2):172–191.

Webster, M., and Katzenberg, M. A., 2009   Dietary Reconstruction and Human Adaptation in West Central Chihuahua. *In* Celebrating Jane Holden Kelley and her Work. M. Kemrer, ed. pp. 67–96. New Mexico Archaeological Council Special Publication 5.

White, C. D., 1993   Isotopic Determination of Seasonality in Diet and Death from Nubian Mummy Hair. Journal of Archaeological Science 20(6):657–666.

White, C. D., and Armelagos, G. J., 1997   Osteopenia and Stable Isotope Ratios in Bone Collagen of Nubian Female Mummies. American Journal of Physical Anthropology 103(2):185–199.

White, C. D., Price, T. D., and Longstaffe, F. J., 2007   Residential Histories of the Human Sacrifices at the Moon Pyramid, Teotihuacan. Ancient Mesoamerica 18:159–172.

White, C. D., and Schwarcz, H. P., 1994   Temporal Trends in Stable Isotopes for Nubian Mummy Tissues. American Journal of Physical Anthropology 93:165–187.

Williams, J., 2005   Investigating Diet and Dietary Change Using the Stable Isotopes of Carbon and Nitrogen in Mummified Tissues from Puruchuco-Huaquerones, Peru. Ph.D. Thesis. Department of Archaeology, University of Calgary.

CHAPTER 7

An Epidemiological Approach to Paleopathology

Jesper L. Boldsen and George R. Milner

INTRODUCTION

Paleoepidemiology is applicable to the part of paleopathology that pertains to disease in a population context. Such studies are usually oriented toward documenting changes in disease experience that accompanied major transitions in human existence, most notably the development of agricultural economies, complex sociopolitical organizations, widespread intergroup contact, and greater population size and density. Collectively, these investigations cover everything from our distant hunter-gatherer ancestors to the inhabitants of relatively recent nation states, such as those of medieval Europe.

Studies that are paleoepidemiological in nature have less to do with the other important emphasis of paleopathology: the identification of specific diseases that were present in past populations. That work is often, although not always, oriented toward skeletal lesions in single individuals, or just a handful of them. For the most part, it is undertaken by matching the physical characteristics of archaeological skeletons to clinical cases from the 19th century onward.

The two principal objectives of paleopathology are obviously related to one another. One is useful for identifying the range of diseases—specifically those identifiable in bones and teeth—that were present at a particular place and time. The other is concerned with estimating how common these pathological conditions were in groups of people that differed by geographical origin, time period, sex, age, socioeconomic position, residential location, and the like. The later is what paleoepidemiological research is all about.

A Companion to Paleopathology, First Edition. Edited by Anne L. Grauer.
© 2012 John Wiley & Sons, Ltd. Published 2016 by John Wiley & Sons, Ltd.

Characterizations of the disease experience of past populations, or subsets of them, often focus on pathological lesions identified in two or more skeletal samples. They feature statements such as one group defined by age, sex, social status, or time horizon was sicker than another such group. All such comparisons, however, are not truly paleoepidemiological in nature since more must be done than simply count skeletal or dental lesions. By themselves, frequencies are nothing more than inventories of what can be observed in a collection of bones. In fact, what are examined— the skeletons—are not really what are of most interest. That might sound odd in a volume dedicated to skeletal remains, but it accurately reflects the fact that we are not primarily concerned with dead people. We are instead concerned with the lives they led before they died. That is, our interest is in characterizing conditions in past communities, not describing skeletons from cemeteries. The skeletons are a select sample of all people of each age who ever lived in the populations of interest; they are, in fact, the individuals who for whatever reason failed to survive long enough to make it to the next age-interval.

At a bare minimum, paleoepidemiological investigations must pay attention to the origin and representativeness of skeletal samples, the association between diseases and lesions (sensitivity and specificity), the selective effect of mortality in the formation of cemetery assemblages (the osteological paradox), the causes and consequences of disease in various cultural and natural settings, and the specific details of local conditions that impinged on health. The immediate objective is to reconstruct the lives of people using data derived from those who died (a mortality sample). That is done in large part through the age distributions of people with and without specific kinds of skeletal lesions. The ultimate goals are to further our understanding of the history of specific diseases, with an emphasis on host–pathogen evolution, as well as the relationships between disease load, population growth, and cultural evolution.

DISEASES, LESIONS, AND POPULATIONS

In order to avoid endless repetition, we take some liberties with wording from here on out. "Disease" is used as a shorthand for many different maladies that reduced life expectancy, and it pertains to what someone suffered from when alive. "Health," sometimes spoken of in opposition to disease, is a fuzzy concept as it can range from good to bad. At one end of the spectrum are those people who are so unhealthy they are about to die, hence enter the mortality sample that might be obtained in some future archaeological excavation of a cemetery. The term "lesion" is used, without further elaboration, for any physical sign of a pathological process on bones or teeth. Such a sign might be definitively associated with a specific pathological condition, such as tuberculosis or leprosy, or only consistent with it. Many only indicate a person had suffered from an illness of unknown origin.

It is, in fact, usually difficult to sort out precisely what led to the pathological conditions seen in archaeological skeletons. One such example is porotic hyperostosis related to anemia that, in turn, is likely to have come about through a combination of poor nutrition and infection (e.g., intestinal parasites), or in some populations an underlying genetic condition. Identifying the root cause of a pathological condition involves considerations of both exposure, such as to a particular pathogen, and a host's susceptibility and response to it. The former varies among populations and subsets of them, and even the latter do not remain constant throughout an individual's lifetime. Paleoepidemiological analyses, therefore, must take into account the nature of specific local contexts as well as the age-at-death of individual skeletons.

Diseases

In a paleoepidemiological context, disease is an identifiable condition that increases the risk of dying. There are two critical points here. First, it must be identifiable in skeletal remains. As many pathological conditions do not affect bones or teeth, it is hard to determine if they were indeed once present. Doing so could involve analyses of mummified soft tissue, preserved microorganisms, descriptions of symptoms in old documents, or ancient DNA, but not simply skeletons. As far as human remains are concerned, preserved soft tissue can be ignored for the most part, as it would be unlikely that there would be enough systematically examined mummies to do anything of paleoepidemiological consequence. Second, the pathological condition must have had an effect on well-being, thus on mortality. That does not mean it was an immediate cause of death, or even contributed to it in some reasonably direct way. All that is necessary is that the pathological condition is associated, no matter how indirectly, with a greater risk of dying. In this respect we differ from Waldron (1991) who includes as diseases of paleopathological interest conditions that increase the risk of death and those that ostensibly do not.

It will be readily understood that contracting diseases such as tuberculosis or experiencing severe trauma affects the risk of dying from that point onward. Yet even pathological processes as seemingly inconsequential as tooth decay had a measurable effect on the risk of death (Boldsen 1997a). Such conditions have an important role in paleoepidemiological analyses, even though they have not attracted the same level of attention as some of the great scourges of humankind, such as tuberculosis and syphilis, with the notable exception of the developmental defects known as enamel hypoplasia. In fact, when reconstructing how people lived and quantitatively assessing the risks they faced, the rather commonplace maladies are arguably as important, if not more so, than major diseases. That is simply a matter of the limited size of most archaeological samples, the infrequent occurrence of many pathological conditions in human populations, and the problems associated with recognizing skeletal lesions attributable to specific diseases. Dental caries, in contrast, are reasonably common in most past populations, teeth preserve rather well, and tooth decay and abscesses are easy to identify. The same could be said for enamel hypoplasia and bone lesions such as nonspecific periostitis (or osteomyelitis).

As a general rule, any impairment of function experienced by a member of a traditional society would affect the risk of dying, no matter how slight that increase might have been. Infirmities among adults, in particular, were important as they would have had a disproportionate effect on a family's well-being, or even that of the community as a whole, in societies where largely self-sufficient households were the primary economic units. Even simple fractures interfere with a person's ability to keep up with a mobile hunter-gatherer group or to do heavy field work in an agricultural society. The effects of such conditions might be difficult to identify, given generally small sample sizes and problems in accurately estimating age and consistently recognizing pathological conditions in skeletons. That does not alter the fact that, for much of human existence, any disability was likely to have had both immediate and long-term consequences.

As sufferers of disease tend to die at younger ages than nonsufferers, reasonably accurate and precise age estimates are essential in skeletal analyses. Obtaining them is, at present, easier for juveniles than for adults, although new developments in age estimation promise to increase the accuracy and precision of age estimates throughout the entirety of adulthood (see Chapter 15, by Milner and Boldsen, in this volume). That is good news as much hinges on the disease experience of adults in past societies because when men or women fell ill or were maimed, entire households were likely to suffer.

Lesions

Skeletal lesions, as they are a consequence of disease, can be viewed as risk factors for death in the sense of the term as it is used in standard epidemiological studies. As such, the lesions are indicative of differences among people. In fact, a particular kind of lesion shared by everyone would be unimportant from a paleoepidemiological perspective since it would be uninformative about variation among individuals. The picture is complicated by the fact that not everyone who suffers from a particular disease has a recognizable skeletal manifestation of it. It is also true that all people with a certain kind of lesion did not have the disease that is most often associated with it. So, identifications of specific diseases, sometimes labeled differential diagnosis, are inherently probabilistic in nature.

It should be recognized that differential diagnosis, by itself, has little to do with paleoepidemiology. It is part of the other principal emphasis of paleopathology—the identification of particular diseases in the past—not the quantification of their effect on groups of people defined in various ways. Where differential diagnosis comes in useful is in the identification of the specific diseases (i.e., pathogens) that affected individual populations in the past. It is not useful for identifying the prevalence of those diseases in the communities. That is because the procedure is intended to identify the most likely candidate for a particular suite of lesions on one or more skeletons, so it rests heavily on lesions considered to be "classic" in terms of their configuration, size, and distribution.

Lesions suitable for paleoepidemiological analyses can include those attributable to specific pathological conditions, including diseases such as leprosy, and others whose origin cannot be determined either reliably or consistently. The latter include the proliferation of woven bone commonly called periostitis that results from a number of infectious organisms as well as trauma. For the purposes of paleoepidemiological analyses, lesions can range from sites of bone resorption or proliferation that were active at the time of death to those that were fully healed. Once again, the concern is with signs of pathological conditions that were associated, no matter how indirectly, with an increased risk of dying. For example, evidence of developmental upsets in childhood, such as enamel hypoplasia, or a completely healed fracture can be of interest even though the people with them survived the events that caused the observable alterations to normal dental or skeletal structure. Their presence could indicate individuals who experienced poor living conditions or participated in hazardous activities associated with some decrease in life expectancy.

Populations

Not all skeletal collections are of equal value for paleoepidemiological investigations. The skeletons collectively must be part of a population, a term that is used loosely in the literature. First of all, there must be some biological and cultural coherency to the skeletal collection; that is, the sample should be more than just a group of skeletons from a broadly defined time period, cultural stage, or archaeological culture. In general, the skeletons should be drawn from one or more well-documented cemeteries associated with single communities. Sometimes skeletons might be derived from only part of a community, so the sample would not be representative of the whole population. Here sample adequacy is defined by cultural context, as well as the number of skeletons and the preservation of bones. For example, skeletons from a pauper's graveyard might provide a glimpse of what life was like for part of a community, specifically people who were among the most disadvantaged socially and economically. But they would not be representative of society as a whole, no matter how many skeletons might be available for examination. Thus, one of the first steps in a paleoepidemiological study is to characterize the nature of the skeletal samples in terms of their cultural context. Without doing so, conclusions drawn from skeletons have little significance as they are uninterpretable.

For the most part, cemetery samples are largely made up of the skeletons of individuals who lived much or all of their lives in the local community or, at most, the nearby region (e.g., a medieval monastery serving as a hospital or hostel). To be sure, the extent to which that statement is true varies according to the kinds of communities that are examined, which is why establishing the cultural context is so very important. One would expect that a cemetery in a medieval European city would include individuals from more diverse origins than a burial ground for a Neolithic village. Nevertheless, as a general rule a cemetery sample reflects local conditions. As they are indicative of what happened locally, one cannot automatically assume that skeletons from a single site are representative of what took place across a broader region among their culturally similar contemporaries. For example, populations hard-hit by war—including communities in small-scale societies, such as a 700-year-old village in Illinois where at least one-third of the adults died in a series of ambushes (Milner et al. 1991)—cannot be considered typical of conditions that prevailed during more peaceful times.

In all but mass graves holding fatalities from epidemics, famine, other natural disasters, or war, the skeletons from archaeological sites accumulated over some period of time, often several centuries or more. So there is a temporal dimension, typically spanning multiple generations, to an archaeological population. Most cemeteries, therefore, only provide a time-averaged picture of the period during which the skeletons accumulated. If living conditions changed enough during the course of cemetery use to affect disease experience, then any signature obtained from bones would be muddied, perhaps to the point of being indecipherable. Interpretations are not automatically easier for skeletons from burial grounds used for only a generation or so. These deaths could have been affected by stretches of especially good or bad years that affected wild food or crop yields, as well as periods when high-mortality epidemic diseases repeatedly broke out or were absent. Skeletons in such cemeteries might provide an excellent picture of what happened during narrow windows of time, but simultaneously present a distorted perspective on general conditions over the long run.

As local conditions were what had the greatest effect on the lives of people throughout most of human existence, the immediate cultural and natural settings need to be established, mainly with archaeological materials. General labels such as "hunter-gatherer" and "village agriculturalist" tell us little about the specific groups of people that produced the skeletons from individual cemeteries, including what led to a greater or lesser exposure to pathogens, activity-related injury, and food short-falls. That is because such categories encompass considerable diversity in what people actually did on a daily basis, where they lived, how often they moved, what they ate, and how much contact they had with neighboring groups, all of which affect disease and trauma experience.

Disease–lesion relationships

Interpreting skeletal data would be considerably simplified if two conditions held true: first, there was a one-to-one relationship between diseases and their corresponding lesions, and, second, there was a reasonably straightforward means of going from lesion frequencies in mortality samples to disease prevalence in living populations. Unfortunately, neither is true. All we can really be sure about is what is seen in the skeletons, but, once again, the bones are not really what are of greatest interest. Instead, the reason all this work is done is to characterize what life was like in the distant past, and how disease experience was related to cultural and natural settings that varied greatly from one place and time to another.

Sensitivity and specificity It is widely recognized that paleopathological analyses are complicated by an inexact relationship between the diseases of interest and skeletal lesions, an issue discussed by Ortner (1991), among others. These particular problems are summarized by sensitivity and specificity. Sensitivity refers to the individuals with a particular disease who test positive (true positive). For skeletons, it means the sick people who also had a recognizable skeletal marker of that disease. Specificity pertains to the individuals without the disease who are recognized as being disease free (true negative). Thus, they are conditioned probabilities:

$$\text{Sensitivity} = \text{true positive}/\text{all sick};$$
$$\text{Specificity} = \text{true negative}/\text{all non-sick}.$$

The concepts of sensitivity and specificity were first developed as tools in screening epidemiology. In screening programs the aim is similar to what a paleopathologist wants to estimate: the individual probability that the person (skeleton) did indeed suffer from the disease under study. In fact, sensitivity and specificity can help paleoepidemiologists define the lesions of relevance for any specific disease because such lesions are characterized by sensitivity $> (1 - \text{specificity})$. The concepts of positive and negative predictive values, also discussed below, are related to sensitivity and specificity; however, they draw on estimates of population (sample) prevalence of the disease to be calculated.

Paleopathologists face two problems with making sense of lesions visible on archae-ological skeletons. First, not all diseases that can affect bone do so in a recognizable manner in all cases. Not all individuals who suffer from such a disease actually develop the bony lesions, and not all of those lesions are sufficiently distinctive to be considered reliable markers of that particular disease. So in archaeological samples, two groups of skeletons lack lesions: those from people who never had the disease (true negative), and others from sick individuals who did not develop clear skeletal signs of the illness (false negative). It should be noted that an additional complication is being ignored here: the confounding effect of unequal bone preservation. One cannot assume the absence of a lesion means the same thing in fragmentary skeletons as it does in intact ones. Second, lesions can be mistakenly attributed to a particular disease, such as tuberculosis or leprosy, when the individual did not suffer from it at all, but instead had something else entirely. Once again, skeletons with similar lesions fall into two groups: those from sufferers of the disease with lesions (true positive), and those with lesions caused by some other disease (false positive).

Statistically speaking, sensitivity and specificity are related to one another: when one goes up, the other usually goes down. If one wanted to increase the chances that individuals with particular kinds of skeletal lesions were identified as suffering from a particular disease, more inclusive criteria could be used to identify markers of that disease (in this instance, the form and location of skeletal lesions). That might be done by including examples of pathological skeletal involvement that depart from those that are considered "classic" expressions of the lesions associated with the disease. Perhaps they might be thought of as early stages prior to the development of full-blown "classic" forms of the lesions. Doing so, however, comes at a cost as the chance of misclassification is simultaneously increased (false positive). Of course, limiting oneself solely to "classic" lesions is no solution as it verges on certainty that some cases go unrecognized, perhaps the great majority of them.

Sensitivity and specificity describe important aspects of the natural history of a disease. The manifestations of any particular disease, including how many exposed people break out with clinically recognizable signs of it, do not remain fixed over time and space. When dealing with deep time, as in archaeological samples, one must also be concerned with continually evolving host–pathogen relationships. So estimating the effect of any particular disease on a past community is by no means a simple matter. Historically, sensitivity (the probability that a sufferer of the disease actually had the lesion used as a marker of that disease) is more stable than specificity (the probability that an individual who did not suffer from the disease did not show the lesion). As what we observe—the skeletal lesions—are the result of diverse pathological processes, sensitivity and, particularly, specificity are dependent on the prevalence of other diseases capable of producing lesions that mimic those of the disease of prime interest.

Selective mortality   There is an additional twist to the tale. While it might seem unnecessary to state the obvious—cemeteries are full of dead people—the implications of that statement represent a major difficulty for paleoepidemiologists. The problem stems from wanting to say something meaningful about living populations when only mortality samples are available.

Perhaps this problem can be most easily visualized by imagining yourself in the local hospital's Emergency Room. If you monitored the front door long enough, taking note of new admissions, it would be possible to get a good idea of the range of severe conditions suffered by the members of your community (and many others that are not so troubling). But forming some idea about their prevalence would be another matter entirely. For that one would need to know the population at risk and estimate the risk of developing the conditions of interest (e.g., being exposed to a pathogen and developing symptoms of the infection, or performing certain activities that might lead to breaking a bone).

It is widely, but not universally, accepted that the path from skeletal lesion frequencies to the prevalence of particular pathological conditions in past populations is tortuous, and the way forward is by no means fully worked out. This issue—often referred to as the "osteological paradox" after the title of the article that introduced the subject to paleopathologists (Wood et al. 1992)—has sparked hot debate about the extent of the problem and its effect on the interpretation of skeletal data (Goodman 1993; Jackes 1993; Saunders and Hoppa 1993; Byers 1994; Cohen 1994, 1997; Wood and Milner 1994; Goodman and Martin 2002; Steckel and Rose 2002b; Wright and Yoder 2003; Milner et al. 2008). Discussion of this topic has even cropped up independently, underscoring a growing concern over what can be done with pathological lesions visible on archaeological skeletons (Waldron 1991, 1994, 1996; Bird 1996). In fact, some aspects of the controversy over the meaning of skeletal lesion frequencies were raised before the osteological paradox article appeared (Cook and Buikstra 1979; Cook 1981; Guagliardo 1982; Palkovich 1985; Ortner 1991; Waldron 1991).

The problem boils down to two simple issues. First, populations are heterogeneous insofar as they are composed of people who vary in their individual frailty. Second, mortality is selective with respect to which individuals are most likely to die at a given age. Neither point should excite controversy; in fact, they make a good deal of common sense.

The original discussion of heterogeneous frailty owed much to work by Vaupel and colleagues (1979, 1985a, 1985b), among others (e.g., Weiss 1990). Everyone recognizes that not all people of a given age are equally fit, so some individuals experience a greater risk of dying than other members of their cohort. That is, some people are more likely to enter the mortality sample at any particular age than others. In many instances, perhaps even the great majority of them, the causes of heterogeneous frailty remain hidden. They go unrecognized by whatever indicators one chooses to use when measuring whether a person appears to be in good shape or not. The difficulties one might experience with living patients in that respect are exacerbated for osteologists, since they are severely limited by what can be seen in the bones and teeth of people who died long ago. Nevertheless, it is still true that those who die at a given age tend to be the sickest or most poorly nourished members of their cohorts, or the people who by necessity or choice routinely engage in dangerous activities. Cautious souls in superb shape certainly die for any number of reasons, including in accidents, whereas sick people of the same age can survive only to die at some later date. Yet deaths in any age group are still weighted toward people who were ill or regularly engaged in hazardous behavior, rather than those who were just fine and did little that would expose them to major trauma.

It follows that skeletal samples are biased pictures of once-living populations. That is true even in the unlikely event that a completely excavated cemetery held every community member who died over some period of time, bones were uniformly well preserved, no skeletons were lost when later graves were dug, and

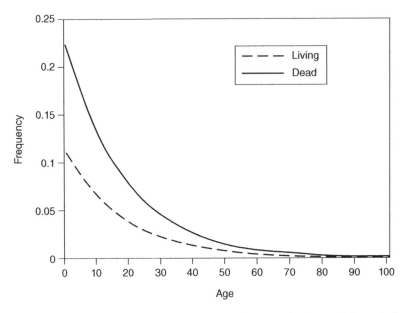

Figure 7.1   A deliberately simplified example where of 10,000 births, 1,000 people had a lesion and the remainder did not. Individuals with the lesion experienced a constant mortality rate of 0.1, whereas the others had a rate of 0.05. The lines show the frequencies of individuals at each age with the lesion in the cemetery sample (solid line) and living population (broken line). Throughout life, there were proportionately more individuals with the lesion at each age in the cemetery (dead) than there were in the community (alive) from which the dead were drawn.

archaeologists did not miss any of them. In other words, all people must die at some point, but not everyone of the same age is equally likely to do so. As skeletons are weighted toward individuals who experienced the greatest risk of dying at each age, a cemetery sample is not the same as what one would observe if transported magically back in time to visit the living members of the community. Because sick people are preferentially weeded out of the population at each age, skeletons with lesions exceed the individuals with exactly the same skeletal lesions in the original population who were alive at any single age. At least that will occur if the lesions are associated with an increased risk of dying (if that were not true, they would uninteresting to the paleopathologist), the bones are well preserved, and all lesions are recognizable without error. The second and third conditions, of course, are rarely met, especially the last, as it is related to sensitivity and specificity.

The point that people with lesions are more common among the dead than the living is shown in Figure 7.1. In this instance, the example is simplified to illustrate the main point; as discussed elsewhere in this paper, the actual situation is considerably more complex. Here it is assumed that of 10,000 births (the original population), 1,000 individuals had a lesion and 9,000 did not. In this very strange population the risk of dying does not change with age; that is, it remains fixed throughout the entirety of the potential lifespan. The constant mortality rates for the two groups are 0.1 for individuals born with a lesion, and 0.05 for those without one. Note that this condition cannot be acquired after birth, all individuals with the condition associated with higher mortality have a distinctive lesion, and all lesions are classified without error. If the situation were so easy, paleopathologists would not have to go through such great contortions when interpreting what archaeological skeletons might have to tell us about life in the distant past.

The solid line in Figure 7.1 illustrates the age-specific frequency (prevalence) of individuals with the lesion among the dead. It is what would be seen in skeletons from the cemetery. The dashed curve shows the frequency of lesions among the living. That is what would be observed if it was possible to conduct thorough clinical examinations of all community members. At all ages, the frequency of lesions is twice as high among the dead as among the living. Yet in spite of that disparity at each age, once all of the original 10,000 individuals had died, the total frequency of the lesion would still be 10 percent among the skeletons, just as it was among the individuals on the day of their birth.

Even catastrophic death assemblages do not provide an easy way around the problem. Archaeologically minded readers might object by citing the example of Pompeii. Yet even there the unfortunate victims cannot be considered a representative cross-section of Pompeii's population the day before the eruption, which devastated the city. Much like what takes place in natural disasters during modern times, the people who stayed behind, in aggregate, were not the same as the rest of the population that had the opportunity and possessed the wherewithal to depart in a timely fashion. The selective effect of mortality even left a detectible osteological signature in skeletons that filled a medieval English plague pit, which is rather remarkable considering the overall high death rate associated with the Black Death (DeWitte and Wood 2008). That is, even the plague was not marked by indiscriminate deaths.

As the principal points of the osteological paradox—heterogeneous frailty and selective mortality—are consistent with everyday experience, it is surprising that it sparked such immediate and vehement opposition among osteologists when the issue was first raised in the early 1990s. Perhaps that is because the paper (Wood et al. 1992), conceived as methodological in nature, was interpreted by some as a "pro-state and pro-civilization" assault on conventional wisdom that emphasized the negative effects on health of the transition from a hunting-and-gathering past to the urban present (Cohen 1992:359). A more legitimate concern, one without polemical overtones, focused on what could be done if skeletal lesion frequencies cannot be interpreted in a reasonably straightforward fashion.

There were, and still are, two practical problems that have to be tackled before skeletons can contribute to a quantitatively based understanding of the disease experience of past populations. The first is a need for better age estimates, especially to eliminate biases in adult estimates and to deal effectively with the elderly. They are necessary to measure the selective effects of mortality across the entire human lifespan. The second pertains to devising ways of using mortality samples (what we observe) to learn about living populations (what we would like to observe, but cannot because the people are long dead). These two problems have proved to be particularly vexing.

APPROACHES TO PALEOEPIDEMIOLOGICAL UNDERSTANDINGS OF THE PAST

Current practice

Most studies of archaeological samples over the past few decades rely on tabulations of lesion frequencies in variously defined groups of skeletons. Such frequencies are still often considered the equivalent of disease prevalence in once-living populations, after due allowances are made for the widely recognized problems posed by sensitivity and specificity. In conventional studies, conclusions are thought to follow quite naturally from counts of lesions, so samples with the highest lesion frequencies are believed to come from the sickest populations. Such inferences necessitate the typically unstated assumption of a direct and proportional relationship between lesions in the dead and the living. So, following this logic, when lesion frequencies in skeletal samples increased (or decreased) over time, such as during the move from hunting-and-gathering to agricultural economies, the numbers of sick people in the original populations were correspondingly higher (or lower).

Whatever the merits of summarizing what is observed in skeletal collections, tabulations of lesion frequencies cannot be considered paleoepidemiology. That is even true when skeletal samples are divided into broadly defined age groups, such as juvenile and adult (with the latter perhaps separated into young, middle, and old). Aggregating data into such long age-intervals entails a loss of information and, worse, can be misleading (Baker and Pearson 2006).

Boldsen (1997a), for example, found that the frequency of active dental disease was equal among adult males and females in medieval Danish skeletons from Tirup, although there was a sharp contrast in age-specific frequencies of dental disease between the sexes. In males, the frequency of active dental disease at death did not change significantly with age. For women, however, hardly any young (<30 years) and old (>55 years) adults died with active dental disease, and most of the women who died around 40 years had active dental disease. These findings indicate there were significant differences between males and females in the risk factors associated with dental disease, and the additional risk of dying associated with active dental disease was not the same for males as compared to females throughout the lifespan. The similarity between male and female lesion frequencies when data were aggregated into a single adult age group missed entirely the more interesting story: age-specific risks of developing dental disease and dying with it were different for the two sexes. The key issue here is "dying with" not "dying from" a particular condition. The lesions, regardless of their origin, are only used as an indicator of an increased risk of dying. They need not be the actual cause of death to tell us something useful about conditions in the past.

Diagnoses are probabilistic

The most efficient means of tackling simultaneously the problems posed by sensitivity, specificity, and selective mortality are not yet fully developed. Nevertheless, it is worth illustrating several aspects of these issues to underscore the importance of abandoning simple counts of lesions as the primary means of characterizing the disease experience of past populations. Several studies point us toward paleoepidemiology's future. It is

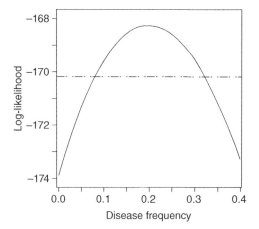

Figure 7.2   Somewhere between 8.1 and 32.3 percent of the individuals in the once-living population suffered from a particular disease, as estimated from the following information. The lesion is known to be present in 80 percent of individuals with the disease of interest, and it is present in 30 percent of people without that disease; and the lesion is found in 100 of 250 excavated skeletons. The horizontal line is the 95 percent confidence interval cutoff line, so the disease frequency interval is demarcated by the part of the log-likelihood curve that lies above the line.

important to recognize at the outset that since the relationship between diseases and skeletal lesions is probabilistic, paleoepidemiology is likewise probabilistic in nature.

Perhaps it is easiest to start by imagining a certain kind of lesion that commonly occurs in skeletal samples and is associated with a specific disease. To begin with, one could estimate the number of people who had the disease by assuming it had the same effect on people in the past as it does today in terms of producing skeletal lesions. In this example, let us assume that studies of historical or clinical sources indicate that 80 percent of sufferers are likely to have the lesion, while 30 percent of individuals without the disease also have it (some misclassification always occurs, even in the best of circumstances). In our archaeological sample, lesions are found on 100 of 250 skeletons. What we are interested in knowing is the number of people in the original population who had the disease. That can be solved by the maximum likelihood principle (Boldsen et al. 2002): 20 percent were sufferers while the remaining 80 percent were disease free (95% limits for sufferers are 8.1 to 32.3%). These estimates have been derived from the likelihood function illustrated in Figure 7.2. The parabola-like curve shows the likelihood of our observation (100 out of 250 having the lesion) as a function of the assumed prevalence of the disease in the population from which the skeletons were derived. The horizontal line is the cutoff line for the 95 percent confidence interval. The confidence interval is defined by the part of the curve that lies above the horizontal line.

This example, of course, requires a greater leap of faith than we might be willing to make in terms of the constancy between modern and ancient times in the occurrence of skeletal lesions in sufferers and nonsufferers of the disease of interest. As mentioned previously, the physical manifestations of disease cannot be expected to remain the same from one situation and time horizon to another. Nevertheless, the use of modern and historical data remains a useful starting point for a first approximation of disease prevalence in the past.

Taking the example one step further, we also want to know the probability a skeleton with a lesion was from a person who actually had the disease (positive predictive value, *PPV*), and a skeleton without a lesion did not have it (negative predictive value, *NPV*). That can be determined when sensitivity (*sen* = 0.8), specificity (*spe* = 1 – 0.3 = 0.7), and the population disease frequency (*p* = 0.2) can be estimated, as in the above example.

$$PPV = \frac{p \cdot sen}{p \cdot sen + (1-p) \cdot (1-spe)} = \frac{0.2 \cdot 0.8}{0.2 \cdot 0.8 + 0.8 \cdot 0.3} = 0.4$$

$$NPV = \frac{(1-p) \cdot spe}{p \cdot (1-sen) + (1-p) \cdot spe} = \frac{0.8 \cdot 0.7}{0.2 \cdot 0.2 + 0.8 \cdot 0.7} = 0.93$$

In this hypothetical example, only 40 percent of the skeletons with lesions came from people who suffered from the disease. This number corresponds to the positive predictive value (PPV). And 7 percent of those lacking lesions had the disease, corresponding to 1 minus the negative predictive value (NPV). So it would be hard to tell who had and who did not have the disease of interest. Nevertheless, the sample prevalence can be estimated as 20 percent using the original 250 skeletons, as follows: $((100 \cdot 0.4) + (150 \cdot 0.07))/250$. That figure, however, is not entirely accurate when it comes to population prevalence as this example, while it addresses the issue of sensitivity and specificity, does nothing whatsoever to take into take into account selective mortality.

The principal point is that it is possible to estimate disease prevalence without knowing exactly which skeletons were from individuals who suffered from the disease of interest. Doing so represents a departure from the traditional approach where skeletons with particular kinds of lesions are summed under the assumption that they approximate the number of people who were sick, once due allowance is made for the inexact relationship between diseases and lesions. Estimating population prevalence is important because it will never be possible to identify from lesions alone how many skeletons were from individuals who suffered from the diseases that are of great concern to paleopathologists, including tuberculosis, the treponemal infections, and leprosy. The problem stems from the fact that not all lesions are "classic" expressions of the diseases, and even all lesions that conform to "classic" forms are not always the result of the diseases of interest.

It follows that, from an analytical perspective, osteologists should be mostly interested in characterizing populations, not individuals, in terms of disease experience. Perhaps more than anything else, this issue—estimating disease prevalence in populations without knowing precisely which individuals (skeletons) had the disease—separates paleoepidemiology from the standard approach that focuses first on specific cases, and only later on all of them considered collectively.

Leprosy in medieval Europe

Considering the interpretive difficulties posed by low sensitivity and specificity, selective mortality, biased age estimates, and imperfect and unequal bone preservation, one might ask whether it will ever be possible to estimate disease prevalence in the

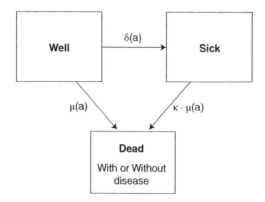

Figure 7.3   A Well–Sick–Dead model derived from Usher (2000) that is of potential use to paleoepidemiologists. Individuals can undergo a transition from Well to Sick, and from both Well and Sick to Dead. The Dead category, of course, can be counted in an archaeological sample; the Well and Sick have to be estimated. The model is interesting because parameter values that can be derived from it are descriptive of life in the past. Much information is contained in the κ-parameter, as it describes the relative mortality risk experienced by people with a given lesion compared to people without it.

distant past. That is, can we go beyond simple frequencies of observed lesions, which provide little in the way of useful information about disease experience in past populations? The outlook is, in fact, unexpectedly bright. While considerable work remains to be done, progress has been made in recent years toward reaching this ambitious goal (Usher 2000; Boldsen 2001, 2005a, 2005b, 2008; DeWitte and Wood 2008; DeWitte and Bekvalac 2010; also Lynnerup and Boldsen, Chapter 25 in this volume).

One such research effort has focused on leprosy in medieval Europe (Boldsen 2001, 2005a, 2005b, 2008; Boldsen and Mollerup 2006). The means of measuring the effects of leprosy on medieval European populations, using archaeological skeletal samples, is based on Usher's (2000) model featuring three states: well, sick, and dead (Figure 7.3). There are likewise three transitions between these categories. A Gompertz mortality model, which fits archaeological samples reasonably well, characterizes the transition from well to dead. The well to sick transition is approximated by historical information, specifically Norwegian data from 1850 to 1920, the Pre-Antibiotic Era (Irgens and Skjærven 1985). The last transition, from sick to dead, is estimated through a proportional hazard model using the proportionality factor κ; this is a standard approach in survival analysis. The proportionality factor, κ, can be interpreted directly as the mortality rate ratio. For all diseases and paleopathologically relevant lesions, κ > 1 since people with a disease (or lesion) experienced higher mortality than those without it.

All the gyrations necessary to estimate disease prevalence in past populations are not the end of the story. That is, they do not provide a full appraisal of the effects of various pathological conditions on past communities. To couch these data in human terms—what various pathological conditions, considered singly or in aggregate, might have meant to people on the ground—osteologists must to return to whatever information can be assembled on local cultural and natural conditions. That is because the ultimate objectives of paleoepidemiological studies are to understand what life was like in the past and to identify the challenges that stemmed from the disease experience of groups defined in various ways, including sex, age, social position, occupation, and residence location.

To illustrate how quantitative assessments from lesion frequencies to disease prevalence can be translated into the life experiences of past people, it is worth returning to the leprosy example, specifically the medieval Danish village of Tirup. Overall, leprosy cut two years off the life expectancy of Tirup's inhabitants. For the unfortunates who had leprosy, they died, on average, seven years before they would have done so had they not suffered from that disease. People with leprosy lived, on average, ten years with the disease, much of which time was spent disabled. In an economy where hard work was a fact of daily life, and there were few if any alternatives to physically demanding labor, these people were certainly a drain on community resources. Leprosy likely reduced the productive potential of the village population by some 10 percent, as estimated from the combined effect of a shorter lifespan and years spent disabled.

Taken together, these numbers indicate leprosy placed a great burden on the community. That is because individual households had to be largely self-sufficient with regard to the provisioning of food and other essentials, including clothing and shelter, and they possessed only a limited capacity to produce surpluses for distribution to needy neighbors. Perhaps the villagers' experience with leprosy, plus other debilitating ailments that were an inevitable outcome of a hard life, was one reason why Tirup eventually disappeared without a trace during the Middle Ages. All that remained was a long-forgotten churchyard, which was discovered and excavated some 600 years later (Boldsen 1984). The community simply could not sustain itself over the long run with its economic and reproductive potential so greatly diminished; in fact, it is a testament to the villagers' resiliency that Tirup lasted for as long as it did. The reduced productive capacity so essential for survival, especially in hard times, put the community at risk when it was eventually hit by famines and the Black Death that repeatedly hammered northern Europe during the 14th century. The village did not last to the end of that century, although the population did not die out altogether. The survivors of the Black Death appear to have moved, as a result of a political decision made outside of the community, to a neighboring bigger village. Other villages in the area, all depopulated, were also consolidated at that time (Kieffer-Olsen et al. 1986). So although paleoepidemiological investigations tell us much about the village population and the hardships they faced, Tirup's final disappearance was brought about by decisions made by politically powerful figures in the region. Tirup's abandonment underscores how information gleaned from skeletons must be embedded in detailed information on local cultural contexts to understand what took place in individual communities.

General patterns in human history

Regardless of what can be gleaned from Tirup's skeletons, a community that failed to survive cannot be considered representative of all of rural medieval Denmark, since many others struggled successfully through the dark years of the 14th century. To return to an earlier point, cemeteries are filled with people whose lives were lived locally, and that is precisely why each site has a unique tale to tell. The villagers of Tirup, more precisely their skeletons, arguably provide us with an excellent view of life for some members of the medieval rural Danish population. The picture is incomplete, however: many other communities were presumably better off, as they managed to make it through the hard times, with many lasting to the present day.

It could be argued that what is of most interest are not individual communities, or small numbers of them, but general patterns in disease experience beginning with our hunting-and-gathering ancestors and extending up to comparatively recent complex societies (e.g., medieval Denmark). Indeed, it is precisely an interest in the biological costs of major transitions in ways of life that serves as the impetus for much of the paleopathological and paleodemographic research undertaken since the 1970s (e.g., Acsádi and Nemeskéri 1970; Cohen and Armelagos 1984; Cohen 1989; Steckel and Rose 2002a). There are, in fact, good reasons why we should be concerned with general trends. But no matter how you look at them, cemetery samples are still largely a product of local conditions. So lumping skeletons together into stages (hunter-gatherer or village agriculturalist) or by broadly defined geographical area, time horizon, and culture period (medieval Scandinavians) yields samples that are difficult to interpret. That is because whatever results one might obtain are heavily dependent on the cemetery samples that happen to be examined, and they, above all else, are a product of what took place in and around specific communities. Only when samples are very large can it be argued, although not entirely convincingly, that local effects are swamped by generally prevailing tendencies to the point where a fuzzy picture of widespread conditions emerges.

We hasten to add that we are also interested in general trends, but prefer to build on a paleoepidemiological approach that features a contextually nuanced understanding of the individual communities (cemeteries) that happen to be studied. After all, archaeological excavations are anything but random, and the samples thereby produced cannot automatically be assumed to be typical of a particular time horizon, cultural stage, or archaeological period (culture). Only through the accumulation of investigations done on a site-by-site basis, with considerable attention focused on local circumstances, can broader generalizations be reached through comparative analyses of multiple contextually secure samples. That is a much harder and longer process than merely treating cemetery samples as somehow representative of higher-order groupings, such as hunter-gatherer or village agriculturalist, but it is one that will lead to a truer picture of the past in all of its rich complexity.

Again using Tirup as an example, many aspects of everyday life in this medieval village were very likely similar to what their Neolithic ancestors experienced. Yet Tirup was also part of a complex, state-level society, whereas Danish Neolithic villages were parts of relatively acephalous tribal societies. The frequency of intercommunity interactions, the numbers of people involved, and the distances covered by at least some members of Danish medieval society must have played a role in patterns of disease spread, hence the lives of local villagers in places like Tirup. In fact, had the Tirup cemetery been in use only a few hundred years earlier than it was, during a time when there were only five market towns in present-day Denmark instead of the 70 such towns when the village was in existence, the pattern of mortality would have been much different (Boldsen 1997b). Communication, in particular, altered the disease experience for small, Tirup-like communities, setting them apart from similarly sized communities a few hundred to several thousand years earlier.

Conclusion

Perhaps the most important points are as follows. Making sense of skeletal lesions in archaeological samples is not easy, as sensitivity, specificity, and selective mortality all come into play. Fortunately, promising progress is being made in terms of estimating the prevalence of pathological conditions in past populations. The entire process of doing so, accounting for sensitivity, specificity, and selective mortality, means that paleoepidemiology is inherently probabilistic in nature. Although it is possible to estimate the population prevalence of a disease, identifying precisely which skeletons belonged to individuals who actually suffered from the disease of interest is much more difficult. Once estimates of disease prevalence are obtained, it is then necessary to estimate its effect on the population. That extra step, which requires a return to archaeological materials (when appropriate, also historical sources), is what provides us with an understanding of the local conditions that had such a great effect on the health of the community. But however much we focus on local communities where lives were actually lived, it must not be forgotten that these people did not exist in splendid isolation. They were always part a larger social fabric that involved connections among neighboring and distant settlements. The nature of those interactions—how frequent they were, the distances covered, and the number of people involved—had an effect on disease experience and mortality at the local level, especially in organizationally complex societies such as medieval Denmark.

The identification of general trends in disease experience associated with major transitions in ways of life has been, and will continue to be, one of the principal reasons to examine ancient skeletons. To get to that point, however, paleoepidemiological studies should first focus on contextually well-characterized communities for the simple reason that what is seen in skeletons mostly reflects local cultural and natural conditions. Only later should separate community-based studies be incorporated as part of broader comparative studies designed to identify what might well turn out to be common, but by no means universal, effects of shifts in economic and sociopolitical systems, settlement size and longevity, and population size and density.

REFERENCES

Acsádi, G., and Nemeskéri, J., 1970   History of Human Life Span and Mortality. Budapest: Akadémiai Kiadó.

Baker, J., and Pearson, O. M., 2006   Statistical Methods for Bioarchaeology: Applications of Age-Adjustment and Logistic Regression to Comparisons of Skeletal Populations with Differing Age-Structures. Journal of Archaeological Science 33:218–226.

Bird, J., 1996   Prevalence Studies in Skeletal Populations. International Journal of Osteoarchaeology 6:320.

Boldsen, J. L., 1984   Bygholm. Department of Theoretical Statistics, University of Aarhus.

Boldsen, J. L., 1997a   Estimating Patterns of Disease and Mortality in a Medieval Danish Village. In Integrating Archaeological Demography: Multidisciplinary Approaches to Prehistoric Population. Richard R. Paine, ed. pp. 229–241. Carbondale: Center for Archaeological Investigations, Southern Illinois University.

Boldsen, J. L., 1997b   Patterns of Childhood Mortality in Medieval Scandinavia. Revista di Antropologia 74:147–159.

Boldsen, J. L., 2001   Epidemiological Approach to the Paleopathological Diagnosis of Leprosy. American Journal of Physical Anthropology 115:380–387.

Boldsen, J. L., 2005a   Leprosy and Mortality in the Medieval Danish Village of Tirup. American Journal of Physical Anthropology 126:159–168.

Boldsen, J. L., 2005b   Testing Conditional Independence in Diagnostic Palaeoepidemiology. American Journal of Physical Anthropology 128:586–592.

Boldsen, J. L., 2008   Leprosy in the Early Medieval Lauchheim Community. American Journal of Physical Anthropology 135:301–310.

Boldsen, J. L., Milner, G. R., Konigsberg, L.W., and Wood, J. W., 2002   Transition Analysis: A New Method for Estimating Age from Skeletons. In Paleodemography: Age Distributions from Skeletal Samples. Robert D. Hoppa, and James W. Vaupel, eds. pp. 73–106. Cambridge: Cambridge University Press.

Boldsen, J. L., and Mollerup, L., 2006   Outside St. Jørgen: Leprosy in the Medieval Danish City of Odense. American Journal of Physical Anthropology 130:344–351.

Byers, S. N., 1994   On Stress and Stature in the "Osteological Paradox." Current Anthopology 35:282–284.

Cohen, M. N., 1989   Health and the Rise of Civilization. New Haven: Yale University Press.

Cohen, M. N., 1992   Comments. Current Anthropology 33:358–359.

Cohen, M. N., 1994   The Osteological Paradox Reconsidered. Current Anthropology 35:629–631.

Cohen, M. N., 1997   Does Paleopathology Measure Community Health? A Rebuttal of "The Osteological Paradox" and Its Implication for World History. In Integrating Archaeological Demography: Multidisciplinary Approaches to Prehistoric Population. R. R. Paine, ed. pp. 242–260. Occasional Paper 24. Carbondale: Center for Archaeological Investigations, Southern Illinois University.

Cohen, M. N., and Armelagos, G. J., eds., 1984   Paleopathology at the Origins of Agriculture. Orlando, FL: Academic Press.

Cook, D. C., 1981   Mortality, Age-Structure and Status in the Interpretation of Stress Indicators in Prehistoric Skeletons: A Dental Example from the Lower Illinois Valley. In The Archaeology of Death. R. Chapman, I. Kinnes, and K. Randsborg, eds. pp. 133–144. Cambridge: Cambridge University Press.

Cook, D. C., and Buikstra, J. E., 1979   Health and Differential Survival in Prehistoric Populations: Prenatal Dental Defects. American Journal of Physical Anthropology 51:649–664.

DeWitte, S. N., and Bekvalac, J., 2010   Oral Health and Frailty in the Medieval English Cemetery of St Mary Graces. American Journal of Physical Anthropology 142:341–354.

DeWitte, S. N., and Wood, J. W., 2008   Selectivity of Black Death Mortality with Respect to Preexisting Health. Proceedings of the National Academy of Sciences 105:1436–1441.

Goodman, A. H., 1993   On the Interpretation of Health from Skeletal Remains. Current Anthropology 34:281–288.

Goodman, A. H., and Martin, D. L., 2002   Reconstructing Health Profiles from Skeletal Remains. In The Backbone of History: Health and Nutrition in the Western Hemisphere. R. H. Steckel, and J. C. Rose, eds. pp. 11–60. Cambridge: Cambridge University Press.

Guagliardo, M. F., 1982   Tooth Crown Size Differences Between Age Groups: A Possible New Indicator of Stress in Skeletal Samples. American Journal of Physical Anthropology 58:383–389.

Irgens, L. M., and Skjærven, R., 1985   Secular Trends in Age at Onset, Sex Ratio, and Type Index in Leprosy Observed During Declining Incidence Rates. American Journal of Epidemiology 122:695–705.

Jackes, M. K., 1993   On Paradox and Osteology. Current Anthropology 34:434–439.

Kieffer–Olsen, J., Boldsen, J. L., and Pentz, P., 1986   En Nyfunden Kirke ved Bygholm. Vejle Amts Årbog 1986:24–51.

Milner, G. R., Anderson, E., and Smith, V. G., 1991   Warfare in Late Prehistoric West-Central Illinois. American Antiquity 56:581–603.

Milner, G. R., Wood, J. W., and Boldsen, J. L., 2008   Advances in Paleodemography. *In* Biological Anthropology of the Human Skeleton, 2nd edition. M. A. Katzenberg, and S. R. Saunders, eds. pp. 561–600. New York: Wiley-Liss.

Ortner, D. J., 1991   Theoretical and Methodological Issues in Paleopathology. *In* Human Paleopathology: Current Syntheses and Future Options. D. J. Ortner, and A. C. Aufderheide, eds. pp. 5–11. Washington, DC: Smithsonian Institution Press.

Palkovich, A. M., 1985   Interpreting Prehistoric Morbidity Incidence and Mortality Risk: Nutritional Stress at Arroyo Hondo Pueblo, New Mexico. *In* Health and Disease in the Prehistoric Southwest. C. F. Merbs, and R. J. Miller, eds. pp. 128–138. Anthropological Research Papers 34. Tempe: Arizona State University.

Saunders, S. R., and Hoppa, R. D., 1993   Growth Deficit in Survivors and Non-Survivors: Biological Mortality Bias in Sub-Adult Skeletal Samples. Yearbook of Physical Anthropology 36:127–151.

Steckel, R. H., and Rose, J. C., eds., 2002a   The Backbone of History: Health and Nutrition in the Western Hemisphere. Cambridge: Cambridge University Press.

Steckel, R. H., and Rose, J. C., 2002b   Patterns of Health in the Western Hemisphere. *In* The Backbone of History: Health and Nutrition in the Western Hemisphere. R. M. Steckel and J. C. Rose, eds. pp. 563–579. Cambridge: Cambridge University Press.

Usher, B. M., 2000   A Multistate Model of Health and Mortality for Paleodemography: The Tirup Cemetery. Ph.D. dissertation, Pennsylvania State University.

Vaupel, J. W., Manton, K. G., Stallard, E., 1979   The Impact of Heterogeneity in Individual Frailty on the Dynamics of Mortality. Demography 16:439–454.

Vaupel, J. W., and Yashin, A. I., 1985a   Heterogeneity's Ruses: Some Surprising Effects of Selection on Population Dynamics. The American Statistician 39:176–185.

Vaupel, J. W., and Yashin, A. I., 1985b   The Deviant Dynamics of Death in Heterogeneous Populations. Sociological Methodology 15:179–211.

Waldron, T., 1991   Rates for the Job. Measures of Disease Frequency in Palaeopathology. International Journal of Osteoarchaeology 1:17–25.

Waldron, T., 1994   Counting the Dead: The Epidemiology of Skeletal Populations. Chichester: Wiley.

Waldron, T., 1996   Prevalence Studies in Skeletal Populations: A Reply. International Journal of Osteoarchaeology 6:320–322.

Weiss, K. M., 1990   The Biodemography of Variation in Human Frailty. Demography 27:185–206.

Wood, J. W., and Milner, G. R., 1994   Reply. Current Anthropology 35:631–637.

Wood, J. W., Milner, G. R., Harpending, H. C., and Weiss, K. M., 1992   The Osteological Paradox: Problems of Inferring Prehistoric Health from Skeletal Samples. Current Anthropology 33:343–370.

Wright, L. E., and Yoder, C. J., 2003   Recent Progress in Bioarchaeology: Approaches to the Osteological Paradox. Journal of Archaeological Research 11:43–70.

CHAPTER 8

The Promise, the Problems, and the Future of DNA Analysis in Paleopathology Studies

Mark Spigelman, Dong Hoon Shin, and Gila Kahila Bar Gal

INTRODUCTION

Paleopathology is the study of ancient *in vivo* lesions and diseases identified in the archaeological skeletal remains (Bartosiewicz 2008). Various agents, such as microbial, viral, as well as genetic disorders, can cause diseases that create pathological lesions. Most developmental pathologies are caused by known mutations, whereas infective agents cause various acquired pathologies. Among the bone pathologies, there are some abnormalities that are specific to a known cause and are considered diagnostic on purely morphological grounds. For instance, paleopathologists can attribute or diagnose several specific bacterial infections due to the unique characteristic deformation they cause. These include syphilis (*Treponema pallidum*), yaws, (*Treponema pallidum* ssp. *pertenue*), tuberculosis (*Mycobacterium tuberculosis*), brucellosis, and Hansen's disease (leprosy; *Mycobacterium leprae*). In other instances, the presence of skeletal lesions may only allow the researcher to assert that the individual suffered from nonspecific osteitis or osteomyelitis. In addition to diseases that cause morphological changes, there are many diseases affecting human and animals that cause epidemic outbreaks, but upon death leave no evidence of their presence on the

A Companion to Paleopathology, First Edition. Edited by Anne L. Grauer.
© 2012 John Wiley & Sons, Ltd. Published 2016 by John Wiley & Sons, Ltd.

bones. Importantly, however, the absence of pathological lesions does not mean that all trace of the microorganisms has been lost—biomolecules can be preserved. If the microorganisms were present in the bloodstream (septicemia) at the time of death, then some residual organic material (DNA, lipids, proteins) of these organisms may remain in the ancient bones of the subject postmortem. Hence, it is possible that the organic remains from the pathogen will be available for analysis along with that from the host. Diseases that may be diagnosed in this manner include tuberculosis, Hansen's disease, brucellosis, malaria, and bubonic plague, as well as viruses such as hepatitis B and influenza (Herrmann 2003; Roberts 2008 and Stone 2009).

The study of ancient biomolecules of human and faunal archaeological remains aims to gain an understanding of both health and disease in the past. The study of "ancient DNA" (aDNA) is a fairly recent development, marrying molecular biology with archaeology. The study is based on the capability to amplify DNA that has suffered autolytic or diagenic degenerative processes through the use of Polymerase Chain Reaction (PCR) (Herrmann 1994). The advance of PCR, developed by Mullis and Faloona (Mullis 1987), has given us the ability to amplify any residual fragmented DNA, either from the host or from any organism, that may have afflicted the host at time of death.

Research into pathogens, especially microbial organisms, is also a developing focus within the scientific field of paleopathology. It enables us to progress from making morphological assumptions about the presence of a particular disease, to at times scientifically confirming its presence. The first aDNA studies focused on the genetic characterization of single or few historical and/or archaeological remains (Higuchi 1984; Hänni 1994; Krings 1997). Technological developments, especially new methods in molecular biology, combined with experience gained over many years of working on specimens from diverse sites, has altered the focus of the field. Now researchers explore disease on a population level, and seek to understand specific characteristics of pathogens, and the interactions between host and pathogen.

Contrary to Robert's assertions (see Chapter 24 in this volume), the first paper on molecular paleopathology was published in 1993 (Spigelman and Lemma1993), focusing on four specimens showing morphological changes attributed to tuberculosis (TB). The diagnosis was reconfirmed with newer technology using three of the same specimens in Spigelman et al. (2003). The following year, Salo et al. (1994) published a study of TB in an Andean mummy. The next organism identified in an archaeological sample was Hansen's disease (leprosy) (*Mycobacterium leprae*). This was found within a bone from Israel (Rafi 1994). Once the feasibility and validity of this work was established, a number of other organisms were found to retain viable DNA in human remains including *Leishmania*, *Yersinia pestis* (Drancourt et al. 1998; Zink 2006), influenza virus (Taubenberger et al. 1997 and 2005), *Trypanosoma cruzi* (Chagas disease; Aufderheide 2006), hepatitis-B virus (Klein 2007), *Plasmodium falciparum* (malaria; Nerlich 2008), *Helicobacter pylori* (Castillo-Rojas 2008), *Brucella* (Mutolo 2006), *Ascaris* (Oh 2010) and *Schistosoma* (Matheson et al. 2009).

The study of ancient diseases is important, as it can have a direct impact on the understanding of modern diseases and their cures. For example, the reconstruction of the 1918 Spanish influenza RNA genome, recovered from an archived formalin-fixed lung autopsy specimen, and from frozen unfixed lung tissues from an Alaskan influenza victim buried in permafrost in November of 1918, provided insight into the nature and origin of this pathogen (Taubenberger 1997; Reid 1999). The comparison of the 1918 pandemic virus with contemporary human influenza H1N1 viruses indicates that the 1918 pandemic virus had the ability to replicate in the absence of trypsin, which may be one of the reasons why the virus had a uniquely high-virulence phenotype (Tumpey 2005). This information is of use in designing management programs to prevent other influenza epidemics and developing future vaccines.

The DNA in ancient specimens is significantly degraded. Degradation of DNA has been shown to be dependent upon the sample's preservation state and not on the antiquity of the sample (Höss and Pääbo 1993). If DNA is preserved and extracted from ancient specimens, it is usually present in extremely small quantities and low quality (Pääbo 2004; Willerslev 2005). Hence, the nature of the aDNA extracted from historical and archaeological remains demands the use of strict methods in order to limit further degradation and to prevent contamination with contemporary DNA. Any contaminating organism that may come in contact with the remains since death will leave its mark and can be amplified parallel to the amplification of the pathogen or host. However, most pathogenic bacteria have unique features, and thus can be easily identified from other organisms. The study of aDNA, mainly infectious disease, must therefore follow careful protocol to insure authenticity.

STANDARDS FOR aDNA PALEOPATHOLOGY STUDIES

The main criticism of aDNA research is the authenticity of the results (Roberts and Ingham 2008; Stone et al. 2009; Wilbur et al. 2009). Roberts and Manchester (1995) declare that, "Perhaps in another 150 years there will be some major advances in this field." However, despite skepticism, research continued in the belief that aDNA could provide a window to the past, presenting unique quantitative data about the genetic links between extant and extinct species (Yang 1997), and making it possible to analyze evolutionary change at the molecular level.

Most aDNA studies amplify eukaryotic DNA, while molecular paleopathology studies are based on the amplification of prokaryotic DNA. The different target requires slightly different approaches; thus some criticisms common for eukaryotic studies (Cooper 2000) are not relevant in paleopathology studies. A rebuttal to the issues and criticism of aDNA studies was published by Donoghue and colleagues (Donoghue et al. 2009). For instance, the idea that contamination, especially in TB studies, is caused from particles floating in the air has been proven false (Witt 2009). Likewise, the concern that there could be carry-over from instruments analyzing DNA from previous experiments was found to be unjustified, particularly when all blank controls were negative. Another serious criticism has been the potential contamination of samples by contemporary DNA. As discussed previously, an important protocol is to carry out aDNA research in a laboratory dedicated only to ancient studies, not modern molecules. To our knowledge, in all paleopathological publications the authors clearly outline the precautions taken to prevent contamination and false results.

The demand for independent confirmation of every single study, as proposed by Cooper and Poinar (2000), is understandable but not always possible or realistic.

Replicating another researcher's work is less preferable and less likely to be funded, than conducting one's own research unless the study, at the outset, was designed to be collaborative. We therefore believe that replication of results can, at times, be carried out in the same laboratory following special conditions such as separation in time and reagents. Another useful method of corroboration is the verification of aDNA findings by the detection of species-specific mycobacterial cell-wall lipid biomarkers (Donoghue 1998; Minnikin 2002; Hershkovitz 2008; Redman 2009; Donoghue 2009). This method relies on a different technological method, does not use amplification, and overcomes some of the potential problems with independent verification. Furthermore, the work can be completed using the same extraction of the DNA.

A further criticism is the lack of cloning to support the authenticity of the results. (Roberts and Ingham 2008; Wilbur 2009). It is important to stress that during cloning, errors of vector (base pair mutations) can occur spontaneously (Taylor 2009). When sequencing clones, polymorphisms may also be present due to replication errors, errors introduced by *Taq* polymerase (particularly with repetitive loci, (Tindall 1988), and due to genuine template heterogeneity. Moreover, cloning requires an additional facility, as the clones can be a source of contamination since each bacterium contains numerous copies of the molecule of interest. Cloning is a useful tool and we certainly use it, but it must not be assumed that failure to clone puts the results in doubt. Direct sequencing of a specific amplification using constrained primer sets of the targeted gene may be a better strategy, minimizing the possibility of error. Overall, we believe that it is the responsibility of the researcher to convince the reader that the published results are reliable and that the study was conducted using all precautions in order to prevent contamination and amplify authentic results.

Following a Careful Protocol

The collecting and sampling of specimens must be carried out taking great precautions to avoid contamination. This can be accomplished using gloves, sterile tools and sterile collection tubes (Spigelman 1996). While demanding "sterile" procedures throughout an entire excavation, as suggested by Roberts and Ingham (2008), might circumvent contamination, employing an appropriate sampling strategy in the field. Adopting protocols which remove suspected contemporary DNA contamination in the laboratory appears to be a more practical and feasible solution (Taylor 2009). Collecting specimens from all excavators who had contact with the specimens under investigation is also advised. In the case of mummies, endoscopically taking internal biopsies insures sterility and reduces the chances for contamination.

The collected specimens need to be packed and stored to prevent further degradation. For example, samples should not be stored in closed plastic bags, as this can encourage bacterial growth, especially if the temperature rises. The Iceman, found in 1991 in the Tyrolean Alps, exemplifies this point. One of the samples collected from the body contained fungal spores. While it is possible that the fungal spores were ancient specimens surviving on the hay padding in the leather boots, it is also possible that the spores represent recent contamination which proliferated due to the storage of the mummy at room temperature (Gams 1994).

It is also imperative to evaluate and control for the macro- and microenvironment of the sample. Archaeologists and/or curators can provide important information concerning pre- and post-excavation processes, such as exposure to sunlight, moisture conditions, treatment with preservatives, and even the use of X-rays or CT imaging; all of which can damage the residual DNA in a specimen. Even the physical distribution of material to different laboratories can cause an unexpected hazard, such as exposure to ionizing radiation, which is known to reduce the amount of amplifiable DNA (Götherström 1995; Grieshaber 2008). Specimens can be exposed to radiation if sent by airmail. On flights, photographic pouches should be considered.

Understanding the way specimens are treated post-excavation will help in the sample-selection process and might circumvent the waste of resources on experiments doomed to fail. An example is the Hamann Todd collection of relatively modern bones from people known to suffer from tuberculosis. In spite of many attempts, no TB DNA has ever been found in these bones. The reason probably lies in the preparation and storage of the collection. Modern human bones are usually very greasy, making their handling by the curator unpleasant. To remove the greasiness the collection was boiled in a mixture of chemicals (Hershkovitz personal communication). This would cause TB DNA to leach out and be disposed of with the chemical broth. Sadly, this collection is unlikely to ever yield TB DNA.

A similar experience occurred with a search for the DNA of *Helicobacter pylori* and papilloma virus from the 18th century Hunterian collection at the Royal College of Surgeons of England in London. The collection includes pathological preparations of stomach ulcers and cervical cancers that are morphologically well preserved. According to the curator, specimens have been kept in alcohol. Unfortunately, despite all attempts using various methods for extraction and amplification, Spigelman and colleagues failed to amplify any DNA from the specimens (Spigelman et al 2001). Seeking an explanation, it was discovered that the specimens had been preserved in jars with clean alcohol and without any lipids; a procedure which usually causes lipids from soft tissues to leach out, and which results in the alcohol becoming cloudy. We believe that the 18th-century curators replaced the cloudy alcohol with clear alcohol (perhaps several times), and unknowingly discarded the DNA of the pathogen and the host along with the cloudy alcohol.

For samples derived from archaeological remains, general information concerning the antiquity of the specimens, climate fluctuation and macroenvironmental conditions, and the composition of the soil, are very important in determining the potential preservation of the DNA in the samples. Similarly, microenvironmental influences are critical to understanding DNA preservation. For instance, preservation of microbial and/or host DNA is not identical along the entire human skeleton. This is often due to the body's tendency to decompose at different rates in different anatomical areas. Bacteria, the body's decomposers, are effective against host cells but the human body does not contain any enzymes capable of destroying a bacterial cell wall. In 1985, Ginsburg noted that there is no enzyme present in a mammalian host that can destroy a bacterial cell wall, unless the cell wall is first deacetylated. Thus, there is a difference between the degradation of prokaryotic cells (such as bacteria) and eukaryotic cells (which make up the human body) and so it is possible that preservation of host DNA in one area of a specimen will be poor, but bacterial DNA will be preserved, or vice versa.

Difficulties in amplification of aDNA can be encountered if there is too little sample available to permit repetitive sampling. For example, testing for septicemia, it is possible that a single negative biopsy will be misleading. We insist on several negative DNA extractions and amplifications from the same specimen (from different areas of the same bone and various skeletal remains) before coming to the conclusion that there is no microbial DNA in the specimen. Since enzymatic, physical and chemical processes that begin after death of the organism are the cause of degradation, quantification of organic biomolecules in the remains can be used as a marker of molecular diagenesis. The degree of amino acid racemization in a specimen is taken as a measure of the extent of DNA preservation of the eukaryotic cell (Poinar 1999), but prokaryotic DNA preservation may act in a different manner, which must be taken into consideration during its quantification.

Maintaining a clean lab is essential. Hence, research should be conducted in a clean laboratory where no work with modern DNA is carried out (Cooper 2000; Capelli 2005; Donoghue 2008). There are a number of essential precautions that researchers must take:

1. physical separation of pre- and post-PCR procedures in two different laboratories;
2. carrying out all procedures under UV-irradiated hoods with sterile equipment to prevent contamination;
3. strict protocols to prevent introduction of modern DNA to the aDNA samples;
4. the use of negative controls;
5. replication of DNA studies of all specimens to confirm authenticity of results (preferably with replication being carried out in different laboratories);
6. assessment of sequence data to confirm the results;
7. observation of an inverse relationship between fragment size and PCR efficiency.

Lastly, we would stress that with the study of microbial disease it is important to establish the health status of team members prior to the extraction. As an example, our laboratory work is on ancient tuberculosis, so chest X-rays are mandatory for all members. In the case of hepatitis-B, a blood test to check on carrier status is also mandatory (Shin, personal communication). It is important to realize that in order to prove authenticity there must be good evidence that proper sampling and contamination-prevention procedures were carried out, otherwise no amount of laboratory expertise will be of use.

DISEASES OF THE PAST AND aDNA STUDIES

Using the analysis of DNA and other biomolecules can shed light on human diseases in the past, the relationship between humans and their environment, cultural behavior, and the relationship between the host and the parasitic organism. Furthermore, it can provide information on diet, traumatic events, genetic mutations and epidemics (Bartosiewicz 2008). Given the potential to provide an evolutionary perspective, aDNA studies impact our understanding and prediction of modern diseases, whether genetically based or appearing as re-emerging infectious diseases.

Pathological conditions linked to inherited diseases help shed light on the origin of mutations and may elucidate the environmental or population pressures that caused their occurrence. In 1995, Filon and colleagues identified the beta thalassemia mutation which probably caused the pathological lesion recognizable on the frontal bone of an 8-year-old child from the archaeological site of Akhziv, Israel (Filon et al. 1995). Similarly, Marchetti (1996) identified the K-RAS mutation directly associated with a tumor present in Ferrante I of Aragon, King of Naples. Currently, a specimen from a Korean mummy, a female dated to the 15th century, is being examined for suspicion of cherubism (Shin, Hershkowitz and Spigelman personal communication). Though the morphological findings suggest the presence of cherubism, a rare autosomal dominant disease caused by a mutation of the *SH3BP2* gene, it is a condition most often found in males.

Different agents cause infectious diseases. Because the biological characterizations and life cycles of each agent differ, the survival of the evidence of infection in the historical remains differs as well. Further complicating the host–pathogen relationship is the fact that the host reaction, reflecting its immune system and/or the genetic profiles of specific receptors towards each pathogen, is also different. This leads to the responses of hosts being different even when the same pathogen is present. In addition, there are differences between pathogens themselves, with some displaying greater capacity to resist stress. *E. coli*, for instance, when subjected to stress collect their DNA and surround it with a crystalline protein, taking on a circular shape and converting their metabolism to one capable of repairing DNA. This can occur without spore formation, and is called biocrystallization of DNA (Wolf 1999).

Another factor to consider is the difference in chemical composition and permeability between prokaryotes, such as the mycobacteria and actinomycetes, and eukaryotic cells. Mycobacteria, which include the causative organisms of tuberculosis and leprosy, produce cell walls of unusually low permeability, which contribute to their resistance to therapeutic agents. The DNA of *M. tuberculosis* and *M. leprae* is more stable than mammalian DNA. In addition, the bacterial DNA is contained within a lipid-rich, resistant and persistent cell wall, and the bacterium has the ability to develop a pseudocapsule for added protection (Spigelman and Donoghe 2003; Donoghue 2004). This may help explain why pathogens might still be viable in a host well after death and long after the host's DNA has lost its activity. This has even led to infections among workers in postmortem rooms.

Environmental factors are equally important to consider and understand. The postmortem instability of the nucleic acids is central for the degradation of the DNA. Environmental factors such as temperature and climate changes have a direct influence on the long-term survival of aDNA molecules (Lindahl 1993; Hofreiter 2001; Smith 2001; Willerslev 2004). Small fragments of DNA (100–500 bp) will survive for no more than 10 kyr in temperate regions and for a maximum of 100 kyr at higher latitudes owing to hydrolytic damage (Poinar 1996; Smith 2001). However, other factors, such as rapid desiccation and high salt concentrations may prolong DNA survival (Lindahl 1993). Researchers have found that specimens buried beneath the Mediterranean Sea (a salty environment) yielded DNA preserved for 9,000 years (Hershkovitz 2008).

Understanding the postmortem conditions and treatment of the body, the bone, or the specimen is equally important. While often critical to the diagnostic process, using noninvasive sampling techniques such as X-ray, CAT scan and/or magnetic resonance imaging (MRI) can damage residual DNA in the remains. X-ray techniques, for instance, have the highest irradiation energy and are likely to damage fragile aDNA

(Grieshaber 2008). If imaging of the pathological lesion is conducted, especially using an X-ray, sampling the specimens prior to irradiation is recommended. The use of imaging techniques with lower radiation (CAT scans, for instance) or no radiation (MRI) is thus advisable.

The different biology and life cycles among the various pathogens that cause infection directly influence research strategies. Bacterial infections which produce macroscopically recognizable alteration to bone tissue can be studied using both morphological and molecular methodologies. The link between these two method-ologies, however, is not always simple or straightforward. Syphilis for example, is an infectious disease that is difficult to study through aDNA analysis due to the structure of the causative microorganisms (*Treponema pallidum*) and stages of the disease in the human body. The bone changes identified on skeletal remains generally reflect the tertiary stage of the disease, by which time there are few or no treponemal organisms present. The bone changes are technically caused by the host's immune response, not directly by the presence of the pathogen itself. Only occasionally, if the secondary stage of syphilis is prolonged, or the host is a child with congenital syphilis, is it possible to find the microbial DNA in the specimen.

In their review, von Hunnius and colleagues (von Hunnius 2007) described the study of ancient syphilis and noted that thus far only Kolman and colleagues (Kolman 1999) have been successful in the endeavor to amplify a fragment of the bacteria, while other attempts have failed (Bouwman 2005; Barnes 2006). What has not been men-tioned in these papers is the bacteriological information on *Treponema pallidum*. *T. pallidum* is a fragile Gram-negative bacterium, highly sensitive to temperature changes and which lyses quite easily. Thus, after death of the host, the pathogen's DNA is not protected by its cell wall and is exposed to destruction. While the absence of *T. pallidum* in the archaeological record may indicate that the disease was not present in antiquity, it is equally important to consider that the genetic makeup of pathogens (and hosts) does not remain static. Hence, the "absence" of the pathogen might actually be due to the pathogen's change since its first appearance in epidemic form. In the late 15th century it appears that the bacterium was quite virulent, killed the host within a few years, and caused an enormous epidemic. At the Diet of Worms in 1495, the Emperor Maximilian described syphilis as a new disease "*posen plattern*" ravaging his empire—"one that never occurred before nor been heard of within memory" (Bentley 1985). This is not the illness described in the 18th century as one taking 20 years or more to reach its final stage. Hence, it is possible that this may be due to genetic change in the bacteria itself, or due to increased immunity of the host. *T. Pallidum* might very well be present far back in antiquity but "masked" by our *a priori* assump-tions of its genetic structure and its inability to survive long after the host has died.

The study of tuberculosis is challenging as well. Paleopathological analysis suggests that tuberculosis was present in early hominids. It was once believed that humans acquired tuberculosis from animals after domestication. However, recent findings on the evolution of the *Mycobacterium tuberculosis* (MTB) complex, the group of bacteria that cause tuberculosis, indicate that *M. tuberculosis* and *M. bovis* separated from a common ancestor and that *M. tuberculosis* has more ancestral characteristics than originally believed. This reinforces the evidence that *M. tuberculosis* could not have arisen from *M. bovis* (Brosch 2002; Baker 2004; Donoghue 2004).

Most studies in paleopathology conducted on the molecular level have identified TB among historical specimens, and thus do not address the issue of origin of the pathogen. The identification, for instance, of *M. tuberculosis* complex infection in a bison from North America dated to 17,000 B.C. indicates the presence of the bacterium, but the species could only be distinguished as a member of the MTB complex due to its level of preservation (Rothschild 2001). Subsequent morphological and molecular studies of skeletal remains from a woman and infant buried together at the archaeological site of Atlitt-Yam located offshore from Haifa, Israel in the Mediterranean Sea and dated to the Pre-Pottery Neolithic C Period (9,250–8,160 years B.P.) identified tuberculosis. The molecular studies identified *Mycobacterium tuberculosis* DNA from five genetic loci. The DNA was sequenced and typed. The *M. tuberculosis* DNA was of a TbD1 deleted lineage, as are many of the MTB lineages found in the world today. Detection of *M. tuberculosis*-specific mycolic acid lipid biomarkers by high-performance liquid chromatography further confirmed these cases of human tuberculosis (Hershkovitz 2008). In another instance, the collection of mummies found in the 18th century Va'c church in Hungary provided excellent church records detailing the time of death, occupation, family linkage and, in some cases, actual mode of death of the individuals in the sample. The aDNA studies together with histological studies of the lesions in the lung biopsies identified TB in 63 percent of all specimens (n=176 individuals) (Fletcher 2003a; Fletcher 2003b). A more comprehensive review of the microbiological aspects of TB can be found in Spigelman and Donoghue (2003).

Hansen's disease, or leprosy, is another complex disease. The causal organism of leprosy is *Mycobacterium leprae* which can be harbored in the noses of healthy carriers. The organism primarily infects peripheral nerves, and then may spread to skin and bones. How it affects the host depends on the immune response and ranges from a slowly developing form with few bacilli, termed tuberculoid leprosy, to a more serious disease, lepromatous leprosy, where there are extremely high numbers of bacilli and extensive tissue destruction. In order to study Hansen's disease on the molecular level in archaeological or mummified specimens, we need to sample the bone that carries the bacterium. It is important to realize that the characteristic hand and foot changes commonly associated with the presence of the disease are due to secondary infection from recurrent ulceration. Thus, only in cases of the rarer lepromatous leprosy will *M.leprae* DNA be found in the bones of the fingers and toes. In some cases the bones are missing due to successful surgical amputation carried out as treatment (Walker 2009).

The bacterium was first identified in a metatarsal bone from Palestine dated to 600 A.D. (Rafi 1994). Later, Haas (2000) found the DNA in two skulls from Southern Germany that displayed the common osseous changes associated with the disease. Similarly, a skull from a 13th–14th century Orkney, Scotland tested positive for *M. leprae* DNA using PCR primers based on a 153-bp locus in RLEP1, and was further confirmed by sequencing (Taylor 2000).

The ability to identify the DNA of Hansen's disease helped to resolve a paleopathological argument about an unusual lesion found in a 1500-year-old skeleton from Israel (Hershkovitz 1992; Hershkovitz 1993; Manchester 1993). Both Madura foot (a fungal infection) and Hansen's disease were suggested as possible causes of the lesion in this specimen. A sample was examined for the presence of ancient microbial DNA in an attempt to determine the responsible microorganisms. *Mycobacterium leprae* DNA was detected, indicating that leprosy was the underlying cause of the pathology, but the possibility of a mixed or secondary fungal infection could not be excluded (Spigelman 2001).

Determining the presence of a specific disease can also begin to answer larger questions about the past. For instance, the finding of a body in the Hinon Valley in Jerusalem from the Second Temple Period that had been sealed in a locus since burial and not reburied in an ossuary, as was the custom in those days, was particularly perplexing. Many possible answers were considered, one of which was that the individual died of a disease that made his family unwilling to rebury for fear of contamination. Simultaneous exploration into the potential presence of both Hansen's disease and TB in different collaborating laboratories led to the discovery of the first cases of co-infection with *M. leprae* and *M. tuberculosis* (Donoghue 2005; Matheson 2009). This specimen is also the earliest known case of leprosy that has been confirmed by aDNA. Co-infection has also been shown to be present in other samples from Egypt, Hungary, Poland and Israel, leading researchers to support the argument that TB was actually responsible for ending the medieval leprosy epidemic in Europe (Donoghue 2005).

Brucellosis is one of the world's major zoonotic pathogens, and is responsible for enormous economic losses and human morbidity in endemic areas (Boschiroli 2001). The causative agent of the disease was first identified in the late 1800s by David Bruce, who isolated the bacteria among British troops during an epidemic on the island of Malta. Today the disease is common in developing nations and is often transmitted to humans through unpasteurized milk. Common zoonotic hosts for the *Brucella* bacteria are mainly Bovidae such as goats (*Brucella melitensis*) and cows (*Brucella abortus*), but other forms are known such as *Brucella suis* from pigs, and *Brucella canis* from dogs. Interestingly, these four species (goat, cow, dog and pig), together with sheep, are the earliest animals that were domesticated by humans (Garrard 1996; Horwitz and Den Driesch 1999; Peters 1999). The close association between animals and humans provided an easy path for transmission of the disease. Although brucellosis is not a sustainable disease in humans, the source of human infection resides in domestic or wild animal reservoirs. Routes of infection are multiple: food-borne, occupational or recreational and travel. New *Brucella* strains or species may emerge or existing *Brucella* species may adapt to changing social, cultural, travel and agricultural environments (Boschiroli et al. 2001; Godfroid et al. 2005). Therefore, at least from the Neolithic Period to the present, an evolutionary race between different strains of *Brucella* and the host immune systems has been taking place. At least one case of brucellosis DNA has been found in the human remains from Atlitt Yam an early Neolithic village in Israel, and stands as one of the first cases where domesticated cattle remains were discovered (Kahila Bar Gal personal communication). In addition, *Brucella* aDNA has been demonstrated in an adult female with vertebral lytic lesions from Iron Age Siberia dated to 360–170 B.C. (Bendrey 2008). *Brucella* spp. was also identified on skeletal remains of two adolescent males from the ancient Albanian city of Butrint, dated to the 11th–13th century A.D. Anthropologists identified severe circular lytic lesions on the thoracic vertebrae, as well as porosity of the ribs. Molecular analysis based on the IS6501 insertion element and Bcsp31 gene confirmed that it was caused by *Brucella* (Mutolo 2009).

The history and evolution of the Bubonic plague has captivated researchers' interest for years. In 1894, Alexandre Yersin identified the bacterium now known as *Yersinia pestis* (Yersin 1894) as the agent of the bubonic plague. Recently, the specific DNA for *Y. pestis* was amplified from 16th- and 18th-century human teeth believed to be from

French plague victims (Drancourt 1998) and from 14th-century French victims of Black Death (Raoult 2000). Based on these findings a study was conducted on 108 teeth from 61 specimens excavated from European burial sites believed to contain plague victims dating from the late 13th to 17th centuries. Although several different sets of primers were used, including those previously documented to yield positive results on ancient DNA extracts, the study failed to amplify *Y. pestis*. Other bacterial DNA sequences from the original samples were obtained from the extracted DNA in the hope that fatal bacteraemia in historical epidemics can be better understood. The results await independent corroboration (Thomas 2004), as does the controversial identification of plague in Athens (Papagrigorakis 2006; Shapiro 2006; Littman 2009).

Viruses are also difficult to recognize and interpret paleopathologically. With the development of advanced technology, attempts have been made to study viral pathogenic agents and a new field has developed: paleovirology. Thus far, the study of has been largely limited to endogenous retroviruses that can be directly identified from their remnants in host genomes (Greenwood 2001; Calvignac 2008). Exceptions are the studies of the human T cell lymphotropic virus type 1 (HTLV-1) (Li et al. 1999), the human papilloma virus (HPV-18) (Fornaciari et al. 2003), influenza virus (Taubenberger 1997; Reid 1999; Taubenberger 2005), the hepatitis B virus (Klein et al. 2007) and a small fragment of a virus belonging to the Anelloviridae (Bedarida et al. 2011).

CHARACTERIZATION OF THE HOST GENETIC PROFILE

Co-evolution between the host and the parasite genome is well documented. The susceptibility to developing infections such as tuberculosis is often determined by the genetic makeup of the host. Studying the genetic profile of human populations dated to the periods before the advent of antibiotics and intravenous drug injection provides a potential opportunity for finding significant genetic changes influencing host susceptibility and resistance to an infection. A well-known example is resistance/susceptibility to HIV. It was found that the CCR5 receptor (a CC-chemokine receptor) mediates leukocyte chemotactic responses and supports an efficient host response to viral and potentially microbial challenges (Galvani 2003; Cohn 2006; Glass 2006; Thio 2007). In the case of HIV, a deletion here prevents the entry of the HIV virus into the cell, and the host is therefore resistant to the disease (Dean 1996). Studies of ancient archaeological samples indicate that the mutation was found in historical human populations in Europe (Stephens 1998; Hummel 2005; Lidén 2006; Zawicki 2008). Mummy remains from the Va'c church in Hungary, dated to 1731–1792 A.D. that were found to be TB-positive (Fletcher 2003a; Fletcher 2003b) were tested for this mutation but were shown to not carry it (Kahila Bar-Gal personal communication).

The SLC11A1 gene, formerly NRAMP1, is another example of a gene where certain mutations are linked to the resistance of several infections (including *Mycobacterium tuberculosis, M. bovis, M. leprae, M. lepraemurium, Salmonella typhimurium* and *Leishmania donovani*) in contemporary human populations around the world. A study was embarked upon to identify the known mutation among ancient specimens, predominantly in the mummies from the Va'c church, (Matheson personal communication). The aim of the study was to determine if those individuals who tested positive for TB and were believed to have died from the disease, had a different genetic profile from those who had TB (such as a Ghon lesion) and were cured or never contracted the disease. Three specimens which were not infected with TB were found to be homozygote carrying the specific allele (A (CA)$_5$ TG (CA)$_5$ TG (CA)$_9$C). This allele was also found among three heterozygote specimens positive to TB (Matheson and Spigelman personal communication). Comparison of the mutations and the frequencies between the modern and ancient human populations may help elucidate pattern of genetic change and the subsequent consequences.

The study of the co-evolution of the host and pathogen is essential to the study of re-emerging diseases, which are increasing worldwide due to globalization and environmental changes. An understanding of the processes and patterns of genetic change, and the subsequent host–pathogen relationship, might lead to management programs preventing the spread of a disease.

THE FUTURE OF DNA IN PALEOPATHOLOGY RESEARCH

Future studies integrating research in ancient DNA with paleopathology will be directly affected by the development and advancement of technologies in the field of molecular biology, as well as by the questions asked by archaeologists, paleopathologists, microbiologists, and medical practitioners. The development of the PCR assay will be key to the field of aDNA research. The new technology, which allows for the study of whole genomes instead of gene fragments, will dramatically change the thrust of research. The two widely used assays are whole genome amplification (WGA) and single nucleotide polymorphism (SNP). Each screens the entire genome but uses different technology and applies a different approach.

Single Nucleotide Polymorphism (SNP)

A single nucleotide may be changed (substitution), removed (deletion) or added (insertion) to a polynucleotide sequence. The nucleotide change can cause a change in the amino acid coded, and in the case of insertion or deletion SNPs (InDels) may shift translation frame (Vignal 2002; Yue 2006; Väli 2008). SNPs are highly conserved throughout evolution and almost all common SNPs have only two alleles, which can be used as markers to differentiate between species, populations and individuals. Millions of SNPs have been catalogued in the human genome; many of them are normal variations in the genome. Others, however, are responsible for disease such as sickle cell disease, and can affect the human response to pathogens, chemicals, drugs, vaccines, and other agents.

An SNP array (hundreds of thousands of probes that are arrayed on a small chip, allowing for many SNPs to be interrogated simultaneously (Rapley 2004)) is a useful tool for studying the whole genome. The most important application of the SNP array is in determining disease susceptibility, and consequently, in pharmacogenomics, by measuring the efficacy of drug therapies specifically for the individual. As each individual has many SNPs that together create a unique DNA sequence, SNP-based genetic linkage analysis could be performed to map disease loci, and hence determine the presence of disease susceptibility genes within an individual. The Affymetrix

Human SNP 5.0 GeneChip performs genome-wide assays that can genotype over 500,000 human SNPs at once (Affymetrix 2007). Using the human SNP database, the origin of Siberian peoples referred to as "The Yakuts" was studied from 58 mummified frozen bodies dated to the 15th–19th centuries, excavated from Yakutia (Crubézy 2010). Due to the small amplicon amplified around the SNP (50–100 bp), the assay was found to be very suitable for aDNA studies. For example, in our studies SNP analysis was found to be a useful method of identifying animal species, especially when the archaeological remains could not be identified using common morphological methods. Another example is the use of the SNP assay to detect mutations associated with drug resistance to rifampin (RMP) and isoniazid (INH) among clinical isolates of *M. tuberculosis* (Ereqat et al. 2011).

Whole Genome Amplification (WGA)

One of the major breakthroughs in genomic sequencing technology occurred in 2005, with a report in *Nature* by researchers at 454 Life Sciences of a massively parallel genome sequencing method (Margulies 2005). The advanced technology provides a robust and accurate method of amplifying nanogram quantities of material. The small amplicon sequences along the genome of the organism, overall millions of nucleotides, are overlapped to determine the complete sequence of the genome. The method has revolutionized aDNA studies, successfully generating more than 13 million bp of DNA from a woolly mammoth (Poinar et al. 2006) and a million bp from a Neanderthal fossil (Green 2006). Recently, the 454 method was used on 18 remains of hair from mammoths that had died thousands of years ago (Gilbert 2007). The mammoth hair samples were sufficient for generating the entire mitochondrial DNA (mtDNA) genome sequence for each of the 18 mammoths, even though some samples had as little as 0.2 g of hair available and some had been stored at room temperature for over 200 years (Gilbert 2007). Currently, we are examining the WGA of two pathogens (*M. tuberculosis* and *hepatitisB*) isolated from ancient samples. Comparison of the ancient pathogens with the sequence of the modern pathogens causing the same disease will shed light on the evolution of the pathogen and the cause of its virulence.

The study of ancient molecular palaeopathology in human and animal remains is a developing but important field. The results of these types of paleopathological studies have implications in understanding human and animal evolution, predicting emerging and re-emerging diseases, and the future management of modern diseases. Establishment of databases available to the public will surely generate support for future research.

REFERENCES

Aufderheide, A. C., Salo, W., Madden, M., Streitz, J., Buikstra, J., Guhl, J., Arriaza, B., Renier, C., Wittmers, L. E., Jr., Fornaciari, G., and Allison, M. A., 2004   9,000-year record of Chagas' disease. Proceedings of the National Academy of Sciences, USA. 101(7):2034–2039.

Baker, L., Brown, T., Maiden, M. C., and Drobniewski, F., 2004   Silent Nucleotide Polymorphisms and a Phylogeny for *Mycobacterium tuberculosis*. Emerging Infectious Diseases 10:1568–1577.

Barnes, I., and Thomas, M. G., 2006   Evaluating Bacterial Pathogen DNA Preservation in Museum Osteological Collections. Proceedings of the Royal Society B 273:635–653.

Bartosiewicz, L., 2008   Taphonomy and Palaeopathology in Archaeozoology. Geobios 41: 69–77.

Bedarida, S., Dutour, O., Buzhilova, A. P., de Micco, P., and Biagini, P. 2011   Identification of Viral DNA (Anelloviridae) in a 200-year-old Dental Pulp Sample (Napoleon's Great Army, Kaliningrad, 1812). Infection, Genetics and Evolution 11:358–362.

Bendrey, R., Taylor, G. M., Bouwman, A. S., and Cassidy, J. P., 2008   Suspected Bacterial Disease in Two Archaeological Horse Skeletons From Southern England: Palaeopathological and Biomolecular Studies. Journal of Archaeological Science 35:1581–1590.

Bentley, G. W., 1985   Shakespeare and the New Disease: The Dramatic Function of Syphilis. Peter Lang, New York.

Boschiroli, M. L., Foulongne, V., and O'Callaghan, D., 2001   Brucellosis: A Worldwide Zoonosis. Current Opinion in Microbiology 4:58–64.

Bouwman, A. S., and Brown, T. A., 2005   The Limits of Biomolecular Paleopathology: Ancient DNA Cannot Be Used to Study Venereal Syphilis. Journal of Archaeological Science 32:703–713.

Brosch, R., Gordon, S. V., Marmiesse, M., Brodin, P., Buchrieser, C., Eiglmeier, K., Garnier, T., Gutierrez, C., Hewinson, G., Kremer, K., Parsons, L. M., Pym, A. S., Samper, S., van Soolingen, D., and Cole, S. T., 2002   A New Evolutionary Scenario for the *Mycobacterium tuberculosis* Complex. Proceedings of the National Academy of Science USA 99:3684–3689.

Calvignac, S., Terme, J. M., Hensley, S. M., Jalinot, P., Greenwood, A. D., and Hänni, C., 2008   Ancient DNA Identification of Early 20th Century Simian T-cell Leukemia Virus Type 1. Molecular Biology and Evolution 25:1093–1098.

Capelli, C., and Tschentscher, F., 2005   Protocols for Ancient DNA Typing. *In* Forensic DNA Typing Protocols. A. Carracedo, ed., pp. 265–278. Totowa, NJ: Humana Press.

Cohn, S. K., and Weaver, L. T., 2006   The Black Death and AIDS: CCR5-Δ32 in Genetics and History QJM: An International Journal of Medicine 99:497–503.

Cooper, A., and Poinar, H. N., 2000   Ancient DNA: Do It Right or Not at All. Science 289:1139.

Castillo-Rojas G., Cerbón, M. A., and López-Vidal, Y., 2008   Presence of *Helicobacter pylori* in a Mexican Pre-Columbian Mummy. BMC Microbiology 8:119. doi:10.1186/1471-2180-8-119.

Crubézy, E., Sylvain, A., Keyser, C., Bouakaze, C., Bodner, M., Gibert, M., Röck, A., Parson, W., Alexeev, A., and Ludes, B., 2010   Human Evolution in Siberia: From Frozen Bodies to Ancient DNA. Evolutionary Biology 10:25.

Cui-Ying, L., and Shi-Feng, Y., 2006   A Novel Mutation in the SH3BP2 Gene Causes Cherubism: Case Report. BMC Medical Genetics 7:84.

Dean, M., Carrington, M., Winkler, C., Huttley, G. A., Smith, M.W., Allikmets, R., Goedert, J. J., Buchbinder, S. P., Vittinghoff, E., Gomperts, E., Donfield, S., Vlahov, D., Aslow, R. K., Saah, A., Rinaldo, C., Detels, R., and O'Brien, S. J., 1996   Genetic Restriction of HIV-1 Infection and Progression to AIDS by a Deletion Allele of the CKR5 Structural Gene. Science 273:1856–1862.

Donoghue, H. D., 2008   Palaeomicrobiology of Tuberculosis. *In* Paleomicrobiology: Past Human Infections. D. Raoult, ed., pp. 75–97. Berlin: Springer Verlag.

Donoghue, H. D., Hershkovitz, I., Minnikin, D. E., Besra, G. S., Lee, O. Y., Galili, E., Greenblatt, C. L., Lemma, E., Spigelman, M., and Kahila Bar-Gal, G., 2009   Biomolecular Archaeology of Ancient Tuberculosis: Response to "Deficiencies and Challenges in the Study of Ancient Tuberculosis DNA" by Wilbur et al. 2009. Journal of Archaelogical Science 36:2797–2804.

Donoghue, H.D., Marcsik, A., Matheson, C., Vernon, K., Nuorala, E., Molto, J., Greenblatt, C., and Spigelman, M., 2005   Co-Infection of *Mycobacterium tuberculosis* and *Mycobacterium leprae* in Human Archaeological Samples – A Possible Explanation for the Historical Decline of Leprosy. Proceedings of The Royal Society: Biological Sciences. Proceedings B 272:389–394.

Donoghue, H. D., Spigelman, M., Greenblatt, C., Lev-Maor, G., Kahila Bar-Gal, G., Matheson, C., Vernon, K., Nerlich, A. G., and Zink, A. R., 2004   Tuberculosis: From

Prehistory to Robert Koch, As Revealed By Ancient DNA. Lancet Infectious Diseases 4:584–592.

Donoghue, H. D., Spigelman, M., Zias, J., Gernaey-Child, A. M., and Minnikin, D. E., 1998 *Mycobacterium tuberculosis* Complex DNA in Calcified Pleura From Remains 1400 Years Old. Letters in Applied Microbiology 27:265–269.

Drancourt, M., Aboudharam, G., Signoli, M., Dutour O., and Raoult, D., 1998 Detection of 400-Year-Old *Yersinia pestis* DNA in Human Dental Pulp: An Approach to the Diagnosis of Ancient Septicemia. Proceedings of the National Academy of Sciences of the USA 95: 12637–12640.

Ereqat S., Kahila Bar-Gal, G., Nasereddin, A., Azmi, K., Qaddomi, S. E., Greenblatt, C. L., Spigelman, M., and Abdeen, Z., 2011 Rapid Differentiation of *Mycobacterium tuberculosis* and *M. bovis* by High-Resolution Melt Curve Analysis. Journal Of Clinical Microbiology, Vol. 48(11):4269–4272.

Filon, D., Faerman, M., Smith, P., and Oppenheim, A., 1995 Sequence Analysis Reveals a Beta-Thalassemia Mutation in the DNA of Skeletal Remains from the Archaeological Site of Akhziv, Israel. Nature Genetics 9:365–368.

Fletcher, H. A., Donoghue, H. D., Holton, J., Pap, I., and Spigelman, M., 2003a Widespread Occurrence of *Mycobacterium tuberculosis* DNA from 18th–19th Century Hungarians. American Journal of Physical Anthropology 12:144–152.

Fletcher, H. A., Donoghue, H. D., Taylor, G. M., van der Zanden, A. G., and Spigelman, M., 2003b Molecular Analysis of *Mycobacterium tuberculosis* DNA From a Family of 18th Century Hungarians. Microbiology 149:143–151.

Fornaciari, G., Zavaglia, K., Giusti, L., Vultaggio, C., and Ciranni, R., 2003 Human papillomavirus in a 16th-century mummy. Lancet 362:1160.

Galvani, A. P., and Slatkin, M., 2003 Evaluating Plague and Smallpox as Historical Selective Pressures for the CCR5-Δ32 HIV-1 Resistance Allele. Proceedings of the National Academy of Sciences of the United Sates of America 100:15276-15279.

Gams, W., and Stalpers, J. A., 1994 Has the Prehistoric Ice-Man Contributed to the Preservation of Living Fungal Spores? FEMS Microbiology Letters 120:9–10.

Garrard, A., Colledge, S., and Martin, L., 1996 The Emergence of Crop Cultivation and Caprine Herding in the "Marginal Zone" of the Southern Levant. *In* The Origins and Spread of Agriculture and Pastoralism in Eurasia. D. Harris, ed., pp. 204–226. London: UCL Press.

Gilbert, M. T., Tomsho, L. P., Rendulic, S., Packard, M., Drautz, D. I., Sher, A., Tikhonov, A., Dalen, L., Kuznetsova, T., Kosintsev, P., Campos, P. F., Higham, T., Collins, M. J., Wilson, A. S., Shidlovskiy, F., Buigues, B., Ericson, P. G., Germonpre, M., Gotherstrom, A., Iacumin, P., Nikolaev, V., Nowak-Kemp, M., Willerslev, E., Knight, J. R., Irzyk, G. P., Perbost, C. S., Fredrikson, K. M., Harkins, T. T., Sheridan, S., Miller, W., and Schuster, S. C., 2007 Whole-Genome Shotgun Sequencing of Mitochondria from Ancient Hair Shafts. Science 317: 1927–1930.

Glass, W. G., McDermott, D. H., Lim, J. K., Lekhong, S., Yu, S. F., Frank, W. A., Pape, J., Cheshier, R. C., and Murphy, P. M., 2006 CCR5 Deficiency Increases Risk of Symptomatic West Nile Virus Infection. Journal of Experimental Medicine 23:35–40.

Godfroid, J., Cloeckaert, A., Liautard, J. P., Kohler, S., Fretin, D., Walravens, K., B. Garin-Bastuji, B., and Letesson, J. J., 2005 From the Discovery of the Malta Fever's Agent to the Discovery of a Marine Mammal Reservoir, Brucellosis Has Continuously Been a Re-Emerging Zoonosis. Veterinary Research 36:313–326.

Götherström, A., Fischer, C., and Linden, K., 1995 X-Raying Ancient Bone: A Destructive Method in Connection with DNA Analysis. Laborativ Arkeologi 8:26–28.

Green, R. E., Krause, J., Ptak, S. E., Briggs, A. W., Ronan, M. T., Simons, J. F., Du, L., Egholm, M., Rothberg, J. M., Paunovic, M., and Paabo, S., 2006 Analysis of One Million Base Pairs of Neanderthal DNA. Nature 444:330–336.

Greenblatt, C. L., Schlein, Y., and Schnur, L. F., 1985 Leishmaniasis in Israel and Vicinity *In* Leishmaniasis. K. P. Chang, and R. S. Bray, eds, pp. 415–426. Amsterdam: Elsevier Science Publishers.

Greenwood, A. D., Lee, F., Capelli, C., DeSalle, R., Tikhonov, A., Marx, P. A., and MacPhee, R. D. E., 2001   Evolution of Endogenous Retrovirus-like Elements of the Woolly Mammoth (*Mammuthus primigenius*) and its Relatives. Molecular Biology and Evolution 18:840–847.

Grieshaber, B. M., Osborne, D. L., Doubleday, A. F., and Kaestle, F. A., 2008   A Pilot Study Into the Effects of X-ray and Computed Tomography Exposure on the Amplification of DNA From Bone. Journal of Archaeological Science 35:681–687.

Haas, C. J., Zink, A., Pálfi, G., Szeimies, U., and Nerlich, A. G., 2000   Detection of Leprosy in Ancient Human Skeletal Remains by Molecular Identification of *Mycobacterium leprae*. American Society for Clinical Pathology 114:428–436.

Hänni, C., Laudet, V., Stehelin, D., and Taberlet, P., 1994   Tracking the Origins of the Cave Bear (*Ursus spelaeus*) by Mitochondrial DNA Sequencing. Proceedings of the National Academy of Sciences, USA 91:12336–12340.

Herrmann, B., and Hummel S., 1994   Ancient DNA: Recovery and Analysis of Genetic Material from Paleontological Archaeological, Museum, Medical and Forensic Specimens. New York: Springer Verlag.

Herrmann, B., and Hummel, S., 2003   Ancient DNA Can Identify Disease Elements *In* Emerging Pathogens: Archaeology, Ecology and Evolution Of Infectious Disease. C. Greenblatt, and M. Spigelman, eds. pp. 143–149. Oxford: Oxford University Press.

Hershkovitz, I., Donoghue, H. D., Minnikin, D. E., Besra, G. S., Lee, O. Y.-C. Gernaey, A. M., Galili, E., Eshed, V., Greenblatt, C. L., Lemma, E., Kahila Bar-Gal, G., and Spigelman, M., 2008   Detection and Molecular Characterization of 9000-Year-Old *Mycobacterium tuberculosis* from a Neolithic Settlement in the Eastern Mediterranean. PLoS ONE 3:e3426.

Hershkovitz, I., Spiers, M., and Arensberg, B., 1993   Leprosy or Madura Foot? The Ambiguous Nature of Infectious Disease in Paleopathology: Reply to Dr Manchester. American Journal of Physical Anthropology 91:251–253.

Hershkovitz, I., Spiers, M., Katznelson, A. A., and Arensberg, B., 1992   Unusual Pathological Conditions in the Lower Extremities of a Skeleton From Ancient Israel. American Journal of Physical Anthropology 88:23–26.

Higuchi, R. G., Bowman, B., Freiberger, M., Ryder, O. A., and Wilson, A. C., 1984   DNA Sequences From the Quagga, An Extinct Member of the Horse Family. Nature 312: 282–284.

Hofreiter, M., Serre, D., Poinar, H. N., Kuch, M., and Pääbo. S., 2001   Ancient DNA. Nature Reviews Genetics 2:353–360.

Horwitz, L. K., Tchernov, E., Ducos, P., Becker, C., Von, A., Den Driesch, L. M., and Garrard, A., 1999   Animal Domestication in the Southern Levant. Paleorient 25:63–80.

Hummel, S., Schmidt, D., Kremeyer, B., Herrmann, B., and Oppermann, M., 2005   Detection of the CCR5-D32 HIV Resistance Gene in Bronze Age Skeletons. Genes and Immunity 6:371–374.

Klein, A., Spigelman M., Grant, P., Pappo, O., Kim, M. J., Shin, D. H., and Shouval, D., 2007   Tracing Hepatitis B Virus DNA Back to the 16th Century in a Korean Mummy. Hepatology 46:648A.

Kolman, C. J., Centurion-Lara, A., Lukehart, S. A., Owsley, D. W., and Tuross, N., 1999   Identification of *Treponema pallidum* subspecies *pallidum* in a 200-Year-Old Skeletal Specimen. Journal of Infectious Diseases 180:2060–2063.

Krings, M., Stone, A., Schmitz, R. W., Krainitzki, H., Stoneking, M., and Pääbo, S., 1997   Neanderthal DNA Sequences and the Origin of Modern Humans. Cell 90:19–30.

Li, H. C., Fujiyoshi, T., Lou, H., Yashiki, S. Sonoda, S., Cartier, L., Nunez, L., Munoz, I., Horai, S., and Tajima, K. 1999   The presence of ancient human T-cell lymphotropic virus type I provirus DNA in an Andean mummy. Nature Medicine 5:1428–1432.

Lidén, K., Linderholm, A., and Gotherstrom, A., 2006   Pushing It Back: Dating the CCR5- Δ32 bp Deletion to the Mesolithic in Sweden and Its Implications for the Meso/Neo Transition. Documenta Praehistorica 633:577.572.

Lindahl, T., 1993   Instability and Decay of the Primary Structure of DNA. Nature 362:709–715.

Littman, R. J., and Litt, M., 2009   The Plague of Athens: Epidemiology and Paleopathology. Mount Sinai Journal of Medicine 76:456–467.

Manchester, K., 1993   Unusual Pathological Condition in the Lower Extremities of a Skeleton From Ancient Israel. American Journal of Physical Anthropology 91:249–250.

Marchetti, A., Pellegrini, S., Bevilacqua, G., and Fornaciari, G., 1996   K-RAS Mutation in the Tumour of Ferrante I of Aragon, King of Naples. The Lancet 347:1272.

Margulies, M., Egholm, M., Altman, W. E., Attiya, S., Bader, J. S., Bemben, L. A., Berka, J., Braverman, M. S., Chen, Y., Chen, Z., Dewell, S. B., Du, L., Fierro, J. M., Gomes, X. V., Godwin, B. C., He, W., Helgesen, S., Ho, C. H., Irzyk, G. P., Jando, S. C., Alenquer, M. L. I., Jarvie, T. P., Jirage, K. B., Kim, J.-B. Knight, J. R., Lanza, J. R., Leamon, J. H., Lefkowitz, S. M., Lei, M., Li, J., Lohman, K. L., Lu, H., Makhijani, V. B., McDade, K. E., McKenna, M. P., Myers, E. W., Nickerson, E., Nobile, J. R., Plant, R., Puc, B. P., Ronan, M. T., Roth, G. T., Sarkis, G. J., Simons, J. F., Simpson, J. W., Srinivasan, M., and Karrie, R., 2005   Genome Sequencing in Microfabricated High-Density Picolitre Reactors. Nature 437:376–380.

Matheson, C. D., Vernon, K. K., Lahti, A., Fratpietro, R., Spigelman, M., Gibson, S., Greenblatt, C. L., and Donoghue, H. D., 2009   Molecular Exploration of the First-Century Tomb of the Shroud in Akeldama, Jerusalem. PLoS One 4:e8319.

Minnikin, D. E., Kremer, L., Dover, L. G., and Besra. G. S., 2002   The Methyl-Branched Fortifications of *Mycobacterium tuberculosis*. Chemistry & Biology 9:545–553.

Mullis, K. B., 1987   Specific Synthesis of DNA In Vitro Via a Polymerase-Catalyzed Chain Reaction. Methods in Enzymology 155:335–350.

Mutolo, M. J., 2006   Molecular Identification of Pathogens in Ancient Skeletal Remains From Butrint and Diaporit, Albania. PhD Dissertation. Michigan State University.

Nerlich, A. G., Schraut, S., Dittrich, S., Jelinek, T., and Zink, A. R., 2008   *Plasmodium falciparum* in Ancient Egypt. Emerging Infectious Diseases 8:1317–1319.

Oh, C. S., Seo, M., Chai, J. Y., Lee, S. J., Kim, M. J., Park, J. B., and Shin, D. H., 2010   Amplification and Sequencing of *Trichuris trichiura* Ancient DNA Extracted From Archaeological Sediments. Journal of Archaeological Science 37:1269–1273.

Pääbo, S., Poinar, H., Serre, D., Jaenicke-Després, V., Hebler, J., Rohland, N., Kuch, M., Krause, J., Vigilant and L., Hofreiter, M., 2004   Genetic Analyses From Ancient DNA. Annual Review of Genetics 38:645–679.

Papagrigorakis, M. J., Yapijakis, C., Synodinos, P. N., and Baziotopoulou-Valavani, E., 2006   DNA Examination of Ancient Dental Pulp Indicates Typhoid Fever as a Probable Cause of the Plague of Athens. International Journal of Infectious Diseases 10:206–214.

Peters, J., Helmer, D., Von Den Driesch, A., and Sana-Segui, M., 1999   Early Animal Husbandry in the Northern Levant. Paleorient 25:27–57.

Poinar, H. N., and Stankiewicz, B. A., 1999   Protein Preservation and DNA Retrieval From Ancient Tissues. Proceedings of the National Academy of Sciences, USA 96:8426–8431.

Poinar, H. N., Höss, M., Bada, J. L., and Pääbo, S., 1996   Amino Acid Racemization and the Preservation of Ancient DNA. Science 272:864–866.

Rafi, A., Spigelman, M., Stanford, J., Donoghue, H., Lemma, E., and Zias, J., 1994   DNA of Mycobacterium leprae detected by PCR in ancient bone. International Journal of Osteoarchaeology 4(4):287–290.

Raoult, D., Aboudharam, G., Crubezy, E., Larrouy, G., Ludes, B., and Drancourt, M., 2000   Molecular Identification By "suicide PCR" of *Yersinia pestis* as the Agent of Medieval Black Death. Proceedings of the National Academy of Sciences, USA 97:12800–12803.

Rapley, R., and Harbron, S., 2004   Molecular Analysis and Genome Discovery. Chichester: John Wiley & Sons.

Redman, J., Shaw, M., Mallet, A., Santos, A., Roberts, C., Gernaey, A., and Minnikin, D., 2009   Mycocerosic Acid Biomarkers for the Diagnosis of Tuberculosis in the Coimbra Skeletal Collection. Tuberculosis 89:267–277.

Reid, A. H., Fanning, T. G., Hultin, J. V., and Taubenberg, J. K., 1999   Origin and Evolution of the 1918 "Spanish" Influenza Virus Hemagglutinin Gene. Proceedings of the National Academy of Sciences, USA 96:1651–1656.

Rice, D., 2005   Craniofacial Anomalies: From Development to Molecular Pathogenesis. Current Molecular Medicine 5:699–722.

Roberts, C. A., and Ingham, S., 2008   Using Ancient DNA Analysis in Palaeopathology: A Critical Analysis of Published Papers, With Recommendations For Future Work. International Journal of Osteoarchaeology 18:600–613.

Roberts, C. A., and Manchester, K., 1995   The Archaeology of Disease. Ithaca, NY: Cornell University Press.

Roberts, C. A., Pfister, L. A., and Mays, S., 2009   Letter to the Editor: Was Tuberculosis Present in Homo Erectus in Turkey? American Journal of Physical Anthopology 139:442–444.

Rothschild, B. M., Martin, L. D., Lev, G., Bercovier, H., Kahila Bar-Gal, G., Greenblatt, C., Donoghue, H., Spigelman, M., and Brittain, D., 2001   Mycobacterium tuberculosis Complex DNA from an Extinct Bison Dated 17,000 Years before the Present. Clinical Infectious Diseases 33:305–311.

Salo, W. L., Aufderheide, A. C., Buikstra, J., and Holcomb, T. A., 1994   Identification of Mycobacterium tuberculosis DNA in a Pre-Columbian Peruvian Mummy. Proceedings of the National Academy of Sciences, USA 91:2091–2094.

Shapiro, B., Rambaut, A., and Gilbert, T. P., 2006   No Proof That Typhoid Caused the Plague of Athens. International Journal of Infectious Diseases 10:334–335.

Smith, C. I., Chamberlain, A. T., Riley, M. S., Cooper, A., Stringer, C. B., and Collins, M. J., 2001   Neanderthal DNA: Not Just Old But Old and Cold? Nature 10:771–772.

Spigelman, M., 1996   The Archaeologist and Ancient Bio-Molecules: Field Sampling Strategies to Enhance Recovery. Papers from the Institute of Archaeology 7:69–74.

Spigelman, M., 2001   Unusual Pathological Condition in the Lower Extremities of a Skeleton From Ancient Israel. American Journal of Physical Anthropology 114:92–93.

Spigelman, M., Barnes, I., Holton, J., Vaira, D., and Thomas. M., 2001a   Long-Term DNA Survival in Ethanol-Preserved Archival Material. Annals of the Royal College of Surgeons of England 83:283–284.

Spigelman, M., and Donoghue, H. D., 2003   Paleobacteriology With Special Reference to Pathogenic Mycobacteria. In Emerging Pathogens: Archaeology, Ecology and Evolution of Infectious Disease. S. M. Greenblatt, ed. pp. 175–188. Oxford: Oxford University Press.

Spigelman, M., and Lemma, E., 1993   The Use of Polymerase Chain Reaction to Detect Mycobacterium tuberculosis in Ancient Skeletons. International Journal of Osteoarchaeology 3:137–143.

Spigelman, M., Matheson, C., Lev, M., Greenblatt, C., and Donoghue, H. D., 2003   Confirmation of the Presence of Mycobacterium tuberculosis-Complex Specific DNA in Three Archaeological Specimens. International Journal of Osteoarchaeology 12:393–401.

Stephens, J. C., Reich, D. E., Goldstein, D. B., Shin, H. D., Smith, M. W., Carrington, M., Winkler, C., Huttley, G. A., Allikmets, R., Schriml, L., Gerrard, B., Malasky, M., Ramos, M. D., Morlot, S., Tzetis, M., Oddoux, C., di Giovine, F. S., Nasioulas, G., Chandler, D., Aseev, M., Hanson, M., Kalaydjieva, L., Glavac, D., Gasparini, P., Kanavakis, E., Claustres, M., Kambouris, M., Ostrer, H., Duff, G., Baranov, V., Sibul, H., Metspalu, A., Goldman, D., Martin, N., Duffy, D., Schmidtke, J., Estivill, X., O'Brien, S. J., and Dean, M., 1998   Dating the Origin of the CCR5-Delta32 AIDS-Resistance Allele by the Coalescence of Haplotypes. American Journal of Human Genetics 62:1507–1515.

Stone, A. C., Wilbur, A. K., Buikstra, J. E., and Roberts, C. A., 2009   Tuberculosis and Leprosy in Perspective. American Journal of Physical Anthropology 140:66–94.

Taubenberger, J. K., Reid, A. H., Krafft, A. E., Bijwaard, K. E., and Fanning, T. G., 1997   Initial Genetic Characterization of the 1918 "Spanish" Influenza Virus. Science 275:1793–1796.

Taubenberger, J. K., Reid, A. H., Lourens, R. M., Wang, R., Jin, G., and Fanning, T. G., 2005   Characterization of the 1918 Influenza Virus Polymerase Genes. Nature 437:889–893.

Taylor, G. M., Mays S. A., and J. F. Huggett 2009 Ancient DNA (aDNA) Studies of Man and Microbes: General Similarities, Specific Differences. International Journal of Osteoarchaeology 20(6):747–751.

Taylor, G. M., Widdison, S., Brown, I. N., and Young, D., 2000 A Mediaeval Case of Lepromatous Leprosy from 13–14th Century Orkney. Scotland Journal of Archaeological Science 27:1133–1138.

Thio, C. L., Astemborski, J., Bashirova, A., Mosbruger, T., Greer, S., Witt, M. D., Goedert, J. J., Hilgartner, M., Majeske, A., O'Brien, S. J., Thomas, D. L., and Carrington, M., 2007 Genetic Protection Against Hepatitis B Virus Conferred by CCR5-Δ32: Evidence That CCR5 Contributes to Viral Persistence. Journal of Virology 81:441–445.

Thomas, M., Gilbert, P., Cuccui, J., White, W., Lynnerup, N., Titball, R. W., Cooper, A., and Prentice, M. B., 2004 Absence of *Yersinia pestis*-Specific DNA in Human Teeth From Five European Excavations of Putative Plague Victims. Microbiology 150:341–354.

Tindall, K. R., and Kunkel, T. A., 1988 Fidelity of DNA Synthesis by the *Thermus aquaticus* DNA Polymerase. Biochemistry 27:6008–6013.

Tumpey, T. M., Basler, C. F., Aguilar, P. V., Zeng, H., Solórzano, A., Swayne, D. E., Cox, N. J., Katz, J. M., Taubenberger, J. K., Palese, P., and García-Sastre, A., 2005 Characterization of the Reconstructed 1918 Spanish Influenza Pandemic Virus. Science 310:77–80.

Väli, Ü., Brandström, M., Johansson, M., and Ellegren, H., 2008 Insertion–Deletion Polymorphisms (Indels) As Genetic Markers in Natural Populations. BMC Genetics 9:8.

Vignal, A., Milan, D., Sancristobal, M., and Eggen, A., 2002 A Review on SNP and Other Types of Molecular Markers and Their Use in Animal Genetics. Genetics 34:275–305.

von Hunnius, T. E., Yang, D., Eng, B., Waye, J. S., and Saunders, S. R., 2007 Digging Deeper Into the Limits of Ancient DNA Research on Syphilis. Journal of Archaeological Science 34:2091–2100.

Walker, D., 2009 The Treatment of Leprosy in 19th-Century London: A Case Study from St Marylebone Cemetery. International Journal of Osteoarchaeology 19:364–374.

Wilbur, A. K., Bouwman, A. S., Stone, A. C., Roberts, C. A., Pfister, L. A., Buikstra, J. E., and Brown, T. A., 2009 Deficiencies and Challenges in the Study of Ancient Tuberculosis DNA. Journal of Archaeological Science 36:1990–1997.

Willerslev, E., and Cooper, A., 2005 Ancient DNA. Proceedings in Biological Science 272:3–16.

Willerslev, E., Hansen, A. J., and Poinar, H. N., 2004 Isolation of Nucleic Acids and Cultures From Ice and Permafrost. Trends in Ecology and Evolution 19:141–147.

Witt, N., Rodger, G., Vandesompele, J., Benes, V., Zumla, A., Rook, G. A., and Huggett, J. F. 2009 An Assessment of Air as a Source of DNA Contamination Encountered When Performing PCR. Journal of Biomolecular Techniques 20:236–240.

Wolf, S. G., Frenkiel, D., Arad, T., Finkel, S. E., Kolter, R., and Minsky, A., 1999 DNA Protection By Stress-Induced Biocrystallization. Nature 400:83–86.

Yang, H., 1997 Ancient DNA From Pleistocene Fossils: Preservation, Recovery, and Utility of Ancient Genetic Information for Quaternary Research. Quaternary Science Reviews 16: 1145–1161.

Yersin, A., 1894 La Peste Bubonique à Hong Kong. Annales de l'Institut Pasteur 8: 662–667.

Yue, P., and Moult, J. 2006 Identification and Analysis of Deleterious Human SNPs. Journal of Molecular Biology 356:1263–1274.

Zawicki, P., and Witas, H. W., 2008 HIV-1 Protecting CCR5-Δ32 Allele in Medieval Poland. Infection, Genetics and Evolution 8:146–151.

Zink, A., and Nerlich, A. G., 2003 Molecular Analyses of the "Pharaohs": Feasibility of Molecular Studies in Ancient Egyptian Material. American Journal of Physical Anthropology 121:109–111.

Zink, A. R., Spigelman, M., Schraut, B., Greenblatt, C. L., Nerlich, A. G., and Donoghue, H. D., 2006 Leishmaniasis in Ancient Egypt and Upper Nubia. Emerging Infectious Diseases 12:1616–1617.

CHAPTER 9

The Analysis and Interpretation of Mummified Remains

Michael R. Zimmerman

INTRODUCTION

Paleopathological studies, by adding the crucial dimension of time, improve our understanding of the evolution of diseases and their role in human biological and social history. Information on ancient disease patterns can be obtained from ancient pathological material and several other sources. Our concept of the distribution of disease in populations in the past is based primarily on historic records, such as Egyptian medical papyri and studies of the role of epidemic diseases in the decimation of the indigenous populations after European colonization of the New World and Polynesia. We also have accounts of the great plagues of medieval Europe and epidemics in the more recent past. Works of art, including such diverse media as paintings, pottery effigies, figurines and religious statuary, and figures and faces on coins often give information on disease.

The examination of skeletal material and mummies yields the most reliable paleopathological information. However, the study of ancient (and indeed modern) disease is not without its pitfalls. Skeletonization may be an erratic process and pathological changes may be obscured or obliterated by the vagaries of preservation. Pseudopathological changes can be produced by erosive forces or anthropophagic animals, and most diseases leave little or no direct mark on the bones. Lines of arrested growth (Harris lines) in the long bones have been suggested as an index of morbidity in a given population, modern or ancient. As Harris lines may disappear later in life and could result from trivial subclinical infections, they are useful only in a general sense.

A Companion to Paleopathology, First Edition. Edited by Anne L. Grauer.
© 2012 John Wiley & Sons, Ltd. Published 2016 by John Wiley & Sons, Ltd.

Mummies, defined as bodies preserved either naturally (by freezing, drying, or tanning), or artificially, hold a much greater potential for paleopathological examination. The preservation of tissues postmortem is based on inactivation of proteolytic enzymes in invading bacteria and fungi and in the tissues themselves. In the modern laboratory this process is called fixation and is accomplished by immersion of the tissue in a liquid chemical fixative, usually some variant of formaldehyde. Rapid fixation results in good preservation. In contrast, mummification is based on enzyme inactivation by heat or cold, usually combined with desiccation. The process is often erratic and anything but rapid, classical sources describing a 70 day period of mummification in ancient Egypt. The inevitable result is a degree of tissue destruction, autolysis, which may obscure pathological change or even underlying tissue structure.

Naturally preserved bodies have been found in bogs in Europe and Britain and in arctic and arid areas. While the majority of paleopathological work has been conducted on Egyptian mummies, extensive work has been conducted on material from North and South America, including Alaska and the Aleutian Islands, the southwestern United States, Kentucky, Peru and Chile. Mummified bodies of Scythians from the steppes of Russia have also been studied (Artamonov 1965). Most recently, 2,000- to 4,000-year-old mummies have been found in central China—interestingly, of Caucasian appearance (Mallory and Mair 2000). The Guanche of the Canary Islands mummified their dead in a fashion similar to that of ancient Egypt (Brothwell et al. 1969), leading G. Elliot-Smith to ascribe this practice, mistakenly, to heliocentrism (the belief that culture radiated out from Egypt). However, the native peoples of the Torres Strait, between Australia and New Guinea, also practiced mummification in an indigenous style, providing evidence in repudiation of heliocentrism.

Relatively standard medical postmortem studies can be performed on mummies. While loss of the features of color and consistency that are so useful in examining fresh tissue limits the interpretation of gross pathological change, the tissues can be rehydrated and sections prepared for microscopic examination. Rehydration of desiccated tissue is based method originally developed by Sir Marc Armand Ruffer (1921). In spite of a number of limitations and potential biases, paleopathological diagnoses of diseases found within mummified remains can be made with a considerable degree of confidence and accuracy.

THE HISTORY OF MUMMY PALEOPATHOLOGY

The earliest period of interest in mummies might well be called "the use and abuse of mummies." Mummies were powdered into drugs, exhibited as "unrollings" for the paying public, and burned for warmth by workmen during the construction of the Egyptian railroad system. A review of this history is provided by Aufderheide (2003) in the first chapter of his comprehensive book on mummies. Scientific studies were initiated by the work of Ruffer and Elliot-Smith in pre-World War I Egypt. Ruffer (1927) introduced the use of modern techniques such as microscopy into the field. Unfortunately, photomicroscopy was not available to him and we are left only with sketches of his findings. However, his method of rehydration is still in use in its original or a modified form. Sporadic studies followed until the recent resurgence of interest in the examination of mummies, with studies begun in the mid-20th century by Allison and coworkers on Peruvian mummies, the Paleopathology Association, and A. Rosalie David's reactivation of the study of mummies in the Manchester Museum that had been originated in 1910 by Margaret Murray. In 1967, Brothwell and Sandison published *Diseases in Antiquity*. Chapters in this book focused on diseases of specific organ systems, including the skin, respiratory, gastrointestinal, genitourinary, reproductive and vascular systems. A little later, Sandison (1970) published a comprehensive chapter on mummies. Other, less thorough reviews have followed (Janssens 1980; Hart 1983; Zimmerman 2001, 2004). There have been many recent reports specifically focusing on histopathology identified in mummies (see Tyson 1997; Aufderheide and Rodriquez-Martin 1998; Cockburn et al. 1998; Aufderheide 2003).

METHODS AND TECHNIQUES USED TO EXAMINE MUMMIFIED REMAINS

Paleopathology is, of necessity, multidisciplinary. The examination of mummified bodies requires the collaborative efforts of specialists from many fields, including physical and cultural anthropology, archeology, history, demography, epidemiology, genetics, pathology, orthopedics, radiology, obstetrics, hematology, parasitology, dentistry, orthodontics and others. The actual dissection of a mummy takes one or two days and is usually followed by a year or more of laboratory work, followed in turn by a comprehensive effort to focus the study on the interface of the physical, social, biological and medical sciences, with publication of the findings as an eventual result. Radiographic techniques, including tomography and xerography, tell much about bodies before they are removed from their coffins or have been unwrapped. Histological techniques, both light and electron microscopy, demonstrate the remarkable preservation of normal and diseased structures. Studies of bone reveal evidence of age, sex, disease, nutrition and trauma. The dentist can help assess age, nutrition, disease and racial characteristics. Most importantly, the value of anatomists should be recognized—not only can they identify dubious structures, but they can also direct the dissection so that the anatomic detail can be preserved for display, enhancing the scientific and museological value of rare paleopathological material.

In addition to light microscopy and scanning and transmission electron microscopy, mummified tissues (and bones) can be studied by a number of other techniques, including chemical analyses, and paleoserology (examination for blood type and antibodies against specific disease organisms). Microbiological studies have not been useful, as viable pathogens have not been cultured from paleopathological material, although organisms can be identified histologically, since viruses can be visualized by electron microscopy. Any organisms cultured have always been postmortem contaminants.

The advent of computerized tomography (CT) scanning and other non-invasive techniques has allowed for more thorough examination of mummies either before or instead of traditional autopsy procedures (see Chapter 18 in this volume by Wanek et al. for more discussion on imaging techniques in paleopathology). CT scanning is particularly useful, although distortions in anatomy due to the efforts of ancient embalmers, position of the mummy, or taphonomic changes may be confusing.

Magnetic resonance imaging (MRI) had not been useful until recently, as the technique depends upon the activation of water molecules, a scarce item in fully mummified bodies. CT scanning can be followed by endoscopically guided biopsies of sites of interest, thereby avoiding the somewhat destructive nature of a full autopsy examination.

The analysis of mummified material for ancient DNA (aDNA) has received much recent interest. Although there are technical problems related to the survival of aDNA and contamination by modern DNA in the field or laboratory (Zimmerman 2007), applications can determine genetic relationships, such as the remarkable clarification of the relationships among the members of King Tutankhamun's family (Hawass et al. 2010). Rasmussen et al. (2010) report determination of a genome sequence from 4,000 year-old hair of a Greenland Paleo-Eskimo, providing evidence indicating that the Arctic New World was populated before modern Native Americans and Inuit. DNA studies also hold the potential for sex determination, which is otherwise still based on anatomic features such as robusticity of the forehead, chin, eyebrows, mastoid processes and muscle attachments, and the angle of the sciatic notch.

Determining when the individual lived and the age at death of the individual are important aspects of the examination of mummies. Dating places the individual in historical context and allows for insight into diseases and conditions within particular groups and/or in particular environments. Radiocarbon (^{14}C) dating, the gold standard for determining the antiquity of a specimen, may be supplemented by electron spin resonance, dendrochronology, and evaluation of tattoos or mummification styles. At the individual level, determining age at death is essential because many diseases occur in specific age ranges, and age determination can be a critical factor in differential diagnosis. Determining age at death requires gross evaluation of the skeleton and viscera, hand-wrist radiographs, bone histology, the evaluation of dental changes and, in special cases, amino acid racemization.

Chemical analysis of bone is also a promising technique, as it allows for dietary reconstruction. Determination of stable isotopes ratios for carbon and nitrogen have been used in determining the relative levels of meat and vegetable matter in ancient diets, and in determining whether food sources were marine or terrestrial. Trace elements are also studied. Strontium can give us an idea of how much meat was eaten, and lead levels indicate either domestic or industrial exposure to this toxic material. For example, it has been found in skeletons recovered from plantations in the southern United States that the owners had much higher lead levels than their slaves, due to lead used in pewter tableware (Aufderheide et al. 1981).

EXPERIMENTAL APPROACHES TO MUMMIFICATION

There have been two experimental approaches to the study of the effects of mummification. In the first, tissues from recently deceased individuals were removed during hospital-based autopsies, desiccated, rehydrated and examined by light microscopy (Zimmerman 1972, 1977a), while the second was a replication of ancient Egyptian mummification in an effort to determine whether the appearance of the tissues of Egyptian mummies is due to the effects of the Egyptian mummification process alone or is modified by the passage of the several millennia following the labors of the ancient embalmers (Brier and Wade 1997; Zimmerman, Brier and Wade 1998).

In the first study, tissues were obtained from adult human cadavers undergoing postmortem examination, all within 24 hours after death. Specimens were bisected, with one half processed immediately as controls and the other half dried in an oven and then rehydrated. Comparison between the fresh and the mummified tissues revealed variable preservation of histological detail. Cytoplasmic structures were discernable, particularly with a variety of specific stains, but nuclei were poorly preserved in normal tissues. An important point was that the large hyperchromatic nuclei of malignant cells were much better preserved than the smaller, normochromatic nuclei of benign cells. Changes in the tissues generally resembled those seen in human mummies.

The second major study was an attempt to replicate Egyptian mummification. The donor, an elderly male, had died of heart failure and a clinically diagnosed massive myocardial infarction. As the intention was to replicate a first-class mummification of the period at which the embalmers' art was at its peak, a royal mummification of the eighteenth dynasty was attempted, using a replica handmade embalming board, copper, bronze, and obsidian tools, with evisceration through a small left lower quadrant abdominal incision.

The heart and brain were not examined in this study. In antiquity, the heart was left in the body, as part of funerary ritual, and the brain, the function of which was unknown to the ancient Egyptians, was discarded after extraction through the cribriform plate of the ethmoid bone or the foramen magnum. In this study, a hook was inserted through the nose in order to pulverize and remove the brain. Linen packets of natron, a naturally occurring compound of sodium carbonate and bicarbonate that had been imported for the study from the Wadi Natrun in Egypt, were used to pack the thoracic and abdominal cavities. A layer of natron was placed under the body on the board, and the remainder of the natron was used to cover the body, incidentally making clear the necessity for a wide embalming board. The internal organs removed in this procedure, consisting of lungs, stomach, liver, spleen, pancreas, intestines, and kidneys, were placed in ceramic dishes, covered with natron, and placed at the corners of the embalming board. The cadaver and organs were kept at 115 F and 28–30 percent humidity for 35 days in a room of dimensions 4.3 × 2.2 meters.

Specimens of the mummified organs were examined by the pathologist in this study (MRZ), who intentionally was kept unaware of the clinical history of the subject. The general state of preservation of the organs in this experimental mummy was slightly better than but comparable to that of actual 1,800–3,200-year-old Egyptian mummies. It was also possible to make presumptive diagnoses in this individual of several pathological processes, including pulmonary edema, pulmonary thromboemboli, hepatic fibrosis, and chronic pyelonephritis, the first three being related to the death of this individual due to chronic congestive heart failure, as was documented clinically. Pathological changes affecting these organ systems in Egyptian mummies have previously been reported (Bruetsch 1959; Tapp et al. 1975; Reyman et al. 1977; Zimmerman 1977b, 1990, 1993; Cockburn et al. 1998), and such diagnoses can now be viewed with confidence in both the efficacy of ancient Egyptian mummification practices and our diagnostic ability in examining ancient human remains.

What Have We Learned From Mummies?

The examination of mummies has two goals: the diagnosis of diseases in individual mummies and the fitting of those disorders into a picture of the health status of a given ancient population; and providing information on the evolution of diseases. Diseases in a number of broad categories diagnosed in mummies have helped in both these efforts, although many of the diagnoses offered in these studies of mummies have been presumptive, a necessary limitation considering the poor preservation usually encountered, and relatively few definitive microscopic diagnoses have been made. An abbreviated list of conditions found in human mummified remains is given in Table 9.1 towards the end of this section.

A number of congenital skeletal deformities have been seen in mummies. The diagnoses were made by gross or radiological examination, with microscopic confirmation being unnecessary. In a survey of Egyptian mummies in museums in Great Britain and Europe, Gray (1973) noted several cases of spina bifida, a pathological failure of closure of the sacrum. Clubfoot was diagnosed in a XIIth-Dynasty mummy and in the mummy of the Pharaoh Siptah by Rowling in 1961, although he noted that contemporary portraits of the Pharaoh did not show this abnormality. Subsequent studies have led to the conclusion that this lesion is probably due to poliomyelitis. Clubfoot has recently been diagnosed in Amenhotep III and Tutankhamun, with the latter also showing oligodactyly (absence of a toe phalanx). These deformities, resulting in difficulty in walking, account for numerous depictions of Tutankhamun being seated in activities that normally require an upright posture, as well as the presence of numerous walking sticks, several showing wear, in Tutankhamun's tomb (Hawass et al. 2010). A rare condition, alkaptonuric arthritis, was diagnosed radiologically in several mummies in the 1960s. This hereditary disorder results in a distinctive calcification of the intervertebral cartilages. It has become apparent that the increased radiodensity of these cartilages is seen in many Egyptian mummies as an artifact of mummification rather than a real pathological condition.

Traumatic injuries have also been reported in mummies. Post and Donner (1972) documented radiologically a case of frostbite in a Pre-Columbian mummy of northern Chile. The pattern of loss of the toes was characteristic, but no histological confirmation was attempted. Fractures have been noted in Peruvian, Alaskan and Egyptian mummies, but only in one was there histological confirmation, with hemorrhage seen in a skull fracture in an Alaskan mummy (Zimmerman and Smith 1975). Confoundingly, some fractures are postmortem or "embalmer's fractures", but Gray (1973) was able to demonstrate in his study of Egyptian mummies radiological evidence of healing in some cases. Gray also X-rayed the mummy of an elderly Egyptian man whose hand had been amputated during his youth; his mummy was fitted with an artificial hand by the embalmers.

Aspiration of foreign material has been seen in two mummies. A Peruvian mummy of 950 A.D. was found to have aspirated a molar tooth, which was firmly impacted in a left bronchus (Allison et al. 1974a). Pneumonia distal to the obstruction was demonstrated microscopically. It was an interesting coincidence that at the same time an elderly patient died of the same process in the same hospital in which the mummy was being studied. Similarly, An Inuit mummy was found to have died due to aspirating material from the roof of her house when she was trapped because of a landslide or earthquake (Zimmerman and Smith, 1975).

Calculus (stone formation) has been found in the gall bladder and urinary tract, including the kidney, urinary bladder and ureter of mummified remains (Steinbock 1990). Microscopic confirmation for these diagnoses is unnecessary.

There have been many reports of infectious and inflammatory processes in mummified remains. Infectious disease is categorized by the etiologic agents, i.e. viruses, bacteria, fungi, protozoan and metazoan parasites, and insects. Two viral diseases that have been noted in ancient Egyptian material, poliomyelitis and small-pox, do show diagnostic lesions grossly and on light microscopy. The manifestations of poliomyelitis are due to unopposed contraction of unaffected muscles, which results in characteristic deformities. As early as 1900, Mitchell noted a shortened leg in an elderly male of 3,700 B.C. as indicative of this condition, and Ruffer reported a characteristic deformity of the foot of the Pharaoh Siptah of the XIXth Dynasty. X-ray examination of the Pharaoh in the 1970s showed an overall shortening of the entire right leg and atrophy of the soft tissues, indicating the presence of a neuromuscular disease in childhood, almost certainly poliomyelitis. Histological examination has not been done, and current Egyptian government policy limits examination of the royal mummies, but one would expect the pattern of muscle atrophy to be well preserved and to substantiate this diagnosis. While infection by the polio virus is almost universal in underdeveloped countries, the clinical disease is rare. Infection usually occurs early in life, when maternal antibodies prevent the development of the disease, allowing the virus to set up a commensal relationship in the intestinal tract of the host. With improved sanitation, the result of infection by the virus is more likely to be disease than a commensal state. The finding of this disease in a Pharaoh can be attributed to the isolation of the Royal Family and provides historical verification of this concept of the pathogenesis of poliomyelitis.

In 1967, Ruffer and Ferguson reported a smallpox-like eruption on the mummy of Ramses V, and noted the characteristic microscopic features of the vesicles of smallpox in a XXth-Dynasty mummy, attesting to the antiquity of this disease.

Bacterial infections have also been found in mummies. One of the most common of the bacterial infections, and a major cause of death in the Pre-Antibiotic Era, is pneumonia. The antiquity of this disease is well attested by reports in mummies from Egypt (Shaw 1938), the Aleutian Islands, (Zimmerman et al. 1971, 1981) and Peru (Allison et al. 1974). These studies indicate that the microscopic features of pneumonia can be preserved by mummification. Pneumonia can be complicated by the entry of microorganisms into the bloodstream and the development of metastatic abscesses in other organs. Such abscesses were noted in one of the Aleut mummies in the heart, lungs, and kidneys (Zimmerman et al. 1971). The heart valves can become affected, and healed endocarditis has been reported in an Egyptian mummy and an Eskimo mummy. A skin infection, Carrion's disease, caused by *Bartonella bacilliformis*, has been demonstrated in a Peruvian mummy (Allison et al. 1974b). In all these cases, attempts to culture the organisms have been unsuccessful.

The question of the origin of syphilis has always aroused much controversy (see Chapter 26 by Cook and Powell in this volume, for more discussion). Although characteristic skeletal lesions have been described in ancient material from both sides of the Atlantic, not a single case of syphilis has ever been diagnosed in a mummy.

In contrast, the antiquity of tuberculosis has been well documented in mummies. Allison et al. (1973) were the first to stain and identify *Mycobacterium tuberculosis* in a mummy of a Peruvian child of 700 A.D. The characteristic lesions of tuberculosis present in this child (necrosis, spinal deformity, psoas abscess, etc.) had been amply demonstrated in other mummies, but not the organisms. A review by Morse (1964) establishes the presence of tuberculosis in ancient Egypt, and a case report by Long (1964) documents caseation necrosis and scoliosis. The mummy of an Egyptian child showed scoliosis due to osseous tuberculosis, with death due to pulmonary hemorrhage (Zimmerman 1979). As the pathogenesis of tuberculosis in a young child is infection from an much older adult, this finding confirms the archeological evidence of extended multigenerational households in ancient Egypt, a pattern that persists today (Nahas 1969). Of interest is the failure of Elliot-Smith (1908) to find any examples of tuberculosis in 6,000 Pre-Dynastic Nubian skeletons and mummies, suggesting the Dynastic period for the onset of human tuberculosis in the Nile Valley, although recent molecular evidence indicates a much older date for the evolution of human tuberculosis (Stone et al. 2009).

No specific fungal infections have been identified in mummies. An Inuit mummy from St. Lawrence Island had a chronic granulomatous process affecting the lymphatic system, consistent with histoplasmosis, but only a contaminating fungus, *Candida albicans*, was identified (Zimmerman and Smith 1979).

For protozoan parasites our evidence is mostly indirect. A 9,000-year history of Chagas disease has been documented by DNA probes of Chilean mummies (Aufderheide et al. 2004). Heizer and Napton (1969) found Charcot–Leyden crystals in coprolites from Lovelock Cave, Nevada. These crystals are seen in diarrheal states, particularly in cases of amebiasis. The cysts of *Entamoeba histolytica* were seen by Pizzi and Schenone in their report on a 450-year-old Peruvian mummy (1954). Ruffer (1921) detected enlarged spleens in two Egyptian mummies and diagnosed malaria. Microscopic examination of the spleens was unsatisfactory because of fungal contamination of the tissue. Angel (1966) noted a geographic overlap between malaria and porotic hyperostosis of the skull that fits in well with current thinking on the protective effects of hereditary anemias against malaria. DNA studies have determined that Tutankhamun suffered from falciparum malaria, the most serious form of the disease (Hawass et al. 2010).

We are on much firmer ground in diagnosing metazoan infections and infestations. Parasitic worms and their ova remain well preserved for millennia (see Chapter 10 in this volume by Dittmar et al. for further discussion), and the characteristic ova of *Ascaris lumbricoides*, *Schistosoma hematobium*, and *Taenia solium* have been reported in Egyptian mummies (Reyman et al. 1977). The ova of *Ascaris*, *Trichuris*, and fish tapeworm have been seen in European bog bodies dated to 600 B.C. and 500 A.D. (Jones and Nicholson 1988) and cysticercosis, a complication of pork tapeworm infestation, has been seen in an Egyptian mummy (Bruschi et al. 2006). Parasitic infestations have also been seen in the New World. Adult hookworms, *Ancylostoma duodenale*, have been seen in the small intestine of a 1,000-year-old Peruvian mummy. Pizzi and Schenone also found the ova of *Trichuris trichiura* in the mummy with amebiasis noted above. The ova of *Enterobius vermicularis*, the pinworm, have been found in 10,000-year-old western American coprolites (Fry and Moore 1969), as have ova of *Acanthocephala*, the thorny-headed worm (Moore et al. 1969). The latter probably represents an accidental infestation of the individual by this rodent parasite, as does the finding of a fish trematode, *Cryptocotyle lingua*, in the St. Lawrence Eskimo woman's mummy. The scalp of an Aleut mummy in the collection of the Harvard's Peabody Museum contained numerous ova and adult headlice (*Pediculus humanus capitis*), attached to the hair and visible to the naked eye. The scalp itself showed excoriations consistent with such an infestation. The rehydrated lice, examined by scanning electron microscopy, showed extraordinary preservation and no essential difference from modern lice, providing clear evidence of the antiquity and ubiquity of pediculosis capitis (Zimmerman et al. 1981).

Osteomyelitis has been seen in Peruvian mummies and in PUM-II, the second mummy of the University of Pennyslvania Museum series, who had a periostitis of the right fibula that had resulted in swelling of the entire leg (Cockburn et al. 1975). Histological study showed no specific changes and the suggested diagnosis was that of a secondary change due to some chronic process, perhaps varicose veins.

The skin is often well preserved, and careful examination can be rewarding (Leslie et al. 2006). A rare skin condition, subcorneal pustular dermatosis, was reported in a 3,200-year-old Egyptian mummy (Zimmerman and Clark 1976). This condition was not described in modern patients until 1956. One of the lessons of paleopathology is that diseases may exist long before their modern clinical diagnosis.

Dental and middle ear disease have also long been part of the human condition. Periodontal disease and caries have been noted in Pharaohs and fellahs. These conditions can lead to infection of the middle ear and mastoid sinuses (Flohr and Schultz 2009), as Wells (1964) vividly dramatized in his reconstruction of the death of a Rhodesian Neanderthal afflicted with caries and mastoiditis. Perforated eardrums have also been seen in an Egyptian mummy (Lynn and Benitez 1974). The location and shape of the perforations were consistent with an episode of acute otitis media occurring sometime during life. Similar changes are seen in modern patients with persistent perforations due to middle ear disease.

The degenerative process most commonly seen in mummies is osteoarthritis. Twenty percent of Egyptian mummies X-rayed by Gray (1973) showed arthritic changes. Arthritis has been diagnosed in Coptic and Nubian mummies (Armelagos 1969), with the longer-lived Christian population showing an increased incidence, as would be expected in this degenerative disorder. Its presence in the hot dry climate of Egypt and Nubia belies the folk attribution of the disease to damp climates. X-rays of Ramses II revealed severe osteoarthritis in his hips (Harris and Wente 1980) and the disease has been found in Peruvian mummies (Appelboom and Struyven 1999) and in a crew member of the sunken 16th-century Swedish ship *Vasa* (During et al. 1994) completing the documentation of its geographic and social ubiquity. In marked contrast, only rare cases of gout have been identified, one example being that of the Spanish Emperor Charles V (Ordi et al. 2006), as discussed below.

A more life-threatening disorder, atherosclerosis, has been very well documented by historic evidence and the finding of atherosclerosis in many Egyptian mummies. Atherosclerosis was diagnosed in the aorta of the Pharaoh Merneptah (Ruffer 1921) and Ruffer found involvement of all arteries, large and small, to be very common in the hundreds of mummies he examined. Long (1931) demonstrated not only coronary artery disease but also myocardial fibrosis and renal arteriolar sclerosis (a change associated with hypertension) and Shaw (1938) noted involvement of the superior mesenteric artery in the mummy of Har-Mose. More recent studies have been similar.

Harris saw calcification of leg arteries in Amenhotep II and Ramses II radiologically and an Egyptian mummy in the collection of the University of Pennsylvania Museum (PUM-II) had severe atherosclerosis of the aorta and diffuse arteriolar sclerosis on microscopic examination of his organs. Gray, in his 1973 radiological survey, diagnosed atherosclerosis in only 4 of 88 adult mummies, reflecting the inadequacy of standard radiological evaluation for this disease, but a more recent study employing computerized tomography (CT) scanning, a more sensitive technology, identified a much higher incidence of the disorder, affecting 9 of 22 mummies in the Cairo Museum (Allam et al. 2009). Atherosclerosis has also been seen in the New World mummies from the Aleutian Islands and St. Lawrence Island. This condition has been found in only a few of the many Peruvian mummies studied, suggesting a possible lower incidence in ancient Peru or shorter lifespans.

The diagnosis of the various disturbances of circulation resulting from atherosclerosis (thrombosis, embolization, myocardial or pulmonary infarction, etc.) has not been made in any mummy. This is surprising in view of the historic and anatomic evidence of atherosclerosis noted above, but may be the result of a problem in preservation. Infarcted tissue is autolyzed *in situ*, and as such would not be distinguishable from postmortem autolysis (Zimmerman 1978).

Another common degenerative process is the accumulation of foreign material, particularly in the lungs. This condition has been noted in mummies from all areas. The deposition of carbon pigment (anthracosis) has been seen in mummies from Egypt, Alaska and the Canary Islands. The combination of carbon and silica particles (anthracosilicosis) was seen in the lungs of Egyptian mummies PUM-II (Cockburn et al. 1975), ROM-I (Reyman et al. 1977) and Nekht-Ankh, a XII Dynasty male (Tapp et al. 1975). The findings of anthracosis have been attributed to life-long exposure to open fires, for heating and cooking, while the silicosis is probably due to inhalation during the sandstorms common to Egypt. Several mummies have shown damage to the lungs as a result of these exposures. Emphysematous changes have been noted on microscopic examination of the lungs of the Alaskan and Egyptian mummies.

Rare elements can also accumulate in the tissues. Charles Francis Hall, an Arctic explorer, died and was buried in Greenland in 1871. An autopsy on the frozen body was performed a century later (Paddock et al. 1970), and hair and nail samples showed high levels of arsenic. The presumption is that he was poisoned by his crew after a winter trapped in the ice.

There have only been a few tissue diagnoses of tumors in mummified remains. Benign tumors in Egyptian mummies include a few skin tumors (Sandison 1967; Zimmerman 1981) and a sacral nerve sheath tumor (Strouhal and Němečková 2004). A few other examples have been reported (Kramar et al.1983; Horne 1986; Ortner and Aufderheide 1991; Jonsdottir et al. 2003; Ortner 2003) including a chest wall lipoma in the mummy of a 14 year-old Chilean girl (Gerszten and Allison 1991).

Although the gross diagnosis of malignant tumors of various types has been made in skeletal material and in the skeletons of mummies, until recently there had never been a microscopic diagnosis of a malignant tumor in ancient material. Gray (1973) specifically noted the total absence of any radiological evidence of malignancy in his survey of 133 mummies. Possible reasons for this absence of cancer in mummies include poor preservation, early age at death and the rarity of cancer in antiquity. The literature contains only three reports of histologically confirmed malignancies. One is an extremely rare skeletal muscle cancer in a Peruvian child's mummy (Gerszten and Allison 1991). The other two are colorectal carcinomas. This diagnosis in the mummy of Ferrante I of Aragon, a 13th-century King of Naples (Fornaciari 1993) was confirmed by a DNA test for a specific oncogenetic mutation (Thillaud 2006). The other case was a histological diagnosis in a ca. 200 A.D. mummy from the Dakhleh Oasis, Egypt (Zimmerman 2003). While several biochemical tests for cancer have been developed recently, such as prostate specific antigen (PSA) and carcinoembryonic antigen (CEA), a significant incidence of false positive results has limited these tests in modern medicine to post-treatment follow-up, and histology remains the ultimate criterion for the diagnosis of cancer.

It has been suggested that the short lifespan of individuals in antiquity precluded the development of cancer. Although this statistical construct is true, individuals in ancient Egypt did live long enough to develop such diseases as atherosclerosis, Paget's disease of bone, and osteoporosis. It must also be remembered that, in modern populations, bone tumors primarily affect the young. Another explanation for the lack of tumors in ancient remains is that tumors might not be well preserved. The difficulty of examining mummies for evidence of cancer has been reviewed by Halperin (2004), but the experimental study noted above (Zimmerman 1977a) indicates that the features of malignant cells are favorable to preservation by mummification. In an ancient society lacking surgical intervention, evidence of cancer should remain in all cases. The virtual absence of malignancies must be interpreted as indicating their rarity in antiquity and indicate that carcinogenic factors are limited to societies affected by modern industrialization and tobacco usage. See Table 9.1 for an abbreviated list of conditions found in human mummified remains.

The Future of Mummy Studies

As in all other branches of scientific and medical investigation, the study of mummies will be facilitated by the development of new technology (Reyman et al. 1998). The past century has seen the progression of radiological study from relatively basic X-rays (Harris and Wente 1980) to sophisticated CT analyses (Wong 1981; Chemm and Rühli 2005; Allam 2009; O'Brien et al. 2009; Hawass 2010), resulting in nondestructive examinations, including CT-guided endoscopic biopsies, yielding a high level of diagnoses, such as the determination of the cause of death of the Iceman 5,300 years B.P., a stone arrowhead having lacerated his subclavian artery (Pernter et al. 2007).

Future advances in paleoserology may allow the detection of antibodies to pathogenic microorganisms, while improvements in DNA detection (Roberts and Ingham 2008) are already expanding our knowledge of the history and evolution of diseases including rheumatoid arthritis (Fontecchio et al. 2006) and infections such as plague (Bianucci et al. 2008a), malaria (Nerlich et al. 2008, Hawass et al. 2010) and influenza (Tumpey et al. 2005). Pathogenic DNA from the 1918 flu virus has been recovered from remains buried in permafrost (Taubenberger et al. 2005) and DNA diagnostic of tuberculosis has been recovered from a Pre-Columbian Peruvian mummy (Salo et al. 1994), confirming the presence of this disease on both sides of the Atlantic. Immunological tests have also provided information on parasitic diseases such as malaria (Bianucci et al. 2008b).

Table 9.1 Abbreviated list of conditions found in human mummified remains

Disease classification	Diagnoses	Site	Reference
Trauma	Accidental inhumation	Alaska	Zimmerman and Smith 1975
	Arrow wound	Italy	Pertner et al. 2007
	Frostbite	Peru	Post and Donner 1972
Circulatory abnormalities	Cardiovascular disease	Egypt	Long 1931
	Coronary artery disease	Egypt	Bruetsch 1959
Infection and infestation	Carrion's disease	Peru	Allison et al. 1974b
	Chagas disease	Chile	Aufderheide et al. 1974
	Cystercercosis	Egypt	Bruschi 2006
	Lice	Alaska	Zimmerman et al. 1971
	Malaria	Egypt	Bianucci et al. 2008b
	Malaria	Egypt	Hawass et al. 2010
	Malaria	Egypt	Nerlich et al. 2008
	Malaria	Egypt	Ruffer 1921
	Malaria, schistosomiasis, cirrhosis	Egypt	Cockburn et al. 1975
	Pinworm	SW United States	Fry and Moore 1969
	Pneumonia	Chile	Allison et al. 1974a
	Pneumonia	Alaska	Zimmerman et al. 1981
	Smallpox	Egypt	Ruffer and Ferguson 1967
	Trichuriasis	Peru	Pizzi and Schenone 1954
	Trichuriasis, ascariasis	Britain	Jones and Nicholson 1988
	Tuberculosis	Chile	Allison et al. 1973
	Tuberculosis	Egypt	Morse et al. 1964
	Tuberculosis	Peru	Salo et al. 1994
	Tuberculosis	Egypt	Zimmerman 1979
Metabolic disorders	Anthracosis	Canary Islands	Brothwell et al. 1969
	Arsenic poisoning	Canada	Paddock et al. 1970
	Gout	Spain	Ordi et al. 2006
	Sand pneumoconiosis	Egypt	Tapp et al. 1975
Neoplastic disease	Adenocarcinoma	Italy	Fornaciari 1993
	Carcinoma, rectum	Egypt	Zimmerman 2003
	Histiocytoma	Egypt	Zimmerman 1981
	Leiomyoma	Europe	Kramar et al. 1983
	Lipoma	Chile	Gerszten and Allison 1991
	Meningioma	Alaska	Jonsdottir et al. 2003
	Rhabdomyosarcoma	Chile	Gerszten and Allison 1991
Miscellaneous	Subcorneal pustular dermatosis	Egypt	Zimmerman and Clark 1976

Enhancements in nuclear magnetic resonance technology have allowed the examination of mummies without the need for rehydration (Munnemman et al. 2007; Rühli et al. 2007a), including the use of portable equipment (Rühli et al. 2007b). Gas chromatography mass spectrometry has been used in the study of ancient Egyptian embalming materials (Buckley et al. 2004; Tchapla et al. 2004).

Even the smallest of specimens can be studied with these exotic techniques. Remarkably, DNA has been extracted from hair shafts of Siberian mummies (Amory et al. 2006) and a preserved finger from the body of the Holy Roman Emperor of Spain, Charles V, examined by scanning electron microscopy, confirmed the diagnosis made in the 16th century of severe gout (Ordi et al. 2006). Beyond the mere establishment of a diagnosis, Charles V resigned because of this disabling illness and his son, Phillip II, sent the Spanish Armada to its destruction, changing the course of history.

Such advances can also facilitate a look back at studies in the past. Detailed imaging of a Peruvian mummy taken to Belgium in 1840 has revealed previously undiagnosed sacroiliac osteoarthritis (Appleboom and Struyven 1999). The University of Manchester's KNH Centre for Biomedical Egyptology and the Natural History Museum, London, have begun a re-examination of the 20,000 specimens collected in the early-20th-century Archeological Survey of Nubia (Elliot-Smith and Jones 1908).

In a 21st-century world where privacy has been "blown to bits" (Abelson et al. 2008), a caution is raised by Markel (2010) regarding issues of privacy for ancient historical figures. Should they have the same rights as deceased modern individuals? He asks if we need to develop rules for mummy studies and if and how they have the potential to change our approach to threatening diseases such as influenza, as well as our understanding of the past. Today, research on mummified remains is conducted with the same dignity and scientific purpose as hospital-based postmortem examinations of the recently deceased. The results of these studies have given these ancient individuals importance and relevance to modern scientists and to the general public.

Paleopathological study of mummies thus expands our knowledge of the life stories and fate of ancient individuals, their relationship to others, ancient migrations, the evolution of disease, and the role of ancient disease in human evolution and social history, as well as applications to modern medicine and implications for health in the modern world.

REFERENCES

Abelson, H., Ledeen, K., and Lewis, H., 2008   Blown to Bits: Your Life, Liberty, and Happiness After the Digital Explosion. Boston: Pearson Education.

Allam, A. H., Thompson, R. C., Wann, S. L., Miyamoto, M. I., and Thomas, G. S.,   2009   Computed Tomographic Assessment of Atherosclerosis in Ancient Egyptian Mummies. Journal of the American Medical Association 302:2091–2094.

Allison, M. J., Mendoza, D., and Pezzia, A., 1973   Documentation of a Case of Tuberculosis in Pre-Columbian America. American Review of Respiratory Diseases 107:985–991.

Allison, M. J., Pezzia, A., Gerszten, E., Giffler, R. F., and Mendoza, D., 1974a   Aspiration Pneumonia due to Teeth – 950 AD and 1973 AD. Southern Medical Journal 67:479–483.

Allison, M. J., Pezzia, A., Gerszten, E., and Mendoza, D., 1974b   A Case of Carrion's Disease Associated with Human Sacrifice from the Huari Culture of Southern Peru. American Journal of Physical Anthropology 41:295–300.

Amory, S., Keyser, C., Crubezy, E., and Ludes, B., 2006   STR Typing of Ancient DNA Material Extracted from Hair Shafts of Siberian Mummies. Forensic Science International 166:218–229.

Angel, J. L., 1966   Porotic Hyperostosis, Anemias, Malarias and Marshes in the Prehistoric Eastern Mediterranean. Science 153:760–763.

Appelboom, T., and Struyven, J., 1999   Medical Imaging of the Peruvian Mummy Rascar Capac. Lancet 354:2153.

Armelagos, G. J., 1969   Disease in Ancient Nubia. Science 163:255–259.

Artamonov, M. I., 1965   Frozen Tombs of the Scythians. Scientific American 212(5):101–109.

Aufderheide, A. C., 2003   The Scientific Study of Mummies. Cambridge: Cambridge University Press.

Aufderheide, A. C., and Rodriquez-Martin, C., 1998   The Cambridge Encyclopedia of Human Paleopathology. Cambridge: Cambridge University Press.

Aufderheide, A. C., Salo, W., Madden, M., Streitz, J., Buikstra, J. E., Guhl, F., Arriaza, B., Renier, C., Wittmers, L. E., Jr., Fornaciari, G., and Allison. M., 2004   A 9,000-Year Record of Chagas' Disease. Proceedings of the National Academy of Sciences, USA 101:1034–2039.

Aufderheide, A. C., Neiman, F. D., Wittmers, L. E., Jr., and Rapp, G., Jr., 1981   Lead in Bone II: Skeletal Lead Content as an Indicator of Lifetime Lead Ingestion and the Social Correlates in an Archaeological Population. American Journal of Physical Anthropology 55(3):285–291.

Bianucci, R., Rahalison, L., Massa, E. R., Peluso, A., Ferroglio, E., and Signoli, M., 2008a   Technical Note: a Rapid Diagnostic Test Detects Plague in Ancient Human Remains: An Example of the Interaction between Archeological and Biological Approaches (Southwestern France, 16th–18th Centuries). American Journal of Physical Anthropology 136:361–367.

Bianucci, R., Mattutino, G., Lallo, R., Charlier, P., Jouin-Spriet, H., Peluso, A., Higham, T., Torre, C., and Massa, E. R., 2008b   Immunological Evidence of *Plasmodium falciparum* Infection in an Egyptian Child Mummy from the Early Dynastic Period. Journal of Archeological Science 35:1880–1885.

Brier, R., and Wade, R. S., 1997   The Use of Natron in Human Mummification: A Modern Experiment. *Zeitschrift für Ägyptische Sprache und Altertumskunde* 124:89–100.

Brothwell, D. R., and Sandison, A. T., eds., 1967   Diseases in Antiquity. Springfield, IL: C.C. Thomas.

Brothwell, D. R., Sandison, A. T., and Gray, P. H. K., 1969   Human Biological Observations on a Guanche Mummy with Anthracosis. American Journal of Physical Anthropology 30:333–347.

Bruetsch, W. L., 1959   The Earliest Record of Sudden Death Possibly Due to Atherosclerotic Coronary Occlusion. Circulation 20:438–441.

Buckley, S. A., Clark, K. A., and Evershed, R. P., 2004   Complex Organic Balms of Pharaonic Animal Mummies. Nature 431:294–299.

Chhem, R. K., and Rühli, F., 2004   Paleoradiology of Mummies. Canadian Association of Radiologists Journal, Special Issue: 55:No. 4.

Cockburn, T. A., Barraco, R. A., Reyman, T. A., and Peck, W. H., 1975   Autopsy of an Egyptian Mummy. Science 187:1155–1160.

Cockburn, T. A., Cockburn, E., and Reyman, T. A., eds., 1998   Mummies, Disease and Ancient Cultures. 2nd edition. Cambridge: Cambridge University Press.

During, E., Zimmerman, M. R., Kricun, M. E., and Rydberg, J., 1994   Helmsman's Elbow: An Occupational Disease of the 17th Century. Journal of Paleopathology 6:19–27, 1994.

Elliot-Smith, G., and Jones, F. W., 1908   Anatomical Report. Archaeological Survey of Nubia Bulletin 2:29–54.

Flohr, S., and Schultz, M., 2009 Mastoiditis – Paleopathological Evidence of a Rarely Reported Disease. American Journal of Physical Anthropology 138:266–273.

Fontecchio, G., Ventura, L., Azzarone, R., Fioroni, M. A., Fornaciari, G., and Papola, F., 2006 Letter: HLA-DRB Genotyping of an Italian Mummy from the 16th Century with Signs of Rheumatoid Arthritis. Annals of the Rheumatic Diseases 65:1676–1677.

Fornaciari, G., 1993 Adenocarcinoma in the Mummy of Ferrante I of Aragon, King of Naples (1431–1494). Paleopathology Club Newsletter 83:5–8.

Fry, G. F., and Moore, J. G., 1969 *Enterobius vermicularis*: 10,000 Year Old Human Infection. Science 166:1620.

Gerszten, E., and Allison, M., 1991 Human Soft Tissue Tumors in Paleopathology. *In* Human Paleopathology: Current Syntheses and Future Options. D. J. Ortner and A. C. Aufderheide, eds. pp. 260–277. Washington, DC: Smithsonian Institution Press.

Gray, P. H.K., 1973 The Radiography of Mummies of Ancient Egyptians. Journal of Human Evolution 2:51–53.

Halperin, E. C., 2004 Paleo-Oncology: The Role of Ancient Remains in the Study of Cancer. Perspectives in Biology and Medicine 47:1–14.

Harris, J. E, and Wente, E. F., eds., 1980 An X-Ray Atlas of the Royal Mummies. Chicago: The University of Chicago Press.

Hart, G. D., ed., 1983 Disease in Ancient Man: An International Symposium. Toronto: Clarke Irwin.

Hawass Z., Gad, Y. Z., Ismail, S., Khairat, R., Fathalla, D., Hasan, N., Ahmed, A., Elleithy, H., Ball, M., Gaballah, F., Wasef, S., Fateen, M., Amer, H., Gostner, P., Selim, A., Zink, A., and Pusch, C. M., 2010 Ancestry and Pathology in King Tutankhamun's Family. Journal of the American Medical Association 303:638–647.

Heizer, R. F., and Napton, L. K., 1969 Biological and Cultural Evidence from Prehistoric Human Coprolites. Science 165:563–568.

Horne, P. D., 1986 Case No. 23: Angiokeratoma Circumscriptum. Paleopathology Club Newsletter 27:1.

Janssens, P. A., 1970 Paleopathology: Diseases and Injuries of Prehistoric Man. London: Baker.

Jones, A. and Nicholson, C., 1988 Recent Finds of *Trichuris* and *Ascaris* Ova from Britain. Paleopathology Newsletter 62:5–6.

Jonsdottir, B., Ortner, D. J., and Frohlich, B., 2003 Probable Destructive Meningioma in an Archaeological Adult Male Skull from Alaska. American Journal of Physical Anthropology 122:232–239.

Kramar, C., Baud, C.-A., and Lagier, R., 1983 Presumed Calcified Leiomyoma of the Uterus: Morphological and Chemical Studies of a Calcified Mass Dating from the Neolithic. Archives of Pathology and Laboratory Medicine 107:91–93.

Leslie, K. S., Levell, N. J., and Dove, S. L. B., 2006 Cutaneous Findings Visible on the Skin of Mummies in the British Museum. Journal of Visual Communication in Medicine. 28: 156–162.

Long, A. R., 1931 Cardiovascular Disease: Report of a Case Three Thousand Years Ago. Archives of Pathology 12:92–94.

Long, E. R., 1964 Tuberculosis and Leprosy. Some Correlations in Retrospect. Lancet 84:395–400.

Lynn, G. E., and Benitez, J. T., 1974 Temporal Bone Preservation in a 2600-Year-Old Egyptian Mummy. Science 183:200–202.

Mallory, J. P., and Mair, V. H., 2000 The Tarim Mummies. New York: Thames and Hudson.

Markel, H., 2010 King Tutankhamun, Modern Medical Science and the Expanding Boundaries of Historical Inquiry. Journal of the American Medical Association 303:667–668.

Masetti, M., Locci, M. T., Ciranni, R., and Fornaciari, G., 2006 Short Report: Cystercercosis in an Egyptian Mummy of the Late Ptolemaic Period. American Journal of Tropical Medicine and Hygiene 74:598–599.

Moore, J. G., Fry, G. F., and Englert, E., Jr., 1969   Thorny-headed Worm Infection in North American Prehistoric Man. Science 163:1324–1325.

Morse, D., Brothwell, D. R., and Ucko, P. J., 1964   Tuberculosis in Ancient Egypt. American Review of Respiratory Diseases 90:524–541.

Munnemann, K., Boni, T., Colacicco, G., Blumich, B., and Ruhli, F., 2007   Noninvasive ^{1}H and [23]Na Nuclear Magnetic Resonance of Ancient Egyptian Mummified Tissue. Magnetic Resonance Imaging 25:1341–1345.

Murray, M. A., 1910   The Tomb of the Two Brothers. Manchester: Sherratt and Hughes.

Nahas, M. K., 1969   The Family in the Arab World. In Peoples and Cultures of the Middle East. A. Shiloh, ed. pp. 174–204. New York: Random House.

Nerlich, A. G., Schraut, B., Dittrich, S., Jelinek, T., and Zink, A. R., 2008   Plasmodium falciparum in Ancient Egypt. Emerging Infectious Diseases 14: 1317–1319.

O'Brien J. J., Battista, J. J., Romagnoli, C., and Chhem, R. K., 2009   CT Imaging of Human Mummies: A Critical Review of the Literature (1979–2005). International Journal of Osteoarchology 19:90–98.

Ordi, J., Alonso, P. L., de Zuleta, J., Esteban, J., Velasco, M., Mas, E., Campo, E., and Fernando, P. L., 2006   Occasional Notes: The Severe Gout of Holy Roman Emperor Charles V. New England Journal of Medicine 355:516–520.

Ortner, D. J., 2003   Identification of Pathological Conditions in Human Skeletal Remains, 2nd edition. San Diego: Academic Press.

Ortner, D. J., and Aufderheide, A. C., eds., 1991   Human Paleopathology: Current Syntheses and Future Options. Washington, DC: Smithsonian Institution Press.

Paddock, F.K., Loomis, C. C., and Perkons, A. K., 1970   An Inquest on the Death of Charles Francis Hall. New England Journal of Medicine 282:784–786.

Pernter, P., Gostner, P., Vigi, E. E., and Rühli, F. J., 2007   Radiologic Proof for the Iceman's Cause of Death (ca. 5,300 BP). Journal of Archeological Science 34:1784–1786.

Pizzi, T., and Schenone, H., 1954   Hallazgo de huevos de Trichuris trichiura en Contenido intestinal de un cuerpo arqueológico Incaico. [The Finding of Trichuris trichiura Eggs in the Intestinal Contents of an Inca Body.] Bol Chileno de Parasitologá 9:73–75.

Post, P. W., and Donner, D. D., 1972   Frostbite in a Pre-Columbian Mummy. American Journal of Physical Anthropology 37(2):187–191.

Rasmussen, M., et al., 2010   Ancient Human Genome Sequence of an Extinct Palaeo-Eskimo. Nature 463:757–762.

Reyman, T. A., Nielsen, H., Thuesen, I., Notman, D. N. H., Reinhard, K. J., Tapp, E., and Waldron, T., 1998   New Investigative Techniques. In Mummies, Disease and Ancient Cultures. A. Cockburn, E. Cockburn, and T. A. Reyman, eds. pp. 353–394. Cambridge: Cambridge University Press.

Reyman, T. A., Zimmerman, M., and Lewin, P. K., 1977   Autopsy of an Egyptian Mummy: Histopathologic Investigation. Canadian Medical Association Journal 117:470–471.

Roberts, C. A., and Ingham, S., 2008   Using Ancient DNA Analysis in Paleopathology: A Critical Analysis of Published Papers with Recommendations for Future Work. International Journal of Osteoarchology 18:600–613.

Rowling, J. T., 1961   Pathological Changes in Mummies. Proceedings of the Royal Society of Medicine 54:409–415.

Ruffer, M. A., 1921   Studies in the Paleopathology of Egypt. Chicago: University of Chicago Press.

Ruffer, M. A., and Ferguson, A. R., 1967   An Eruption Resembling that of Variola in the Skin of a Mummy of the Twentieth Dynasty (1200–1100 B.C.). In Diseases in Antiquity. D. R. Brothwell and A.T. Sandison, eds. pp. 346–348. Springfield, IL: C. C. Thomas.

Rühli, F. J., Böni, T., Perlo, J., Casanova, F., Baias, M., Egarter, E., and Blumich, B., 2007a   Non-invasive Spatial Tissue Discrimination in Ancient Mummies and Bones by Portable Nuclear Magnetic Resonance. Journal of Cultural Heritage 8:257–263.

Rühli, F. J., von Waldenburg, H., Nielles-Vallespi, S., Böni, T., and Speier, P., 2007b   Clinical Magnetic Resonance Imaging of Ancient Dry Mummies without Rehydration. Journal of the American Medical Association 298:2618–2620.

Salo, W. L., Aufderheide, A. C., Buikstra, J. E., and Holcomb, T. A., 1994   Identification of *Mycobacterium tuberculosis* DNA in a Pre-Columbian Peruvian Mummy. Proceedings of the National Academy of Sciences USA 91:2091–2094.

Sandison, A. T., 1967   Diseases of the Skin. *In* Diseases in Antiquity, D. R. Brothwell and A. T. Sandison, eds. pp. 449–463. Springfield, IL: C. C. Thomas.

Sandison, A. T., 1970   The Study of Mummified and Dried Human Tissues. In: Science in Archeology. D. R. Brothwell and E. Higgs, eds. pp. 490–502. New York: Praeger.

Shaw, A. F. B., 1938   A Histological Study of the Mummy of Har-mose, the Singer of the Eighteenth Dynasty (Circa 1490 B.C.). Journal of Pathology and Bacteriology 47:115–123.

Steinbock, R. T., 1990   A Review of Ancient Calcified Soft Tissues and Organic Concretions, with Particular Reference to Renal and Urinary Bladder Stones and Gallstones. [Abstract]. *In* Papers on Paleopathology Presented at the Eighth European Members Meeting of the Paleopathology Association, E. Cockburn, ed. p. 20.Cambridge, England, 19–22 September 1990.

Stone, A. C., Wilbur, A. K., Buikstra, J. E., and Roberts, C. A., 2009   Tuberculosis and Leprosy in Perspective. Yearbook of Physical Anthropology 52:66–94.

Strouhal, E., and Nemeckova, A., 2004   Paleopathological Find of a Sacral Neurilemmoma from Ancient Egypt. American Journal of Physical Anthropology 125:320–328.

Tapp, E. A., Curry, A. and Anfield, C.,1975   Sand Pneumoconiosis in an Egyptian Mummy. British Medical Journal 2:276.

Taubenberger, J. K., Reid, A. H., Lourens, R. M., Wang, R., Jin, G., and Fanning, T. G., 2005   Characterization of the 1918 Influenza Virus Polymerase Genes. Nature 437:889–893.

Tchapla, A., Mejanelle, P., Bleton, J., and Goursaud, S., 2004   Characterization of Embalming Materials of a Mummy of the Ptolemaic Era. Comparison with Balms of Mummies of Different Eras. Journal of Separation Science 27:217–234.

Thillaud, P. L., 2006   Paleopathologie du Cancer, Continuite ou Rupture? Bulletin du Cancer 93:767–773.

Tumpey, T. M., Basler, C. F., Aguilar, P. V., Zeng, H., Solórzano, A., Swayne, D. E., Cox, N. J., Katz, J. M., Taubenberger, J. K., Palese, P., and García-Sastre, A., 2005   Characterization of the Reconstructed 1918 Spanish Influenza Pandemic Virus. Science 310:77–80.

Tyson, R., ed., 1997   Human Paleopathology and Related Subjects: An International Bibliography. San Diego: San Diego Museum of Man.

Wells, C., 1964   Bones, Bodies and Diseases: Evidence of Disease and Abnormality in Early Man. London: Thames and Hudson.

Wong, P. A., 1981   Computed Tomography in Paleopathology: Technique and Case Study. American Journal of Physical Anthropology 55:101–110.

Zimmerman, M. R., 1972   Histologic Examination of Experimentally Mummified Tissue. American Journal of Physical Anthropology 37:271–280.

Zimmerman, M. R., 1976   Rehydration of Accidentally Desiccated Pathologic Specimens. Laboratory Medicine 7:13–17.

Zimmerman, M. R., 1977a   An Experimental Study of Mummification Pertinent to the Antiquity of Cancer. Cancer 40:1358–1362.

Zimmerman, M. R., 1977b   The Mummies of the Tomb of Nebwenenef: Paleopathology and Archeology. Journal of the American Research Center in Egypt 14:33–36.

Zimmerman, M. R., 1978   The Mummified Heart: A Problem in Medicolegal Diagnosis. Journal of Forensic Sciences 23:750–753.

Zimmerman, M. R., 1979   Pulmonary and Osseous Tuberculosis in an Egyptian Mummy. Bulletin of the New York Academy of Medicine 55:604–608.

Zimmerman, M. R., 1981   A Possible Histiocytoma in an Egyptian Mummy. Archives of Dermatology 117:364–365.

Zimmerman, M. R., 1990   The Paleopathology of the Liver. Annals of Clinical Laboratory Science 20:301–306.

Zimmerman, M. R., 1993   The Paleopathology of the Cardiovascular System. Texas Heart Institute Journal 20:252–257.

Zimmerman, M. R., 2001   The Study of Preserved Human Tissue. *In* Handbook of Archeological Sciences. D. R. Brothwell and A. M. Pollard, eds. pp. 249–257. Chichester: John Wiley.

Zimmerman, M. R., 2003   Histology of Rectal Carcinoma. *In* The Scientific Study of Mummies. A.C. Aufderheide, ed. p. 373 Cambridge: Cambridge University Press.

Zimmerman, M. R., 2004   Paleopathology and the Study of Ancient Remains. *In* Encyclopedia of Medical Anthropology: Health and Illness in the World's Cultures. C. R. Ember and M. Ember, eds. pp. 49–58. New York: Kluwer/Plenum, New York.

Zimmerman, M. R., 2007   Report to Assassination Records Review Board, National Archives, Washington, DC, Dec. 14, 1999. Excerpted in Reclaiming History: The Assassination of President John Fitzgerald Kennedy. V. Bugliosi, ed. Endnotes, pp. 424–425. New York: Norton.

Zimmerman, M. R., Brier, R., and Wade, R. S., 1998   20th Century Replication of an Egyptian Mummy – Implications for Paleopathology. American Journal of Physical Anthropology 107:417–420.

Zimmerman, M. R., and Clark, W. H., Jr., 1976   A Possible Case of Subcorneal Pustular Dermatosis in an Egyptian Mummy. Archives of Dermatology 112:204–205.

Zimmerman, M. R., and Smith, G. S., 1975   A Probable Case of Accidental Inhumation of 1,600 Years Ago. Bulletin of the New York Academy of Medicine 51:828–837, 1975.

Zimmerman, M. R., Trinkaus, E., LeMay, M., Aufderheide, A. C., Marrocco, G. R., Ortel, R. W., Benitez, J., Laughlin, W. S., Schultes, R. E., and Coughlin, E. A., 1981   The Paleopathology of an Aleutian Mummy. Archives of Pathology and Laboratory Medicine 105:638–641.

Zimmerman, M. R., Yeatman, G., Sprinz, H., and Titterington, W. P., 1971   Examination of an Aleutian Mummy. Bulletin of the New York Academy of Medicine 47:80–103.

CHAPTER **10** The Study
of Parasites
Through Time:
Archaeoparasitology
and Paleoparasitology

*Katharina Dittmar, Adauto
Araújo, and Karl J. Reinhard*

INTRODUCTION

Two scientific fields have been pivotal in starting and advancing the study of parasite remains through time: archaeology and parasitology. Together, they have contributed major methodological and theoretical advances to the recovery and interpretation of parasitological data from times past. Specifically, the onset of the field of paleopathology proved to be a catalyzing factor in the early days of the field, as it not only brought attention to the study of pathology manifested in human skeletal remains, but also to other, understudied biological remains from archaeological sites, such as pollen, coprolites, latrine soils, and/or food remains. It was only a matter of time before conclusive findings of parasite remains at archaeological sites were recovered, and since then the study of parasite remains has gained appreciation and independence as a scientific field. The development of special collecting and survey techniques rapidly followed, leading to new possibilities through the rise of advanced DNA-sequencing techniques. Ever since its inception, the field has embraced an interdisciplinary approach, most notably collaborating with the fields of parasitology, (bio)archaeology, anthropology, paleopathology, ecology, and evolutionary biology.

A Companion to Paleopathology, First Edition. Edited by Anne L. Grauer.
© 2012 John Wiley & Sons, Ltd. Published 2016 by John Wiley & Sons, Ltd.

Trailblazers in Archaeological Parasitology

The beginning of the study of parasitological remains is firmly tied to biological remains (mummies) at archaeological sites. What we now consider the first parasitological study was undertaken on Egyptian mummies by Ruffer (1910), who reported the presence of blood fluke eggs (*Schistosoma haematobium*). This find changed the landscape of archaeological studies, and brought parasitological data into the focus of mummy research. Within this framework, parasitology was largely seen as providing physical (causal) evidence to explain distinct pathology pertaining to particular individuals (see, for instance, Brothwell and Sandison 1967).

Some researchers, however, started to pursue other avenues of parasitological investigations in archaeological settings. These studies were not predicated on the recovery of human remains, but rather, followed Taylor's (1955) recognition of the value of parasite examination of latrine sediments in interpreting living conditions of populations (rather than the singular health of an individual). Despite these important developments, it was not until the 1970s and 1980s that the field had a unified theoretical framework or standard methods. Since then four active research regions have emerged worldwide, each with its distinct approach.

United Kingdom   Building on previous pioneering work by Pike (1975), Sandison (1967) and Taylor (1955), Andrew Jones established parasitological studies in archaeological contexts. In particular, Jones modified parasitological techniques to recover helminth eggs from mineralized coprolites and latrine samples ranging from very low to very high egg quantities (Jones 1982a, 1985). Most importantly, by comparing quantities between archaeological deposits, he was able to document variation associated with different site functions. He also showed that parasite eggs were part of the normal medieval urban background fauna and, thus, exposure to parasite infection was an unavoidable aspect of medieval life (Jones et al. 1988a, 1988b). His theoretical framework emerged from environmental archaeology as practiced at the York Archaeological Trust. Their integrative approach sought to combine diverse biological data sets to establish solid, holistic interpretations from archaeological deposits (Jones 1982b). In particular, the group combined soil analysis, archaeopalynology, archaeobotany, zooarchaeology, human remains, archaeoentomology, malacology, parasitology, and other fields. In the late 1980s Jones's work inspired paleoparasitological research in Japan by Akira Matsui and his colleagues Masaaki and Masako Kanehara, addressing issues of site use, transhumance, medicinal plant use, and sanitation. Their work resulted in the identification of ascarid roundworms, whipworms, Yokogawa flukes, Chinese liver flukes and beef or pork tapeworms in Japanese latrine sediments and coprolites (see review by Matsui et al. 2003).

Germany   In Germany, paleoparasitology began with the work of Bernd Herrmann on medieval latrines. Hermann and colleagues developed a quantitative analysis of medieval villages to trace the distribution of parasite species. Theoretically, Herrmann and Schulz (1986) defined the interpretive variables at play that defined parasite egg spectra in latrine sediments. These included the social group or groups that used specific latrines, the demographic make-up of those groups, differential egg production between species, and effects of soil chemistry and decay organism on egg preservation (Hermann 1985). Hermann's work drew the attention of other researchers in Europe, among them most prominently Horst Aspöck from Austria. Using a multitude of techniques, Aspöck and colleagues confirmed the antiquity (2,000 to 5,000 years ago) of taeniid tapeworms, ascarid and whipworm, *Dicrocoelium* flukes and *Fasciola hepatica* flukes in Europe (Aspöck et al. 1995, 1996, 1999; Aspöck 2000).

South America   Researchers in the Americas started prolific work with a broad range of material, including coprolites, mummies, and latrines. In Brazil, Luiz Fernando Ferreira initiated the field through the Fundação Oswaldo Cruz (FIOCRUZ) in 1978. With his colleagues, he evaluated the parasitological theories of heirloom and souvenir parasites presented by Sprent (1969), and summarized later by Kliks (1983) with paleoparasitological data. According to theory, heirloom parasites include species that co-evolved with their human hosts (such as hookworms). Heirloom parasites tend to be human-specific, or nearly so. Souvenir parasites are generalist parasites that infect many host species. They opportunistically infected humans after the evolution of the genus *Homo* as humans adapted to different environments. A variety of zoonotic parasites are included in the souvenir category.

Theoretical advances were also made through work in South America. Previously, the theories of Fonseca (1972) and Manter (1967) argued that environmental conditions of the Arctic would prevent the introduction of pathogens into the New World human migrations. Therefore heirloom parasites such as hookworm, whipworm and ascarid roundworms would not be found in prehistoric New World archaeological sites. However, the FIOCRUZ researchers discovered ample evidence of hookworm and whipworm infection in Brazil by analyzing mummies and coprolites (Araújo et al. 1981; Ferreira et al. 1980; Araújo et al. 2008). Thus, the hypothesis of a Beringian cold filter was cast into doubt, making either parasite survival likely, or invoking a scenario of alternative, non-Beringian migration routes (see discussion below).

Methodologically, the FIOCRUZ research addressed the effects of desiccation and mummification on parasite remains (Araújo et al. 1981; Bastos et al. 1996). This taphonomic research included experimental desiccation of modern parasite eggs to determine whether or not diagnosis from dehydrated remains could be based on egg measurements. Research found little taphonomic alteration for whipworm developmental stages, but desiccation altered the shape of hookworm eggs. This makes micrometer measurement an unreliable diagnostic technique for dehydrated hookworm species, as measurement might not reflect true sizes, potentially leading to misidentification.

North America   Research in North America also saw the use of different data sources. In the western states of the USA, coprolite analysis was the research focus of prehistoric parasitism. In the eastern states, the emphasis was on latrines from historic sites – a natural division driven largely by preservation conditions for biological material. Samuels (1965) was first to recover parasite eggs from coprolites, and subsequent research highlighted the importance of coprolite studies (Fry and Hall 1969; Heizer and Napton 1969; Moore et al. 1969). These works subsequently gained the interest of the larger parasitological community (Schmidt 1971), which set the stage for new developments in the 1970s and 1980s.

Coprolite research was conducted at several universities, including UC Berkeley, Washington State University, University of Utah, Northern Arizona University, University of Colorado, and Texas A&M University (Reinhard 2006). The commonality of training at these universities was that students simultaneously studied palynology, archaeobotany, zooarchaeology and parasitology. This period also saw the formal development of methods for dietary and environmental analysis of coprolites. Parasitological data were interpreted in an interdisciplinary and quantitative context of diet, environment, and culture (Reinhard et al. 1985, 1987; Reinhard 1988; Reinhard and Bryant 2008).

Analysis of latrine sediments for parasites began in North America (Reinhard et al. 1985), largely inspired by the earlier works of Jones (UK) and Herrmann (Germany). Similar to coprolite analyses, parasite remains were interpreted in the context of diet and environment. However, historic archaeology also provided socio-economic contexts for interpretation of parasitism by class and ethnicity.

Faulkner and Patton (Faulkner et al. 1989) were trailblazing archaeological parasitologists who published extensively on the ancient parasitology of the southeast USA. They discovered *A. lumbricoides, E. vermicularis*, hookworms, fleas, and *Giardia* protozoan cysts in coprolites from Big Bone Cave, Tennessee, USA. The cysts were identified as *Giardia* using an indirect immunofluorescent antibody test. The coprolites dated to 2,177 +/- 145 yrs ago.

In the 1980s, researchers from the four geographic regions began to communicate, resulting in a renewed worldwide interest in the field. Methods from all four major regions were summarized by Reinhard et al. (1986), setting the stage for the growth of the field in the 1990s.

ADVANCES IN ARCHAEOLOGICAL PARASITOLOGY

In the 1990s, research on ancient parasites took on a global focus and started to address a diversity of topics including paleoepidemiology, taphonomy, and methods (Araújo et al. 2003). Major efforts were made to place results into a cultural context. Particularly prolific data came from extensive research in the Andes, South America. Santoro and colleagues, for instance, summarized the parasitological impact of Inca Empire expansion in the Lluta Valley of Chile (Santoro et al. 2003). They showed that the settlement pattern imposed on indigenous farmers increased "crowd diseases," as well as exposure to fecal-borne pathogens. Reinhard and Buikstra (2003) showed that standard epidemiological expectations of modern parasites could be discovered in ancient louse distribution in mummies. This paper was critical in showing that paleoepidemiological data from a large series of mummies produced infection patterns that follow parasite ecology principles. Dittmar and colleagues (2003) analyzed ectoparasites (parasites that inhabit the "outside" of an organism) for aDNA from animal mummies from southern Peru, showing that combining morphological examination with molecular study provides a reliable basis for species identification. Similarly, Iñiguez and coworkers (Iñiguez et al. 2003, 2006) published an overview of molecular approaches to pinworm DNA. This work ultimately led to the definition of different pinworm populations in the prehistoric Americas, reflecting multiple migrations of human populations, each bringing its own pinworm variant. Taphonomy was addressed by Reinhard and Urban (2003), who described taphonomic alteration of fish tapeworm eggs from Chinchorro mummies in Chile. Arriaza et al. (2010) then explored the pathoecology of Chinchorro parasitism by linking natural climate variation to tapeworm prevalence. The Chinchorro existed during a time of cyclical El Niño events. The results point to oscillations of tapeworm infections related to the appearance and disappearance of the El Niño phenomenon, as differences in the temperature of marine currents favored fish species infected by the parasite consumed by the Chinchorro.

Outside of the Andes, investigation shows that Chagas disease spread far from the Andes in ancient times. Chagas disease is the clinical manifestation of infection with *Trypanosoma cruzi*, a protozoan transmitted by triatomine vectors (Insecta). Chagas disease has a longstanding fascination for parasitologists. The clinical manifestations of the disease, megacolon and megaesophagous, were described in the 1980s (Rothammer et al. 1985). Soon, procedures were developed in several labs to recover DNA from *Trypanosoma cruzi*, the causative agent (Aufderheide et al. 2004; Ferreira et al. 2000; Fornaciari et al. 1992; Guhl et al. 1997,1999). Data from these papers show that *Trypanosoma cruzi* DNA can be reliably recovered from mummified tissue. This line of research culminated in the work of Aufderheide et al. (2004), who analysed 283 mummies for trypanosome DNA, and were able to establish that Chagas disease had been a constant human infection in the Andes since 9,000 years ago. Importantly, this discovery revised the conventional wisdom that Chagas became a health threat only after animal domestication and sedentary life style was established. Aufderheide and his colleagues showed that sylvatic infection was common in the most ancient Andean hunter-gatherers.

Chagas disease was not restricted to the western Andean cordillera region, as shown by recent work (Reinhard et al. 2003). Although it was believed that human infection by *T. cruzi* appeared in the Bolivian highlands, it was hypothesized that prehistoric people in Brazil and other parts of the Americas could also be infected by the parasite. That was proved in North America, where a partially mummified body was found on the Texas–Coahuila border and dated to 1,200 years before present. The mummy exhibited an extreme case of megacolon (Reinhard et al. 2003). DNA verification of trypanosome infection has been accomplished, but is as yet unpublished. In the Brazilian lowland (Minas Gerais) ancient human remains (900 to 7,000 B.P.) were also tested for *T. cruzi* infection. One of the mummified bodies showed a mass of feces obstructing the intestinal tract, suggesting another case of megacolon. Both mummies were positive to ancient *T. cruzi* DNA (Fernandes et al. 2009). Lima and colleagues (2008) found molecular evidence for the presence of *T. cruzi* I in a mummy from Brazil, representing a genotype currently absent from the area. Linked to extant data, this points to microclimate shifts, and changing epidemiological profiles of the pathogen. The paleoepidemiology and pathoecology of Chagas disease was recently reviewed by Araújo and colleagues (2009).

Parasitologists recently began analyzing material from the southernmost frontier of South America. This work highlighted the value of parasitological data in modeling aspects of human adaptation. Studies in Patagonia are focused on documenting pathoecological transitions before and after contact with the European colonists

(Fugassa and Guichón 2005). Before contact, Fugassa and colleagues (2006) documented that zoonotic parasites from wild animals were most common in indigenous groups, variation existing between hunter-gatherers in different Patagonian environments. In contrast, in historic contexts, human-specific fecal-borne infection with *Ascaris lumbricoides* was most common. As discussed by Leles et al. (2009) this was the pattern for all of the Americas. Prehistoric groups differed from each other in subsistence strategy in different areas ranging from coast to grasslands (Borrero 1999; Guichón et al. 2006). Human coprolites contained parasite eggs from the ingestion of animals infected with parasites. Some of these parasites can infect humans, but others pass harmlessly through the human intestinal tract (Fugassa et al. 2008b, Fugassa et al. 2010).

In general, the epidemiology of human groups in Patagonia is quite different from that of prehistoric humans from other regions. It appears that widely diverse and unusual parasites became part of the Patagonian epidemiology. Analyzing the available parasite data for Patagonia, Fugassa (2006) and Fugassa and Guichón (2005), describe different paleoepidemiological scenarios presenting risk factors for humans in diverse ecosystems. This required synthesizing anthropological, archaeological, historical, and ethnographic knowledge of the region. They concluded that parasitism was influenced by many factors including group size, mobility, tropism, the ecosystem occupied, food preparation, and hygiene. These were isolated in a dichotomous risk table indicating probability of infection by parasites (Fugassa and Guichón 2008).

Parasites from Asian archaeological sites have also been explored in the past. Matsui et al. (2003) explored how parasite eggs from archaeological sites have changed the interpretation of architectural features and the immigration of foreign officials to Japan. Broader issues in Japanese archaeology have also been addressed, including the introduction of rice agriculture and its association with *A. lumbricoides* infection. Han et al. (2003) provide a summary of parasites that were present in ancient Korea. Building on Matsui et al. (2003) and Han et al. (2003), Asian colleagues are also developing new techniques to investigate and recover parasites from archaeological deposits. For instance, Seo, Shin, and colleagues (Seo et al. 2007; Shin et al. 2009) have extensive procedures to address issues of taphonomy. Through careful excavation combined with highly refined analysis of degradation, superb data regarding the prevalence of parasites in mummies was obtained (Shin et al. 2009). Through exacting field and lab methods, Shin and colleagues are defining some of the conditions that promote preservation and degradation of parasite eggs in burial contexts.

Researchers in France currently dominate archaeological parasitology in Europe. These labs focus on French, Belgian, German and Swiss archaeological sites (Bouchet et al. 1996; Bouchet et al. 2002; Bouchet et al. 2003a). Studies include analysis of parasitological remains over long periods, focusing on single horizons at several sites or on single time periods at one site (Bouchet et al., 2003b; Loreille and Bouchet, 2003). For example, Rocha et al. (2006) examined parasitism from Gallo-Roman times to nearly modern times at the "Place d'Armes" in Namur, Belgium. Animal parasites, such as *Oxyuris equi*, as well as common human parasites were recovered. Zoonotic parasites such as beef tapeworm, fish tapeworm, and sheep liver fluke were also present. Le Bailly et al. (2005) document fish tapeworm infection at Neolithic sites in Germany and Switzerland. This suggests that the parasite was well established in Europe by the Neolitic era. Their work includes the surprising find of *Dioctophymale* sp. (giant kidney worm) in human coprolites dating to nearly 3,400 years B.C. (Le Bailly et al. 2003). This parasite is most commonly found in the kidneys of carnivores that eat frogs, fish, or other intermediate hosts. Because the location of the coprolite finds is near a lake, it is suggested that ancient humans ate freshwater fish and frogs without completely cooking them, and so became infected.

Large overviews of evidence of single parasite species have also greatly contributed to our knowledge of the past. Le Bailly and Bouchet (2010) summarized the evidence of lancet fluke infection, which is remarkably common in ancient times and relatively rare today. They document that the parasite was indigenous to Western Europe before humans arrived. The archaeological evidence shows a low prevalence until medieval times and its transfer to the Americas by the 16th century. This work details how a parasite species can move from one continent to another with human migration accompanied by livestock. The earliest human parasites were documented in Paleolithic coprolites. The stellar work by Bouchet et al. (1996), which documented *Ascaris lumbricoides* in Paleolithic humans, is an example of how a single find can stand conventional wisdom on its head. At the time of publication, the origin of *A. lumbricoides* was thought to be tied to pig domestication. This discovery showed that it was a human parasite thousands of years before pigs were presumably domesticated.

Pathoecology

Reinhard (1974) coined and published the term "pathoecology" in context of epidemiology. As applied by Reinhard, pathoecology is the retrospective analysis of environmental data that reveal the environmental influences, both social and natural, on the development of disease conditions. It was not until 2003, that pathoecology was applied to parasite remains from archaeological sites by Martinson et al. (2003), as the study of biotic, abiotic, and cultural elements of disease. This term would later be incorporated into the field as a major theoretical and interpretive parasitological framework (Reinhard 2007, 2008; Reinhard and Bryant 2008). Reinhard and Bryant formalized the term and defined its theoretical base based on Cockburn (1967; 1971) and Pavlovsky (1966).

Cockburn defined the evolution of infectious disease in detail by describing the influence of human social evolution on a variety of pathogen species. This provided the basis for hypothesis development for a variety of pathogens in various archaeological settings. Pavlovsky introduced the nidus concept to explain specific disease bursts in finite social space. As applied to bioarchaeology by Reinhard and Bryant, nidus was defined as follows:

> The nidus is a geographic or other special area containing pathogens, vectors, reservoir hosts and recipient hosts that can be used to predict infections based on one's knowledge of ecological factors related to infection. Ecological factors include the presence of vectors, reservoir hosts, humans, and external environment favorable for the transmission of parasites. An individual nidus therefore reflects the limits of transmission of a given parasite or pathogen within specific areas of interaction: bedbugs in a bedroom, for example. Thus, a nidus is a focus of infection. A nidus can be as confined as a single room containing a bed and with access to the room by rodents

carrying plague-infected fleas. However, a nidus can also be as large as the community and its surrounding area in which there is a transmission of hookworms.(Reinhard and Bryant 2008:207–208).

Based on these definitions, the case was made that any prehistoric habitation was a system of overlapping nidi, and by analyzing social use of space and behavior by ancient people, one can identify potential nidi and the infection threats they contained. As an example, Reinhard argues in his analysis of Ancestral Puebloans who occupied Chacoan greathouses (Reinhard 2008) that use of *kivas* developed discrete nidi that promoted pinworm infection. Development of granaries and food processing foci may have enhanced the cycle of tapeworms associated with grain beetles by establishment of nidi for its transmission and a mechanism by which their cysts were added to food. Irrigation agriculture may have established hookworms in large nidi outside of the village.

Pathoecologists are also exploring other details of paleopathology. Miranda Chaves and Reinhard (2006) developed a theoretical fault-tree analysis system for interpreting pollen evidence regarding use of medicinal plants. This resulted in the identification of a cluster of medicinal plants used to treat symptoms of parasitism. In historic archaeology, the development of sanitation and social aspects of parasite control have become a theme. Most recently, Fisher et al. (2007) summarize the emergence and control of parasitism in Albany, New York from the social perspective.

Long-Standing Issues in Archaeological Parasitology

Before archaeological parasitology existed, the parasitological literature was replete with "just-so" stories about the origins of parasites. These stories were based on the best guesses of the origins of parasites based on their known geographic distributions. Most researchers considered these as hypotheses to be tested. However, some stories became reified to axioms by repeated publication in the literature and repeated retelling in the lecture halls. Consequently, some just-so stories became impediments to advancing archaeological parasitology. One of these stories related to the origin of hookworms.

The conventional wisdom regarding hookworm was that it could not have been a prehistoric New World parasite. It was assumed that hookworm species arrived with enslaved Africans and European colonists. Beginning in the 1970s and 1980s, hookworm was found in prehistoric mummies and/or coprolites in North America and South America. The "hookworm debate" began with Ferreira and colleagues' findings of hookworm eggs in Brazil prior to the European conquest (Ferreira et al. 1980, 1983; Araújo et al. 1981, 1988). Initially, there was sharp opposition to these results (Kliks 1983), suggesting that Ferreira and colleagues had confused animal coprolites with human coprolites. This argument was countered when Ferreira and colleagues replicated the hookworm (and whipworm) finds in a prehistoric human mummy (Ferreira et al. 1983). Summaries by Reinhard (1990, 1992a) provided a comprehensive review of similar evidence from South America, including the documentation of earlier finds by Allison and his colleagues (1974). Just when one would have thought that the hookworm debate was satisfactorily resolved,

Fuller (1997) questioned the validity of hookworm finds in Tennessee, Peru, and Brazil. She argued that the current distribution of parasite species should be considered in diagnosing ancient finds. She also suggested that the adult hookworms found attached to the intestinal wall of a Peruvian mummy were in reality pinworm larvae. According to Fuller, it was impossible to find hookworm embryonated eggs and larvae in certain archaeological contexts, because of limits imposed on development by hookworm biology. Two responses written by key researchers systematically addressed every issue brought up by Fuller, finally laying this debate to rest (Faulkner and Patton 2001; Reinhard et al. 2001; Sianto et al. 2005). The authors refuted Fuller's perceptions of hookworm morphology and biology by literature review and experimental evidence. Based on this foundation in biological fact, they were able to defend all hookworm finds.

In the years between the Kliks and Fuller papers, the debate among parasitologists shifted to the mechanism by which hookworms arrived in the New World. Hawdon and Johnston (1996) proposed that the phenomenon of hypobiosis could have been at play. Hypobiosis refers to the ability of some parasites (i.e. hookworms) to suspend development in host tissues until the external environment to the host is optimal for successful embryonation of eggs. Hawdon and Johnston suggested that the parasites survived in migrating humans through hypobiosis until the humans encountered optimal conditions for the extracorporeal stage of hookworm life cycle. Ferreira and Araújo (1996) argued that hypobiosis does not explain other parasites such as whipworm, which arrived in the New World without hypobiotic capabilities. The latest, contribution suggests that multiple parasites, including hookworm, arrived with non-Beringian migrations (Araújo et al. 2008).

Another important and ongoing debate within paleopathology (discussed by Kozlowski and Witas, Chapter 22 of this volume, and Ortner, Chapter 14 of this volume) relates to the etiology of porotic hyperostosis (porous lesions often accompanied by diploic thickening of the parietals and/or superior orbital walls of the cranium) in the archaeological record (Carlson et al. 1974; El-Najjar et al. 1976; Walker 1985, 1986; Reinhard 1992b, 2007; Stodder and Martin 1992; Holland and O'Brien 1997; Reinhard and Bryant 2008; Walker et al. 2009). This pathology is thought to be the skeletal manifestation of chronic iron-deficiency anemia. One cause for porotic hyperostosis may be parasitic infestation and activity in the host (Reinhard 2008; Reinhard and Bryant 2008; Walker et al. 2009). Blood-sucking helminths and diarrhea-causing protozoa are most associated with anemia in the modern world.

Initially, El-Najjar et al. (1976) presented a maize-dependency hypothesis to explain a statistically significant higher rate of porotic hyperostosis among Anasazi skeletal series from "canyon bottom" versus "sage plain" environments. In essence, they suggested that relative to sage plain populations, Anasazi populations in "canyon bottoms" consumed more maize and less meat than "sage plain" Anasazi, thus resulting in chronic iron deficiency. They also assumed that human parasitism among Anasazi peoples was uncommon and consequently did not contribute to iron-deficiency anemia. These dietary and environmental reconstructions were based on methods and perspectives prevalent in the 1970s that predated the wide application of pollen analysis, flotation, and refined methods for coprolite analysis.

During the 1980s, authors began to challenge the maize-dependency hypothesis as applied to the Anasazi (Walker 1985). Walker (1985) suggested that a series of pathoecological factors caused porotic hyperostosis, including weaning practices and parasitic diarrheal diseases. Fink (1985) presented details of Anasazi life such as communal life, lack of knowledge of the germ theory, and cramped living conditions that promoted infectious diseases.

Towards the end of the 1980s, and into the 1990s, the balance of opinion shifted towards parasitism as an etiological factor for iron deficiency and porotic hyperostosis. Reinhard (1988, 1992b, 1996; Reinhard et al. 1987) discovered that parasitism, far from being rare among Anasazi, was an unavoidable aspect of Anasazi life. The relation between parasitism and porotic hyperostosis was demonstrated statistically (Reinhard 1992b). Finally, in 1997 Holland and O'Brien suggested that infectious disease played a greater role than diet in the etiology of porotic hyperostosis.

Walker et al. (2009) offered a summation of the debate and presented an integrated idea that insufficient diet combined with parasitic disease exacerbated by population aggregation caused anemia and porotic hyperostosis. Reinhard (2008) and Reinhard and Bryant (2008) illustrate this with case examples from Ancestral Pueblos. Reinhard (2008) showed that the three parasites most hazardous to infants and pregnant mothers, hookworm, *Giardia lamblia* and *Entamoeba histolytica*, became established at some Ancestral Pueblos and are an obvious parasitic cause of anemia. Reinhard and Bryant (2008) using environmental and archaeological data show that during the Great Drought of A.D. 1150–1450 the Ancestral Pueblo region of Canyon de Chelly witnessed several destructive changes. These were: first, population aggregation; second, fouling of water sources with hookworm, *G. lamblia* and *E. histolytica*; and third, a decline in food quality.

DIVERSIFICATION OF TECHNIQUES

Insights into the presence and composition of parasite fauna through time can be obtained by either relying on direct specimen evidence from fossil or archaeological material, or by identifying or sequencing of parasite DNA from the samples. Early work in the field benefited from traditional parasitological diagnostic techniques, which were modified and refined to suit the unique needs and challenges associated with historical and ancient material (Callen and Cameron 1960; Reinhard et al. 1986; Reinhard 1988, 1992a). Unfortunately, preservation and taphonomic issues often limit the resolution and applicability of these techniques. Therefore, other possibilities were pursued to verify the presence of, and exposure to parasites in archaeological material. One such avenue is the use of immunological techniques. Here, one detects the presence of antigens or antibodies in a sample. If antigens are targeted, a direct test of the physical presence of a parasite or pathogen is conducted. Testing for the presence of antibodies verifies exposure to a parasite or pathogen. Because both antigens, and antibodies are essentially proteins, they may be better suited to survive long term than DNA. Deelder and colleagues (1999) pioneered this approach for archeological material, and applied immunological techniques (enzyme-linked immunosorbent assay; ELISA) with some samples, verifying the exposure of ancient Egyptians to *Schistosoma* sp. Similarly, Gonçalves and colleagues (2002, 2004), Le Bailly et al. (2008) and Mitchell et al. (2008) used ELISA techniques. Gonçalves and colleagues found evidence of *Giardia intestinalis* at archeological sites, whereas

Mitchell and colleagues verified the presence of dysentery in two medieval latrines in Israel. While these techniques are useful, caution has to be exerted, as it is possible that positive reactions are triggered by non-specific elements present in the sample. Therefore, other, more direct methods are used, if possible.

In the past two decades, ancient DNA (aDNA) research has inspired the use and development of sequence-based methods in paleoparasitology. Ancient DNA is postmortem-preserved genetic information, and like any other biological material it is subject to degradation and contamination (see Spigelman et al., Chapter 8 of this volume, for further discussion on the analysis of DNA in paleopathology). Any molecular approaches should therefore be subjected to well-established criteria of authenticity (Pääbo et al. 2004).

Currently, aDNA evidence exists for a variety of parasites (Dittmar 2009). So far, all molecular-based paleoparasitological research has relied on short individual gene sequences, which target particular species. With the development of multiplex DNA techniques has come the opportunity to increase the length of obtainable sequence fragments, using only small amounts of starting template (Krause et al. 2006). Retrieving longer contiguous data from any single gene, and (or) data from a variety of genes is important for a more stringent identification of a parasite of interest. Therefore, another trend in aDNA research is to amplify the total DNA of a sample, and subject it to a high-throughput technique called "pyrosequencing" (Margulies et al. 2006). Adapting these new techniques to samples from animal middens, latrine soils or any other kind of paleoparasitological evidence holds the promise of directly sequencing and identifying parasite DNA, which might go unnoticed if one uses traditional sedimentation methods to recover parasite eggs (Dittmar 2009). Genomic approaches can help us understand how evolutionary factors have influenced the variation of life-history traits. This is particularly interesting for parasites, as it becomes clear that the interplay of different suites of genes may result in the same phenotype. The comparison of genome data from ancient and extant parasites can therefore inform us about the evolutionary trajectory of such variation on a large scale.

Both immunological and ancient DNA techniques require the preservation of molecules. Although the recovery of these molecules has been attempted from fossilized samples, none of the results are credible. However, it recently became possible to visualize endosymbionts and parasites in fossils using neutron, x-ray computed tomography, as well as synchotron-computer-tomography. Pioneering work has been done by Neumann and colleagues, who work on the visualization of amber insects, as well as endosymbionts and endoparasites (symbionts or parasites that live in tissues or cavities of an organism) preserved in fossil specimens (Neumann and Wisshak 2009; Neumann et al. 2009). These early records of parasitic association are useful in the context of understanding the age of parasitism in certain animal phyla.

HUMAN–PARASITE INTERACTIONS IN THE CONTEXT OF EVOLUTION AND ENVIRONMENT

Paleoepidemiology, paleoecology, and phylogeny have all made contributions to the understanding of human parasitism through evolutionary time. Most recently, Dittmar (2009) summarized the development of the study of parasites in ancient material pointing to trends and perspectives in the field. This overview of recent developments highlights how these results are able to broaden the scope of the field, placing paleoparasitology and archaeoparasitology at the interface of a wide array of studies, including parasitology, climate change, human evolution, and evolutionary processes. Lambrecht (1980) pioneered this approach with his reconstruction of African sleeping sickness as an evolutionary pressure on early hominid populations. Since Lambrecht, several approaches have emerged and recent examples of these are summarized below.

Nozais (2003) used current knowledge of infectious diseases to reconstruct the origin and dispersion of some parasites in the Old World, discussing human parasite infections of African origin that were introduced into Europe. Sallares (2002) analyzed the history of malaria in Rome. His analysis of social and natural factors affecting ancient malaria transmission is exhaustive. Deforestation, road construction, military campaigns, nutrition, agriculture, and other subjects are identified as contributing to the establishment of malaria in Rome. Montenegro et al. (2006) similarly related the paleobiogeography of parasites to ancient human migrations, realizing that the life-cycle requirements and presence of hookworm in the ancient Americas could be explained by paleoclimate modeling. He and his colleagues formalized a coastal migration in pre-Clovis times that would have allowed for the transfer of hookworms from Asia to the Americas. Araújo et al. (2008) further developed this model using several parasite species found in the ancient Americas. With increased human mobility and climate change, long-distance transfer of parasites has become a focal point for epidemiologists. Analysis of Chinese immigrant latrines in California showed that long-distance transfer of parasites with immigration is common, and occurred with liver flukes in the late 19th century (Reinhard et al. 2008).

Zink et al. (2006) recovered aDNA from bone marrow samples from Nubia and Egypt. Based on the analysis of these for aDNA of *Leishmania donovani*, they concluded that visceral leishmaniasis had its origin in Sudan. The molecular biology evidence shows that leishmaniasis was present in early Christian-period Nubia. It infected ancient Egyptians through their trade with Nubia during the Middle Kingdom.

Several researchers studied ectoparasites, particularly human head lice, in archeological material to elucidate human migrations (Sadler 1990). Records of head lice, pubic lice and body lice in ancient material mainly stem from the Old World (Mumcuoglu and Zias 1988; Capasso and DiTota 1998; Kenward 2001; Mumcuoglu et al. 2003). Horne (1979) documented head lice infestation by gross examination and scanning electron microscopy in a 200-year-old Aleutian mummy. Based on these records, it was assumed for a long time that head louse infections in the Americas were a result of the arrival of the Spanish *conquistas*. However, Araujo et al. (2000) and Martinson et al. (2003) reported the presence of head lice from archaeological material in Brazil and Peru, respectively. Raoult et al. (2008) amplified DNA from Pre-Conquista Peruvian head lice, showing their affiliation with a globally distributed louse phylotype A, thus verifying that several phylotypes of head lice have coexisted for centuries in humans, and supporting the claim that type A lice were present in the Americas before the time of Columbus.

Two authors in particular looked at the broader pictures of parasitism and human evolution. Hoberg (2006) analyzed the phylogeny of the more common tapeworms of human: *Taenia saginata*, *T. solium*, and *T. asiatica*. He concludes that *T. saginata* and *T. asiatica* represent independent spatially and temporally distinct trajectories; evolutionary lineages that separated from a common ancestor between 0.78 and 1.71 million years ago in Africa or Eurasia. Ashford (2000) uses the spectra of parasites in humans and apes to attempt to understand the ecological origins of humans by analyzing ecological associations of human specialist parasites. His analysis reveals some interesting trends. Of some 40 "specialist" parasites species, most cluster in sub-Saharan Africa, which shows that most host–parasite associations for humans came from a long period of evolutionary time in Africa. A second group of associations comes from the Paleoarctic and/or Orient regions, which shows that humans continued to evolve for a long period outside of Africa.

FOSSILS AND NON-HUMAN EVIDENCE OF PARASITISM

The beginning of the study of parasite remains is firmly tied to archeological studies. This however, limits our focus to a human context, and creates a bias towards a lesser appreciation of fossil, animal, and plant parasite records. Parasites pervade all domains of life. They have been very successful throughout evolutionary history, as directly evidenced by fossil parasites. Due to several randomizing factors influencing preservation, these fossils finds are anecdotal at best, and are almost always removed from the context of the host. The majority of parasite fossils stems from invertebrate-rich amber deposits of the Cretaceous (145.5–65.5 million years ago), and the Oligocene (33.9–23 million years ago) (Weitschat and Wichard 1998; Poinar and Milki 2001; Kobbert 2005; Poinar and Poinar 2007). Examples are amber-preserved bloodsucking ectoparasites, such as fleas (as reviewed in Whiting et al. 2008), mites (Witalinski 2000), and ticks (de la Fuente 2003; Poinar and Brown 2003). A more recent subfossil record relates to the recovery of a nit glued to a human hair from a cave sediment sample dated at ca. 10,000 years ago (Araújo et al. 2000). Because of the lack of host records for most parasite fossils, we can only venture an educated guess about their preferred habitats. Most of these parasite records are from ectoparasites. Endoparasites are rarely preserved. Exceptions to this rule are the Lower Carboniferous fossil crinoids infested with an endosymbiotic tabulate coral (*Cladochonus* sp.) (Neumann et al. 2009).

Very rarely, both host and parasite have fossilized together, thus providing direct evidence for ancient host associations. Examples are only known from invertebrate hosts, such as entomopathogenic mermithid nematodes from the Mid-Tertiary (Poinar 1984), parasitic fungi on the cuticle of a mosquito, or trypanosomatid flagellates in an Early Cretaceous biting midge (Poinar and Poinar 2005). Frequent signs of paleoparasitism come from trace evidence. In this case, the parasite is not preserved, but has inflicted visible pathology that fossilized with its host. If these pathologies are similar to the damage done by known extant parasites, they should be considered as the inflicting cause. Credible records of early parasitic relationships in metazoans concern pathological evidence from vermiform animals on specimens of Ordovician graptolites (Conway Morris 1981), and 520 MYA Lower Cambrian brachiopods (Bassett et al. 2004). Other evidence comes from gall-producing parasitoid arthropods on plants, such as seen on Cretaceous plant fossils from Israel (Krassilov 2008), which have contributed to our understanding of how endophytic parasite communities evolved in basal angiosperms. Yet evidence of life's constant exposure to parasites may come from a completely extant data source – namely animal genomes. Fossilized "genetic scars" from the invasion of ancient (and recent) endogenous retroviruses (ERVs) are visible all over the genomes of birds, reptiles, amphibians, fish, and mammals (Tarlington et al. 2006). Most of these invaders have been silenced over long evolutionary timeframes, but recent studies on koalas found an endogenizing virus that entered the koala germline ca. 200 years ago, with the ability still to produce infectious virus.

There are many opportunities to extend paleoparasitological research to animals, especially considering the development of new molecular approaches. Notable sources are animal mummies, which are preserved in permafrost, and in Egyptian, and South American archaeological sites. For instance, Dittmar et al. (2003), reported the presence of fleas from guinea pig and dog mummies of the Chiribaya Culture (Peru). The above mentioned sources of paleoparasitological research mostly pertain to domesticated animals, yet so far unexplored research opportunities exist for wild animals, such as pack rat, caribou, or shell middens, owl pellets, bird-, or bat-guano (Insoll and Hutchins 2005; Dittmar 2009).

CONCLUSION

The evolutionary fate of each organism is forged by reciprocal interactions with its local and global environment. Through time every organism undergoes ecological and evolutionary changes across space. Without the opportunity to directly sample the past, we can only estimate the dynamics of these changes by observing the present patterns of genetic diversity, geographical distribution or ecological parameters. Historical and ancient evidence of parasites however, can provide direct access to additional information in time and space, thus facilitating the understanding of parasite dynamics in biological networks. One of the primary goals of paleo- and archaeoparasitological studies is to connect glimpses into the past to the present. These important records can provide clues to a variety of constantly changing processes, such as transmission cycles, host-use, co-evolution, feedback relationships between diet and infection levels, or impact of human activity on infection cycles. Most parasites require certain conditions for their successful development, and fluctuation in temperatures may permit or drive parasites to migrate and colonize areas other than their extant distribution. Results from paleo- and archaeoparasitological studies can also further our understanding of the influence of global and local climate change on the transmission of parasitic diseases.

REFERENCES

Allison, M. J., Pezzia, A., Hasegawa, I., and Gerszten, E., 1974 A Case of Hookworm Infestation in a Precolumbian American. American Journal of Physical Anthropology 41:103–106.
Araújo, A., Ferreira, L. F., and Confalonieri, U., 1981 A Contribution to the Study of Helminth Findings in Archaeological Material in Brazil. Revista Brasileira de Biologia 41:873–881.

Araújo, A., Ferreira, L. F., Confalonieri, U., and Chame M., 1988   Hookworms and the Peopling of America. Cadernos de Saúde Pública 4(2):226–233.

Araújo, A., Ferreira, L. F., Guidon, N., Freire, N. M. S., Reinhard, K. J., and Dittmar, K., 2000   Ten Thousand Years of Head Lice Infection. Parasitology Today 16:269.

Araújo, A., Jansen, A. M., Bouchet, F., Reinhard, K., and Ferreira, L. F., 2003   Parasitism, the Diversity of Life, and Paleoparasitology. Memórias do Instituto Oswaldo Cruz 98 (Suppl.1):5–11.

Araújo, A., Reinhard, K. J., Ferreira, L. F., and Gardner, S. L., 2008   Parasites as Probes for Prehistoric Human Migrations? Trends in Parasitology 24:112–115.

Araújo, A., Jansen, A., Reinhard, K. J., and Ferreira, L. F., 2009   Paleoparasitology of Chagas Disease: a Review. Memórias do Instituto Oswaldo Cruz 104:9–16.

Arriaza, B. T., Reinhard, K. J., Araújo, A., Orellana, N. C., and Standen, V. G., 2010   Possible Influence of the ENSO Phenomenon on the Pathoecology of Diphyllobothriasis and Anisakiasis in Ancient Chinchorro Populations. Memórias do Instituto Oswaldo Cruz 105:66–72.

Ashford, R. W., 2000   Parasites as Indicators of Human Biology and Evolution. Journal of Medical Microbiology 49:770–771.

Aspöck, H., 2000   Paläoparasitologie: Zeugen der Vergangenheit. Nova Acta Leopoldina 83:159–181.

Aspöck, H., Auer, H., and Picher, O., 1995   The Mummy from the Hauslabjoch: A Medical Parasitology Perspective. Alpe Adria Microbiology Journal 2:105–114.

Aspöck, H., Auer, H., and Picher, O., 1996   Trichuris trichiura Eggs in the Neolithic Glacier–Mummy from the Alps. Parasitology Today 12:255–256.

Aspöck H., Auer H., and Picher O., 1999   Parasites and Parasitic Diseases in Prehistoric Human Populations in Central Europe. Helminthologia 36:139–145.

Aufderheide A. C., Salo, W., Madden, M., Streitz, J., Buikstra, J., Guhl, F., Arriaza, B., Renier, C., Wittmers, L. E., Fornaciari, G., and Allison, M., 2004   A 9,000–Year Record of Chagas' Disease. Proceedings of the National Academy of Sciences of the USA 101(7):2034–2039.

Bassett, M. G., Popov, L. E., and Holmer, L. E., 2004   The Oldest–Known Metazoan Parasite? Journal of Paleontology, 78:1214–1216.

Bastos, O. M., Araújo, A., Ferreira, L. F., Santoro, A., Wincker, P., and Morel, C. M., 1996   Experimental Paleoparasitology: Identification of T. cruzi DNA in Desiccated Mouse Tissue. Paleopathol News 94:5–8.

Borrero, L. A., 1999   Human Dispersal and Climatic Conditions During Late Pleistocene Times in Fuego–Patagonia, Quaternary International 53–54:93–99.

Bouchet, F., Baffier, D., Girard, M., Morel, P., Paicheler, J. C., and David, F., 1996   Paléoparasitologie en Contexte Pléistocène Premières Observations à la Grande Grotte d'Arcy–sur–Cure (Yonne), France. Comptes Rendus des Seances de l'Academie des Sciences. Série III. Sciences de la Vie 319(2):147–151.

Bouchet, F., Guidon, N., Dittmar, K., Harter, S., Ferreira, L. F., Miranda Chaves, S. A., Reinhard, K., and Araújo, A., 2003b   Parasite Remains in Archaeological Sites. Memórias do Instituto Oswaldo Cruz 98 Suppl 1:47–52.

Bouchet, F., Harter, S., and Le Bailly, M., (2003a)   The State of the Art of Paleoparasitological Research in the Old World. Memórias do Instituto Oswaldo Cruz 98 Suppl 1:95–101.

Bouchet, F., Harter, S., Paicheler, J. C., Araújo, A., and Ferreira, L. F., 2002   First Recovery of Schistosoma mansoni Eggs from a Latrine in Europe (15th–16th centuries). Journal of Parasitology 88(2):404–405.

Brothwell, D. R., and Sandison A. T., eds., 1967   Diseases in Antiquity. Springfield, IL: C. C. Thomas.

Callen, E. O., and Cameron T. W. M., 1960   A Prehistoric Diet Revealed in Coprolites. New Scientist 8:35–40.

Capasso, L., and Di Tota, G., 1998   Lice Buried Under the Ashes of Herculaneum. Lancet 351(9107):992.

Carlson, D., Armelagos, G., and Van Gerven, D., 1974   Factors Influencing the Etiology of Cribra Orbitalia in Prehistoric Nubia. Journal of Human Evolution 3(3):405–410.

Cockburn, T. A., 1967   Infectious Diseases, Their Evolution and Eradication. Springfield, IL: C. C. Thomas.

Cockburn, T. A., 1971   Infectious Diseases in Ancient Populations. Current Anthropology 12:45–62.

Conway, M. S., 1981   Parasites and the Fossil Record. Parasitology, 82:489–509.

Deelder, A. M., Miller, R. L., de Jonge, N., and Krijger, F. W., 1990   Detection of Schistosome Antigen in Mummies. Lancet 335(8691):724–725.

Dittmar, K., Mamat, U., Whiting, M., Goldmann, T., Reinhard, K., and Guillen, S., 2003   Techniques of DNA-Studies on Prehispanic Ectoparasites (*Pulex* sp., Pulicidae, Siphonaptera) From Animal Mummies of the Chiribaya Culture, Southern Peru. Memórias do Instituto Oswaldo Cruz 98(Suppl 1):53–58.

Dittmar, K., 2009   Old Parasites for a New World: The Future of Paleoparasitological Research. The Journal of Parasitology 95:215–221.

El–Najjar, M. Y., Ryan, D. J., Turner, C. G., and Lozoff, B., 1976   The Etiology of Porotic Hyperostosis Among the Prehistoric and Historic Anasazi Indians of Southwestern United States. American Journal of Physical Anthropology 44:477–487.

Faulkner, C. T., and Patton, S., 2001   Pre-Columbian Hookworm Evidence From Tennessee: A Response To Fuller (1997). Medical Anthropology, 20:92–96.

Faulkner, C. T., Patton, S., and Johnson, S. S., 1989   Prehistoric Parasitism in Tennessee: Evidence From the Analysis of Desiccated Fecal Material Collected from Big Bone Cave, Van Buren County, Tennessee. Journal of Parasitology 75:461–463.

Fernandes, A., Iñiguez, A. M., Lima, V. S., Souza, S. M., Ferreira, L. F., Vicente, A. C., and Jansen, A. M., 2009   Pre-Columbian Chagas Disease in Brazil: *Trypanosoma cruzi* I in the Archaeological Remains of a Human in Peruaçu Valley, Minas Gerais, Brazil. Memórias do Instituto Oswaldo Cruz 103:514–516.

Ferreira, L. F., de Araújo, A. J., and Confalonieri, U. E., 1980   Finding of Helminth Eggs in Human Coprolites from Unai, Minas Gerais, Brazil. Transactions of the Royal Society of Tropical Medicine and Hygiene 76:798–800.

Ferreira, L. F., de Araújo, A. J., and Confalonieri, U. E., 1983   The Finding of Helminth Eggs in a Brazilian Mummy. Transactions of the Royal Society of Tropical Medicine and Hygiene 77(1):65–67.

Ferreira, L. F., and de Araujo, A. J., 1996   On Hookworms in the Americas and Trans-Pacific Contact. Parasitology Today 12:454–455.

Ferreira, L. F., Britto, C., Cardoso, M. A., Fernandes, O., Reinhard, K., and de Araújo, A. J., 2000   Paleoparasitology of Chagas Disease Revealed by Infected Tissues From Chilean Mummies. Acta Tropica 75:79–84.

Fink, T. M., 1985   Tuberculosis and Anemia in a Pueblo II–III (ca. AD 900–1300) Anasazi Child From New Mexico. *In* Health and Disease in the Prehistoric Southwest. C. F. Merbs and R. J. Miller, eds. Arizona State University, Anthropological Research Papers 34.

Fisher, C. L., Reinhard, K. J., Kirk, M., and DiVirgilio, J., 2007   Privies and Parasites: The Archaeology of Health Conditions in Albany, New York. Historical Archaeology 41(4): 172–197.

Fonseca, Filho O., 1972   Parasitismo e Migrações Humanas Pré–históricas. Mauro Familiar. Editor, Rio de Janeiro, 446pp.

Fornaciari, G., Castagna, M., Viacava, P., Tognetti, A., Bevilacqua, G., and Segura, E. L., 1992   Chagas' Disease in a Peruvian Inca Mummy. Lancet 339:128–129.

Fry, G. F., and Hall, H. J., 1969   Parasitological Examination of Prehistoric Human Coprolites From Utah. Proceedings of Utah Academy of Sciences, Arts, and Letters, 46: part 2, 102–105.

Fuente, J. de la, 2003   The Fossil Record and the Origin of Ticks (Acari: Parasitiformes:Ixodida). Experimental and Applied Acarology, 29:331–344.

Fugassa, M. H., Araújo, A., and Guichon, R. A., 2006   Quantitative Paleoparasitology Applied to Archaeological Sediments. Memorias do Instituto Oswaldo Cruz 101 Suppl 2:29–33.

Fugassa, M. H., Beltrame, M. O., Sardella, N. H., Civalero, M. T., and Aschero, C., 2010   Paleoparasitological Results From Coprolites Dated at the Pleistocene–Holocene Transition as Source of Paleoecological Evidences in Patagonia. Journal of Archaeological Science 37:880–4.

Fugassa, M. H., and Guichon, R. A., 2005   Análisis Paleoparasitológico de Coprolitos Hallados en Sitios Arqueológicos de Patagonia Austral: Definiciones y Perspectivas, Magallania 33:13–9.

Fugassa, M. H., and Guichón, R. A., 2008   Modelos Paleoepidemiológicos Para el Holoceno Patagónico, 7th Jornadas de Arqueología de la Patagonia. 21–25 de Abril de 2008, Ushuaia.

Fugassa, M. H., Sardella, N. H., Taglioretti, V., Reinhard, K., and Araújo, A., 2008   Eimeriid Oocysts from Archaeological Samples in Patagonia, Argentina. Journal of Parasitology 94(6):1418–20.

Fuller, K., 1997   Hookworm: Not a Pre–Columbian Pathogen. Medical Anthropology 17:297–308.

Gonçalves, M. L. C., Araújo, A., Duarte, R., da Silva, J. P., Reinhard, K., Bouchet, F., and Ferreira, L. F., 2002   Detection of Giardia duodenalis Antigen in Coprolites Using a Commercially Available Enzyme-Linked Immunosorbent Assay. Transactions of the Royal Society of Tropical Medicine and Hygiene 96(6):640–643.

Gonçalves, M. L. C., Silva, V. L., Andrade, C. M., Reinhard, K. R., Le Bailly, M., Bouchet, F., Ferreira, L. F., and Araújo, A., 2004   Amoebiasis Distribution in the Past: First Steps Using a Immunoassay Technique. Transactions of the Royal Society of Tropical Medicine and Hygiene 98:88–91.

Guhl, F., Jaramillo, C., Vallejo, G. A., Yockteng, R., Cardenas-Arroyo, F., Fornaciari, G., Arriaza, B., and Aufderheide, A. C., 1999   Isolation of Trypanosoma cruzi DNA in 4.000–year–old Mummified Human Tissue from Northern Chile. American Journal of Physical Anthropology 108:401–407.

Guhl, F., Jaramillo, C., Yockteng, R., Vallejo, GA., and Cardenas–Arroyo, F., 1997   Trypanosoma cruzi DNA in Human Mummies. Lancet 349:1370.

Guichon, R. A., Suby, J. A., Casali, R., and Fugassa, M. H., 2006   Health at the Time of Native–European Contact in Southern Patagonia: First Steps, Results, and Prospects. Memorias do Instituto Oswaldo Cruz 101 Suppl 2:97–105.

Han, E. T., Guk, S. M., Kim, J. L., Jeong, H. J., Kim, S. N., and Chai, J. Y., 2003   Detection of Parasite Eggs from Archaeological Excavations in the Republic of Korea. Memórias do Instituto Oswaldo Cruz 98(Suppl. 1):123–126.

Hawdon, J. M., and Johnston, S. A., 1996   Hookworms in the Americas: An Alternative to Trans-Pacific Contact. Parasitology Today, 12:72–74.

Heizer, R. F., and Napton, L. K., 1969   Biological and Cultural Evidence From Prehistoric Human Coprolites. Science, 165:563–568.

Herrmann, B., 1985   Parasitologisch-Epidemiologische Auswertungen Mittelalterliche Kloaken. Zeitschrift für Archäologie des Mittelalters 13:131–161.

Herrmann, B., and Schulz, U., 1986   Parasitologische Untersuchungen eines Spätmittelalterlich–Frühneuzeitlichen Kloakeninhaltes aus der Fronerei auf dem Schrangen in Lübeck. Lübecker Schriften zur Archäologie und Kulturgeschichte 12:167–172.

Hoberg, E. P., 2006   Phylogeny of Taenia: Species Definitions and Origins of Human Parasites. Parasitology International 55(Suppl):s23–s30.

Holland, T. D., and O'Brien, M. J., 1997   Parasites, Porotic Hyperostosis, and the Implications of Changing Perspectives. American Antiquity 62(2):183–193.

Horne, P. D., 1979   Head Lice from an Aleutian Mummy. Paleopathology Newsletter 25:7–8.

Iñiguez, A. M., Reinhard, K. J., Araujo, A., Ferreira, L. F., and Vicente, A. C. P., 2003   Enterobius vermicularis: Ancient DNA From North and South American Human Coprolites. Memórias do Instituto Oswaldo Cruz 98 (Suppl.1):67–69.

Iñiguez, A. M., Reinhard, K., Gonçalves, M. L. C., Ferreira, L. F., Araújo, A., and Vicente, A. C. P., 2006   SL1 RNA Gene Recovery From *Enterobius vermicularis* Ancient DNA in Pre–Columbian Human Coprolites. International Journal for Parasitology 36(13):1419–1425.

Insoll, T., and Hutchins, E., 2005   The Archaeology of Disease: Mollusks as Potential Disease Indicators in Bahrain. World Archaeology 37(4):579–588.

Jones, A. K. G., 1982a   Recent Finds of Intestinal Parasite Ova at York, England. Papers on Paleopathology, 4th European Members Meeting. Middelburg, Antwerpen, p. 7.

Jones, A. K. G., 1982b   Human Parasite Remains: Prospects for a Quantitative Approach. *In* Environmental Archaeology in the Urban Context. A. R. Hall and H. K. Kenward, eds. pp. 66–70. The Council for British Archaeology, Research Report No. 1.

Jones, A. K. G., 1985   Trichurid Ova in Archaeological Deposits: Their Value as Indicators of Ancient Feces. *In* Paleobiological Investigations: Research Design, Methods and Data Analysis. N. J. R. Fieller, D. D. Gilbertson and N. G. A. Ralph, eds. pp. 105–114. BAR International Series 266. Oxford:British Archaeological Reports.

Jones, A. K. G., and Nicholson, C., 1988a   Recent Finds of *Trichuris* and *Ascaris* Ova From Britain. Paleopatol News 62:5–6.

Jones, A. K. G., Hutchinson, A. R., and Nicholson, C., 1988b   The Worms of Roman Horses and Other Finds of Intestinal Parasite Eggs from Unpromising Deposits. Antiquity 62:275–276.

Kenward, H., 2001   Pubic Lice in Roman and Medieval Britain. Trends in Parasitology 17(4):167–168.

Kliks, M. M., 1983   Paleoparasitology: On the Origins and Impact of Human–Helminth Relationships. *In* Human Ecology and Infectious Disease. N. A. Croll and J. H. Cross (eds.), pp. 291–313. Academic Press.

Kobbert, M. J., 2005   Bernstein – Fenster in die Urzeit. Goettingen.

Krassilov, V., 2008   Mine and Gall Predation As Top–Down Regulation in Plant–Insect Systems from the Cretaceous of Negev, Israel. Paleogeography, Paleoclimatology, and Paleoecology, 261:261–269.

Krause, J., Dear, P. H., Pollack, J. L., Slatkin, M., Spriggs, H., Barnes, I., Lister, A. M., Ebersberger, I., Pääbo, and S., Hofreiter, M., 2006   Multiplex Amplification of the Mammoth Mitochondrial Genome and the Evolution of Elephantidae. Nature 439:734–737.

Lambrecht, F. L., 1980   Paleoecology of Tsetse Flies and Sleeping Sickness in Africa. Proceedings of the American Philosophical Society 124(5):367–85.

Le Bailly, M., and Bouchet, F., 2010   Ancient Dicrocoeliosis: Occurrence, Distribution and Migration. Acta Tropica 115:175–80.

Le Bailly, M., Gonçalves, M. L. C., Harter-Lailheugue, S., Prodéo, F., Araujo, A., and Bouchet, F., 2008   New Finding of *Giardia intestinalis* (Eukaryote, Metamonad) in Old World Archaeological Site Using Immunofluorescence and Enzyme–Linked Immunosorbent Assays. Memórias do Instituto Oswaldo Cruz 103:298–300.

Le Bailly, M., Leuzinger, U., and Bouchet, F., 2003   Dioctophymidae Eggs in Coprolites From Neolithic Site of Arbon–Bleiche 3 (Switzerland). Journal of Parasitology 89:1073–6.

Le Bailly, M., Leuzinger, U., Schlichtherle, H., and Bouchet, F., 2005   *Diphyllobothrium*: Neolithic parasite? The Journal of Parasitology 91:957–959.

Leles, D., Araújo, A., Vicente, A. C., Iniguez, A. M., 2009   Molecular Diagnosis of Ascariasis From Human Feces and Description of a New *Ascaris* sp. Genotype in Brazil. Veterinary Parasitology 163:167–170.

Lima, V. S., Iniguez, A. M., Otsuki, K., Fernando Ferreira, L., Araújo, A., Vicente, A. C., and Jansen, A. M., 2008   Chagas Disease in Ancient Hunter-Gatherer Populations, Brazil. Emerging Infectious Diseases 14:1001–1002.

Loreille, O., and Bouchet, F., 2003   Evolution of Ascariasis in Humans and Pigs: A Multi–Disciplinary Approach. Memorias do Instituto Oswaldo Cruz 98 Suppl 1:39–46.

Manter, H. W., 1967   Some Aspects of the Geographical Distribution of Parasites. The Journal of Parasitology, 53:2–9.

Matsui, A., Kanehara, M., and Kanehara, M., 2003   Paleoparasitology in Japan—Discovery of Toilet features. Memorias do Instituto Oswaldo Cruz 98 Suppl 1:127–136.

Margulies, M., Egholm, M., Altman, W. E. et al., 2006   Genome Sequencing in Microfabricated High-Density Picolitre Reactors. Nature 437:376–380.

Martinson, E., Reinhard, K. J., Buikstra, J. E., and Dittmar,K., 2003   Pathoecology of Chiribaya Parasitism. Memorias do Instituto do Oswaldo Cruz 98:195–205.

Miranda Chaves, S. A., and Reinhard, K. J., 2006   Critical Analysis of Coprolite Evidence of Medicinal Plant Use in Piaui, Brazil. Paleogeography, Paleoclimatology, Paleoecology 237(1):110–118.

Mitchell, P. D., Stern, E., and Tepper, Y., 2008   Dysentery in the Crusader Kingdom of Jerusalem: An ELISA Analysis of Two Medieval Latrines in the City of Acre (Israel). Journal of Archaeological Science 35(7):1849–1853.

Montenegro, A., Araújo, A., Hetherington, R., Ferreira, L. F., Weaver, B., and Eby, M., 2006   Parasites, Paleoclimate, and the Peopling of the Americas: Using the Hookworm to Time the Clovis Migration. Current Anthropology 47(1):193–200.

Moore, J. G., Fry, G. F., and Englert E., Jr., 1969   Thorny-Headed Worm Infection in North American Prehistoric Man. Science 163:1324–1325.

Mumcuoglu, Y. K., and Zias, J., 1988   Head Lice, *Pediculus humanus capitis* (Anoplura, Pediculidae) from Hair Combs Excavated in Israel and Dated From the First Century BC to the Eighth Century AD. Journal of Medical Entomology 25(3):545–547.

Mumcuoglu, Y. K., Zias, J., Tarshis, M., Lavi, M., and Stiebel. G. D., 2003   Body Louse Remains Found in Textiles Excavated at Masada, Israel. Journal of Medical Entomology 40(4):585–587.

Neumann, C., and Wisshak, M., 2009   Gastropod Parasitism on Late Cretaceous to Early Paleocene Holasteroid Echinoids – Evidence From *Oichnus halo* isp. n. Palaeogeography, Palaeoclimatology, Palaeoecology 284:115–119.

Neumann, C., Kardjilov, N., and Hilger, A., 2009   Endosymbionts of Fossil *Echinodermata* Using Neutron and X–ray Computed Tomography. *In* BENSC Experimental Reports 2008. A. Rodig, A. Brandt, and H. A. Graf, eds. pp. 622:252. Berlin: Berichte des Helmholtz-Zentrums HMI-B.

Nozais, J. P., 2003   The Origin and Dispersion of Human Parasitic Diseases in the Old World (Africa, Europe and Madagascar). Memórias do Instituto Oswaldo Cruz 98(Suppl 1):13–19.

Pääbo, S., Poinar, H., Serre, D., Jaenicke–Després, V., Hebler, J., Rohland, N., Kuch, M., Krause, J., Vigilant, L., and Hofreiter, M., 2004   Genetic Analyses From Ancient DNA. Annual Review of Genetics 38:645–679.

Pavlovsky, E. N., 1966   Natural Nidality of Transmissible Diseases with Special Reference to the Landscape Ecology of Zooanthroponoses. F. K. Plous, Jr. trans. Urbana: University of Illinois Press.

Pike, A. W., 1975   Parasite Eggs. *In* Excavations in Southampton. C. Platt and R. Coleman-Smith, eds. pp. 347–348. Leicester: Leicester University Press.

Poinar, G. O., 1984   Fossil Evidence of Nematode Parasitism. Revue Nematologique, 7(2): 201–203.

Poinar, G. O., and Brown, A. E., 2003   A New Genus of Hard Ticks in Cretaceous Burmese Amber (Acari: Ixodida: Ixodidae). Systematic Parasitology, 54:199–205.

Poinar, G. O., and Milki, R., 2001   Lebanese Amber. Corvallis, Oregon: Oregon State University Press.

Poinar, G. O., and Poinar, R., 2005   Fossil Evidence of Insect Pathogens. Journal of Invertebrate Pathology, 89:243–250.

Poinar, G. O., and Poinar, R., 2007   What Bugged the Dinosaurs?: Insects, Disease, and Death in the Cretaceous. Princeton: Princeton University Press.

Raoult, D., Reed, D. L., Dittmar, K., Kirchman, J. J., Rolain, J. M., Guillen, S., and Light. J. E., 2008   Molecular Identification of Lice from Pre–Columbian Mummies. Journal of Infectious Diseases 197(4):535–543.

Reinhard, K. J., 1988   Cultural Ecology of Prehistoric Parasitism on the Colorado Plateau as Evidenced By Coprology. American Journal of Physical Anthropology 77:355–366.

Reinhard, K. J., 1990   Archaeoparasitology in North America. American Journal of Physical Anthropology 82:145–162.

Reinhard, K. J., 1992a   Parasitology as an Interpretive Tool in Archaeology. American Antiquity 57:231–245.

Reinhard, K. J., 1992b   Patterns of Diet, Parasitism, and Anemia in Prehistoric West North America. *In* Diet, Demography, and Disease: Changing Perspectives on Anemia. P. Stuart-Macadam, and S. Kent, eds. pp. 219–258. Aldine de Gruyter: New York,

Reinhard, K. J., 1996   Parasite Ecology of Two Anasazi Villages. *In* Case Studies in Environmental Archaeology. E. J. Reitz, L. A. Newson, and S. J. Scudder, eds. Plenum Press: New York.

Reinhard, K. J., 2006   A Coprological View of Anasazi Cannibalism. American Scientist 94:254–262.

Reinhard, K. J., 2007   Pathoecology of Two Anasazi Villages. *In* Case Studies in Environmental Archaeology, 2nd edition. E. J. Reitz, ed. pp. 191–210. Plenum Press: New York.

Reinhard, K. J., 2008   Parasite Pathoecology of Chacoan Great Houses: The Healthiest and Wormiest Ancestral Puebloan. *In* Chaco's Northern Prodigies: Salmon, Aztec, and the Ascendancy of the Middle San Juan Region after AD 1100. P. F. Reed, ed. pp. 86–95. Salt Lake City: University of Utah Press.

Reinhard, K. J., Ambler, J. R., and McGuffie, M., 1985   Diet and Parasitism at Dust Devil Cave. American Antiquity 50(4):819–824.

Reinhard, K. J., Anderson, G. A., and Hevly, R. H., 1987   Helminth Remains From Prehistoric Coprolites on the Colorado Plateau. Journal of Parasitology 73:630–639.

Reinhard, K. J., Araújo, A., Ferreira, L. F., and Coimbra, C., 2001   American Hookworm Antiquity. Medical Anthropology 20:96–101.

Reinhard, K. J., Araújo, A., Ferreira, L. F., and Herrmann, B., 1986   Recovery of Parasite Remains From Coprolites and Latrines: Aspects of Paleoparasitological Technique. Homo 37:217–239.

Reinhard, K. J., Araújo, A., Sianto, L., Costello, J. G., and Swope, K., 2008   Chinese Liver Flukes in Latrine Sediments from Wong Nim's Property, San Bernardino, California: Archaeoparasitology of the Caltrans District Headquarters. Journal of Parasitology 94(1):300–303.

Reinhard, K. J., and Bryant, V. M., 2008   Pathoecology and the Future of Coprolite Studies. *In* Reanalysis and Reinterpretation in Southwestern Bioarchaeology. A. L. W. Stodder, ed. pp. 199–216. Tempe: Arizona State University Press.

Reinhard, K. J., and Buikstra, J., 2003   Louse Infestation of the Chiribaya Culture, Southern Peru: Variation in Prevalence by Age and Sex. Memorias do Instituto do Oswaldo Cruz 98:173–179.

Reinhard, K. J., Fink, M., and Skiles, J., 2003   A Case of Megacolon in Rio Grande Valley as a Possible Case of Chagas Disease. Memorias do Instituto do Oswaldo Cruz 98:165–172.

Reinhard, K. J, and Urban, O., 2003   Diagnosing Ancient Diphyllobothriasis from Chinchorro Mummies. Memorias do Instituto do Oswaldo Cruz 98:191–193.

Reinhard, K. R., 1974   Relation of Climate to Epidemiology of Infectious Disease Among Arctic Populations. Alaska Medicine 16:25–30.

Rocha, G. C., Harter-Lailheugue, S., Le Bailly, M., Araújo, A., Ferreira, L. F., da Serra-Freire, N. M., and Bouchet, F., 2006   Paleoparasitological Remains Revealed by Seven Historic Contexts from "Place d'Armes", Namur, Belgium. Memórias do Instituto Oswaldo Cruz 101:43–52.

Rothhammer, F., Allison, M. J., Nuñez, L., Staden, V., and Arriza, B., 1985   Chagas Disease in Pre-Columbian South America. American Journal of Physical Anthropology 68:495–498.

Ruffer, M. A., 1910   Note On the Presence of *Bilharzia haematobia* in Egyptian Mummies of the Twentieth Dynasty (1250–1000 B.C.). British Medical Journal 1:16.

Sadler, J. P., 1990   Records of Ectoparasites on Humans and Sheep from Viking-Age Deposits in the Former Western Settlement of Greenland. Journal of Medical Entomology 27(4): 628–631.

Sallares, R., 2002   Malaria and Rome: A History of Malaria in Ancient Italy. Oxford University Press: Oxford, England.

Samuels, R., 1965   Parasitological Study of Long-Dried Fecal Samples. *In* Contributions of the Wetherill Mesa Archaeological Project. D. Osborne, and B. S. Katz, eds. pp. 175–179. Society for American Archaeology, Memoirs, 19.

Sandison, A. T., 1967   Parasitic Diseases. *In* Diseases in Antiquity. D. Brothwell, and A. T. Sandison, eds. pp. 178–183. Springfield, IL: C. C. Thomas.

Santoro, C., Vinton, S. D., and Reinhard, K., 2003   Inca Expansion and Parasitism in the Lluta Valley: Preliminary Data. Memorias do Instituto do Oswaldo Cruz 98:161–163.

Schmidt, G. D., 1971   Acanthocephalan Infections of Man, With Two New Records. Journal of Parasitology 57:582–584.

Seo, M., Guk, S. M., Kim, J., Chai, J. Y., Bok, G. D., Park, S. S., Oh, C. S., Kim, M. J., Yi, Y. S., Shin, M. H., Kang, I. U., and Shin, D. H., 2007   Paleoparasitological Report on the Stool From a Medieval Child Mummy in Yangju, Korea. The Journal of Parasitology 93:598–592.

Shin, D. H., Lim, D. S., Choi, K. J., Oh, C. S., Kim, M. J., Lee, I. S., Kim, S. B., Shin, J. E., Bok, G. D., Chai, J. Y., and Seo. M., 2009   Scanning Electron Microscope Study of Ancient Parasite Eggs Recovered From Korean Mummies of the Joseon Dynasty. The Journal of Parasitology 95(1):137–145.

Sianto, L., Reinhard, K. J., Gonçalves, M. L. C., and Araújo. A., 2005   The Finding of *Echinostoma* (Trematoda: Digenea) and Hookworm Eggs in Coprolites Collected from a Brazilian Mummified Body Dated of 600–1,200 Years Before Present. Journal of Parasitology 91:972–975.

Sprent, J. F. A., 1969   Evolutionary Aspects of Immunity of Zooparasitic Infections. *In* Immunity to Parasitic Animals. G. J. Jackson, ed. pp. 13–64. New York: Appleton.

Stodder, A. W., and Martin, D. L., 1992   Health and Disease in the Southwest Before and After Spanish Conquest. *In* Disease and Demography in the Americas. J. W. Verano, and D. H. Ubelaker, eds. pp. 55–73. Wiley: New York.

Tarlington, R. E., Meers, J., and Young, P. R., 2006   Retroviral Invasion of the Koala Genome. Nature 442:79–81.

Taylor, E. L., 1955   Parasitic Helminths in Mediaeval Remains. Veterinary Record 67:216–218.

Walker, P. L., 1985   Anemia Among Prehistoric Indians of the American Southwest. *In* Health and Disease in the Prehistoric Southwest. C. F. Merbs, and R. J. Miller, eds. pp. 139–154. Tempe: Arizona State University Anthropological Research Papers 34.

Walker, P. L., 1986   Porotic Hyperostosis in a Marine-Dependent California Indian Population. American Journal of Physical Anthropology 69:345–354.

Walker, P. L., Bathurst, R. R., Richman, R., Gjerdrum, T., and Andrushko, V. A., 2009   The Causes of Porotic Hyperostosis and Cribra Orbitalia: A Reappraisal of the Iron-Deficiency-Anemia Hypothesis. American Journal of Physical Anthropology 139:109–125.

Weischat, W., and Wichard, W., 1998   Atlas der Planzen und Tiere im Baltischen Bernstein. Germany.

Whiting, M. F., Whiting, A. S, Hastriter, M., and Dittmar, K., 2008   A Molecular Phylogeny of Fleas (Insecta: Siphonaptera): Origins and Host Associations. Cladistics 24:677–707.

Witalinski, W., 2000   *Aclerogamasus stenocornis* sp. n., A Fossil Mite From The Baltic Amber (Acari: Gamasida: Parasitidae). Genus 11(4):619–626.

Zink, A. R., Spigelman, M., Schraut, B., Greenblatt, C. L., Nerlich, A. G., and Donoghue, H. D., 2006   Leishmaniasis in Ancient Egypt and Upper Nubia. Emerging Infectious Diseases 12(10):1616–1617.

11 More Than Just Mad Cows: Exploring Human–Animal Relationships Through Animal Paleopathology

Beth Upex and Keith Dobney

INTRODUCTION

From the earliest hunters, through the advent of domestication to the most recent outbreak of swine flu, animals and humans have had a long and complex relationship. Anthropology and archaeology (more specifically the disciplines of zooarchaeology and paleopathology) provide a key temporal framework for the exploration and understanding of this relationship and the wide range of questions that are encompassed by it. Paleopathological studies in ancient human and nonhuman species can be broadly defined as the study of past health, disease and injury, through the analysis of calcified tissue, primarily bones and teeth. However, compared to human paleopathology, animal paleopathology has received little or no serious attention within the wider fields of archaeology and even zooarchaeology. The reasons for this are varied (and discussed in more detail below), but are principally related to the more complex, intensive taphonomic processes that animal bone assemblages often undergo compared to human remains. As we hope this brief chapter will highlight, the study of animal paleopathology has the potential to add a great deal of information to our understanding of the past. Whilst perhaps the most obvious area of study relates to the spread and management of human diseases, this is just the "tip of the iceberg".

A Companion to Paleopathology, First Edition. Edited by Anne L. Grauer.
© 2012 John Wiley & Sons, Ltd. Published 2016 by John Wiley & Sons, Ltd.

Animal Paleopathology: A Brief History

The history of animal paleopathological studies is considerably shorter and less glamorous than its illustrious cousin, human paleopathology, with many of the early publications focusing upon simple descriptive case studies. For example, Harcourt (1967), an experienced veterinarian, published the first detailed description of osteoarthritis in a Romano-British dog skeleton. However, the potential of animal paleopathology for exploring broader questions about human–animal relationships in the past was first recognized by the pioneer of the field, Don Brothwell. He undertook and published the first review of pathological conditions affecting animal bones and teeth, which focused upon Pleistocene mammals (Brothwell 1969). Soon afterwards, Harcourt (1971) published what was probably the first systematic study of pathology present in zooarchaeological material (O'Connor 2008:166), whilst Siegel (1976) provided one of the first attempts at describing specimens, identifying conditions and classifying disorders in disarticulated archaeological animal bones. There followed the publication of a seminal and (to this day) unprecedented scholarly account of animal paleopathology by Baker and Brothwell (1980), which provides the benchmark for the subject. Their book contains a series of chapters dedicated to the discussion of a wide variety of pathological conditions likely to be found (and identified) in the bones and teeth of animals from archaeological and paleontological sites, in addition to proposed classifications of disorders (e.g. joint arthropathies) that are still widely used today. The textbook became (and still remains) the only book dedicated to this subject. Tragically, the publisher commissioned only one print run and, due to initial slow sales, the book was remaindered, with spare copies being subsequently pulped.

More recent works in animal paleopathology are either review-style papers that investigate specific pathological conditions such as joint arthropathies (Groot 2005; Joannsen 2005; Bartosiewicz 2008b; O'Connor 2008), fractures (Urdrescu 2005; Bartosiewicz 2008a; Groot 2008) and dental conditions (Ervynck and Dobney 1999; Dobney and Ervynck 2000, 2002; Teegan 2005), or are site-based pathology reports where some (but not all) merely describe rather than interpret the evidence (e.g. Brothwell 1995; Fabis 2004; Csippan and Daroczi-Szabo, 2008; Daroczi-Szabo 2008; Miklikova 2008). For a field with so much potential, very little has been realized since the publication of the Baker and Brothwell volume over thirty years ago, with O'Connor (2000:98) describing the field of animal paleopathology as "an inchoate discipline, pursued by a relatively small number of analysts."

The development of the discipline has been slow and hindered by a plethora of methodological and taphonomic problems. This has contributed to the field of animal pathology being relegated to an "interesting" but rather "quirky" category of zooarchaeology, with the majority of publications limited to little more than "just-so stories," descriptions of "interesting or unusual" specimens, but lacking detailed discussion, interpretation or occasionally even differential diagnosis. The "stamp-collecting" approach has further led to a lack of synthesis and, rather more worryingly, a lack of basic quantitative information on the occurrence of animal pathology in the past. However, this situation has developed into a "catch 22," since the publication of interesting or unusual specimens and basic prevalence rates is an essential prerequisite to the development and publication of larger syntheses and more detailed analyses.

In an effort to remedy the situation in 1999, the International Council for Archaeozoology (ICAZ) set up the Animal Pathology Working Group (APWG) with the aims of creating more integration between paleopathological data and other forms of zooarchaeological evidence, standardizing recording practices and focusing on geographic and diachronic trends, and understanding the underlying biological processes and consequences of different types of pathology.

Years after the APWG was set up, there is a growing awareness of both the potential of animal paleopathology studies and the need for systematic and quantitative recording. However, the lack of standardization in recording and documenting pathological conditions remains one of the biggest problems facing the advancement of the discipline. Unlike the guides and protocols established in human paleopathology (e.g. Buikstra and Uberlaker 1994; Brickley and McKinley 2004), there remains a general lack of guidelines and consistency regarding the recording and documentation of pathology in nonhuman material, although this balance is slowly being redressed (e.g. Brothwell et al. 1996; Bartosiewicz et al. 1997; Dobney and Ervynck 1998; Vann and Thomas 2006). The APWG has now published three collections of papers resulting from conferences; two monographs (Davis et al. 2005 and Miklikova and Thomas 2008), and a collection of papers published in the journal *Veterinarija and Zootechnika* (2008). General volumes on paleopathology are also beginning to include sections on animal paleopathology (e.g. Grupe et al. 2008, and this volume), although the analysis of animal remains is still often allocated less space, and occasionally viewed purely as a means of furthering the understanding of human diseases, rather than a valid branch of paleopathology in itself.

PROBLEMS

The field of animal paleopathology is hampered by variety of problems, some of which are common to those working on human remains, but also some more specific to nonhuman material. Thomas and Mainland (2005) highlight the problem common to all paleopathological studies—that of multiple etiologies, i.e. the often ubiquitous nonspecific nature of bony changes and the difficulties of identifying specific disease states. Specific taphonomic factors associated with the disarticulation, butchery, consumption and final disposal of animal bones further complicate the situation. For example, investigating the impact of a poorly healed long-bone fracture on the joints of the opposing limb, or identifying conditions such as tuberculosis, which rely on the presence of several features across the skeleton, are compromised when the skeletal material is severely disarticulated and co-mingled.

More fundamental is the general lack of research into the impact of disease on the skeletons and teeth of modern animal populations, largely due to the (at least in managed domestic populations) remedial action or slaughter of animals early on in the disease process. As a result, pathological conditions rarely advance to the point where diagnostic changes to calcified tissue become apparent. Consequently, there remains a lack of primary modern comparative material which can be used to study the impact of chronic pathological conditions on the bones and teeth of animals. As a further limitation, modern veterinary literature tends to focus on the identification of soft tissue conditions and the extent to which they are treatable. The majority of datasets that do exist are from studies that would not be considered ethically acceptable today. However, they provide some of the most important comparative data for animal paleopathological studies (e.g. Mellanby 1929; McCance et al. 1961; Suckling et al. 1983, 1986).

As previously mentioned, complex predepositional taphonomic processes that animal bone assemblages undergo add to the complexity of identification and interpretation of skeletal and dental pathology. Complications associated with food selection, processing, cooking and disposal deeply impact our understanding of pathology in faunal remains (Bartosiewicz 2008c). Similarly, burial biases (e.g. the unlikelihood of sick animals being included in the food chain, and the subsequent use of varied disposal methods such as separate (complete) burial, cremation, or feeding of these remains to other animals such as dogs and pigs) render the survival of some forms of hard-tissue pathology more likely than others. For example, joint arthropathies, leading to the formation of dense compact bone, are far more likely to survive complex taphonomic processes than fragile newly developed woven bone relating to new breaks or infections. On the other hand, articular ends of bones are less dense, and these key areas are often the target of butchery practices associated with disarticulation of the skeleton, as well as the most likely region to be severely gnawed by dogs and other carnivores.

As a further complicating issue, the sheer number and range of vertebrate taxa potentially represented in zooarchaeological assemblages (mammals, birds, reptiles, amphibians and fish), along with the presence of wild and domestic versions of some species (and potentially even different varieties of domesticates), render the identification of pathological conditions difficult, particularly with regard to identifying abnormal versus normal development. There is a huge degree of variation across the animal kingdom and features which are considered "normal" for some species are considered "abnormal" for others. For example, commonly observed coarse pitting in the skulls of pinnipeds (seals) is considered normal, yet if seen in primates could only be interpreted as pathological (Baker and Brothwell 1980:192).

WHAT CAN WE LEARN FROM ANIMAL PALEOPATHOLOGY?

Given the problems inherent to the study of vertebrate paleopathology, it is not surprising that the development of the field has been slow and tentative. Animal paleopathology, however, provides a little-explored window into a better understanding of past communities, cultural attitudes, economies, agricultural practices, environmental change and animal management. It also, importantly, provides insight into human health and disease.

Human health and disease

Whilst the exploitation of animals over the last 10,000 years has led to massive advancements in agriculture and transport, it has come at a significant demographic cost to human populations. Even with the development of modern vaccines and medical intervention, outbreaks of zoonotic diseases remain fatal, with their impact being felt across the globe. According to the World Health Organization (see Pandemic H1N1 2009–Update 108: http://www.who.int/csr/don/2010_07_09/en/index.html), the most recent zoonotic disease pandemic of Swine Flu, (H1N1) affected 214 countries with over 18,311 confirmed deaths. Although causing worldwide concern (and some hysteria), these are small numbers compared with the AIDS epidemic sweeping Africa, the great influenza pandemic of the early 20th century, or the early-mid 14th century depopulation of western Europe by the "Great Plague" or "Black Death." There can be little doubt that the transmission of infectious zoonotic diseases between humans and animals played a dramatic (and perhaps pivotal) role in the evolution and development of past societies.

Tuberculosis   In terms of understanding the impact of animals on past human health, tuberculosis is probably the most studied infectious disease that directly affects calcified tissue (see Chapter 24 by Roberts, this volume, for discussion of this disease in human populations). According to Roberts and Buikstra (2003), tuberculosis is responsible for around 5,000 deaths per day, and of all the infectious diseases it is the most common cause of deaths in adults. Tuberculosis is an enzootic disease in mammals, but occurs most commonly in cattle, pigs and carnivores. Many species have specific versions of the bacteria, for example *Mycobacterium tuberculosis* in humans, *Mycobacterium avium* in birds and *Mycobacterium bovis* in cattle, although many of these mycobacteria are naturally transmissible to other hosts.

The phylogenetic relationship of these bacteria is still unclear. It was originally assumed that *M. tuberculosis* evolved directly from *M. bovis*, most likely during the early domestication of cattle (Haas and Haas 1996; Diamond 1997; Stead 1997; Wolfe et al. 2007). However, more recent genomic studies challenge this assumption, and demonstrate that it is unlikely that *M. tuberculosis* arose from a bovine strain (Gordon et al. 1999; Brosch et al. 2001; Gutacker et al. 2006). It has also been suggested that *M. bovis* may be an example of a reverse zoonosis, i.e. the spread of the human strain of *M. tuberculosis* to cattle, which then evolved into *M. bovis* (Gibbons 2008). However, whilst it is probable that the most recent common ancestor of *M. bovis* was a human-adapted strain of the disease, it appears unlikely that the founding strain of tuberculosis (TB) in cattle was derived directly from humans. The ancestor of the disease in cattle is more likely to have been another animal-adapted strain (Smith et al. 2009). Although the new genetic studies suggest a more complex and deeper evolutionary trajectory for the human and bovine forms, it is still largely assumed that the first transmissions of the disease occurred in the Old World, as cattle were first domesticated in the Levant and Mesopotamia in the 8th millennium B.C. However, smaller Bovidae species such as sheep and goats were domesticated even earlier and, although less susceptible to TB than cattle, may have also played a significant role as a disease vector to humans in these early pastoralist societies.

Primary TB infection in cattle is often via the respiratory tract (95 percent of cases) and occasionally through the skin. The crowded conditions often seen on dairy farms today make dairy cattle far more susceptible to this form of the disease than beef cattle, with the disease most commonly being spread to humans through the consumption of infected milk products (Ortner 2003). Studies in the early 20th century by Nocard and Leclainche in 1903 and Eber in 1932 (Lignereux and Peters 1999) showed that bovine tuberculosis was highly prevalent in cattle populations in Europe, with 17.5 percent to 40 percent of animals in France and Germany (between the beginning of the 20th century and 1927) being slaughtered due to infection. Lignereux and Peters (1999) provide a detailed synthesis of the prevalence and occurrence of lesions in some modern domestic animal populations, with bone lesions on the ribs, vertebra and sternum observed in 5–9 percent of infected cattle at the beginning of the 20th century. Sheep and goats showed a 0.001–0.2 percent occurrence in Germany at the end of the 19th century, and in pigs, 2–5 percent of the population was infected with TB at end of the 20th century. In the case of pigs it was also noted that a 2–5 percent population infection rate was considerably higher than in earlier periods, perhaps due to new farming methods which supplemented pigs' diets with milk (Lignereux and Peters 1999). As in human populations, only a small percentage of TB-infected animals show evidence of bony changes, suggesting that the identified prevalence rates from archaeological remains represent a fraction of the true extent of frequency of this condition in the past.

In spite of the amount of research conducted on TB in humans, little work has been carried out on the identification of the condition in animals. Lignereux and Peters (1999) blame this situation on the lack of pathognomonic lesions caused by TB and a lack of modern comparative material, which makes accurate diagnosis especially difficult in disarticulated remains. An exception to this are claims for the presence of tuberculosis in a Pleistocene bison, arguably identified through both DNA and macroscopic analysis (Rothschild and Martin 2001; 2003; 2006). Rothchild and Laub (2006) assert that tuberculosis was a disease of pandemic proportion in Pleistocene Mastodon populations, contributing to their eventual extinction. Many researchers, however, remain highly skeptical of these conclusions, questioning the likelihood that aDNA of mycobacterial pathogens could preserve in the bones and teeth of fossils in a form which could be readily replicated and sequenced. In fact the only systematic study of aDNA bacterial pathogen survival in recent museum specimens (known to have suffered from infectious disease) obtained "no reproducible evidence of surviving pathogen DNA, despite the use of extraction and PCR-amplification methods determined to be highly sensitive" (Barnes and Thomas 2005:645). Clearly, interpretations of disease in the past based on such data should be treated with a high degree of caution.

Alternatively, two more rigorous studies into ancient TB (and the potential cross infection of cattle and humans strain) exist. Taylor et al. (1999) studied two human skeletons from medieval London, both of which had skeletal evidence of tuberculosis. Using aDNA analysis they claimed that both individuals had suffered from *M. tuberculosis* (the human strain); this result was not unexpected as both individuals came from urban areas with assumed minimal contact with cattle. Mays (2001) investigated TB in humans from Wharram Percy (a medieval rural farming community), where it was argued (given the rural, farm-based economy of the site) that some of the individuals affected by tuberculosis might have been suffering from the bovine strain. However, the subsequent analysis indicated that the infection was due to *M. tuberculosis* and not, as expected, *M. bovis*. Three cattle ribs associated with the site, which displayed evidence of periostitis, were also tested for mycobacterium strains. All came back negative for tuberculosis. Mays (2001) suggest that this may be the result of taphonomic processes, for example cooking, or, alternatively, that the lesions were the result of a different infectious disease.

Other infections   Other than tuberculosis, there are very few zoonotic infections that are currently identifiable from animal bone assemblages. Although other zoonoses surely existed in the past, such as rabies, ebola, plague, and anthrax, they leave no physical traces in dry bones. However, in 1967, Lewis reported that nematode infections in mustelid skulls can cause perforations (Baker and Brothwell 1980:180). This raises the possibility of finding pathological evidence of these infections in archaeological material. An excessive parasitic load carried by domestic or wild animals, causing severe malnutrition, as seen in animal models studied by Suckling et al. (1983; 1986), might also be inferred from enamel hypoplasia (a deficiency in the thickness and quality of dental enamel formed during crown development), or growth retardation in young animals. However, it would be unwise to leap to the assumption that parasitic infestation was the cause of enamel hypoplasia in archaeological assemblages without the added presence of large numbers of related and preserved parasite ova (which is highly unlikely).

The only other zoonotic disease that has been explored in the animal paleopathology literature is brucellosis (caused by the bacteria *Brucella abortus*), a condition that primarily affects the respiratory tract of cattle but can have a series of debilitating effects on humans if cross-infection occurs. According to Baker and Brothwell (1980), the condition often affects the joints in horses and cattle, but does not produce any bony changes that make the condition diagnosable. However, it does occasionally affect the lumbar and cervical vertebral bodies, causing irregular bone erosion, as well as extensive periosteal new bone formation (Baker and Brothwell 1980). Unfortunately, separating the condition from bony lesions produced by other diseases (such as TB) is difficult. For example, an Iron Age horse skeleton excavated from Basingstoke (UK), displayed evidence of a systemic infection which was suggested to be caused by either tuberculosis or brucellosis, or both (Bendrey 2008a). The application of aDNA techniques to this skeleton (along with another articulated Iron Age horse skeleton from Kent (UK) displaying similar lesions), failed to recover equine aDNA or specific pathogen DNA related to either disease (Bendrey et al. 2008).

The impact of domestication on human health

Although animals are an important reservoir for human infection and disease, using their fossil remains to understand the impact and spread of specific zoonotic infections remains fraught with difficulties. Regardless, zooarchaeological analysis can shed some light on the general health and incidence of disease in human populations. The advent of animal domestication, whilst bringing many advantages to humans, also brought many (often deleterious) changes to patterns of human health and disease. Horwitz and Smith (2000) studied the link between domestication and the prevalence of zoonotic pathologies in the Levant. Their study highlighted the link between domestication and the spread of disease by demonstrating that animals that have been domesticated for the longest period share more diseases with humans than other species with shorter histories of domestication. For example, dogs—the first domesticated species—share 65 diseases with humans, whereas camels (far more recently domesticated) share just seven (McNeill 1976).

Apart from the rise in zoonotic diseases associated with the advent of more complex human-animal relationships, there also appears to be a concomitant increase in the frequency and intensity of general stress indicators, such as enamel hypoplasia, that can shed light on the domestication process itself. Several studies have been carried out investigating changes in the levels of physiological stress between hunter-gatherer societies and sedentary farming communities from the Levant (Smith et al. 1984), Ecuador (Ubelaker 1984), Peru and Chile (Allison 1984) and Egypt and Nubia (Starling and Stock 2007). All of these studies noted that the occurrence of enamel hypoplasia in humans rose with increased agricultural intensification, implicating the linked increase of factors such as disease and malnutrition with this major biocultural transition.

The impact of domestication on the animals themselves is also visible in the vertebrate faunal record. Work by Ervynck et al. (2001) used simple frequencies of dental enamel defects to assess the domesticated status of pigs from the Neolithic site of Çayönü Tepesi in south-eastern Anatolia, Turkey, a site long accepted to be one of the earliest foci of pig domestication in western Asia. The analysis of linear enamel hypoplasia (in tandem with biometrical evidence, principally measurements of molar tooth crowns) showed a gradual increase in the frequency of hypoplastic defects over time, interpreted as a gradual increase in physiological stress in wild boar and pigs directly associated with the domestication process. The results of this study were further supported by Dobney et al. (2004, 2005, 2007a; Rowley-Conwy and Dobney 2007) in *Sus* remains across Eurasia. They identified consistently high (and varying) frequencies of enamel hypoplasia in archaeological domestic pig samples, compared with much lower values in modern and archaeological wild boar populations from the same geographic locations. They interpreted the high frequency of enamel hypoplasia identified in Neolithic domestic pig populations as the direct consequence of animal husbandry pressures, and the wide variation often observed between them to reflect differences in animal husbandry regimes and ecological conditions

HUMAN EXPLOITATION OF ANIMALS

Animal paleopathology can offer far more than just understanding the impact of animals on human health and disease. In fact, perhaps more detailed and interesting research questions can be asked when the study is reversed to focus on the impact of humans on animal health and disease. Animals have been exploited by humans throughout history and whilst their most obvious use is as a food resource, there are multiple other ways in which animals are utilized, many of which can be inferred by the physical impact on the skeleton.

Traction
Investigation of joint arthropathies in cattle, as evidence for their use as traction beasts, has been one of the most intensely studied topics in animal paleopathology. Bartosiewicz et al. (1997) published the first attempt at systematically classifying and interpreting joint arthropathies in the metapodials and phalanges of cattle and their use in identifying animals used for traction. By studying the skeletonized remains of

eighteen modern Romanian oxen used extensively for traction, comparative data from young cattle not used for traction, and a Roman cattle sample from Belgium, this research provided a systematic baseline for identifying plow oxen in the zooarchaeological record. Their scoring system has subsequently been applied to a number of zooarchaeological cattle studies (De Cupere et al. 2000; Telldahl 2005; Johannsen 2005; Vann 2008; Thomas 2008). Whilst there are obvious problems comparing the presence of joint arthropathies between young and old animals (given that such conditions are both activity and age related phenomena), this type of study on modern material is still essential to the development of standard interpretative baselines for the study of archaeological material.

Whilst the protocol of Bartosiescwiz et al. (1997) has enabled a more robust interpretative framework for better understanding of joint arthropathies in the context of the zooarchaeological evidence for traction, careful consideration of the evidence is still required to avoid over-simplistic interpretation of the data. For example, Telldahl (2005) interprets an increase in cattle arthropathies through time as evidence for the adoption of a deeper (and therefore heavier) type of plow. Whilst this is perhaps a reasonable assumption to make, there are also several other unexplored and equally valid explanations for such a pattern, such as variation in age profiles and cattle body weight between the periods studied.

The need for careful interpretation and consideration of all the available evidence is clearly highlighted by the study of lower limb pathology in 780 adult cattle metapodia from medieval Dudley Castle in England by Thomas (2008). In this study the methodology of Bartosiescwiz et al. (1997) was systematically applied to investigate the occurrence of joint pathologies, in conjunction with standard zooarchaeological analyses and the study of extensive historical documentation about the castle. The dramatic increase seen in lower limb joint arthropathies in the mid-late 14th century could quite simply have been interpreted as evidence of the more intensive use of cattle for traction. However, a careful examination of the associated historical evidence revealed that out-breaks of plague in 14th-century Britain had reduced the population to such an extent that large tracts of land were turned over to pasture and that the use of cattle for plowing was consequentially dramatically decreased.

A chronological study of demographic and biometric evidence from the cattle remains at Dudley Castle also revealed that in the mid-late 14th century cattle were being slaughtered at an older age and that there had been an increase in body size from the earlier period. Although the heavy plow scenario provides an obvious explanation for the arthropathy data, the historical and zooarchaeological evidence here strongly implies that the increase in cattle arthropathies could equally have been linked to the longer life span and heavier body weight of the cattle in the mid-late 14th century.

With careful interpretation and consideration of the broader archaeological evidence, the pathological analysis of zooarchaeological remains can provide intriguing glimpses into the use of cattle for traction. Fabis (2005) studied two complete cattle skeletons found buried together in a pit in a Baden Culture cemetery in Hungry. Based on epiphyseal fusion, dental wear, and metric analysis, both animals were identified as castrated males, with the older being over eight years old and the younger between three and a half to four years old. The older animal displayed evidence of vertebral/sacral pathology as well as broadening of the trochlea in the metapodalis. Interestingly, this occurred on the lateral trochlea of the right metatarsal and on the medial trochlea of the left metatarsal, indicating that when moving forward the animals hind legs were leaning to the right, as would be the case if it was harnessed in a pair and always on the left hand side of the cart. The younger animal, whilst showing no evidence of pathology in the lower limbs or spine, showed marked rotation of the spinous processes, which Fabis (2005) suggested was due to the dorsal muscles on the right of the animal being worked more intensely, and that, given the young age of the animal, this abnormal stress altered its shape. Based on the evidence of the pathologies from both cattle it was concluded the two animals must have been a draught team, with the older individual always harnessed on the left and the younger on the right. This paper provides a good example of the degree of interpretation that is possible when complete animals skeletons are available for study. One fact that has been rather overlooked in the literature but is highlighted by this study is that there are clearly other osseous changes associated with the use of cattle for traction (such as rotation of the spine) and that it is essential that broader pathologies be thoroughly investigated.

Riding

In addition to the impact draught cattle had on early transport and the intensification of early agriculture, the domestication and exploitation of the horse for riding and transportation revolutionized the development and formation of later societies. The difficulties of separating age-related arthropathies from work-related, stress-induced arthropathies are well documented, and in horses there have been several studies that have attempted to identify pathologies related specifically to riding and traction versus those that are purely age-related.

Bendrey (2007) discusses the ossification of ligaments between metapodials, and documents their occurrence in modern, wild, unworked horse populations along with their increasing prevalence with age. A later study by the same author (Bendrey 2008b) focused on modern populations with known life histories and proposed a simple recording system for scoring the ossification of the nuchal ligament. The author suggests that while this musculoskeletal stress marker is clearly age-related, it may also be caused by excessive fast riding, since it has a higher prevalence in racehorses. These studies all provide essential modern baselines for the interpretation of archaeological data.

Using this modern comparative framework, Bendrey et al. (2009) notes the extensive ossification of the nuchal ligament in a horse skeleton from Tuva Republic in Central Asia. This evidence was compounded by the presence of lesions on the occipital bone, interpreted tentatively as the first archaeologically recorded case of poll evil identified in the cranial bones—a condition previously only suggested from lesions in the atlas (Baker and Brothwell 1980; Bendrey et al. 2009). Poll evil results from the inflammation and possible rupture of the supra-atlantal bursa (the small sac that provides a cushion between the dorsal aspect of the atlas and the nuchal ligament) typically following trauma to the area. The authors again conclude that the presence of nuchal ligament ossification could be evidence for the over-riding of the Tuva horse, although it is unclear to what extent which came first—the infection, the ligament ossification, or if both pathologies could be a product of the same causal factor(s).

Some of the most conclusive evidence for the potential of paleopathology to identify horse riding derives from a detailed study of the incidence of pathology in groups of modern wild and ridden horses. By comparing modern animals used for riding with those that have never been ridden, Levine et al. (2000) identified several key pathologies (located in the caudal, thoracic and lumbar vertebrae) in ridden horses which were absent in wild horses. The identifying features of these so-called vertebral riding pathologies are osteophyte formation, overriding or impinging spinous processes, horizontal fissures through the caudal epiphyses, and new bone formation on and around the articular surfaces (Levine et al. 2000:127). Using comparative data from these modern populations, Levine et al. (2005) later investigated the occurrence of vertebral riding pathologies in six Iron Age horse skeletons recovered from Scythian cemeteries in the Altai Mountains, Siberia. All of the animals were found in contexts that suggest that they were used for riding and a pathological investigation revealed that each animal had a degree of abnormal caudal thoracic vertebral pathology as defined above. Levine et al. (2005) compared these data, and that from the Eneolithic site of Botai (where there was no evidence of vertebral pathology), with the two modern populations of ridden and wild horses discussed previously. Their conclusions suggest that the horses from Botai were actually wild—or at least unridden. This fits with the earlier conclusions reached by Levine (1999) based on a detailed analysis of the population structure from this site.

It is clear the use of modern baselines as an interpretative framework for archaeological material is essential, particularly with regard to conditions (such as joint arthropathies) that are also related to age. As the above studies demonstrate, with the careful study and interpretation of the wider archaeological context, and comparison with modern analogs, paleopathological interpretations can shed light on the use and manipulation of animals by humans.

ANIMAL HUSBANDRY PRACTICES

Along with providing details on how humans were exploiting animals, the study of animal paleopathology can also provide information on how they were housing, feeding and managing them. This is neatly demonstrated by Von den Dreisch et al. (2004), who used paleopathology to make inferences about the breeding, feeding and housing of primates from the Saitic-Ptolemaic animal necropolis of Hermopolis Magna (Tuna el-Gebel, Middle Egypt). A total of 245 mummified primates were recovered from this site, although it is thought that this is just a fraction of the true number of animals present. Detailed pathological examination of the mummified remains identified a vast range of pathological conditions, ranging from minor dental conditions such as abnormal wear, to metabolic conditions such as rickets and osteoporosis. When the chronological prevalence of pathologies was studied, there was a clear increase in metabolic conditions from 15 percent to 46 percent, indicating a steady decline in the health of these primates through time. The types of pathologies seen in the animals suggests that they were affected by chronic malnutrition and inadequate housing (as indicated by the occurrence of rickets), combined with the likely effects of inbreeding. The authors suggest that many of these sociable animals would have suffered pain, potentially making them aggressive towards their keepers leading to isolation and even higher levels of stress.

This study provides a fascinating glimpse into the ways in which careful paleopathological analysis can be used to make more detailed inferences about human management, control and husbandry of animals in the past. While this is obviously a rare and special dataset, there are other potential pathological conditions affecting domestic livestock remains that can also provide important evidence for husbandry practices. For example, Whitwell (1996) suggested that stalling horses for long periods might cause abnormalities in their cervical vertebra on the basis that horses kept in open pasture spend much of their time grazing with their heads down, whereas stabled animals spend the majority of their time with their heads up, eating from mangers/hayracks or looking over doors. However, this suggestion remains untested.

An intriguing and enigmatic pathology (with unknown etiology), identified in archaeological caprine and cattle remains, has also been tentatively interpreted as relating to husbandry practice. The condition has been described as a "half cigar shaped expansion of cortical bone on the medial or lateral ridges of the proximal anterior shaft" of caprine metatarsals (Brothwell et al. 2005:75). Whilst the condition has been most commonly noted on sheep distal limb bones (especially metapodials), there is an example noted on a cattle metatarsal from the city of Lincoln (Dobney et al. 1996). In almost all cases, the pathology appears as a ridge or series of bony protrusions on the proximal upper third of the shaft (Dobney et al. 1996).

O'Connor (1984:42), describing these bony protrusions on sheep metapodials from post-medieval York, suggests that this form of abnormality may have been the result of severe bruising, potentially the result of animals being hobbled (that is, being incapacitated by a device used to limit mobility), which could have placed prolonged pressure on various bones, leading to bleeding and new bone formation. However, Brothwell et al. (2005) suggest hobbling is too simplistic an explanation for the condition, given the variation in nature, location and extent of these defects in the populations observed. They also noted that the condition was identified in a modern red deer, an animal that was certainly never hobbled. Dobney et al. (1996:43) suggest a variety of alternative causes including a change in gait, potentially as the result of chronic pain and discomfort; prolonged penning on hard surfaces leading to jarring of joints, and additional physical stress or mechanical loading associated with animals kept on rough pasture (noting that the condition is seen in Herdwick sheep kept on rough upland pastures). Vann and Grimm (2010) investigated the condition in an archaeological assemblage of sheep from a post-medieval site at Tiverton in Devon and although they also were unable to identify a definitive cause for the defect, they do note that the area where the animals grazed would certainly have been similar to rough upland pasture. Further studies are clearly needed of this condition in archaeological and modern populations in order to investigate more thoroughly the etiology behind its occurrence. However, it clearly has potential to shed light on a variety of animal husbandry factors such as grazing terrain or direct human control or both.

Another enigmatic pathology, long thought to be directly linked with the impact of human control of livestock, is so-called "penning elbow," again recorded principally in caprines. This condition, which produces osteophytes of bone on the lateral sides of the proximal radius and distal humerus (leading eventually to severe immobility of the joint), has often been interpreted as caused by trauma related to overcrowding, rough handling or penning (Baker and Brothwell 1980:127). Dobney et al. (2007b:184) investigated the occurrence of the condition in the very large Anglo-Saxon animal bone assemblage from Flixborough (North Lincolnshire) and noted an increase in prevalence from 2.9 percent in the late 7th to early 8th century, to a maximum of

10.1 percent in the late 8th to 9th centuries, before decreasing again to 2.8 percent in the late 10th to early 11th centuries A.D. Although these significant shifts in low-grade trauma to sheep coincided with major changes in the economy and sociopolitical identity of the site, no further insight into the actual causal factors involved in "penning elbow" were forthcoming. As with the condition on metapodials discussed above, and for so many of these potentially informative, low-grade, pathological conditions, more comparative data from modern populations needs to be studied before full etiologies can be revealed.

Several other pathological lesions have the potential to provide information on the conditions in which animals were kept, although there has been little research into either their causes or prevalence in zooarchaeological populations. Conditions such as rickets, which is commonly used to infer poor living conditions and nutrition in human populations, is almost never discussed in the animal bone record. Baker and Brothwell (1980:49), however, claim that it is seen fairly commonly in the archaeological record, and briefly discuss ten sites from the UK where the condition has been noted. Interestingly, of the ten sites they discuss, only four are in the north and they suggest that the prevalence of the condition in the south may relate to animals being kept indoors and reared for beef, while in the north they were raised outdoors for milk or draught purposes. Other conditions, such as infections of the phalanges in ungulates, and spavin (osteoarthritis leading to the fusion of the tarsal bones to the metatarsal) in horses, can potentially be related to the housing and use of animals. Caprines, for example, are exceptionally prone to foot infections when kept on damp or boggy pastures, and spavin has been linked to repeated percussion to the joint by keeping or riding animals on hard ground, or by activities such as jumping (Baker and Brothwell 1980:117, 123–126).

CULTURAL ATTIDUDES TO ANIMALS

Care and treatment

The study of animal health can allow archaeologists access to a range of otherwise unknowable sociocultural aspects of past human populations regarding the treatment, care and attitudes towards animals. For instance, there have been several paleo-pathological studies that have attempted to explore the treatment and healing of pathological conditions in animals. Perhaps the issue that resonates most deeply with us today is the care and attention lavished upon pets. The concept of keeping pets is not new—many animals such as cats and dogs were kept, or tolerated, in past times as working animals for hunting, catching pests, assisting with herding and guarding property. MacKinnon (2010) provides an interesting study into pathological conditions present in Mediterranean Roman dog skeletons. This synthesis indentified only thirteen out of two hundred site assemblages with evidence of dog pathology, although MacKinnon notes that this is probably more to do with lack of reporting or recording of pathological conditions than the actual prevalence of pathology. Of particular interest in this study is the identification of "toy breeds" of dog. One such specimen from Yasmina (Carthage) had lost almost all of its teeth, and suffered from extensive osteoarthritis, a dislocated right femur and spondylosis deformans. MacKinnon suggests that these conditions would have caused the animal great discomfort, yet the animal clearly survived and was mobile for a considerable time, given the degree of bone remodeling present. The paleopathological evidence suggests that this animal would have been well cared for and, based on the high ^{15}N values recorded in its bone, was probably fed a diet of ground meat that could be eaten without teeth.

Other studies have looked at the prevalence rates of fractures to understand the relationships between humans and their animals. Groot (2008) carried out a systematic analysis of fractures in domestic animals at the Roman site of Tiel-Passewaaij in the Netherlands. Of all the animals studied, dogs showed the highest incidence of trauma-related fractures. Groot (2008) concluded that this was multi-factoral, relating to both the dog's close relationship with humans (making them more likely to get kicked or abused), but also their role in Roman societies as working animals. Working dogs are far more likely to be injured as the result of being kicked by cattle, or injured during the course of a hunt, compared with other domesticated species. Udrescu and Van Neer (2005) also explored fracture frequencies, this time in domestic ungulates. They highlight several cases cited in the literature where human intervention has been suggested (e.g. Grant 1975:404). They conclude that in the case of small and medium sized mammals such as sheep and goats, well-healed and aligned fractures can occur naturally with little or no human intervention (as they do in wild species such as deer), whereas in large mammals, healing would be slow and unlikely without human intervention. Such direct intervention (in the form of a bandage and splint) is suggested by Grimm (2008) in the healing of a possible greenstick fracture in a cattle metacarpal from medieval Emdem (Germany).

In the case of limb fractures, large animals, such as horses, over seven months of age are often euthanized rather than treated, even in modern populations (Udrescu and Van Neer 2005). However, this has not always the case, as demonstrated by Antikas (2008) who reports the excavation of a horse from a cemetery dating from 600 to 300 B.C. near Sindos, Greece. The animal was buried along with human remains and a cart, and the extensive lower limb arthropathies present were interpreted as evidence of its use as a pack or carthorse. Of particular interest was the compound fracture to the animal's left metacarpal. The fracture was well-healed, although at an angle, and showed extensive evidence of osteomylitis. Antikas (2008) suggests that the animal's forelimb had been immobilized and press-bandaged for at least 3 to 4 months, and that the misalignment of the limb may have been due to the animal returning to work before the break had fully healed. Other examples of healed metapodial fractures in horses are recorded in the literature. Baker and Brothwell (1980:89) briefly mention a well-healed horse metatarsal from Skedemosse (Denmark) and suggest that the animal was not stressed or compelled to move about much during its recuperation period. However, Udrescu and Van Neer (2005) conclude that in archaeological populations injured animals such as cattle and horses are more likely to be have been killed and eaten than have had any attempt at healing made.

Selective breeding

One animal pathology that has been subject to a degree of comparative investigation is the presence of cranial perforations in bovids. Brothwell et al. (1996) provided the first systematic analysis and differential diagnosis of perforations that appear commonly on the posterior (cranial) portion of the skull as rounded, single or multiple holes into the sinus cavity. After a detailed differential diagnosis that included possible parasites, infection and tumors, Brothwell et al. concluded that there were two possible causes; a congenital (nonpathological) abnormality, or reoccurring pressure (possibly due to yoking) via the horns or head. Manaseryan et al. (1999) and Baxter (2002) went on to identify the occurrence of the condition in ancient *Bison bonasus* and *Bos primigenius* skulls. The presence of cranial perforations in wild bovidae species led to the rejection of the traction hypothesis and a consensus that congenital defects are the most likely cause. The etiology of the perforations was investigated further using computer tomography (Llado et al. 2008). However, the study produced inconclusive results, except to suggest that the lack of association with age or sex in cattle made a congenital defect a more likely cause than yoking. The final conclusive evidence that the perforations were of a congenital origin came from Fabis and Thomas (2009), who identified cranial perforations in a modern domestic pig skull. This study addressed the issue of why perforations occur in some species and not others and concluded that the size of the frontal sinus is a likely key factor in their expression. They went on to suggest that the condition is probably linked to a recessive gene and that further study of this condition could be used to investigate selective breeding and genetic bottlenecks in archaeological cattle populations.

Along with achieving the desired size, physical conformation and other phenotypic attributes required in a domestic animal, selective breeding can also have undesirable consequences in the form of detrimental congenital defects. Baker and Brothwell (1980:40) highlight an example of Swedish Highland Cattle, which were selectively bred to be white. However, this led to an associated increase in gonadal hypoplasia that caused 17 percent of all male animals to be infertile, combined with a high number of sterile females due to malformation of the reproductive tracts. Another example is congenital hip dysplasia, a condition often seen in Labradors and Alsatian dogs, although to the best of our knowledge is currently unrecorded in archaeological populations. One form of congenital defect that is potentially visible in archaeological populations is congenital dwarfism. Dexter cattle are an example of congenital dwarfism deliberately bred into populations, with all short-legged Dexters displaying the condition. However, this intensive selective breeding can have disastrous consequences, as when two affected animals are bred producing "bulldog" calves. These calves display gross developmental defects and are naturally aborted before they reach full term. Baker and Brothwell (1980:41) note a dyschondroplastic calf humerus from the early Neolithic site of Knap of Howar, Orkney, and from the Anglo-Saxon site of Flixbrough a possible example of a "bulldog" calf has been identified (Dobney unpublished data and Brothwell personal communication).

One final form of congenital defect that has been recorded in the archaeological record are crested chickens. In living birds, the crest is often a large tuft of feathers, rendering the breeds very distinctive. Osteologically, the birds show a congenital cerebral hernia, causing the frontal bones of the skull to be bulbous and protrude. Brothwell (1979) identified this condition in Roman chicken skulls from Britain, and almost thirty years later Teegen (2008) identified this condition from another Roman site in Trier. The zooarchaeological incidence of crested fowl appears to increase by the 16th and 17th centuries, and it is suggested that these birds were being specifically bred (Teegen 2008).

CLIMATE AND ENVIRONMENT

Recently, investigations into animal populations have begun to explore what paleopathology can tell us about the local environment and climate. Enamel hypoplasia has been the main focus of these studies and it has been used to demonstrate a range of climatic and environmental factors. Franz-Odendaal (2004) investigated differences in the occurrence of enamel hypoplasia in wild and captive giraffes (*Giraffa camelopardalis*). Her results showed that wild giraffes exhibited no signs of enamel hypoplasia, suggesting that wild giraffes suffer considerably lower levels of systemic stress than their captive counterparts. Collet and Teaford (2008) investigated the relationship between external environmental conditions and enamel hypoplasia prevalence in *Cebus* monkeys. They explored the association between higher frequencies of enamel hypoplasia and increased environmental stress. Their study of 38 *Cebus* monkeys from 15 locations across Brazil, correlated the frequency of enamel hypoplasia with the amount of rainfall occurring in each region. As water availability is associated with high levels of bio-availability and subsequently increased access to nutritional resources, areas with high levels of rainfall were considered to be less environmentally stressful. They inferred that monkeys from these areas would show lower levels of enamel hypoplasia occurrence. This hypothesis was supported by their results, which demonstrated that *Cebus* monkeys from savannah and scrubland exhibited significantly higher levels of enamel hypoplasia than animals from rainforest or coastal regions. Lastly, Guatelli-Steinberg and Benderlioglu (2006) investigated nutritional stress in an isolated population of Rhesus monkeys (*Macaca mulatta*) on the island of Caya Santiago. The animals were reportedly in poor health until a feeding program was instituted in 1956, and enamel hypoplasia frequencies were calculated for animals from before and after the feeding program was introduced. There was a statistically significant decrease in enamel hypoplasia occurrence after 1956, clearly linking the occurrence of enamel hypoplasia with levels of nutrition and food availability.

The archaeozoological study of enamel hypoplasia has focused on the investigation of seasonal periods of stress. A series of studies investigating enamel hypoplasia in Pleistocene and early Holocene bison (Niven 2000; Niven et al. 2004; Byerly 2007) suggest that enamel hypoplasia in ancient bison can be potentially related to a range of regular seasonal stresses including post-rut nutritional deficiencies, weaning, cold-season forage quality, availability and seasonally dependent mineral deficits. Kierdorf et al. (2006), however, questioned the fundamental data underpinning this work, suggesting that the observations of enamel hypoplasia were actually developmental defects in coronal cementum.

As previously mentioned, archaeological pigs have perhaps been the most intensively studied species in terms of developmental enamel defects (Ervynck et al. 2001; Dobney et al. 2004, 2005, 2007a; Rowley-Conwy and Dobney 2007). The chronology of enamel hypoplasia occurrence on the molar teeth from both wild and domestic *Sus* across temperate Eurasia clearly points to seasonal physiological stressors impacting on enamel development, specifically related to nutritional deficiencies principally associated with winter.

Finally, investigations into the impact of environmental variation in modern caprines by Balasse et al. (2010) compared two contrasting environmental regions in Kenya, one semi-arid and one more mesic. Their data clearly demonstrates the potential for enamel hypoplasia to be used to investigate climate and environmental variation in archaeological caprine populations.

THE FUTURE OF ANIMAL PALEOPATHOLOGY

The study of animal paleopathology has a great deal to offer archaeologists. As this chapter has demonstrated, animal paleopathology can be used to explore a wide range of archaeological questions that cannot be investigated through other avenues. Although the field has advanced over the last ten years from a relative backwater of zooarchaeology (relegated to the collection of purely "interesting specimens"), to a scientific discipline in its own right, in many ways the subject is still in its infancy. With over 30 years since the publication of Baker and Brothwell's seminal book on animal paleopathology, the vast majority of studies are still little more than "just-so stories," with limited and speculative interpretations and conclusions. Rarely is research grounded in a comparative veterinary or clinical framework. Yet the potential for pathological conditions (identified in zooarchaeological material) to significantly advance our understanding of the human past, is becoming more apparent.

The rapid development and application of biomolecular approaches and techniques to the study of zooarchaeological remains is perhaps one of the most exciting, yet still controversial and experimental areas of study. Whilst, there are problems with both the preservation and integrity of ancient DNA, particularly with regards ancient pathogens, the extraction and identification of the DNA of infectious organisms from the fossil record remains promising.

No one technique can provide all the answers. The real future of the field of animal paleopathology lies in adopting integrated, broad-scale, populational approaches that may include a range of techniques. The accurate and systematic recording of animal pathology needs to become published as standard in zooarchaeological reports, moving the field away from the publication of purely "interesting specimens" and allowing true comparative (geographic and temporal) syntheses of the prevalence and occurrence of pathological conditions to be undertaken. This clearly needs to be coupled with more intensive research into the etiology and prevalence of conditions in modern comparative populations with known life histories, so that pathological conditions can be accurately identified and related to known causal factors. The development of these approaches to animal paleopathological research is essential if it is to be truly recognized as a discipline that can make valid contributions to understanding the "bigger picture" of human-animal relationships in the past.

REFERENCES

Allison, M., 1984   Palaeopathology in Peruvian and Chilean Populations. *In* Palaeopathology at the Origins of Agriculture. M. N. Cohen, and G. J. Armelagos, eds. pp. 515–530. New York: Academic Press.

Antikas, T. G., 2008   They Didn't Shoot Horses: Fracture Management in a Horse of the 5th century BC from Sindos, Central Macedonia, Greece. Veterinarija & Zootechnika 42:24–27.

Baker, J., and Brothwell. D., 1980   Animal Diseases in Archaeology. London: Academic Press.

Balasse, M., Upex, B., and Ambrose, S., 2010   The Influence of Environmental Factors on Enamel Hypoplasia in Domestic Sheep and Goats in Southern Kenya, Masailand. *In* Tracking Down the Past: Ethnohistory Meets Archaeozoology. G. Grupe, G. McGlynn, and J. Peters, eds. Documenta Archaeobiologiae 7:17–37.

Barnes, I., and Thomas, M., 2005   Evaluating Bacterial Pathogen DNA Preservation in Museum Osteological Collections. Proceedings of the Royal Society (Biology):273 (1587):645–653.

Bartosiewicz, L., 2008a   Description, Diagnosis and the Use of Published Data in Animal Palaeopathology: A Case Study Using Fractures. Veterinarija Ir Zootechnika, 41:12–24.

Bartosiewicz, L., 2008b   Bone Structure and Function in Draft Cattle. Documenta Archaeobiologiae: Limping Together Through the Ages, Joint Afflictions and Bone Infections, 6:153–164.

Bartosiewicz, L., 2008c   Taphonomy and Palaeopathology in Archaeozoology. Geobios, 41:69–77.

Bartosiewicz, L., Neer, W. V., and Lentacker, A., 1997   Draught Cattle: Their Osteological Identification and History. Annales du Musée Royal de l'Afrique Centrale.

Baxter, I. L., 2002   Occipital Perforations in a Late Neolithic Probable Aurochs (*Bos primigenius bojanus*) Cranium from Letchworth, Hertfordshire, UK. International Journal of Osteoarchaeology, 12:142–143.

Bendrey, R., 2007   Ossification of the Interosseous Ligaments Between the Metapodials in Horses: A New Recording Methodology and Preliminary Study. International Journal of Osteoarchaeology, 17:207–213.

Bendrey, R., 2008a   A Possible Case of Tuberculsis or Brucellosis in an Iron Age Horse Skeleton from Viables Farm, Basingstoke, England. Proceedings of the Second ICAZ Animal Pathology Working Group Conference., BAR International Series 1844, 19–26.

Bendrey, R., 2008b   An Analysis of Factors Affecting the Development of an Equid Cranial Enthesopathy. Veterinarija and Zootechnika 41:25–32.

Bendrey, R., Cassidy, J. P., Bokovenko, N., Lepetz, S., and Zaitseva, G. I., 2009   A Possible Case of 'Poll–Evil' in an Early Scythian Horse Skull from Arzhan 1, Tuva Republic, Central Asia. International Journal of Osteoarchaeology, published on line 3 Sept 2009. doi. 10.1002/oa.1099.

Bendrey, R., Taylor, G. M., Bouwman, A. S., and Cassidy, J. P., 2008   Suspected Bacterial Disease in Two Archaeological Horse Skeletons From Southern England: Palaeopathological and Biomolecular Studies. Journal of Archaeological Science 35:1581–1590.

Brickley, M., and McKinley, J. I., eds., 2004   Guidelines to the Standards for Recording Human Remains. Institute for Field Archaeologists Paper No. 7. Southampton: British Association for Biological Anthropology and Osteoarchaeology, Institute for Field Archaeologists, and the Department of Archaeology University of Southampton.

Brosch, R., Pym, A. S., Gordon, S. V., and Cole, S. T., 2001   The Evolution of Mycobacterial Pathogenicity: Clues From Comparative Genomics. Trends in Microbiology 9:452–458.

Brothwell, D., 1969   The Palaeopathology of Pleistocene and More Recent Mammals. *In* Science in Archaeology. D. Brothwell, and E. Higgs, eds. pp. 310–314. London: Thames and Hudson.

Brothwell, D., 1979   Roman Evidence of a Crested Form of Domestic Fowl, as Indicated by a Skull Showing Associated Cerebral Hernia. Journal of Archaeological Science 6:291–293.

Brothwell D., 1995   The Special Animal Pathology. *In* Danebury, An Iron Age Hillfort in Hampshire, 6: A Hillfort Community in Perspective. B. Cunliffe, ed. pp. 207–233. York: CBA Research Report 102.

Brothwell, D., Dobney, K., and Ervynck, A., 1996   On the Causes of Perforations in Archaeological Domestic Cattle Skulls. International Journal of Osteoarchaeology 6:471–487.

Brothwell, D., Dobney, K., and Jaques, D., 2005   Abnormal Sheep Metatarsals: A Problem in Aetiology and Historical Geography. Diet and health in Past Animal Populations:

Current Research and Future Directions, Proceedings of the 9th ICAZ Conference, Durham. J. Davies, M. Fabis, I. Mainland, M. P. Richards, and R. Thomas, eds. Oxford: Oxbow Books.

Buikstra, J. E., and Ubelaker, D., 1994   Standards for Data Collection from Human Skeletal Remains: Proceedings of a Seminar at the Field Museum of Natural History. Fayetteville: Arkansas Archaeological Survey Press.

Byerly, R. M., 2007   Palaeopathology in late Pleistocene and Early Holocene Central Plains Bison: Dental Enamel Hypoplasia, Fluoride Toxicosis and the Archaeological Record. Journal of Archaeological Science 34:1847–1858.

Collet, M. B., and Teaford, M. F., 2008   Ecological Stress and Linear Enamel Hypoplasia in Cebus. American Journal of Physical Anthropology 46(suppl.):78–79.

Csippan, P., and Daroczi-Szabo, M., 2008   Animal Disease from Medieval Buda. Proceedings of the Second ICAZ Animal Pathology Working Group Conference, BAR International Series 1844:74–79.

Daroczi-Szabo, M., 2008   Animal Disease at a Celtic–Roman Villa in Hungry. Proceedings of the Second ICAZ Animal Pathology Working Group Conference, BAR International Series 1844:57–62.

Davies, J., Fabis, M., Mainland, I., Richards, M., and Thomas, R., 2005   Diet and Health in Past Animal Populations: Current Research and Future Directions, Proceedings of the 9th ICAZ Conference, Durham. Oxford: Oxbow Books.

De Cupere, B., Lentacker, A., Neer, W. V., Waelkens, M., and Verslype, L., 2000   Osteological Evidence for the Draught Exploitation of Cattle: First Applications of a New Methodology. International Journal of Osteoarchaeology 10:254–267.

Diamond, J., 1997   Guns, Germs, and Steel: The Fates of Human Societies. London: W. W. Norton & Company.

Dobney, K. M., Anezaki, T., Hongo, H., Matsui, A., Yamazaki, K., Ervynck, A., Albarella, U., and Rowley-Conwy, P., 2005   The Transition From Wild Boar to Domestic Pig as Illustrated by Dental Enamel Effects (LEH): A Japanese Case Study Including the Site of Torihama. Torihama Shell Midden Papers 4, 5:51–78.

Dobney, K. M., and Ervynck, A., 1998   A Protocol for Recording Linear Enamel Hypoplasia on Archaeological Pig Teeth. International Journal of Osteoarchaeology 8:263–273.

Dobney, K., and Ervynck, A., 2000   Interpreting Developmental Stress in Archaeological Pigs: The Chronology of Linear Enamel Hypoplasia. Journal of Archaeological Science 27:597–607.

Dobney, K., Ervynck, A., Albarella, U., and Rowley-Conwy, P., 2004   The Chronology and Frequency of a Stress Marker (Linear Enamel Hypoplasia) in Recent and Archaeological Populations of Sus scrofa in North-West Europe, and the Effects of Early Domestication. Journal of Zoology 264:197–208.

Dobney, K., Ervynck, A., Albarella, U., and Rowley-Conwey, P., 2007a   The Transition From Wild Boar to Domestic Pig in Eurasia, Illustrated by a Tooth Developmental Defect and Biometric Data. In Pigs and Humans: 10,000 Years of Interaction. U. Albarella, K. Dobney, A. Ervynck, and P. Rowley-Conwey, eds. pp. 57–82. Oxford: Oxford University Press.

Dobney, K., Jaques, D., et al., 2007b   Farmers, Monks and Aristocrats: The Environmental Archaeology of Anglo–Saxon Flixbrough. Excavations at Flixbrough 3. Oxford: Oxbow Books.

Dobney, K., Ervynck, A., and La Ferla, B., 2002   Assessment and Further Development of the Recording and Interpretation of Linear Enamel Hypoplasia in Archaeological Pig Populations. Environmental Archaeology 7:35–46.

Dobney, K. M., Jaques, D., et al., 1996   Of Butchers and Breeds. Report on Vertebrate Remains From Various Sites in the City of Lincoln. In Lincoln Archaeological Studies. Lincoln: City of Lincoln Archaeological Unit.

Ervynck, A., and Dobney, K., 1999   Lining Pp on the M1: A Tooth Defect as a Bio-Indicator for Environment and Husbandry in Ancient Pigs. Environmental Archaeology 4:1–8.

Ervynck, A., Dobney, K., and Hongo, H., 2001   Born Free? New Evidence of the Status of Pigs at Neolithic çayönü Tepesi Southeastern Anatolia, Turkey. Palaeorient 27:47–73.

Fabis, M., 2004   Palaeopathology of Findings among Archaeofaunal Remains of Small Seminar Site in Nitra. Acta Vet Bruno 73:55–58.

Fabis, M., 2005   Pathological Alteration of Cattle Skeletons – Evidence for the Draught Exploitation of Animals? In Diet and Health in Past Animal Populations: Current Research and Future Directions, Proceedings of the 9th ICAZ Conference, Durham. J. Davies, M. Fabis, I. Mainland, M. P. Richards, and R. Thomas, eds. pp. 58–67. Oxford: Oxbow Books.

Fabis, M., and Thomas, R., 2009   Not Just Cattle: Cranial Perforations Revisited. International Journal of Osteoarchaeology, published online doi: 10.1002/oa. 1133.

Franz-Odendaal, T., 2004   Enamel Hypoplasia Provides New Insights into Early Systemic Stress in Wild and Captive Giraffes (Giraffa camelopardalis). Journal of Zoology 263:197–206.

Gordon, S. V., et al., 1999   Identification of Variable Regions in the Genomes of Tubercle Bacilli Using Bacterial Artificial Chromosome Arrays. Molecular Microbiology 32:643–655.

Gibbons, A., 2008   Tuberculosis Jumped From Humans to Cows, Not Vice Versa. Science 608.

Grant, A., 1975   The Animal Bones. In Excavations at Porchester Castle, vol. I. Roman. B. Cunliffe, ed. pp. 378–406. London: The Society of Antiquaries of London.

Grimm, J. M., 2008   Break a Leg: Animal Health and Welfare in Medieval Emden, Germany. Veterinarija Ir Zootechnika 41:49–59.

Groot, M., 2005   Palaeopathological Evidence for Draught Cattle on a Roman Site in the Netherlands. In Diet and Health in Past Animal Populations: Current Research and Future Directions, Proceedings of the 9th ICAZ Conference, Durham. J. Davies, M. Fabis, I. Mainland, M. P. Richards, and R. Thomas, eds. pp. 52–58. Oxford: Oxbow Books.

Groot, M., 2008   Understanding Past Human–Animal Relationships Through the Analysis of Fractures: A Case Study From a Roman Site in the Netherlands. Proceedings of the Second ICAZ Animal Pathology Working Group Conference. BAR International Series 1844:40–50.

Grupe, G., McGlynn, G., and Peters, J., eds., 2008   Limping Together Though The Ages: Joint Afflictions and Bone Infections. Documenta Archaeobiologiae.

Guatelli–Steinberg, D., and Benderlioglu, Z., 2006   Brief Communication: Linear Enamel Hypoplasia and the Shift From Irregular to Regular Provisioning in Cayo Santiago Rhesus Monkeys (Macaca mulatta). American Journal of Physical Anthropology 131:416–419.

Gutacker, M., et al., 2006   Single-Nucleotide Polymorphism–Based Population Genetic Analysis of Mycobacterium tuberculosis Strains from Four Geographic Sites. Journal of Infectious Diseases 193:121–128.

Haas, F., and Haas, S., 1996   The Origins of Tuberculosis and the Notion of its Contagiousness. In Tuberculosis. W. N. Rom, and S. Garay, eds., pp. 3–19. London: Little, Brown and Co.

Harcourt, R. A., 1967   Osteoarthritis in a Romano–British Dog. Journal of Small Animal Practice 8:521–522.

Harcourt, R. A., 1971   The Palaeopathology of Animal Skeletal Remains. Veterinary Record 89:267–272.

Horwitz, L. K., and Smith, P., 2000   The Contribution of Animal Domestication to the Spread of Zoonoses: A case Study From the Southern Levant. Anthropozoologica 31:77–84.

Johannsen, N., 2005   Palaeopathology and Neolithic Cattle Traction: Methodological Issues and Archaeological Perspectives. In Diet and Health in Past Animal Populations: Current Research and Future Directions, Proceedings of the 9th ICAZ Conference, Durham. J. Davies, M. Fabis, I. Mainland, M. P. Richards, and R. Thomas, eds. pp. 39–51. Oxford: Oxbow Books.

Kierdorf, H., Zeiler, J., and Kierdorf, U., 2006   Problems and Pitfalls in the Diagnosis of Linear Enamel Hypoplasia in the Cheek Teeth of Cattle. Journal of Archaeological Science 33:1690–1695.

Levine, M. A., 1999   Botai and the Origins of Horse Domestication. Journal of Anthropological Archaeology 18:29–78.

Levine, M. A., 2005   Domestication and Early History of the Horse. *In* The Domestic Horse: The Origins and Management of its Behavior, D. S. Mills, and S. M. McDonnell, eds. pp. 5–22. Cambridge: Cambridge University Press.

Levine, M. A., Bailey, G. N., Whitwell, K. E., and Jeffcott, L. B., 2000   Palaeopathology and Horse Domestication: The Case of Some Iron Age Horses From the Altai Mountains, Siberia. *In* Human Ecodynamics. G. Bailey, R. Charles, and N. Winder, eds. pp. 123–133. Oxford: Oxbow Books.

Levine, M. A., Whitwell, K. E., and Jeffcott., L. B., 2005   Abnormal Thoracic Vertebrae and the Evolution of Horse Husbandry. Archaeofauna 14:93–109.

Lignereux, Y., and Peters, J., 1999   Elements for the Retrospective Diagnosis of Tuberculosis on Animals Bones from Archaeological Sites. *In* Tuberculosis Past and Present. G. Palfi, O. Dutor, J. Deak, and I. Hutas, eds. pp. 339–349. Szeged: Golden Book Publishers Ltd.

Llado, E., Gaitero, L., M. Pumarola, M., and Saña, M., 2008   Perforations in Archaeological Neolithic Cattle Skulls: A New Methodological Approximation for Their Study and Explanation. Veterinarija and Zootechnika 43:58–61.

MacKinnon, M., 2010   "Sick As A Dog": Zooarchaeological Evidence for Pet Dog Health and Welfare in the Roman World. World Archaeology 42:290–309.

Manaseryan, N., Dobney, K., and Ervynck, A., 1999   On the Causes of Perforations in Archaeological Cattle Skulls: New Evidence. International Journal of Osteoarchaeology 17:514–523.

Mays, S., 2001   Palaeopathological and Biomolecular Study of Tuberculosis in a Medieval Skeletal Collection From England. American Journal of Physical Anthropology 114:298–311.

Mays, S., 2005   Tuberculosis as a Zoonotic Disease in Antiquity. *In* Diet and Health in Past Animal Populations: Current Research and Future Directions, Proceedings of the 9th ICAZ Conference, Durham. J. Davies, M. Fabis, I. Mainland, M. P. Richards, and R. Thomas, eds. pp. 125–134. Oxford: Oxbow Books.

McCance, R. A., Ford, E. H. R. and Brown, W. A. B., 1961   Severe Undernutrition in Growing and Adult Animals 7. Development of the Skull, Jaws and Teeth in Pigs. British Journal of Nutrition 15:213–224.

McNeill, W. H., 1976   Plagues and People. New York: Anchor Press.

Mellanby, L. M., 1929   Diet and Teeth: An Experimental Study. Part 1, Dental Structure in Dogs. *In* Medical Research Council Special Report Series 104. London: H.M.SO.

Miklikova, Z., 2008   Skeletal Alterations of Animal Remains From the Early Medieval Settlement of Bajc, South–West Slovakia. *In* Proceedings of the Second ICAZ Animal Pathology Working Group Conference, BAR International Series 1844:63–73.

Miklikova, Z., and Thomas, R., eds., 2008   Current Research in Animal Pathology. BAR International Series 1814.

Niven, L., 2000   Enamel Hypoplasia in Bison: Paleoecological Implications for Modelling Hunter-Gatherer Procurement and Processing on Northwestern Plains. *In* Assessing Season of Capture Age and Sex of Archaeofaunas: Recent Work. A. Pike-Tay, ed. pp. 101–112.

Niven, L., Egeland, C. P., and Todd, L. C., 2004   An Inter-Site Comparison of Enamel Hypoplasia in Bison: Implications for Paleoecology and Modelling Late Plains Archaic Subsistence. Journal of Archaeological Science 31:1783–1794.

Norcard, E., and Leclaiche, E., 1903   Les Maladies Microbiennes des Animaux. 3rd edition, vol. 1. Paris: Masson et Cei.

O'Connor, T. P., 1984   Selected Groups of Bones from Skeldergate and Walmgate. *In* The Archaeology of York: The Animal Bones. York: York Archaeological Trust and the Council for British Archaeology.

O'Connor, T. P., 2000   The Archaeology of Animal Bones. London: Sutton Publishing.

O'Connor, T. P., 2008   On the Differential Diagnosis of Arthropathy in Bovids. Documenta Archaeobiologiae: Limping Together Through the Ages, Joint Afflictions and Bone Infections 6:165–186.

Ortner, D., 2003   The Identification of Pathological Conditions in Human Skeletal Remains. London: Academic Press.

Roberts, C. A., and Buikstra, J. E., 2003   The Bioarchaeology of Tuberculosis: A Global Perspective on a Reemerging Disease. Gainesville: University Press of Florida.

Roberts, C. A., and Manchester, K., 2005   Archaeology of Disease. Stroud: Sutton Publishing.

Rothschild, B. M., and Laub, R., 2006   Hyperdisease in the Late Pleistocene: Validation of an Early 20th Century Hypothesis. Naturwissenschaften 93:555–564.

Rothschild, B. M., and Martin, L. D., 2003   Frequency of Pathology in a Large Natural Sample From Natural Trap Cave with Special Remarks on Erosive Disease in the Pleistocene. Reumatismo 55:58–65.

Rothschild, B. M., and Martin, L.D., 2006   Did Ice-Age Bovids Spread Tuberculosis? Naturwissenschaften 93:565–569.

Rothschild, B. M., Martin, L. D., and Lev, G., 2001   *Mycobacterium tuberculosis*-Complex DNA From an Extinct Bison Dated 17,000 Years BP. Clinical Infectious Diseases 33:305–311.

Rowley–Conwy, P., and Dobney, K., 2007   Wild Boar and Domestic Pigs in Mesolithic and Neolithic Southern Scandinavia. In Pigs and Humans: 10,000 Years of Interaction. U. Albarella, K. Dobney, A. Ervynck, and P. Rowley-Conwy, eds. pp. 131–155. Oxford: Oxford University Press.

Siegel, J., 1976   Animal Pathologies – Possibilities and Problems Journal of Archaeological Science 3:349–384.

Smith, P., Bar-Yosef, O., and Sillen, A., 1984   Archaeological and Skeletal Evidence for Dietary Change During the Late Pleistocene / Early Holocene in the Levant. In Palaeopathology at the Origins of Agriculture. M. N. Cohen and G. J. Armelagos, eds. pp. 101–127. New York: Academic Press.

Smith, N. H., Hewinson, R.G., Kremer, K., Brosch, R., and Gordon, S.V., 2009   Myths and Misconceptions: The Origin and Evolution of *Mycobacterium tuberculosis*. Nature Reviews Microbiology 7:537–544.

Starling, A. P., and Stock, J. T., 2007   Dental Indicators of Health and Stress in Early Egyptian and Nubian Agriculturalists: A Difficult Transition and Gradual Recovery. Journal of Physical Anthropology 134:520– 528.

Stead, W. W., 1997   The Origin and Erratic Global Spread of Tuberculosis: How the Past Explains the Present and is the Key to the Future. Clinics in Chest Medicine 18:65–77.

Sucking, G., Elliott, D. C., and Thurley, D.C., 1983   The Production of Developmental Defects of Enamel in the Incisor Teeth of Penned Sheep Resulting From Inducted Parasitism. Archives of Oral Biology 28:393–399.

Suckling, G., Elliott, D. C. and Thurley, D. C., 1986   The Macroscopic Appearance and Associated Histological Changes in the Enamel Origin of Hypoplastic Lesions of Sheep Incisor Teeth Resulting from Induced Parasitism. Archives of Oral Biology 31:427–439.

Taylor, G. M., Goyal, M., Legge, A. J., Shaw, R. J., and Young, D., 1999   Genotypic Analysis of *Mycobacterium tuberculosis* From Medieval Human Remains. Microbiology 145:899–904.

Teegan, W. R., 2005   Linear Enamel Hypoplasia in Medieval Pigs From Germany. In Diet and Health in Past Animal Populations: Current Research and Future Directions, Proceedings of the 9th ICAZ Conference, Durham. J. Davies, M. Fabis, I. Mainland, M. P. Richards, and R. Thomas, eds. pp. 89–92. Oxford: Oxbow Books.

Teegan, W. R., 2008   A Crested Fowl From Late Antique Augusta Treverorum / Trier. Documenta Archaeobiologiae: Limping Together Though The Ages. Joint Afflictions and Bone Infections 6:204–207.

Telldahl, Y., 2005   Can Palaeopathology Be Used as Evidence for Draught Animals? In Diet and Health in Past Animal Populations: Current Research and Future Directions, Proceedings of the 9th ICAZ Conference, Durham. J. Davies, M. Fabis, I. Mainland, M. P. Richards, and R. Thomas, eds. pp. 63–67. Oxford: Oxbow Books.

Thomas, R., 2008   Diachronic Trends in Lower Limb Pathologies in Later Medieval and Post–Medieval Cattle From Britain. Documenta Archaeobiologiae: Limping Together Though the Ages. Joint Afflictions and Bone Infections 6:187–201.

Thomas, R., and Mainland, I., 2005   Introduction: Animal Health and Diet–Current Perspectives and Future Directions. *In* Diet and Health in Past Animal Populations: Current Research and Future Directions, Proceedings of the 9th ICAZ Conference, Durham. J. Davies, M. Fabis, I. Mainland, M. P. Richards, and R. Thomas, eds. pp. 1–7. Oxford: Oxbow Books.

Ubelaker, D., 1984   Prehistoric Human Biology of Ecuador: Possible Temporal Trends and Cultural Correlations. *In* Palaeopathology at the Origins of Agriculture. M. N. Cohen and G. J. Armelagos, eds. pp. 515–530. New York: Academic Press.

Urdrescu, M., and Neer, W. V., 2005   Looking for Human Therapeutic Intervention in the Healing of Fractures of Domestic Animals. *In* Diet and Health in Past Animal Populations: Current Research and Future Directions, Proceedings of the 9th ICAZ Conference, Durham. J. Davies, M. Fabis, I. Mainland, M. P. Richards, and R. Thomas, eds. pp. 24–33. Oxford: Oxbow Books.

Vann, S., 2008   Animal Palaeopathology at Two Roman Sites in Southern Britain. Proceedings of the Second ICAZ Animal Pathology Working Group Conference., BAR International Series 1844:27–37.

Vann, S., and Grimm, J., 2010   Post-Medieval Sheep (*Ovis aries*) Metapodia from Southern Britain. Journal of Archaeological Science 37:1532–1542.

Vann, S., and Thomas, R., 2006   Humans, Other Animals and Disease: a Comparative Approach Towards the Development of a Standardised Recording Protocol from Animal Pathology. Internet Archaeology 20.

Von den Driesch, A., Kessler, D., and Peters, J., 2004   Mummified Baboons and other Primates from the Saitic-Ptolemaic Animal Necropolis of Tuna el-Gebel, Middle Egypt. Documenta Archaeobiologicae: Conservation Policy and Current Research 2:231–278.

Whitwell, K. E., 1996   Hyperion Revisited: Observations on the Life and Mortal Remains of an Equine legend. Proceedings Association of Veterinary Teachers and Research Workers Conference, Scarborough, April 1996. p. 38.

Wolfe, N. D., Dunavan, C. P., and Diamond, J., 2007   Origins of Major Human Infectious Diseases. Nature 447:279–283.

CHAPTER **12** How Does
The History
of Paleopathology
Predict its Future?

*Mary Lucas Powell and Della
Collins Cook*

Those who cannot remember the past are condemned to repeat it.
George Santayana, *The Life of Reason*

As we sit gazing into our gleaming crystal skull, searching for clues to the future
of paleopathology, we confess to a certain unease about our ability to forecast
accurately the course of our discipline. Recent advances in the reconstruction of
ancient natural and cultural environments (e.g., Roberts 2003; Chapter 6 by
Katzenberg, this volume; and Chapter 4 by Buzon, this volume), in nondestructive
imaging technologies (e.g., Chhem and Brothwell 2008), and in the recovery and
analysis of ancient microbial DNA (e.g., Chapter 8 by Spigelman et al., this volume)
and publication of the genomes of venereal syphilis (von Hunnius et al. 2006; and
Chapter 26 by Cook and Powell, this volume), tuberculosis (Roberts and Buikstra
2003; and Chapter 24 by Roberts, this volume), leprosy (Chapter 25 by Lynnerup
and Boldsen, this volume), malaria (Hawass et al. 2010, Chapter 10 by Dittmar,
this volume), and other infectious diseases have revolutionized our attempts to
reconstruct patterns of health and disease in the past. However, unless our
interpretations of the data from these marvelous new technical innovations are
based on a solid understanding of the synergistic interaction of human biology and
behavior and key stressors (microbial and otherwise), we may learn little that is new
about the endless human struggle against 'the thousand natural shocks that flesh is
heir to'. Drawing upon our time spent studying the history of paleopathology and

A Companion to Paleopathology, First Edition. Edited by Anne L. Grauer.
© 2012 John Wiley & Sons, Ltd. Published 2016 by John Wiley & Sons, Ltd.

mindful of George Santayana's dictum, we offer five lessons to be learned from our discipline's diverse and sometimes murky history.

FIVE LESSONS FROM PALEOPATHOLOGY'S PAST

Who's an idiot? Or, what, exactly, is "pathological"?   In 1866, Langdon Down noted in his "Observations on an ethnic classification of idiots" that certain features common in crania from Ethiopia, the Malay peninsula, and various Native American tribes were common in mentally deficient Britons (Down 1866). He proposed a typology of developmental conditions in which each earlier racial stage appeared as an atavism among Europeans. His "Mongolian idiocy," known today as Down syndrome, is the sole survivor of his diagnostic typology. Paleontologists no longer see Europeans as the pinnacle of evolution, geneticists have rejected Down's equation of deviance from European cranial morphology with "pathological" stigmata, and students of biological anthropology today are (we hope) familiar with the well-documented world-wide patterns of human morphological variation and are explicitly taught to reject the earlier, pejorative interpretations of regional variations. However, the recent continuing controversy over the Flores hominid (Jacob et al. 2006; Obendorf et al. 2008) reminds us that paying equal attention to regional and temporal patterns and to dysmorphology is still essential for the evaluation of atypical human remains. We applaud the fascinating literature that the unfortunately named Hobbit is generating!

"The farmer and the cowman should be friends" ("Oklahoma!" Rodgers and Hammerstein, 1943), or "Let a thousand flowers bloom; let a hundred schools of thought contend." (Chairman Mao). Don't be misled by false oppositions   In a recent review, Armelagos and van Gerven (2003) charge that paleopathology has not advanced today meaningfully beyond "the old questions of differential diagnosis" (Armelagos and van Gerven 2003:60). They posit that "The traditional approach of paleopathology had been the diagnosis of specific diseases such as tuberculosis, leprosy, and syphilis" (2003:58–59), ignoring the hundreds of publications during the 20th century on trauma, dental pathology, congenital and developmental disorders and other topics (for examples see Buikstra and Cook 1980; Ubelaker 1982; Ortner and Aufderheide 1991; Roberts and Manchester 1995, 2005; Aufderheide and Rodríguez-Martín 1998; Ortner 2003, among others). They view this traditional approach as "inherently limited [because] bones and teeth do not often respond with the kind of specificity necessary for a clinical diagnostic approach to all diseases." (Armelagos and van Gerven 2003:59). In opposition to their straw man, Armelagos and van Gerven favor the study of "Responses such as trauma, patterns of growth and development, periosteal inflammation, enamel hypoplasia, and differential mortality [because these] can be used to ask a host of interesting questions" (Armelagos and van Gerven 2003:59), unlike the outmoded focus on "the diagnosis of individual cases", which in their view obviously deals only with *un*interesting questions. It must be said that their view of the paleopathological literature during the second half of the 20th century seems sadly narrow, missing as it does the significant expansion of interest to broad-based population studies that incorporate archaeological, ecological, ethnographic, historical

and clinical data interpreted within a broad theoretical framework (Owsley and Jantz 1994; Webb 1995; Cucina and Tiesler 2003; Roberts 2003; Merbs 2005; Lukacs and Largespada 2006; Cook and Powell 2006; Watson et al. 2010).

We agree with Armelagos and van Gerven that the most productive studies of human health and disease in the past should examine patterns of pathology (whether specific or nonspecific) "by age, sex, and environmental (cultural and natural setting) setting." However, the "osteological paradox" (Wood et al. 1992) notwithstanding, not all diseases have equal effects on morbidity and mortality, and failure to attempt differential diagnosis of skeletal pathology *in appropriate cases* limits interpretations of nonspecific stress indicators as such to epidemiologically nonspecific conclusions about differing patterns of prevalence in different populations. For example, the three specific infectious diseases named by Armelagos and van Gerven as examples of paleopathology's old-fashioned preoccupation with clinical diagnosis—tuberculosis, leprosy, and syphilis—all produce "periosteal inflammation" among their clinical manifestations. However, they may also produce distinctive pathognomonic lesions and can clearly shape "differential mortality" in quite different ways (Powell 1992; Aufderheide and Rodríguez-Martín 1998; Ortner 2003). Simply recording observations of "periosteal inflammation" in an individual or a population yields no information on whether one is seeing the effects of an infectious disease that produces relatively high morbidity and low mortality (e.g., syphilis) and which carries culturally-bound pejorative social connotations and negative demographic consequences (Quetel 1990), or one that is characterized by a pattern of low morbidity and very high mortality, with cultural interpretations ranging from sexual allure to disgust (e.g., tuberculosis) (Roberts and Buikstra 2003). A third pattern is also possible—one with variable morbidity and moderate mortality but associated with horrific social consequences (e.g., leprosy [Hansen's disease]) (Boldsen 2001; and see Chapter 25 by Lynnerup and Boldsen, this volume).

When we began our careers in paleopathology in the early 1970s, the diagnosis of specific infectious diseases in ancient human remains relied upon careful point-by-point matching of macroscopic, microscopic, and radiographic details of observed skeletal pathology with differential models based on clinical and epidemiological data. Today, these methods are augmented (but certainly not superseded) by standardized protocols for the recovery, analysis, and interpretation of molecular evidence (metabolites and DNA from ancient pathogens) for tuberculosis (Roberts and Buikstra 2003), Hansen's disease (Rafi et al. 1999), Chagas disease (Guhl et al. 1999), malaria (Hawass et al. 2010), and other infectious diseases. Molecular identification of treponemal disease in ancient human remains has proved far more difficult (Ortner et al. 1992; Kolman et al. 1999; Von Hunnius et al. 2006), because of the ephemeral character of the spirochetes relative to the more durable *Mycobacterium* spp and other pathogenic organisms.

For paleopathology to claim scientific status among the subdisciplines of anthropology, it must "develop and test hypotheses", as Armelagos and van Gerven exhort. However, hypotheses are only as sound as the data used to construct them, and in our view the ideal paleopathological analysis of an ancient population sample would *integrate*, not reject, clinically informed descriptive studies of skeletal pathology with relevant ecological and cultural data in the construction and testing of these hypotheses. The "systematic environmental forces" which Armelagos and van Gerven see as responsible for the "patterns of stress response evidenced in ancient populations" (Armelagos and van Gerven 2003:59) indeed comprise two points of the classic epidemiological triangle: "the environment" and "the stressor" (or "pathogen"). But the third point, the human population, functions simultaneously within unique biological *and* cultural contexts. As anthropologists we argue that understanding the cultural factors that affect the diseases and disorders visible on ancient bodies is a crucial counterpart to the biology and ecology of the human hosts and their stressors. One aspect of the incorporation of this social dimension is the accurate identification of pathological entities that carry specific social connotations. If we limit ourselves *only* to "nonspecific" observations and disdain from attempting appropriate differential diagnoses, which could expand our understanding, we voluntarily limit the scope of our reconstructions of life in the past, to no real purpose that we can imagine.

Contra Mies van der Rohe, less is not *always more: one bone can't tell you everything*
Until the middle of the 20th century, physical anthropology was obsessed with craniology, so much so that older collections largely consist of adult skulls. The early paleopathologists lamented this bias, and those of us who work today with older collections still fume at the myopia of our predecessors.

In 1995, Rothschild and Rothschild published a set of six diagnostic criteria ("SPIRAL") that promised to distinguish bone lesions of venereal syphilis from those of endemic syphilis (bejel) and yaws. They employed data from three skeletal series: venereal syphilis was represented by cases selected from the late 19th- and early 20th-century Hamann–Todd Collection, a medical autopsy skeletal series curated at the Cleveland Museum of Natural History (total N = 2,069); the model for bejel was a 18th- and 19th-century Bedouin archaeological series (N = 40) from the site of Be'er Sheva Tel in Israel; and a late 17th-century pre-European contact archaeological series from the Gognga Gun Beach site in Guam (N = 213) provided the model for yaws.

The Rothschild and Rothschild SPIRAL method, however, includes several serious flaws, including "part to whole" statistical comparisons: their analytical sample for venereal syphilis was "limited to those individuals with diagnosed syphilis" within the very large Hamann–Todd series, while the other two samples are small archaeological series with no previous clinical diagnoses (i.e., they lack the 100 percent prevalence rate for syphilis that defined the Hamann–Todd subsample). A second problem is that all examples of "periostitis" in the two small series are attributed to treponematosis; there is no discussion of alternative causes such as trauma, osteomyelitis, or other infectious diseases. The most pervasive flaw in SPIRAL is its exclusive concentration on *one* bone of the post-cranial skeleton: three of the six diagnostic criteria focus solely on tibial lesions, with relevant pathology of the cranium, face, and teeth are completely ignored.

In their desire to provide a "quick and easy" diagnostic model, Rothschild and Rothschild ignore the complexity of disease manifestations and omit precisely those categories of pathological evidence that have been advanced as most useful in distinguishing treponemal disease (whether venereal syphilis or the nonvenereal syndromes) from the numerous other causes of periosteal proliferative lesions in the clinical and paleopathology literature over the past century (summarized in Hackett 1976; Aufderheide and Rodríguez-Martín 1998; Ortner 2003).

As T. Aidan Cockburn warned us in 1963, "Avoid emotional attachment to one's ideas.": the most carefully constructed diagnostic models may be overturned by new data   In their 2003 treatise, *The Bioarchaeology of Tuberculosis, a Global View on a Reemerging Disease*, Roberts and Buikstra summarize the intense debate over the natural history of that disease in the Americas, beginning with Aleš Hrdlička's 1909 study of tuberculosis on Indian reservations. Hrdlička's rejection of Pre-Columbian TB was based on his careful evaluation of archaeological and historic evidence in the light of his research on tuberculosis among living Indian tribes. Half a century later, the eminent paleopathologist and physician T. Aidan Cockburn agreed with Hrdlička's assessment, noting that "the Americas before Columbus were very thinly populated" (Cockburn 1963:89), lacking the dense population concentrations where tuberculosis flourished in recent times. Physician and amateur archaeologist Dan F. Morse (1967) concurred, noting that the ancient skeletal lesions did not match current clinical models for bone involvement and furthermore, there were no herds of domesticated bovids to provide reservoirs for the disease prior to European colonization. Both Morse and Cockburn had gained considerable clinical experience with the modern forms of tuberculosis, and their diagnostic models, like Hrdlička's, incorporated current clinical and epidemiological features regarding lesions, case prevalence, base population size, and the widely accepted evolutionary scenario that saw tuberculosis as a zoonotic disease.

Two decades later, Janet McGrath (1988)'s computer simulation, focused on data from multiple Pre-Columbian archaeological sites in the lower Illinois River Valley, attempted to reconcile the growing body of paleopathological evidence from securely dated Pre-Columbian contexts in North and South America with current clinical disease models. Her simulations were well-constructed, informed by the latest archaeological, epidemiological, and clinical evidence, but the ancient evidence failed to match her clinically-based criteria for a prehistoric New World form of the disease.

These carefully considered conclusions collectively incorporated more than 100 years of clinical, epidemiological, and archaeological research, *and yet today we know that they all were wrong*: the Pre-Columbian presence of tuberculosis has been unequivocally demonstrated at numerous sites in both North and South America, as chronicled by Roberts and Buikstra (2003). Other research, stimulated by recent advances in technology—microscopic imaging of actual ancient mycobacterial pathogens in ancient human tissues and the molecular identification of both their metabolic products and their aDNA, has revealed many surprises. The greatest of these is the destruction of the model of *M. bovis* as the zoonotic ancestor of the human disease (Zink et al. 2003). We are bemused at the reluctance of many contributors to medical pedagogy and medical history to give up this outworn model. But then, it was only in the late twentieth century that "mongolism," a name that enshrines Dr Langdon Down's racialist model for his eponymous syndrome, was final abandoned.

Don't put all of your eggs in one basket. (Mother Goose)   Proposals to separate pathological bones from all other bones in the collections of ancient human remains go back at least to Roy L. Moodie (1880–1923), pioneer of our field, whose requests were rebuffed by all the archaeologists with whom he corresponded (Wellcome Museum Archives). Perhaps the refusals were motivated by the critical desire to preserve the context of archaeological specimens. Creating universal registry collections of paleopathological specimens (i.e., gathering all specimens into one place) is an equally bad idea, since these unique specimens could all be destroyed by one disaster, as happened during WW II with the destruction by a single bomb of much of the Hunterian Museum pathology collection. However, 21st-century digital technology makes possible the creation of "virtual registries". This would document specimens located all over the world. The data, not the specimens themselves, would be available from one source. Examination of these data should be, *ideally*, the prelude to examination of the actual physical specimens.

CONCLUSIONS

Everything that has happened once can happen again.
(Jacobo Timerman, *Prisoner Without a Name, Cell Without a Number*)

Nature abhors a vacuum.
(François Rabelais, *Gargantua and Pantagruel*)

In his 1963 treatise on *The Evolution and Eradication of Infectious Diseases*, the eminent paleopathologist T. Aidan Cockburn wrote "we can look forward with confidence to a considerable degree of freedom from infectious diseases at a time not too far in the future. Indeed, if the present pace of research and the present increase in the world's wealth continue ... it seems reasonable to anticipate within some measurable time, such as 100 years, all the major infections will have disappeared." (Cockburn 1963:150). At the same time, he cautioned, "This desirable goal will not be easily reached, for the difficulties are many, and unpleasant surprises are inevitable ... And even as we are successfully eliminating one set of infections, new ones will almost certainly appear, for we live in a world swarming with potential pathogens in many forms. Evolution is not merely something that happened in the past; it is an essential part of both the present and the future, so that out of all the microorganisms that are continually seeking to invade our bodies, one that is favored by changing conditions will occasionally succeed." (Cockburn 1963:150).

Four years later, the U.S. Surgeon General, William H. Stewart, assured an invited audience of state and territorial public health officials at the White House in Washington DC that it was "time to close the book" on infectious diseases because of the phenomenal success of new antimicrobial drugs and vaccines (quoted in Garrett 1994:33). Of course, these were early days in the antibiotic war on infectious diseases—before the amazing ability of some pathogenic microbes to evolve resistance to currently effective drugs was widely recognized (except by microbiologists).

Today, more than a half century later, we see that it is the *second* part of Cockburn's prescient remarks that prevails. Stewart's optimism was challenged within two decades by the resurgence of tuberculosis and the appearance of drug-resistant strains of various bacterial pathogens (e.g., *Staphylococcus aureus*) in the large cities of the most economically and scientifically advanced nations (Garrett 1994; Fauci 2003; Roberts and Buikstra 2003), as well as the emergence of new infectious diseases from exotic locales (Pendergast 2010), thus illustrating the truth of the two aphorisms that head this section.

What do these new medical threats have to do with paleopathology, you ask? They remind us that *life is not static*: microbes, parasites, and vectors evolve in response to natural selection, typically at a much faster rate than their host populations because of the very short lifespans of these tiny organisms, and the documented human evolutionary responses to infectious diseases, such as the hemoglobin variations that offer protection against malaria, are fairly limited but also reasonably effective. In the past century we have seen important mutations in some common infectious diseases, such as influenza, tuberculosis, and some staphylococcal and streptococcal illnesses. All three pathogens have the capacity to greatly affect patterns of mortality (which would be identified by bioarchaeologists of the future through mortuary analyses). Of the three, only tuberculosis leaves distinctive skeletal evidence, but today it is also detectable in dry bone series by molecular assay techniques searching for mycolic metabolites and aDNA. A greater number of bacterial and even viral pathogens may eventually yield their secrets to molecular paleopathology. The emergence of a new infectious disease at one point in time may meaningfully affect the prevalence—and virulence—of concurrent infections, as HIV did for tuberculosis and a host of other opportunistic infections in the late 20th century. We should not assume that a similar synchronous synergy never existed in the past, or that it will never appear again.

Rapid air travel between continents means that victims of acute infectious diseases who in earlier times would not have survived a lengthy voyage by sea or trek overland may now inadvertently introduce novel infectious pathogens to previously "virgin soil" populations (Garrett 1994; Pendergast 2010, Muehlenbein et al. 2010), to the bewilderment and frustration of clinicians and future paleopathologists alike. And just as infectious diseases may change through time (and not only in the past) due to combinations of natural and cultural factors (Ewald 1994), other human ailments may also change in manifestation or prevalence, due to cultural factors. Will the current "epidemic" of obesity in the U.S. result in a higher prevalence of DISH (diffuse idiopathic skeletal hyperostosis) in those population segments who combine poor dietary and activity habits with higher genetic risks for this disorder? Will paleopathologists of the late 21st century and beyond wonder at the increased number of amputated feet and legs in historic skeletal series from recent times, the legacy of widespread uncontrolled diabetes? And will changes in global climate patterns and international travel exacerbate the spread of "tropical" diseases such as South American leishmaniasis and Chagas disease and African zoonotic infections, by expanding the range of both pathogens and vectors? One thing is certain: our students, and *their* students after them, will face interesting new challenges, as well as some unresolved old questions. Paleopathology will never become a dull endeavor.

CODA

We have seen the future and it is not us. (with apologies to Lincoln Jeffers)

Or

Never trust anyone over 30. (Jack Weinberger)

When we began our professional training in anthropology some four decades ago, analytical techniques such as electron microscopy and nondestructive three-dimensional imaging were just beginning to be employed, with molecular paleopathology lying far in the future. The first intimations of widespread repatriation of human skeletal remains had appeared, but the perpetual availability of securely curated museum skeletal collections was largely unquestioned. Much has changed now. The destruction of important regional collections is far advanced in the United States, South Africa (Legassick and Rassool 2000) and New Zealand. However, while "repatriation" appears largely as an Anglophone colonial issue, in other countries such as Portugal, Greece, and throughout Europe and most of Latin America, there remains a commitment to long-term curation of human remains. This will, no doubt, yield valuable scientific information on health and conditions of life in the past. (For a cogent discussion of the current legislation affecting repatriation in the US and in Canada, see Buikstra (2006) and Webb (1995), who presents a concise overview of this issue regarding Australian Aboriginal human remains.)

Four decades ago, anthropologically based studies of large population samples were proliferating, with emphasis on investigating diachronic changes in health and disease within the context of cultural and ecological changes. The then "New Archaeology" played a key role in providing that context. The professionalization of paleopathology emphasized training in cultural anthropology, archaeological theory and methods, human biology, and the specialized fields of pathology, epidemiology, and medical geography, as the era of interdisciplinary research replaced the solitary scientist who "knew everything there was to know". And women began to enter the field of paleopathology in growing numbers, a development dear to our own hearts.

We have both devoted a large part of our professional lives to the curation of older collections. A particularly interesting consequence to this era of repatriation is that older collections have become more valuable to our profession. Cadaver collections from the recent past have become an important resource for study on the natural history of untreated infectious diseases, and for the rigorous truth-grounding of conventional techniques (for example, see Redman et al. 2009 and Alves Cardoso and Henderson 2010). The relative costs and benefits of destructive analysis have shifted in response to the productive new paradigms provided by stable isotope chemistry, histology, and ancient DNA analysis. It is vitally important that curators attend to potential new uses while documenting previous ones. A case in point is the likelihood that both old high-dose radiography as well as the new tomographic imaging technologies damage ancient DNA (Grieshaber et al. 2008).

So what can we say about the future of paleopathology, based on its past history?

We expect that it will be surprising. It will be (we hope) broadly integrative of essential aspects of cultural and biological anthropology, clinical medicine, genetic research, and current issues in medical ethics. We hope that it will experience a return to a close association with the history of medicine, a field that is undergoing its own renewal (e.g. Quetel 1990; Waller 2009). And we are boldly optimistic that the next generations of paleopathologists will discover "secrets of the past" that we only dreamed of unlocking. As human beings, we are all infinitely capable of enlightenment, and the future of paleopathology is, in a sense, just beginning.

REFERENCES

Alves Cardoso, F., and Henderson, C. Y., 2010   Enthesopathy Formation in the Humerus: Data From Known Age-at-Death and Known Occupation Skeletal Collections. American Journal of Physical Anthropology 141(4):550–60.

Armelagos, G. J., and van Gerven, D. P., 2003   A Century of Skeletal Biology and Paleopathology: Contrasts, Contradictions, and Conflicts. American Anthropologist 105 (1):53–64.

Aufderheide, A. C., and Rodríguez-Martín, M., 1998   The Cambridge Encyclopedia of Human Paleopathology. Cambridge: Cambridge University Press.

Boldsen, J. E., 2001   Epidemiological Approach to the Paleopathological Diagnosis of Leprosy. American Journal of Physical Anthropology 115(4):380–387.

Buikstra, J. E., 2006   Repatriation and Bioarchaeology: Challenges and Opportunities. In Bioarchaeology: the Contextual Analysis of Human Remains. J. E. Buikstra, and L. A. Beck, eds. pp. 389–415. Amsterdam: Academic Press.

Buikstra, J. E., and Cook, D. C., 1980   Paleopathology: An American Account. Annual Review of Anthropology 9:433–470.

Chhem, R. K., and Brothwell, D. R., eds., 2008   Paleoradiology. Imaging Mummies and Fossils. Springer:New York.

Cockburn, T. A., 1963   The Evolution and Eradication of Infectious Diseases. Baltimore: Johns Hopkins University Press.

Cook, D. C., and Powell, M. L., 2005   Piecing the Puzzle Together: North American Treponematosis in Overview. In The Myth of Syphilis: the Natural History of Treponematosis in North America. M. L. Powell, and D. C. Cook, eds. pp. 442–479. Gainesville: University Press of Florida.

Cook, D. C., and Powell, M. L., 2006   The Evolution of American Paleopathology. In Bioarchaeology: the Contextual Analysis of Human Remains. J. E. Buikstra, and L. A. Beck, eds. pp. 281–322. Amsterdam: Academic Press.

Cucina, A., and Tiesler, V., 2003   Dental Caries and Antemortem Tooth Loss in the Northern Peten Area, Mexico: A Biocultural Perspective on Social Status Differences Among the Classic Maya. American Journal of Physical Anthropology 122(1):1–10.

Down, J. L. H., 1866   Observations on an Ethnic Classification of Idiots. London Hospital Clinical Lectures and Reports 3:259–62. Reprinted in Down's Syndrome (Mongolism): a Reference Bibliography. R. F. Vollman, ed. Washington, DC: USDHEW–PHS–NIH. 1969.

Ewald, P. W., 1994   Evolution of Infectious Disease. Oxford: Oxford University Press.

Fauci, A. S., 2005   Emerging and Reemerging Infectious Diseases: The Perpetual Challenge. Academic Medicine 80(12):1079–1085.

Garrett, L., 1994   The Coming Plague: Newly Emerging Diseases in a World Out of Balance. New York: Farrar, Straus and Giroux.

Guhl, F., Jaramillo, C., Vallejo, G. A., Yockteng, R., Cárdenas–Arroyo, F., Arriaza, B., and Aufderheide A. C., 1999   Isolation of Trypanosoma cruzi DNA in 4000-Year-Old Mummified Human Tissue From Northern Chile. American Journal of Physical Anthropology 108:401–407.

Grieshaber, B. M., Osborne, D. L., Doubleday, A. F., and Kaestle, F. A., 2008   A Pilot Study Into the Effects of X–Ray and Computed Tomography Exposure on the Amplification of DNA From Bone. Journal of Archaeological Science 35:681–687.

Hackett, C. J., 1976   Diagnostic Criteria of Syphilis, Yaws and Treponarid (Treponematoses) and Some Other Diseases in Dry Bones (for Use in Osteo–Archaeology). In Sitzungsberichte der Heidelberger Akademie der Wissenschaften Mathematisch-naturwissenschaftliche Klasse, Abhandlung 4. Berlin: Springer–Verlag.

Hawass, Z., Gad, Y. Z., Ismail, S., Khairat, R., Fathalla, D., Hasan, N., Ahmed, A., Elleithy, H., Ball, M., Gaballah, F., Wasef, S., Fateen, M., Amer, H., Glstner, P., Selim, A., Zink, A., and

Pusch, C. M., 2010   Ancestry and Pathology in King Tutankhamun's Family. Journal of the American Medical Association 303(7):638–647.

Hrdlička, A., 1909   Tuberculosis Among Certain Indian Tribes of the United States. Washington, DC: U.S. Government Printing Office.

Jacob, T., Indriati, E., Soejono, R. P., Hsü, K., Frayer, D. W., Eckhardt, R. B., Kuperavage, A. J., Thorne, A., and Henneberg, M., 2006   Pygmoid Australomelanesian *Homo sapiens* Skeletal Remains From Liang Bua, Flores: Population Affinities and Pathological Abnormalities. Proceedings of the National Academy of Sciences, USA 103(36):13421–13426.

Kolman, C. J., Centurion-Lara, A., Lukehart, S. A., Owsley, D. W., and Tuross, N., 1999   Identification of *Treponema pallidum* Subspecies *pallidum* in a 200-year-old Skeletal Specimen. Journal of Infectious Diseases 180:2060–2063.

Legassick, M., and Rassool, C., 2000   Skeletons in the Cupboard: South African Museums and the Trade in Human Remains, 1907–1917. South African Museum, Cape Town.

Lukacs, J. R., and Largaespada, L. L., 2006   Explaining Sex Differences in Dental Caries Prevalence: Saliva, Hormones, and "Life-History" Etiologies. American Journal of Human Biology 18(4):540–555.

McGrath, J., 1988   Social Networks of Disease Spread in the Lower Illinois Valley: A Simulation Approach. American Journal of Physical Anthropology 77:483–496.

Merbs, C. F., 2005   A New World of Infectious Disease. Yearbook of Physical Anthropology 35(S15):3–42.

Muehlenbein, M. P., Martinez, L. A., Lemke, A. A., Ambu, L., Nathan, S., Alsisto, S., and Sakong, R., 2010   Unhealthy Travelers Present Challenges to Sustainable Primate Ecotourism. Travel Medicine and Infectious Disease 8: 169–175.

Morse, D. F., 1967   Tuberculosis. *In* Diseases in Antiquity. D. Brothwell, and A. T. Sandison, eds. pp. 249–271. Springfield IL: C. C. Thomas.

Obendorf, P. J., Oxnard, C. E., and Kefford, B. J., 2008   Are the Small Human-Like Fossils Found on Flores Human Endemic Cretins? Proceedings of the Royal Society B (Biological Sciences) 275(1640):1287–96.

Ortner, D. J., 2003   Identification of Pathological Conditions in Human Skeletal Remains. 2nd edition. Amsterdam: Academic Press.

Ortner, D. J., and Aufderheide, A. C., eds., 1991   Human Paleopathology: Current Syntheses and Future Options. Washington, DC: Smithsonian Institution Press.

Ortner, D. J., Tuross, N., and Stix, A. I., 1992   New Approaches to the Study of Disease in Archaeological Populations. Human Biology 64:337–360.

Owsley, D. W., and Jantz, R. L., 1994   Skeletal Biology in the Great Plains: Migration, Warfare, Health and Subsistence. Washington, DC: Smithsonian Institution Press.

Pendergrast, M., 2010   Inside the Outbreaks, the Medical Detectives of the Epidemic Intelligence Service. New York: Houghton Mifflin Harcourt.

Powell, M. L., 1992   Endemic Treponematosis and Tuberculosis in the Prehistoric Southeastern United States: the Biological Costs of Chronic Endemic Disease. *In* Human Paleopathology: Current Syntheses and Future Options. D. J. Ortner, and A. C. Aufderheide, eds. pp. 173–180. Washington, DC: Smithsonian Institution Press.

Quetel, C., 1990   History of Syphilis. Baltimore: Johns Hopkins University Press.

Rafi, A., Spigelman, M., Stanford, J., Lemma, E., Donoghue, H, and Zias, J., 1994   DNA of *Mycobacterium leprae* Detected by PCR in Ancient Bone. International Journal of Osteoarchaeology 5:287–290.

Redman, J. E., Shaw, M. J., Mallet, A. I., Santos, A. L., Roberts, C. A., Gernaey, A. M., and Minnikin, D. E., 2009   Mycocerosic Acid Biomarkers for the Diagnosis of Tuberculosis in the Coimbra Skeletal Collection. Tuberculosis 89(4):267–77.

Roberts, C. A., 2003   Health and Disease in Britain: from Prehistory to the Present Day. Stroud: Sutton Publishing.

Roberts, C. A., and Buikstra, J. E., 2003   The Bioarchaeology of Tuberculosis. A Global View on a Reemerging Disease. Gainesville: University Press of Florida.

Roberts, C. A., and Manchester, K., 1995   The Archaeology of Disease. 2nd edition. Stroud: Sutton Publishing.

Roberts, C. A., and Manchester, K., 2005   The Archaeology of Disease. 3rd edition. Stroud: Sutton Publishing.

Rothschild, B. M., and Rothschild, C., 1995   Treponemal Disease Revisited: Skeletal Discriminators for Yaws, Bejel, and Venereal Syphilis. Clinical Infectious Diseases 20: 1402–1408.

Stewart, W. H., 1967   "A Mandate for State Action", Presented at the Association of State and Territorial Health Officers, Washington DC, December 4, 1967.

Ubelaker, D. H., 1982   The Development of American Paleopathology. In A History of American Physical Anthropology 1930–1980. F. Spencer, ed. pp. 337–356. New York: Academic Press.

Von Hunnius, T. E., Roberts, C. A., Boylston, A., and Saunders, S. R., 2006   Histological Identification of Syphilis in Pre–Columbian England. American Journal of Physical Anthropology 129(4):559–566.

Waller, J., 2009   The Dancing Plague. Naperville: Sourcebooks.

Watson, J. T., Fields, M., Martin, D. L., 2010   Introduction of Agriculture and Its Effects on Women's Oral Health. American Journal of Human Biology. 22(1):92–102.

Webb, S., 1995   Palaeopathology of Aboriginal Australia: Health and Disease Across a Hunter-Gatherer Continent. Cambridge: Cambridge University Press.

Wood, J. M., Milner, G. R., Harpending, H. C., and Weiss, K. M., 1992   The Osteological Paradox: Problems of Inferring Prehistoric Health from Skeletal Samples. Current Anthropology 33(4):343–370.

Zink, A. R., Sola, C., Reischl, U., Grabner, W., Rastogi, N., Wolf, H., and Nerlich, A. G., 2003   Characterization of Mycobacterium Tuberculosis-Complex DNAs From Egyptian Mummies by Spoligotyping. Journal of Clinical Microbiology 41(1):359–67.

PART II   Methods and Techniques of Inquiry

13   A Knowledge of Bone at the Cellular (Histological) Level is Essential to Paleopathology

Bruce D. Ragsdale and
Larisa M. Lehmer

INTRODUCTION

An anthropologist deduces a species from a bone, a face from a skull, and life habits from a skeleton; the paleopathologist deduces disease from dry bone abnormalities by reference to medical museum specimens, pathology texts, and roentgenologic literature. But these "knowns" are classic, extreme and rare. In most instances determining the etiology of a bone lesion is extremely difficult. Bone, like all tissues, has few ways of reacting, so different causes may produce similar results. However, recognizing successive gradual pathologic stages from normal states may reveal the mechanisms of disease. Cells are the effectors of change, and an understanding of them is a necessary foundation for diagnosis. So, knowledge of skeletal dynamics in growth and adaptation is fundamental (Johnson 1966).

This chapter will review the influences that drive the cellular activities that modify mineralized tissue. A dry bone from antiquity is like a fossilized footprint left in mud. Just as the foot is obviously of more interest than the print, so, too, in paleopathology the disease is more interesting than the resulting lesion. One's frame of mind must be to mentally "put the soft tissue that abutted solid substance back on the specimen," to find clues to the host response that modified bone, and then to deduce the

A Companion to Paleopathology, First Edition. Edited by Anne L. Grauer.
© 2012 John Wiley & Sons, Ltd. Published 2016 by John Wiley & Sons, Ltd.

mechanisms that produced the change. Through this process one might then determine the category of disease provoking the skeletal change, and only then, take a stab at a specific diagnosis.

BASIC PRINCIPLES

Disease is the alteration of living tissue that jeopardizes "ease" and survival. The causes of skeletal disease are aberrations of normal growth, development, and maintenance, rather than the interjection of something totally new, which limits the number of ways in which bone can react. Consequently, "morphologic residua" of different diseases significantly overlap. Even the Basic Categories of disease (discussed below, and see Table 13.2) overlap to some extent, since inflammation includes repair, trauma is associated with circulatory changes, and metabolic diseases may call forth mechanical compensation.

The sequence from normal, through adaptation, to sickness and death is a continuum. However, most of the skeletal alterations that paleopathologists encounter are the result of long-standing disease, which may or may not have led to the death of the individual. Hence, using modern diseases, which are intercepted by medical professionals comparatively early, as reference standards for diagnosing diseases of the past, is problematic as they bear little resemblance to the untreated end stage of the process (Ragsdale 1997). While autopsy specimens are more likely to represent advanced expressions of disease, the morphologic expression is generally modified by therapeutic intervention. Ortner and Putschar's (1981) work is of greater relevance for paleopathological diagnosis, since many of its illustrations are cases collected in the 19th century or come from Dr. Putschar's experiences in underdeveloped countries (Ragsdale 1996). Unfortunately, further complicating the diagnostic process is the fact that well-defined structural change may appear in many diseases, and a particular disease may produce more than one morphologic expression. Only in the presence of a few extreme abnormalities can relatively specific diagnoses be made. Obviously, this casts doubt on the diagnostic specificity of a small slice of dry bone studied histologically.

Further complicating our ability to diagnose the presence of specific disease from dry bone specimens is the fact that pathologic morphology is linked to only three aberrations: circulation, metabolic factors, and mechanical stress. These three influences act on the bone fabric of collagen bundles impregnated with mineral crystals through three cell types. Only osteoblasts produce significant osteoid matrix (bone fabric). Only osteoclasts can significantly resorb (remove) bone (and do so much faster than osteoblasts create bone, so osteopenia is the usual initial expression of a bone reaction). Osteocytes can both resorb and form bone, but only to a limited degree along the walls of their lacunae where they maintain bone, serve as mechanical sensors, and control calcium flux. These three cells are normally active in the daily remodeling and rejuvenation of the living skeleton. No other cell types directly modify bone structure. Pathologic anatomy results when local (pathological) conditions stimulate abnormal activity of these cells. An understanding of these conditions is a necessary background to diagnose skeletal disease, and is facilitated by keeping a few basic principles in mind (Table 13.1).

Table 13.1   The twelve basic principles of orthopedic pathology

1. There are seven Basic Categories of disease.
2. There are only three influences which govern cell activities: circulatory dynamics, metabolic factors, and mechanical forces.
3. Only osteoblasts make bone, are favored by elevated pH, passive hyperemia (venous congestion and edema), negative bioelectric charge, and growth hormone. Only osteoclasts delete bone, favored by low pH, active hyperemia (increased arterial input), positive bioelectric charge, and parathormone and work 100 times faster than osteoblasts.
4. Bone formation proceeds on surfaces; what is lost is gone forever, but persisting surfaces can be added to.
5. The "endosteum," if the term must be retained in adults, consists of the mature fat cells that abut internal bone surfaces and are capable of modulating into osteoblasts.
6. The interface between periosteum and cortex along growing bones can be a plane of bone formation (as in additions to diaphyseal diameter) or resorption (as in metaphyseal cutback); these same activities can be reawakened in disease as periosteal new bone or cortical lysis.
7. Turnover and remodeling proceed, even in the presence of disease. In face of disease, the mechanically most important (i.e., the stressed) bone components are preserved the longest and may indeed be strengthened.
8. Fully differentiated mesenchymal cells, e.g., "fibroblasts" and adipocytes of soft tissue and of the marrow, are not irrevocably differentiated cells, but retain a capacity to modulate into other functional cell types as conditions dictate. Fat, fascia, and muscle contribute to "periosteal reactions."
9. All cartilage formed, except for articular, ear/nasal/airway and sarcomatous cartilage, is doomed to resorption (e.g., epiphyseal and fracture callus cartilage).
10. Bone is an enclosed space bound by cortex and internally baffled by cancellous modules. This shields vessels from the mechanical compressive forces operative in soft tissue but makes bone's blood supply vulnerable, especially to compressive obliteration when anything is added to the marrow space.
11. Age epochs modify likely sites of disease, reactive potential and therefore morphologic expression.
12. Benign tumors may stabilize in size, but even so, by strategic position, can incapacitate or kill; malignant tumors inexorably grow and by definition, can metastasize.

CATEGORIES OF DISEASE

Factors that influence the speed with which bone lesions evolve leave clues that distinguish chronic from acute skeletal disease. The tempo of the process is a potentially important factor in diagnosis and can be discerned by "morphologic analysis," i.e., attention to the character of margins, periosteal reactions, and presence or absence of regional remodeling effects (Ragsdale 1993a). The location of solitary lesions is also helpful in diagnosis, and the character and distribution of multifocal lesions are similarly important in narrowing a differential diagnosis.

In 1996, Ragsdale and co-workers proposed using seven basic disease categories as part of the diagnosis of paleopathological conditions. These can be easily remembered by use of the acronym VITAMIN (Table 13.2) (Ragsdale and Miller 1996). Approaching differential diagnosis by looking for clues as to which of the Seven Basic

Table 13.2 The seven basic categories of disease

I	V	Vascular
II	I	Innervation/Mechanical
III	T	Trauma/Repair
IV	A	Anomaly
V	M	Metabolic
VI	I	Inflammatory/Immune
VII	N	Neoplastic

Categories of Disease is represented helps minimize the overwhelming bewilderment that can result from attempting to choose from several hundred individual entities that can afflict bone. Thus, paleopathologists should redirect their enthusiasm for diagnosing specific diseases in human skeletal remains, to deducing the less ambitious (but more often correct) classification by disease category. This will enhance the comparability of data within paleopathology.

Vascular disturbances

There are two basic vascular mechanisms which can impact bone. The first is active hyperemia (increased arterial blood flow), which increases tissue oxygen tension, fueling mitochondria-rich osteoclasts that engage in bone resorption. The second is passive hyperemia (slow flow in patulous veins) which creates back pressure and increases permeability, leaking the edematous protein substrate that osteoblasts use to produce matrix. Due to the physics of streaming potentials, these contrasting vascular flows are characterized by opposite local electrical field effects (MacGinitie et al. 1993). Charged solutes in blood moving past fixed charges in vessel walls create energy fields similar to an electromagnet. Active hyperemia engenders a positive electrical environment, favoring osteoclasts, which work best in a slightly acidic pH. Passive hyperemia evokes a net negative field and a basic environment favoring osteoblasts.

Examples of hyperemic conditions include:

Active hyperemia: Enlarged cortical penetration ports of vessels near any inflammation; lytic phase of early acute osteomyelitis; juxta-articular osteolysis in rheumatoid arthritis; pencil deformity (concentric atrophy) of ainhum; loss of neuroregulation in dysvascular responses (Sudeck's atrophy, post-traumatic osteolysis) and leprosy; increased blood flow demanded by metastatic carcinoma.

Passive hyperemia: Dense medullary edema around a Brodie's abscess supplies protein substrates for marginal sclerosis; back pressure in varicose veins or distal to an atriovenous fistula evokes an edematous periosteal reaction; calvarial hyperostosis due to invasive meningioma clogs venous outflow (Huggins et al. 1981); and pachydermohyperostosis (Uehlinger disease) is its systemic manifestation.

Biphasic hyperemias: Circulation is initially active and later passive: acute osteomyelitis transitioning into chronic; diabetic osteopathy; Charcot joint; melorheostosis; hypertrophic pulmonary osteoarthropathy; Paget's disease.

Importantly, the mechanical, hematopoietic, and fat reserve functions of bone are dependent on an adequate blood supply, i.e., viability. Injury or disease elicits a vascular response. There are three noteworthy peculiarities of the vascular supply to bone which directly impact pathological processes in bone.

The first is that bone is a completely rigid compartment. The blood vessels in bone are totally protected from the mechanical compressive forces operative in soft tissue and have, therefore, thin walls. This confers a susceptibility to hydrostatic compression by anything that is added to the marrow space such as edema, pus, fat cell swelling, histiocytes filled with abnormal lipid or mucopolysaccharide, tumor, etc. Cessation of flow leads to ischemia and necrosis. At the end of a bone and in cuboidal bones this is "avascular (aseptic) necrosis." Conversely, ischemic necrosis in metaphyses or diaphyses is a "bone infarct." This can be seen with storage diseases (Gaucher's disease), vasculitis, and sickle cell disease. Bone infarcts are uncommon in the modern era and likely less so in antiquity.

Ischemic aseptic necrosis is most commonly found today in the femoral head, a precariously supplied spheroid with narrower inlet (the neck) than equatorial girth. Anything that swells fat cells (e.g., alcoholism, steroid therapy, genetically caused storage diseases like Gaucher's) will compress the only thing that can yield in the closed space defined by the cortex-marrow vessels: blood vessels. Hip dislocation and femoral neck fracture simply transect vessels, vasculitis occludes them with thrombus, and sickle cells clog them. Avascular necrosis of the femoral head in a child (Legg-Calvé-Perthes) follows a growth spurt that is unaccompanied by sufficient vascular remodeling. Whatever the cause, the attempt at repair of a dead femoral head segment begins with removal of dead cancellous bone along a reactive interface which cuts away the necrotic segment, followed by collapse and eventual severe osteoarthritis due to the resultant shape change and head fragmentation.

The second peculiarity of the vascular supply to bone is that only one-third to one-half of the vascular network in bone is open at any given time. Blood is shunted from regions of lesser to greater need by chemo-sensors (the glomoid system) that act over neural pathways to modify arterial caliber (Johnson 1971). The periosteal, osseous and medullary vascular "fields" constantly shift (take turns) from moment to moment. This is regulated by the angioglomic system that is commonly upset in disease. Therefore, circulatory disturbances are regional, affecting bone segments or bones served by the same vascular trunk.

Thirdly, the capacity of the venous system in bone is 6–8 times that of the arterial system. The shared vascular connections between muscle and bone play an important part in blood movement. The loss of bone density that accompanies disuse and muscle paralysis is in part due to the loss of the muscular "pump's" assistance to venous return. This results in progressive distention of intraosseous sinusoids caused by osteoclastic removal of bone. A further effect of disuse is the removal of the mechanical stimulus for bone formation while the lysis of physiologic turnover proceeds.

Innervation and mechanical disease

Mechanical factors operate through bioelectric signals proportional to use or disuse. Bone is a composite, crystalline substance. Hydroxyapatite hexagonal prisms solidify the periodic organic crystal fabric of collagen. Electric charge generated by pressure on crystals is known as piezoelectricity. Whether affecting a solitary trabeculum or an entire longbone, deformation (i.e., slight bending), results in a charge separation in the crystalline fabric. When thus loaded, concavities become sites of negative charge favoring osteoblasts that bolster the site with new bone. Convexities are the site of unburdening, positive bioelectric charge, and osteoclasia. In this way, bone is optimally "repositioned" under the load.

Bone adapts to its mechanical environment during life. The proposition that differences in morphology can be used to investigate differences in past mechanical environments is widely accepted among paleoanthropologists and bioarchaeologists (Ruff 2005), but has important limitations (see Chapter 29 by Jurmain et al., this volume, and Chapter 28 by Waldron, this volume, for further discussion). Examples should come readily to mind: hypertrophy of bone and/or alterations in shape due to specific occupations, sports activities or other specialized activities; "flowing" hyperostosis along vertebral bodies on the opposite side of handedness; relative loss of bone on the convex side of a scoliotic curve; loss of bone following denervation, and inactivity of senescence.

A negative feedback control mechanism involving bioelectricity has been proposed as the basis for changes in bone structure. Here, extrinsic force is translated by the osseous transducer (bone fabric) into a proportional electrical command signal, eliciting a cellular response that creates a structural change to resist the extrinsic force. In the feedback model presented by Lanyon et al. (1982), strain (not stress) on bone is the stimulus for the feedback loop causing bone formation. The new bone reduces strain in the area of formation, and the loop ends at what Lanyon et al. termed the "optimum customary strain level," or equilibrium. Decreased strain on an area would, by this same hypothesis, lead to bone loss, thereby increasing the strain on any single area of bone and ultimately leading back to equilibrium. The impact of strain will differ with location within a single bone and within the skeleton (Carter 1984), due to age (Currey et al. 2002), disease state, hormonal status (Lanyon and Skerry 2001) and genetics. The type and tempo of strain may also play an important role.

Trauma and repair

This category includes both accidental and intentionally caused changes to the skeletal system, both of which provide important information about the culture and environment in which the individual lived. Trauma includes fractures, dislocations, posttraumatic deformity, cuts, borings and scrapes to bone (e.g. incomplete trephination or scalping, mutilation), amputations, decapitations or other dismemberments of living individuals, cranial deformation, cauterization, thermal injuries, and others (Ortner 2003, and see Chapter 20 by Judd and Redfern, this volume).

Repair is an accelerated recapitulation of normal growth. The stages and rate of repair are constant (though attenuated in calvaria and exaggerated in ribs) so that the elapsed time since the fracture can be determined from roentgenographic and histologic details. The three phases of fracture healing are circulatory (initial active hyperemia supporting osteolysis of injured matrix), metabolic (passive hyperemia supporting osteoblastic synthesis of internal and external callus) and mechanical (prolonged remodeling under mechano-piezoelectric guidance). Most fractures heal with good function in the absence of medical care. Healing with a false joint (pseudoarthrosis or neoarthrosis) may mimic a tumor or infection. Surviving cells in injured muscle may modulate to bone, ossifying to contribute to an exuberant callus that eventually becomes a bony prominence (exostosis).

Pathologic fracture caused by an underlying pathological condition is relegated to the category containing the responsible condition (tumor = Neoplastic; osteoporosis = Metabolic; osteogenesis imperfecta = Anomaly; etc.) rather than in the Trauma category.

Anomalies (misgrowths)

Anomalies, or "misgrowths," are local distortions of shape and size that result from skewed growth due to malfunction of cells or organelles (for more discussion of these conditions see Chapter 21 by Barnes, this volume) Generalized developmental failure produces dwarfism; the several types of epiphyseal and metaphyseal dwarfism (whose deformed bone ends resemble osteoarthritis) must be separated from achondroplastic dwarfism. Localized failures produce cleft palate, clubfoot, and congenital dislocation of the hip. Abnormal organelles affecting mucopolysaccharide function skew growth-plate and articular cartilage development to produce distorted dwarfism. Osteoblasts that lack a normal genetic recipe for collagen produce a thin-boned skeleton (osteogenesis imperfecta) that suffers multiple fractures and has excessive osteocyte lacunae per unit area of bone appreciable histologically. Incompetent osteoclasts in Albers-Schonberg disease (osteoporosis; marble bone disease) leave most bone that is produced, resulting in lack of metaphyseal cutback and retention of enchondral bone leaving scant space for hematopoietic marrow, which takes up residence in an enlarged liver and spleen.

The metabolic category

Metabolic disease has to do with abnormal production, mineralization, or maintenance of the matrix of bone, i.e. the collagen fabric impregnated with mucopolysaccharide and crystals (see Chapter 22 by Kozlowski and Witas, this volume). So metabolic abnormalities are generalized throughout the skeleton with bilaterally symmetrical zones of maximal intensity, and where normal growth and remodeling are most rapid at the time of the disease. They result from an activation or suppression of cell activity and include nutritional inadequacy, vitamin deficiency, hormonal variations, metabolic errors, and poisons.

Important concepts concerning metabolic disorders include:

1. Bone is a composite material (like steel-reinforced concrete), comprised of synthesized matrix (osteoid) and mineral.
2. The simple rules, reactions, rates and investigative methods of inorganic chemistry do not apply in complex living biological systems and will not provide all-inclusive explanations.
3. Matrix made in abnormal genetic, metabolic nutritional, drug therapy, etc., or circulatory environments, fails to mineralize properly because it is *inherently* abnormal, not because of the popularly held notion that calcium is less available. (If the heart is beating, the serum calcium is sufficient to mineralize properly formed osteoid.)

4. Phosphate has more to do with the normal process of matrix building and mineralization than calcium. Understanding this is particularly important to understanding rickets, the various types of osteomalacia, and renal osteodystrophy.
5. Bones cannot become "decalcified" in the body, as they can if interred in an acidic environment. There is always total removal of the organic component (matrix) as well as mineral. This requires cellular activity and only osteoclasts and (to a much lesser degree) osteocytes possess this capability. Thus, the term "osteopenia" is preferred over "demineralization" when describing skeletal x-rays that show rarefaction.
6. Skeletal radiographs reflect a momentary summation of a dynamic interplay between simultaneous bone formation and bone removal ("turnover"). Osteoclasia and osteoblastic repair are always coupled, but one or the other may predominate in a given disease state at a given time period, and thus dominate the radiograph.

The problem of identifying the trigger for metabolic disorders is difficult in the context of paleopathological studies where skeletal remains and their archaeological context are our only sources of evidence (Walker et al. 2009). For instance, the effects of vitamin deficiency are dependent on intensity, duration, and age of the individual. Similarly, intermittent inadequate nutrition is registered simply as growth-arrest lines, while continuous malnutrition might lead to stunted stature. Recognizing the presence of these conditions in the archaeological record does not clearly lead to a precise cause. Adding to the complexity is a tendency for conditions to mimic one another. Acute infantile scurvy masquerades as a metaphyseal fracture with most of the cancellous bone missing; during healing, much of the bone is encased in a shell of new periosteal bone around organizing hematoma which will remodel to normal in later years (hematomas themselves do not "ossify": they organize). Osteomalacia will appear on roentgenograms as an osteopenia (the unmineralized osteoid will have been lost *in situ*), associated with bilaterally symmetrical incomplete cortical fractures (infractions); prolonged osteomalacia starting at an early age results in bowing deformities. Excesses of thyroxin, insulin, and cortisone all produce osteopenia without distinguishing features, but are unlikely to be seen in Stone-Age populations with a shorter lifespan.

Inflammation and immunity

Here the primary cause of bone change is an immune reaction to outside challenges such as bacteria, fungi, parasites, and viruses; or to changes in the immune system caused by failure of the body to recognize its own constituents. Many of these conditions are acute and leave no marks on the skeleton (e.g., pediculosis, typhus, malaria); some leave highly characteristic changes (e.g., leprosy, tuberculosis, treponemal diseases); some leave nonspecific changes related to the immune system malfunction over an extended period of time (e.g., sterile subacute polyostotic osteomyelitis).

The inflammatory process is an exaggeration of the normal disposal mechanism for everyday cell turnover. However, it is activated by the immune system rather than an inherent abnormality in circulation or vasoregulatory systems characterizing the conditions in the *Vascular* category. Cells within the human body are continually shed and renewed at varying rates. Infection damages normal cells and accelerates this

Table 13.3   The three fundamental inflammatory patterns

	Septic	*Granulomatous*	*Angiitic*
Cell	Polymorphonuclear	Histiocyte	Plasma cell
Response	Exudation	Migration inhibition	Perivascular proliferation & cuffing
Effect	Liquefaction	Caseation	Infarction
Focus	Abscess	Tubercle	Gumma
Host response	Great	Moderate	Low
Course	Fast	Slow	Late

turnover; the heightened response is pathological inflammation. In bone, the character, quality, quantity, and distribution of lytic change and reactive bone are indications of the tempo and character of the inflammatory fight with a foreign invader. They are also largely determined by the predominant cell type recruited and the attendant increased blood supply.

There are three fundamental inflammatory patterns: septic, granulomatous, and angiitic (Table 13.3). Septic inflammation can be suppurative (e.g., triggered by staphylococcus) with rapid production of pus that fills the marrow cavity, strips off the periosteum and isolates a dead cortex (sequestrum) within the reparative shell of new periosteal bone (involucrum). A less violent septic reaction, i.e., a bone abscess, is walled off by a dense border of bone. Important to diagnosis is the fact that extensive local repair may mimic sequelae of trauma or a neoplasm. Granulomatous infection (e.g. tuberculosis) entails the slow accumulation of rounded aggregates (granulomas) of macrophages and lymphocytes around a necrotic center that enlarges as satellite granulomas spread out with gradual bone destruction and scant if any, repair. This result can mimic neoplasm. Angiitic (or "vasculitic") inflammation (e.g., syphilis) is mediated by antibody-producing cells (lymphocytes and plasma cells) accumulating around blood vessels (angiitis); vascular thrombosis results in dead tissue (gumma) in bone and soft tissue. Angiitis is widespread over a region and in periosteum provokes bilaterally symmetrical reactions. Different still is "toxic" inflammation (e.g., rheumatoid arthritis), beginning as a fluid effusion that then becomes a low-grade inflammation dominated by the humoral side of the immune system (antibodies). Sustained active hyperemia, which stimulates osteoclasts, results in erosion of cortex at the joint capsule attachment, and erosion of subchondral bone that shares the boosted vascular supply of synovium. "Reactive" inflammation of the soft tissues and within bone to viruses (e.g., smallpox) or worms (e.g., echinococcus) is mild, nonspecific, and usually asymptomatic.

However, not all reactions are triggered by "outside" invaders. One's microbial flora includes pathogens held in check by a variety of commensal organisms and saprophytes. Infection may result from a break in the epithelial surface by malnutrition (inadequate cell shedding), exhaustion (inadequate secretions), or cuts and bites. But even germ injection does not ensure disease, for the core of the immune system, the reticuloendothelial system, thrives on a continuous diet of germs. Disease results only when the number and vigor of the organisms permit successful competition with cells for essential nutrients, enzymes, or space. The augmented turnover of injured cells drives inflammation.

The influence of the portal of entry is illustrated by the variety of diseases that can result from the same streptococcus. For instance, streptococcus entering from the skin produces erysipelas; entering via the mouth, septic sore throat and mastoiditis; from the gut, phlegmonous enteritis; venereally, puerperal fever; from the lung, hemorrhagic pneumonia; and from inoculation as by penetrating trauma, endocarditis and osteomyelitis. Furthermore, streptococci can provoke granulomatous inflammation (the Aschoff nodules of rheumatic fever), angiitic inflammation (polyarteritis and Schonlein-Henoch purpura), and collagen vascular disease (autoimmune nephritis). The lesson is clear: the disease morphology is determined by the reaction, not the specific trigger (organism).

The relevance of this to paleopathology is obvious. Through human history, specific diseases (e.g., Lucio's leprosy (Roverano et al. 2000; Choon and Tey 2009) and "malignant" or "galloping" syphilis (Cripps and Curtis 1967; Watson et al. 2004)) have burst on the scene with aggressive, rapidly progressive, even fatal expressions, later evolving into chronic illnesses with very different skeletal effects and individual case tempo. This suggests that we are naïve when we use particular modern clinical infections to model the expected histologic structures found in ancient dry bone specimens.

Neoplasms (Tumors)

Neoplasms are overgrowths of cells that commonly arise from antecedent accelerated cell proliferation related to normal growth and development or a sustained reactive state (Johnson 1953; and see Chapter 23 by Brothwell in this volume). Benign tumors, generally solitary, may achieve dramatic proportions or stabilize at a lesser size. They can weaken a bone to the point of fracture but remain a local problem.

Malignancy is characterized by inexorable local growth and metastases. Malignant bone neoplasms, arising there or coming to bone from a visceral organ are rare, and rarer still in antiquity with the lower turnover rate, the slower growth associated with diminished stature, and shorter lifespan. Primary bone, and joint sarcomas in particular, should be rare since the annual number diagnosed in the USA is only about 2,650 as of 2010 (American Cancer Society 2010). However, the age distribution of bone tumors derived from modern surgical pathology series is irrelevant to paleopathology since, benign or malignant, the endpoint in the paleopathological specimen was the death of the individual.

Neoplasms are named for the cells of which they are composed (e.g., lymphocytes—lymphoma, vessels—angioma) or their products (e.g., osteoid, cartilage, fibrous tissue, fat, etc.). Primary bone tumors emphasize the proliferative and functional capabilities in one or more of the three skeletogenic cell types (Johnson et al. 2000): chondroblast, osteoblast and osteoclast. Particular types of primary bone neoplasms occur in that portion of a bone where the normal cells, from which they arise, are most abundant and active (Figure 13.1). Therefore, roentgenograms signal the probability of a specific diagnosis by showing the site of involvement (e.g., osteosarcomas in metaphyses; chondroblastomas in epiphyses) (Madewell et al. 1981). A list of somatic tissues, in the order of normal cell turnover rate, is also a list of the relative likelihood of neoplasms. Since neither cells nor unmineralized matrices usually remain in paleopathological specimens, most neoplasms will be perceived as large holes, some

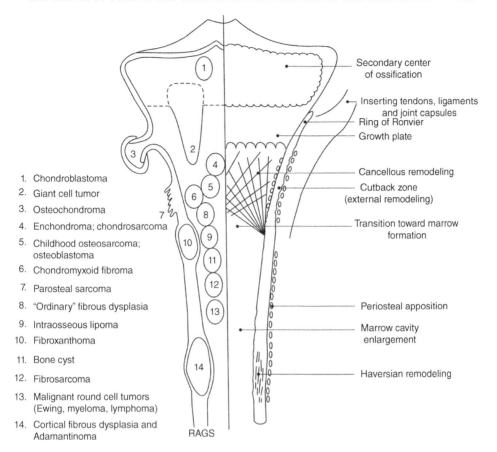

1. Chondroblastoma
2. Giant cell tumor
3. Osteochondroma
4. Enchondroma; chondrosarcoma
5. Childhood osteosarcoma; osteoblastoma
6. Chondromyxoid fibroma
7. Parosteal sarcoma
8. "Ordinary" fibrous dysplasia
9. Intraosseous lipoma
10. Fibroxanthoma
11. Bone cyst
12. Fibrosarcoma
13. Malignant round cell tumors (Ewing, myeloma, lymphoma)
14. Cortical fibrous dysplasia and Adamantinoma

Secondary center of ossification
Inserting tendons, ligaments and joint capsules
Ring of Ronvier
Growth plate
Cancellous remodeling
Cutback zone (external remodeling)
Transition toward marrow formation
Periosteal apposition
Marrow cavity enlargement
Haversian remodeling

RAGS

Figure 13.1 Typical bone tumor locations as they relate to events of normal growth and development.

bordered by reactive bone. The character of periosteal reactions reflect the speed of the growth (Ragsdale 1993a). Some tumors have an internal margin of dense bone indicating slow growth.

Metastatic cancer (lytic, blastic, or mixed), frequently affects the skeleton. Due to greater blood supply of hematopoietic marrow compared to fatty marrow, more cancer cells are delivered to it, and so it is no wonder that the spine (80 percent), proximal femur (40 percent), ribs and sternum (25 percent), skull and pelvis (25 percent), humerus and shoulder girdle (7 percent) are common sites for metastasis seen in autopsy cases (Ortner and Putschar 1981). Tumors arising from the breast, prostate, thyroid, lung, and kidney possess a special propensity to spread to bone (Coleman 1997), but using modern statistics on the prevalence of various cancers to understand the past must be undertaken cautiously. For instance, since 1990 there has been a precipitous decline in gastric cancer and an exponential rise in lung cancer, both attributed to altered lifestyles. The AIDS epidemic is afflicting large numbers of individuals with new or rare expressions of "old" neoplasms such as lymphoma (e.g., primary CNS involvement). Clearly, large changes in disease incidence are possible over short time intervals and in different cultures (Ragsdale 1995).

DIAGNOSTIC ACCURACY

Once having deduced the disease category, one can reasonably attempt to extend the diagnostic effort to a specific disease entity and/or chronologic stage in the morphologic spectrum of that entity. Despite the advances in diagnostic tools (e.g., CT and MRI scanning, DNA, molecular identification of pathogens, histologic study), diagnostic accuracy remains questionable (see Chapter 18 by Wanek et al. in this volume, and Chapter 8 by Spigelman et al. in this volume, for further discussions on these issues). In a series of four workshops held at annual meetings of the Paleopathology Association (PPA), Miller et al. (1996) compared participants' accuracy in specific versus categorical diagnosis of a series of 20 macerated, modern dry bone skeletal disease specimens of known diagnosis. Diagnosis by disease category (43 percent) was statistically more accurate than specific diagnosis (29 percent). Although, as the authors state, "the conditions did not approximate those found in most field or laboratory situations" and used modern specimens, the study serves to highlight "the difficulty in diagnosing specific conditions on the basis of dry bone and specimen radiographs alone" (Miller et al. 1996: 224). Rather than simply asserting "expert opinion" concerning the diagnosis of a specimen, recording a detailed descriptive account and the possible disease *category* ensures a maximum of comparable data that can be enhanced over time. Clearly, the descriptive/categorical approach enhances the narrative of ancient epidemiology by removing both the technological and single-expert biases that currently limit the scope of comparative analysis in paleopathology.

In sum, even though the mechanisms operative in skeletal disease are variations of normal biology, and bone has a limited number of ways in which it can respond, each of the seven basic categories provoke skeletal lesions that may be diagnostic in dry bone specimens. However, because some disease conditions of antiquity may have been eliminated by evolutionary selection with no modern counterpart, paleopathologists should be cautious when diagnosing specific diseases in skeletal populations.

HISTOLOGY IN PALEOPATHOLOGY

Various methods for preparing ground thin sections for light microscopic study, and thus eventually for paleopathological investigation, can be found in the literature (Maat et al. 2001; Schultz 2001; Beauchesne and Saunders 2006; Von Hunnis et al. 2006). Preparation involves embedding samples in plastic resin, sawing a thin slice and polishing it even thinner by hand or on a glass surface with diamond paste abrasive. Maat et al. (2001) provide a revised method which produces images of equivalent quality to samples prepared using embedding media. But only relatively dense and intact specimens hold up to the physical demands of the manual grinding procedure (Beauchesne and Saunders 2006).

The promise
A disease unfolds as a movie; the specimen is a single frame. But bone, like a single frame, can represent a record of past events. Therefore, each specimen provides clues to earlier frames in the show. Through the use of roentgenograms, stereo- and light

microscopy and scanning electron microscopy, rates of bone destruction and formation can be viewed. Applying microscopic techniques allows an even higher "resolution" to identify osseous changes (Flohr and Schultz 2009). However, this approach may only "make larger" what one does not understand grossly, or worse, lead to "tunnel vision," whereby the minute is explored at the expense of the "big picture." It has long been known that "the gross anatomy (as corroborated by radiographs) is often a safer guide to a correct clinical conception of the disease than the variable and uncertain structure of a small piece of tissue" (Ewing 1922).

As mentioned previously, bone has a limited number of ways in which it can react to the presence of disease (production or destruction of bone, or a combination of the two), and varying perhaps in the type of bone laid down, (e.g. woven or lamellar). These distinctions are not always observable at the macroscopic level and may be better identified by histological and ultrastructural techniques. By providing a simple description of the activity on bone surfaces, such as type of bone laid down and the location (endosteal, cortical or periosteal), important information can be added to other lines of evidence and improve a macroscopic differential diagnosis (Klepinger 1983). For instance, pockmarks created by Howship's lacunae on a surface is evidence of a rapid resorptive process due to an active pathological condition as in a giant cell tumor of bone (Figure 13.2). They can be seen with sidelight under a powerful dissecting microscope, but better still using SEM (Figure 13.2, upper right) or thin-section microscopy (Flohr and Schultz 2009). After maceration though, chemically or archeologically over time, all reactive matrix (osteoid) superficial to the mineralizing front will be lost. A "micro" snapshot of the formative surface, however, will appear pock-marked with osteocyte lacunae where osteoblasts have settled into lacunae as osteocytes (Figure 13.2, lower right).

There are a number of other ways in which histological analysis informs paleopathology. The histological analysis of bone cross-sections (e.g., cortical area, cortical thickness) have been shown to be powerful tools in the assessment of nutrition, growth patterns, and adult bone loss (Martin and Armegalos 1985). Although uniform procedures for the collection and reporting of data are wanting, standardized terms for reporting are available in detail (Parfitt 1987).

The use of histology to estimate age-at-death, understand taphonomic processes, and diagnose disease in human skeletal and mummified remains is also extremely promising (Schultz 2001).

The presence of pseudopathology can be determined with a good working knowledge of the microscopic structure of bone, including the histogenesis and growth of bone and the potential and/or presence of decomposition and diagenesis. For instance, long bones of South African diamond miners who died between 1897 and 1900, were macroscopically diagnosed with periostitis. Subsequent histologic sectioning of the bones revealed no pathological changes to the structure of the bone, suggesting that these bones were normal (Van Der Merwe et al. 2010). Schultz et al. (2007), after conducting a histological examination of bones from a child buried at the Early Bronze Age cemetery, argue that the macroscopic changes initially attributed to periosteal reaction were actually caused by post-mortem taphonomic changes. These, and other findings have led Van Der Merwe et al. (2010) to declare that 'periostitis' is likely over-diagnosed in archaeologically derived samples.

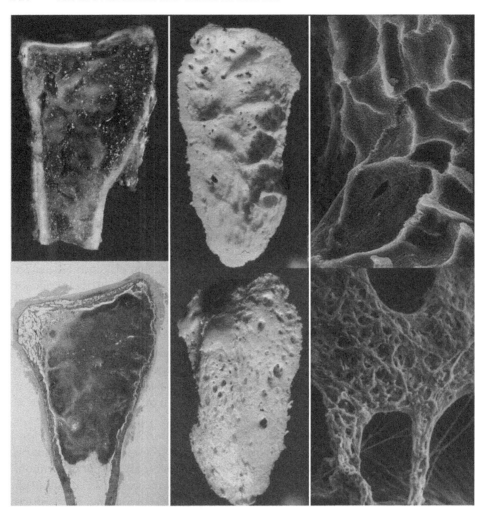

Figure 13.2 Giant cell tumor of bone, distal radius. [Left column]: gross view of longitudinally saw-cut specimen (top) and large format histological section (bottom) with the new ridged shell-type periosteal reaction that replaced original cortex along the right side of the tumor. [Center column]: macerated specimens of the ridged shell-type periosteal reaction, endosteal surface (top) periosteal surface (bottom). [Right column]: scanning electron micrographs – Endosteal surface (top) displaying overlapping Howship's lacunae, exposing an elliptical osteocyte lacuna and above that, a somewhat erosion-resistant cement line (1300x). The periosteal surface (bottom) pockmarked by shallow depressions where osteoblasts were settling in as osteocytes beneath a layer of unmineralized osteoid, removed during maceration (520x).

Histopathology makes its most definite contributions when soft tissue is present and can be rehydrated. Sabbatani and Fiorino (2009) reviewed paleopathological studies performed in mummified tissues, with reference to infectious diseases. Through correlation of histological and biomolecular information, the authors explored dietary and hygiene conditions of ancient populations. Similarly, histological analysis of soft tissue in a 2,500-year-old Egyptian mummy enabled specific diagnosis of a benign neurilemmoma (Strouhal and Německová 2009). However, Zweifel et al. (2009), through the use of meta-analysis, emphasize that minimum publication standards for paleopathologic studies are necessary to improve evidence-based research in paleopathology.

The pitfalls

Despite growing awareness of diagnostic and analytical potential of histological analysis, its application within paleopathology and bioarchaeology is rare (Von Hunnis 2009). The reason for this is undoubtedly pragmatic, in part, as the cost of the equipment and materials, the preparation time, and the methodological knowledge required for the preparation of thin sections can be daunting (Beauchesne and Saunders 2006). Serving as further obstacles are the facts that the biologic basis of histomorphological features must be clearly understood in order to interpret histological data, and a comprehensive understanding of diagenetic processes that affect bone preservation is needed. Much of the biological and taphonomic knowledge required for the analysis of bone thin sections can be acquired through detailed literature reviews and experiential learning. Basic medical knowledge is necessary if histopathological evidence is to be competently interpreted (Schultz 2001). We dispute the claim that experience and familiarity with bone biology and histology can be swiftly gained (Beauchesne and Saunders 2006).

The availability of a reference collection of well-diagnosed specimens for comparison is also necessary for developing reliable histopathologic conclusions (Schultz 2001). With histology being used rarely, scant comparative paleohistopathological data is available, and much of it is presented with the enthusiasm and certitude attendant with new techniques.

Further confounding histopathological research within paleopathology is the fact that external (i.e., the macroscopic) preservation of a bone does not always correlate to internal (i.e. microscopic) preservation. A bone appearing in good condition macroscopically might have undergone substantial diagenetic change. Diagenesis operates through soil and water, plant roots, fungi, algae, bacteria, protozoa, and arthropods and their larvae, causing microscopic fungal and/or microorganism intrusion, focal destruction by microorganisms (Wedl canals) (hackett 1981), and the presence of foreign materials (e.g., soil particulates, crystals) within lacunae and/or haversian systems. Thus microscopic control is advisable as a precursor to extraction studies (Schmidt-Schultz and Schultz 2007) whose goals might be to isolate particular proteins or fragments of DNA.

Thus, a thorough differential diagnosis is essential in any paleopathological assessment (Roberts et al. 2009). The reliability of a final diagnosis increases as all diagnostics (macroscopic, radiological, endoscopic, and light and scanning electron microscopic techniques) point in the same direction. None of these is more important than the others.

Terminology: Gross and Microscopic

Paleopathology is essentially a morphologic study. Though, fundamentally, it has many points in common with modern pathologic anatomy, it must be considered different from the latter due to the peculiar nature of the materials investigated.

Despite application of rigorous scientific criteria, sometimes no diagnosis can be made and only descriptive findings will result. This should not be interpreted as an inaccurate examination, or evidence of insufficient technical or diagnostic capacity. Indeed, the healthy trend today is towards a meticulous description rather than the stipulation of a likely but not thoroughly proven diagnosis. Descriptive criteria in pathological anatomy are well defined and can be applied with great accuracy in paleopathology. In 1981, Sweet et al. standardized descriptive terminology for the macroscopic (gross and radiologic) features of bone lesions in a three-part article series. Most of the terminology was applicable to neoplastic and inflammatory lesions. Their assertion was that attention to the three parameters of *margins, periosteal reactions,* and *matrix patterns,* as disclosed in plain films, permitted a diagnostic accuracy in excess of 90 percent for bone tumors. Furthermore, combinations of periosteal alterations, margins, and density changes can help refine an "Inflammatory Category" diagnosis into one of the three patterns of skeletal inflammation: septic, granulomatous, or angiitic. Post-traumatic and some metabolic (e.g., hyperparathyroid bone disease) changes are also succinctly described with these terms. The authors emphasized that for accurate description, and as a permanent record of a specimen, specimen radiographs are indispensable, since choice of descriptive terms in part relies on radiographic appearances. Though directed at today's pathologists, radiologists, and orthopedic surgeons, the terminology therein defined can be advocated without alteration for use in paleopathologic descriptions. An update of these concepts was subsequently published (Ragsdale 1993b).

Histomorphometry terminology is a much simpler matter than a standard terminology for gross and microscopic paleopathology. A review of even a dozen papers finds a host of terms used in dry bone microscopy which are unfamiliar to the modern-day orthopedic pathologist. The following is a sample:

- *Faserfilz-osteon*
- Fissural gap
- *Grenzstreifen* (= *grenzlinie*)
- Pin-like or spicular structures
- Plate-like proliferation
- Newly built bone
- Net-like bone formation
- Trabecular type of lesion
- Porotic type of lesion
- Polsters/padding
- Radiating trabeculae

How work employing these terms can be assisted by textbooks and papers describing modern-day orthopedic pathology is questionable. Schultz states that "… in the paleohistopathology of dry bones, the diagnostic criteria can be quite different from those used in recent pathology" (Schultz 2001:120), especially where soft tissue and cells are the major basis for diagnosis. Therefore, "the classification of diseases investigated by paleohistopathologists does not necessarily conform in all ways to that used by recent pathologists." (Schultz 2001:121). This contradicts the accepted paleopathological paradigm of "biomedical clinical analogy," where it is assumed that the same pathologic changes and signs used to diagnose today's patient can be applied as diagnostic criteria in the interpretation of ancient material (Von Hunnis et al. 2006).

There is a need for greater scientific rigor, especially for data that will be compared and used by different investigators (Fulcheri et al. 1994). The approach of agreeing on a few experts to set the terms, and then agreeing on common usage in the major journals in the field, will likely help paleopathology, as it has in other fields. The first concession, of course, is that investigators must agree that their pet terms may not become the accepted terms and may need to be sacrificed in the interest of universality. With conscientious authors and reviewers working alongside diligent, strong-willed editors, rules will become habits, and manuscripts using uncanonized terminology will go unpublished.

TACKLING HISTOLOGIC SPECIFICITY

Soft tissues and cells that have modified the osseous fabric during life, and which are the main basis for modern histologic diagnoses, are gone and cannot be studied in archaeological skeletal remains. All that remains is the architecture of the cortical, compact, and spongy bone, and any new bone additions. It has been declared that there is almost always a pattern in architectural elements of the cortical, compact and spongy bone substances associated with a particular disease that enable the paleopathologist to make relatively reliable diagnoses (Schultz 2001). This is contested by others (Weston 2009, and Chapter 27 this volume). Previous work has demonstrated a lack of diagnostic characteristics (qualitative or quantitative) in the macroscopic and radiographic appearance of periosteal reactions (Weston 2008). According to Hackett (1974), a diagnostic criterion of a disease is a change that by itself always indicates with confidence the presence of that disease. In the specimens examined by Weston (2008), it was shown that Schultz's (2001) characteristic traits of *grenzstreifen*, polsters, and sinuous lacunae were not characteristic of specific diseases and are extremely variable in shape and distribution, and highly dependent on the part of the section investigated (Van Der Merwe et al. 2010). It was proposed that these three traits are simply manifestations of general inflammatory processes (Weston 2008). That there are specific histomorphological features for each disease is implausible (Van Der Merwe et al. 2010). Hence, as with the macroscopic and radiographic characteristics of periosteal new bone, it appears that the microscopic characteristics are similar regardless of the disease etiology, and thus should not be relied on alone to provide disease diagnoses (Weston 2008).

Unlike the macroscopic/radiologic differences in bone changes due to suppurative, granulomatous, or angiitic inflammation, on a histological level infectious changes in bone are most likely very similar regardless of the specific condition that caused the osteomyelitis. For example, Schultz states that "Unfortunately, bone lesions caused by syphilis are frequently very similar to alterations caused by nonspecific bone diseases, particularly in long bones" (Schultz 2001:126). However, Van Der Merwe contests that it was only in conjunction with macroscopic investigation and a clear description of the distribution pattern of the lesions across a skeleton that a case of osteomyelitis could be attributed to treponematosis (Van Der Merwe et al. 2010).

Of further concern is Schultz's (2001) creation of only three groups of micro-scopically "proliferate patterns" bone change: hemorrhagic, inflammatory, and tumorous. While this assertion may be correct, approaching the task of diagnosis in this limited fashion may dissuade the researcher from thinking of changes caused by vascular disturbance (e.g., infarct), trauma and repair, anomalies (osteopetrosis), metabolic disease (fluorosis), and neuromechanical (osteophytosis) processes. Diagnoses become oversimplified and potentially wrong.

The Cutting Edge

Tens of thousands of histological slides of normal bone and radiologic studies of some skeletal diseases are housed at the Anatomical Division of the National Museum of Health and Medicine (NMHM) in Washington DC. Theses slides are from the former research collection of the Orthopedic Pathology Department of the Armed Forces Institute of Pathology (AFIP). Particularly relevant to paleopathologists interested in histopathology are the stained large-format histological glass slides and ground sections of bone cut from plastic embedded, undecalcified samples. Researcher access to the collection (which is currently unavailable due to the planned relocation of the facility) is expected to resume in 2012. In addition, senior author Ragsdale's personal collection of large glass slides assembled while at the AFIP may soon be available for qualifying scholars to study.

The primary methods for creating whole-mount slides are outlined in the out-of-print 1968 AFIP Histological Staining Methods Manual (Luna 1968). Hematoxylin and eosin (H & E) stains predominate in the whole-mount collection, though Masson's trichrome and other special stains were commonly used. All processing of large-bone specimens was done by hand whereas automatic processors handled the smaller pieces (AFIP 1949).

Large glass slides of these modern-day specimens with known diagnoses have much to contribute. Soft tissue structures and cellularity responsible for the adjacent mineralized fabric changes are present in these preparations, as seen in Figure 13.2 (lower left), unlike in a dry bone sample. The disease *is* the soft tissue and cellularity; bone is merely a bystander modified by it. The goal should be to find mineralized matrix clues to specific subjacent soft tissue/cellular compo-nents. One would hope (but we doubt) that details in hard-tissue histomorphology can be reliably linked to immediately adjacent soft tissue/marrow content and then this "experience" and morphologic standards taken back to dry bone studies.

The growing number of histopathological citations in paleopathological publications indicates an interest in the regular use of histopathology and the need for a standardization of methods, descriptive terms, and reporting. More work is required in the field, particularly in characterizing the histopathological features of pathological bone from macerated specimens with known diagnoses (Weston 2009). Specimens donated to the Anthropology Department at Arizona State University in Tempe by Dr. Ragsdale were designed to meet this need. Macerates are on file with photographic documentation of what the surrounding soft tissue looked like. Figure 13.3 is an

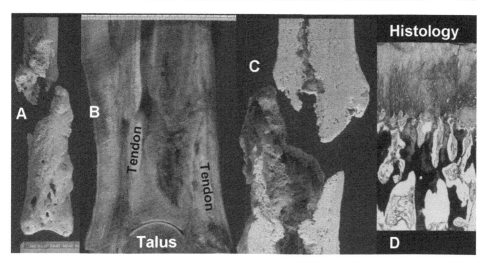

Figure 13.3   Infected non-union of distal tibia. External (A) and internal (C) views of proximal and distal macerated tibia segments flank the saw-cut fresh specimen (B). The gap shown in "A" and "C" was occupied by a gristle-like fibrocartilage band that triggered enchondral ossification where it contacted bone as an attempt toward bony union (D). Would the histopathology of dry bone "C" predict this pathologic enchondral ossification?

example of a post-traumatic cartilagenous non-union gap in dry bone segments, flanking the longitudinally saw cut fresh specimen. The right panel is a trichrome histological view of the blue-stained fibrocartilage that filled the gap, and the enchondral ossification working to delete it from below.

A review of the literature indicates that fewer studies have used scanning electron microscopy (Weston 2009). Other techniques, such as fluorescence microscopy (e.g., for the detection of the vestiges of postmortem or intravitam fungi), phase contrast microscopy, and interferential contrast microscopy (e.g., examination of soil-living microorganisms), are also useful in paleopathology (Schultz 2001).

Confocal microscopy has also been adapted. The confocal laser scanning microscope (CLSM) is a relatively new and advanced microscopic imagining method which is rarely used in anthropological research, although numerous clinical and basic science research projects have shown its value to study bone growth, bone micro-architecture and 3D bone morphometry (Papageorgopoulou et al. 2009). The advantage of CLSM over conventional histological imaging is three-dimensional visualization, making it possible to evaluate the osteocyte distribution and the architecture of the canalicular network. CLSM users claim to differentiate pre-existing microdamage sustained *in vivo* from microdamage acquired through the processing of the sample (Papageorgopoulou et al. 2009).

Ancient bones in a good preservation state, ascertained by microscopic techniques, conserve extracellular matrix proteins and other macromolecules over thousands of years (Schmidt-Schultz and Schultz 2007). DNA analysis provides enormous potential and reliability in the identification of human remains (see Chapter 8 by Spigelman et al., this volume, for more detailed discussion on this topic). Analysis of degraded

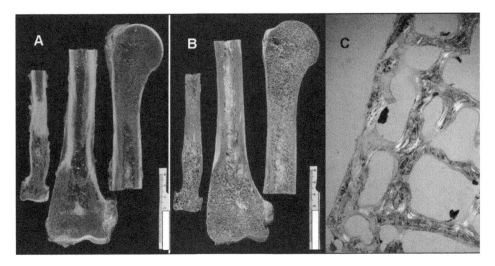

Figure 13.4 Sickle cell disease. A: Saw cut fresh autopsy bones of the adult male who died in sickle cell crisis (Faerman et al. 2000). Proximal radius (left) and distal humerus (center) have old sclerotic residua of bone infarcts appearing white against the hyperplastic red marrow filling the bones. A recent infarct appears white in the distal diaphyseal marrow of the proximal humeral segment. B: Same specimens after chemical (papain) maceration display larger than normal cancellous modules. C: Ground section at extreme right viewed under polarized light (courtesy of Michael Schultz, MD; 40x).

DNA in forensic and archaeological specimens is still hampered by methodological difficulties and the question of the authenticity of DNA isolated from human remains. Trace amounts of highly damaged DNA in forensic and archaeological samples require the application of the extremely sensitive polymerase chain reaction (PCR) method, which is prone to contamination even deep in the specimen. However, this may be overcome, or at least predicted, by applying a set of clear precautions and controls which are enhanced by careful microscopic evaluation of the material (Cooper 1997).

Differential diagnosis of anemias in dry skeletal remains is difficult using traditional anthropological methodology. However, microscopic examination as part of preliminary screening will facilitate the molecular detection of hemoglobinopathies in archaeological specimens with suspected anemia. In anemia, cancellous-type modules are clearly carved out from antecedent compact bone to afford additional space in which to house hyperplastic marrow as in the fatal sickle cell disease case presented in Figure 13.4. Trabeculae of cancellous bone become thin and relatively long as cancellous modules of the marrow space are enlarged. In a study of a modern autopsy bone sample from a documented case of sickle cell anemia, Faerman et al. (2000) demonstrated the power of DNA analysis applied to the ß-globin gene, mtDNA sequences, Y-chromosome DNA polymorphisms, and sex identification in the creation of a genetic portrait that was corroborated by microscopic examination of bone structure, and historical and medical records (Faerman et al. 2000).

Conclusion

Explaining the dynamic pathobiology of bone lesions must address the only factors influencing the presence of lesions: vascular, neuromechanical, and metabolic factors. These three influences operate to variable extents in the seven basic categories of disease which are capable of leaving recognizable hallmarks in dry bone specimens. Just as fossilized footprint is of interest for they tell us about the creature that made them the alteration of solid bone substance is the result of the action of ephemeral soft tissue elements acting on bone surfaces, internal and external. Understanding this dynamic interface is the essence of studying the mechanisms of disease. The analysis of a dry bone, either grossly or at the microscopic level, must be done with an eye that can "see" what is no longer there, the soft tissue that would have occupied its "holes" and covered its "bumps." We assert that only through detailed descriptions and diagnoses to general disease categories, will a stronger methodological basis for comparative research in paleopathology be reached.

REFERENCES

AFIP (Armed Forces Institute of Pathology), 1949   Annual Report. Washington DC: AFIP Historical Archives National Museum of Health and Medicine. AFIP Home Page. http://www.afip.org/index.html

American Cancer Society, 2010   What are the key statistics about bone cancer? Electronic document http://www.cancer.org/Cancer/BoneCancer/DetailedGuide/bone-cancer-key-statistics .

Beauchesne, P., and Saunders, S., 2006   A Test of the Revised Frost's "Rapid Manual Method" for the Preparation of Bone Thin Sections. International Journal of Osteoarchaeology 16:82–87.

Carter, D. R., 1984   Mechanical Loading Histories and Cortical Bone Remodeling. Calcified Tissue International 36(1):19–24.

Choon, S. E., and Tey, K. E., 2009   Lucio's Phenomenon: A Report of Three Cases Seen in Johor, Malaysia. International Journal of Dermatology 48(9):984–8.

Coleman, R. E., 1997   Skeletal Complications of Malignancy. Cancer 80:1588–1594.

Cooper, A., 1997   Reply to Stoneking: Ancient DNA – How Do You Really Know When You Have it? American Journal of Human Genetics 60:1126–1135.

Cripps D. J., and Curtis, A. C., 1967   Syphilis Maligna Praecox. Syphilis of the Great Epidemic? An Historical Review. Archives of Internal Medicine 119(4):411–418.

Currey, J. D., Rho, J. Y., Zioupos, P., and Pharr, G. M., 2002   Microstructural Elasticity and Regional Heterogeneity in Human Femoral Bone of Various Ages Examined by Nano-Indentation. Journal of Biomechanics 35(2):189–198.

Ewing, J., 1992   A review and classification of bone sarcomas. Archives of Surgery 4:485–533.

Faerman, M., Nebel, A., Filon, D., Thomas, M. G., Bradman, N., Ragsdale, B. D., Schultz, M., and Oppenheim, A., 2000   From a Dry Bone to a Genetic Portrait: a Case Study of Sickle Cell Anemia. American Journal of Physical Anthropology 111:153–163.

Flohr, S., and Schultz, M., 2009   Mastoiditis – Paleopathological Evidence of a Rarely Reported Disease. American Journal of Physical Anthropology 138:266–273.

Fulcheri, E., Massa, E. R., and Badini, A., 1994   Proposal for the Use of SNOMED Coding System in Paleopathology. Paleopathology Newsletter 13:6–10.

Hackett, C. J., 1974   Possible Treponemal Changes in a Tasmanian Skull. Man 9(3):436–43.

Hackett, C. J., 1981   Microscopical Focal Destruction (Tunnels) in Excavated Human Bones. Medical Science Law 21:243–265.

Huggins, T. J., Ragsdale, B. D., Schnapf, D. J. S., Madewell, J. E., and Youngblood, L., 1981   Radiologic–Pathologic Correlation from the Armed Forces Institute of Pathology: Calvarial Invasion by Meningioma. Radiology 141:709–713.

Johnson, L. C., 1953   A General Theory of Bone Tumors. Bulletin of New York Academic Medicine 29(2):164–171.

Johnson, L. C., 1966   Kinetics of Skeletal Remodeling. In Birth Defects: Structural Organization of the Skeleton. R. Milch, ed. pp. 66–142. New York: National Foundation: March of Dimes.

Johnson, L. C., 1971   Discussion – General Biology of the Skeleton. In Craniofacial Growth in Man. R. E. Moyers, and W. M. Krogman, eds. New York: Pergamon Press.

Johnson, L. C., Vinh, T. N., and Sweet, D. E., 2000   Bone Tumor Dynamics: an Orthopedic Pathology Perspective. Seminar of Musculoskeletal Radiology 4(1):1–15.

Klepinger, L. L., 1983   Differential Diagnosis in Paleopathology and the Concept of Disease Evolution. Medical Anthropology 7:73–77.

Lanyon, L. E., and Skerry, T., 2001   Postmenopausal Osteoporosis as a Failure of Bone's Adaptation to Functional Loading: a Hypothesis. Journal of Bone and Mineral Research 16(11):1937–1947.

Lanyon, L. E., O'Connell, A., and MacFie, H., 1982   The Influence of Strain-Rate on Adaptive Bone Remodeling. Journal of Biomechanics 15:767–781.

Luna, L. G., 1968   Manual of Histologic Staining Methods of the AFIP. New York: McGraw-Hill.

Maat, G. J. R., Van Den Bos, R. P. M., and Aarents, M. J., 2001   Manual Preparation of Ground Sections for the Microscopy of Bone Tissue: Update and Modification of Frost's "Rapid Manual Method." International Journal of Osteoarchaeology 11(5):366–374.

MacGinitie, L. A., Wu, D. D., and Cochran, G. V., 1993   Streaming Potentials in Healing, Remodeling and Intact Cortical Bone. Journal of Bone and Mineral Research 8(11): 1323–1335.

Madewell, J. E., Ragsdale, B. D., and Sweet, D. E., 1981   Radiologic and Pathologic Analysis of Solitary Bone Lesions. Part I: Internal Margins. Radiologic Clinics of North America 19:715–748.

Martin, D. L., and Armelagos, G. J., 1985   Skeletal Remodeling and Mineralization as Indicators of Health: an Example from Prehistoric Sudanese Nubia. Journal of Human Evolution 14:527–537.

Miller, E., Ragsdale, B. D., and Ortner, D. J., 1996   Accuracy in Dry Bone Diagnosis: a Comment on Palaeopathological Methods. International Journal of Osteoarchaeology 6:221–229.

Ortner, D. J., 2003   Identification of Pathological Conditions in Human Skeletal Remains, Second Edition. Amsterdam: Academic Press.

Ortner, D. J., and Putschar, W. G. J., 1981   Metastatic Tumors: Pathology. In Identification of Pathological Conditions in Human Skeletal Remains. pp. 391–398. Washington, DC: Smithsonian Institution Press.

Papageorgopoulou, C., Kuhn, G., Ziegler, U., and Ruhli, F. J., 2009   Diagnostic Morphometric Applicability of Confocal Laser Scanning Microscopy in Osteoarchaeology. International Journal of Ostoearchaeology 20(6):708–718.

Parfitt, A. M., 1987   Bone Histomorphometry: Standardization of Nomenclature, Symbols, and Units. In Report of the ASBMR Histomorphometry Nomenclature Committee. Journal of Bone and Mineral Research 2(6):595–610.

Ragsdale, B. D., 1993a   Morphologic Analysis of Skeletal Lesions: Correlation of Imaging Studies and Pathologic Findings. Advances in Pathologic and Laboratory Medicine 6:445–490.

Ragsdale, B. D., 1993b   Polymorphic Fibro-osseous Lesions of Bone: An Almost Site-Specific Diagnostic Problem of the Proximal Femur. Human Pathology 24:505–512.

Ragsdale, B. D., 1995   Comment – Appearance of Cancer in Defleshed Human Bones. Paleopathology Newsletter 90:7–8.

Ragsdale, B. D., 1997   The Irrelevance of Contemporary Orthopedic Pathology to Specimens from Antiquity, Continued. Paleopathology News Letter 97:8–12.

Ragsdale, B.D., and Miller, E., 1996   Workshop A. Skeletal Disease Workshop VIII: Several of the Seven Basic Categories of Disease. *In* Papers on Paleopathology Presented at the 23rd Annual Meeting of the Paleopathology Association, Durham, North Carolina. Cockburn, E., ed., p. 1. Detroit: Paleopathology Association.

Roberts, C. A., Pfister, L. A., and Mays, S., 2009   Letter to the Editor: Was Tuberculosis Present in *Homo erectus* in Turkey? American Journal of Physical Anthropology 139: 442–444.

Roverano, S., Paira, S., and Somma, F., 2000   Lucio's Phenomenon: Report of Two cases and Review of the Literature. Journal of Clinical Rheumatology 6(4):210–213.

Ruff, C. B., 2005   Mechanical Determinants of Bone Form: Insights from Skeletal Remains. Journal of Musculoskeletal Neuronal Interactions 5:202–212.

Sabbatani, S., and Fiorino, S., 2009   Contribution of Palaeopathology to Defining the Pathocoenosis of Infectious Diseases: Part II. Le Infezioni in Medicina 17(1):47–63.

Schmidt-Schultz, T. H., and Schultz, M., 2007   Well Preserved Non-Collagenous Extracellular Matrix Proteins in Ancient Human Bone and Teeth. International Journal of Osteoarchaeology 17:91–99.

Schultz, M., 2001   Paleohistopathology of Bone: A New Approach to the Study of Ancient Diseases. Yearbook of Physical Anthropology 44:106–147.

Schultz, M., Timme, U., and Schmidt-Schultz, T. H., 2007   Infancy and Childhood in the Pre-Columbian North American Southwest: First Results of the Palaeopathological Investigation of the Skeletons from the Grasshopper Pueblo, Arizona. International Journal of Osteoarchaeology 17:369–379.

Strouhal, E., and Němečková, A., 2009   Palaeopathological Diagnosis After 2500 Years: The Case of Imakhetkherresnet, Sister of Priest Iufaa. Prague Medical Report 110(2):102–113.

Sweet, D. E., Madewell, J. E., and Ragsdale, B. D., 1981   Radiologic and Pathologic Analysis of Solitary Bone Lesions: Part III: Matrix Patterns. Radiologic Clinics of North America 19:785–814.

Van Der Merwe, A. E., Maat, G. J. R., and Steyn, M., 2010   Ossified Hematomas and Infectious Bone Changes on the Anterior Tibia: Histomorphological Features as an Aid for Accurate Diagnosis. International Journal of Osteoarchaeology 20:227–239.

Von Hunnis, T., 2009   Using Microscopy to Improve a Diagnosis: an Isolated Case of Tuberculosis-Induced Hypertrophic Osteopathy in Archaeological Dog Remains. International Journal of Osteoarchaeology 19:397–405.

Von Hunnis, T., Roberts, C. A., and Saunders, S. R., 2006   Histological Identification of Syphilis in Pre-Columbian England. American Journal of Physical Anthropology 129:559–566.

Walker P. L., Bathurst, R. R., Richman, R., Gjerdrum, T., and Andrushko, V. A., 2009   The Causes of Porotic Hyperostosis and Cribra Orbitalia: A Reappraisal of the Iron-Deficiency-Anemia Hypothesis. American Journal of Physical Anthropology 139:109–125.

Watson, K. M., White, J.M., Salisbury, J. R., and Creamer, D., 2004   Lues Maligna. Clinical Experimental Dermatology 29(6):625–627.

Weston, D. A., 2008   Investigating the Specificity of Periosteal Reactions in Pathology Museum Specimens. American Journal of Physical Anthropology 137:48–59.

Weston, D. A., 2009   Brief Communication: Paleohistopathological Analysis of Pathology Museum Specimens: Can Periosteal Reaction Microstructure Explain Lesion Etiology? American Journal of Physical Anthropology 140:186–193.

Zweifel, L., Büni, T., and Rühli, F. J., 2009   Evidence-Based Palaeopathology: Meta-Analysis of PubMed-Listed Scientific Studies on Ancient Egyptian Mummies. HOMO Journal of Comparative Human Biology 60(5):405–427.

CHAPTER **14** # Differential Diagnosis and Issues in Disease Classification

Donald J. Ortner

INTRODUCTION

In Chapter VI of Lewis Carroll's *Through the Looking Glass* there is a delightful comment "'When I use a word' Humpty Dumpty said, in a rather scornful tone,' it means just what I choose it to mean—neither more nor less.'" Defining words as we choose may work well in Carroll's Wonderland, but it does not work in scientific research, discourse, or publication. The names we apply to the abnormalities encountered in human paleopathology, and how we classify these abnormalities, often make a big difference between confusion and understanding. Indeed, a basic component of any scientific discipline is rigor in defining and using terms.

Equally, classification is fundamental to all the biological sciences. Classification provides a tool to show potential relationships between different animals, plants and phenomena. In paleopathology, it aids in establishing the relationship between different diseases, as well as their cause and pathogenesis. However, we need to remember that classificatory categories are artificial mental constructs that, although often convenient and critical, can interfere with our understanding of the disease process (pathogenesis). An example of this is the erosive arthropathies such as the seronegative spondyloarthropathies (see Chapter 28 by Waldron in this volume for more detailed discussion of these conditions). These arthropathies are appropriately classified with the other types of arthritis, including osteoarthritis, as joint disorders. The cause of the erosive arthropathies remains a subject of intense research, but the

A Companion to Paleopathology, First Edition. Edited by Anne L. Grauer.
© 2012 John Wiley & Sons, Ltd. Published 2016 by John Wiley & Sons, Ltd.

general prevailing understanding is that there is a genetic factor that is not expressed until there is a triggering agent. The triggering agent for the erosive arthropathies is thought to be infectious pathogens. This hypothesis is fairly well established with inflammatory bowel disease and Reiter's syndrome (Resnick 2002; Ortner 2003:580). Are these joint disorders better classified as infectious diseases? The current classification is unlikely to change but in trying to understand the pathological process, one needs to avoid the inherent mental traps of a classificatory system that inevitably reflects an oversimplification of the complex factors that affect the expression of disease.

In the 1960s one of my mentors, J. Lawrence Angel, published a paper on what he called "porotic hyperostosis" and linked this term to the skeletal manifestations of thalassemia (Angel 1966), one of the disorders he encountered in his research on the bioarchaeology of the Eastern Mediterranean. The term simply means porous, abnormal bone formation, and could be applied to any of several skeletal disorders that stimulate this abnormality including anemia, infection, cancer, scurvy and rickets. Angel, of course, understood that a diagnosis of anemia in archaeological human skeletal remains depends on abnormal enlargement of the marrow (marrow hyperplasia) to accommodate the increased demand for hematopoietic tissue. In the skull vault this is expressed as porosity caused by enlargement of diploë at the expense, usually, of the outer table, although the inner table may be affected in severe cases. Angel assumed that all who used the term porotic hyperostosis would understand this as well.

Unfortunately the term often became linked exclusively to anemia in research conducted and published on archaeological human skeletal samples. The distinction is often not made between cranial porosity caused by marrow hyperplasia, and porosity, potentially caused by other disorders. This has resulted in abnormalities being diagnosed as anemia when other diagnostic options are more probable. It also means that much of the literature on the prevalence of anemia in archaeological human skeletal samples is likely to overestimate the true prevalence of the skeletal disorder in antiquity.

One of the objectives of this chapter is to provide the reader with some basic descriptive tools to use in analyzing skeletal disorders encountered in human archaeological remains. Correct description of the disorders encountered also provides the basis for classification. It aids in identifying the relationships that may exist between different expressions of skeletal disorders, and can serve as the basis for diagnosis of skeletal abnormalities. Examples will be included about how basic information is used in classifying disorders and in determining specific diagnosis in cases when possible. The reader should understand that this chapter cannot provide a comprehensive review of all the diagnostic challenges confronting a researcher analyzing an archaeological human skeletal sample for evidence of paleopathology. There are numerous sources in the medical literature that are available for consultation (e.g. Resnick 2002) as well as general reference works on human paleopathology (e.g., Aufderheide and Rodriguez-Martin 1998; Ortner 2003; Roberts and Manchester 2005; Pinhasi and Mays 2008; Chhem and Brothwell 2008; Waldron 2009). These sources provide a more complete source of information on diagnostic options when confronted by skeletal abnormality in an archaeological skeleton. There are also important reference sources on specific diseases in paleopathology (e.g., Roberts and Buikstra 2003; Powell and Cook 2005; Brickley and Ives 2008).

Basics of Skeletal Paleopathology

Bone cells and their function, as well as endochondral and intramembraneous bone development are reviewed elsewhere in this book (see Chapter 13 by Ragsdale and Lahmer in this volume, and Chapter 5 by Gosman in this volume). These topics are also presented in other reference works on skeletal paleopathology (e.g., Ortner and Turner-Walker 2003; Tuross 2003) and in medical sources (e.g., Manolagas 2000; Resnick et al. 2002), and will not be reviewed in detail in this chapter. However, quite basically, two bone cells, osteoblasts, which form bone, and osteoclasts, which destroy bone, produce the bone reactions associated with skeletal disorders. In some disorders abnormal bone is formed, in other disorders bone is destroyed. However, a combination of both, as occurs, for example, in Paget's disease, is a common expression of skeletal disease. Careful attention to the characteristics of the abnormal features created by the bone cells provides critical information basic to description and diagnosis of all skeletal disorders. Complication arises because the expression of a specific skeletal disorder is variable in terms of the speed of the bone cell action and the mechanism by which the cells are activated or not activated.

Very rapid abnormal bone formation produces woven bone, which tends to be poorly organized both at the microscopic and gross levels. At the other end of the abnormal bone formation continuum, relatively slow bone formation produces dense compact bone. Woven bone is a temporary tissue that, with long term survival, is gradually remodeled into compact bone. In aggressive, bone-forming sarcomas (primary cancer of bone), tumor bone may be poorly organized woven bone, although dense compact bone may be found in some primary bone sarcomas.

Similarly, there is a continuum in the expression of pathological bone destruction from very slow destruction to very rapid and aggressive bone destruction. The anatomical and radiological features of destruction will be discussed in greater detail later in this chapter. An additional complicating factor is the need to distinguish between bone destruction and a failure to form. These two expressions of abnormal absence of bone will also be discussed in greater detail later in this chapter.

In human skeletal paleopathology, the first step in differential diagnosis is to distinguish between the normal and the abnormal. What may be quite normal bone structure in an infant may be abnormal in a child or adult. For example, woven bone (Figure 14.1a) is associated with very rapid growth and characterizes all the bone formed *in utero* and most of the bone formed in infancy, because bone growth is relatively much faster during this phase of development. The presence of woven bone after about one to two years of age is always abnormal with the exception of the alveolar sockets where the surface consists of woven bone throughout life as long as teeth are present. Similarly the cortical porosity regularly encountered in the metaphysis of long bones during growth (Figure 14.1b) is usually a normal result of the modeling taking place as the metaphysis of the lengthening bone is reduced in diameter by action of osteoclasts and osteoblasts to form the diaphysis. This growth–modeling process exposes the vascular channels in the compact bone of the metaphysis leaving the fine holes apparent on the new bone surface. These holes are filled during normal growth. A common pathological expression in response to infection and other causes of inflammation is cortical porosity that may occur at the same location on

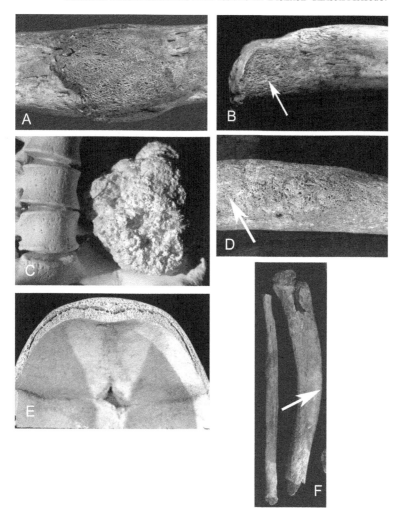

Figure 14.1   (A)Woven bone fracture callus in the diaphysis of an adult right humerus
(NMNH 364816-32). (B) Cutback zone (arrow) in the proximal metaphysis of the right tibia
in a 4-year-old child from an archaeological site in the American southwest (NMNH
326194). (C) Osteosarcoma arising on the left ilium in a modern 25-year-old male (Institute
of Pathological Anatomy, University of Zurich 389/69). Note the poor organization of the
bone. (D) Bone-forming lesion of the tibial shaft in an adult male from a postcontact
archeological site in Canaveral, Florida (NMNH 377457). Note the area of porosity in the
central area of the lesion and the compact bone at the margins of the lesion (arrow).
(E) Abnormal rachitic bone separated from earlier spongy bone by fine spicules of trabecular
bone connecting the two areas of bone. Section through the skull vault of a child 2 years of
age with rickets (Federal Pathologic-Anatomy Museum, Vienna, Austria catalog no. 5251).
(F) Abnormal compact bone formation on the medial and anterior surfaces (arrow) of the
right tibia in a case of treponematosis. Male, 30 years old, from a prehistoric site in Virginia
(USA NMNH 385788).

a bone in a person who died during growth. One has to distinguish between normal and abnormal porosity utilizing the combined context of both location and age.

In the following paragraphs I will explore some of the basic principles of classification and differential diagnosis in human skeletal paleopathology, but will also emphasize the need for careful use of terminology in describing the abnormalities one encounters in human skeletal remains. In working with students and colleagues I share a basic principle learned from my mentors in pathology, i.e., carefully describe the abnormalities you encounter, including the type of abnormal bone and the distribution of abnormalities throughout the skeleton, before attempting differential diagnosis.

Another principle that I emphasize (and is carefully discussed in detail in Ragsdale and Lehmer, Chapter 13 in this volume) is to avoid overdiagnosis, i.e., avoid a degree of specificity that is not justified by the evidence. An example of this tendency is being overspecific in the diagnosis of cancer. Sarcoma and carcinoma are the two types of cancer and both cause bone lesions. Sarcomas are uncommon and derive from connective tissues (bone, cartilage, muscle, fat etc.). Carcinoma develops in epithelial tissue that forms the surface of skin and organs, and is the most common cause of cancer affecting bone. There are some carcinomas which produce lesions permitting specific diagnosis, such as the mossy, bone-forming lesions occurring in the pelvic area of some cases of prostate carcinoma. However, most carcinomas that metastasize to bone are destructive and have similar dry bone manifestations. Identifying the organ from which the cancer originated is often impossible even in modern clinical contexts. In such situations the appropriate strategy is to identify the lesions as probably caused by cancer and not attempt a more specific diagnosis.

Sarcomas originating in bone or cartilage usually form tissue that is poorly organized and can develop into large lesions (Figure 14.1c). Sarcomas originating in soft tissues can metastasize to bone and tend to be destructive, with lesions that are indistinguishable from most carcinomas. Clearly, this illustrates the need to avoid unwarranted specificity, certainly in the diagnosis of cancer.

DESCRIPTION IN HUMAN SKELETAL PALEOPATHOLOGY

Terminology for bone abnormalities

There are four basic abnormal conditions that occur in human skeletal paleopathology: abnormal bone formation, abnormal absence of bone, abnormal size and abnormal shape. However, there are subcategories within each of these basic conditions that provide additional information about the pathological process (pathogenesis) which produced the abnormal condition.

Abnormal bone formation   All abnormal bone formation is the result of action by osteoblasts. Woven bone is the most common type of abnormal bone formed as part of a pathological process. Its expression ranges from very poorly organized, coarse woven bone seen in some types of primary bone cancer, to well organized fine woven bone apparent, for example, in the early stage of callus formation following bone fracture (see Figure 14.1c). Woven bone, as the name implies, has spaces within the tissue that can be confused with porosity formed by vascular holes penetrating through cortical bone. Many of the spaces in woven bone are fairly superficial and do not penetrate through the entire cortex. However, vascular holes are also present in woven bone since the stimulus to form woven bone is often associated with an inflammatory process. In these instances, blood vessels are present to provide pathways for blood cells to respond to the pathology present at the site of woven bone formation. Vascular conditions associated with inflammation also create conditions that directly stimulate bone formation.

Woven bone is usually a temporary tissue, at least in relatively chronic conditions where the patient survives for months or years after the lesion was formed. If survival extends beyond a few months after a woven bone lesion is formed, it will gradually be remodeled into compact bone. One often encounters woven bone in which the fine spaces within the bone are being filled and the woven bone itself is being transformed into compact bone (Figure 14.1d). The presence of un-remodeled woven bone lesions in an archaeological human skeleton indicates that the pathological process stimulating the lesion was active at the time the person died.

Woven bone can occur as a solitary lesion but it often is multifocal and usually develops with well-defined borders relative to adjacent normal bone. A good example of this expression of woven bone is the lesions seen on the long bones of some skeletons affected by treponematosis. Woven bone can also occur as one or more sheets of porous bone very loosely attached to underlying bone. This gives rise to the laminated appearance in radiographs, for example, of some cases of infection, rickets (Figure 14.1e) and also Ewing's sarcoma, a primary cancer of bone.

Although compact bone lesions can be created as woven bone lesions remodel into compact bone, they can also be the primary form of some abnormalities which develop slowly. Benign tumors such as osteomas are examples of abnormal compact bone development without an initial phase of woven bone. Trauma induced activation of the periosteum can also result in compact bone lumps on a long bone. Similarly chronic infectious diseases such as treponematosis can stimulate compact bone formation, as is seen in some cases of saber shin abnormality in this disorder (Figure 14.1f).

Abnormal bone formation is more often associated with chronic conditions. In contrast, bone destruction tends to be a response to acute, aggressive disorders. There are also disorders in which both types of lesions occur in the same skeleton and, indeed, in the same lesion. An example of the latter is the classic caries sicca lesion seen in treponematosis that will be reviewed in the section on destructive lesions of bone (see also Cook and Powell, Chapter 26 in this volume). In the skull, the lesion begins as a circular focus of porosity (Hackett 1976). In a later stage bone begins to form at the margins of the lesion, creating, in cross section, a crater-like lesion with a central destructive focus surrounded by reactive bone formation. Although the type and distribution of lesions seen in treponematosis are virtually pathognomonic for that disorder, the combination of bone destruction and formation within a single lesion is not an uncommon feature in other types of skeletal disorder.

Another form of abnormal bone formation occurs in some types of infant and childhood anemia when there is an abnormal increase in demand for hematopoietic marrow. This type of abnormal bone formation is also accompanied by abnormal bone destruction, as marrow space enlarges at the expense of cortical bone. In the postcranial skeleton, this results in decreased thickness of cortex. In the cranium, the diploë, composed of spongy bone, enlarges primarily at the expense of the outer table of the skull, creating the porous surface apparent in some types of anemia. Skull vault porosity, accompanied by marrow hyperplasia, is usually associated with one of the anemias, but it can occur in other disorders such as rickets (see Figure 14.1e).

Abnormal absence of bone   Lesions associated with an abnormal absence of bone can occur either from a failure to form bone or from a destructive process in which normal existing bone is destroyed. An example of abnormal absence of bone, caused by failure to form, is a defect associated with an epidermal inclusion cyst (Figure 14.2a). These cysts begin to form during embryological development when epidermal tissue remains in the embryonic tissue precursor of bone. The cyst continues to proliferate and prevents bone formation since bone cannot form in epidermal tissue. Other examples of abnormal absence of bone are apparent where there is a central focus of infection (e.g., a granuloma) that creates vascular conditions stimulating marginal bone

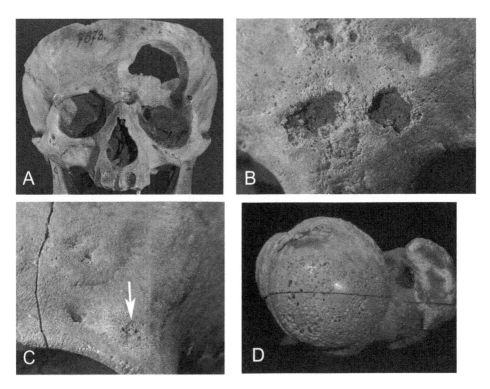

Figure 14.2   (A)Lesion characterized by a failure to form bone resulting from the presence of a congenital epidermoid cyst. Note the compact bone margins of the lesion indicative of a benign condition; modern 61-year–old male (Department of Pathology, University of Strasbourg, France catalog no. 7878). (B) Destructive lesions with reactive bone formation at the margins of the lesion in a case of treponematosis. Frontal bone from an adult male about 35 years of age from a medieval site at the Hull Magistrate's Court, Hull, England (burial no. 1216). (C) Early-stage porous destructive bone lesion in a case of treponematosis. Frontal bone from an adult male about 35 years of age from a medieval site at the Hull Magistrate's Court, Hull, England (burial no. 1216). (D) Porosity and eburnation of subchondral bone in the left femoral head in a modern adult femur (Museum of Man, San Diego, California catalog no. 1981-30-755).

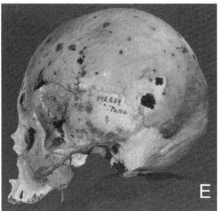

Figure 14.2 (*cont'd*) (E) Multiple "punched out" lytic lesions of the cranium in an adult female skull about 35 years of age from an archaeological site near Caudivilla, Peru dated between 500 and 1530 A.D. The destructive lesions were caused by multiple myeloma. Note the large lesion in the right central portion of the figure with lobulated margins that probably represents the coalescence of several smaller lesions. Note also that there is no evidence of reactive bone formation associated with any of the lesions. (NMNH 242559.) (F) Aggressive bone destruction in the right anterior temporal, sphenoid and parietal bones in a probable case of metastatic carcinoma. Note the very ragged and poorly defined margins of the lesion. Adult female 17th century from Varangerfjord, Norway (NMNH 241876.)

formation but prevents bone formation in the immediate area occupied by the focus. This is a fairly common feature in treponematosis (Figure 14.2b).

The overwhelming majority of lesions in which there is an abnormal absence of bone are destructive disorders in which some of the normal bone of the skeleton is deleted. All destruction of bone is the result of action by osteoclasts. Destructive processes may occur at one or more specific sites with clearly demarcated lesions where bone has been destroyed, or there may be a generalized rarefaction as occurs in osteoporosis and osteopenia. Although cortical bone is affected in osteoporosis and osteopenia, the greatest relative bone loss occurs in trabecular bone.

Bone destruction can exhibit a considerable range of morphological features from clusters of fine porosity penetrating through cortical bone to large areas of bone destruction with poorly defined margins. Porosity occurs in several disorders including treponematosis, rickets, scurvy and some types of anemia. However, the distribution of porosity tends to be quite different. In treponematosis porosity is most commonly seen in the skull vault as a circular cluster of fine holes (see Figure 14.2c). This lesion is the initial stage for what will become the more typical caries sicca lesion of the skull vault, with its crater-like bone formation (see Figure 14.2b) and fine lines radiating from the center of the lesion (Hackett 1976; and see Chapter 26 by Cook and Powell, this volume, for further discussion of treponematosis).

In rickets, the porosity is generally apparent in the subchondral surfaces of the growth plate where small and circular areas (like pores) of the subchondral osteoid are not mineralized (Ortner and Mays 1998). The unmineralized osteoid is lost postmortem, leaving the small holes. Elsewhere in the skeleton a common abnormal feature in rickets are "streamers" of bone left between narrow (usually less than 1mm wide) and long areas of unmineralized osteoid (which is lost postmortem along with the other nonmineralized proteins of the body). These "streamers", or struts, occur at the growing ends of bone and on the surface of the cranial vault. The strut-like mineralized structures demonstrate the attempt by the body to maintain as much biomechanical function as possible. This abnormality is particularly apparent in the costochondral end of the ribs.

Scurvy, caused by a deficiency in dietary vitamin C, can stimulate porous lesions of the cortex and porous hypertrophic bone formation. However, the pathogenesis is fundamentally different in this disorder. Vitamin C is an important cofactor in the formation of collagen, one of the major proteins in connective tissue including bone, but also in the supportive sheath around blood vessels. In children with scurvy, new blood vessels formed during growth are weaker and easily ruptured with even minor trauma. Blood outside the vascular system is an irritant and invokes an inflammatory response by the immune system. Bleeding occurs in several areas of the skeleton but is most easily identified in the skull where chronic bleeding can occur from even mild trauma to the skull vault but is more typically associated with rupturing caused by normal muscle action during chewing. This gives rise to a virtually pathognomonic distribution of porous lesions particularly in the bone underlying the temporalis muscle, which, along with the masseter muscle, is the muscle used in chewing. The arterial branches supplying the temporalis muscle lie between the temporalis and the underlying greater wing of the sphenoid and the anterior portion of the temporal bone. Porous lesions develop as part of the reaction to chronic bleeding at this site. These lesions are almost always bilateral, as one would expect if the cause is bleeding associated with chewing (see Chapter 22 by Kozlowski and Witas in this volume for further discussion on metabolic and endocrine disorders).

Porosity of the skull vault and/or orbital roof is also apparent in some cases of anemia. When this type of porosity occurs in combination with marrow enlargement caused by hyperplasia, a likely, although not certain, cause is anemia. If marrow enlargement is not present, the diagnostic options include infection, metabolic disorders and cancer.

Bone destruction can involve the formation of clusters of fine holes (2 millimeters or less) as occurs in the early stage of caries sicca in treponematosis (Figure 14.2c), but also in scurvy as a vascular reaction to an overlying hematoma. Fine holes also occur in the subchondral bone of the joints in some cases of arthritis. These holes are often, but not exclusively, associated with eburnation (Figure 14.2d). Larger holes usually occur in bone involvement when they are caused by cancer, particularly the carcinomas. As we have seen in the case of epidermal inclusion cysts, they can also occur in other skeletal disorders. The size of a destructive focus may be the result of a single lesion that enlarges. However, large destructive lesions may also be the result of smaller tumors coalescing to form a larger destructive lesion.

With all holes, the margins of the destructive process provide important information about the pathogenesis of bone destruction. In general, the slower the process, the more clearly defined the margin of a destructive lesion will be. For example, the benign epidermal inclusion cyst discussed earlier (Figure 14.2a) has a well-defined margin lined with compact bone. In a radiograph, the margin appears dense or sclerotic, which is indicative of a disorder that either was always or had become a slow destructive process or possibly a more destructive process that is in the healing phase.

In destructive lesions there is a continuous spectrum from a well-defined sclerotic margin to a margin where the border between the destructive process and normal bone is difficult, if not impossible, to define. Multiple myeloma (plasmacytoma) is an example of cancer in which the margins of the destructive foci in bone are well defined (Figure 14.2e). The lytic lesions tend to be small with the primary focus in marrow where the plasma cells normally occur. These lesions can coalesce to form larger, lobulated lesions with very sharply defined, purely lytic, "punched out" margins. Sclerosis does not occur in lesions caused by multiple myeloma. Most of the bone destruction associated with carcinoma is aggressively destructive with poorly defined margins since the process involves relentless growth until the death of the patient (Figure 14.2f) (see Chapter 23 by Brothwell in this volume, for further discussion of tumors).

With abnormal absence of bone, the two major factors relevant to diagnosis are the size and morphology of the hole and the margins of the lesion. Careful attention to these two features will usually permit assignment to at least a general category of disorder.

Abnormal bone size

Both abnormal bone size and abnormal bone shape tend to be associated with disorders in development (see Chapter 21 by Barnes in this volume for more detailed discussion of these disorders). One of the main categories of bone disorder is the dysplasias for which a genetic defect is the cause. There are hundreds of these disorders and differential diagnosis is likely to be challenging, if not impossible, in some cases encountered in archaeological human remains. Fortunately, most of the dysplasias are very rare disorders. It is also important to stress that many of the disorders resulting in abnormal bone size are also associated with abnormal bone shape. An example of this is seen in the most common dysplasia, achondroplastic dwarfism. As the name implies the problem is a deficiency of cartilage production. Although the most noticeable effect of achondroplasia is a substantial reduction in long bone length (impeded endochondral ossification), this disorder also affects the shape of bones including the skull. Here, the skull vault grows normally but the skull base does not, giving rise to the midfacial depression. The reason for this abnormality is that endochondral growth occurs in the base of the skull during development, whereas the skull vault is entirely formed by periosteal development (intramembranous ossification) in which cartilage plays no role. The long bones also exhibit both abnormal size and shape. In the fetal femur, shown in Figure 14.3a, the bone is obviously abnormally short with a flared metaphysis that is disproportionately larger than expected on the basis of the length of the diaphysis. The large size of the metaphysis is the result of relatively normal circumferential intramembranous bone formation, while growth in length is greatly diminished.

There are other disorders beside the dysplasias that cause either abnormal bone size and/or shape. Infectious disorders such as polio and osteomyelitis can affect growth, including reduced size. But increased growth in length can also be stimulated by the disorder (Figure 14.3b). One of the endocrine disorders, endemic cretinism caused by a deficiency in thyroid hormone resulting from a dietary lack of iodine, causes skeletal abnormalities that are similar to the dwarfism of achondroplasia. Trauma can also disrupt growth resulting in abnormal size and shape.

Another of the growth disturbances is associated with a defect of the pituitary gland: the master gland in human physiology. There are complex and important interactions with other glands including the thyroid. In the context of abnormal bone size, the pituitary gland secretes growth hormone, a deficiency of which results in pituitary dwarfism. In this disorder both endochondral and intramembranous ossification are diminished. Because both types of bone formation are affected, the proportions of the body, including the skeleton, are normal; just the size is affected.

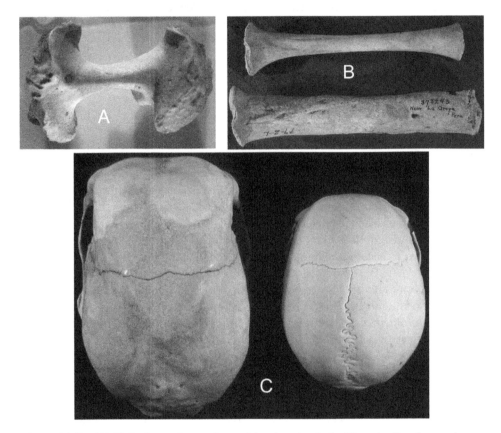

Figure 14.3   (A) Right femur from a fetus with achondroplasia. Note the flared metaphyses that are abnormally wide relative to the length of the femur. Growth in width is minimally affected in achondroplasia so that the margins of the metaphyses are relatively normal in size. (From the Wellcome Museum, Royal College of Surgeons of England, London catalog no. S59.3.) (B) Posterior views of the left (lower) and right tibias of a child about 9 years of age with osteomyelitis of the left tibia from an archaeological site near La Otoya, Peru dated between 1400 and 1530 A.D. Note the greater length of the abnormal tibia in which the growth process was stimulated by chronic inflammation (NMNH 378243). (C) Abnormal size apparent in the cranium in the left of the figure compared with a normal cranium in the right portion of the figure. The abnormal size is caused by a defect in the pituitary gland which secreted excessive growth hormone during development. (Photo courtesy of Dr. Takao Suzuki, Shinjyuku-ku, Tokyo, Japan.)

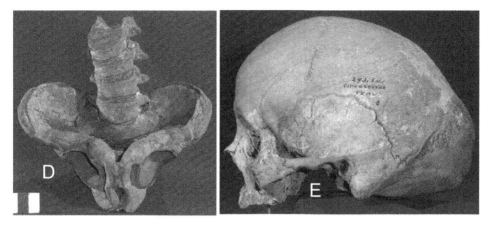

Figure 14.3   (*cont'd*) (D) Severe deformity in a case of rickets resulting from a deficiency in vitamin D in a male about 18 years of age at the time of death. Note the inward collapse of the pelvis caused by the weight of the body on the femoral heads. (Department of Pathology, University of Strasbourg, France catalog number 7664d.) (E) Left lateral view of a cranium with premature fusion of the sagittal suture resulting in an elongated cranium. Adult male from a pre-Columbian site in Cinco Cerros region of Peru (NMNH 293841).

Gigantism occurs if the pituitary secretes too much growth hormone during development or continues to secrete the hormone after growth should have ceased (Figure 14.3c).

Abnormal bone shape   As noted in the previous section, abnormal bone shape often accompanies abnormal bone size. However, there are some disorders in which abnormal shape is the most noticeable abnormality. One of these is the result of vitamin D deficiency, which causes rickets in subadults and osteomalacia in adults. The skeletal deformities resulting from these disorders can be quite dramatic, since the basic problem is that bone protein formed either during growth or as the result of the continuous process of remodeling (which continues throughout life) does not completely mineralize and becomes flexible (Figure 14.3d). If the dietary deficiency is corrected, the bone protein will mineralize preserving the abnormal shape of the bone in some cases. An important observation in identifying vitamin D deficiency is the interosseous line of the tibia. Normally this line is straight in both the anterior–posterior and medio–lateral axes. In rickets and osteomalacia the line is curved. There are other disorders that can result in abnormal shape such as poor alignment of a long bone after fracture. The fact that fracture is very rarely bilateral and the effect of vitamin D deficiency is bilateral helps in differentiating between the two disorders.

Growth disturbances are another cause of abnormal bone shape. An example of this is premature fusion of the skull sutures (craniostenosis) during growth of the skull. This disorder results in abnormal shape that is determined by which sutures fuse prematurely and when the fusion takes place. One manifestation of craniostenosis is early childhood premature fusion of the sagittal suture that results in an abnormally lengthened skull (Figure 14.3e).

CLASSIFICATION AND DIAGNOSIS IN HUMAN SKELETAL PALEOPATHOLOGY

Classification

Classification is basic to all aspects of biology including paleopathology. Through classification we seek to show the relationship between different types of biological organisms and phenomena such as disease. The problem with biological and pathological classificatory systems is that the organisms and phenomena being classified are complex. The criteria used in classification do not always permit unambiguous assignment to a single category. There is no good way to avoid this limitation in a classificatory system. However, an important objective in creating and using a classificatory system in skeletal paleopathology is to minimize ambiguity in assigning a specific disorder to a category in the system. Even with a generally useful classificatory system, it is important to remember that the system is an artificial construct. Care is needed to ensure that use of classificatory categories does not result in oversimplification in our understanding of the disease process.

In pathology there are two fundamental criteria in disease classification. One criterion is the cause of the disease; the other is pathogenesis of the disorder. Classification in pathology may be based on either or both of these factors. For most infectious diseases, cause is the primary factor in classification. An exception to this is the infectious disease actinomycosis which is caused by a bacterium. The suffix "mycosis" implies that the disease is caused by a fungus, which is clearly not the case. However, the pathogenesis of the disease is similar to other, mycotic, infections giving rise to the use of a potentially misleading suffix.

In contrast, assignment to the category of metabolic disorders is largely based on pathogenesis. Metabolic disorders involve disturbances in the formation and mineralization of osteoid, the major component of which is collagen. Vitamin, mineral or protein–calorie deficiency is often associated with these disorders. Similarly, pathogenesis is the major factor in the category of the erosive arthropathies even though there is convincing evidence that an infectious organism may be the triggering agent in an individual with the genetic potential for the disorder (Resnick 2002).

There are several options in developing a classificatory system for disease that depend on the relative importance of cause and pathogenesis. The late Lent Johnson, M.D., the legendary American orthopaedic pathologist, vigorously promoted a system that had seven basic categories of disease: anomaly, trauma and repair, inflammatory-immune, circulatory (vascular), metabolic, neuromechanical and neoplastic disorders. This is the system adopted and carefully explained by Ragsdale and Lehmer in Chapter 13 in this volume. However, any review of the table of contents in reference sources on orthopaedic pathology or skeletal radiology reveals the difficulty in establishing categories of skeletal diseases. This is, in part, because of the complexity of different diseases and the fact that the criteria used in classification overlap in some disorders. For example, Resnick's monumental work on skeletal radiology (Resnick 2002) has divided the discussion of skeletal disease into seventeen sections. In a major reference work on human paleopathology Aufderheide and Rodríguez-Martín (1998) divide their discussion of disease into thirteen parts.

When the late Walter G. J. Putschar, M.D. and I were organizing the contents of the first edition of *Identification of Pathological Conditions in Human Skeletal*

Remains (Ortner and Putschar 1981), we consulted some of the recently published reference works on orthopedic pathology. At that time we utilized twelve general categories that were used in both editions (Ortner and Putschar 1981; Ortner 2003): trauma, infectious diseases, circulatory disorders, reticuloendothelial and hematopoietic disorders (disorders of white and red blood cells and the tissues that make them), metabolic disorders, endocrine disorders, congenital and neuromechanical disorders, dysplasias, tumors and tumor-like disorders, joint disorders, dental and jaw disorders and miscellaneous disorders. We recognized that assignment of specific diseases to one of these categories was somewhat arbitrary, at least in a few cases, and that a problem in assigning some specific diseases to one of these categories was possible.

Fortunately, for most specific diseases assignment to a general category of pathology does not pose a problem. We used the general category of "miscellaneous disorders" for some of the diseases where the cause was unknown. An example of the problem that can develop is the classification of iron deficiency anemia. For many years the porous enlargement of diploë that occurs in some archaeological skeletal samples was attributed to this disease (see Ortner 2003). Walker et al. have challenged this possibility, arguing that the cause is megaloblastic anemia which is the result of a deficiency in vitamin B_{12} and/or folic acid (Walker et al. 2009). But the question remains, is an anemia caused by a dietary deficiency to be classified as a hematopoietic disease because of its pathogenesis, or would it be better to broaden our definition of metabolic disorders to include other vitamin-deficiency diseases? The important point to be made about classification of disease is the need to understand the pathogenesis and, where possible, the cause of the disorder and not let the assignment to a specific category of disease obscure our understanding of the basic biology of disease.

Diagnosis and research design in human skeletal paleopathology

In any research program on human skeletal paleopathology you will encounter cases of bone abnormality that do not fit into one of the diagnostic categories. However, the commonest types of bone abnormality, such as trauma, arthritis and infection, are often relatively easy to identify. Greater specificity is frequently more challenging, and those analyzing archaeological skeletons for evidence of disease need to avoid diagnosis to a level of specificity that is not justified by the evidence. Of course, as our knowledge and experience grow, it becomes possible to be more specific, and that is a major objective of reference works on paleopathology and the training that is now available at many centers of higher education.

An important component of differential diagnosis is to be aware of the most common skeletal disorders that can produce a given type and distribution of skeletal lesions. It is often possible to eliminate many, if not all, of the alternative diagnostic options for any given abnormality. The basis for this is careful evaluation of the type and distribution of skeletal lesions. But the first step in this evaluation is using the correct descriptive tools to determine the type and distribution of lesions present in a case of skeletal paleopathology.

A good example of this type of research is the analysis of metabolic disorders conducted by Brickley and Ives (2008). Metabolic diseases that receive emphasis in this study include rickets, scurvy in subadults and osteopenia/osteoporosis. The criteria for the first two of these disorders are relatively well known and, with experience, relatively easy to diagnose. Osteopenia/osteoporosis, which causes generalized loss of bone mass, is a more challenging issue in paleopathology. Diagnosis requires some measure of generalized loss of bone mass. This is challenging, particularly for osteopenia which is the less severe manifestation of this disorder. Taphonomic factors have to be considered as a possible cause of reduced bone mass. But even assuming that the diagnosis can be made, the cause of the disorder is much more difficult to determine since there are several diseases that can result in this abnormality.

Rickets and scurvy are both associated with malnutrition and provide insight into an important component of health in archaeological human skeletal populations. Osteopenia/osteoporosis can be caused by severe malnutrition but it is also caused by hormonal changes associated with aging and other disorders as well. If the disorder occurs in several archaeological human skeletons from a single site that have an estimated age at death of less than 40 years, the probability that severe malnutrition is the cause greatly increases. If this can be determined, the observation provides additional support for the presence of severe malnutrition and the next step in the research is to determine why this is the case and what is the significance of this evidence.

Another strategy in designing research in human skeletal paleopathology is to pursue a different type of problem. For example, a design that attempts to determine if skeletal sample A is healthier than skeletal sample B. The first thing that must be done is to decide what criteria are to be used to determine the relative health of each of the skeletal samples. One option is to define the prevalence of all the disorders encountered in each of the skeletal samples and create some method of establishing their relative significance. It is possible to do this but determining the diagnosis of all the potential disorders one might encounter in a skeletal sample requires a level of knowledge and experience that is difficult to achieve. Also, since different observers are likely to have different opinions about at least some of the diagnoses, comparison of prevalence data between samples analyzed by various scientists may not be effective. One possible solution is for a single observer to collect all the data needed for the research design. However, the time and cost factors in doing this create significant barriers.

A strategy for minimizing interobserver error is being utilized in the Global History of Health Project (global.sbs.ohio-state.edu/project_overview.htm) based at Ohio State University. This research has been under way for several years and has provided important insight about the relative health of different human societies and the factors that influenced the health of the people represented by the skeletal samples (Steckel and Rose 2002). In this research project, seven relatively general categories of skeletal abnormalities (linear enamel hypoplasia, stature, anemia, dental deficiency, degenerative joint disease, trauma and infection) are defined and used to create an index of health for different skeletal samples and their associated populations (e.g., Steckel and Rose 2002).

The variables used in this project are relatively easy to evaluate in part because the researchers utilize general categories of data that minimize interobserver error. The assumption is made that the variables provide a range of data indicative of general health of the sample. Because of this very basic research design, it is possible have many observers collect data that can be combined in the general study. This provides the option of creating large databases that address one of the major limitations of research in human skeletal paleopathology. An additional benefit is that it compensates at least partially for the inadequacy of sample sizes associated with most archaeological skeletal samples.

DISCUSSION AND CONCLUSIONS

In any research involving human skeletal paleopathology, the fundamental question is: What is the meaning and significance of the data? A very general response to what we have learned in human skeletal paleopathology is that knowing the time-depth and geographical range of various disorders provides at least a basic knowledge about what diseases our ancestors had to confront. However, even this basic information requires accurate diagnosis of the abnormalities encountered in the study of an archaeological skeletal sample and this is difficult in some cases if not impossible. Miller et al. (1996) provide evidence that the error rate in diagnosis increases significantly as one tries to move from a general category to specific diagnosis (e.g., a diagnosis of syphilis relative to the more general category of infection). The data for this study were based on observations made during workshops in which there were limitations on the time available for diagnosis of the abnormalities presented during the workshop. Nevertheless, this observation provides a note of caution about the design of research in human skeletal paleopathology. All skeletal abnormalities are indicative of disorder but the impact of various indicators of disease varies from insignificant to probable cause of severe disability and death. One of the important questions that is central to human paleopathology is the significance of disease in a given human society. It is unlikely that any human society did not have troublesome encounters with disease, particularly infectious disease. We know from recent history the social devastation that accompanied epidemic diseases such as the influenza pandemic that struck in 1918, just as World War I was ending, when millions of people died. During the medieval period, plague killed millions in Europe and caused major social and economic disruption. One of the tasks of paleopathology is to use the evidence of disease in archaeological human skeletal samples to extend our knowledge of the impact of disease on human societies well back into antiquity (Ortner 2009).

To accomplish this objective will require careful development of our knowledge regarding human disease and its effect on the human skeleton. However, it is equally important to recognize the limitations of diagnosis and avoid the temptation of overextending our data and ascribing specific diagnoses where this is not possible or where the potential for error is significant.

Furthermore, some assignment of significance needs to be attached to the presence of skeletal abnormalities in a skeletal sample. For example, the presence of button osteomas has no significance relative to the ability of the affected person to function effectively in the biocultural environment in which the person lives. Some infectious diseases that can affect the skeleton, such as brucellosis, have, in most cases, very little impact on the individual's ability to perform effectively and live a normal lifespan. Other disorders such as tuberculosis or carcinoma are much more serious, and often affect the ability to function and shorten life expectancy.

Fortunately, most of the abnormalities encountered in archaeological samples can be classified to at least one of the major categories of skeletal disorder e.g., trauma, infection, arthritis, tumor. More specific diagnosis within these major categories requires careful attention to the type and distribution of abnormalities and an awareness of the diagnostic options and limitations associated with these expressions of disease. What I have tried to accomplish in this chapter is to highlight an approach to diagnosis and classification of skeletal disease in archaeological human burials that emphasizes careful attention to the type and distribution of skeletal abnormalities. The first step in this analysis is careful use of terminology that minimizes the potential of masking the pathological processes that caused the skeletal abnormality. The descriptive terminology presented earlier provides the basic terms necessary to do this and enables other researchers to conduct an independent evaluation of the diagnostic options.

Acknowledgments

I owe a great debt to the many colleagues and students who have participated in the short courses and workshops with which I have been involved during the past forty years. Their enthusiasm for learning about skeletal disease was always an inspiration for me to try and find new ways of communicating the essentials of human skeletal paleopathology. Many have gone on to highly successful careers as researchers and teachers who have added much to our knowledge of paleopathology.

REFERENCES

Angel J. L., 1966   Porotic Hyperostosis, Anemias, Malarias, and Marshes in the Prehistoric Eastern Mediterranean. Science 153:760–763.

Aufderheide, A., and Rodríguez-Martín, C., 1998   The Cambridge Encyclopedia of Human Paleopathology. Cambridge: Cambridge University Press.

Brickley, M., and Ives, R., 2008   The Bioarchaeology of Metabolic Bone Disease. Amsterdam: Elsevier.

Chhem, R., and Brothwell, D., 2008   Paleoradiology. Berlin: Springer.

Hackett, C., 1976   Diagnostic Criteria of Syphilis, Yaws and Treponarid (Treponaematoses) and of Some Other Diseases in Dry Bones. In Sitzungsberichte der Heidelberger Akademie der Wissenschaften Mathematisch-naturwissenschaftliche Klasse, Abhandlung 4. Berlin: Springer.

Manolagas, S. C., 2000   Birth and Death of Bone Cells: Basic Regulatory Mechanisms and Implications for the Pathogenesis and Treatment of Osteoporosis. Endocrine Reviews 21: 115–137.

Miller, E., Ragsdale, B. D., and Ortner, D., 1996   Accuracy in Dry Bone Diagnosis: A Comment on Palaeopathological Methodology. International Journal of Osteoarchaeology 6:221–229.

Ortner, D. J., 2003   Identification of Pathological Conditions in Human Skeletal Remains. 2nd Edition. Amsterdam: Academic Press.

Ortner, D. J., 2009   Issues in Paleopathology and Possible Strategies for Dealing With Them. Anthropologischer Anzeiger, 67, 323–340.

Ortner, D. J., and Mays, S., 1998   Dry-Bone Manifestations of Rickets in Infancy and Early Childhood. International Journal of Osteoarchaeology 8:45–55.

Ortner, D. J., and Putschar, W. J. G., 1981   Identification of Pathological Conditions in Human Skeletal Remains. Washington, DC: Smithsonian Institution Press.

Ortner, D. J., and Turner-Walker, G., 2003   The Biology of Skeletal Tissues. In Identification of Pathological Conditions in Human Skeletal Remains. 2nd Edition. D. J.Ortner, ed. pp. 11–35. Amsterdam: Academic Press.

Pinhasi, R., and Mays, S. eds., 2008   Advances in Human Palaeopathology. Hoboken, NJ: John Wiley and Sons.

Powell, M. L., and Cook, D. C., eds., 2005   The Myth of Syphilis: The Natural History of Treponematosis in North America. Gainesville: University Press of Florida.

Resnick, D. J., ed., 2002   Diagnosis of Bone and Joint Disorders. 4th Edition. Philadelphia: W. B. Saunders.

Resnick, D. J., Manolagas, S. C., and Fallon, M. D., 2002   Histogenesis, Anatomy and Physiology of Bone. *In* Diagnosis of Bone and Joint Disorders. 4th edition. D. J. Resnick, ed. pp. 647–687. Philadelphia: W. B. Saunders.

Roberts, C. A., and Buikstra, J. E., 2003   The Bioarchaeology of Tuberculosis: A Global View on a Reemerging Disease. Gainesville: University Press of Florida.

Roberts, C. A., and Manchester, K., 2005   The Archaeology of Disease. 3rd edition. Gloucester: Sutton Publishing.

Steckel, R. H., and Rose, R. C., 2002   The Backbone of History. Cambridge: Cambridge University Press.

Tuross, N., 2003   Recent Advances in Bone, Dentin and Enamel Biochemistry. *In* Identification of Pathological Conditions in Human Skeletal Remains. 2nd edition. D. J., Ortner, ed. pp. 65–72. Amsterdam: Academic Press.

Waldron, T., 2009   Palaeopathology. Cambridge: Cambridge University Press.

Walker, P. L., Bathurst, R. R., Richman, R., Gjerdrum, T., and Andrushko, V. A., 2009   The Causes of Porotic Hyperostosis and Cribra Orbitalia: A Reappraisal of the Iron-Deficiency-Anemia Hypothesis. American Journal of Physical Anthropology 139:109–125.

CHAPTER **15** Estimating Age
and Sex from
the Skeleton, a
Paleopathological
Perspective

*George R. Milner
and Jesper L. Boldsen*

INTRODUCTION

Estimating age and sex are among a paleopathologist's first and most important tasks. That information, along with sample size, cultural context, and the physical condition of bones, plays a big part in determining what can be inferred from human remains from archaeological sites. Age and sex, for example, are critical components of identifying the specific diseases that were present in the past because many pathological conditions favor one segment of a population over another. For much the same reason, classifying samples by age and sex is essential when determining which segments of communities were hardest hit by diseases or injuries, and when assessing the biological costs of different ways of life.

Much has been written on age and sex estimation, but here we focus narrowly on several issues of immediate concern to paleopathologists. General approaches to the problem of extracting that information from skeletons are discussed, not the specific details of how it is done. Readily visible or measurable aspects of the skeleton are emphasized because those characteristics can be recorded quickly and do not require destructive analyses. Both are highly desirable, as large samples are needed to say something meaningful about past populations, as distinct from specific individuals,

A Companion to Paleopathology, First Edition. Edited by Anne L. Grauer.
© 2012 John Wiley & Sons, Ltd. Published 2016 by John Wiley & Sons, Ltd.

and destroying bones for thin sections (to estimate age by counting osteons) and genetic analyses (to establish sex) is often not possible.

Concerns about estimate accuracy, precision, and reliability are only the beginning of the issues paleopathologists face when trying to establish how skeletal lesions were distributed in past populations (Milner et al. 2008). Even in ideal circumstances where age and sex estimates are largely correct, one still has to contend with the effects of selective mortality when making sense of lesion frequencies in mortality samples. Osteologists want to know about life in the past, but they only have skeletons to look at. That is a problem because the individuals who died at each age were a biased sample of all people ever alive at that particular age because the phenomena of interest—diseases, trauma, and poor nutrition—increase the risk of dying (Wood et al. 1992). The interpretive difficulties posed by cemetery samples from archaeological cemeteries cannot be resolved without better estimates of sex and age, with the latter being a special concern. There is a fuller discussion of selective mortality in Chapter 7 of this volume.

AGE ESTIMATION

All osteologists would agree that age estimation is more of a problem for adult skeletons than it is for juvenile remains. That is because estimates of age into the late twenties are based on highly constrained developmental processes. For the rest of adulthood, estimates are derived from skeletal characteristics that can be loosely characterized as degenerative in nature, and the skeletal changes that accompany advancing age occur later in some individuals than in others. Variability in the appearance of these degenerative changes eventually becomes so great that the uppermost category in conventional estimation procedures is typically open-ended (e.g., 50+ years). The use of such terminal intervals is an honest admission of the limitations of existing methods, but recent work has shown that such a shortcoming is not a necessary feature of age-estimation procedures (Boldsen et al. 2002; Weise et al. 2009; Milner 2010).

Approaches to age estimation

Different approaches to age estimation can be summarized by considering how estimates are obtained and the nature of the age intervals assigned to skeletons. What is commonly done in everyday practice, of course, is a bit of a moving target, since what are regarded as best practices change over time.

Age estimates can be based on an investigator's best judgment or a fixed procedure that focuses attention on changes in one or more anatomical structures, as documented in a particular reference sample. Many years ago, subjective appraisals were the norm, but since the mid-20th century, beginning with Todd's (1920) work with the pubic symphysis, much work has focused on the definition of sequences of stages characterizing age-related alterations in a few skeletal features. This work moved the field from the realm of individual judgment to procedures that could be duplicated and verified; provided systematic documentation of several age-related changes in the skeleton; and helped ensure greater comparability of results. Expert assessments, therefore, have fallen out of favor over the past half-century. But that is by no means the end of the story.

The use of fixed estimation procedures certainly reduces the subjective element, but does not eliminate it altogether. Observers must have a firm grasp of the considerable morphological variability normally found in the skeleton to match specimens to a particular stage in a sequence. Experience plays a large part in being able to classify anatomical structures consistently and with little error relative to the published stage descriptions and illustrations (hence age intervals). So the observer's role has not been removed through the introduction of various formal classifications of the age-related changes that occur in several parts of the skeleton. Furthermore, experience-based assessments were never truly left behind, as osteologists commonly use their best judgment when reconciling divergent results from different parts of the skeleton. So the pubic symphysis, sacroiliac joint (iliac auricular surface), sternal ends of ribs, and cranial sutures, to name several commonly used age indicators, are relied upon to a greater or lesser extent depending on a researcher's training and prior experience, bone preservation, and the unique characteristics of each skeleton. How that balancing act is performed in daily practice is not the same for all osteologists. It is also likely to differ from one skeleton to the next, as the perceived importance of various age indicators is affected by pathological conditions and postmortem damage that alter the appearance of bones.

For juvenile skeletons, age estimates are mostly based on tooth development and eruption, epiphyseal appearance and union, and bone dimensions. The late-fusing iliac crest and the clavicle's medial epiphysis can be useful into the twenties. It has long been recognized that tooth formation and eruption are less severely affected by poor nutrition and infectious diseases, which upset normal development, than bone growth and maturation (e.g., Lewis and Garn 1960; Ubelaker 1987). Teeth are also generally well preserved and easily recognizable—important factors when dealing with archaeological remains. So for several reasons teeth are usually the preferred means of estimating age, and there are several ways of doing so. Charts depict various points in the development of the entire dentition, individual teeth can be categorized by developmental stage, the lengths of incomplete teeth can be measured, and erupted teeth can be identified or simply counted (e.g., Moorrees et al. 1963a, 1963b; Townsend and Hammel 1990; Liversidge and Molleson 1999; Ubelaker 1999; AlQahtani et al. 2010). As long as skeletons are well-preserved and reasonably complete, age ranges can span several months for infants up to a few years for teenagers. Such short intervals are generally accurate, as they often encompass the true age.

The picture is not nearly so rosy for most of adulthood. That is because even long intervals that cover as much as several decades (poor precision) often fail to encompass the true age (poor accuracy). For adults, the focus on a set of fixed procedures has had one important, if unintended, consequence: estimates tend to be based on a mere handful of anatomical structures. The parts of the skeleton most often used to estimate the ages of adults include the pubic symphysis (Todd 1920; McKern and Stewart 1957; Meindl et al. 1985a; Katz and Suchey 1986), sacroiliac joint (Lovejoy et al. 1985a; Buckberry and Chamberlain 2002), cranial sutures (Todd and Lyon 1924; Meindl and Lovejoy 1985), and sternal ends of ribs (İşcan et al. 1984). They can be divided in one way or another into several stages, with the ages assigned to them based on the age distributions of the original reference samples. Occasionally, non-overlapping age intervals follow each other in succession, such as Todd's (1920) sequence of ten pubic symphysis stages. Age ranges in most procedures, however,

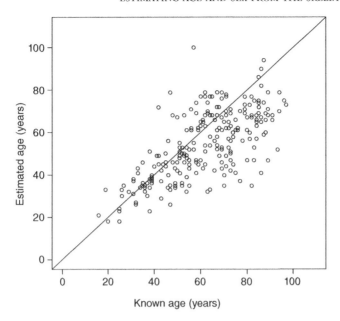

Figure 15.1 A comparison of actual and estimated ages for 239 modern American (Bass Collection) skeletons examined by Milner. Ages were estimated with the current Transition Analysis program, and they are based on pelvic and cranial characteristics, as well as a uniform prior distribution. The diagonal line indicates a perfect match between actual and estimated ages. Note that a large fraction of the sample is over 50 years, a part of the lifespan commonly handled through the use of a terminal open-ended age interval.

overlap one another, including those of the often-used six Suchey–Brooks pubic symphysis stages (Katz and Suchey 1986). The latter are far more realistic as they better capture the variability that actually exists in the appearance of skeletons from individuals of exactly the same chronological age.

Age estimates can be based on single anatomical structures, such as the pubic symphysis, or several in combination. Multiple skeletal characteristics are usually used because it is thought they yield, on average, more accurate estimates than just one such feature. A composite age estimate can be obtained by balancing available information, putting the most emphasis on what seems to be the most reliable indicator(s) in individual skeletons. That is, the pubic symphysis is generally considered more accurate than cranial sutures, but sometimes it might not be used because the joint looks anomalous or has been altered by trauma. So an individual's judgment, once again, comes into play. Fixed procedures for combining different sources of information can also be employed, notably the Complex or Multifactorial methods mostly used in Europe and North America, respectively (WEA 1980; Lovejoy et al. 1985b). In recently developed Transition Analysis, the age distributions of various cranial and pelvic characteristics are combined, yielding an overall estimate tailored to the specific mix of traits found in each skeleton (Boldsen et al. 2002). That is a more sound way to proceed, but even so the resulting estimates are not without problems (Milner 2010) (Figure 15.1).

Some problems with age estimation

Several difficulties with age estimation apply to both juveniles and adults. For paleopathological purposes, however, the error associated with juvenile estimates is usually small enough to be of little consequence, at least relative to other problems arising from the use of small, incomplete, and biased archaeological samples. The most important difficulties pertain to adults, because standard estimation methods tend to feature poor accuracy, precision, and replicability; biased estimates; and an inability to deal with the upper part of the lifespan (Bocquet-Appel and Masset 1982; Konigsberg and Frankenberg 1992, 1994; Konigsberg et al. 1997; Hoppa and Vaupel 2002a; Milner et al. 2008). Collectively, they can derail the process of making sense of disease patterning in adult skeletons.

Conventional adult age-estimation procedures, including those that rely on the pubic symphysis, sacroiliac joint, ribs, and cranial sutures, tend to yield biased results (Murray and Murray 1991; Saunders et al. 1992; Aiello and Molleson 1993; Dudar et al. 1993; Galera et al. 1998; Oettlé and Steyn 2000; Schmitt 2004; Mulhern and Jones 2005; Martrille et al. 2007; Hens et al. 2008; Passalacqua 2010). Often, the ages of individuals in their twenties are overestimated. Perhaps more important is a systematic underestimation of the ages of people from roughly their fifties onward, because discrepancies between estimated and actual ages tend to be greater in old people. The combined effect means too many individuals in archaeological samples supposedly died in their thirties or forties, since there is a tendency for young adults to be overestimated while the opposite is true for old people. Few people are thought to have survived past their fiftieth birthday.

Biased estimates result from the fact that the age structure of the reference sample (the skeletons used in developing the procedure) has an effect on the estimated ages of archaeological skeletons. It is useful to think of this issue, first brought to the attention of osteologists by Bocquet-Appel and Masset (1982), in Bayesian terms as the effect of a prior distribution on age estimates (Konigsberg and Frankenberg 1992, 1994; Hoppa and Vaupel 2002b; Caussinus and Courgeau 2010). Osteologists were initially skeptical of Bocquet-Appel and Masset's (1982) argument, if not downright hostile to it, partly because the article was provocatively titled "Farewell to Paleodemography." Several decades later, there is widespread, if not universal, agreement that this issue is important and must be addressed (e.g., Hoppa and Vaupel 2002a). The difficulty can be illustrated with McKern and Stewart's (1957) classification of age-related changes in the male pubic symphysis. Their approach—the symphyseal face is divided into separately scored "components" that are combined to arrive at a final age estimate—was an important contribution to the discipline. But the sample only consisted of soldiers who died in the Korean War. The age distribution for any trait present in the sample would appear to span some fraction of early adulthood, even if the trait was actually more typical of middle or old age in the American population of the early 1950s. Most reference collections, of course, are not as badly skewed as the Korean War sample, but that does not eliminate the effect of their age distributions on estimates produced for archaeological skeletons. Note that this issue has nothing whatsoever to do with how well a researcher scores various age-related skeletal characteristics. In fact, if traits are recorded properly, one is assured of getting biased results (doing a bad job is not an attractive option because the outcome would likely be even worse and entirely unpredictable). So, the age ranges and measures of central tendency and dispersion that are published as part of standard methods, while indicating a progression from young to old, are influenced by the age structure of the samples used when the procedures were initially developed.

It is commonly believed that multiple age indicators are better than single features. Age-estimate inaccuracy, however, is not automatically corrected by having several skeletal markers of age, as in the Multifactorial and Complex methods (WEA 1980; Lovejoy et al. 1985b). Such procedures are only as good as the methods upon which they are based. One cannot assume that the error in one procedure will simply cancel out the error in another. In fact, as most standard procedures underestimate the ages of old people (those beyond about 50 years), combining the results of several methods, most or all of which err in the same direction, would not lead to any real improvement in accuracy.

A second troublesome feature of conventional methods is the use of open-ended terminal intervals. Little can be said about diseases in the terminal age group, except that certain pathological conditions were present in old people. The inability to say much about people who lived longer than about 50 years might at first glance seem of little consequence because so few individuals survived that long, to judge from paleodemographic studies. That argument, however, does not take into account the tendency to underestimate age in the upper part of the lifespan, so our perceptions about adult life expectancy in the distant past are likely to be wide of the mark. In fact, there are studies where enough individuals apparently lived into the sixth decade and beyond to be consistent with the mortality experience of small-scale societies from the recent past (Milner et al. 1989; Paine 1989). Fortunately, new methods are being developed that allow the ages of old people to be estimated (Boldsen et al. 2002; Weise et al. 2009). Transition Analysis, for example, has produced estimates for two elderly people, one an early-modern-period Dane and the other a Pre-Columbian Maya Indian, that closely approximate their ages as documented in records or glyphs, respectively (Boldsen et al. 2002; Buikstra et al. 2006). So with appropriate procedures, old people can be identified in archaeological samples, if they are indeed present.

A third issue of concern is the use of fixed-length age intervals when initially estimating the ages of skeletons, and then grouping them to form the categories used when comparing lesion frequencies in separate samples. Perhaps the most obvious problem is that many published age ranges, especially for middle to old age, are quite wide, sometimes spanning much of adulthood. Examples include high stages in the Suchey–Brooks pubic symphysis procedure (Katz and Suchey 1986), where these stages encompass many decades. Long intervals are desirable as they accommodate the actual age distribution of people associated with the morphological stages. Yet, they can be so long that little real value is gained by using them. That said, they are still preferable to overly short intervals, such as the Todd (1920) pubic symphysis and the Meindl et al. (1985a) "modal" sacroiliac joint ages based on Todd's original work. Tight ranges imply a level of precision that is not justified given the variation that exists in the rates of change in the skeletal features commonly used for age-estimation purposes. It follows that the true ages of many individuals, perhaps even the majority of them, fall outside short intervals. Perhaps it is useful to think of an ideal age range as splitting the difference between being short enough to say something useful, but not overly so, as that risks having many individuals with the trait not being included within the interval.

The trouble with published age intervals, however, goes deeper than accuracy (do intervals encompass the true age?) and precision (are intervals too short or long to be accurate or useful?). An interval of fixed length implies that one can be equally confident that all individuals assigned to it actually belong there. In essence, that means the age-informative characteristics of all individuals in a category—say 45–50 years for one of the late Todd (1920) stages—are the same. Practical experience shows that in some skeletons there is a great deal of consistency in how age-informative markers change with advancing age; however, in others that does not occur. That is as true of single anatomical structures as it is for the entire skeleton. It follows that osteologists can be more certain that some skeletons truly fall in a particular age interval than other skeletons, even if they end up in the same category. For skeletons with reasonably consistent indicators of age, one can have more confidence in the resulting age estimates than for those with markers that point in different directions. The use of fixed intervals, however, carries with it the implication that one is equally certain about the ages of all skeletons assigned to a particular age category.

Despite several troublesome issues, the outlook is not as bleak as it is sometimes portrayed. Improvements in age estimates, however, require that three issues are addressed simultaneously: the osteological (better age indicators) and mathematical (better analytical procedures) foundation of this work must be coupled with user-friendly computer programs to implement new methods as they are developed.

Recently, progress has been made on furthering our understanding of the age-related changes in anatomical structures that have long been the centerpiece of adult age estimation: the pubic symphysis and sacroiliac joint (Boldsen et al. 2002; Buckberry and Chamberlain 2002). These complex structures have been broken down into separately scored traits, following the lead of McKern and Stewart (1957) and Gilbert and McKern (1973) who split the pubic symphysis into different components for recording purposes. By doing so, it is possible to characterize interindividual variation and differences in rates of change in specific age-informative traits better than can be done with a sequence of stages based on anatomical structures in their entirety.

New analytical methods have been introduced, like Transition Analysis and, more recently, Calibrated Expert Inference; the first one is accompanied by a computer program to facilitate its use (Boldsen et al. 2002; Kimmerle et al. 2008; Konigsberg et al. 2008; DiGangi et al. 2009; Weise et al. 2009). In these two procedures, age-interval lengths can vary from one skeleton to the next, based on the mix of characteristics that happens to be present (Boldsen et al. 2002; Weise et al. 2009). So uncertainty about age is quantified on a case-by-case basis for each skeleton.

The prior distribution issue is not fully addressed in all new procedures, as pointed out by Boldsen et al. (2002) and Caussinus and Courgeau (2010). That said, maximum-likelihood age estimates calculated using Boldsen and coworkers' (2002) computerized version of Transition Analysis and two different prior distributions, one uniform and the other from seventeenth-century Danish rural parish records, are essentially the same up to about 40 years (Milner 2010). Beyond that point, there is usually a difference of no more than five years. This discrepancy, while still troubling, pales beside the problems posed by lengthy age estimates, especially in middle to old

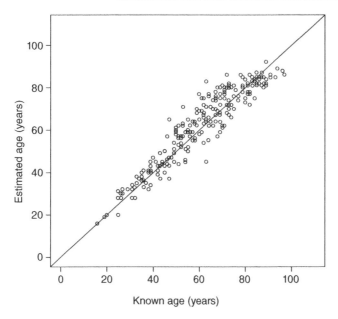

Figure 15.2 A comparison of the actual and estimated ages of the same 239 modern American (Bass Collection) skeletons, with the estimates being experience-based impressions based on a wide array of characteristics distributed throughout the skeleton. The diagonal line indicates a perfect match between actual and estimated ages. These results indicate there is much more that is informative about age than is captured by conventional methods of scoring skeletal structures that emphasize the pelvis, cranium, and ribs.

age, and poor accuracy, most notably for the cranial sutures. Disappointing accuracy and precision underscore the inadequacy of the pelvic and cranial structures that have traditionally served as the principal means of estimating the age of adults. In fact, no published method based on the skeletal structures commonly used to estimate age during adulthood yields particularly satisfactory results, and it is not likely that further elaborations of scoring procedures for those particular anatomical features will materially improve the situation.

Fortunately, a wider array of skeletal characteristics promises to improve estimate accuracy and precision, judging from experience-based assessments of age undertaken in blind trials (Weise et al. 2009; Milner 2010) (Figure 15.2). Individually, these traits contribute little to age estimates (typically only "young" versus "old" for the trait in question); however, collectively they yield surprisingly good results. That is especially true when employing a procedure developed by Weise and coworkers (2009) that accommodates the importance of estimating the overall mortality distribution before obtaining individual ages-at-death, as discussed in Hoppa and Vaupel (2002a, 2002b). It is always difficult to predict the future, but improvements in age estimation for adults will probably combine new age-informative characteristics and procedures that explicitly address analytical problems that have long plagued paleodemography. The full implementation of those procedures, in turn, requires the development of user-friendly computer programs.

Sex Estimation

The situation for sex estimation is rather the reverse of that for age—it is principally a problem with juveniles, not adults. The skeletons of people older than roughly 15 years can usually be assessed fairly accurately, as long as bones are reasonably well preserved, although there is still room for improvements in methods. For immature skeletons, new procedures are periodically proposed, but initial enthusiasm soon wears thin following examinations of additional samples by different researchers. At present, no means of estimating the sex of juveniles is sufficiently reliable to be of any real use in paleopathological analyses, and the outlook for such a procedure remains bleak.

There is, of course, the possibility of using genetic markers to identify an individual's sex. In the foreseeable future the value of ancient DNA (aDNA) for this purpose is limited by high cost, lengthy analysis time, poor preservation, and the possibility of contamination. However useful aDNA might be for specific, high-profile cases— skeletons that are especially interesting because of their cultural context or, for paleo-pathologists, particular suites of bony lesions—it is not feasible for the large samples required for population-oriented research. The latter includes paleodemographic studies as well as paleopathological analyses that assess the disease experience of past communities.

At the outset, it is necessary to clarify the use of two terms: gender and sex (Walker and Cook 1998). Gender is sometimes substituted for sex when referring to the physical characteristics of skeletons. Yet doing so is misleading as it blurs a useful distinction. Osteologists observe sexually dimorphic skeletal characteristics that in large part reflect an individual's genetic make-up. Gender, in contrast, is culturally specified, as it pertains to the roles a person assumes and how that individual is perceived by fellow community members. A male's skeleton, for example, might be found with artifacts normally reserved for females, so the sex and gender distinction is important in reconstructing the life history of that person.

Approaches to sex estimation

Estimates of sex are based on either visual assessments or measurements of one or more bones. Single measurements capture variation in size, and some of them, including the diameters of the humeral and femoral heads, permit a reasonably good separation of the sexes. Simple ratios (e.g., the Ischio-Pubic Index) or complex multivariate analyses have the capacity to characterize shape as well as size. The computer program, Fordisc, intended for forensic applications, is perhaps the most notable example of multivariate analyses (Jantz and Ousley 2005). The complexities of skeletal structures tend to be better characterized by many measurements than just a few of them. But not all dimensions of one or more bones are equally efficacious in discriminating between the sexes, so simply adding one measurement after another does not guarantee a correspondingly better outcome.

Skeletal indicators of sex are often described in terms of classic male or female characteristics, although everyone realizes that for each skeletal trait there is a continuum from one extreme to the other. Measurement-based work obviously accommodates variation in size and shape. There are also several classifications of selected anatomical structures, which feature categories ranked from gracile to robust, that are designed to help with the visual assessment of shape (Hoshi 1962; WEA 1980; Buikstra and Ubelaker 1994; Walker 2008).

With regard to accuracy, when skeletons are relatively complete, the judgments of experienced osteologists perform about as well as measurements. So if accuracy was the sole concern, it would make little difference if an osteologist simply looked at bones or measured them. Yet quantitative methods have one great advantage: they permit estimates of how certain one might be about the sex of skeletons. While that is also possible with categorical data, as has been done for several cranial features (Walker 2008), usually uncertainty in visual inspections is denoted by appending one or two question marks to male or female. The score a particular skeleton would receive, of course, is liable to vary from one observer to the next, depending on each researcher's experience.

Some problems with sex estimation

Although it is computationally possible to obtain estimates of the probability a given skeleton is one sex or the other, they rarely play a part in decision-making based on bone measurements. It is instead common practice to use a sectioning point to separate skeletons classified as males from those considered females, or to specify a middle ground that encompasses individuals of unknown sex. Doing so, however, is not particularly desirable as potentially important information is lost when allocating skeletons to two or three groups (male, female, and unknown). Misclassifications are inevitable—particularly when using a single sectioning point to distinguish males from females—as overlapping distributions in size and shape mean there is never a clear separation between the sexes.

Being able to calculate a numerical measure of uncertainty, however, is not the end of the story. That is true for assessments based on measurements, which are by far the most common, and classifications of traits in the gracile to robust spectrum (e.g., Walker 2008). There are two issues of special concern. For probabilities of being a male or female to have any validity, the study skeletons must be drawn from the same population that gave rise to the reference sample used when developing the sex-estimation procedure. Incidentally, the same applies to sectioning points that divide the sexes. That is because the world's populations display considerable variation in body size, limb and trunk proportions, skeletal robustness, and degree of sexual dimorphism in bones. Finding appropriate reference samples is difficult for skeletons from recent times, such as forensic cases. Doing so for archaeological skeletons is probably impossible, since living conditions were considerably different from those experienced by the modern populations represented by known-sex reference samples. The best that can be done is to get the closest match possible, which will generally be in terms of general ancestry and country of origin.

One must also know the sex composition of the archaeological sample being investigated: that is, one needs to know the ratio of females to males (Albanese et al. 2005; Milner and Boldsen 2012). Take, for example, a situation where decisions about sex are made on the basis of one measurement; one might use the diameter of the humeral head, as it does a good job of distinguishing between the sexes based on

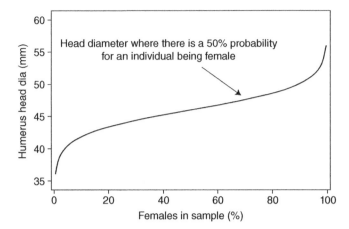

Figure 15.3   Humeral head diameters (mm) that correspond to an equal probability of a skeleton being female or male are not the same in study samples with different sex ratios. When females are disproportionately represented in the sample of skeletons, perhaps because they came from an excavated convent cemetery, one can expect that large bones are more likely to be from females than they would be in a sample that had few women but many men.

size. If the archaeological sample happened to be heavily weighted toward females, perhaps because the cemetery was associated with a medieval convent, then even large bones will likely be from women. These big women would be misclassified as men unless probabilities (or a sectioning point, see Figure 15.3) were shifted upward to take into account the composition of the archaeological sample. The reverse can also occur, such as when a male-dominated sample is found in a battlefield mass grave. It is rarely acknowledged that the process of using measurements, as well as categorical data such as the classification of cranial features used by Walker (2008), is predicated upon the assumption that one is equally likely to have a male as a female skeleton in the sample being investigated (Milner and Boldsen 2012). By far the greatest difficulties arise when the archaeological skeletons are overwhelmingly one sex or the other, such as the convent or battlefield examples.

PALEOPATHOLOGICAL IMPLICATIONS

Paleopathologists might well ask what all of the above means in terms of their day-to-day work. Of particular interest is what can be done at present, not at some point in the future, when the major bumps in age and sex-estimation procedures might, perhaps, be smoothed over.

For adults, the experienced eye, or measurements from suitable reference samples, tend to yield accurate estimates of sex, at least for reasonably intact skeletons. The qualifier about completeness is important, as paleopathologists necessarily rely heavily on skeletons that are well preserved. After all, it is easiest to detect subtle signs of bone resorption or proliferation on such specimens. So it is often the case that the skeletons of greatest paleopathological interest are also those where one has the most confidence about sex. While it is usually possible to correctly assess the sex of well-preserved skeletons, several researchers have noted a tendency to misclassify young males as females and old females as males (Meindl et al. 1985b; Milner and Smith 1990; Walker 1995). To the extent this occurs (and we have seen, but not quantified, a slight tendency in those directions in skeletons of different ancestries, especially with regard to cranial morphology), one must be careful about accepting what might initially appear to be higher female mortality early in the reproductive period and greater male longevity toward the end of the lifespan, should either one of them be seen in an archaeological sample.

It would be useful to know the probability that a particular skeleton was one sex or the other. That often will be problematic given the need for proper reference collections; furthermore, the likely composition of the archaeological sample also influences such figures. Nevertheless, even crude measures of one's certainty about sex can be important when a question arises about the pathological conditions visible in a particular skeleton, such as those attributable to a disease or injury that tends to be associated with one sex more so than the other.

As there are no effective means of estimating the sex of juveniles from simply examining or measuring bones, paleopathologists must combine the skeletons of boys and girls. That eliminates the possibility of determining whether the skeletons of one sex had more pathological lesions than the other. One might be tempted to count markers of childhood ill-health that are visible in adult skeletons, such as enamel hypoplasia and radiopaque Harris lines, and then assume that the sex with the greatest number of such markers experienced worse conditions early in life. Such a conclusion, however, ignores the confounding effects of selective mortality. The straightforward explanation would be that the sex with the fewest markers of childhood ill-health was, on average, the healthier one. Yet it could also be true that many sick children of one sex died more or less immediately, whereas hard-hit members of the other sex tended to survive, only to die later as adults still carrying signs of their earlier disease experience. Extrapolating backward from adults to identify differences in the lives of boys and girls is not at all a straightforward process (Wood et al. 1992).

In contrast, age estimates for juveniles are generally adequate for most paleopathological purposes. More precisely, estimates tend to be satisfactory when based on dental development and, to a lesser extent, epiphyseal union. As growth faltering is a common consequence of poor nutrition and a high disease-load, bone dimensions such as long bone lengths confound age with living conditions. Estimating age from the sizes of bones is further complicated to the extent that individuals suffering from chronic conditions that retard growth were preferentially winnowed out of the population at an early age. But because teeth are so commonly used to estimate the ages of juvenile skeletons, estimate inaccuracy is not terribly significant in most instances for the simple reason that archaeological samples tend to be small, so there is a limit to what can be realistically done with them. An under-representation of the youngest skeletons, particularly infants, because of mortuary customs (burial elsewhere) or poor preservation (loss of tiny and fragile bones) is generally a greater problem when reconstructing age-at-death distributions from the human remains in excavated cemeteries.

Age estimates for adults are another matter altogether. Biased estimates produced by conventional methods mean that the categories often used for comparative purposes are of questionable value. For example, when tallying skeletons with particular kinds of lesions, one might divide adults into "young," "middle," and "old." In fact, Waldron (1991) has noted that intervals of 10-year duration are about as narrow as paleopathologists can hope to achieve, with the last one typically being open-ended. The ages assigned to whatever categories might be used, of course, will vary from one study to the next according to research design, investigator preference, and sample size, but something along the lines of 20–35, 35–50, and 50+ years would not be unusual. Young adults can often be identified by slight to moderate tooth wear, the relatively late fusion of the clavicle's medial epiphysis, and alterations in the pubic symphysis, including the progressive loss of a ridge-and-furrow (billowed) surface and ventral rampart formation. An open-ended terminal category encompassing the upper end of the lifespan acknowledges the limitations of standard methods, as little can be done to separate old skeletons from very old ones. The biases in age estimates, as discussed above, mean a separation into young, middle, and old adult will result in too many individuals being placed in the first two groups, especially the middle one, and not enough in the last. The same would be true of other age categories one might use with archaeological samples, such as the 10-year intervals commonly employed in paleodemographic studies. On the bright side, reasonable (although not perfect) approximations of age for the period when most adults in traditional societies tended to die can be obtained from the existing computerized version of Transition Analysis (Boldsen et al. 2002; Milner 2010) (Figure 15.1). At least that is true for the population as a whole, although estimates for individual skeletons can be far off the mark. Fortunately, new methods are being developed that yield more accurate and precise age estimates (e.g, Weise et al. 2009) (Figure 15.2).

Whatever the advantages of grouping skeletons into several age categories, such as "juvenile," "young adult," "middle adult," and "old adult," such coarse intervals are still not a very good way to proceed, even in the ideal situation where one has unbiased age-at-death estimates. Differences in disease experience among separate groups of skeletons can be obscured when samples are divided into age intervals that span many years, if not decades.

It would be far better to estimate the age-specific risk of having the pathological condition of interest. For example, in a rural medieval Danish sample no statistically discernible difference was found between males and females in terms of dental caries when they were treated simply as adults (Boldsen 1997). Based on the aggregated data, one could reasonably infer that men and women had the same (miserable) experience with tooth decay. That would be the conclusion reached in most paleo-pathological studies as an "adult" category is often used when summarizing lesion frequencies. The same criticism, of course, would apply to putting all juveniles in a single "subadult" group.

A very different picture, however, was obtained when men and women from this medieval village were compared in terms of age-specific risks of experiencing active dental decay (Boldsen 1997). The two sexes were not at all alike in terms of age-specific probabilities of having bad teeth, with women being particularly susceptible to tooth decay toward the end of their reproductive period. The principal point is that looking at age-specific risks of having pathological conditions is a much better means of characterizing disease experience than simply lumping skeletons into broadly defined categories, such as "juvenile" and "adult." That is even true when those groups are broken down into categories such as "young," "middle," and "old" adult, which still span many years. Not only is information lost, but conclusions can be utterly wrong, as would have happened in the Danish example had the analysis stopped with frequencies derived from aggregated data.

CONCLUSION

Despite problems that have long beset paleodemography, the outlook for sex and age estimation is bright. At least one can be optimistic as long as several issues are tackled.

With regard to sex estimation, assessments of adult skeletons are already reasonably accurate, although it would often be useful to know the probability that a skeleton was from an individual of a particular sex. The effect of the composition of archaeological samples on sex estimates is largely unexplored, and there are sites where that would be of great importance. Fortunately, situations where skeletal samples might be heavily weighted toward one sex or the other can often be detected by paying close attention to the archaeological context. Although it would be highly desirable to be able to estimate the sex of juveniles, we do not see that there is a clear path toward that objective using bones alone.

Present methods, despite room for improvement, are sufficient for estimating the ages of juveniles for most paleopathological purposes. When it comes to adults, however, the situation is much different. We probably have gone as far as is possible with the current emphasis on just a few parts of the skeleton, most notably the pubic symphysis, sacroiliac joint, cranial sutures, and ribs. There is reason to believe that it is worth looking at a broader array of age-informative features distributed throughout the skeleton, as long as there is a simultaneous development of the necessary analytical procedures to use them to good advantage. More work must be done to account for the effect of the reference sample's age composition on estimates generated for archaeological skeletons. In short, the research effort must proceed on three fronts. First, new skeletal traits have to be defined, and their age distributions documented. Second, the mathematical means must be developed to produce unbiased estimates of age. Third, computer programs are required to facilitate the use of computationally intensive procedures.

Paleopathologists need not wait until all of these issues are fully resolved, as that will take some time. Methods are now in place, or will be shortly, that are of sufficient accuracy to allow the age-specific risk of various diseases or trauma-related conditions to be estimated. That would be a welcome change from a long-standing, and rather complacent, reliance on procedures that yield biased adult age estimates and the use of broad age categories that obscure the disease patterns of interest.

REFERENCES

Aiello, L. C., and Molleson, T., 1993 Are Microscopic Ageing Techniques More Accurate Than Macroscopic Ageing Techniques? Journal of Archaeological Sciences 20:689–704.

Albanese, J., Cardoso, H. F. V., and Saunders, S.R., 2005 Universal Methodology for Developing Univariate Sample-Specific Sex Determination Methods: An Example Using the Epicondylar Breadth of the Humerus. Journal of Forensic Sciences 32:143–152.

AlQahtani, S. J., Hector, M. P., and Liversidge, H. M., 2010   The London Atlas of Human Tooth Development and Eruption. American Journal of Physical Anthropology 142: 481–490.

Bocquet-Appel, J.-P., and Masset, C., 1982   Farewell to Paleodemography. Journal of Human Evolution 12:353–360.

Boldsen, J. L., 1997   Estimating Patterns of Disease and Mortality in a Medieval Danish Village. In Integrating Archaeological Demography: Multidisciplinary Approaches to Prehistoric Population. Richard R. Paine, ed. pp. 229–241. Carbondale: Center for Archaeological Investigations, Southern Illinois University.

Boldsen, J. L., Milner, G. R., Konigsberg, L.W., and Wood, J. W., 2002   Transition Analysis: A New Method for Estimating Age from Skeletons. In Paleodemography: Age Distributions from Skeletal Samples. R. D. Hoppa, and J. W. Vaupel, eds. pp. 73–106. Cambridge: Cambridge University Press.

Buckberry, J. L., and Chamberlain, A. T., 2002   Age Estimation From the Auricular Surface of the Ilium: A Revised Method. American Journal of Physical Anthropology 119: 231–239.

Buikstra, J. E., and Ubelaker, D. H., eds., 1994   Standards for Data Collection from Human Skeletal Remains. Research Series 44. Arkansas Archeological Survey, Fayetteville, AR.

Buikstra, J. E., Milner, G. R., and Boldsen, J. L., 2006   Janaab' Pakal: The Age-at-Death Controversy Re-visited. In Janaab' Pakal of Palenque: Reconstructing the Life and Death of a Maya Ruler. V. Tiesler, and A. Cucina, eds. pp. 48–59. Tucson: University of Arizona Press.

Caussinus, H., and Courgeau, D., 2010   Estimating Age Without Measuring It: A New Method in Paleodemography. Population-E 65:117–144.

DiGangi, E. A., Bethard, J. D., Kimmerle, E. H., and Konigsberg, L. W., 2009   A New Method for Estimating Age-at-Death From the First Rib. American Journal of Physical Anthropology 138:164–176.

Dudar, J. C., Pfeiffer, S., and Saunders, S. R., 1993   Evaluation of Morphological and Histological Adult Skeletal Age-at-Death Estimation Techniques Using Ribs. Journal of Forensic Sciences 38:677–685.

Galera, V., Ubelaker, D. H., and Hayek, L. C., 1998   Comparison of Macroscopic Cranial Methods of Age Estimation Applied to Skeletons from the Terry Collection. Journal of Forensic Sciences 43:933–939.

Gilbert, B. M., and McKern, T. W., 1973   A Method for Aging the Female Os Pubis. American Journal of Physical Anthropology 38:31–38.

Hens, S. M., Rastelli, E., and Belcastro, G., 2008   Age Estimation from the Human Os Coxa: A Test on a Documented Italian Collection. Journal of Forensic Sciences 53:1040–1043.

Hoppa, R. D., and Vaupel, J. W., eds., 2002a   Paleodemography: Age Distributions from Skeletal Samples. Cambridge: Cambridge University Press.

Hoppa, R. D., and Vaupel, J. W., 2002b   The Rostock Manifesto for Paleodemography: The Way from Stage to Age. In Paleodemography: Age Distributions from Skeletal Samples. R. D. Hoppa and J. W. Vaupel, eds. pp. 1–8. Cambridge: Cambridge University Press.

Hoshi, H., 1962   Sex Difference in the Shape of the Mastoid Process in Norma Occiptalis and Its Importance to the Sex Determination of the Human Skull. Okajimas Folia Anatomica Japonica 38:309–313.

İşcan, M. Y., Loth, S. R., and Wright, R. K., 1984   Estimation from the Rib by Phase Analysis: White Males. Journal of Forensic Sciences 29:1094–1104.

Jantz, R. L., and Ousley, S. D., 2005   FORDISC 3.0 Personal Computer Forensic Discriminate Functions. Knoxville: Forensic Anthropology Center, University of Tennessee.

Katz, D., and Suchey, J. M., 1986   Age Determination of the Male Os Pubis. American Journal of Physical Anthropology 69:427–435.

Kimmerle, E. H., Konigsberg, L. W., Jantz, R. L., and Baraybar, J. P., 2008   Analysis of Age-at-Death Estimation Through the Use of Pubic Symphyseal Data. Journal of Forensic Sciences 53:558–568.

Konigsberg, L. W., and Frankenberg, S. R., 1992   Estimation of Age Structure in Anthropological Demography. American Journal of Physical Anthropology 89:235–256.

Konigsberg, L. W., and Frankenberg, S. R., 1994   Paleodemography: "Not quite dead". Evolutionary Anthropology 3:92–105.

Konigsberg, L. W., Frankenberg, S. R., and Walker, R. B., 1997   Regress What on What? Paleodemographic Age Estimation as a Calibration Problem. *In* Integrating Archaeological Demography: Multidisciplinary Approaches to Prehistoric Population. R. R. Paine, ed. pp. 64–88. Occasional Paper 24. Carbondale: Center for Archaeological Investigations, Southern Illinois University.

Konigsberg, L. W., Herrmann, N. P., Wescott, D. J., and Kimmerle, E. H., 2008   Estimation and Evidence in Forensic Anthropology: Age-at-Death. Journal of Forensic Sciences 53:541–557.

Lewis, A. B., and Garn, S. M., 1960   The Relationship Between Tooth Formation and Other Maturational Factors. The Angle Orthodontist 30:70–77.

Liversidge, H. M., and Molleson, T. I., 1999   Developing Permanent Tooth Length as an Estimate of Age. Journal of Forensic Sciences 44:917–920.

Lovejoy, C. O., Meindl, R. S., Pryzbeck, T. R., and Mensforth, R. P., 1985a   Chronological Metamorphosis of the Auricular Surface of the Ilium: A New Method for the Determination of Adult Skeletal Age at Death. American Journal of Physical Anthropology 68:15–28.

Lovejoy, C. O., Meindl, R. S., Mensforth, R. P., and Barton, T. J., 1985b   Multifactorial Determination of Skeletal Age at Death: A Method and Blind Tests of Its Accuracy. American Journal of Physical Anthropology 68:1–14.

Martrille, L., Ubelaker, D. H., Cattaneo, C., Seguret, F., Tremblay, M., and Baccino, E., 2007   Comparison of Four Skeletal Methods for the Estimation of Age at Death on White and Black Adults. Journal of Forensic Sciences 52:302–307.

McKern, T. W., and Stewart, T. D., 1957   Skeletal Age Changes in Young American Males. Technical Report EP-45. Natick, MA: Quartermaster Research and Development Command.

Meindl, R. S., and Lovejoy, C. O., 1985   Ectocranial Suture Closure: A Revised Method for the Determination of Skeletal Age at Death Based on the Lateral-Anterior Sutures. American Journal of Physical Anthropology 68:57–66.

Meindl, R. S., Lovejoy, C. O., Mensforth, R. P., and Walker, R. A., 1985a   A Revised Method of Age Determination Using the Os Pubis, with a Review and Tests of Accuracy of Other Current Methods of Pubic Symphyseal Aging. American Journal of Physical Anthropology 68:29–38.

Meindl, R. S., Lovejoy, C. O., Mensforth, R. P., and Don Carlos, L., 1985b   Accuracy and Direction of Error in the Sexing of the Skeleton: Implications for Paleodemography. American Journal of Physical Anthropology 68:79–85.

Milner, G. R., 2010   Transition Analysis and Subjective Assessments of Age in Adult Skeletons. *In* ADBOU 1992–2009: Forskiningsresultater. J. L. Boldsen, and P. Tarp, eds. pp. 15–27. Odense, DK: ADBOU, Syddansk Universitet.

Milner, G. R., and Boldsen, J. L., 2012   Humeral and Femoral Head Diameters in Recent White American Skeletons. Journal of Forensic Sciences, in press.

Milner, G. R., and Smith, V. G., 1990   Oneota Human Skeletal Remains. *In* Archaeological Investigations at the Morton Village and Norris Farms 36 Cemetery, S. K. Santure, A. D. Harn, and D. Esarey, eds. pp. 111–148. Reports of Investigations 45. Springfield: Illinois State Museum.

Milner, G. R., Humpf, D. H., and Harpending, H. C., 1989   Pattern Matching of Age-at-Death Distributions in Paleodemographic Analysis. American Journal of Physical Anthropology 80:49–58.

Milner, G. R., Wood, J. W., and Boldsen, J. L., 2008   Advances in Paleodemography. *In* Biological Anthropology of the Human Skeleton, 2nd edition. M. A. Katzenberg, and S. R. Saunders, eds. pp. 561–600. New York: Wiley-Liss.

Moorrees, C. F. A., Fanning, E. A., and Hunt, E. E., 1963a   Age Variation of Formation Stages for Ten Permanent Teeth. Journal of Dental Research 42:1490–1502.

Moorrees, C. F. A., Fanning, E. A., and Hunt, E. E., 1963b   Formation and Resorption of Three Deciduous Teeth in Children. American Journal of Physical Anthropology 21:205–213.

Mulhern, D. M., and Jones, E. B., 2005   Test of Revised Method of Age Estimation from the Auricular Surface of the Ilium. American Journal of Physical Anthropology 126:61–65.

Murray, K. A., and Murray, T., 1991   A Test of the Auricular Surface Aging Technique. Journal of Forensic Sciences 36:1162–1169.

Oettlé, A. C., and Steyn, M., 2000   Age Estimation from Sternal Ends of Ribs by Phase Analysis in South African Blacks. Journal of Forensic Sciences 45:1071–1079.

Paine, R., 1989   Model Life Table Fitting by Maximum Likelihood Estimation: A Procedure to Reconstruct Paleodemographic Characteristics from Skeletal Age Distributions. American Journal of Physical Anthropology 79:51–61.

Passalacqua, N. V., 2010   The Utility of the Samworth and Gowland Age-at-Death "Look-Up" Tables in Forensic Anthropology. Journal of Forensic Sciences 55:482–487.

Saunders, S. R., Fitzgerald, C., Rogers, T., Dudar, C., and McKillop, H., 1992   A Test of Several Methods of Skeletal Age Estimation Using a Documented Archaeological Sample. Canadian Society of Forensic Sciences Journal 25:97–118.

Schmitt, A., 2004   Age-At-Death Assessment Using the Os Pubis and the Auricular Surface of the Ilium: A Test on an Identified Asian Sample. International Journal of Osteoarchaeology 14:1–6.

Todd, T. W., 1920   Age Changes in the Pubic Bone. I. The Male White Pubis. American Journal of Physical Anthropology 3:285–334.

Todd, T. W., and Lyon, D. W., 1924   Endocranial Suture Closure, Its Progress and Age Relationship: Part I. Adult Males of White Stock. American Journal of Physical Anthropology 7:325–384.

Townsend, N., and Hammel, E. A., 1990   Age Estimation from the Number of Teeth Erupted in Young Children: An Aid to Demographic Surveys. Demography 27:165–174.

Ubelaker, D. H., 1987   Estimating Age at Death from Immature Human Skeletons: An Overview. Journal of Forensic Sciences 32:1254–1263.

Ubelaker, D. H., 1999   Human Skeletal Remains: Excavation, Analysis, Interpretation. 3rd edition. Manuals on Archeology 2. Washington, DC: Taraxacum.

Waldron, T., 1991   Rates for the Job. Measures of Disease Frequency in Palaeopathology. International Journal of Osteoarchaeology 1:17–25.

Walker, P. L., 1995   Problems of Preservation and Sexism in Sexing: Some Lessons from Historical Collections for Palaeodemographers. In Grave Reflections: Portraying the Past through Cemetery Studies. S. R. Saunders, and A. Herring, eds. pp. 31–47. Toronto: Canadian Scholars' Press.

Walker, P. L., 2008   Sexing Skulls Using Discriminant Function Analysis of Visually Assessed Traits. American Journal of Physical Anthropology 136:39–50.

Walker, P. L., and Cook, D. C., 1998   Gender and Sex: Vive la Différence. American Journal of Physical Anthropology 106:255–259.

Weise, S., Boldsen, J. L., Gampe, J., and Milner, G. R., 2009   Calibrated Expert Inference and the Construction of Unbiased Paleodemographic Mortality Profiles. American Journal of Physical Anthropology Supplement 48:269 (abstract).

Wood, J. W., Milner, G. R., Harpending, H. C., and Weiss, K. M., 1992   The Osteological Paradox: Problems of Inferring Prehistoric Health from Skeletal Samples. Current Anthropology 33:343–370.

WEA (Workshop of European Anthropologists), 1980   Recommendations for Age and Sex Diagnoses of Skeletons. Journal of Human Evolution 9:517–549.

CHAPTER 16

The Relationship Between Paleopathology and the Clinical Sciences

Simon Mays

Human paleopathology may be considered to have two principal aims. Firstly, it seeks to trace the history of particular diseases in human populations. In this role it may be used in combination with written sources, or for some conditions it may be the sole source of evidence. This may be termed the medicohistorical approach. Secondly, paleopathological data may be harnessed to address wider archaeological, historical or other questions about the human past, normally by relating patterns of disease to cultural, social or environmental variables. This may be termed the biocultural approach. In this contribution, I shall consider the relationship between clinical sciences and paleopathology, as it pertains to both medicohistorical and biocultural paleopathological research. The first part of the chapter concentrates on issues concerning disease diagnosis, the second on the balance between qualitative and quantitative research in paleopathology.

Diagnostic Issues in Paleopathology

The baseline for diagnosis in paleopathology is bone changes in modern or recent individuals with diseases that were accurately diagnosed using criteria independent of the bone changes. A number of laboratory manuals detailing the changes wrought by various diseases have been prepared with the aim of enabling diagnosis of paleopathological specimens (Steinbock 1976; Zimmerman and Kelley 1982; Ortner

A Companion to Paleopathology, First Edition. Edited by Anne L. Grauer.
© 2012 John Wiley & Sons, Ltd. Published 2016 by John Wiley & Sons, Ltd.

and Putschar 1985; Aufderheide and Rodriguez-Martin 1998; Ortner 2003; Mann and Hunt 2005). Manuals such as these mainly use as their reference material specimens from medical pathology museums. This material has the advantage that much of it was gathered prior to the advent of effective antibiotic and other treatments, so that disease would have progressed essentially unchecked, except by the body's natural defenses. They thus provide evidence for advanced cases of disease that are likely to resemble those that we might expect to encounter in archaeological populations. Because many specimens were kept as dry bone preparations, they enable direct morphological comparison with archaeological material. However, the use of medical pathology museum specimens as a source of baseline data for paleopathological diagnosis has some drawbacks. Globally, the number of specimens representing any one disease is fairly small. In addition, specimens were often selected for retention because they were unusual or spectacular manifestations of disease. Therefore, the full spectrum of skeletal manifestations of a disease is unlikely to be represented. Lastly, relying on the accuracy of past researchers' diagnoses may also be problematic

A second source of baseline data on skeletal disease for paleopathology is radiological imaging data (see Chapter 18 by Wanek et al., in this volume for more technical discussion of paleoimaging techniques), particularly radiographic images, from living patients with known diseases. Radiological images of paleopathological specimens enable direct comparison of lesion morphology with that visualized in living patients. Because radiography has been used for medical imaging of lesions for more than 100 years, there is a substantial published corpus of radiographic images in the medical literature depicting cases of disease whose progress was unhindered by modern treatments. The published radiographic record generally displays a greater range of manifestations of a given disease than do medical pathology museum specimens, and therefore gives a more realistic idea of the variety of skeletal changes that may be wrought by a given disease. In addition, the recognition of some diseases (for example many arthropathies and some metabolic diseases) has only taken place in the latter half of the 20th century, after many medical museums ceased collecting dry bone specimens. For these conditions, radiographic and other medical imaging data are fundamental for deriving a baseline for interpretation of paleopathological cases.

Until about the 1960s it was commonly thought that diagnostic radiography was beyond the grasp of the paleopathologist and that interpretation of radiographic images of archaeological specimens should only be undertaken by professional radiologists (Wells 1963; Sandison 1968). This attitude reflected the status of paleopathology at that time as a minor adjunct of the clinical disciplines. However, it has since matured into a field with its own academic traditions, training and skills base. Radiography is a core skill within the discipline, and laboratory manuals in paleopathology routinely include radiographic as well as dry bone images of disease.

Radiographic interpretation in paleopathology presents its own specific problems. Postdepositional artifacts may alter the radiographic picture. Soil ingress may cause fluffy areas of radiodensity, and if severe may prevent the production of diagnostically useful images. Postdepositional uptake of lead from coffins or artifacts may mimic the effects of diseases such as osteopetrosis (Molleson et al. 1998). Soil erosion may lead to radio-lucencies that might be mistaken for the effects of disease. Radiographs of archaeological bones should always be examined in conjunction with the specimen itself.

For some bone diseases, clinical radiographic diagnostic criteria can be directly applied to paleopathological cases. For example, the radiographic picture of Paget's disease of bone (PDB) is highly diagnostic, and in clinical practice is the mainstay for diagnosis (Walsh 2004). PDB is a focal disease of bone remodeling. There is initially increased bone resorption, so that lysis is the predominant process. This provokes an osteoblastic response which results in the rapid formation of chaotically organized new bone. After a variable period, osteoclastic activity slows so that the balance shifts in favor of bone formation and the bones become heavy and sclerotic. Radiographically, in early phase disease there is localized radiolucency. The patchy rarification and sclerosis of mid-phase disease produces a mottled appearance on X-ray and there is trabecular coarsening and loss of normal corticomedullary distinction. In long bones disease generally begins in subchondral bone and advances along the shaft, with a sharp, V-shaped boundary between Pagetic and normal bone. In late phase disease, the V-shaped demarcation between normal and abnormal bone in long bones is lost and sclerosis predominates (Mirra et al. 1995a, b; Smith et al. 2002). These features may be clearly visualized in radiographs of dry bone specimens (Figure 16.1) enabling diagnosis.

For other diseases some clinical radiographic criteria may not be very useful for dry bone specimens, so that paleopathologists need to apply them selectively. For example, in rickets, most clinical radiographic diagnostic features relate to bending deformities, metaphyseal thickening, concavity of metaphyseal subchondral bone and "fraying" of bone beneath the epiphyseal growth plate (Thacher et al. 2000; Pettifor et al. 2003:555–557), which are readily visible to gross inspection in the dry bone. Radiographic diagnostic criteria for rickets in paleopathology emphasize aspects such as coarsening / thinning of the trabecular structure and loss of corticomedullary distinction that are not visible grossly in intact specimens (Mays et al. 2006).

For many diseases with skeletal manifestations, clinical radiographic diagnostic criteria relate partly to aspects that cannot be visualized in radiographs of dry bones. For example, in clinical medicine, joint space narrowing (taken to represent loss of articular cartilage), marginal osteophyte formation, and the presence of subchondral cysts and sclerosis are important indicators of osteoarthritis (Rogers et al. 1990). Clearly, the first of these criteria cannot be evaluated in dry bones. The other criteria could be visualized by radiographing the specimen, but in practice this is not usually the way osteoarthritis is diagnosed in paleopathology. Marginal osteophytes can be seen grossly, so radiography is redundant. In paleopathology, joint-surface porosis and eburnation, features that are scored by visual examination of the specimen, are taken as indicators of osteoarthritis. Porosis and eburnation correspond to the subchondral cysts and sclerosis visualized radiographically in living patients—pores normally communicate with subchondral cysts and eburnated bone is sclerotic (Rogers and Waldron 1995:44; and Chapter 28 by Waldron in this volume). For the diagnosis of osteoarthritis, clinical radiographic criteria have been selectively used to derive paleopathological diagnostic criteria based on visual examination of specimens.

There are a number of advantages in formulating dry bone diagnostic criteria in paleopathology. Firstly, although the link with clinical diagnostic criteria is retained, dry bone diagnostic indicators are much more suited to the circumstances of paleopathology. Large numbers of skeletons can be rapidly assessed visually, whereas it would clearly be impractical routinely to radiograph every bone. In addition, bony

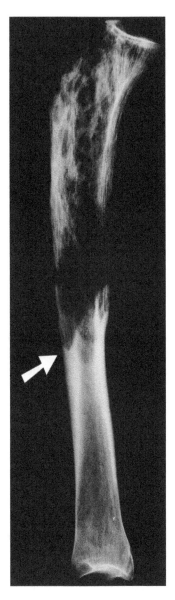

Figure 16.1 Lateral radiograph of a tibia from a medieval case of Paget's disease of bone (PDB). The diseased proximal part of the bone has a mottled appearance and shows loss of corticomedullary distinction. The advancing front of Pagetic bone (arrowed) is V-shaped. (Pinhasi and Mays, 2008, Figure 11.7.)

changes are much more readily apparent to visual inspection than they are to medical imaging techniques. A density change of about 40 percent is needed before alteration is evident on plain-film radiography (Ortner 1991). Many lesions will therefore be invisible radiographically even when they are clearly apparent in the gross specimen. For example, in a study of osteoarthritis at the knee, Rogers et al. (1990) found that only two of the 24 dry-bone specimens they examined were abnormal radiographically, whereas 16 showed obvious bony changes characteristic of the disease upon visual assessment. These are powerful arguments for the development, where possible, of dry bone diagnostic criteria rather than relying solely on clinical radiographic ones.

Many bone diseases lead to microscopic alterations visible in histological section. In the clinical sciences, histopathology is a well-established research tool for studying the etiology, pathogenesis and treatment of bone disease, normally using biopsy from the iliac crest (Recker 2008). Histopathology has also assisted in diagnosis in clinical practice but its role today as a diagnostic tool in minor. It is an unpleasant, invasive procedure. Although there some exceptions (e.g. renal osteodystrophy—Balon and Bren 2000) for most skeletal diseases, laboratory tests, medical imaging and patient history are sufficient to secure a diagnosis without recourse to bone histology. For example, although the histological picture of Paget's disease of bone is highly diagnostic, bone biopsy is not normally carried out as other findings, particularly the radiographic picture, are adequate for diagnosis (Walsh 2004; Siris and Roodman 2008). However, because the battery of diagnostic techniques available to paleopathologists is more restricted, bone histopathology has a potentially greater role to play (see Chapter 13 by Ragsdale and Lehmer in this volume).

Clinical histopathologists study both structural and kinetic variables (Recker 2008). Kinetic variables relate to rates of bone formation and resorption, and are normally measured by administering a labeling agent to the experimental subject before biopsy. Clearly, kinetic features cannot be measured in paleopathology. Some structural features of bone histology commonly used in health sciences research cannot be analyzed in dry bones. These include measures of the amount of osteoid and populations of osteoblasts and osteoclasts, and the presence of abnormal cells or tissue in marrow spaces. However, structural features of mineralized tissue can be investigated in dry bones. This means that, as with radiography, diagnostic features used in biomedical histology can be selectively applied to paleopathological specimens.

Although almost no pathological changes that can be seen microscopically in archaeological bone are specific to a particular disease (Turner-Walker and Mays 2008), microscopic examination is nevertheless useful to support inferences made from gross examination and other techniques. It is particularly helpful for aiding paleopathological diagnosis of metabolic diseases of bone (Brickley and Ives 2008). For example, unlike in clinical medicine, histology is useful for the identification of Paget's disease of bone in paleopathology. This is especially so when postdepositional damage to specimens prevents the production of adequate radiographic views or when changes are too slight for radiographic views to be diagnostic. Features such as a disorganized histological structure characterized by multiple cement lines separating small islands of bone tissue—the so-called mosaic structure—are normally preserved in archaeological specimens showing PDB (Figure 16.2).

The above discussion relates to lesion-based paleopathology, but some diseases may produce no diagnostically useful skeletal lesions but rather changes in the quantity or density of bone mineral. For example, osteoporosis leads to progressive, age-related bone loss from the endosteal envelope causing rarification of the trabecular structure and thinning of cortical bone. The gold standard for monitoring osteoporosis in living populations is dual X-ray absorptiometry (DXA). DXA uses an X-ray source that emits two beams at different energy levels. In living patients this enables differing attenuation of the beam due to bone and soft tissue to be calculated, and hence a

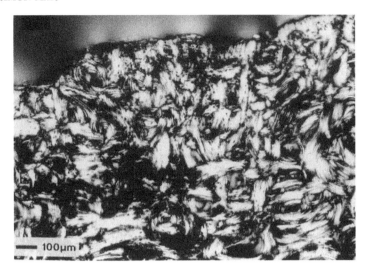

Figure 16.2 A histological section of bone from the same individual as Figure 16.1. The randomly orientated, fragmented lamellar pattern is typical of PDB. (Mays and Turner-Walker, 1999, Figure 17b.)

measure of bone mineral density (BMD) to be produced (Lees et al. 1998). Plotting BMD versus age enables inter-population comparisons of age-related loss of BMD. Because archaeological specimens lack marrow and soft tissue, absolute BMD values cannot be compared with those in living subjects. However, provided that significant diagenetic changes in BMD have not occurred in archaeological specimens, comparison of age-related patterns between living and skeletal populations can be made (discussion in Mays 2008). Another method used to study osteoporosis in clinical work is radiogrammetry—measurement of cortical thickness—particularly in the metacarpals. This has the advantage for paleopathological work that provided specimens showing soil erosion are excluded; results are closely comparable to data from living subjects (Mays 2008).

For accurate paleopathological diagnosis, it is vital that diagnostic criteria have a secure basis which derives ultimately from clinical medicine. The history of the paleopathological interpretation of porotic hyperostosis (pitting and thickening) of the orbital roofs and cranial vault (Figures 16.3a and 16.3b) illustrates both the advantages of close collaboration between paleopathology and the clinical sciences, and some of the problems that may arise when paleopathology drifts away from clinical diagnostic criteria.

Porotic hyperostosis (PH) has been noted in crania from archaeological sites since the 19th century (Virchow 1874; Welcker 1888). It is a common lesion, particularly in children. In some skeletal populations more than half of all individuals may show it. Because of this, paleopathologists and physical anthropologists naturally began to cast around for explanations of its occurrence. Early interpretations were rather speculative—for example it was seen as a racial marker or as a consequence of artificial cranial deformation. In the 1920s, clinicians began to describe radiographic changes in the cranial bones in cases of congenital anemias, including thalassemia and sickle cell anemia (Cooley et al. 1927; Moore 1929). These changes included hyperplasia of

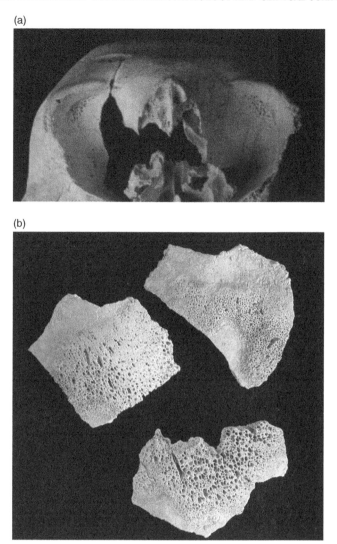

Figure 16.3   (a) Porotic hyperostosis of the orbital roofs. (b) Porotic hyperostosis on the ectocranial surface of skull vault fragments.

the diploë (which is composed of red, hemopoietic marrow in children) and consequent resorption of the outer table. The hyperplasia of the red marrow is an erythropoietic response to anemia. Clinicians in the 1920s and 30s who were aware of the paleopathological cases of porotic hyperostosis (although it was not called that then—the term was coined by Larry Angel in 1966), noted the resemblance between the radiographic changes seen in patients with congenital anemias and radiographs of ancient crania with grossly visible lesions of porotic hyperostosis (Moore 1929; Vogt and Diamond 1930; Feingold and Case 1933). In the dry specimen, the thickening of the cranial bones is due to diploic hyperplasia, and the surface pitting to resorption of the outer table revealing the trabeculae beneath. These and subsequent workers (e.g., Angel 1964, 1966) interpreted such changes in archaeological crania as suggestive of congenital anemias, such as thalassemia, in geographical regions where those conditions were prevalent.

It was realized that the high prevalence and wide geographic spread of porotic hyperostosis in early populations meant that, in most populations, congenital hemolytic anemias were not likely to be the most important cause. Beginning in the 1950s, sporadic reports began to appear in the clinical literature of cranial changes in chronic iron-deficiency anemia in children. These alterations were similar to those previously reported in crania in congenital anemias—diploic hyperplasia and resorption of the outer table (Eng 1958; Britton et al. 1960; Burko et al. 1960; Shahidi and Diamond 1960; Moseley 1961; Aksoy et al. 1966; Lanzkowski 1968; Agarwal et al. 1970; Reiman et al. 1976). The cases described were mainly in children. In the adult, bone marrow is mainly of the fatty yellow type rather than the hemopoietic red type. In adults a need for increased red blood cell production is normally met by reconversion of yellow to red marrow, so in acquired anemias there is usually no hyperplasia of the marrow space (Stuart-Macadam 1985). Iron deficiency anemia is the most widespread type (Bothwell 1995; Brugnara 2003). Therefore iron-deficiency anemia appeared to be both a plausible cause of porotic hyperostosis, and one which was commensurate with the high frequency of the lesion in early populations. This explanation for porotic hyperostosis in ancient crania was first suggested by a clinician, John Moseley, in the mid-1960s (Moseley 1965). Since then it has been enthusiastically adopted by paleopathologists, so that until recently porotic hyperostosis was almost synonymous with iron-deficiency anemia in paleopathology.

However there are some difficulties with this. Firstly, the clinical literature on cranial changes in anemia is primarily concerned with alterations to the skull vault. There is very little on the orbital roof lesions, which are usually more prevalent than vault changes in paleopopulations (e.g. Stuart-Macadam 1991; Steckel et al. 2002). Methodological work in paleopathology has produced important evidence that lesions in the two areas share a common cause. Studying British skeletal material, Stuart-Macadam (1989) noted that vault lesions rarely occurred without accompanying lesions of the orbits. Gross, radiographic and microscopic examination showed morphological similarity between lesions in the two locations; indeed as the bone in the orbital plates is continuous with that on the rest of the frontal bone it would be perverse to argue for different etiology in the two locations. The thinness of the orbital lamina means that hyperplasia of the diploë may more readily result in surface porosity here than in cranial vault bones (Wapler et al. 2004).

There is a second and more serious problem. Early paleopathologists (e.g., Williams 1929; Hamperl and Weiss 1955) were often careful to evaluate marrow hyperplasia when studying porotic hyperostosis. This can be done grossly or microscopically at postdepositional breaks or in cut sections, or else radiographically. But over time there developed a divergence between the paleopathology literature and the clinical literature on the nature of the cranial bone changes in anemia. The latter continued to emphasize the "hyperostosis" aspect (i.e. the diploic hyperplasia); the paleopathology community increasingly emphasized the "porotic" aspect (i.e. the pitting on orbital roof and vault surfaces), probably because it is easily observed and scored in dry skulls. In general, there was a tendency in paleopathology toward equating pitting in the orbital roofs or cranial vault bones with anemia. This is problematic as porosity of the orbital roofs or cranial vault may occur in a variety of other conditions such as rickets

(Ortner and Mays 1998) or scurvy (Ortner 2003:385–386), and these conditions may also lead to thickening of the cranial bones, albeit not by marrow hyperplasia. That this may lead to serious problems of interpretation has been demonstrated by Wapler et al. (2004). They recorded pitting in the orbital roofs in archaeological material from the Sudan, and found that in only 54 percent of cases was pitting *in vivo* associated with diploic hyperplasia and hence suggestive of anemia.

Although there has recently been a refocusing on diploic hyperplasia rather than superficial pitting in the analysis of porotic hyperostosis, questions have been raised among paleopathologists concerning the types of anemia that may lead to this. Since the 1960s in paleopathology, iron-deficiency anemia has been accepted as the most common cause of porotic hyperostosis. However Walker et al. (2009), from a reading of the biomedical literature, argue that iron deficiency leads to a depression rather than elevation of erythrocyte production, and hence could not lead to hyperplasia of bone marrow. They contend that megaloblastic anemia (due to deficiency in vitamin B_{12} and / or folic acid) will trigger marrow hyperplasia and is hence a more likely cause of hyperplasia of the diploë. There are some difficulties with Walker et al's arguments. Firstly, not all the biomedical literature supports their view that marrow hyperplasia does not occur in iron-deficiency anemia (e.g. Cavill 2002). Secondly, as cited above, there is clinical evidence that diploic hyperplasia can occur in patients with iron deficiency anemia, and Walker et al. (2009) do not fully engage with this. However, the clinical literature is sparse and much is in the form of case studies, so the prevalence of bone changes in the various acquired anemias is rather unclear (see Chapter 22 by Kozlowski and Witas in this volume, for further discussion on metabolic and endocrine disorders). In order to shed further light on the problem, further clinical studies, on large numbers of patients are needed. This may come through MRI imaging. In living patients this enables active, hemopoietic red marrow to be distinguished from yellow fatty marrow, as well as facilitating measurements of marrow thickness. A clinical paper (Yildirim et al. 2005) demonstrated reconversion of yellow to red marrow in the cranial bones of adults with acquired anemia (mainly iron-deficiency anemia), which would seem to demonstrate that, *contra* Walker et al. (2009), iron-deficiency anemia can elicit an erythropoietic response. Further work of this nature is needed to resolve the relationship between porotic hyperostosis and anemia.

The example of porotic hyperostosis demonstrates the difficulties that arise in paleopathological interpretation of lesions if paleopathology drifts away from the clinical evidence or if the clinical basis linking lesion to disease is slender. Problems in interpretation may also arise even when paleopathologists manage to diagnose lesions correctly. This may occur when paleopathologists attempt to link the diseases they are seeing to risk factors for those diseases that have been identified in clinical epidemiological studies. For example, the clinical literature tells us that mechanical loading at a joint is a risk factor in the development of osteoarthritis. This has led many paleopathologists to interpret patterning in osteoarthritic changes in skeletal series in terms of differences in activity patterns in past populations. However, close evaluation of the clinical literature shows that the evidence linking osteoarthritis to the strenuousness of a person's occupation or habitual activities is rather inconsistent, and many other risk factors have also been identified (Jurmain 1999; Weiss and Jurmain 2007; see Chapter 29 by Jurmain et al. in this volume). Another case in point is diffuse idiopathic skeletal hyperostosis (DISH). There is modern epidemiological evidence that obesity is a risk factor for DISH. This has led some to argue that it can be used to monitor rates of obesity in past populations. Again, however, the link with obesity is not an invariable finding in modern epidemiological studies, there are other risk factors which have also been identified (and indeed the cause of the disease remains unknown), and is little evidence to suggest that the link with obesity is strong enough for the frequency of DISH to be used as an indicator of obesity in past populations (Mays 2006). These examples suggest that sometimes paleopathologists have a tendency to seize upon risk factors for diseases identified in skeletal remains that we think may enable us to say interesting things about past populations, and we tend to interpret disease prevalence in those terms. In a kind of wishful thinking, we may give insufficient consideration to clinical evidence that tends to undermine these links and calls into question simple interpretations.

A movement away from specific diagnosis—the rise of stress indicators in paleopathology

Because the aim in clinical practice is to give a prognosis and to instigate effective treatment, diagnostic accuracy is paramount. In paleopathology applied to medico-historical questions, accuracy of diagnosis is also a key issue. When the aim is the biocultural study of disease in past societies, accuracy of diagnosis may be less imperative if the ultimate aim is not to study a particular disease but to obtain a more general picture of disease burden in a population. A key aspect of this latter approach has been the paleopathological conception of nonspecific stress.

Taking their cue from the physiological work of Selye and others (reviewed in Goodman et al. 1988; Bush 1991), paleopathologists have defined stress as physiological disruption resulting from adverse environmental circumstances (Huss-Ashmore et al. 1982). The degree of disruption to an organism will depend both upon the duration and the severity of the stressor, and on the adequacy of the organism's response (Huss-Ashmore et al. 1982). External stressors which may provoke a physiological response in skeletal tissues include disease, poor nutrition, psychological stress or excessive heat or cold (Gilbert 1985). Stress indicators in skeletal populations include pathological lesions, growth disruptions and mortality.

In biocultural studies in paleopathology there are two major pragmatic reasons for the use of research strategies aimed at evaluating stress rather than diagnosing particular diseases. Firstly, because of the limited response of bone to disease, it is often difficult to arrive at a specific diagnosis on the basis of skeletal lesions. Secondly, the prevalence of most specific diseases which can be identified in skeletal populations is low (< 5 percent). This means that unless very large series are examined, the statistical power of the study is likely to be limited. Taking an approach which effectively involves counting lesions, rather than trying to identify specific diseases, bypasses these problems.

Among the lesions most often used as stress indicators in paleopathology are periosteal new bone formation, porotic hyperostosis of the orbital roof / cranial vault, and dental enamel hypoplasias. These are selected to a great extent simply because they are common in skeletal remains, and so provide useful amounts of data with which to work. As well as studies of stress in single populations or regions, some continental or wider scale works have also been attempted. Two volumes edited by Mark Cohen and his collaborators, *Paleopathology at the Origins of Agriculture*

(Cohen and Armelagos, 1984) and *Ancient Health: Skeletal Indicators of Agricultural and Economic Intensification* (Cohen and Crane-Kramer 2007) explored the health impacts arising from the transition to agriculture in different parts of the world. Most of the contributors to these volumes analyzed commonly occurring stress indicators such as those described above. Results indicated that in general skeletal lesions became more frequent, suggesting to the editors a decline in health with the adoption of agriculture.

A difficulty with the *Paleopathology at the Origins of Agriculture* and *Ancient Health: Skeletal Indicators of Agricultural and Economic Intensification* volumes is that different authors collected different types of data in different ways, meaning different datasets were not precisely comparable (see Chapter 19 by Stodder in this volume, for further discussion of issues involving data collection). In an attempt to overcome this, Steckel and his collaborators initiated two continental-scale projects to look at skeletal health which involved systematic scoring of stress indicators in a very large number of skeletons. The first, begun in 1988, is "Health and Nutrition in the Western Hemisphere." This involved recording seven commonly occurring stress indicators (stature, porotic hyperostosis, dental enamel hypoplasias, periosteal new bone formation, trauma, dental disease, and degenerative joint disease) in over 12,000 skeletons from 65 locations in the Americas. The material ranged in date from about 5000 B.C. to A.D. 1900. At each location, the scores for each stress indicator were combined, together with terms to adjust for the age at death of the individual and the severity of lesions, to yield an overall health index. This could theoretically range from zero (most severe expression of all indicators) to 100 (complete absence of lesions or signs of deficiency). The results from this project (Steckel and Rose 2002) suggested a long-term temporal decline in health in the pre-contact New World, and supported the New World findings of contributors to the Cohen and Armelagos (1984) and Cohen and Crane-Kramer (2007) volumes in suggesting that the transition from hunter-gatherer to agricultural subsistence was marked by deterioration in skeletal health.

A similar project was initiated in Europe by Steckel and colleagues in 2001 (Steckel 2003). The "History of Health in Europe" project involved calculating health indices from over 15,000 skeletons. Although results have yet to be fully published, it appears that European material generally has health indices about 10 points lower than those found in the Western Hemisphere. There were also some temporal trends. For example the health index was higher in Classical Antiquity than in the preceding or succeeding periods (Steckel et al. 2009).

Projects such as "Health and Nutrition in the Western Hemisphere," and the "History of Health in Europe," explicitly owe their rationale not to research designs from clinical medicine but to those from the social sciences. The skeletal health index is analogous to health measures (such as mortality, morbidity, and stature) used for living populations by social scientists, particularly economists and economic historians, to measure national and personal well-being (Steckel 2006).

Although it facilitates quantitative analyses, and speeds data-capture and analysis, there are also drawbacks to treating lesions in a generic fashion as stress indicators. Given that a particular stress indicator does not, by its very nature, have a single cause, frequency figures combine data on conditions with rather different causes. Data on prevalence of stress indicators is therefore rather heterogeneous and may be difficult to interpret. The degree to which this causes problems depends upon the nature of the lesions being recorded, and the way in which the data are analyzed. If the causes of a particular type of lesion of interest have much in common, then the problems may not be very great. For example, current consensus seems to be that infectious disease is likely to be the most frequent cause of periosteal new bone formation in premodern populations (Ortner 2003:209). If this is correct, then lesions consisting of periosteal new bone can be used as an indicator of skeletal infection in past populations (see Chapter 27 by Weston in this volume, for further discussion on the interpretations of this lesion in skeletal populations). Transmission of most infections is favored by dense aggregation of populations, poor hygiene and pollution of water supplies, factors which, prior to the introduction of piped sewerage and modern notions of hygiene, tended to be found together. Although one needs to keep in mind nuances of interpretation (e.g., Wood et al. 1992), it seems reasonable to use prevalence of periosteal new bone lesions as an index of these types of living conditions in past populations (e.g., Grauer 1993; Mays 2010a:213–215).

There are more problems of interpretation when a lesion has a more heterogeneous range of causes. As we have seen, porotic hyperostosis of the orbital roofs and cranial vault may have a variety of causes, including childhood anemia, rickets and scurvy. These in turn have rather different risk factors which include gut parasites and inadequate iron or vitamin B_{12} intake (anemia), inadequate exposure of the skin to sunlight (rickets), and inadequate dietary intake of vitamin C (scurvy). Different populations might have similar rates of porotic hyperostosis and hence similar rates of "stress." However the lesions in different populations might have arisen due to different diseases which might imply very different things about lifestyle.

Paleopathologically, porotic hyperostosis due to anemia, rickets, and scurvy can be often be distinguished, both by careful study of the morphology of the orbital / cranial vault lesions themselves and by examination of the rest of the skeleton for accompanying lesions (Ortner and Mays 1998; Ortner et al. 2001; Ortner 2003:370–375, 384–393). Although lesions with potentially heterogeneous causes, such as porotic hyperostosis, can simply be treated as stress indicators, merely to consider them as such and not attempt diagnosis of cause in individual cases risks losing much of the data that makes paleopathology interesting. This problem is amplified when many different types of lesions are combined to produce a single index intended to give some kind of overview of the skeletal health of a population.

QUALITATIVE VERSUS QUANTITATIVE WORK IN PALEOPATHOLOGY

In the medical sciences, the single case study has been a recognized means of presenting clinical evidence for at least 500 years (DeBakey and DeBakey 1983). From the beginnings of human paleopathology in the 19th century, until about the 1930s, the case study was its archetypal publication (Armelagos 1997). This was in keeping with its status at the time as an adjunct of the medical sciences. Today, research in all sciences has a basis in quantitative analysis of data. In paleopathology, the introduction of the paleoepidemiological approach to ancient disease, in which the population rather than the individual is the object of interest, is widely credited (e.g. Roney 1959; Angel 1981; Armelagos 1997) to the landmark study in 1930 by Earnest Hooton (in collaboration with several clinicians) on the osteology of the Indians of Pecos Pueblo in the American Southwest (Hooton 1930). The decades which followed saw the publication of other population studies in paleopathology. Some of these publications took the form of descriptive population studies of disease frequency (e.g. Stewart 1931; Goldstein 1957; Roney 1959), others were quantitative methodological studies (e.g. Stewart 1953, 1956), and a few combined pathology and cultural data in early attempts at biocultural studies (e.g. Angel 1946). However, such works were generally the exception rather than the norm. The development of paleopathology toward a quantitative science was slow. Until the 1960s, paleopathology was mainly carried out by physicians with no formal training in anthropology or archaeology and who wrote reports on individual skeletons recognized as abnormal by archaeologists (Jarcho 1966). This tended to isolate medically trained paleopathologists from the archaeological data which would have given context to their findings, and helped reinforce the case study as the dominant form in paleopathology.

There was an upsurge in quantitative paleopathology during the 1960s. In the USA this was facilitated by the rise of the so-called "new archaeology." A cornerstone of the new archaeology was a problem-orientated, hypothesis-testing approach involving quantitative analysis of archaeological data. Its understanding of culture was as the extrasomatic means of adaptation of the human organism (White 1959:8). Under this paradigm, humans were seen as surviving not through biological adaptation or cultural adaptation but through biocultural adaptation (Blakely 1977). Paleopathological data had the potential to provide important insights into biocultural adaptation in past societies, and their quantitative analysis had the potential to facilitate the testing of biocultural hypotheses. In addition, and in contrast to earlier periods, from the 1960s most paleopathologists working in the U.S.A. were trained in anthropology rather than medicine (Cook and Powell 2006), a field in which the population rather than the individual is normally the focus.

From the late 1960s, more attention begun to be paid to the methodological problems raised by quantifying disease in skeletal remains. Although attempts at recording commonly occurring lesions in a systematic way have a long history in paleopathology (e.g., Welcker 1888), from the 1960s onward, there was an increased interest in contriving schemes which expressed the degree of severity of bone changes in a repeatable manner which facilitated quantitative analysis of results. Early efforts include that of Nathan and Haas (1966) for cribra orbitalia and Sager (1969) for osteoarthritis and vertebral osteophytosis. The importance of distinguishing, where possible, between active lesions, indicating a disease process affecting the bone at time of death, and old, healed lesions also started to be recognized (e.g. Mensforth et al. 1978). It also began to be more widely realized that clinical schemes which express the prevalence of diseases with respect to numbers of individuals observed were not on their own adequate for the fragmentary and incomplete skeletons encountered by paleopathologists, and that recording systems based on the number of observable bones or parts of bones were needed (e.g., Lovejoy and Heiple 1981). Paleopathological scoring schemes for common conditions are now a routine feature in osteology laboratory manuals (e.g., Brothwell 1981; Buikstra and Ubelaker 1994; Brickley and McKinley 2004), and existing recording methodologies continue to be evaluated and improved, and new ones suggested (e.g., Waldron and Rogers 1991 for osteoarthritis, Judd 2002 for fractures, and Stuart-Macadam 1985 and Jacobi and Danforth 2002 for porotic hyperostosis).

Although the new archaeology of the 1960s was also influential to archaeology in Britain, British paleopathology remained firmly bound to the medical sciences for another 20 years. Nevertheless, some workers in Britain, and elsewhere in Europe, were already taking a quantitative approach to ancient disease. In part this may have been a response to the large size of many European assemblages of human remains, which made case-by-case descriptions of frequently occurring pathologies in osteological reports impractical and in part forced a summary, quantitative approach. Even when quantitative, much of this work was not problem-orientated, but was descriptive in nature, the aim being to characterize the populations under study rather than too address broader questions (e.g. Gejvall 1960; Brothwell 1961; Bourke 1967). There were some exceptions, where attempts were made to investigate archaeological or historical questions. Examples include Inglemark's classic study of the distribution of blade injuries in bones from a mass grave at a medieval battle site (Inglemark 1939), and Acsádi et al.'s. (1962) attempt to relate frequency of skeletal lesions to social status in medieval Hungary. Other quantitative problem-orientated work in Europe was directed at medical or medicohistorical questions. For example, Lindblom (1951) examined osteophytes in vertebral bodies of lumbar vertebrae in prehistoric skeletons from Sweden and compared them with radiographs of the lumbar area in living Swedes. He found a difference in osteophyte distribution which he ascribed to differences in lifestyle. In Denmark, Møller-Christensen (1961) studied large numbers of skeletons from a medieval leper hospital and documented the nature and frequency of leprous changes seen in them. This work showed both an appreciation of the need for rigorous quantification of data and made attempts to classify lesions into standardized grades of severity. The aim of this work was to accurately characterize the bone changes of leprosy, and the data capture methodology directly echoed that used in clinical studies of living patients (Møller-Christensen 1974).

The strong ties between medicine and paleopathology in Britain may be one reason why biocultural approaches, which had become commonplace in North America from the early 1970s, were slow to gain acceptance (see Chapter 4 by Buzon in this volume, for further discussion of the bioarchaeological approach in paleopathology). As late as 1991, Bush and Zvelebil still felt able to characterize the biocultural approach as a "new area" which had "yet to become established" in Britain or elsewhere in Europe (Bush and Zvelebil, 1991:5). It was only during the 1990s, when postgraduate courses teaching paleopathology within an explicitly archaeological framework became more widely available at British universities (Roberts 2006), that the biocultural approach was fully established. However even today, population based paleopathological studies remain more frequent in the U.S.A. than in the U.K. (Mays 2010b).

The traditionally strong links between the clinical sciences and paleopathology in the U.K. have also resulted in subtle differences between population-based studies in the U.K. and the U.S.A. Firstly, data relating to articles published in selected major journals 2001–2007 (Mays 2010b) shows that about one-third of population studies from U.K.-based authors had a strong medicohistorical element involving comparison of patterns seen in ancient populations with those in modern clinical studies. During the same period such a focus was lacking in population studies carried out by U.S. authors published in the same journals, these works being biocultural studies directed at archaeological questions.

Secondly, the last 20 years have seen the application of analytical techniques from clinical epidemiology into paleopathology (see Chapter 15 by Milner and Boldsen in this volume), a trend that appears to have begun in the U.K. These include statistical methods for comparing disease prevalence between populations and for dealing with obvious confounding variables such as age at death (Waldron 1994; Baker and Pearson 2006; Pinhasi and Turner 2008). One important approach is the case-control study, in which individuals showing a disease (the cases) are compared, on some other variable, with controls who do not show the disease but are matched with the cases in potential confounder variables such as age and sex (Gordis, 2009:177–199). In paleopathology they are a means of evaluating associations between different diseases or types of lesions or between these and other aspects of skeletal biology. Case control studies are particularly useful in paleopathology because they are well suited to studying conditions which have low prevalence, as do most specific diseases in skeletal series, and are economical of time and effort. For example, studies of this type have been used to assess links between sacroiliac fusion and DISH and various other conditions (Waldron and Rogers 1990), the association between osteoarthritis at and the hand and the knee (Waldron 1997), and associations between spondylolysis and aspects of vertebral morphology (Ward et al. 2010).

Although simple epidemiological approaches, such as the case-control study design, are of obvious value in paleopathology, some workers have recently attempted to transfer more complex statistical approaches from medical epidemiology to paleoepidemiology. Bayes' theorem has been used to as a diagnostic approach in a clinical setting, and attempts have been made to apply it in paleopathology (e.g., Byers and Roberts 2003). Attempts have also been made to apply the basic clinical concepts of sensitivity and specificity in paleopathological analysis. The sensitivity of a particular sign of disease is the probability that a person suffering from the particular disease in question has a positive score for the lesion. Specificity is the probability that a person who did not suffer from the disease had a negative score for the lesion. Estimates of both sensitivity and specificity of skeletal lesions associated with particular diseases can be used to try and arrive at estimates of the prevalence of specific diseases in earlier populations from the prevalence of bone lesions. Boldsen (2001, 2005, 2008) has pioneered this approach in the paleopathology of leprosy. Although the applications of these analytical methods are interesting and innovative, the estimation of parameters required by the models presents special problems in paleopathology, and it remains to be seen whether these approaches will make a significant impact on the discipline.

Qualitative research in paleopathology—the case study

Despite the advances in quantitative population and methodological studies in paleopathology it, like clinical medicine, retains a significant qualitative, descriptive element. In both paleopathology and clinical sciences this is exemplified by the persistence of the case study as a form of publication.

In the clinical sciences, the case report now makes up only a minor proportion of publications. A study of health science periodicals (Patsopoulos et al. 2005) found that (for articles published in 2001) only 12 per cent were case reports. In addition, the proportion of publications made up of case reports has been declining in the health sciences (Farmer 1999; Fenton et al. 2004; Patsopoulos et al. 2005), with a decreasing number of periodicals accepting them for publication (e.g. Kenyon 2002;

Table 16.1   Citation frequency of case reports versus other types of study in clinical sciences and in paleopathology

	Case reports		Other studies		
	N	Median citation frequency	N	Median citation frequency	Reference
Clinical Sciences	249	1	1493	4–9	Patsopoulos et al. 2005
Palaeopathology	150	1	123	3–4	Mays in press

Notes: The clinical sciences study analyzed citations in 2002–3 to articles published in various medical journals in 2001. The paleopathology study analyzed citations since publication to articles on paleopathology published in the *International Journal of Osteoarchaeology* 1995–2007. Median citation of the studies other than case reports varies depending on the specific type of study.

Fenton et al. 2004; Albrecht et al. 2005; Patsopoulos et al. 2005). Debate has centered over whether the case report still has a value in modern clinical science, where statistical studies of large numbers of patients are seen as the gold standard for evidence-based medicine (e.g. Farmer 1999; Mason 2001; Bydder 2008).

An analogous debate is taking place in paleopathology. As we have seen, from the mid 20th century onward there has been a rise in quantitative epidemiological work in paleopathology. However, commentators have argued that a greater shift in this direction is still needed, and that within this environment the potential of the case study to advance the discipline is limited (Roberts and Manchester 1995:196; Mays 1997; Roberts 2006). Some have gone further, and have questioned whether case reports should continue to have a place at all (Armelagos and van Gerven 2003).

Despite this, and in contrast to the clinical literature, the case report continues to play a prominent role in paleopathology. Analysis of paleopathology publications in selected periodicals over the last 15–20 years demonstrates that 35–40 percent are case reports, with little evidence of a decline during that period (Mays 1997, 2010b).

The impact of different types of study on a discipline can be assessed using citation analysis. In both the clinical sciences, and in paleopathology, recent citation analyses (Patsopoulos et al. 2005; Mays in press) indicate that case studies receive fewer citations than other types of study. However this deficit is less marked in paleopathology (Table 16.1). Case studies appear to have a more influential role in paleopathology than they do in the clinical sciences.

The "impact factor" of a periodical is based on the number of citations in the current year to papers published in that journal in the previous two years, divided by the number of articles published in the journal over that same two-year period (Garfield 1996). Because, in the medical literature, case reports are cited less often than other types of study, the impact factors for medical journals that continue to publish them tend to suffer (Kenyon 2002). Given the emphasis on impact factors in the health sciences generally, this is likely a major factor in the decline in the numbers of clinical science journals which continue to publish case reports, and hence in their general diminution in the clinical literature. Impact factors calculated in this way are less relevant to small disciplines, such as paleopathology, where the numbers of citations over a given period are necessarily few compared with larger disciplines. This, coupled with the observation that in paleopathology the deficit in citation for case reports compared to other types of publication is less than in the clinical sciences, may mean that there is less pressure on journal editors to exclude them on these grounds.

The decline of the case report in the clinical science literature has led some to spring to its defense. Case reports can be a basis for meta-analyses of data (Vandenbroucke 2001), and may also be important in refutation of medical hypotheses (Farmer, 1999). Teaching in medicine has traditionally involved case presentation, so case studies serve an educational purpose (Fenton et al. 2004). Physicians deal with a high volume of patients which increases their chances of encountering unusual cases (Wright and Kourouki, 2000), and reporting of such cases is a means by which those in clinical practice can contribute to international academic discourse (Har-El 1999). Analogous arguments to those above can be made for the case study in paleopathology, and some of these are examined below.

The nature of paleopathology, at least to some extent, necessitates a case-by-case approach if specific diseases rather than nonspecific stress are the topic under investigation. In skeletal assemblages, the low prevalence of most specific diseases, coupled with the fragmentary nature of archaeological skeletons and the small size of most skeletal assemblages may, for many conditions, render problematic the type of quantitative analysis of disease in particular communities that is routine in modern epidemiology. In order to conduct problem-orientated quantitative work on rarer conditions in paleopathology, meta-analysis of data is needed. In paleopathology, case studies provide an important source of data for meta-analyses of particular diseases (e.g. Buikstra 1999; Mays 2010c), and also for more general geographical or thematic syntheses (e.g. Roberts and Cox 2003; Larsen 1997). In addition, case studies may, on occasion, be useful in helping refute medicohistorical hypotheses—for example Columbian hypotheses for the origin of Old World treponemal disease (e.g., Erdal 2006).

Although commentators have often stated that progress in paleopathology involves moving from a descriptive case-study-based approach toward one where the population not the individual is the unit of analysis, reliable diagnosis of disease at the individual level is the foundation of reliable prevalence data on specific diseases in populations, and in laboratory classes we continue to emphasize to our students the importance of rigorous differential diagnosis of individual specimens. Hence, the carefully presented case study has potential value to the discipline, for example by describing conditions not well-covered in textbooks or else presenting lesion descriptions or differential diagnosis in more detail than is possible in more general works. It may also be that paleopathology, being a young and relatively small discipline, has more need than most clinical sciences of case reports dealing with basic diagnostic issues.

Many osteoarchaeologists work not in an academic environment but in a commercial one in which they are employed to produce osteological reports on skeletal series excavated in advance of development. The large number of skeletons they deal with means that they often encounter interesting or unusual cases. Publication of these cases enables contract osteoarchaeologists to participate in international academic discourse, which is otherwise often difficult given the time and other practical constraints of working in a commercial environment. There is evidence that this sector publishes a higher proportion of case studies than do other parts of the profession. For example, among U.K.-based authors, 44 percent of paleopathology case studies published in the period 2001–2007 came from workers employed in contract osteology, 24 percent from those in museums and heritage organizations, and 32 percent from those in universities. For other paleopathology publications (methodological or population studies, or reviews) the figures were 3 percent, 36 percent and 61 percent respectively (Mays 2010b).

Conclusions

Human paleopathology began as a minor adjunct to medical history, but has matured into a fully fledged discipline with its own academic traditions, training and skills-base. The changing relationship with clinical disciplines is reflected in the ways in which the paleopathological approach to diagnosis has altered over recent decades. Diagnostic criteria for specific diseases, whilst having a clinical base, are usually adapted so that they are suited to the circumstances of paleopathology rather than uncritically taken from clinical sources. This particularly involves the crafting of dry-bone criteria for disease diagnosis using a combination of medical pathology museum specimens and medical imaging data on living patients.

Today there appear to be two approaches to diagnosis within paleopathology. On the one hand there is work which emphasizes the value of undertaking as accurate a diagnosis as possible at the individual level. This forms the basis of medicohistorical work and of much biocultural work. Some of the medicohistorical work remains at the level of case reports, whereas more recent studies are population-based and quantitative, investigating the ways in which prevalence and severity of disease have varied over time. Most biocultural work in paleopathology is inherently quantitative. Some of this work emphasizes the quantitative study of lesions rather than of diagnosed cases of specific diseases—the analysis of stress indicators. The rise of the "stress indicators" approach in the last few decades is an indication of the growing independence of paleopathology from clinical medicine. At a theoretical level, this work has an explicit population emphasis, and is focused on disease burden as a whole rather than patterns of particular diseases that are the focus of modern epidemiology. There is normally an explicit biocultural approach in which pathology data are combined with social or environmental information to address questions about past societies. The intellectual allegiance of this work is much more with social sciences than with clinical medicine.

Although the emergence of paleopathology from beneath the shadow of clinical medicine is to be welcomed, the lessening of an emphasis on accurate diagnosis at the level of the individual skeleton, which is part and parcel of this, has been something of a mixed blessing. On the one hand, it is recognition of the fact that we often cannot arrive at firm diagnoses in paleopathology, and it enables us to make use of nondiagnostic lesions. On the other hand, the data it generates are by their very nature rather heterogeneous, and the multiplicity of factors that may lead to lesions commonly used as stress indicators may make it difficult to interpret the data in terms of population lifestyles. One may be left with simply interpreting the data in terms of relative levels of "stress" in paleopopulations but without any very clear idea of what that actually implies in terms of lifestyle differences. The study of differences in prevalences of specific diseases is likely to give a much finer-grained picture of ancient lifestyles and how these varied with social, economic and environmental factors. This potentially enables investigation of more nuanced hypotheses about the past. Whilst acknowledging the value of the "stress indicator" approach, the emphasis in paleopathology should still be on the accurate diagnosis of as many cases of disease as possible at the individual level, and methodological work aimed at the end is fundamental to the discipline. We should only lump together lesions as stress indicators when a more precise diagnosis is not possible.

Most sciences begin as qualitative, descriptive endeavors, but as they mature they move toward a basis in quantitative analyses. Paleopathology appears to be travelling along this timeline, albeit rather slowly. Qualitative work in the discipline persists and largely consists of descriptions of individual cases of disease. They still form a large proportion of the literature—much larger than in clinical sciences, where the case study seems to be an endangered species. It could be suggested that this may reflect the fact that paleopathology is a younger discipline, and perhaps has more need of descriptions of individual cases of disease. However many paleopathologists feel that case studies should have a lesser role than they do at present.

ACKNOWLEDGEMENTS

Thanks to Gordon Turner-Walker for Figure 16.2, which originally appeared in Mays and Turner Walker (1999) *Journal of Paleopathology* Vol. 11. Thanks are also due to the Missouri State Medical Association for supplying a xerox copy of one of their publications.

REFERENCES

Acsádi, G., Harsányi, L., and Nemeskéri, J., 1962   The Population of Zalavár in the Middle Ages. Acta Archaeologica Hungaricae 14:113–148.

Agarwal, K. N., Char, N., and Bhardwaj, O. P., 1970   Roentgenologic Changes in Iron Deficiency Anaemia. American Journal of Roentgenology 110:635–637.

Aksoy, M., Çamli, N., and Erdem, S., 1966   Roentgenographic Bone Changes in Chronic Eron Deficiency Anemia. Blood 27:677–686.

Albrecht, J., Meves, A., and Bigby, M., 2005   Case Reports and Case Series from Lancet Had Significant Impact on Medical Literature. Journal of Clinical Epidemiology 58:1227–1232.

Angel, J. L., 1946   Skeletal Changes in Ancient Greece. American Journal of Physical Anthropology 4:69–97.

Angel, J. L., 1964   Osteoporosis: Thalassemia? American Journal of Physical Anthropology 22:369–374.

Angel, J. L., 1966   Porotic Hyperostosis, Anemias, Malarias, and Marshes in the Prehistoric Eastern Mediterranean. Science 153:760–763.

Angel, J. L., 1981   History and Development of Paleopathology. American Journal of Physical Anthropology 56:509–515.

Armelagos, G. J., 1997   Palaeopathology. *In* The History of Physical Anthropology. F. Spencer, ed. vol. 2. pp. 790–796. New York: Garland.

Armelagos, G. J., and van Gerven, D. P., 2003   A Century of Skeletal Biology and Paleopathology: Contrasts, Contradictions, and Conflicts. American Anthropologist 105: 53–64.

Aufderheide, A. C., and Rodriguez-Martin C., 1998   The Cambridge Encyclopedia of Human Palaeopathology. Cambridge: Cambridge University Press.

Baker, J., and Pearson, O. M., 2006   Statistical Methods for Bioarchaeology: Applications of Age-Adjustment and Logistic Regression to Comparisons of Skeletal Populations with Differing Age-Structures. Journal of Archaeological Science 33:218–226.

Balon, B. P., and Bren, A., 2000   Bone Histomorphometry is Still the Golden Standard for Diagnosing Renal Osteodystrophy. Clinical Nephrology 6:463–469.

Blakely, R. L., 1977   Introduction: Changing Strategies for the Biological Anthropologist. In Biocultural Adaptation in Prehistoric America. R. L. Blakely, ed. pp. 1–9. Southern Anthropological Society Proceedings No. 11. Athens, GA: University of Georgia Press.

Boldsen, J. L., 2001   Epidemiological Approach to the Paleopathological Diagnosis of Leprosy. American Journal of Physical Anthropology 115:380–387.

Boldsen, J. L., 2005   Leprosy in the Early Medieval Lauchheim Community. American Journal of Physical Anthropology 135:301–310.

Boldsen, J. L., 2008   Leprosy and Mortality in the Medieval Danish Village of Tirup. American Journal of Physical Anthropology 126:159–168.

Bothwell, T. H., 1995   Overview and Mechanisms of Iron Regulation. Nutrition Reviews 53:237–245.

Bourke, B., 1967   A Review of the Paleopathology of the Arthritic Diseases. In Diseases in Antiquity. D. R. Brothwell, and A. T. Sandison eds. pp. 352–370.

Brickley, M., and Ives, R., 2008   The Bioarchaeology of Metabolic Disease. London: Academic Press.

Brickley, M., and McKinley, J., 2004   Guidelines to the Standards for Recording Human Remains. Southampton: British Association for Biological Anthropology and Osteoarchaeology / Institute of Field Archaeologists.

Britton, H. A., Canby, J. P., and Kohler, C. M., 1960   Iron-Deficiency Anemia Producing Evidence of Marrow Hyperplasia in the Calvarium. Pediatrics 25:621–628.

Brothwell, D. R., 1961   The Palaeopathology of Early British Man: An Essay on the Problems of Diagnosis and Analysis. Journal of the Royal Anthropological Institute 91: 318–343.

Brothwell, D. R., 1981   Digging Up Bones. 3rd edition. Oxford: Oxford University Press / British Museum (Natural History).

Brugnara, C., 2003   Iron Deficiency and Erythropoiesis: New Diagnostic Approaches. Clinical Chemistry 49:1573–1578.

Buikstra, J. E., 1999   Paleoepidemiology of Tuberculosis in the Americas. In Tuberculosis Past and Present. G. Pálfi, O. Dutour, J. Deák, and I. Hutás eds. pp. 479–494. Budapest: Golden Books / Tuberculosis Foundation.

Buikstra, J. E, and Ubelaker, D. H., 1994   Standards for Data Collection from Human Skeletal Remains. Arkansas Archaeological Survey Research Series No. 44. Fayetteville: Arkansas Archaeological Survey.

Burko, H., Mellins, H. Z., and Watson, J., 1961   Skull Changes in Iron Deficiency Anemia Simulating Congenital Hemolytic Anemia. American Journal of Roentgenology 86:447–452.

Bush, H., 1991   Concepts of Health and Stress. In Health in Past Societies. Biocultural Interpretations of Human Skeletal Remains in Archaeological Contexts. H. Bush, and M. Zvelebil, eds. pp. 11–21. British Archaeological Reports, International Series 567. Oxford: Tempus Reparatum.

Bush, H., and Zvelebil, M., 1991   Pathology and Health in Past Societies: An Introduction. In Health in Past Societies. Biocultural Interpretations of Human Skeletal Remains in Archaeological Contexts. H. Bush, and M. Zvelebil, eds. pp. 3–9. British Archaeological Reports, International Series 567. Oxford: Tempus Reparatum.

Bydder, S., 2008   Ten Years of Radiation Oncology Case Reports. Journal of Medical Imaging and Radiation Oncology 52:527–530.

Byers, S. N., and Roberts, C. A., 2003   Bayes' Theorem in Paleopathological Fiagnosis. American Journal of Physical Anthropology 121:1–9.

Cavill, I., 2002   Erythropoiesis and Iron. Best Practice and Research in Clinical Haematology 15:399–409.

Cohen, M. N., and Armelagos G. J., 1984 Paleopathology at the Origins of Agriculture. London: Academic Press.

Cohen, M. N., and Crane-Kramer, G., 2007 Ancient Health: Skeletal Indicators of Agricultural and Economic Intensification. Gainesville: University Press of Florida.

Cook, D. C., and Powell M. L., 2006 The Revolution of American Paleopathology. *In* Bioarchaeology: The Contextual Analysis of Human Remains. J. E. Buikstra, and L. Beck eds. pp. 281–322. New York: Elsevier.

Cooley, T. B., Witwer, E. R., and Lee, P., 1927 Anemia in Children With Splenomegaly and Peculiar Changes in the Bones Report of Cases. American Journal of Diseases of Children 34:347–363.

DeBakey, L., and DeBakey S., 1983 The Case Report. I. Guidelines for Preparation. International Journal of Cardiology 4:357–364.

Eng, L., 1958 Chronic Iron Deficiency Anaemia With Bone Changes Resembling Cooley's Anaemia. Acta Haematologica 19:263–268.

Erdal, Y. S., 2006 A Pre-Columbian Case of Congenital Syphilis From Anatolia (Nicaea, 13th Century AD). International Journal of Osteoarchaeology 16:16–33.

Farmer, A., 1999 The Demise of the Published Case Report – Is Resuscitation Necessary? British Journal of Psychiatry 134:93–94.

Feingold, B. F., and Case J. T., 1933 Roentgenologic Skull Changes in the Anemias of Childhood. American Journal of Roentgenology and Radium Therapy 29:194–202.

Fenton, J. E., Khoo, S. G., Ahmed, I., Ullah, I., and Shaikh M., 2004 Tackling the Case Report. Auris Nasus Larynx 31:205–207.

Garfield, E., 1996 How Can Impact Factors be Improved? British Medical Journal 313:411–413.

Gejvall, N-G., 1960 Westerhus. Mediaeval Population and Church in the Light of the Skeletal Remains. Lund: Vitterhets Historie och Antikvitets Akademien.

Gilbert, R. L., 1985 Stress, Paleonutrition and Trace Elements. *In* The Analysis of Prehistoric Diets. R. L. Gilbert, and J. M. Mielke, eds. pp. 339–358. London: Academic Press.

Goldstein, M. S., 1957 Skeletal Pathology of Early Indians in Texas. American Journal of Physical Anthropology 15:299–311.

Goodman, A. H, Thomas, R. B., Swedlund, A. C., and Armelagos, G. J., 1988 Biocultural Perspectives on Stress in Prehistoric, Historical, and Contemporary Population Research. Yearbook of Physical Anthropology 31:169–202.

Gordis, L., 2009 Epidemiology. 4th edition. Philadelphia: Saunders Elsevier.

Grauer, A. L., 1993 Patterns of Anaemia and Infection from Mediaeval York, England. American Journal of Physical Anthropology 91:203–213.

Hamperl, H., and Weiss, P., 1955 Über Die Spongiöse Hyperostose an Schädeln aus Alt-Peru. Virchows Archiv für Pathologische Anatomie und Physiologie 327:629–642.

Har-El, G., 1999 Does It Take a Village to Write a Case Report? Otolaryngology – Head and Neck Surgery 120:787–788.

Huss-Ashmore, R., Goodman, A. H., and Armelagos, G. J., 1982 Nutritional Inference From Paleopathology. Advances in Archaeological Method and Theory 5:395–474.

Inglemark, B., 1939 The Skeletons. *In* Armour from the Battle of Wisby, 1361. B. Thordemann, ed. pp. 149–205. Stockholm: Vitterhets Historic och Antikvitets Akademien.

Jacobi, K. P., and Danforth, M. E., 2002 Analysis of Interobserver Scoring Patterns in Porotic Hyperostosis and Cribra Orbitalia. International Journal of Osteoarchaeology 12:248–258.

Jarcho, S., 1966 The Development and Present Condition of Paleopathology in the United States. *In* Human Paleopathology. S. Jarcho, ed. pp. 3–30.

Judd, M. A., 2002 Comparison of Long Bone Trauma Recording Methods. Journal of Archaeological Science 29:1255–1265.

Jurmain, R., 1999 Stories From the Skeleton. Behavioral Reconstruction in Human Osteology. London: Gordon and Breach.

Kenyon, G., 2002 Editorial. Journal of Laryngology and Otology 116:493.

Lanzkowski, P., 1968   Radiological Features of Iron-Deficiency Anemia. American Journal of Diseases in Childhood 116:16–29.

Larsen, C. S., 1997   Bioarchaeology: Interpreting Behaviour from the Human Skeleton. Cambridge: Cambridge University Press.

Lees, B., Banks, L. M., and Stevenson, J. C., 1998   Bone Mass Measurements. *In* Osteoporosis. J. C. Stevenson, and R. Lindsay eds. pp. 137–160. London: Chapman and Hall.

Lindblom, K., 1951   Backache and Its Relation to Ruptures of the Intervertebral Disks. Radiology 57:710–719.

Lovejoy, C. O., and Heiple, K. G., 1981   The Analysis of Fractures in Skeletal populations with an Example from the Libben Site, Ottowa County, Ohio. American Journal of Physical Anthropology, 55(4):529–541.

Mann, R. W., and Hunt, D. R., 2005   Photographic Regional Atlas of Bone Disease: A Guide to Pathologic and Normal Variation in the Human Skeleton. Springfield, IL: C. C. Thomas.

Mason, R. A., 2001   The Case Report – An Endangered Species? Anaesthesia 56:99–102.

Mays, S. A., 1997   A Perspective on Human Osteoarchaeology in Britain. International Journal of Osteoarchaeology 7:600–604.

Mays, S. A., 2006   The Osteology of Monasticism in Mediaeval England. *In* Social Archaeology of Funerary Remains. R. Gowland, and C. Knüsel, eds. pp. 179–189. Oxford: Oxbow Books.

Mays, S. A., 2008   Radiography and Allied Techniques in the Paleopathology of Skeletal Remains. *In* Advances in Human Palaeopathology. R. Pinhasi, and S. A. Mays, eds. pp. 77–100. Chichester: Wiley.

Mays, S. A., 2010a   The Archaeology of Human Bones. 2nd edition. London: Routledge.

Mays, S. A., 2010b   Human Osteoarchaeology in the UK 2001–2007: A Bibliometric Perspective. International Journal of Osteoarchaeology 20:192–204.

Mays, S. A., 2010c   Archaeological Skeletons Support a North-West European Origin for Paget's Disease of Bone. Journal of Bone and Mineral Research 25:1839–1841.

Mays, S. A., 2012   The Impact of Case Reports Relative to Other Types of Publication in Palaeopathology. International Journal of Osteoarchaeology 22:81–85.

Mays, S. A., Brickley, M., and Ives, R., 2006   Skeletal Manifestations of Rickets in Infants and Young Children in an Historic Population from England. American Journal of Physical Anthropology 129:362–374.

Mensforth, R. P., Lovejoy, C. O., Lallo, J. W., and Armelagos, G. J., 1978   The Role of Constitutional Factors, Diet, and Infectious Disease in the Etiology of Porotic Hyperostosis and Periosteal Reactions in Prehistoric Infants and Children. Medical Anthropology 2(1): 1–59.

Mirra, J. M., Brien, E. W., and Tehranzadeh J., 1995a   Paget's Disease of Bone: Review With Emphasis on Radiologic Features. Part I. Skeletal Radiology 24:163–171.

Mirra, J. M., Brien, E. W., and Tehranzadeh J., 1995b   Paget's Disease of Bone: Review With Emphasis on Radiologic Features. Part II. Skeletal Radiology 24:173–184.

Møller-Christensen, V., 1961   Bone Changes in Leprosy. Copenhagen: Munksgaard.

Møller-Christensen, V., 1974   Changes in the Anterior Nasal Spine and the Alveolar Processes in Leprosy. A Clinical Examination. International Journal of Leprosy 42:431–435.

Molleson, T. I., Williams C. T., Cressey G., and Din V. K., 1998   Radiographically Opaque Bones From Lead-Lined Coffins at Christ Church, Spitalfields, London – An Extreme Example of Bone Diagenesis. Bulletin de la Société Geologique de France 169:425–432.

Moore, S., 1929   The Bone Changes in Sickle Cell Anemia With Note on Similar Changes Observed in Skulls of Ancient Mayan Indians. Journal of the Missouri Medical Association 26:561–564.

Moseley, J. E., 1961   Skull Changes in Chronic Iron Deficiency Anemia. American Journal of Roentgenology 85:649–652.

Moseley J. E., 1965   The Paleopathological Riddle of "Symmetrical Osteoporosis". American Journal of Roentgenology 95:135–142.

Nathan, H., and Haas N., 1966   "Cribra Orbitalia." A Bone Condition of the Orbit of Unknown Nature. Israel Journal of Medical Sciences 2:171–191.

Ortner, D. J., 1991   Theoretical and Methodological Issues in Paleopathology. *In* Human Paleopathology: Current Syntheses and Future Options. D. J. Ortner, and A. C. Aufderheide, eds. pp. 5–11. Washington, DC: Smithsonian Institution.

Ortner, D. J., 2003   Identification of Pathological Conditions in Human Skeletal Remains. London: Academic Press.

Ortner, D. J, Butler, W., Cafarella, J., and Milligan, L., 2001   Evidence of Probable Scurvy in Subadults From Archaeological Sites in North America. American Journal of Physical Anthropology 114:343–351.

Ortner, D. J, and Mays, S., 1998   Dry-Bone Manifestations of Rickets in Infancy and Early Childhood. International Journal of Osteoarchaeology 8:45–55.

Ortner, D. J., and Putschar, W. G. J., 1985   Identification of Pathological Conditions in Human Skeletal Remains. Washington: Smithsonian Institution.

Patsopoulos, N. A., Analatos, A. A., and Ioannidis, J. P. A., 2005   Relative Citation Impact of Various Study Designs in the Health Sciences. Journal of the American Medical Association 293:2362–2366.

Pettifor, J. M., 2003   Nutritional Rickets. *In* Pediatric Bone. Biology and Diseases. F. Glorieux, J. Pettifor, and H. Jüppner, eds. pp. 541–565. New York: Academic Press.

Pinhasi, R., and Mays, S., 2008   Advances in Human Palaeopathology. Chichester: Wiley.

Pinhasi, R., and Turner, K., 2008   Epidemiological Approaches in Paleopathology. *In* Advances in Human Palaeopathology. R. Pinhasi, and S. A. Mays eds. pp. 45–56. Chichester: Wiley.

Recker, R. R., 2008   Bone Biopsy and Histomorphometry in Clinical Practice. *In* Primer on the Metabolic Bone Diseases and Disorders of Mineral Metabolism. 7th Edition. C. Rosen, ed. pp. 180–186. Washington, DC: American Society for Bone and Mineral Research.

Reiman, F., Talasli, U., and Gökman, E., 1976   Zur Röntgenologischen Bestimmung der Dicke der Schädelknochen und Ihrer Verbreiterung bei Patienten Mit Schwerer Bluterkrankung und Hyperplasie des Roten Knochenmarks. Fortschritte auf dem Gebiete der Röntgenstrahlen und der Nuklearmedizin 125:540–545.

Roberts, C. A., 2006   A View from Afar: Bioarchaeology in Britain. *In* Bioarchaeology: the Contextual Analysis of Human Remains. J. E. Buikstra, and L. Beck, eds. pp. 417–439. Elsevier: New York.

Roberts, C. A., and Cox M., 2003   Health and Disease in Britain. Stroud: Sutton.

Roberts C. A., and Manchester K., 1995   The Archaeology of Disease. 2nd edition. Stroud: Sutton.

Rogers J., and Waldron T., 1995   A Field Guide to Joint Disease. Chichester: Wiley.

Rogers, J., Watt, I., and Dieppe, P., 1990   Comparison of Visual and Radiographic Detection of Bony Changes at the Knee Joint. British Medical Journal 300:367–368.

Roney, J. G., 1959   Palaeopathology of a Californian Archaeological Site. Bulletin of the History of Medicine 33:97–109.

Sager, P., 1969   Spondylosis Cervicalis. Aarberetning Københavns Universitets Medicinsk-Historiske Museum. Copenhagen: Munksgaard. Cited in Brothwell 1981.

Sandison, A. T., 1968   Pathological Changes in the Skeletons of Earlier Populations Due to Acquired Disease, and Difficulties in Their Interpretation. *In* The Skeletal Biology of Earlier Human Populations. D. R. Brothwell, ed. pp. 205–243. Oxford: Pergamon.

Shahidi, N. T., and Diamond, L. K., 1960   Skull Changes in Infants With Chronic Iron-Deficiency Anemia. New England Journal of Medicine 262:137–139.

Siris, E. S., and Roodman, G. D., 2008   Paget's Disease of Bone. *In* Primer on the Metabolic Bone Diseases and Disorders of Mineral Metabolism. 7th edition. C. Rosen, ed. pp. 335–343. Washington, DC: American Society for Bone and Mineral Research.

Smith, S. E., Murphey, M. D., Motamedi, K., Mulligan, M. E., Resnik, C. S., and Gannon, F. H., 2002   Radiologic Spectrum of Paget Disease of Bone and Its Complications With Pathologic Correlation. RadioGraphics 22:1191–1216.

Steckel, R. H., 2003   Research Project. A History of Health in Europe From the Late Paleolithic Era to the Present. Economics and Human Biology 1:139–142.

Steckel, R. H., 2008   Biology and Culture: Assessing the Quality of Life. In Between Biology and Culture. H. Schutkowski, ed. pp. 67–104. Cambridge: Cambridge University Press.

Steckel, R. H, and Rose J. C., eds., 2002   The Backbone of History. Health and Nutrition in the Western Hemisphere. Cambridge: Cambridge University Press.

Steckel, R. H, Kjellström, A, Rose, J, Larsen, C. S., Walker, P. L., Blondiaux, J., Grupe, G., Maat, G. et al., 2009   Summary Measurement of Health and Wellbeing: The Health Index. American Journal of Physical Anthropology Supplement 48:247

Steinbock, R. T., 1976   Paleopathological Diagnosis and Interpretation. Springfield, IL: C. C. Thomas.

Stewart, T. D., 1931   Incidence of Separate Neural Arch in the Lumbar Vertebrae of Eskimos. American Journal of Physical Anthropology 16:51–62.

Stewart, T. D., 1953   The Age-Incidence of Neural Arch Defects in Alaskan Natives, Considered From the Standpoint of Etiology. Journal of Bone and Joint Surgery 35A: 937–950.

Stewart, T. D., 1956   Examination of the Possibility That Certain Skeletal Characters Predispose to Defects of the Lumbar Neural Arches. Clinical Orthopaedics 8:44–60.

Stuart-Macadam, P., 1985   Porotic Hyperostosis: Representative of a Childhood Condition. American Journal of Physical Anthropology 66:391–398.

Stuart-Macadam, P., 1989   Porotic Hyperostosis: Relationship Between Orbital and Vault Lesions. American Journal of Physical Anthropology 80:187–193.

Stuart-Macadam, P., 1991   Anaemia in Roman Britain: Poundbury Camp. In Health in Past Societies. Biocultural Interpretations of Human Skeletal Remains in Archaeological Contexts. H. Bush, and M. Zvelebil, eds. pp. 101–113. British Archaeological Reports, International Series 567. Oxford: Tempus Reparatum.

Thacher, T. D., Fischer, P. R., Pettifor, J. M., Lawson, J. O., Minister, B. J., and Reading, J. C., 2000   Radiographic Scoring Method for the Assessment of the Severity of Nutritional Rickets. Journal of Tropical Pediatrics 46:132–139.

Turner-Walker, G., and Mays S. A., 2008   Histological Studies on Ancient Bone. In Advances in Human Palaeopathology. R. Pinhasi, and S. A. Mays, eds. pp. 121–146. Chichester: Wiley.

Vandenbroucke, J. P., 2001   In Defense of Case Reports and Case Series. Annals of Internal Medicine 134:330–334.

Virchow, R., 1874   Altpatagonische, Altchilenische und Moderne Pampas Schädel. Verhandlung der Berliner Gesellschaft für Anthropologie, Ethnologie und Urgeschichte 6:51–64.

Vogt, E. C., and Diamond, L. K., 1930   Congenital Anemias, Roentgenologically Considered. American Journal of Roentgenology and Radium Therapy 23:625–630.

Waldron, H. A., 1997   Association Between Osteoarthritis of the Hand and Knee in a Population of Skeletons from London. Annals of the Rheumatic Diseases 56:116–118.

Waldron, T., 1994   Counting the Dead: The Epidemiology of Skeletal Populations. Chichester: Wiley.

Waldron, T., and Rogers, J., 1990   An Epidemiologic Study of Sacroiliac Fusion in Some Human Skeletal Remains. American Journal of Physical Anthropology 83:123–127.

Waldron, T., and Rogers, J., 1991   Inter-Observer Variation in Coding Osteoarthritis in Human Skeletal Remains. International Journal of Osteoarchaeology 1: 49–56.

Walker, P. L., Bathurst, R. R., Richman, R., Gjerdrum, T., and Andrushko, V. A., 2009   The Causes of Porotic Hyperostosis and Cribra Orbitalia: A Reappraisal of the Iron-Deficiency-Anemia Hypothesis. American Journal of Physical Anthropology 139:109–125.

Walsh, J. P., 2004   Paget's Disease of Bone. Medical Journal of Australia 181:262–265.

Wapler, U., Crubézy, E., and Schultz, M., 2004   Is Cribra Orbitalia Synonymous with Anemia? Analysis and Interpretation of Cranial Pathology in Sudan. American Journal of Physical Anthropology 123:333–339.

Ward, C. V., Mays, S. A., Child, S., and Latimer, B., 2010   Lumbar Vertebral Morphology and Isthmic Spondylolysis in a British Mediaeval Population. American Journal of Physical Anthropology 141:273–280.

Weiss, E., and Jurmain, R., 2007   Osteoarthritis Revisited: A Contemporary Review of Aetiology. International Journal of Osteoarchaeology 17:437–450.

Welcker, H., 1888   Cribra Orbitalia, Ein Ethnologische-Diagnostisches Merkmal am Schädel Mehrer Menschenrassen. Archiv für Anthropologie 17:1–18.

Wells, C., 1963   The Radiological Examination of Human Remains. *In* Science in Archaeology. D. R. Brothwell, and E. Higgs, eds. pp. 401–412. London: Thames and Hudson.

White, L. E., 1959   The Evolution of Culture. New York:McGraw-Hill.

Williams, H. U., 1929   Human Paleopathology With Some Original Observations on Symmetrical Osteoporosis of the Skull. Archives of Pathology 7:839–902.

Wood, J. W., Milne, G. R., Harpending, H. C., and Weiss, K. M., 1992   The Osteological Paradox: Problems of Inferring Prehistoric Health From Skeletal Samples. Current Anthropology 33:343–370.

Wright, S. M., and Kouroukis, C., 2000   Capturing Zebras: What To Do With a Reportable Case. Canadian Medical Association Journal 163:429–431.

Yildirim, T., Agildere, A. M., Oguzkurt, L., Barutcu, O., Kizilkilic, O., Kocak, R., and Niron, E. A., 2005   MRI Evaluation of Cranial Bone Marrow Signal Intensity and Thickness in Chronic Anemia. European Journal of Radiology 53:125–130.

Zimmerman, M. R., and Kelley, M. A., 1982   Atlas of Human Paleopathology. New York: Praeger.

CHAPTER 17

Integrating Historical Sources with Paleopathology

Piers D. Mitchell

INTRODUCTION

Historical sources of evidence for life in the past are those that have been written down at the time by people from the communities under study, or those who encountered them. Such texts can tell us directly about health and disease in those times, and can also help us to understand archaeological findings that might otherwise be perplexing to modern researchers. Currently, the majority of research on health and disease in the past focuses on archaeological evidence and modern genetic evidence (Aufderheide and Rodrigues-Martin 1998; Pinhasi and Mays 2008). Textual evidence is widely used by medical historians to study the social consequences of disease in the past, but it is less commonly used to study the disease itself. When palaeopathologists do explore written sources, it is often to study evidence from relatively recent times written in modern languages (Grauer 1995; Herring and Swedlund 2003). The reasons for this include the skills learnt by students of different disciplines, and also traditional interests within different fields. In this chapter, the consequences of these two conflicts are explored, and the debate amongst historians regarding the reliability of past written sources for studying disease is outlined. An approach is then given that allows researchers to assess the degree to which historical sources can reliably inform us about disease in the past, so that we are can use these texts in a safe manner, and not overinterpret the written word over and beyond safe limits.

PRIMARY AND SECONDARY SOURCES

Writing is thought to have first developed in the ancient Middle East by the Sumerians in the late 4th millennium B.C. (Black 2008). Cuneiform text was pressed into damp

A Companion to Paleopathology, First Edition. Edited by Anne L. Grauer.
© 2012 John Wiley & Sons, Ltd. Published 2016 by John Wiley & Sons, Ltd.

clay tablets that were then hardened and these tablets have been preserved thousands of years. Other civilizations have written on stone, papyrus, animal skins (vellum), and paper among other media (Harris 1986). Some of these early languages have been difficult to interpret as they are no longer used today, and so when first found these texts were not understood. However, the discovery of the same text written in different languages by merchants and statesmen involved in foreign policy has allowed specialist linguists to decipher how these languages were put together and what the words meant (Parkinson 1999). Nonetheless, there will always be some words where the modern interpretation or translation is an educated guess rather than one known with certainty. In later times, texts were written in languages that are well known today or even still spoken today. For example, continued study of ancient Greek and Latin in Europe over the past 2,000 years means that ancient texts in these languages are much better understood than ancient Hittite, for example (Kloekhorst 2008). The same principle applies to ancient languages from any part of the world, and must be considered when interpreting words believed to indicate disease in those past populations.

Considerable attention has been paid to making original sources that discuss disease and medicine in the past available for consultation without readers having to travel the world to consult the rare surviving copies in the original manuscript form. Printed editions of these original, handwritten manuscripts have made study much more affordable and accessible. The approach may be to reproduce photocopies of the original handwritten sources, or to print the wording using modern type fonts (Sudhoff and Singer 1925; Rawcliffe 1995; Lo and Cullen 2004; Scurlock and Andersen 2005). Hand-lists summarizing the available editions have also been produced to act as reference tools (Jayawardene 1982; Barrett 2003). Guidance on how to use such sources can be found in works that discuss research methods, methodological issues, and interaction with sources and nonhistorical research fields such as anthropology (Brier 1980; MacDonald 1983; Webster 1983; Weindling 1983).

The term "primary source" is used today to highlight that these records contain information from eyewitnesses or those that lived at the time of an event. This is in contrast to a "secondary source," a term that refers to texts written at a much later time, or modern commentaries on past events, that give the opinion of authors who took their evidence from primary sources. In order to assess the quality and likely reliability of secondary sources, it can be helpful to look at the references and determine how much of the evidence comes from primary sources, and how much is quoted from pre-existing secondary sources. Ideally, the core evidence for a paper studying any aspect of life in the past based on written information should come from genuinely primary sources. Secondary sources can then be incorporated to the paper to give the opinion of others, and allow discussion of whether those opinions are still valid or may be mistaken. Authors who take their primary evidence from secondary sources may be misled by the opinion of those previous authors. For example, previous authors may have never read the original evidence, they may have only seen an unrepresentative selection of the available evidence, or they may have been writing from a particular background that biases their perspective. One example of a research area that has suffered considerably from these problems is the study of crucifixion in Roman times (Maslen and Mitchell 2006). In other words, it is wise never to rely on the opinion of others if you can check the evidence yourself. The more I work in the field, the more I have learnt to become skeptical of everything written, unless the material is well referenced to an original source. In consequence, papers I wrote a decade ago (e.g., Mitchell 1999) would not meet the criteria I am recommending here.

When using primary sources a number of criteria must be satisfied before we can be comfortable that our interpretation is the best it can be. Clearly, primary sources from past populations will be written in languages used by those populations, which will generally be different languages to those used today. The best way to study the original text will be to read it in the original language. Translations can be helpful for students and for those screening original sources for information likely to be of relevance to disease in the past, but translations should not be relied upon for evidence. This is because the translation may not use the best choice of words to describe the terms referring to disease in the past. In general, modern translators of ancient texts are not medical historians or textual palaeopathologists, so may select the wrong medical terms without realizing the importance of their choice. For example, on the Seventh crusade to Egypt in 1248 A.D., French troops were trapped between two canals by the Sultan of Egypt and were running short of food. A nobleman named Jean de Joinville wrote in medieval French of the illness from which he and his troops suffered. The English translation of his chronicle uses the phrase "the camp fever" to refer to his illness (John of Joinville 1955:24). If we were to use this phrase to attempt a differential diagnosis for retrospective diagnosis, we would be considering diseases that cause fever in a stressed army camp, and so include infectious diseases that cause epidemics. However, reading the original medieval French texts shows that the words Joinville actually used to describe the illness were "*la maladie de l'ost*" (John of Joinville 1874:6). This does not translate as camp fever, but rather as "the illness of the host," or the sickness of the army camp. Joinville does not mention a fever, so the differential diagnosis for this illness would be very different. This is an example of inappropriate use of words by a translator who probably did not realize this would matter very much. As it happens, Joinville goes on to give a very detailed description of the symptoms, including the appearance of brown patches on the legs, overgrowth of the gums, and spontaneous bleeding from the nose (John of Joinville 1874:160, 166). These are highly suggestive of scurvy, vitamin C deficiency (Mitchell 2004:185). Without checking the original source in the original language it would be easy to be misled by the translation. In fact, too many myths about disease in the past have been perpetuated by authors uncritically accepting past opinions, and not reassessing the sources themselves (Mitchell 2002, 2010). This is why it is always a good idea when writing an article to include the original wording in the source language, either in the body of the text if just a few words are translated, or as a footnote for longer passages.

Our understanding of the meaning of words has improved over time in parallel with so many other areas of academic endeavor. In consequence, a translation from a century ago (frequently the best or only translations of some ancient and medieval texts are this old) will be out of date in its interpretation of many words referring to disease. However, even modern dictionaries of ancient languages may describe disease-related words incorrectly, so should be used with appropriate caution. For example, the medieval Latin word *dysenteria* is generally translated as dysentery in medieval Latin–English dictionaries (Niermeyer and Van de Kieft 2002:445).

However, the Latin term *dysenteria* does not directly mean dysentery as we would understand it as a modern biological diagnosis, but rather a range of diseases that caused diarrhea. In the setting of a medieval army during a siege, an outbreak of infective dysentery may well be the cause (Mitchell et al. 2008). However, diarrhea may not only be caused by amoebic and bacterial dysentery, but also cholera, typhoid, viral gastroenteritis, and other causes of loose, watery stools (Farthing 2002). This highlights that the scholars who compile dictionaries of ancient languages are rarely experts in specific medical terminology and medical diagnoses. This is another reason why original sources quoted as translations in the text should always be given in the original language, preferably as a footnote, allowing nonspecialists to read the article smoothly while still allowing experts in the source language to check the original wording to satisfy themselves that the original text has been interpreted correctly.

WHAT IS DISEASE?

It might seem obvious to us what disease is, but it is actually a concept that will vary quite considerably in meaning for different people in different parts of the world living at different times during history. One very broad and thought-provoking definition that I find helpful is that of Charles Rosenberg, who describes disease as "at once a biological event, a generation-specific repertoire of verbal constructs reflecting medicine's intellectual and institutional history, an occasion of potential legitimization for public policy, an aspect of social role and individual—intrapsychic—identity, a sanction for cultural values, and a structuring element in doctor and patient interactions" (Rosenberg and Golden 1992:xiii). I like to simplify this definition by considering two factors: the modern biological diagnosis and the social diagnosis.

The modern biological diagnosis is one based upon current scientific understanding of the causes of disease (Mitchell, in press (a)). This allows us to explore disease in the past without the cultural–social overlay and expectations of past populations. The modern biological diagnosis will clearly include many diseases that people in the past had no idea existed, but will also omit diseases that do exist but have yet to be identified. As such, a modern biological diagnosis reflects a snapshot of disease as we understand it at the time of writing about it, but will inevitably become out of date for future generations who make new scientific discoveries. Most palaeopathologists automatically attempt to determine a modern biological diagnosis when evaluating human skeletal remains, mummies or other archaeological material for evidence of disease in the past.

The social diagnosis is the label given to a particular condition as past civilizations may have perceived it (Mitchell, in press (a)). This may occasionally be the same as the modern biological diagnosis, but will often be something quite different. The social diagnoses used in the past are fascinating and comprise a perfectly valid research field. However, it is wise to keep social diagnosis distinct from attempts to identify the modern biological diagnosis, as they are not interchangeable. If hundreds of years ago a population wrote of a disease in a specific individual, mentioning a range of quite distinctive signs and symptoms that we would commonly see today in people with leprosy, it is not unreasonable to suggest that the modern biological diagnosis for that disease is leprosy (Mitchell 2000). However, in the absence of a description of those distinctive signs and symptoms, we should not assume that all (or even the majority) of cases labeled with the term, or social diagnosis, in the past, did indeed have leprosy. It is only when the term is repeatedly linked with a similar range of distinctive signs and symptoms that we can have confidence that the social diagnosis in that time-period is routinely referring to the same disease. Furthermore, the social diagnosis may represent just one component or one presentation of a larger modern biological diagnosis, or it may group together many modern biological diagnoses under one term.

Some past social diagnoses may not be conditions we would regard as diseases in the modern sense at all, and some modern diseases may not have been regarded as disease to past populations. We might regard hallucinations and hearing voices as a mental illness today, whereas in some society in the past it may have been regarded as a gift from God (MacDermot 1971). Many mistakes have been made trying to infer modern biological diagnoses from social-diagnostic terms recorded in the past. Criticism of such an approach has been vigorous (Paterson 1998; Arrizabalaga 2002; Hays 2007; Metcalfe 2007; Karenberg 2009) as the meaning of a term may have changed over time, diagnostic criteria may have evolved, and the expertise of individuals within society applying that label may change. Similarly, it may have become socially more or less profitable or desirable to use one label as opposed to another for a similar disease. For example, if two societies used the same diagnostic criteria to come to a particular social diagnosis for a disease that was equally common in both societies, but diagnosis in one society resulted in financial support from the state while in the other it resulted in imprisonment or execution, the frequency of such a social diagnosis might be very different in the two populations. So, while the actual disease may have been equally common in both communities, it would not appear to be so to modern researchers relying on textual records from those societies.

For all these reasons diagnostic labels applied in the past without record of the actual disease signs and symptoms are very difficult to assess and compare reliably. This social diagnosis can be compared in its social context, but we should avoid assuming that the modern biological diagnosis is reliably linked to it. The social diagnosis will vary depending upon which members of a past society were regarded as the right people to make such diagnoses, what their background was, and what the consequences of allocating that diagnostic label was. It is only when we understand the social function of that diagnostic term in a past civilization that we can start to evaluate how best to use records in a safe manner to improve our understanding of health in past societies. When studying textual sources for evidence of disease in the past, it is of key importance that we remember that the social diagnosis and modern biological diagnosis are not the same things.

THE CUNNINGHAM DEBATE

Andrew Cunningham holds strong views on the potential to determine a modern biological diagnosis based on textual sources from past civilizations, and so the controversy can be referred to as the Cunningham debate. Cunningham argues that since we cannot reliably equate a modern biological diagnosis to an ancient social diagnosis, there is no way we can identify the modern biological diagnosis from written historical sources (Cunningham 2002). He argues that a modern biological diagnosis is actually irrelevant to the past population involved, as they did not see disease in the way we see it today, and in 500 years time our view of medicine will again seem just as out of date as medicine advances further. He feels that it is much more reliable, and more relevant, to study the social diagnosis of disease, how past populations viewed illness, how it affected their society, and what they did to try to cure it.

There is much wisdom in his argument, and the debate has helped clarify what is important when researching this area. The social diagnosis is a valuable topic to study, but I would argue that we should not be limited to this. Medical historians in Britain, America and Australia come mostly from a background in history, and not from the study of medicine, in contrast to mainland Europe. Lack of training in medicine means that there is less interest in obtaining a modern biological diagnosis, and also greater technical difficulty in understanding the meaning of symptoms and signs recorded in historical sources. This is in no way a criticism, but merely an explanation as to why many medical historians from English-speaking countries are content to focus on the social diagnosis. However, those researchers with a background in medicine and paleopathology are just as keen to understand ancient disease as they are the people affected by the disease. It is as interesting, and as academically valid, to research past microorganisms as past humans.

Individuals from one background with a long tradition of what can and cannot be done, or what is or is not a valid topic of study, can find it difficult to cross the boundary to another field of study. This has been noted for many disciplines: Stephen Jay Gould termed this phenomenon "the misconceived gap" (Gould 2003). Individuals who have been trained in both the humanities and the sciences generally feel more comfortable crossing the boundaries between fields. Also, teams from diverse backgrounds can combine their skills to do so. I would argue that individuals trained in just one field should concentrate their efforts within that field to ensure their interpretation and conclusions are robust and reliable, and that individuals trained in several fields, or teams comprised of experts from several fields, should tackle interdisciplinary topics in order to advance our knowledge of the past as a whole. In this way, when distinctive symptoms and signs are recorded that are either pathognomonic or highly indicative of a particular disease, we can then determine a probable modern biological diagnosis. If we avoid the social diagnosis, and instead rely on symptoms and signs from eyewitness sources, many of the difficulties raised by Cunningham are resolved (Mitchell in press (a)).

IDENTIFYING DISEASE IN PRIMARY TEXTUAL SOURCES

Having ensured that at least one member of the research team can read the original languages of the texts under study, and that symptoms and signs of disease, rather than past social-diagnostic labels, are being sought out, the full range of available source texts must be searched for evidence of disease. Basing opinions on one vivid, but potentially unrepresentative, text may lead to the wrong conclusions. Only by reading as many relevant sources as possible that have survived to modern times can we minimize the risk of mistakes. The range of sources that might record information on past diseases includes eye-witness histories and chronicles, epics and tales, biographies, personal letters and diaries, legal documents and wills, religious proclamations, customs documents, state registers of births and deaths, and hospital records. The strength of evidence found within past texts will vary depending upon who wrote them, why they were written, for whom they were written, and exactly when they were written. A thorough knowledge of the literary history of that population is needed in order to understand this context. For example, when the pox (syphilis) spread across Europe in the late 1490s and early 1500s, each country blamed the disease on the country that appeared to have spread it to them. Just because the Italians referred to the pox as the French Disease, does not necessarily mean the disease originated there. The French themselves blamed the Spanish, as cases were reported there before any appeared in France (Arrizabalaga 1997), and there is reasonable evidence now to suggest the Spanish acquired it from overseas (Harper 2008).

The nature of sources studied, and the social genre of those sources, is important to guide interpretation of their content. If we want to know today which diseases exist in different countries the logical place to start would be a medical book or article. However, the descriptions of diseases in past societies may not be reflected in the medical texts of those societies. It is noticeable that in the medieval Middle East and Europe many passages in medical works were copied from earlier medical texts dating over the previous thousand years all the way back to Greek and Roman times (Alvarez-Milan 2000). The medieval practice of word-for-word copying of ancient or classical passages suggests that descriptions of diseases, medical treatments, or operations were not always evolving and advancing as we might expect if they were being used in daily practice. It seems likely that an operation described centuries before was being included in a medical text out of completeness, to show the range of scholarly reading of the medical author, and may never have been performed or even witnessed by that author (Savage-Smith 2000).

The tradition of copying earlier authoritative medical texts without explicitly acknowledging them was perpetuated for many hundreds of years after the medieval period. Medical authors writing about the plague in Tudor England mentioned how miasma (bad air) coming up from the earth drove subterranean animals such as rats, moles and snakes to the surface. Skeyne (1568:10) wrote, *"as quhan the Moudeuart and Serpent leauis the Eird beand moleftit be the Vapore contenit within the bowells of the famin."* Three decades later Lodge (1603: chap.3, p.1) wrote *"it is a signe that the temperament of the air is altered. And when rats, moules, and other creatures, (accustomed to live under ground) forsake their holes and habitations, it is a token of corruption in the same."* While these might appear to be good eyewitness descriptions of dead rats as found in modern epidemics of bubonic plague, they were actually repeating sections almost word for word from the Canon of Medicine by the 10th-century A.D. Arabic physician Avicenna (Avicenna 1490: vol.2, liber 4, fen.1, tract. 4, cap.3). The Latin translation of his text which would have been available to both Skeyne and Lodge states how, *"& videas reptilia generata ex putredine iam multiplicari. & de eis que fignificant illud eft ut videas mures et animalia que habitant fub terra fugere ad fuperficiem terre & exire manifefte."* This was not an eyewitness anecdote vividly portraying a plague outbreak in 16th-century England, but merely repetition of a description from the Middle East six centuries before. In other words, just because a disease or medical treatment is found in an ancient medical text it does not necessary mean that disease or treatment reflects medical practice at that time. Only by compar-ing the entries in all texts of that civilization, and their sources, can we determine what was in use and what was a relic of earlier times. In consequence, it is important that assessments of disease in past societies are not just based on medical texts, but also on nonmedical sources that happen to describe medical information in passing.

Once interesting descriptions of ancient disease have been identified, they can only be interpreted to determine the modern biological diagnosis in the context of knowledge of those symptoms and signs of disease as we comprehend it today. Symptoms are the changes caused by disease that the sick person experiences and describes, and signs are the changes that another person (such as a medical practitioner) may notice. Clearly our current knowledge of medicine and disease is not perfect and is an evolving research field in its own right. However, comparing past disease with disease as we understand it today is the best we can do at any time. This allows us to identify in ancient texts examples of classic diagnostic symptoms and signs that are known in modern times to be found in particular diseases that we can confidently identify today. Because so many diseases share similar symptoms, it can be helpful to include a medically qualified member in the team to construct a differential diagnosis based on the symptoms and signs, modern knowledge of disease epidemiology, climate favorable to certain diseases today, and other factors that may help in making a modern biological diagnosis.

When King Richard I of England and King Philippe of France took part in the Third Crusade to the eastern Mediterranean in 1191, they both fell ill at the siege of the city of Acre shortly after their arrival. Various eyewitness chroniclers who were with these kings at the siege mentioned the more obvious symptoms and signs. The social diagnosis for the kings is recorded with the medieval Latin term *arnaldia*, and the medieval French term *leonardie* (Wagner and Mitchell in press). Neither of these two terms has been found in any medieval medical texts, and the only manuscripts that survive today that mention these words refer to the kings on the Third Crusade. When a word is found only once in the literature it is known as a hapax. It might have been a genuine word of the time, or a miscopy of the manuscript due to scruffy handwriting. However, the actual symptoms are much more illuminating for our coming to a modern biological diagnosis than social diagnostic terms of unknown meaning. William the Breton, the French king's chaplain, wrote how after a fever, Philippe "was so badly afflicted with illness, so that he lost the nails of his hands and feet as well as the hair and most of the surface of the skin" (*unde et tanta infirmitate gravatus est, quod et ungues manuum et pedum et capillos et fere omnem cutis superficiem amisit*) (William the Breton 1878:70). Roger of Howden was a royal clerk in the English court who accompanied the English king on the crusade. Again an eyewitness, he wrote how both Richard and Philippe appeared to have the same disease and they both lost all their hair (Roger of Howden 1870:113). Multiple eyewitnesses describing what appear to be genuine symptoms and signs are a good start from which to explore a modern biological diagnosis. In modern times, loss of hair, nails and skin after a high fever is known to occur in a variety of severe infectious diseases. In some cases this is due to toxins produced by the infective microorganism, but in many diseases the body shuts down nonessential functions and concentrates all its efforts on fighting the life-threatening illness (Amagai 2000; Sperling 2001). In children this bodily response results in temporary cessation of grown in bones (leading to Harris lines) and teeth (leading to linear enamel hypoplasia), while in adults it can lead to the hair and nails falling out and the skin peeling. This is referred to by the medical term anagen effluvium (Sperling 2001). While there is insufficient detail to determine the modern biological diagnosis for the

Table 17.1 Twelve pitfalls in retrospective diagnosis that can result in errors

1. Too little information surviving in written texts to allow a diagnosis.
2. Material inspected is not representative of the full range of texts originally created.
3. Inability to detect that an apparently eye-witness record was copied from earlier manuscripts.
4. Limited understanding of cultural background of text by researcher.
5. Basing conclusions upon translations of others, not reading original languages.
6. Limited awareness of signs and symptoms required to enable a plausible diagnosis.
7. Failure to maintain an open mind, so that symptoms incompatible with author's theory are not discussed.
8. Failure to think outside range of current diseases in existence, so that symptoms incompatible with any modern diagnosis are not discussed.
9. Inability to consider that more than one concurrent diagnosis was present.
10. Inability to consider that diseases may mutate and change over the centuries.
11. Assuming the diagnosis must be a disease that still exists in modern times.
12. Overstating the probability that the author's diagnosis is the right one.

Table 17.2 Aspects of a source that optimizes reliability in retrospective diagnosis

1. Testimony from eyewitness.
2. Vivid description of the signs and/or symptoms of disease.
3. Both the nature of lesion and its physical location is described.
4. Minimal or no evidence for description mimicking medical views of period.
5. Description of one or more virtually diagnostic symptoms/signs.
6. Plausible epidemiological observations recorded in the text.

infectious disease that caused their illnesses, the diagnosis of anagen effluvium secondary to an infectious disease is reasonable (Wagner and Mitchell in press).

To summarize, in order to safely use written texts in ancient languages to study disease in the past and determine a modern biological diagnosis, the research team needs to be able to read the texts in their original languages, have a good knowledge of the literature of the period under study, search as large a proportion of surviving texts from that period as possible, recognise the symptoms and signs of those diseases as understood by modern science, disregard the ancient diagnostic labels used in the past, and base their attempted modern biological diagnosis on ancient eye-witness descriptions of symptoms and on signs of disease that we can safely interpret today as diagnostic of particular illnesses (Tables 17.1 and 17.2).

ILLUSTRATIONS IN HISTORIC DOCUMENTS

Drawings and pictures in old manuscripts do appear at first glance to be a potentially good source of information on past disease. While such illustrations may occasionally be highly illuminating, as always, we have to be really careful when interpreting them. In many cases the person illustrating a manuscript would not have been the person who wrote the text of that manuscript, but a professional illustrator who may have possessed no medical knowledge whatsoever. Gaps may have been left in the text for these images, or they may be drawn on separate sheets to be bound together with the pages of text. These illustrators may have never seen the disease they were attempting to depict, and created their illustrations from the description in the text. Every time that manuscript was copied the illustrations would change slightly, until after a few generations the pictures looked very little like the original. If only the last version of this manuscript survives to modern times, all the early versions having been lost, the illustrations may bear little resemblance to the disease itself, the intended disease, or even the earlier misguided efforts to represent it (McKinney 1965; Hall 1996; Jones 1998).

The quality of illustrations from a modern perspective will also vary considerably depending upon the century it was created and the artistic skill and development of that civilization. For example, medieval European art from the 12th century was not as realistic as renaissance art from the 16th century, as the technical abilities of painters had evolved over the intervening four hundred years (Humphreys 1995). Furthermore, every culture varies in its perceptions of what is beautiful, what is handsome, and what the ideal person should look like. Eye location on the face, head size or shape, muscle bulk, and relative limb length may all vary between different cultures. In consequence, different artists from different cultures may depict the same person in very different ways. It is important that such variation is not misinterpreted as a disease, and similarly that a true disease is not missed because it is mistakenly perceived to be the style of the culture.

Bearing this in mind, we must critically evaluate every illustration, for technical accuracy, relationship with the adjacent text, plausibility, and typical style for that culture, in order to determine how accurate or helpful it may be in understanding the disease process described in that text. Quite frequently, it is the text rather than the illustrations that help in determining the modern biological diagnosis, especially for earlier time-periods.

WRITTEN EVIDENCE TO OPTIMIZE ARCHAEOLOGICAL INTERPRETATIONS

While written sources can record evidence for past disease, they can also be of use for biological anthropologists when interpreting the findings of archaeological excavations. Understanding the social context within which human skeletal remains were interred may explain confusing burial patterns. For instance, recovering skeletal remains only of babies may represent the presence of infanticide, a maternity hospital disposing of stillbirths, a high neonatal mortality due to difficult births or congenital defects, or social decisions to bury infants away from other individuals. While determining the reasons for the patterns can be enhanced through the incorporation of demographic analyses (Faerman et al. 1998; Mays and Faerman 2002), understanding the written sources of the period can explain attitudes towards young children in that society (Jackson 2002; Koskenniemi 2009). Archaeologically recovering children of a broader age-range buried together may represent separate burial of those not yet reaching adulthood, or the presence of an orphanage. Adult males buried without females may indicate the presence of a military base, a monastic community, a single-sex hospital, or single-sex prison. Females interred without males might signify a nunnery, a brothel, a single-sex hospital or prison, or even the disposed corpses of a serial killer.

A high prevalence of a particular disease might indicate that a cemetery or institution was reserved for individuals thought by their community to have that disease. This can, incidentally, help reconcile the concept of the social diagnosis and modern biological diagnosis. One good example of this is the leprosarium in medieval Europe, where a high proportion of individuals in leprosarium cemeteries have been found to have skeletal changes indicative of leprosy (Møller-Christensen 1978; Magilton et al. 2008). At various times hospitals have been set up to specialize in particular conditions such as venereal diseases, eye diseases, gynecological diseases and cancers (Saunders 1961; Siena 2004), so study of their cemeteries will be considerably easier if textual sources describing the institutional focus are known beforehand. The finding of cut marks or saw marks on bone might indicate that a community practiced complex social funerary practices associated with honoring their ancestors, that individuals had undergone surgical procedures, that torture was being practiced, autopsy to determine cause of death, or postmortem anatomical dissection for teaching medical students (Donnelly and Diehl 2006; Mitchell 2006; Mitchell in press (b)). This highlights how differentiating such man-made changes to bone can be much simplified when textual sources are available.

SUGGESTIONS FOR FUTURE DEVELOPMENT

Many university courses that teach palaeopathology do so exclusively from an anthropological and archaeological perspective. Great attention is devoted to how to recover skeletal remains and then analyze them in order to understand the health of each individual and the population as a whole. However, historical sources that provide evidence for disease in the past are just as valid a research source as bones, mummies or latrine soil. Yet few palaeopathology courses also provide training in the use of historical texts in order to study disease in the past. For example, the master's course on the archaeology and ancient history of disease I took 20 years ago at University College London in the UK has discontinued the history component to focus on skeletal and dental bioarchaeology. Such courses could be profitably linked with others that teach languages and historical context so that a new generation of palaeopathologists can be trained in both fields. Once this takes place, they will have little fear of Gould's "misconceived gap", and will feel comfortable tackling those interdisciplinary topics that interest them.

Increased interaction and debate between palaeopathologists and historians, particularly medical historians, could help to break down the current misconceived gap between these two fields. Presentation at each other's conferences more would also facilitate such interaction. This would not only strengthen interdisciplinary collaboration between fields, but perhaps more importantly help to break down the current perceptions in both fields as to what is, and what is not, a worthy topic to study. It is not enough to know that a topic could be investigated if no one has the interest to get on and do it. The knock-on effect of this would be that funding bodies might become more broad-minded regarding what is a topic worthy of funding. So long as the individual or team has the necessary skills and an exciting topic, funding should not be denied for the reason that the interdisciplinary nature of the project is mistakenly thought to make integrating the textual and anthropological evidence impossible. Funding bodies are just as limited by the misconceived gap as everyone else.

CONCLUSION

While paleopathology is dominated by the study of human skeletal remains, there are other ways to investigate disease in the past. Textual evidence is clearly limited to those time-periods and societies who chose to write something down, but this still spans several thousand years. In many regions such as Europe, the Middle East and Asia, the vast majority of human skeletal remains that have ever been excavated are from civilizations for which written sources survive. If we fail to use these sources to the best of our ability we not only run the risk of misinterpreting the human skeletal remains we study, but we also remain ignorant of the unique evidence for disease recorded in those texts.

REFERENCES

Alvarez-Milan, C., 2000   Practice Versus Theory: Tenth-Century Case Histories From the Islamic Middle East. Social History of Medicine 13:293–306.

Amagai, M., 2000   Toxin in Bullous Impetigo and Staphylococcal Scalded-Skin Syndrome Targets Desmoglein. Nature Medicine 6:1275–1277.

Arrizabalaga, J., 1997   The Great Pox: The French Disease in Renaissance Europe. New Haven: Yale University Press.

Arrizabalaga, J., 2002   Problematizing Retrospective Diagnosis in the History of Disease. Asclepio 54:51–70.

Aufderheide, A. C., and Rodrigues-Martin, C., 1998   The Cambridge Encyclopedia of Human Paleopathology. Cambridge: Cambridge University Press.

Avicenna, 1490   Canon Medicinae. B. Locatellus ed. Venice: O. Scotus.

Barrett, F. A., 2003   Foreign Primary Sources for Medical Geography and Geographical Medicine. Toronto: York University Press.

Black, J. E., 2008   Texts From Ur, Kept in the Iraq Museum and the British Museum. Messina: Di.Sc.A.M.

Brier, G. H., 1980   History of medicine. In A Guide to the Culture of Science, Technology and Medicine. P. T. Durbon, ed. pp.121–194. New York: Free Press.

Cunningham, A., 2002   Identifying Disease in the Past: Cutting The Gordian Knot. Asclepio 54:13–34.

Donnelly, M., and Diehl, D., 2006   Eat Thy Neighbour: a History of Cannibalism. Stroud: Sutton.

Faerman, M., Bar-Gal, G. K., Filon, D., Greenblatt, C. L., Stager, L., Oppenheim, A., Smith, P., 1998   Determining the Sex of Infanticide Victims From the Late Roman Era Through Ancient DNA Analysis. Journal of Archaeological Science 25:861–865.

Farthing, M. J., 2002   Tropical Malabsorption. Seminars in Gastrointestinal Disease 13:221–231.

Gould, S. J., 2003   The Hedgehog, the Fox, and the Magister's Pox: Mending and Minding the Misconceived Gap Between Science and the Humanities. London: Jonathan Cape.

Grauer, A. L., ed. 1995   Bodies of Evidence: Reconstructing History Through Skeletal Analysis. New York: Wiley-Liss.

Hall, B. S., 1996   The Didactic and the Elegant: Some Thoughts on Scientific and Technological Illustrations in the Middle Ages and Renaissance. In Picturing Knowledge: Historical and Philosophical Problems Concerning the Use of Art in Science. B. S. Baigrie, ed. pp.3–39. Toronto: University of Toronto Press.

Harper, K. N., Ocampo, P. S., Steiner, B. M., George, R. W., Silverman, M. S., Bolotin, S., Pillay, A., Saunders, N. J., and Armelagos, G. J., 2008   On the Origin of the Treponematoses: a Phylogenetic Approach. PLoS Neglected Tropical Diseases 2(1): e148.

Harris, R., 1986   The Origin of Writing. London: Duckworth.

Hays, J. N., 2007   Historians and Epidemics: Simple Questions, Complex Answers. *In* Plague and the End of Antiquity: The Pandemic of 541–750. L. K. Little, ed. pp.33–56. Cambridge: Cambridge University Press.

Herring A., Swedlund A. C., eds., 2003   Human Biologists in the Archives: Demography, Health, Nutrition and Genetics in Historical Populations. Cambridge: Cambridge University Press.

Humphreys, H. N., 1995   The Illuminated Books of the Middle Ages: an Account of the Development and Progress of the Art of Illumination. London: Bracken.

Jackson, M., ed., 2002   Infanticide: Historical Perspectives on Child Murder and Concealment, 1550–2000. Aldershot: Ashgate.

Jayawardene, S. A., 1982   Reference Books for the Historian of Science: a Hand List. London: Science Museum.

John of Joinville. 1874   Histoire de Saint Louis. N. de Wailly, ed. Paris: Librairie de Firmin Didot.

John of Joinville. 1955   The Life of Saint Louis. N. de Wailly, ed., and R. Hague, trans. London: Sheed and Ward.

Jones, P. M., 1998   Medieval Medicine in Illuminated Manuscripts. London: British Library.

Karenberg, A., 2009   Retrospective Diagnosis: Use and Abuse in Medical Historiography. Prague Medical Report 110:140–145.

Kloekhorst, A., 2008   Etymological Dictionary of the Hittite Inherited Lexicon. Leiden: Brill.

Koskenniemi, E. 2009   The Exposure of Infants Among Jews and Christians in Antiquity. Sheffield: Sheffield Phoenix.

Lo, V., and Cullen, C., eds., 2005   Medieval Chinese Medicine: the Dunhuang Medical Manuscripts. London: Routledge Curzon.

Lodge, T., 1603   A Treatise of the Plague: Containing the Nature, Signes, and Accidents of the Same. London: Edward White.

MacDermot, V., 1971   The Cult of the Seer in the Ancient Middle East: a Contribution to Current Research on Hallucinations Drawn from Coptic and Other Texts. London: Wellcome Institute of the History of Medicine.

MacDonald, M., 1983   Anthropological Perspectives on the History of Science and Medicine. *In* Information sources in the History of Science and Medicine. P. Corsi, and P. Weindling, eds. London: Butterworth.

Mackinney, L. C., 1965   Medical Illustrations in Medieval Manuscripts. London: Wellcome.

Magilton, J., Lee, A., and Boylston, A., eds. 2008   Lepers Outside the Gate: Excavations at the Cemetery of the Hospital of St James and St Mary Magdalene, Chichester, 1986–7 and 1993. York: Council for British Archaeology.

Maslen, M., and Mitchell, P. D., 2006   Medical Theories on the Cause of Death in Crucifixion. Journal of the Royal Society of Medicine 99:185–88.

Mays, S., and Faerman, M., 2001   Sex Identification in Some Putative Infanticide Victims from Roman Britain Using Ancient DNA. Journal of Archaeological Science 28:555–559.

Metcalfe, N. H., 2007   A Description of the Methods Used to Obtain Information on Ancient Disease and Medicine and of How the Evidence has Survived. Postgraduate Medical Journal 83:655–658.

Mitchell, P. D., 1999 The Integration of the Palaeopathology and Medical History of the Crusades. International Journal of Osteoarchaeology 9:333–343.

Mitchell, P. D., 2000   An Evaluation of the Leprosy of King Baldwin IV of Jerusalem in the Context of the Mediaeval World. *Appendix in* The Leper King and his Heirs: Baldwin IV and the Crusader Kingdom of Jerusalem. B. Hamilton, pp.245–258. Cambridge: Cambridge University Press.

Mitchell, P. D., 2002   The Myth of the Spread of Leprosy with the Crusades. *In* The Past and Present of Leprosy. C. Roberts, K. Manchester, and M. Lewis, eds. pp.175–181. Oxford: Archaeopress.

Mitchell, P. D., 2004   Medicine in the Crusades: Warfare, Wounds and the Medieval Surgeon. Cambridge: Cambridge University Press.

Mitchell, P. D., 2006   The Torture of Military Captives During the Crusades to the Medieval Middle East. *In* Noble Ideals and Bloody Realities: Warfare in the Middle Ages, 378–1492. N. Christie and M. Yazigi, eds. pp. 97–118. Leiden: E. J. Brill.

Mitchell, P. D., 2010   The Spread of Disease with the Crusades. *In* Between Text and Patient: The Medical Enterprise in Medieval and Early Modern Europe. B. Nance and E. F. Glaze, eds. pp. 309–330. Florence: Sismel.

Mitchell, P. D., in press (a)   The Use of Historical Texts for Investigating Disease in the Past. International Journal of Paleopathology.

Mitchell, P.D., ed. in press (b)   Anatomical Dissection in Enlightenment Britain and Beyond: Autopsy, Pathology and Display. Farnham: Ashgate.

Mitchell, P. D., Stern, E., and Tepper, Y., 2008   Dysentery in the Crusader Kingdom of Jerusalem: an ELISA Analysis of Two Medieval Latrines in the City of Acre (Israel). Journal of Archaeological Science 35:1849–1853.

Møller-Christensen,V., 1978   Leprosy Changes of the Skull. Odense: Odense University Press

Niermeyer, J. F., and Van De Kieft, C., 2002   Mediae Latinitatis Lexicon Minus. Leiden: E. J. Brill.

Parkinson, R. B., 1999   Cracking Codes: the Rosetta Stone and Decipherment. London: British Museum Press.

Patterson, J. T., 1998   How Do We Write the History of Disease? Health and History 1:5–29.

Pinhasi, R., and Mays, S. A., 2008   Advances in Human Palaeopathology. Chichester: John Wiley and Sons.

Rawcliffe, C., ed., 1995   Sources for the History of Medicine in Medieval England. Kalamazoo: Western Michigan University.

Roger of Howden, 1870   Chronica. W. Stubbs ed., Rolls series 51. London: Longmans, Green, Reader and Dyer.

Rosenberg, C. E., and Golden, J., eds. 1992   Framing Disease: Studies in Cultural History. New Brunswick: Rutgers University Press.

Saunders, C.J.G., 1961   The Bristol Eye Hospital, Founded 1810. Bristol: Board of Governors of the United Bristol Hospitals.

Savage-Smith, E., 2000   The Practice of Medicine in Islamic Lands: Myth and Reality. Social History of Medicine 13:307–321.

Scurlock, J., and Andersen, B. R., 2005   Diagnoses in Assyrian and Babylonian Medicine: Ancient Sources, Translations, and Modern Medical Analyses. Urbana: University of Illinois.

Siena, K. P., 2004   Venereal Disease, Hospitals and the Urban Poor: London's "Foul Wards", 1600–1800. Rochester: University of Rochester Press.

Skeyne, G., 1860   Ane Breve Descriptioun of the Pest, 1568, *In* Tracts by Dr Gilbert Skeyne, Medicinar to his Majesty. W. F. Skene, ed. Edinburgh: Bannatyne Club.

Sperling, L. C., 2001   Hair and Systemic Disease. Dermatology Clinics 19:711–26.

Sudhoff, K., and Singer, C., eds., 1925   The Earliest Printed Literature on Syphilis, Being Ten Tractates from the Years 1495–1498. Florence: Rhier and Co.

Wagner, T. G., and Mitchell, P. D., in press.   The Illnesses of King Richard and King Philippe on the Third Crusade: an Understanding of Leonardie and Arnaldia. Crusades.

Webster, C., 1983   The Historiography of Medicine. *In* Information Sources in the History of Science and Medicine. P. Corsi, and P. Weindling, eds. pp. 29–43. London: Butterworth.

Weindling, P., 1983   Research Methods and Sources. *In* Information Sources in the History of Science and Medicine. P. Corsi, and P. Weindling, eds. pp. 173–194. London: Butterworth.

William the Breton, 1878   De Gestis Philippi Augusti. M.-J. J. Brial, ed. Paris: Recueil des historiens des Gaules et de la France SS 17.

CHAPTER 18

Fundamentals of Paleoimaging Techniques: Bridging the Gap Between Physicists and Paleopathologists

Johann Wanek, Christina Papageorgopoulou, and Frank Rühli

INTRODUCTION

Paleopathology is one of the most challenging fields of skeletal biology today. It aims to reconstruct the history, evolution and geography of disease. It also aims to study the effect of diseases on human development and to investigate interactions between diseases, cultural practices and social status. Of special interest are the comparisons of prevalence of diseases between past and modern populations, as well as the adaptability of past populations to certain environments and the effects of environment on modern populations. The major contribution of this scientific field rests on the evolutionary perspective that it offers. Paleopathology has the potential to offer valuable information on our evolutionary process. However, reconstructing the history of disease is not easy and requires the employment of many methods for accurate analyses (Aufderheide 2009; Lynnerup 2007). For this reason, the field of paleopathology necessitates the

A Companion to Paleopathology, First Edition. Edited by Anne L. Grauer.
© 2012 John Wiley & Sons, Ltd. Published 2016 by John Wiley & Sons, Ltd.

interdisciplinary cooperation of physicians, and many specialists such as medical physicists, anthropologists, archaeologists and historians.

One of the golden standards in paleopathological investigation is the radiological examination of a specimen using both planar X-ray and more advanced methods. Radiological examination complements the macroscopic analysis, since it offers a quick, relatively nondestructive observation of the specimen (though X-rays can damage ancient DNA). Computed tomography (CT) represents the gold standard of X-ray based 3-D imaging systems. More sophisticated methods of analysis, such as nuclear magnetic resonance imaging (NMR) offer new diagnostic possibilities, both in modern medicine and in paleopathology.

There are many diseases affecting bone that can be recognized radiologically. These include degenerative diseases, trauma and fractures, congenital malformations, metabolic diseases, as well as infectious diseases such as tuberculosis and syphilis. In most cases, the radiological examination contributes significantly to the final diagnosis, especially when coupled with other histological and biomolecular methods of investigation. For mummified remains, CT and MRI examination may contribute to the identification of additional diseases present in soft tissues.

The application of imaging techniques is not without its theoretical problems. For instance, not all macroscopic bone changes are recognizable through imaging techniques. Subtle changes to the periosteal surfaces of bone may go unnoticed using planar X-ray imaging. Alternatively, not all conditions noted using imaging techniques will be visible to the naked eye. Diploic and/or cortical bone changes may go undetected with macroscopic investigation, but become evident when imaging techniques are employed. Hence, differential diagnosis ought to use as many approaches as possible in order to thoroughly evaluate the tissue under investigation. Putting this dictate into practice, however, is not easy. Limitations to the use of imaging technology include the availability of the instruments, the portability of the instruments, the technical expertise of the user and image analyst, the financial cost of the equipment, and the acquisition of permission allowing the specimen to undergo examination.

Lastly, in almost all instances, it is virtually impossible (financially and methodologically) to take images of every bone of every individual under investigation within a population. Hence, a substantial bias is created, whereby the frequency of conditions within individuals and between populations cannot be assessed. As an example, in the assessment of the presence and frequency of trauma within and between populations, one researcher may evaluate the frequency of fractures or trauma using macroscopic evaluation. This researcher would look for indications of callus formation and/or significant changes to the morphology of a bone. Another researcher, choosing to evaluate the same population, might have all the resources and permissions necessary to examine the population using imaging techniques. As bone has the capability to "repair" itself (that is, that calluses caused by fracture can become completely incorporated into the bone cortex), remnants of trauma might not be invisible to the researcher viewing bone macroscopically, but are clearly evident to the researcher using sophisticated imaging techniques. The results obtain by these two researchers, quite obviously, will differ significantly.

In spite of these issues, there has been a revolution in imagining techniques in recent years which has greatly improved the accuracy and the precision of diagnosis. Applying these new methods to paleopathological research is essential. This chapter aims to present an overview of current imaging techniques used in the medical field today, and to demonstrate their applicability to paleopathology. A brief history, as well as the basic principles of each method will be offered to provide fundamental knowledge. Advantages and disadvantages between the methods over other diagnostic methods will also be discussed.

TECHNICAL ASPECTS OF IMAGE QUALITY OF ANCIENT REMAINS

In general, image quality is influenced by several variables, such as the physical and biochemical properties of the sample and the X-ray source energy. It is also influenced by interaction processes like absorption and scattering of the X-rays. For instance, mummified tissue is often composed of microcracks and calcifications which tend to exacerbate X-ray scattering. Unfortunately, in these instances image contrast is decreased along with the ability to differentiate soft tissue layers with similar attenuation (see below for further discussion of attenuation).

An important image quality parameter is called contrast resolution (CR). This is the ability to detect small differences in the attenuation coefficient of adjacent structures (Farr and Allisy-Roberts 2004). Contrast resolution is a crucial parameter in the imaging of dehydrated tissue due to the decreased thickness of soft tissue layers. The contrast between a structure and its surroundings is only detectable if the foreground signal is about 3–5 times greater than the noise in the image (Farr and Allisy-Roberts 2004). The CR can be enhanced by maximizing the number of pixels (picture element) of small structures in the image, but optimal spatial and contrast resolutions are not achievable simultaneously, except by delivering an unacceptable dose of radiation to a living subject (Farr and Allisy-Roberts 2004). While the dose to ancient remains can theoretically be higher, it is wise to follow the ALARA (As Low As Reasonably Achievable) principle, as ancient DNA can be damaged by radiation.

Image quality and CR is also often compromised by quantum noise, such statistical variations in the number of X-ray photons in each pixel (Farr and Allisy-Roberts 2004). One trade-off to reduce noise is to increase the number of photons absorbed per voxel by increasing the slice thickness or the pixel size (a voxel, derived from "volume pixel" is the smallest three-dimensional component of a three-dimensional image or scan). However, such solutions will impair spatial resolution. Noise may also be reduced by increasing the exposure via the product of tube current and exposure time (mAs). This method, however, once again results in an increased dose of radiation to the subject.

Noisy images might be avoided by making the signal to noise ratio (SNR) as high as possible in order to distinguish small structures. For instance, increasing the number of photons (e.g. higher tube current-time product) leads to an increased SNR as the square root of the quantity of photons. In practice, the SNR of a radiographic exposure is about 300:1 (Farr and Allisy-Roberts 2000). Successful discrimination of different tissue types often requires a high contrast to noise ratio (CNR).

Adding to the complexity of image quality is the occasional presence of materials of high density and atomic number in the specimen. These materials might be funerary accessories or body adornments. For instance, amulets were used in funerary ritual in ancient Egypt and were made of high-density materials such as stone or fayence

(Budge 1925). Similarly, beads, weapons, and body adornments have been amongst the dead in many cultures through time and across the globe. Creating useful images for paleopathological investigation is therefore more complex than most researchers realize. With mummified remains and dry bone specimens varying greatly in their composition, creating successful images requires technological sophistication of the instruments and materials. It also requires substantial expertise and training on the part of the imaging technicians.

IONIZING IMAGING TECHNIQUES

Conventional x-ray technique

The discovery of X-rays by Roentgen in November 1895, led to the nondestructive investigation of mummified remains using planar X-ray imaging. Some of the earliest paleoradiological studies were carried out on human and cat mummies the following year (Koenig 1896). Later, the subjects included Egyptian and Peruvian mummies. Today, using planar X-ray imaging is common. Researchers view it as an essential component in the process towards determining the age-at-death of the subject (especially in mummified remains) and for differential diagnosis.

X-rays are highly energetic electromagnetic waves, which consist of photons (waves and particles without charge and mass) in the energy range from 1 keV to 250 keV. This is the range in the electromagnetic spectrum between ultraviolet light and gamma rays. Because X-rays have a wavelength which is much shorter, and a frequency much higher, than visible light they have great penetration power.

As X-rays travel through matter their intensity is attenuated; this key fact can be utilized to obtain an image. X-ray attenuation depends on different factors: material density and the atomic number (Z) of the tissue, and the energy of the X-ray photons. Image collection can be performed by different detectors, notably photographic film, which has been used since the discovery of X-rays. However, the use of film is characterized by poor absorption and low sensitivity to photons, so nowadays digital X-ray imaging systems are often preferred. Digital detectors offer a higher sensitivity due to increased photon absorption, and enable images to be immediately available for examination. The dilemma, however, is that when high-energy X-ray photons are used to image mummified material the photons often fail to be absorbed. This leads to poor visual discrimination between compact tissue layers.

Pros and cons of conventional x-ray imaging

Pros:

- nondestructive to tissue on macro and cellular level
- little or no sample preparation is required
- access to the machines is accessible and affordable to most researchers

Cons:

- limited contrast resolution
- resolution depends on the field of view (FOV)

- blurring of images and contrast reduction common
- limited soft tissue discrimination with high energy X-ray photons
- increased risk of aDNA destruction with low energy X-ray photons
- use of radiation

Computed tomography and dual energy computed tomography

Computed tomography (CT) also uses X-rays for imaging, but is more sophisticated than conventional X-ray imaging. Computed Tomography was founded in 1970, by Sir Godfrey Hounsfield (Hounsfield 1973) and Allan McLeod Cormack, and was used in neuroradiology. The first sophisticated multislice CT scanners (MSCT) were introduced in late 1998 (Goldman 2000), and delivered much more detailed information in comparison to X-ray images, such as cross-sections of organs, and the status of blood vessels, and allowed a better insight into pathogenesis of the body. In contrast to conventional X-ray images, CT has the potential to distinguish between smaller contrast differences, which increases the diagnostic value of this technique.

First-generation CT scanners were equipped with one X-ray tube and one detector, and during the examination of different slices of the body, the source and the detector moved over the patient. Second-generation CT scanners were equipped with one X-ray tube and an array of detectors. In the third generation, a bank of detectors rotated around the patient in opposition to the X-ray source. Now, in the fourth generation, there are over 1000 detectors and only one X-ray source rotating around the patient. CT image production is performed in three phases: during the scanning phase, data is collected; in the reconstruction phase the data are processed and a digital image is produced; finally the greyscale image is built up by a digital-to-analog conversion.

A more recent development in CT scanning, known as dual energy computed tomography (DECT), allows researchers to digitally capture the presence of different materials. This is particularly helpful in the analysis of mummified remains, as many types of materials (such as blood, bone, remnants of organs, and cloth) are often preserved (Figure 18.1).

The first paleopathological application of CT scanning was performed on September 27, 1976 at the Hospital for Sick Children in Toronto, on the preserved and desiccated brain of Nakht, a 14-year-old weaver who died 3.200 years ago in Egypt (Lewin and Harwood-Nash 1977). The results were so promising that further and more extensive images of Egyptian mummies were conducted shortly afterward (Harwood-Nash 1979). The new technique allowed the detection of amulets and diseases of the head and neck region without unwrapping the body. Hoffman and Hudgins (2002) captured images of desiccated brain in female Egyptian mummies from which the intracranial contents had not been removed. CT scans of the skull have also contributed to our understanding of Egyptian mummification procedures, such as the identification of the loss of the ethmoid and sphenoid sinuses caused by excerebration (removal of the brain) (Chan et al. 2008). In addition, CT has been able to differentiate cerebral gray and white matter in high-altitude Argentine mummies (Previgliano 2003). CT has also influenced the investigation of infectious diseases such as leprosy and tuberculosis in skeletal remains. Haas has demonstrated the pathological symptoms

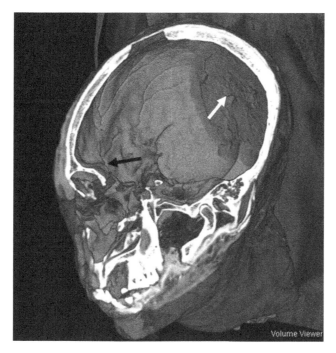

Figure 18.1   Dual-energy computed tomography of a mummy head. Grooves for the middle meningeal artery are visible (black arrow) as well as resin-filled areas in the posterior skull base (white arrow). The DECT scan (SOMATOM Definition Dual Source, Siemens, Germany) illustrates a three-dimensional (3-D) volume rendering of the mummy head (tube voltage 100 kVp, tube current 120 mAs, spatial resolution 0.4 mm, matrix size 512 × 512).

"facies leprosa" of the skull using CT imaging by recognizing extended osteolysis in the mandible and reactive new bone formation (Haas 2000).

Pros and cons of CT imaging

Pros:

- non-destructive 3-D imaging
- little or no sample preparation is required
- extraction of sub-voxel-level details
- large field of view (FOV)
- more sophisticated CTs mean reduced-dose scans

Cons:

- limited contrast resolution (no contrast agents available)
- resolution depends on the field of view (FOV)
- blurring of images and contrast reduction common
- limited soft tissue discrimination of compacted tissue layers
- use of radiation

Micro-computed tomography

While the spatial resolution of clinical CT imaging typically reaches 0.5 mm, X-ray micro-computed tomography (micro-CT), developed in the early 1980s, is able to deliver a spatial resolution 10 to 100 times higher, in the range between 5 and 50 μm (Feldkamp et al. 1989; Kuhn et al. 1990). Such imaging systems have a small field of view (FOV) of about 30 mm, which leads to a small sample size. As with normal CT, two-dimensional X-ray images are integrated to generate a three-dimensional image by using a large data set. Therefore, micro-CT is best suited for applications where small tissue samples are available and higher resolution is required. This technique is also suited for the visualization of the microarchitecture of mummified remains such as tissue or bones (McErlain et al. 2004). Micro-CT can also be used for monitoring density changes over time, such as those occurring during artificial mummification.

Pros- and cons of micro-CT imaging

Pros:

- production of 3D images
- high resolution (5–50 μm)
- investigation of internal structure of mummified samples is possible
- no specific sample preparation needed

Cons:

- use of radiation
- differentiation of dehydrated tissue difficult
- limited sample size (small FOV)

NON-IONIZING IMAGING TECHNIQUES

Magnetic resonance imaging

Magnetic resonance imaging (MRI) is a non-ionizing and noninvasive technique which is widely used in medical diagnoses. It provides a safer environment than X-ray based techniques, and is commonly used to image radiosensitive organs and for soft tissue discrimination. In1983, MRI was attempted on an Egyptian mummy, but proved unsuccessful despite the diversity of pulse sequences and body coils (Notman et al. 1998). Later, however, Piepenbrink et al. (1986) successfully used the technology to image ancient dry soft tissue, but only after invasive rehydration was employed, which had the unfortunate effect of altering the morphology of the sample (Piepenbrink et al. 1986; Sebes et al. 1991). In 2007, Rühli et al. demonstrated that clinical MRI could be applied to visualize historic dry tissue without rehydration using ultrashort echo time pulse sequences (UTE) (Rühli et al. 2007a). To date, MRI of mummified remains continues to suffer from low resolution and weak signal to noise ratio (SNR).

Magnetic resonance imaging uses radio waves and magnetic fields to create an image of a patient, who has been placed in a strong magnetic field. Elementary particles and nuclei with an odd number of protons possess a spin, such as a hydrogen atom ^{1}H in fat or water of a human body. This also applies to other biologically important nuclei, e.g. ^{13}C, ^{14}N, ^{19}F, ^{23}Na and ^{31}P. In 1924, Wolfgang Pauli (1900–1958), an Austrian physicist, discovered the spin and the magnetic moment of atomic nuclei. Protons are randomly orientated and spin about their own axis. Such nuclei produce a magnetic moment like rotating bar magnets. Hydrogen protons will align with the field in one direction or the other in the presence of an external magnetic field.

An important parameter defining the image quality of MRI is the signal-to-noise ratio (SNR). The SNR is defined as the ratio of the receiving signal amplitude from the coil to the amplitude of the noise from the background. Consequently, a high SNR is desirable in clinical practice. The low water content in mummified remains results in a small SNR. However, the SNR can be increased by altering many parameters, such as the field of view, slice thickness, magnetic field strength and the number of acquisitions. Alternatively, a small number of data points in 2-D, called matrix size, helps to improve the SNR. In addition, transmit and receive coils can be small and positioned close to the regions of interest. The scan parameters, such as the repetition time (TR), echo time (TE), or the flip angle, are normally optimized by performing several pre-measurements. Tissues with a very short transverse relaxation time (T2), such as mummified muscles or bone, cannot be detected using conventional sequences because of the rapid decay of the net magnetization M_z (magnetization vector in the z-direction). Therefore ultra-short echo time (UTE) sequences are applied. Bock et al. (2008) provide example parameters which meet the requirements of MRI of dehydrated tissue. These include a field of view (320 × 320), slice thickness (1.25 mm), matrix size (128 × 128), magnetic field strength (3 Tesla), and a flip angle of 14°. Note that doubling the number of acquisitions (averaging) will increase the SNR by square root of two. According to the example parameters, the resolution is given by the FOV/Matrix size = 2.5 mm.

Pros- and cons of MR-imaging in the study of dehydrated tissue
Pros

- Non-invasive imaging
- Non-ionizing imaging
- No radiation and hence safer working environment
- Can be used without rehydration of mummified remains

Cons

- Less likely to be available to researchers and more expensive
- Low signal to noise ratio (SNR) in hydrogen MRI
- Low spatial resolution

- Moderate soft tissue contrast
- No contrast agents useable

Nuclear Magnetic Resonance Imaging

Nuclear Magnetic Resonance (NMR) was discovered in 1946 by Felix Bloch at Stanford University and Edward Purcell at Harvard University. They were later awarded the Nobel Prize in 1952 (Andrew 1983).

In general, imaging of dry tissue using hydrogen MRI suffers from the low content of protons (^{1}H) within the material. Consequently, MRI has not been applied successfully to visualize ancient remains in a diagnostic image quality. For this reason, researchers have been asking for alternative nuclei, which possess an adequate magnetic moment for nuclear magnetic resonance imaging (NMR), such as ^{23}Na or ^{13}C. Previous studies have indicated that the impact of artificial mummification as practiced in ancient Egypt had never been addressed. Egyptian embalmers used natron, a mixture of sodium carbonate and sodium bicarbonate, for mummification. Münnemann investigated the selective enrichment of sodium in the bone and demonstrated the feasibility of nonclinical MRI for visualizing historic dry human tissue (Münnemann et al. 2007). In addition, such investigation potentially shows the spatial distribution of hydrogen (^{1}H) and sodium (^{23}Na) in mummified tissue.

Another interesting way to discriminate ancient tissue was demonstrated by Rühli and colleagues (2007b). This study used the spectroscopic properties of NMR, as demonstrated in structural and conformational analysis in chemistry, biology, medicine, and material science (Rühli et al. 2007b). A representative spectrum of historic natural and artificial mummies with dry and frozen mummified tissue and bone were measured using the NMR-Mouse® (Mobile Universal Surface Explorer). The NMR-Mouse is an open and portable sensor with a receiving coil to detect NMR signals external to the magnet (Eidmann et al. 1996).

The investigations demonstrated that different anatomical layers between soft-tissue and bone and even within bone may be distinguished by superimposing various depth profiles (Rühli et al. 2007a and b).

Similar to magnetic resonance (MR), NMR is based on the spinning phenomena of nuclei in the presence of a static magnetic field. Note that sodium MR imaging has a lower relative sensitivity for an equal number of nuclei and identical magnetic field than hydrogen imaging. An important aspect of NMR, however, is the possibility of performing spectroscopic analysis (Lynnerup 2007) in order define the number and types of atoms in the molecule.

Pros and Cons of NMR
Pros:

- No ionizing radiation
- High resolution
- Non-invasive
- Spectroscopic analysis possible

Cons:

- Expensive
- Time consuming
- Low SNR for nuclei ^{13}C or ^{23}Na and small FOV

CONFOCAL LASER SCANNING MICROSCOPY (CLSM)

Although CLSM is not a radiological technique and requires histological treatment of a sample, we have chosen to discuss it in a broader sense, as it is a digital imaging method capable of producing electronic images and offering high quality 3-D tissue visualization.

The basic concept of confocal microscopy was developed by Marvin Minsky in the mid-1950s (Minsky 1988). However the lack of intense light sources and computer technology did not allowed progress of the technique until the late 1980s, when the necessary advances in computer and laser technology took place. While the first commercial instruments appeared in 1987, modern confocal microscopes have been in use for more than a decade. These microscopes are being employed for routine investigations on molecules, cells, and living tissues, and most recently on human bones mummified tissue of archaeological origin (Maggiano et al. 2010; Papageorgopoulou et al. 2010).

The key feature of confocal microscopy is its ability to acquire in-focus images from selected depths, a process known as optical sectioning. Images are acquired point-by-point and reconstructed digitally with a computer. This is achieved with the use of spatial-filtering techniques with pinholes which eliminate any out-of-focus light. In a conventional wide-field optical epifluorescence microscope, secondary fluorescence emitted by the specimen often occurs through the excited volume and obscures resolution of features that lay in the objective focal plane. This is especially problematic in thicker specimens (greater than 100 micrometers), where the phenomenon is more intense. However, higher z-resolution and reduced out-of-focus blur make confocal pictures crisper and clearer. The number of optical sections defines the z-resolution in the data set. The section thickness together with the x–y-pixel dimension defines the "voxel" size (Ziegler et al. 2010). CLSM predominantly employs fluorescence, but it is also possible to use the transmission mode with conventional contrasting methods, as well as the reflection mode with polarization methods.

Pros and cons of CLSM

Confocal microscopy offers several advantages over conventional wide-field microscopy, most important of which are the production of blur-free images of high quality and the capability to collect serial optical sections from thick specimens (up to 100 μm). Three-dimensional structures in ultracellular levels may be reconstructed using image-processing software. As a result, image parameters such as length, diameter, perimeter, surface and the volume of structures, such as Haversian channels or the osteocyte lacunae, are calculable. The preparation protocols for CLSM are based upon protocols developed for conventional wide-field microscopy and specifically for undecalcified bone histology preparation methods (Maat et al. 2001; Schultz 2001; Papageorgopoulou et al. 2010). CLSM is optimal for manually prepared ground sections (Maat et al. 2001). An advantage of CLSM versus microcomputed tomography (micro-CT), another established method for 3-D analysis of bone, is that CLSM provides an even higher resolution. While conventional desktop micro-CT can visualize cortical and trabecular bone, as well as the parts of the cannular network, it is not sufficient to capture osteocyte lacunae. On the other hand, CLSM is an invasive method and some destruction of the sample is necessary. Additionally, the staining and embedding processes necessitate time for preparation. Finally, the high cost of purchasing and operating a CLSM, in comparison to wide-field microscopes, limits their implementation in smaller laboratories.

The applications of CLSM in paleopathology are many, and have been demonstrated recently by Šefčáková et al. (2001), Maggiano et al. (2006, 2010), and Papageorgopoulou et al. (2010). Papageorgopoulou et al. (2010) for instance, have demonstrated the applicability of CLSM imaging in both archaeological and modern macerated, cremated, and fresh human bone tissue. In a study of bone microdamage, CLSM proved of great value. It was possible to differentiate pre-existing microdamage sustained in vivo from microdamage acquired through the processing of the sample. Maggiano et al. (2010) have demonstrated the advantages of CLSM compared to standard epifluorescence and polarized-light enhanced microscopy, especially in the detection of tetracycline in archaeological bones.

Terahertz Imaging (THz)

With the implementation of novel laser sources, femtosecond laser sources in particular, the THz time domain approach was established. THz-imaging is a non-invasive and non-ionizing diagnostic modality. Due to the larger wavelengths, terahertz radiation provides a lower spatial resolution than CT. The THz time domain approach is based on short, broadband THz pulses which are focused on to the sample and which are detected after transmission. By using a time-delay, the time-dependent electric field is detected. Evaluation of the detected signal provides the frequency-dependent absorption and optical density of the investigated sample. Since water is a strong absorbing media in the THz frequency range (100 GHz–5 THz), THz-imaging allows only close-to-surface tissue differentiation (Fitzgerald et al. 2003) for modern human tissue. However, the extremely low water content in mummified tissue renders THz investigation very promising (see Table 18.1).

Terahertz imaging was successfully performed on various samples, such as a fish mummy, a complete mummified hand, and macerated bone samples (Öhrström et al. 2010). While spatial resolution is strongly limited by diffraction processes and the used wavelength (rendering soft tissue differentiation using low-energy photons impossible), bone and cartilaginous structures can be differentiated from surrounding soft tissues (Öhrström et al. 2010). The resolution was sufficient to identify major anatomical landmarks of the fish mummy and human metacarpal bones. THz imaging also allows simultaneous acquisition of images from the same object at different energies. Such complementary information of the specimen is usable for investigations of structures with different sizes. Hence, THz has great potential for the detection of

Table 18.1 Strengths and weaknesses of THz-imaging of normal and dehydrated tissue

Normal tissue	Dehydrated tissue
Expensive due to the coherent laser source ↓	Expensive due to the coherent laser source ↓
No physical effect due to low energy of about 4 meV ↑	No physical effect ↑
Good spatial resolution near the target resolution (≈ 300 μm)↑	Major morphological structures are identifiable (spatial resolution about 1 mm at 0.3 THz) → Limited signal strength degrades image quality for thicker specimen ↓
Identifying of chemical substances ↑	Identifying of chemical substances ↑
Non ionizing imaging ↑	Non ionizing imaging ↑

↑Good   ↓Poor   →Moderate

Table 18.2 Comparison between CLSM, wide-field microscopy and micro-CT

Imaging aspects	CLSM	Wide-field microscopy	Micro-CT
Sample preparation	Yes	Yes	No (size limitation)
Thickness of specimen	100 μm max.	Small	≈ 20 mm
Image quality [a]	High SNR	Low SNR	High SNR
Dimension of image data	3-D	2-D	3-D
Spatial resolution	≈ 0.5 μm	2–3 μm	5–50 μm

3-D = Three dimensional
SNR = Signal to noise ratio
[a] The signal to noise ratio (SNR) depends on the number of detected photons [N] with $SNR = \sqrt{N_{photon}}$.

hidden objects in historic samples, such as funerary amulets wrapped within mummies, or for the identification of spectral signatures from chemical substances such as embalming fluids (Öhrström et al. 2010).

CONCLUSION

The use of imaging techniques in paleopathology has a relatively long history. Used as noninvasive means to study delicate human remains, X-ray-based imaging such as planar and CT and micro-CT, have offered researchers outstanding diagnostic tools towards the examination of mummified and skeletal remains. Regardless of the technological strides made in imaging techniques over the decades, each method offers benefits and limitations. Choosing which technique to employ is not simple. It requires the assessment and balancing of a great number of variables, ranging from the need to preserve the material under investigation, to the needed quality of the images, to the cost, availability, and needed expertise to use the machines. As a means to assist researchers, information concerning the differences between imaging techniques (see Table 18.2) and the potential for use in paleopathological investigation (see Table 18.3) are offered.

Table 18.3   Imaging modality and its benefit in paleopathology

Modality	Field of view	Approximate spatial resolution	Ionizing radiation	Dry bone	Dry soft tissue	Hair	Teeth
X-Ray	Whole body	~ 0.4 mm	Yes	↑/→	↑	↓	↑
Computed Tomography	Whole body	~ 0.4 mm	Yes	↑/→	↑	↓	↑
Magnetic Resonance Imaging	Whole body	1.5–2.5 mm[b]	No	→	→	↓	↓
Terahertz Imaging	Whole body[a]	~ 1mm at 0.3 THz	No	→	→	↓	→
Micro-Computed Tomography	~ 30 mm	10–50 μm	Yes	↑/→	↑	→	↑
Confocal Laser Scanning Microscopy	0.1–0.6 mm	~ 0.5 μm	No	↑	↑	↑	→[c]

[a] Varies depending on the setup geometry
[b] Dry ancient tissue
[c] Analysing of caries, mineralization/demineralization (Marzuki 2008)
↑ Good   ↓ Poor   → Moderate

REFERENCES

Andrew, E. R., 1983   Perspectives in NMR Imaging. *In* Nuclear Magnetic Resonance (NMR) Imaging. C. L. Partain, A. E. James, F. D. Rollo, and R. R. Price, eds. pp. 7–13. Philadelphia: W. B. Saunders.

Aufderheide, A. C., 2009   Reflections About Bizarre Mummification Practices On Mummies at Egypt's Dakhleh Oasis: A Review, Anthropologischer Anzeiger 67(4):385–390.

Bock, M., Speier, P., Nielles-Vallespin, S., Szimtenings, M., Leotta, K., and Rühli, F., 2008   MRI of an Egyptian Mummy on Clinical 1.5 and 3 T Whole Body Imagers. Proceedings of the 16th Scientific Meeting of the International Society for Magnetic Resonance in Medicine.

Budge, E. A. W., 1925   The Mummy: A Handbook of Egyptian Funerary Archaeology. 2nd Edition. Cambridge: Cambridge University Press.

Chan, S., Elias, J., Hysell, M., and Hallowell, M., 2008   CT of a Ptolemaic Period Mummy from the Ancient Egyptian City of Akhmim. RadioGraphics 28:2023–2032.

Dendy, P. P., and Heaton, B., 2000   Physics For Diagnostic Radiology. 2nd Edition. Bristol: Institute of Physics Publishing.

Eidmann, G., Savelsberg, R., Blumler, P., and Blümich, B., 1996   The NMR MOUSE: A Mobile Universal Surface Explorer. Journal of Magnetic Resonance Series A 122:104–109

Farr, R. F., and Allisy-Roberts, P. J., 2004   Physics for Medical Imaging. London:Elsevier.

Feldkamp, L. A., Goldstein, S. A., Parfitt, A. M., Jesion, G., and Kleerekoper, M., 1989   The Direct Examination of Three-Dimensional Bone Architecture *in Vitro* By Computed Tomography. Journal of Bone and Mineral Research 4(1):3–11.

Fitzgerald, A. J., Berry, E., Zinov'ev, N. N., Homer-Vanniasinkam, S., Miles, R. E., Chamberlain, J. M., and Smith, M. A., 2003   Catalogue of Human Tissue Optical Properties at Terahertz Frequencies. Journal of Biological Physics 29:123–128.

Goldman, L., 2000   Principles of CT and Evolution of CT Technology: CT and US Cross-Sectional Imaging. *In* Categorical Course in Diagnostic Radiology Physics. L. W. Goldman, and J. B. Fowlkes, eds. pp. 33–52. Oak Brook, IL: Radiological Society of North America.

Haas, C. J., Zink, A., Pálfim, G., Szeimies, U., and Nerlich, A. G., 2000   Detection of Leprosy in Ancient Human Skeletal Remains by Molecular Identification of Mycobacterium leprae. Amerian Journal of Clinical Pathology 114:428–436.

Hoffman, H., and Hudgins, P. A., 2002   Head and Skull Base Features of Nine Egyptian Mummies: Evaluation With High-Resolution CT and Reformation Techniques. American Journal of Roentgenology 178:1367–1376.

Hounsfield, G. N., 1973   Computerized Transverse Axial Scanning (Tomography), Part 1, Description of System. British Journal of Radiology 46:1016–1022.

Koenig, W., 1896   14 Photographien von Röntgenstrahlen Aufgenommen im Physikalischen Verein zu Frankfurt a. M. Leipzig: Johann Ambrosius Barth.

Kuhn, J. L., Goldstein, S. A., Feldkamp, L. A., Goulet, R. W., and Jesion, G., 1990   Evaluation Of a Microcomputed Tomography System to Study Trabecular Bone Structure. Journal of Orthopaedic Research 8(6):833–842.

Lewin, P. K., and Harwood-Nash, D. C., 1977   X-Ray Computed Axial Tomography of an Ancient Egyptian Brain. IRCS Medical Sciences 5:78.

Lynnerup, N., 2007   Mummies. Yearbook of Physical Anthropology 45:162–190.

Maat, G. J. R., Van Den Bos, R. P. M., and Aarents, M. J., 2001   Manual Preparation of Ground Sections for the Microscopy of Natural Bone Tissue: Update and Modification of Frost's Rapid Manual Method. International Journal of Osteoarchaeology 11:366–374.

Maggiano, C., Dupras, T., Schultz, M., and Biggerstaff, J., 2006   Spectral Photobleaching Analysis Using Confocal Laser Scanning Microscopy: A Comparison of Modern Archaeological Bone Fluorescence. Molecular Cellular Probes 20:154–162.

Maggiano, C., Dupras, T., Schultz, M., and Biggerstaff, J., 2009   Confocal Laser Scanning Microscopy: A Flexible Tool For Simultaneous Polarization and Three-Dimensional Fluorescence Imaging of Archaeological Compact Bone. Journal of Archaeological Sciences 36(10):2392–2401.

Marzuki, A. F., 2008   Confocal Laser Scanning Microscopy Study of Dentinal Tubules in Dental Caries Stained With Alizarin Red. Archives of Orofacial Sciences 3(1):2–6.

McErlain, D. D., Chhem, R. K., Bohay, R. N., and Holdsworth, D. W., 2004   Micro-Computed Tomography of a 500 Year-Old Tooth: Technical Note. Canadian Association of Radiologists Journal 55(4):242–245.

Minsky, M., 1988   Memoir On Inventing the Confocal Scanning Microscope. Scanning 10:128–138.

Münnemann, K., Böni, T., Colacicco, G., Blümich, B., and Rühli, F., 2007 ^{1}H and ^{23}Na Nuclear Magnetic Resonance Imaging of Ancient Egyptian Human Mummified Tissue. Magnetic Resonance Imaging 25:1341–1345.

Notman, D. N. H., 1983   Nuclear Magnetic Resonance Imaging of an Egyptian Mummy. Paleopathology Newsletter 43(8):257–263.

Öhrström, L., Bitzer, A., Walther, M., and Rühli, F., 2010   Technical Note: Terahertz Imaging of Ancient Mummies and Bone, American Journal of Physical Anthropology 142(3):497–500.

Papageorgopoulou, C., Kuhn, G., Ziegler, U., and Rühli, F., 2010   Diagnostic Morphometric Applicability of Confocal Laser Scanning Microscopy in Osteoarchaeology. International Journal of Osteoarchaeology 20(6):708–718.

Piepenbrink, H., Frahm, J., Haase, A., et al., 1986   Nuclear magnetic resonance imaging of mummified corpses. American Journal of Physical Anthropology 70:27–28.

Previgliano, C. H., Ceruti, C., Reinhard, J., Araoz, F. A., and Diez, J. G., 2003   Radiologic Evaluation of the Llullaillaco Mummies. American Journal of Roentgenology 181:1473–1479.

Rühli, F. J., Von Waldburg, H., Nielles-Vallespin, S., Böni, T., and Speier, P., 2007a   Clinical Magnetic Resonance Imaging of Ancient Dry Human Mummies Without Rehydration. Journal of the American Medical Association 298(22):2618–2620.

Rühli, F. J., Böni, T., Perlo, J., Casanova, F., Baias, M., Egarter, E., and Blümich, B., 2007b Non-Invasive Spatial Tissue Discrimination in Ancient Mummies and Bones in Situ by Portable Nuclear Magnetic Resonance. Journal of Cultural Heritage 8(3):257–263.

Schultz, M., 2001 Paleohistopathology of Bone: A New Approach to the Study of Ancient Diseases. Yearbook of Physical Anthropology 44:106–147.

Sebes, J. I., Langston, J. W., Gavant, M. L., and Rothschild, B. M., 1991 Magnetic Resonance Imaging of Growth Recovery Lines in Fossil Vertebrae. American Journal of Roentgenology 157:415–416.

Šefčáková, A., Strouhal, E., Němečková, A., Thurzo, M., and Staššiková-Štukovská, D., 2001 Case of Metastatic Carcinoma From End of the 8th–early 9th Century Slovakia. American Journal of Physical Anthropology 116:216–229.

Ziegler, U., Greet Bittermann, A., and Hoechli M., 2010 Introduction to Confocal Laser Scanning Microscopy (LEICA). http://www.zmb.uzh.ch/resources/download/CLSM. pdf. Accessed on June 2, 2011.

CHAPTER **19** # Data and Data Analysis Issues in Paleopathology

Ann L.W. Stodder

This chapter addresses some fundamental issues we face in conducting paleopathology research: how we record data, how we interact with and analyze data, and how we store data and make it accessible and comprehensible to other users. The intent here is bring forth some of the challenges and problems specific to paleopathology data that aren't covered in reference books or manuals, and in truth, are infrequently discussed publicly amongst professionals and students alike.

In recent years, data collection on human remains has been endowed with urgency due to repatriation initiatives worldwide (see Lambert, Chapter 2 in this volume, for discussion of repatriation). However, even when assemblages are not slated for reburial, our access to collections can be limited by policy, budgetary restraints, and time. In addition, we increasingly recognize that repeated handling of human skeletal material may jeopardize preservation, especially if the material is particularly fragile. Consequently it is clear that the permanent resource available to paleopathologists is, in fact, the data we generate.

Individuals and institutions have generated data recording protocols for decades, creating a genealogy of sorts since the information is passed between teachers and students, and between researchers in various academic, commercial, and museum contexts. Interestingly, these lineages reflect the evolution of data collection and recording, chronicling the change from handwritten notes on cards, to the creation and use of paper forms, to the development of direct online data entry. Clearly, the types of data we generate (nominal, ordinal, interval) and the values or codes we assign (numeric, alphanumeric, text), as well as the manner in which data are recorded, stored and accessed (spreadsheets, relational databases, paper and electronic archives),

A Companion to Paleopathology, First Edition. Edited by Anne L. Grauer.
© 2012 John Wiley & Sons, Ltd. Published 2016 by John Wiley & Sons, Ltd.

all impact the manner in which we interact with and learn from a body of data. There is no "fits-all-cases" approach to recording paleopathological data, but one principal applies to all research: data collection and management need to be designed with the entire *process* of research in mind, including data collection, data analysis and reporting (Grauer 2008).

The goals of data collection in support of paleopathology research are explicitly tied to the desire to systematically and unambiguously document pathological changes in the skeleton. They are also tied to conservative principals of differential diagnosis and strong inference, along with efforts to quantify the prevalence and frequency of pathological conditions in past populations. There are, however, important (and persistent) issues that we face in trying to meet these goals. These include the acceptance of anecdotal recording procedures, the difficulties surrounding the development of detailed bone inventories, and the development of clear descriptive terminology alongside the use of visual illustrations of observed conditions. In addition, the electronic environment for data entry, storage, and analysis, while affording many opportunities, poses new problems worthy of discussion.

Along with data collection, an equally challenging but rarely addressed component of paleopathology research is data analysis. While students are often trained in the principals of statistics and in the use of statistical software packages, the importance of *exploratory data analysis*, as a stage distinct from significance testing, is often ignored. In order to understand the behavior of a dataset, and as a means of recognizing new patterns and relationships among variables, exploratory data analysis (EDA) is critical to the development of our field. Yet, ironically, it is not a typical component of training researchers.

SYSTEMATIC DATA COLLECTION IN PALEOPATHOLOGY

What is systematic data collection?

Systematic data collection is a premise that a researcher was both thorough and consistent in recording data. Today, this might include looking for the same bony change in every skeleton in the assemblage, recording its presence or absence, and indicating whether the presence or absence could not be determined because of poor preservation or missing elements. This contrasts with what is usually referred to as the "anecdotal approach" where one records the presence of a condition when it is observed, but not its absence on bone elements or instances where the bone element is too damaged to assess. The distinction between these two approaches is critical to paleopathology. Was a particular pathological condition truly absent in the population because the disease and/or condition did not impact any member of the group? Or does the lesion appear to be absent in the population because the skeletal element upon which the lesion might be found was not recovered during the original excavation? Trauma to hand or feet phalanges, for instance, is not commonly discussed in the paleopathological literature. Is this because, after looking at all the phalanges recovered from archaeological sites and assessed in mummified remains that the condition is rare? Or might it be due to the fact that recovering and curating phalanges is particularly difficult due their small size and relatively fragile nature?

It might appear that systematic recording is an obvious component of data collection, or that a systematic method is more easily devised for some kinds of pathological conditions than others. Both assumptions are false. The presence of dental caries serves as an example. Here, one could argue, the recording of quantitative data could be easily developed: the lesion is either present or it isn't. But recording dental caries has proven to be far more problematic over the years. First, like the recording of any other pathological condition, decisions must be made *a priori*. Will recording include determining the presence of caries per individual, per tooth, or per tooth surface? And what other variables might be important to record in order to impart meaning to the data when analysis begins? Indeed, the evolutionary path and pitfalls of systematic recording become evident when we take a close look at the recording of caries within paleopathology over time.

Rose and Burke's review (2006) of the history of dental anthropology and its influence on bioarchaeology asserts that the earliest systematic recording of a pathological condition was Broca's 1879 five-level scale of dental wear, which provided more detail than Mummery's 1870 "qualitative" scoring (Rose and Burke 2006:329). Later, in an article published in the eighth volume of the American Journal of Physical Anthropology, Leigh (1925) compared dental pathology in four archaeological assemblages, informing the reader that "A simple form providing an appropriate notation for the thirty-two teeth in the series was utilized; ample space opposite a number representing each tooth, in connection with a short symbol for each pathological process, enabled accurate and rapid recording. The sex, and the age, estimated to within a decade, were also noted" (1925:181–182). Five years later, Hooton (1930) publishing on the Pecos Pueblo skeletal material, employed systematic recording techniques in an era when reporting "was typically nonsystematic, even anecdotal…" (Cook and Powell 2006:297). Dental pathology data, in particular, were more systematically recorded, although it seems that quite early in the several years long analysis of the Pecos remains Hooton modified the way he recorded dental observations. For instance, Figure 19.1 shows a Peabody Museum Cranial Observations card with data from a skeleton excavated in the 1919 season, early in the Andover Pecos Expedition. The number of teeth with caries (from 1 to 17 or more) was scored as grouped data, as were other cranial and dental conditions such as antemortem tooth loss, tooth wear, and dental morphology.

While the development of these cards moved paleopathological data collection many steps forward in terms of systematic recording, choices made by Hooton directly affected the types of questions that could be posed to and answered by analysis of the data. For instance, without knowledge of which teeth display caries, or which teeth were lost antemortem, an understanding of dental decay in the population is limited. Did individual number 60168 have 1–4 caries on a single tooth? Which two teeth were lost postmortem? Were they molars? If so, with molars being particularly susceptible to caries, the assessment of presence of caries in this individual (and population) is compromised. Current paleopathology literature stipulates that caries development is related to age, sex, upper or lower tooth position and that and that these complexities require separate and careful statistical evaluation (Hillson 2000). Waldron states that since molars preserve better and are also more susceptible to caries, calculating the percentage of teeth with caries is not a useful statistic; the clinically interesting data are to know "what proportion of individuals had dental

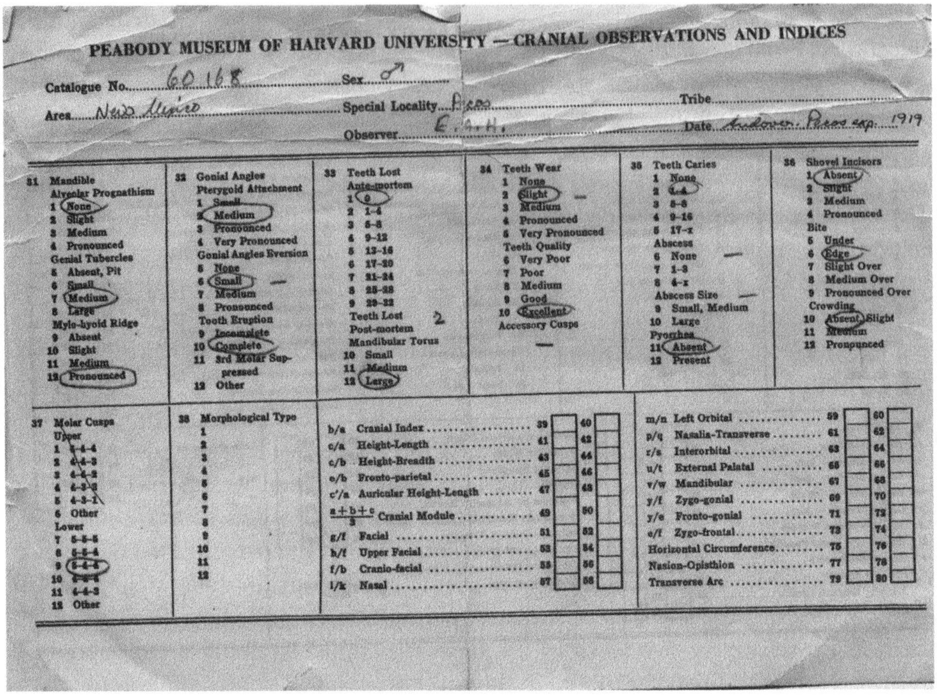
PEABODY MUSEUM OF HARVARD UNIVERSITY — CRANIAL OBSERVATIONS AND INDICES

Catalogue No. 60168 Sex ♂
Area New Mexico Special Locality Pecos Tribe
Observer E.A.H. Date Andover. Pecos exp. 1919

31 Mandible
Alveolar Prognathism
1 None
2 Slight
3 Medium
4 Pronounced
Genial Tubercles
5 Absent, Pit
6 Small
7 Medium
8 Large
Mylo-hyoid Ridge
9 Absent
10 Slight
11 Medium
12 Pronounced

32 Gonial Angles
Pterygoid Attachment
1 Small
2 Medium
3 Pronounced
4 Very Pronounced
Gonial Angles Eversion
5 None
6 Small
7 Medium
8 Pronounced
Tooth Eruption
9 Incomplete
10 Complete
11 3rd Molar Suppressed
12 Other

33 Teeth Lost
Ante-mortem
1 0
2 1–4
3 5–8
4 9–12
5 13–16
6 17–20
7 21–24
8 25–28
9 29–32
Teeth Lost Post-mortem 2
Mandibular Torus
10 Small
11 Medium
12 Large

34 Teeth Wear
1 None
2 Slight
3 Medium
4 Pronounced
5 Very Pronounced
Teeth Quality
6 Very Poor
7 Poor
8 Medium
9 Good
10 Excellent
Accessory Cusps

35 Teeth Caries
1 None
2 1–4
3 5–8
4 9–16
5 17–x
Abscess
6 None
7 1–3
8 4–x
Abscess Size
9 Small, Medium
10 Large
Pyorrhea
11 Absent
12 Present

36 Shovel Incisors
1 Absent
2 Slight
3 Medium
4 Pronounced
Bite
5 Under
6 Edge
7 Slight Over
8 Medium Over
9 Pronounced Over
Crowding
10 Absent, Slight
11 Medium
12 Pronounced

37 Molar Cusps
Upper
1 5-4-4
2 4-4-3
3 4-4-2
4 4-3-3
5 4-3-1
6 Other
Lower
7 5-5-5
8 5-5-4
9 5-4-4
10 4-4-4
11 4-4-3
12 Other

38 Morphological Type
1
2
3
4
5
6
7
8
9
10
11
12

b/a Cranial Index 39 40
c/a Height-Length 41 42
c/b Height-Breadth 43 44
e/b Fronto-parietal 45 46
c′/a Auricular Height-Length 47 48
$\frac{a+b+c}{3}$ Cranial Module 49 50
g/f Facial 51 52
h/f Upper Facial 53 54
f/b Cranio-facial 55 56
l/k Nasal 57 58

m/n Left Orbital 59 60
p/q Nasalia-Transverse 61 62
r/s Interorbital 63 64
u/t External Palatal 65 66
v/w Mandibular 67 68
y/f Zygo-gonial 69 70
y/e Fronto-gonial 71 72
e/f Zygo-frontal 73 74
Horizontal Circumference 75 76
Nasion-Opisthion 77 78
Transverse Arc 79 80

Figure 19.1 A Peabody Museum Cranial Observation card used by Earnest Hooton, with quantification data for antemortem tooth loss, wear, caries, and other dental conditions from a skeleton excavated in the 1919 season of the Andover Pecos Expedition. Courtesy of the Peabody Museum of Archaeology and Ethnology, Harvard University, [Peabody 2011.1.10].

disease, irrespective of the number of teeth affected, although the number of bad teeth may also be recorded separately" (2007:64). And elsewhere in this volume (see Lukacs, Chapter 30 in this volume) we are reminded of the genetic component to caries susceptibility, which suggests additional ways of parsing a larger dataset like that from Pecos.

Interestingly, Hooton altered his data recording cards during the project. The modified card, Cranial Observations D (see Figure 19.2), has only two columns within which to record 1) dental completeness and wear, and 2) "Quality" of the teeth which included caries and abscesses. Gone are the devised scales which provided information on some aspect of quality or quantity pertaining to the teeth. Now the researcher must write specific comments within a single row and column. The dentition is now noted as "complete" or "incomplete", and there is no quantification of antemortem or postmortem tooth loss. Unlike the former version of the card, the number of caries (do we assume this means the number of teeth with caries?) and their relative size are indicated. The ramifications of changing recording procedures are great. On the "plus" side, with the new cards, comparing individuals and creating aggregate data is perhaps made easier. On the "down" side is the fact that comparing data collected using the first card with that of the second is impossible.

The Peabody Museum data cards, including the original version of dental pathology recording shown in Figure 19.1, were used for several decades by researchers at other

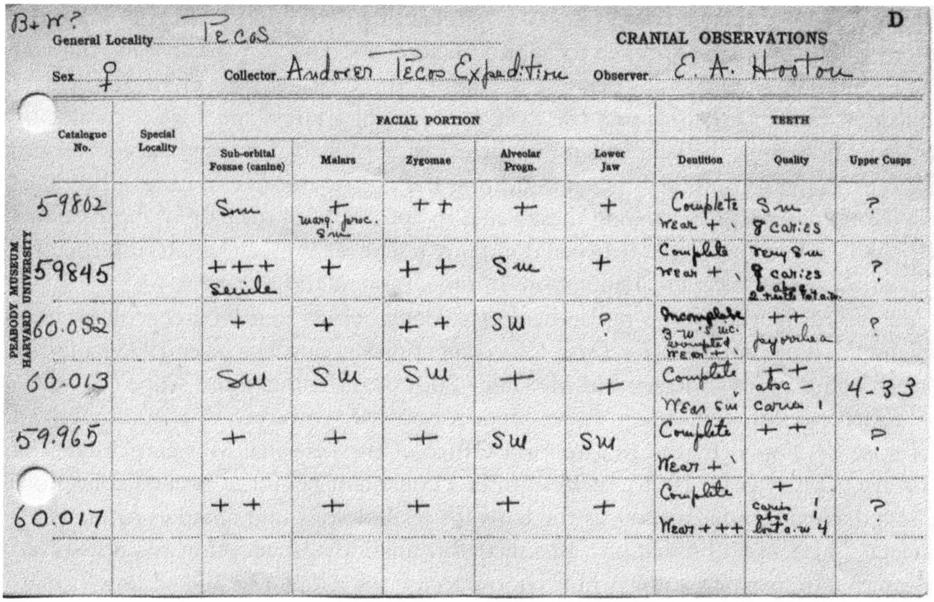
B+W?
General Locality Pecos
CRANIAL OBSERVATIONS
D
Sex ♀
Collector Andover Pecos Expedition
Observer E. A. Hooton

PEABODY MUSEUM HARVARD UNIVERSITY

Catalogue No.	Special Locality	FACIAL PORTION: Sub-orbital Fossae (canine)	Malars	Zygomae	Alveolar Progn.	Lower Jaw	TEETH: Dentition	Quality	Upper Cusps
59802		Sm	+ marg. proc. Sm	++	+	+	Complete Wear +	Sm 8 caries	?
59845		+++ Senile	+	++	Sm	+	Complete Wear +	Very Sm 8 caries	?
60.082		+	+	++	Sm	?	Incomplete	++	?
60.013		Sm	Sm	Sm	+	+	Complete Wear Sm	+	4-3 3
59.965		+	+	+	Sm	Sm	Complete Wear +	++	?
60.017		++	+	+	+	+	Complete Wear +++	+ caries	?

Figure 19.2 The data recording card used by Earnest Hooton for the majority of the Pecos study, with "+" signs replacing scores for dental wear, and caries count replacing group counts. Courtesy of the Peabody Museum of Archaeology and Ethnology, Harvard University, [Peabody ID 2011.1.11].

institutions. Hooton's students, in particular, used the cards elsewhere. George Neuman, for instance, used the Peabody cards at the University of Kentucky Museum of Anthropology in the 1950s and a paper version of the Peabody cards at Indiana University in the 1970s (Della Cook, personal communication). The cards include the 0 to 5 scores for tooth wear, the number of teeth lost antemortem, number of teeth lost postmortem, number of carious teeth, and observations on abscesses (scored from zero to 4 indicating "large – many").

Today the level of specificity advocated for recording dental pathology is far more detailed. It is suggested that observations should be recorded separately for every tooth so the number of teeth lost antemortem, the number of teeth with one or more caries, and the composite dental wear scores are readily obtainable to the researcher. Furthermore, recording this data as second-order rather than first-order (or primary) data is also helpful, as each variable should not have to be calculated using the raw data. This means that, depending on the value given to the primary data, the secondary data becomes relevant to explore or not. For instance, recording the presence and condition of a tooth might be the primary data, while aspects concerning the specific tooth (such as the presence of caries) become secondary data. Using layered data allows the researcher to more readily and efficiently use data to answer questions.

Clearly, issues of specificity and sensitivity in data recording in paleopathology are at the heart of data-collection design. But, given that we are almost always subject to time limitations and that individual research questions demand different kinds of data, how much detail is enough? After all, decisions made at the onset of a project regarding data collection will influence many aspects of future analytical design and potential.

One can always generate composite data by collapsing two or more categories of data, but rarely can one parse grouped data into smaller units. That is, you can lump but you can't split. The ramifications of these decisions can be clearly seen in Hooton's data collection cards. Questions that the early cards might address and answer cannot be tackled when using the later version of the cards. Specificity, while enhanced in some respects on the second cards (such indicating the precise number of caries), becomes lost in other respects. Relegating these issues to "problems of the past" is naïve on our part, as clear tensions persist between our desire to collect as much specific information as possible, and the very real limitations of time, money, and resources.

For more than a decade researchers in paleopathology and bioarchaeology have been guided by *Standards for Data Collection* (Buikstra and Ubelaker 1994) and the British counterpart (Brickley and McKinley 2004) for the minimum scope of data to be collected. Building upon these predecessors, Osteoware©, the computerized software developed by the Repatriation Office of the National Museum of Natural History (Ousley et al. 2005), seeks to offer a consistent platform for data collection. These devised standards serve as the basis for terminology and coding protocols for skeletal and dental observations. But their formulation and adoption has not solved all issues. For instance, some people choose to use the data forms and coding formats provided, while others incorporate some or all of the data protocols into their own forms and procedures. In part, this variation is a product of the different kinds of research that are conducted under the auspices of paleopathology. Research focused on a single condition might be based on specifically crafted data collection protocols designed to document the presence (or absence) of relevant bony changes, and to record important qualitative and/or quantitative aspects of the bone and lesion. They also tend to determine *a priori* the portion of the population that is relevant to the research. At the opposite extreme are projects which charge the osteologist to thoroughly document all kinds of pathological conditions in an assemblage of archaeological remains. In these instances the population might be studied once and then reburied. Large synthetic efforts like The Global History of Health project (Steckel et al. 2006) impose a strict level of comparability as contributors recode extant data which are then transformed into indexes and composite indicators of relative health. All of these kinds of work start with the bone or tooth (or a portion thereof) as the unit of observation, but the *unit of analysis* ranges from the individual skeleton to a local population, to a regional or even larger population, as research goals dictate.

INVENTORY AND OBSERVABILITY

The issue of "How much detail?" is equally central to the topic of the skeletal inventory because this is the basis for quantification of skeletal conditions—the number of elements / individuals *observable* for the condition. Current standards emphasize the importance of skeletal inventory as the initial stage in skeletal analysis, with particular emphasis being placed on "the bones and centers of ossification that are most important in skeletal biological research" (Buikstra and Ubelaker 1994:5). Dedication to developing a detailed skeletal inventory pays off when data analysis begins. As an example, based on the statement: "cribra orbitalia was present in seven individuals in the assemblage of 126 skeletons" one might calculate the frequency of cribra orbitalia in the population as 5.5 percent (7 of 126). But the statement does not specify how many individuals could be observed for the presence or absence of the condition. If only 72 of the individuals represented in the assemblage had (at least) one intact orbital roof, then the frequency nearly doubles to become 9.7 percent (7 of 72). Hence, it is essential to clearly provide the minimum number of observable individuals which served as the denominator for the equation (Waldron 2007: Waldron, Chapter 28 in this volume; and Boldsen and Milner, Chapter 7, in this volume) if we are to meet that most fundamental goal of paleoepidemiology: the "rigorous quantification of disease or lesion frequency" (Pinhasi and Mays 2008:ix).

But creating an inventory is not as straightforward as one might suspect. Many issues requiring decisions and interpretation become quickly evident when one creates an inventory form of one's own or uses the standardized inventory forms currently available. As researchers quickly find, indicating that a bone is present is often an inadequate notation on its own. What exactly does "present" mean? Is the entire bone intact with every surface visible and in good condition? Is the entire bone present, but large sections of the surface has suffered post mortem damage? The list of possible permutations is extensive and not entirely addressed by current standardized forms. For instance, in the *Standards* inventory "missing elements are recorded by leaving the space blank" (Buikstra and Ubelaker 1994:6). The use of blank instead of zero may introduce a level of uncertainty: was this element completely missing or was some of it present just not adequately preserved to record anything, or did the researcher accidentally skip this part of the inventory? Another notable aspect of *Standards* is that there are columns for indicating presence of the left and right scapula, radius, etc., but there is no place on this form to record an element or part of an element of indeterminate side. This carries over to other forms as well, and is problematic given that imperfectly preserved skeletons are quite frequently encountered in many archaeological regions. Not knowing the side of an element should not prevent recording most kinds of pathological lesions, nor do we want to create a situation in which elements are arbitrarily assigned right or left by a conscientious osteologist who does not want to lose data. Efforts to remedy some of these issues, for instance by Brickley and McKinley (2004), have succeeded on some fronts and generated new issues on others.

Explicit documentation of observability for specific conditions is especially crucial when reporting on incomplete or imperfectly preserved skeletal remains (Waldron 2007; Grauer 2008; Pinhasi and Mays 2008). For instance, many researchers have carefully reported the number of frontal bones that are present in their collections. Recording the presence of a frontal bone, however, does not necessarily indicate whether the orbits were present or in good condition. This omission renders the evaluation of the presence of cribra orbitalia (which occurs exclusively on the eye orbits) impossible. Similarly, preservation of anatomical features is also an important factor to record; the maxillary canine may be present, but if the enamel is chipped off then the presence or absence of enamel hypoplasia may not observable. If a specific bone is present, but the cortex is weathered or abraded, then periosteal reaction may not be observable. An all-purpose inventory may not be sufficiently detailed to support accurate quantification of observability and frequency of every condition that impacts the skeleton. The scorbutic rosary, the "enlargement of the ribs adjacent to the costochondral junction" (Brickley and Ives 2008) is a good example. Inventories typically record the number of right and left ribs present, but not whether the costochondral end is present. Inventories might be profitably recrafted to include recording the presence and condition of particular skeletal elements, or regions of elements, in which pathognonomic or diagnostically important lesions supporting a differential diagnosis of known conditions occur (Stodder et al. 2010). Individuals, or individual institutions, choosing to recraft inventory forms, however, brings us back to issues of data sharing and the limitations inherent in creating aggregate data when standardization is absent.

As with the creation of an inventory, documenting the nature and distribution of abnormal bone is of foremost concern in paleopathology, and current texts emphasize the importance of using unambiguous universally understood terminology in description, and of documenting the anatomical locations and extent of bony changes (Buikstra and Ubelaker 1994; Lovell 2000; Ortner 2003, 2008; Roberts and Connell 2004; Grauer 2008). As summarized by Lovell, description and documentation has four components: "1) appearance of the lesions themselves, 2) the location of the lesions within a skeletal element, 3) the skeletal distribution of the lesions in an individual, and 4) the distributions of lesions in a population" (2000:219). Implicit in this elegant précis of paleopathology is the fact that our understanding of prehistoric health is built upon several scales of analysis. Even in discussing only macroscopic analysis, we have the individual lesion, the skeletal element, the individual, the population. The unit of analysis changes throughout the process, and the challenge is to structure the data (and the database) so that it can accommodate all of these stages or scales of investigation. We can create very large databases of coded data, but are they equally useful for differential diagnosis, for the synthetic study of health and disease in a small skeletal assemblage, *and* for a regional paleoepidemiological study?

Once again using *Standards* as an example of the promise and pitfalls of standardized data collection protocol, lists of terms and classifications of bony changes are explicated through the use of a numeric coding system that builds on the inventory data for each element. In fact *Standards* suggests that pathological lesions be recorded at the same time as inventory: "The pathology coding procedure described here is closely integrated with the inventory process. The observer can easily identify and record forms of pathology as the inventory proceeds" (Buikstra and Ubelaker 1994:108). The researcher is provided with carefully devised numeric codes that incorporate the location of a lesion along with the type, scope, and severity of the lesion. In spite of the promise of this approach, the system has been characterized as, "far too cumbersome and restrictive to be of practical use in most cases (especially in contract archaeology). For example, a right ulna with a healed parry fracture would be coded as follows: (1),(3),(9), (4.13), (5.13)" (Roberts and Connell 2004:36). Roberts and Connell suggest that this plethora of codes is distracting and time-consuming. The data entry screen for abnormal bone formation in the Osteoware© program (Figure 19.3) uses check-boxes; in this context the analyst is spared the distraction of the codes. But although they are invisible to the analyst in the lab, the numeric codes are what appear in the actual data tables.

Roberts and Connell point out that numeric codes "do not represent quantitative data" (2004:36), and while they do not elaborate on this, there can be an anti-intuitive dimension to code systems, as the information isn't there; just the coded data. The use of numeric (or other) codes for paleopathology data is rarely addressed in the literature,

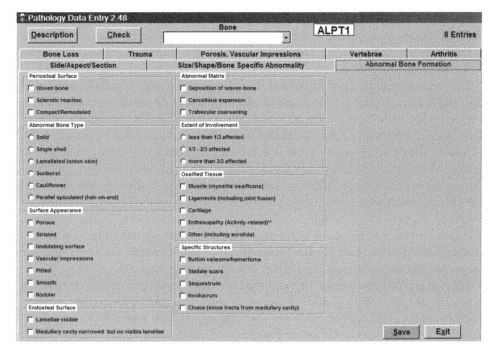

Figure 19.3   Osteoware© data entry screen for abnormal bone formation. Courtesy of the National Museum of Natural History.

although the manner in which we record data is clearly a practice that impacts our work. As with inventories, there is probably no single best way. The data needs of the Global History of Health Project are different from those of a bioarchaeologist writing a report on 60 skeletons from a Neolithic village. The paleoepidemiological study designed to generate estimates of disease based on lesion types in a population involves a different process of data manipulation than estimating disease prevalence in a skeletal assemblage based on the differential diagnosis at the individual level. Numeric data are elegant and make for smaller data files, but these attributes do not necessarily accommodate the process of data manipulation and exploration for all practitioners.

ISSUES OF ANALYSIS: BIG SCIENCE AND LITTLE SCIENCE IN PALEOPATHOLOGY

Efforts in the last couple of decades to compile large amounts of comparable osteological data in systematic formats is part of the maturation process of our field, perhaps akin to what Price (1963) observed in physics in the post WWII boom of large-scale federally funded research known as "Big Science." Biological anthropology joined "Big Science" with the human adaptability studies funded by the International Biological Programme, but expansion in the scale of research specifically involving human skeletal remains has been more recent and episodic. The plethora of synthetic analyses initiated around the time of the Columbian quincentenary (e.g. Verano and Ubelaker 1992, Larsen and Milner 1994, Baker and Kealhofer 1996) exemplify a threshold effect: in the early

1990s there was sufficient data, much of it new, for multiple researchers to approach the "Big" questions about the health impact of European contact on indigenous people of the Americas. However, interobserver issues are mentioned in many, if not most, bioarchaeological contributions.

The most frequently cited limitation to expanding the scale of bioarchaeology / paleopathology research beyond the confines of whatever current project we are writing about (aside from preservation and representativeness) is noncomparability of data. Finding comparable data in published sources involves the search for compatibility not just in *what* data are reported, but in *how* they are reported (raw data, grouped data, descriptive statistics). Often we are faced with using a convention for data reporting that we know to be suboptimal, like reporting frequency data using age groups. These age groups, which might be as broadly defined as "young adult (18–25 years old)" or as specific as "18–21 years old", can mask age-specific trends in disease. Imposing stringent standards to the scope of osteological data collection and publishing coding protocols do provide steps forward from the situation in which abundant data are generated, but meta-analyses are difficult or impossible.

How might paleopathology begin to tackle "big" questions? As summarized by White (2008), several types of database are of interest to paleopathologists, ranging from inventories and catalogs that serve as "finding aids" for research collections, to those with more contextual information and / or actual osteological data. This latter more substantive category presently includes (among others) the databases created by the Smithsonian's Repatriation Laboratory (ROL) (Ousley et al. 2006), the Wellcome Osteological Research Database (WORD) (White 2008), and the Global History of Health project database (Steckel et al. 2006). These differ in terms of how paleopathology data are recorded and accessed. As mentioned above, the Smithsonian's Osteoware© uses data coding protocols based on *Standards* and captures very detailed primary data on individual skeletons. The Wellcome database (WORD) contains archaeological and osteological data. Paleopathological observations (description of lesion location, type of bone, alteration, distribution) are classified by disease category (metabolic, trauma, joint disease, etc.), with descriptions of lesions recorded in free text, not code. The codebook for the Global History of Health project directs participating data providers to enter descriptions of pathologies from a series of options based on location, type, healing status, and size of affected area (Steckel et al. 2006:28). There is a data hierarchy, a limited choice of values for each variable, and the data are comma delimited which suggests that the data are reformatted in a variety of ways for spreadsheets or other formats.

In spite of the promise that databases offer to paleopathology, there are a number of important limitations to address. For instance, at present, accessing data from the large databases discussed above is somewhat restricted. Researchers request data from WORD and the Smithsonian ROL database administrator who runs queries in Standard Query Language and provides the data to the researcher. White describes an in-house use of the WORD database: "the data were exported into Excel spreadsheets that were then manipulated to produce the desired output, especially overall disease prevalences" (White 2008:82). He refers to the future WORD database as "allowing users to build queries using keyword lists, including interpretive groupings and keywords supplied for specialist users ... Expert users would be able to interrogate the database directly themselves" (2008:183). Clearly, we have come far in terms of integrating large volumes of data into more manageable and accessible platforms, with plenty of room for improvement and development.

But what about smaller projects undertaken by a single individual or smaller groups? I would argue that all researchers must be "expert users"—not of the "Big Science" databases, but of our own "Little Science" data. Expertise in this sense means familiarity with the behavior of a dataset as well as all the relevant contextual information. Relational queries and data manipulation are essential tools for exploring and maximizing the usefulness of data. The multifactorial approach that we now seek to employ in paleopathological and paleoepidemiological interpretation increases the importance of a relational, iterative approach to data exploration. This more nuanced approach, however, can be inhibited by reliance on a third party to extract and manage data.

Resistance to using relational databases (also a problem in archaeology as discussed in Keller 2009) may be due in part to the fact that this kind of software doesn't come preinstalled on our computers or prepackaged in larger program suites, as Excel or other spreadsheets tend to do. There is also the perception that relational database software is strictly the domain of experts, too sophisticated for the average user. But this perception is unwarranted, as a number of software programs, such as Dbase, Paradox, FoxPro, and today FileMaker and Access, have been developed to accommodate small scale desktop users. Yes, there are rules to master; there is a learning curve, but these can be effectively used by Dr. or Mr. or Ms. Average User, even if most of us reach a plateau of expertise.

Unquestionably, the standardization of recording represents a critical and essential stage in improving the quality and usability of paleopathology data, and online data entry presents significant advantages. No one would argue with the importance of database design and management expertise for the Big Data resources. But not all research in paleopathology is supported by a download of prevalence data. My concerns about data coding and databases are really about the way we work with data. Research is an iterative process, and the closer we are to our data the better—at least, for the human-scale research that is the core of bioarchaeology. I think that many of us are doing Little Science, and we are doing social science. Increasingly, paleopathology as a component of bioarchaeological inquiry is turning towards a socially and politically (not just biologically) contextualized approach to understanding social processes and agents that drive health patterns within and between communities (see Buzon, Chapter 4 in this volume for further discussion of the bioarchaeological approach). Not only will we benefit from close articulation with the archaeological data, but we also can take to heart some fundamental admonitions on the research process from the archaeological literature.

Data exploration and pattern recognition and are not new concepts, but with the emphasis within educational programs on formal scientific hypothesis testing and statistical confirmation, this initial stage of research is often omitted from training (and practice) in paleopathology and bioarchaeology. Binford's discussion for archaeologists of "Where Do Research Problems Come From?" (by problems he means questions) points to the very fundamental importance of *looking at data*: "Problems derive from the recognition of patterning that is unsuspected, new, and / or not understood" (2001:657). "Theory is not something one brings to data. Theory is developed to explain relational patterns among data that are analytically generated among different observational domains or problem sets" (2001:676). Relational patterns are the very stuff of paleopathology. We operate within a field in which multiple scales of observation, derived from multiple datasets, are explicitly utilized.

Tukey's (1977) seminal text on exploratory data analysis (EDA) presented a series of pencil-and-paper exercises and data treatments (featured in Drennan's 2010 archaeological statistics text and to a lesser degree in Shennan 1997) to be used in understanding the behavior of a dataset. Exploratory data analysis "emphasizes flexible searching for clues and evidence, whereas confirmatory analysis stresses evaluating the available evidence" (Hoaglin et al. 1983:2). EDA includes the recognition that datasets, especially small ones, are often messy; not normally distributed or perfectly representative. In paleopathology, we often use small samples and generate datasets full of holes. Hence, paleopathological (and archaeological) projects do not provide the ideal material for "the critical experiment." In EDA, a set of data is *a batch of numbers*, not a sample, not an assemblage, nor is it a population.

This approach is especially important with small datasets (or batches), where the actual data can reveal some information that is lost in summary statistics. Figure 19.4 presents data on degenerative joint disease (DJD) of the shoulder in adult females from Ancestral Pueblo sites in the Ridges Basin in Southwestern Colorado (Stodder et al. 2010). The skeletal remains were poorly preserved and the DJD data (scored by severity as 0 1, 2, or 3) range widely from 0 – 3 within the sample. The presentation of the average scores and standard deviations by age group is a typically disappointing example of Southwestern paleopathology data, rendering small N's and very few individuals represented in the oldest age group (see Figure 19.4a). Instead of a relatively predictable progression of higher average scores associated with sequentially older age groups, the bar chart is a mess (Figure 19.4b). It might "look better" if we collapse the data into three age groups instead of four, but upon what criteria do we create three from four? And isn't doing this at cross purposes with the intent to look for evidence of progressive physical stress across the lifecourse? The scattergram (Figure 19.4c) shows the individual DJD scores for the glenoid and proximal humerus plotted at the midpoint of the age at death estimate for each individual. Two interesting questions about this community pop out from this humble diagram: Who was the woman with stage 2 and 3 DJD in her shoulder at age 22? And who had *no* DJD at age 54? These are questions about the physical lives of two specific people who stand out, in this measure at least, from their community. The answers lie in the rest of the data from their skeletons, the site and the population. This is the kind of data that can be disposed of in a standard table with apologies for the small sample size. Or we could take another look, because outliers are interesting.

Represented in Figure 19.5 is one of most basic EDA tools, the stem and leaf plot. This is essentially a histogram of digits showing the shape and distribution of numbers in a batch. The purpose of the plot shown here is to contextualize the femur length data for adult females from a prehistoric community, Sacred Ridge (site 5LP245) which seemed to have had markedly short and also markedly tall women (not all represented here due to preservation). The "stem" is the central column of numbers representing the hundreds and tens digits of the maximum femur length (in mm), and the leaves (the ones digits) indicate, on the left, the spread of data points for the eight Sacred Ridge females, and on the right, the 68 other Ancestral Pueblo females from the Mesa Verde / San Juan Basin Region. The regional plot has a definite single peak, but the distribution is certainly not symmetrical (it is skewed)

(a)

Age group	Glenoid (N=20)			Proximal humerus (N=16)		
	N	Avg	SD	N	Avg	SD
YA	3	1.00	1.73	3	0.67	1.15
MA1	6	0.33	0.82	6	1.00	1.09
MA2	9	0.44	0.73	5	0.60	0.55
OA	2	1.50	0.71	2	1.00	1.41

YA = 17–25.9 years; MA1 = 26–39.9; MA2 = 40–49; OA = 50+

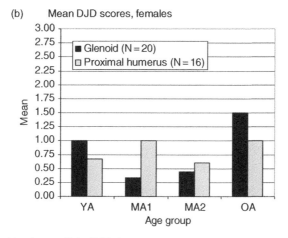

(b)   Mean DJD scores, females

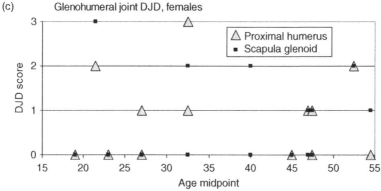

(c)   Glenohumeral joint DJD, females

Figure 19.4   Three presentations of a small dataset, with a) showing the averages and standard deviations of the severity of DJD within each age group; b) presenting the averages as a bar chart; and c) displaying the raw data as a scattergram. Data are from Stodder et al. 2010, courtesy of SWCA Environmental Consultants, Inc.

as there are two outliers at the long (447 mm) and much longer (454) end of the distribution. In contrast, the femur lengths from Sacred Ridge fall mostly in the main cluster of regional data points, and include outliers at the short (333 mm) and long (449 mm) ends. The very short stature (333 mm) of the one individual, which is not a result of dwarfism (Osterholtz 2008), pulls the mean value for this small set of data to 399.68—below the regional data mean 401.66 mm. But even with the

Sacred Ridge (N = 18)		Regional data (except Sacred Ridge) (N = 68)
3	33	
	34	
	35	0
	36	1,4
	37	3,5,8,8,9
3,1	38	1,5,6,7,9
5	39	0,1,2,3,3,4,4,5,6,6,8,9
9,7,5	40	0,0,0,1,1,1,2,3,4,5,5,5,5,7,7.5,8,9,9
6	41	0,0,0,2,2,2,3,3,3,4,6
	42	0,0,1,2,2,3,4,4,5,6
	43	
9	44	7
	45	
	46	
	47	
	48	
	49	
	50	
	51	
	52	
	53	
	54	4
Mean: 399.68		Mean: 401.66
Median: 406.25		Median: 401.5

Figure 19.5 Back-to-back stem and leaf plot, maximum femur length, for ancestral pueblo females.

short outlier in Sacred Ridge and the very tall outlier in the regional sample, the Sacred Ridge median is taller than that of the regional sample. The substantial gap between the femur lengths in the 440 mm range and the next point serves a warning that the 544 mm measurement might be an error, that the individual might be male, or that this person is from a different population. In fact this is the earliest dated individual represented in the plot, and there are a number of other very tall males and females reported for the Basket Maker (pre-Pueblo) period, so it is quite possibly not an error. And while not reaching the extreme of the 544 mm femur, it does appear that the women of Sacred Ridge included some unusually short and some unusually tall individuals.

Like the scattergram and other kinds of visual displays, a stem and leaf plot is a tool for looking at data and recognizing the unique characteristics of datasets. These tools can be especially useful in assessing small datasets where, no matter how torturously data are parsed and reparsed, nonparametric tests do little except to discourage further study (e.g. Cowgill 1977). Hence, this quite simple stage of EDA, coupled with standard descriptive statistics, measures of skewing and kurtosis, quartiles and quantitatively defined "outliers," can reveal behavior in the data that suggest both cautionary notes and promising paths of inquiry that might not otherwise be apparent.

The Future of Data: Issues of Storage

While much of the discussion thus far has focused on the future of data as it pertains to collection and analysis, a considerable amount of thought and discussion ought to be devoted to data storage. The utility of data collection rests not only in its ability to record and reflect aspects of the specimens under examination, and the ability for it to be shared between researchers; it also rests on its ability to be accessible to researchers across time. In spite of strides taken to both facilitate and simplify data collection using computers, most researchers still begin data collection by recording data on paper. In the past, researchers took great efforts to protect their precious pages of information from moisture, mold, and insects. The fruits of these efforts are readily seen as we have Hooton's original data cards, digitized and thus available to use 90 years after they were created. For the past decades, however, with researchers often computerizing their original data sheets, emphasis on protecting the original recording forms has been replaced by efforts to maintain the computerized formats.

There are two issues however, to address with this approach. First, computerized datasets are not necessarily permanent. As many researchers have come to realize, changes in methods of data entry, from the creation of data punch cards to direct data entry and computerized tools rendering touching a keyboard obsolete, have been met with equally amazing changes in software programs and platforms, and in data-storage options. Data computerized using punch cards, based on DOS, and saved on tape, must now be (if possible) migrated into Windows-based or other platforms and moved from one storage medium to another. This migration is not always successful. Hence, while data manipulation and storage appears to be easier to manage today, it might, in fact, be more challenging than ever.

Another issue germane to the topic of storage centers on our efforts to improve standardization. Whether using codes to represent information, or relying on handwritten notations, researchers often need to develop a key (or legend). An examination of the data cards Hooton used to record his observations on the Pecos collection reveals a range of specificity in pathology data collection, which is also typical in large projects today. The Hooton archive includes a much annotated list of his abbreviations and definitions of his codes (a data dictionary). For instance, he used "?" to represent "unknown" and "–" to represent "absent." Observations on lumbar vertebrae appear on the back of the Pelvis Observations cards, with "+", "++", or "+++" indicating degree of exostoses development. As photocopying, rather than printing, became more readily available, it became easier for researchers to include legends on each recording form. This helped insure that data collected in one decade could be used both easily and reliably in the next without having to locate original legends. With the advent of computerized data, however, and especially with the growing emphasis on coding data, the separation of data from the meaning of the codes becomes more common. So now, the onus is on the researcher to successfully migrate files to new operating platforms, to store files on media that can be read by the newest computers (how many computers have you seen recently that have 5¼ inch floppy drives, or 3½ inch drives?), and to ensure that any/all keys and legends are firmly attached to every file. Who said that computerization has made our lives easier?

CONCLUSION

The most valuable things I learned in graduate school: how to get that formalin smell out of my lab coat; paleopathology is an exercise in sampling error; never work with small samples. Alas, I finally had to get a new lab coat (and then another), and I have always studied "small samples" because I work in the regions where people lived in small communities and buried their dead in small groups or singly, in the intimacy of the household. Given that for most of our history as a species people lived in small communities, it seems a bit cowardly to avoid studying their remains, small "Ns" notwithstanding. The close consideration of nuanced and contextualized evidence of intracommunity and intercommunity differences in health and morbidity in Neolithic communities yields some of the baseline evidence of the social, economic, and political processes that drive change in human biocultural adaptation. The way to do this is through careful management of data, use of exploratory tools that clarify the behavior of datasets, as well as the conscientious effort to generate data that will be useful to others and that can be used to build the "big data" that serve larger scale analyses. Truth be told, however, in spite of our best efforts to standardize and systematically record data, and our efforts to familiarize ourselves with datasets small and large, the process of data collection, analysis, and management is not static. It changes with new approaches towards understanding disease processes, the development of new technology, the application of new paradigms and interpretive frameworks, and the availability of material to work with. Transparency in data recording and vigilant curation of our data are essential for the continued growth of knowledge in this dynamic field.

ACKNOWLEDGEMENTS

Many thanks to Della Cook for providing copies of data forms used by Neuman and Gebhard, and to Patricia Kervick, Reference Archivist at the Peabody Museum for assistance with the Pecos data cards, to Chris Dudar for allowing me to participate in the beta testing of Osteoware©, to Mike Pietrusewsky, Donald Ortner, and Charlotte Roberts for help dealing with this amorphous topic. I especially appreciate the many years of discussion with Michele Toomay Douglas, Mary Lucas Powell, and Anne Grauer on the practice of paleopathology. All the heresy herein is my own.

REFERENCES

Baker, B. J., and Kealhofer, L. K., eds., 1996   Disease and Biocultural Frontiers: Native American Adaptation in the Spanish Borderlands. Gainesville: University of Florida Press.

Binford, L. R., 2001   Where Do Research Problems Come From? American Antiquity 66(4):669–678.

Brickley, M., and McKinley, J. I., eds., 2004   Guidelines to the Standards for Recording Human Remains. Institute for Field Archaeologists Paper No. 7. Southampton: British Association for Biological Anthropology and Osteoarchaeology, Institute for Field Archaeologists, and the Department of Archaeology University of Southampton.

Buikstra, J. E., and Ubelaker, D. H., 1994 Standards for Data Collection from Human Skeletal Remains. Arkansas Archeological Survey Research Series No. 44. Fayetteville: Arkansas Archeological Survey.

Cook, D. C., and Powell, M. L., 2006 The Evolution of American Paleopathology. In Bioarchaeology: the Contextual Analysis of Human Remains. J. E. Buikstra, and L. A. Beck, eds., pp. 281–322. New York: Elsevier.

Cowgill, G. L., 1977 The Trouble with Significance Tests and What We Can Do About It. American Antiquity 42:350–368.

Drennan, R. D., 2010 Statistics for Archaeologists: A Common-Sense Approach. 2nd Edition. New York: Springer Science+Business Media.

Grauer, A. L., 2008 Macroscopic Analysis and Data Collection in Paleopathology. In Advances in Human Paleopathology. R. Pinhasi, and S. A. Mays, eds. pp. 57–76. New York: John Wiley & Sons.

Hillson, S., 2000 Dental Pathology. In Biological Anthropology of the Human Skeleton. M. A. Katzenberg and S. R. Saunders, eds. pp. 249–286. New York: Wiley-Liss.

Hillson, S., 2008 The Current State of Dental Decay. In Technique and Application in Dental Anthropology. Irish, J.D., and Nelson, G.C., eds. pp. 111–135. Cambridge: Cambridge University Press.

Hoaglin, D. C., Mosteller, F., and Tukey, J. W., 1983 Understanding Robust and Exploratory Data Analysis. New York: John Wiley & Sons.

Hooton, E. A., 1930 The Indians of Pecos Pueblo: A Study of Their Skeletal Remains. New Haven: Yale University Press.

Keller, A. H., 2009 In Defense of the Database. The SAA Archaeological Record 9(5):26–32.

Larsen, C. S., and Milner, G. R., eds., 1994 In the Wake of Contact: Biological Responses to Conquest. New York: Wiley-Liss.

Leigh, R. W., 1925 Dental Pathology of Indian Tribes of Varied Environmental and Food Conditions. American Journal of Physical Anthropology 8:179–199.

Lovell, N. C., 2000 Paleopathological Description and Diagnosis. In The Biological Anthropology of the Human Skeleton. M. A. Katzenberg, and S. R. Saunders, eds. pp. 217– 248. New York: Wiley-Liss.

Mays S. A., and Pinhasi, R., 2008 Preface. In Advances in Paleopathology. R. Pinhasi, and S. A. Mays, eds. pp. ix–xii. New York: John Wiley & Sons.

Ortner, D. J., 2003 Identification of Pathological Conditions in Human Skeletal Remains. New York: Academic Press.

Ortner, D. J., 2008 Differential Diagnosis of Skeletal Lesions in Infectious Disease. In Advances in Human Paleopathology. R. Pinhasi, and S. A. Mays, eds. pp. 191–214. New York: John Wiley & Sons.

Osterholtz, A. J., 2008 The Long and Short of It: A Case of Diminutive Stature in Prehistoric Ridges Basin. Annual Meeting of the American Association of Physical Anthropologists, Columbus, OH.

Ousley, S. D., Billeck, W. T., and Hollinger, R. E., 2005 Federal Repatriation Legislation and the Role of Physical Anthropology in Repatriation. Yearbook of Physical Anthropology 48:2–32.

Rihan, H. Y., 1930 Appendix I. Dental and Orthodontic Observations on 289 Adult and 53 Immature Skulls from Pecos, New Mexico. In The Indians of Pecos Pueblo. E. A. Hooton, ed. pp. 367–373. New Haven: Yale University Press.

Roberts, C. A., and Connell, B., 2004 Guidance on Recording Paleopathology. In Guidelines to the Standards for Recording Human Remains. Brickley, M., and J. I. McKinley, eds. pp. 34–39. Reading, UK: British Association for Biological Anthropology and Osteoarchaeology, and the Institute of Field Archaeologists. IFA Paper No. 7.

Roberts, C. A., and Cox, M., 2003 Health and Disease in Britain. Stroud: Sutton.

Rose J. C., and Burke, D. L., 2006 The Dentist and the Anthropologist: The Role of Dental Anthropology in North American Bioarchaeology. In Bioarchaeology: The Contextual Analysis of Human Remains. J. E. Buikstra, and L. A. Beck, eds. pp. 323–346. New York: Elsevier.

Shennan, S., 1997   Quantifying Archaeology. 2nd Edition. Edinburgh: Edinburgh University Press.

Steckel, R. H., Larsen, C. S., Sciulli, P. W., and Walker, P. L., 2006   Data Collection Codebook. The Global History of Health Project. http://global.sbs.ohio-state.edu/new_docs/ Codebook-01-24-11-em.pdf. Accessed May 4, 2015.

Steckel, R. H., and Rose, J. C., eds. 2002   The Backbone of History: Health and Nutrition in the Western Hemisphere. New York: Cambridge University Press.

Stodder A. L. W., Mowrer, K., Osterholtz, A. J., and Salisbury, E., 2010   Skeletal Pathology and Anomaly. *In* Animas-La Plata Project Report Series Volume XV: Bioarchaeology. E. Perry, A. L. W. Stodder, and C. Bollong, eds. pp. 89–155. Phoenix: SWCA Environmental Consultants, Inc.,

Tukey, J. W., 1977   Exploratory Data Analysis. Reading, MA : Addison-Wesley.

Verano, J. W., and Ubelaker, D. H., eds. 1992   Disease and Demography in the Americas. Washington, DC: Smithsonian Institution Press.

Waldron, T., 2007   Paleoepidemiology: The Measure of Disease in Past Populations. Walnut Creek, CA: Left Coast Press.

White, W., 2008   Databases. In Advances in Human Paleopathology. R. Pinhasi, and S. Mays, eds. pp. 177–188. New York: John Wiley & Sons.

PART **III** Diseases of the Past: Current Understandings and Controversies

CHAPTER 20 Trauma

Margaret A. Judd
and Rebecca Redfern

INTRODUCTION

"World wide, five million people died as a result of an injury in 2000 ... for each injury death, there are several thousand injury survivors who are left with permanent disabling sequelae" (Peden et al. 2002:2–3). Unlike other disease processes, such as chronic infectious disease, osteoarthritis and neoplasms, injuries are characteristically immediate due to a sudden unexpected event, although current clinical theorizing views most injuries as preventable (Peden et al. 2002). In antiquity, the injury mechanism was comprehensible. For example, Hippocrates observed in *On injuries to the head* (Section 11.3–4) "he who falls from a very high place upon a very hard and blunt object is in most danger of sustaining a fracture and contusion of the bone, and of having it depressed from its natural position." In contrast, the etiology of other diseases was often attributed to supernatural intervention or inexplicable, although superstition pervades explanations for trauma in some modern cultures (Blum et al. 2009). No matter how injury is perceived, the person's life may come to an abrupt end or their quality of life may be drastically altered temporarily or forever (Oakley 2007). In this chapter we: 1) present and discuss how trauma is defined and classified by paleopathologists; 2) provide an overview of how the study of trauma has changed over time; 3) discuss how the field of paleopathology currently undertakes trauma studies, and 4) conclude with recommendations for future consideration.

WHAT IS TRAUMA?

Numerous controversies influence our study of ancient trauma, first and foremost being how trauma is defined and classified and secondly, what lesions and pathological changes are considered to be of traumatic origin, which often diverges with

A Companion to Paleopathology, First Edition. Edited by Anne L. Grauer.
© 2012 John Wiley & Sons, Ltd. Published 2016 by John Wiley & Sons, Ltd.

current clinical and social science research (Kirmayer et al. 2008). It is suggested that these controversies in part stem from the researcher's background and training (archaeological vs. clinical), and the location and type of material they are most familiar with studying (e.g., Europe vs. the Americas, skeletal vs. mummified human remains). These controversies directly affect the presentation and interpretation of results, and therefore our shared past.

In the clinical literature, trauma is defined as "a physical wound or injury, such as a fracture or blow" (Oxford Medical Dictionary 2000:670). The International Classification of Disease published by World Health Organization (Peden et al. 2002) classifies injuries as intentional and unintentional and defines injury as being "caused by acute exposure to physical agents such as mechanical energy, heat, electricity, chemicals, and ionizing radiation in amounts that exceed the threshold of physiological tolerance. In some cases (for example, drowning and frostbite) injuries result from the sudden lack of essential agents such as oxygen or heat" (Baker et al. 1984:1). The absence of soft tissue (in most cases) has shaped our conception of paleotrauma, although there is little consistency. Ortner (2003:120) does not define trauma in his introduction, but instead states four ways in which trauma can affect the skeleton: a partial or incomplete break in a bone; abnormal displacement or dislocation of joints; disruption in nerve and/or blood supply, and artificially induced abnormal shape or contour of bone. In her review of trauma analysis in paleopathology, Lovell (1997:139) provides an uncited definition, "trauma may be defined many ways but conventionally is understood to refer to an injury to living tissue that is caused by a force or mechanism extrinsic to the body." Roberts (1991; 2006; Roberts and Manchester 1995, 2005) has consistently provided a definition reflecting her clinical training – "trauma can be defined as any bodily injury or wound" (Roberts and Manchester 2005:84).

Currently, there is no agreement on what paleotrauma includes and how it should be recorded. This may be because, as Ortner (2003:177) observed, "the variants of trauma ... affect the skeleton in so many ways that a comprehensive review would fill the pages of a substantial book." A review of widely used textbooks shows that considerable variation exists in what conditions are regarded as having a traumatic origin or association, reflecting both changes within the discipline as a whole and the development of forensic anthropology. Steinbock (1976) grouped lesions into five general categories: fractures, crushing injuries, sharp-force injuries, dislocations and Harris lines. Brothwell (1981:119) under the assumption that "there would have been relatively few accidents in early times ... and most injuries probably resulted from intentional blows," classified trauma by the pattern produced by weapons (gross crushing, less extensive fracturing, piercing and cutting), but was one of the few to include blood-stained bones, dental evulsion and trepanation. Ubelaker's (1989) pathological section included fractures (traumatic and pathological), projectile injuries and dislocations. Merbs (1989) discussed fractures, spondylolysis, dental trauma, weapon wounds, dislocations, scalping, surgery, and modification marks (i.e., cut-marks). In Cox and Mays' (2006) edited volume, the chapters by Roberts (2006) and Boylston (2006) included fractures (traumatic and pathological), amputation, trepanation and weapon injuries, but in contrast to other publications, parturition was considered separately (Cox 2006). In *Skeleton Keys*, Schwartz (2007:346) noted that "The term 'trauma' has been extended to incorporate the results of surgery-like or other body-altering behaviors," and cites dental modification/mutilation, foot binding and cranial deformation as examples. Bennike's (2008) overview of trauma grouped fractures, weapon injuries, spondylolysis, dislocations and trepanation. Mays (2010) included myositis ossificans traumatica, dislocations, and scalping but focused on fractures (traumatic and pathological) and sharp-force weapon injuries. Lovell (1997) classified injuries as fractures or dislocations only, with fractures including knife and sword cuts, scalping, projectile points and crushing due to binding. Lovell's (2008) revised scheme included myositis ossificans traumatica and placed bones modified by binding as a separate category, but did not discuss sharp-force or projectile injuries.

In volumes dedicated to paleopathology, these differences persist. Ortner and Putscher (1981) listed fractures, dislocations, deformations, scalping, mutilations, trephinations, traumatic problems arising from pregnancy, and sincipital T-mutilation in their chapter on trauma. Roberts and Manchester (1995; 2005) added myositis ossificans traumatica, spondylolysis and osteochondritis dissecans, as well as groups of injuries (decapitation, domestic violence, infanticide, child abuse, defleshing and cannibalism); amputation and trepanation were subclassified as treatment. Aufderheide and Rodríguez-Martín (1998) more closely reflected the clinical range of trauma with the addition of localized subperiosteal thickenings, strangulation, cauterization, bloodletting and crucifixion, soft-tissue injuries inflicted by others (laceration, stab, and sacrifice victims), accident due to crushing, asphyxia and burn, and finally mutilation. From a paleoradiological perspective, Chhem et al. (2008) limited bone trauma to fracture, an embedded foreign object, trepanation, heteroptopic bone formation and amputation; joint trauma consisted of dislocation, subluxation and diastasis (separation of the fibrocartilage joint). Finally, Mann and Murphy (1990) in their *Regional Atlas of Bone Disease*, designated fractures, Schmorl's nodes and myositis ossificans traumatica as traumatic, while scapular dislocations, trepanation and cranial cut marks were treated separately.

Because physical trauma can result from accidents and intentional acts of violence, caution should be paid to what skeletal changes are used to investigate or support certain hypotheses in archaeological populations since we lack the essential resource available to clinicians—the ability to interview a patient and accurately produce a timeline for these pathologies. For example, Tomoya et al. (2001) reported that a 26-year-old woman who was employed as a golf caddie, presented with myositis ossificans traumatica anteriorly to the right hip joint. The woman had no medical or sport history of the injury, but the clinicians concluded that it had developed because cervix stress was frequently applied in the hip.

It is evident that there is little agreement among paleopathologists about how trauma is defined, classified and by extension what is recorded, thereby narrowing the comparability between analyses. This arises from both how the term trauma is defined and understood, because if the meaning of trauma is applied in its widest sense, including the hard-tissue repercussions of emotional or environmental events (e.g., famine), it reflects the ethos of the bioarchaeological approach (Buikstra 1977), in contrast to the paleopathological approach that embraces the disease process. Consequentially, rather than attempting to generate yet another trauma classification system, the International Classification of Diseases (World

Health Organization 2007) could be considered as a model from which to structure future classifications of trauma.

OVERVIEW OF PAST PALEOTRAUMA RESEARCH

From the broader perspective that includes surgical intervention and cultural modification, ancient trepanation and cranial shaping were at the forefront of interest. Paul Broca's (1867) identification of trepanation on an Incan skull, though not the first account (Aufderheide and Rodríguez-Martín 1998), initiated a worldwide scramble to collect other ancient examples, but more importantly provided the underpinning for the theoretical discourse that followed and continues today (Lucas-Championnière 1878; Arnott et al. 2003). When we adhere to the clinical meaning that equates trauma with injury, anecdotal injuries were increasingly noted by 19th-century anatomists and physicians during their unrolling of Egyptian mummies or in craniometric studies (Pettigrew 1834; Morton 1839). More extensive descriptions were linked with archaeological expeditions. For example, Fouquet's (1896) paleopathological catalogue recorded injuries among 11 skeletons from El-Amrah, Egypt and compared the injury frequency to other Egyptian collections, thus foreshadowing the 20th-century epidemiological approach to paleopathology.

According to Armelagos (2003) and Ortner (2003) a landmark event for paleotrauma, and paleopathology in general, was the Archaeological Survey of Nubia in Egypt and Sudan beginning in 1907. Anatomists observed thousands of individuals and recorded macroscopic pathological lesions, notably injuries. This culminated in a single volume on human remains (Smith and Jones 1910), with an entire chapter devoted to fractures and dislocations (Jones 1910). Jones (1910:294) drew on his medical training to use clinical epidemiology as a means of interpreting the role of the environment and technology in his modal distribution of fractures. He (1910:297) emphasized the importance of the ubiquitous distal ulna fracture and credited Grafton Elliot Smith with attributing it to "fending a blow aimed at the head of the recipient" with the modern Nubian fighting stick the likely weapon; here, Jones made an early reference to gendered violence although unlike the stick-fighting activity he did not actually observe (or at least did not record) any violence against the local females. Fractures were differentiated as being antemortem or perimortem, and made by a blunt or sharp instrument. Of import was a stab wound between two ribs of a mummified individual (Jones 1910:334)—an early observation that fatal (or nonfatal) injury need not affect the bone. Evidence for mass execution and punishment at Shellal (Jones 1908, 1910:100–101, 334–336) included the *in situ* scraps of textile around the neck and/or wrists in addition to perimortem trauma. Finally, in an early reference to injury recidivism, Jones (1910:336) observed that several individuals had accumulated numerous healed injuries and proposed that these men were either soldiers or members of a tribe at war.

Hooton's (1930) epidemiological approach ushered in a new era for paleopathology (Armelagos 2003; Ortner 2003). Through quantification and fracture frequencies he inferred that earlier and later residents at Pecos Pueblo were more prone to intergroup violence (Hooton 1930:315). While Hooton addressed demographic variables within a group, the problem of differential preservation persisted. Lovejoy and Heiple (1981) tackled this issue and included only complete long bones and crania in their fracture analysis of Ohio's Libben people. They calculated an elemental fracture rate (total fractured bones/total bones observed × 100 %), and drawing on Buhr and Cooke's (1959) clinical model they determined that individuals were at greatest injury risk at 10–25 and 45+ years of age, thus introducing aspects of the lifecourse into our understanding of injury.

Lovejoy and Heiple's (1981) systematic fracture recording integrated with comparative clinical studies advocated by Jones's (1910) and the epidemiological approach of Hooton (1930) advanced our ability to more confidently assess injury patterns and treatment within a society, between contemporary communities and over time. Notably, Grauer and Roberts (1996) concluded that similar fracture frequencies among British urban communities placed residents at common injury risks, while the lack of healing deformity as determined from clinical data revealed that treatment was available to the poorest of individuals and was not restricted to those of high status. Similar approaches to trauma analysis have provided insight into a diverse range of social issues and the consequences of technological developments, such as power struggles (Andrushko et al. 2005), military strategies (Šlaus et al. 2010), ecological decline (Walker 1989; Torres-Rouff and Junqueira 2006), fertility rituals (Conlee 2007), peaceful and volatile imperial administrations (Buzon and Richman 2007; Erfan et al. 2009), hazardous occupations (Djurić et al. 2006; Van der Merwe et al. 2010), subsistence change (Domett and Tayles 2006), architectural sophistication (Kilgore et al. 1997) and surgical skill (Mitchell 2006; Redfern 2010). A particularly innovative analysis was that of Berger and Trinkhaus (1995), who proposed that close contact with big game animals was responsible for Neanderthal trauma owing to parallels with injuries sustained by modern rodeo performers.

Our Current Understanding

Since the 1990s, our ability to revisit samples of human remains has undergone a shift in response to changes in laws governing the repatriation and retention of human remains (e.g., Fforde et al. 2002; Ousley et al. 2006). Therefore, it is imperative that data is collected in a manner that allows it to be of use to future researchers. Our ability to "move with the times" and revise interpretations has resulted in increased efforts to improve the diagnosis and recording of different types of trauma, particularly for fractures and weapon injuries, in a range of archaeological periods and populations (Maat and Mastwijk 2000; Glencross and Stuart-Macadam 2001; Knüsel 2005; Lewis 2008; and many others). Perhaps the most popularized example is the ongoing re-examination of King Tutankhamun's mummy in response to new medical technologies (e.g., Hawass 2005; Hawass et al. 2010).

Recording trauma

Guidance on how to record trauma can be found in method documents, such as Buikstra and Ubelaker (1994), Brickley and McKinley (2004) and standards designed for the Global History of Health Project (Steckel et al. 2002), all of which were designed to promote standardization. Other work has focused on how fractures in particular should be recorded. Roberts (2006) and various colleagues (Roberts and Manchester 2005; Grauer and Roberts 1996) provide guidelines on what features and measurements should be documented in macroscopic and radiographic analyses. However, because of differences in funerary treatment, bone preservation and access to imaging facilities, researchers are often unable to fulfill these directions (Prokopec and Halman 1999, 2009, 2010; Redfern 2010). Roberts (2006:347) strongly emphasizes that the bone segment affected must be recorded to provide more detailed information on how the fracture was sustained and also to provide a true prevalence rate calculation. Buikstra and Ubelaker (1994:112–113) enforce the recording of pathology by segment, but little research has been undertaken on how the segment affected is determined, and how that influences data collection and interpretation (Judd 2002a). By correctly locating the fracture to the segment or joint affected and identifying its type (such as transverse, oblique or crush (e.g., Lovell 1997)), our ability to understand how the injury was sustained directly affects how we interpret the injury context and prevalence. The "parry" fracture is a case in point, as reviewed by Judd (2008) and discussed by others, for example, Smith (1996), Alvrus (1999), and Jurmain (2005): parry fractures typically result from a direct blow to the ulna when the arm is raised to shield the face and therefore have been used as indicators of interpersonal violence among past peoples. A more rigorous analysis using the parry signature gleaned from clinical work (transverse fracture on the distal ulna with no radial involvement and minimal unalignment) (Judd 2008) and radiography, aids in differentiating between injuries due to direct force from those obtained by an indirect force, thus facilitating our interpretation of social relations in past societies. Jurmain et al.'s (2009) re-evaluation of a prehistoric skeletal collection from California Bay found that only one of six fractures previously identified as parry trauma met with the criteria, thus greatly reducing the parry fracture frequency and purported victims of interpersonal violence. They concluded that the parry fracture was the least reliable indicator of interpersonal violence and promoted the use of multiple skeletal indicators to infer interpersonal violence.

Focusing on a few key paleoimaging developments, the introduction of portable digital radiography equipment revolutionized our ability to study mummified and skeletalized remains, which for various reasons cannot be removed from their curating institution (e.g., St. Bride's Church, London) or country of excavation (e.g., Egypt, Greece, Syria). This technology requires less training than traditional film radiography, it allows for more images to be taken in a shorter amount of time (Roberts 2010), and because it is a digital resource the acquired images can be more easily archived and disseminated via the internet. Developments in CT and MRI scanning have also improved our diagnostic accuracy in mummified remains, as the acquired images are clearer and can be stored in a digital format. Both forms of imaging produce 2-D cross-sections of tissue and 3-D reconstructions, which better clarifies the state of bone healing and the determination of weapon trajectory. The use of Alicona (2010) 3-D InfiniteFocus imaging microscopes promotes a greater understanding of traumatic injuries, particularly those inflicted by an instrument, because it allows for the creation of a 3-D image that can be quantified, facilitating better comparison to weapon profiles (e.g., Bello and Soligo 2008).

Accidental vs. intentional injury

Determining if a lesion was the product of accidental or intentional violence is reliant on identifying the causative instrument and/or injury mechanism, and integrating contextual evidence (Jurmain 2005:186–188). Some types of trauma, such as sharp-force weapon injuries have a more secure foundation for concluding that they were the result of intentional violence and are clearly differentiated from blunt-force trauma (Jurmain et al. 2009; Jurmain 2005:214–215). For example, injuries sustained by many individuals during the medieval Battles of Towton (England) and Wisby (Sweden) were irrefutably due to sharp weapons (Inglemark 1939; Novak 2000), although it should be noted that sharp-weapons can also produce bone fractures (Knüsel 2005; Novak 2000). However, for fractures it is not such a clear-cut situation. Clinical and forensic literature informs us that we must be cautious in interpreting these data and be aware that many fractures can be caused by an intentional action or accident (Berryman and Haun 1996; Greer and Williams 1999). Moreover, injuries commonly associated with accident, for example those resulting from a fall on an outstretched hand or from a height (Osifo et al. 2010), may be the consequence of an intentional push. In paleotrauma analysis we simply have no way of knowing the ultimate injury mechanism (Judd 2008).

It would seem intuitive that the presence of an embedded foreign object is an unambiguous indicator of intentional violence, but we cannot overlook accident. For example, injuries obtained while cleaning weapons, hunting, or handling weapons by inexperienced individuals are regular occurrences in modern emergency rooms (Krukemeyer et al. 2006). Similarly, injuries involving sharp objects used during domestic or occupational tasks may also produce sharp-force trauma (Hang et al. 2005). While some researchers argue that the direct evidence of an embedded weapon is the only reliable evidence of warfare and proximity of artifacts circumstantial (Jurmain 2005; Jurmain et al. 2009), others counter that we should not expect to see associated weapons, as the fundamental function of the weapon is to maim the essential soft tissue organs rather than hit the bone, and perhaps one of the protagonists removed the offending weapon after the deed is committed (Schulting 2006). Forensic and archaeological experiments have shown just this: the majority of lesions inflicted with a projectile or other sharp weapon, such as an arrow, knife, screwdriver or even gun do not touch the bone (Croft and Ferllini 2007; Smith et al. 2007; Letourneux and Petillon 2008). Furthermore, clinical studies report that the majority of assault injuries are soft tissue lacerations, abrasions, cuts and bruising, while broken bones account for 30 percent or less of the total injuries (Shepherd et al. 1990; Brink et al. 1998). Therefore, weapon-oriented trauma and nonintentional trauma, whether fatal or not, is appreciably underestimated in paleotrauma analysis.

Paleopathology has sought to provide its own methodologies in the absence or limited availability of clinical data, for example, the identification of sharp-force weapon trauma made by instruments no longer in use (Smith et al. 2007; Letourneux and Petillon 2008). The development of forensic anthropology has allowed its techniques to be employed on archaeologically derived material, which has dramatically improved our ability to recognize and classify blunt and sharp-force injuries (Berryman and Haun 1996; Novak 2000; Lewis 2008). This development is important in archaeological periods or locations where weaponry may be absent or recovered in low numbers (Armit et al. 2007; Redfern 2009). Differentiating perimortem injuries from postmortem damage remains problematic, but ongoing experimentation continues to hone differences (Barbian and Sledzik 2008; Wheatley 2008) thus negating inaccurate claims of violence, such as the alleged punctures observed in the Taung skull which were later identified as evidence of bird predation (Berger 2006).

Social and cultural meaning

Paleopathology has increasingly drawn on social science theory, such as feminist (Gilchrist 1999), gender (Gero and Conkey 1995), and age (Sofaer Derevenski 2000), in order to provide more nuanced interpretations. More recently, newer theoretical concepts built on these foundations have given rise to frameworks concerned with identity (Insoll 2007), lifecourse (Hutchinson 2008) and personhood (Fowler 2004). These shifts have greatly enhanced researchers' analyses of individual case studies (Knüsel 2002) as well as larger samples (Stirland 2001), and prompted those working with ancient people to not only critique their application to human remains (e.g., Geller 2005), but also to develop theoretical frameworks specific to human remains. Two significant approaches have come to the fore: the osteobiography approach advanced by Saul (1972; Saul and Saul 1989) and cultivated by Robb (1997; 2002), and that promoted by Sofaer (2006), which attempts to analyze the body in a "material culture" approach. Saul and Saul (1989) encapsulate the osteobiographical approach by posing four questions of the individual or sample under consideration: (1) Who was there? (2) Where did they come from? (3) What happened to them? and (4) What can be said about their way of life? Extending from this is the narrative biography or experienced cultural osteology employed by Robb (2002) to provide new insights into the individual lives, community as a whole and mortuary treatment. Sofaer (2006) identified the academic division between the dead and living body as a false separation of analysis and interpretation (e.g., exemplified in the appended display and discussion of osteological data in site reports), and argued that this arbitrary disconnection does not reflect the true relationship between a person's life, body and underlying skeleton. At first glance, this statement appears self-evident, particularly to paleopathologists. However, we frequently forget that the individuals we study did not have a homogeneous or static life from birth to death. Our greater understanding of skeletal remodeling over the short and long-term (Shaw and Stock 2009) and the interrelationships between immunity, health, culture, environment and society (e.g., Schell 1997; DeWitte 2010) make it very evident that individual lives are shaped by their biological sex, socio-economic status within the community, occupations performed and the environment in which they lived (Molnar 2006; Buzon and Richman 2007). These forces will result in modifications to the underlying skeleton, many of which we can observe, and should not be neglected by trauma studies. For example, young males who died during episodes of warfare or were considered to be warriors may have been interred with this status clearly signified in the burial record (Treherne 1995; Sarauw 2007). However, anthropological studies of the life course reveal that warriorhood may be a transitory stage (Foner and Kertzer 1978; Thomas 1995). Foner and Kertzer (1978:1085) reported that in the Kipsigis tribe of Kenya, the post-infancy male lifecourse is divided into boyhood, warriorhood, and elderhood, with the warrior stage having the highest prestige and personal freedom. Consequentially, survivors who enter other adult status groups may be buried without reference to this previous identity (Sayer 2010).

Trauma studies of interpersonal violence have increased in recent years (Park et al. 2010), perhaps in response to escalating levels of violence in our own communities, motivating paleopathologists to consider this behavior among past peoples, particularly against women and children (Smith 1996; Walker 1997; Walker et al. 1997). Child abuse is especially problematic as most children survive abusive injuries or the ensuing death is not detectible from the skeletal remains, although cases for abuse have been proposed by Blondiaux et al. (2002) in France and by Wheeler et al. (2007) in Egypt. If we look to clinical research for guidance, clinicians have been unsuccessful in their quest for skeletal abuse indicators among adults, although soft tissue facial injuries are profuse (Allen et al. 2007; Brink 2009). We tend to assume that domestic violence was perpetrated by males, when in fact cultural anthropologists observe that active female participation is widespread particularly in same-sex relationships (McClennen 2005) or in polygynous societies, where altercations among co-wives are regarded as unimportant "nonevents" (Burbank 1994). Active female participation in violent acts towards men, children and other women is rarely deliberated by paleopathologists despite a growing body of bioarchaeological research (Guliaev 2003) and mounting contemporary findings (Mechem et al. 1999; Hirschinger et al. 2003). Ethnographic sources not only provide alternatives to consider in our interpretations of trauma but inform the legitimacy of violence within gender and power hierarchies. For example, among the indigenous residents of Mangrove (Australia) fights are mundane occurrences sanctioned by specific rules as to who one does not fight (mothers-in-law, opposite-sex sibling) and which bodily regions (hands, arms, legs) are acceptable to hit with one's fighting stick (Burbank 1994). A paleopathologist studying skeletal remains from this community might therefore attribute the lack of cranial and sharp-force trauma to environmental hazards and accident rather than widespread socially sanctioned interpersonal violence among adults. More importantly, this example illustrates that what we consider to be abusive or deviant behavior is acceptable and expected by contemporary societies elsewhere.

As a discipline we sometimes fail to appreciate that violence is extraordinarily meaningful to the construction and maintenance of identity, gender, age and status (Singer 2006; Sen 2007; Terrell et al. 2008; Graves 2010), and was used by many past populations in religious ceremonies and to enforce social and economic inequalities (Verano 2005; Chacon and Dye 2008). Therefore, we should also be aware of the discrepancy between the meaning of violent acts and the often unpleasant nature of evidence itself. For example, during the Inca Empire children were sacrificed to the Gods, either killed by a blow to the head, strangled or asphyxiated during *capacocha* ceremonies. These children were regarded as messengers or representatives of their communities to the gods, and selection was considered to be a great honor; notably, they were not subject to physical deprivation before being sacrificed (Ceruti 2004).

Paleopathologists have been guilty of making interpretations based on trauma data collected from human remains without first considering the social and cultural meanings of violence in the community under study. This has impacted greatly on how individuals or samples are interpreted within paleopathology (see Jurmain 2005:227–229), but also how these interpretations are perpetuated within the anthropological literature. For example, Chadwick-Hawkes and Wells (1975) asserted that ossified muscle traumas identified on the femora of a young Anglo-Saxon female were the result of a violent rape. This interpretation was incorporated into the wider literature and was not critiqued or questioned until Andrew Reynolds (1988) and later Nicholas Reynolds (2009) demonstrated that this was a classic case of over-interpretation. Nevertheless, the false assertion continues to be included in research (e.g., Arnold 1997). A more famous case is that of Özti the Iceman, an ice-mummy recovered from a glacier at the Italian–Austrian alpine border in 1991, who dates to the Chalcolithic period (3350–3300 B.C.). While the mummy displays evidence of participating in hand-to-hand combat shortly before death, and has an embedded arrow injury in the back, numerous scientists, using trauma and other data, have attempted to reconstruct his identity and the manner and cause of his death; interpretations have ranged from participation in organized violence, human sacrifice and murder, but now the consensus is that he died whilst fleeing from danger (EURAC 2010; Pernter et al. 2007; Nerlich et al. 2009).

Body modification as trauma

As indicated above, not all scholars regard evidence for body modification in life and death as trauma. This may be because some researchers recognize the wider meanings associated with these acts that may not be considered as violent, such as in the construction of gender and status among Italian Iron Age communities (Robb 1997), the maintenance of ethnic identities in periods of sociocultural change through lip piercings and head shaping in the Late Intermediate Period of Chile (e.g., Knudson and Torres-Rouff 2009; Torres-Rouff 2009), or the recognition that the human body can be used to transcend individual identities for wider community purposes, such as traditional Buddhist practices in Tibet which transform human remains into ritual objects (e.g., flutes or bowls) (Malville 2005). Cannibalism epitomizes these differences in approach, as this practice is used as a survival mechanism, to subjugate enemies, and a stage in funerary practices (amongst others, Keenleyside et al. 1997; Degusta 1999; Billman et al. 2000; Beaver 2002; Gottlieb 2007; Lindenbaum 2009). The skeletal markers of cannibalism are similar cross-culturally and temporally (White 1992; Turner and Turner 1998), therefore it is interesting to observe how differently these markers can be presented, interpreted and judged as evidence of the practice (Billman, et al. 2000; Dongoske et al. 2000). It should also be noted that the vast majority of evidence for body modification is presented and discussed as case studies (McNeill 2005), and only in recent years have attempts been made to understand the evidence in its regional, cultural and historical context (Chacon and Mendoza 2007; Torres-Rouff 2009).

Limitations

One hundred years have passed since Smith and Jones's 1910 landmark trauma report and we still struggle with many of the same difficulties specific to trauma that challenged these pioneering paleopathologists. Taphonomy can obscure, mimic or obliterate bone, particularly perimortem trauma where the freshness of the lesion is attractive to scavengers (Lotan 2000; Denys 2002; Berger 2006; Wheatley 2008). We do not know the age at which the fracture occurred and therefore the susceptibility of one age-group to injury cannot be determined unless the fracture was in the process of healing or perimortem (Maat and Huls 2009). The contemporaneity of multiple antemortem injuries is also problematic, although a case for injury recidivism has been argued by Judd (2002b). There is scant evidence for subadult trauma in the paleopathological record (Lewis 2007), although it may be that the reparative resilience of the child's bones masked earlier injuries or that the child survived and entered the sample as an adult. We must also consider that the majority of ancient injuries and accidental fatalities will be undetectable even in the most pristinely preserved skeletons if only soft tissue was involved or as in modern developed and developing nations, the most fatal injuries affecting subadults, such as drowning, burns, fatal stings and poison ingestion, leave no skeletal traces (Baker et al. 1984).

It is often assumed that individuals with violence-related injuries were passive victims. For example, the widely cited site of Jebel Sahaba in Sudan (Wendorf 1968), as being the earliest evidence for warfare based on *some* embedded flakes and multiple burials, is presented by researchers, such as Guilaine and Zammit (2005), as a community massacre even though there is no archaeological evidence that these people were help-less victims. If there were an attack or series of attacks, it may well be that the alleged victims were the initiators of a blood-feud or retaliation. This problematic assignment of "victim" and "assailant" has been dealt with in clinical settings, where injured individuals are nonjudgmentally referred to as "participants" (Shepherd et al. 1990).

While clinical and ethnographic research provides a wealth of information from which to draw interferences and structure hypotheses, it is similarly problematic. The clinical classification of physical trauma is subject to ongoing alterations, because of advances in medical science, the revision of outdated nomenclature, and technological changes in warfare and daily-life, for example, the use of radiographs from 1895 and the introduction of automatic weapons. Clinicians disagree on the origins of minor traumas that also perplex paleopathologists, notably dental avulsion, Harris lines, osteochondritis dissecans, spondylolysis, Schmorl's nodes and myositis ossificans traumatica (Siffert and Katz 1983; Schenck and Goodnight 1996; Bastone et al. 2000; Battie, et al. 2008). Finally, not all injuries or abuse cases are reported or are attended to as a result of proximity, lack of documentation of patient, expense, social inaccessibility, fear or stigma (Mock et al. 1995; Nordberg 1994).

FUTURE CONSIDERATIONS

When considering future directions for paleotrauma research we suggest three key areas: the inclusion of minor trauma; the association between disability, other diseases and injury patterns and prevalence; and the social consequences of traumatic injury. Minor traumas, such as injuries to the dentition, extremities and flat bones of the torso (scapulae, ribs, manubrium and sternum) are often neglected by trauma studies, which predominantly focus on cranial or long-bone injuries. We recognize that such trends are influenced by factors such as funerary strategies, recovery methods employed during excavations and preservation, but clinical studies emphasize the importance of injury patterning to these areas to determine interpersonal vio-lence, accidental or intentional falls, and occupational injury. A small number of studies have shown that important sociocultural inferences can be made from the study of these injuries (e.g., Brickley and Smith 2006). Dental injuries are noticeably absent (Jurmain 2005; Lovell 2008), although they are sometimes included with dental disease (e.g., Ortner, 2003) or occupational modifications (Molnar 2008; Scott and Winn in press). This may be because identification and classification can be problematic and made more troublesome by taphonomic changes (Hinton 1981). Interestingly, the exclusion of dental trauma assessment is not exclusive to paleopathology. Lieger et al. (2009) observed that clinicians rarely differentiate between maxillary or mandibular injuries, and therefore, the true prevalence of dental injury is obscured even in clinical trauma studies. Clinical research shows that the majority of dental injuries are sustained during childhood by falls, play and sports (Bastone et al. 2000; Eyuboglu et al. 2009). In adults, tooth injuries in isolation are in fact seldom pathognomonic of interpersonal violence (Muelleman et al. 1996). Studies such as Holst and Couglan's (2000) detailed assessment of the medieval Towton soldiers and Lukacs's (2007) recent work on dental trauma among ancient Canary Islanders underscore the value of integrating contextual information and evidence for other craniofacial trauma when interpreting these and other direct injuries to the face.

In many trauma studies data collection and interpretation often fails to display or discuss data in relation to other conditions. For example, consider an older female with a Colles' fracture (Mays 2000). Is this evidence for a fall or osteoporosis? Inevitably, the role of classification and access to radiography plays a role in these decisions. We have also neglected to address disabilities and their influence and association with patterns of trauma, as most clearly shown in the study of leprosy and fractures by Judd and Roberts (1998), but also how the co-existence of diseases, such as septic arthritis or tuberculosis, can influence the prevalence and pattern of injuries in individuals (Ferrara and Peterson 2000).

Finally, the emphasis on the populational approach that advocates the health of the group (e.g., Steckel and Rose 2002) to the detriment of the individual depersonalizes trauma, glossing over the devastating impact of the injury on the individual's quality of life and emotional well-being (Mays 2006), and that of their family, friends and community. However, several researchers have provided comprehensive osteobiographies to underscore the debilitating effects of trauma on the individual (Hawkey 1998; Neri and Lancellotti 2004; Mays 2006). Others approach society's response to circumventing more severe impairment through treatment and advances in medical knowledge (Dupras et al. 2010; Redfern 2010).

CONCLUSION

Paleopathologists investigating ancient trauma are presented with an individual's lifetime accumulation of injuries and attempt to explain the injury mechanisms, in addition to the contemporaneity of injuries. As methods of distinguishing and recording ancient injuries increasingly cross forensic, clinical and archaeological boundaries, we might venture to differentiate injury in the clinical and forensic sense, from the broader culturally inclusive meaning of trauma that would include modifications to the body (any tissue) in order to heal, beautify or distinguish. But suppose that we are presented with an individual with a healed amputated forearm? Without documentation and/or other distinguishing funerary features, how do we determine if the amputation was a surgical intervention, an occupational or sports accident, an intentional mutilation received during warfare or a form of punishment with the social debt settled? The fundamental issue of defining paleotrauma is unresolved and the more that we endeavor to categorize lesions that deform, disfigure, disable or destroy the individual the more reasons we find not to—at least from an anthropological perspective. In most cases, paleopathologists will be at a disadvantage, unaware of the cognitive behavior, the individual(s) and the circumstances that initiated the traumatic event. We would be better served to integrate additional lines of contextual evidence and consider how the individual experienced life as a result of their injuries.

REFERENCES

Alicona, 2010   http://www.alicona.com/cms/front_content.php?idcat=11. Accessed June 3, 2011.

Allen, T., Novak, S. A., and Bench, L. L., 2007   Patterns of Injuries: Accident Or Abuse? Violence Against Women 13:802–816.

Alvrus, A., 1999   Fracture Patterns among the Nubians of Semna South, Sudanese Nubia. International Journal of Osteoarchaeology 9:417–429.

Andrushko, V. A., Latham, K. A. S., Grady, D. L., Pastron, A. G., and Walker, P. L., 2005   Bioarchaeological Evidence for Trophy-Taking in Prehistoric Central California. American Journal of Physical Anthropology 127:375–384.

Armelagos, G. J., 2003   Bioarchaeology as Anthropology. In Archaeology is Anthropology. S. Gillespie, and D. Nichols, eds. pp. 27–40. Arlington, VA: Archeological Papers of the American Anthropological Association, no. 13.

Armit, I., Knüsel, C., Robb, J., and Schulting, R., 2007   Warfare and Violence in Prehistoric Europe: An Introduction. In War and Sacrifice. T. Pollard, and I. Banks, eds. pp. 1–10. Studies in the Archaeology of Conflict. Leiden: Brill.

Arnold, C. J., 1997   An Archaeology of the Early Anglo-Saxon Kingdoms. London: Routledge.

Arnott, R., Finger, S., and Smith, C. U. M., 2003   Trepanation. History, Discovery and Theory. Lisse (Netherlands): Swets and Zeitlinger.

Aufderheide, A. C., and Rodríguez-Martín, C., 1998   Cambridge Encyclopedia of Human Paleopathology. Cambridge: Cambridge University Press.

Baker, S. P., O'Neill, B., and Karpf, R. S., 1984   The Injury Fact Book. Lexington, MA: Lexington Books.

Barbian, L. T., and Sledzik, P. S., 2008   Healing Following Cranial Trauma. Journal of Forensic Sciences 53:263–268.

Bastone, E. B., Freer, T. J., and McNamara, J. R., 2000   Epidemiology of Dental Trauma: A Review of the Literature. Australian Dental Journal 45:2–9.

Battie, M. C., Videman, T., Levalahti, E., Gill, K., and Kaprio, J., 2008   Genetic and Environmental Effects on Disc Degeneration by Phenotype and Spinal Level. Spine 33: 2801–2808.

Beaver, D., 2002   Flesh or Fantasy:Cannibalism and the Meanings of Violence. Ethnohistory 49:671–685.

Bello, S. M., and Soligo, C., 2008   A New Method for the Quantitative Analysis of Cutmark Micromorphology. Journal of Archaeological Science 35:1542–1552.

Bennike, P., 2008   Trauma. In Advances in Human Palaeopathology. R. Pinhasi, and S. Mays, eds. pp. 309–328. Chichester: John Wiley & Sons Ltd.

Berger, L. R., 2006   Predatroy Bird Damage to the Taung Type-Skull Of Austalopithecus Africanus Dart 1925. American Journal of Physical Anthropology 131:166–168.

Berger, T. D., and Trinkaus, E., 1995   Patterns of Trauma among the Neandertals. Journal of Archaeological Science 22:841–852.

Berryman, H. E., and Haun, S. J., 1996   Applying Forensic Techniques to Interpret Cranial Fracture Patterns in an Archaeological Specimen. International Journal of Osteoarchaeology 6:2–9.

Billman, B. R., Lambert, P. M., and Leonard, B. L., 2000   Cannibalism, Warfare and Drought in The Mesa Verde Region During the Twelfth Century A.D. American Antiquity 65:145–178.

Blondiaux, G., Blondiaux, J., Secousse, F., Cotten, A., Danze, P., and Flipo, R., 2002   Rickets and Child Abuse: The Case of a Two Year Old Girl From the 4th Century in Lisieux (Normandy). International Journal of Osteoarchaeology 12:209–215.

Blum, L. S., Khan, R., Hyder, A. A., Shahanaj, S., El Arifeen, S., and Baqui., A., 2009   Childhood Drowning in Matlab, Bangladesh: An In-Depth Exploration of Community Perceptions and Practices. Social Science & Medicine 68:1720–1727.

Boylston, A., 2006   Evidence for Weapon-Related Trauma in British Archaeological Samples. In Human Osteology in Archaeology and Forensic Science. M. Cox, and S. Mays, eds. pp. 357–380. Cambridge: Cambridge University Press.

Brickley, M., and McKinley, J. I., eds., 2004   Guidelines to the Standards for Recording Human Remains. Volume Paper No. 7. Reading, UK: IFA BABAO.

Brickley, M., and Smith, M., 2006   Culturally Determined Patterns of Violence: Biological Anthropological Investigations at a Historic Urban Cemetery. American Anthropologist 108:163–177.

Brink, O., 2009   When Violence Strikes the Head, Neck, and Face. Journal of Trauma 67:147–151.

Brink, O., Vesterby, A., and Jensen, J., 1998   Pattern of Injuries due to Interpersonal Violence. Injury 29:705–709.

Broca, P., 1867   Trépanation chez les Incas. Bulletin de la Société d'Anthropologe de Paris 2:403–408.

Brothwell, D. R., 1981   Digging Up Bones. Ithaca, New York: Cornell University Press.

Buhr, A. J., and Cooke, A. M., 1959   Fracture Patterns. The Lancet March 14:531–536.

Buikstra, J. E., 1977   Biocultural Dimensions of Archaeological Study: A Regional Perspective. In Biocultural Adaptation in Prehistoric America. R. Blakely, ed. pp. 67–84. Athens: University of Georgia Press.

Buikstra, J. E., and Ubelaker, D. H., eds., 1994   Standards for Data Collection from Human Skeletal Remains. Fayetteville: Arkansas Archaeological Survey Research Series, Vol. 44.

Burbank, V. K., 1994   Fighting Women: Anger and Aggression in Aboriginal Australia. Berkeley: University of California Press.

Buzon, M., and Richman, R., 2007   Traumatic Injuries and Imperialism: the Effects of Egyptian Colonial Strategies at Tombos in Upper Nubia. American Journal of Physical Anthropology 133:783–791.

Ceruti, C., 2004   Human Bodies as Objects of Dedication at Inca Mountain Shrines (North-Western Argentina). World Archaeology 36:103–122.

Chacon, R. J., and Dye, D. H., eds., 2008   The Taking and Displaying of Human Body Parts as Trophies by Amerindians. New York: Springer.

Chacon, R. J., and Mendoza, R. G., eds., 2007   North American Indigenous Warfare and Ritual Violence. Tucson: University of Arizona Press.

Chadwick-Hawkes, S., and Wells, C., 1975   Crime and Punishment in an Anglo-Saxon Cemetery? Antiquity 49:118–122.

Chhem, R. K., Saab, G., and Brothwell, D. R., 2008   Diagnostic Paleoradiology for Paleopathologists. In Paleoradiology. Imaging Mummies and Fossils. R. K. Chhem and D. R. Brothwell, eds. pp. 73–118. Heidelberg: Springer.

Conlee, C. A., 2007   Decapitation and Rebirth. Current Anthropology 48: 438–446.

Cox, M., 2006   Assessment of Parturition. In Human Osteology in Archaeology and Forensic Science. M. Cox, and S. Mays, eds. pp. 131–142. Cambridge: Cambridge University Press.

Cox, M., and Mays, S., eds., 2006   Human Osteology in Archaeology and Forensic Science. Cambridge: Cambridge University Press.

Croft, A. M., and Ferllini, R., 2007   Macroscopic Characteristics of Screwdriver Trauma. Journal of Forensic Sciences 52:1243–1251.

Degusta, D., 1999   Fijian Cannibalism: Osteological Evidence from Navatu. American Journal of Physical Anthropology 110:215–241.

Denys, C., 2002   Taphonomy and Experimentation. Archaeometry 3:469–484.

DeWitte, S. N., 2010   Sex differentials in Fraility in Medieval England. American Journal of Physical Anthropology 143:285–297.

Djurić, M. P., Roberts, C. A., Rakočević, Z. B., Djonić, D. D., and Lešić, A. R., 2006   Fractures in Late Medieval Skeletal Populations from Serbia. American Journal of Physical Anthropology 130:167–78.

Domett, K. C., and Tayles, N., 2006   Adult Fracture Patterns in Prehistoric Thailand. International Journal of Osteoarchaeology 16:185–99.

Dongoske, K. E., Martin, D. L., and Ferguson, T. J., 2000   Critique of the Claim of Cannibalism at Cowboy Wash. American Antiquity 65:179–190.

Dupras, T. L., Williams, L. J., De Meyer, M., Peeters, C. Depraetere, D., Vanthuyne, B., and Willems, H., 2010.   Evidence of Amputation as Medical Treatment in Ancient Egypt. International Journal of Osteoarchaeology 20:405–23.

Erfan, M., El-Sawaf, A., Soliman, M. A., El-Din, A. S., Kandeel, W. A., El-Banna, R. A. E., and Azab, A., 2009   Cranial Trauma in Ancient Egyptians from the Bahriyah Oasis, Greco-Roman Period. Research Journal of Medicine and Medical Sciences 4:74–84.

EURAC, 2010   The Iceman. Bolzano: Institute for Mummies and the Iceman.

Eyuboglu, O., Yilmaz, Y., Zehir, C., and Sahin, H., 2009   A 6-Year Investigation into Types of Dental Trauma Treated in a Paediatric Dentistry Clinic in Eastern Anatolia Region, Turkey. Dental Traumatology 25:110–114.

Ferrara, M. S., and Peterson, C. L., 2000   Injuries to Athletes with Disabilities: Identifying Injury Patterns. Sports Medicine 30:137–43.

Fforde, C., Hubert, J., and Turnbull, P., eds., 2002   The Dead and Their Possessions. Repatriation in Principle, Policy and Practice. London: Routledge.

Foner, A., and Kertzer, D., 1978   Transitions over the Life Course: Lessons from Age-Set Societies. The American Journal of Sociology 83:1081–1104.

Fouquet, D. M., 1896   Appendice: Note sur les Squelettes d'El-'Amrah. In Recherches sur les Origines de L'Égypte: L'Age de la Pierre et les Métaux. J. De Morgan, ed. pp. 241–270. Paris: Ernest Leroux.

Fowler, C., 2004   The Archaeology of Personhood. An Anthropological Approach. London: Routledge.

Geller, P. L., 2005   Skeletal Analysis and Theoretical Complications. World Archaeology 37: 597–609.

Gero, J. M., and Conkey, W. M., eds., 1995   Engendering Archaeology. Women and Prehistory. Oxford: Blackwell.

Gilchrist, R., 1999   Gender and Archaeology. Contesting the Past. London: Routledge.

Glencross, B., and Stuart-Macadam, P., 2001   Radiographic Clues to Fractures of the Distal Humerus in Archaeological Remains. International Journal of Osteoarchaeology 11: 298–310.

Gottlieb, R. M., 2007   The Reassembly of the Body from Parts: Psychoanalytic Reflections on Death, Resurrection, and Cannibalism. Journal of the American Psychoanalytical Association 55:1217–1251.

Grauer, A. L., and Roberts, C. A., 1996   Paleoepidemiology, Healing, and Possible Treatment of Trauma in the Medieval Cemetery Population of St. Helen-on-the-Walls, York, England. American Journal of Physical Anthropology 100:531–544.

Graves, B. M., 2010   Ritualized Combat as an Indicator of Intrasexual Selection Effects on Male Life History Evolution. American Journal of Human Biology 22:45–49.

Greer, S. E., and Williams, J. M., 1999   Boxer's Fracture: An Indicator of Intentional and Recurrent Injury. American Journal of Emergency Medicine 17:357–360.

Guilaine, J., and Zammit, J., 2005   The Origins of War: Violence in Prehistory. M. Hersey, transl. Malden, MA: Blackwell.

Guliaev, V. I., 2003   Amazons in the Scythia: New Finds at the Middle Don, Southern Russia. World Archaeology 35:112–125.

Hang, H. M., Bach, T. T., and Byass, P., 2005   Unintentional Injuries over a 1-Year Period in a Rural Vietnamese Community: Describing an Iceberg. Public Health 119:466–473.

Hawass, Z., 2005   Press Release-Tutankhamun CT scan. http://drhawass.com/blog/press-release-tutankhamun-ct-scan. Accessed December 7, 2009.

Hawass, Z., Gad, Y. Z., Ismail, S., et al., 2010   Ancestry and Pathology in King Tutankhamun's Family. Journal of the American Medical Association 303:638–647.

Hawkey, D. E., 1998   Disability, Compassion and the Skeletal Record: Using Musculoskeletal Stress Markers (MSM) to Construct an Osteobiography from Early New Mexico. International Journal of Osteoarchaeology 8:326–340.

Hinton, R. J., 1981   Form and Patterning of Anterior Tooth Wear among Aboriginal Human Groups. American Journal of Physical Anthropology 54:555–564.

Hippocrates,   On Injuries to the Head. Section 11. Francis Adams, trans. The Internet Classics Archive. D. C. Stevenson, Web Atomics. http://classics.mit.edu/Hippocrates/headinjur.11.11.html, accessed June 3, 2011.

Hirschinger, N. B., Grisso, J. A., Wallace, D. B. et al., 2003   A Case-Control Study of Female-To-Female Nonintimate Violence in an Urban Area. American Journal of Public Health 93:1098–1103.

Holst, M., and Coughlan, J., 2000   Dental Health and Disease. In Blood Red Roses. V. Fiorato, A. Boylston, and C. Knüsel, eds. pp. 77–89. Oxford: Oxbow Books.

Hooton, E. A., 1930   The Indians of Pecos Pueblo: A Study of their Skeletal Remains. New Haven: Yale University Press.

Hutchinson, E. D., 2008   A Life Course Perspective. In Dimensions of Human Behavior. Third edition. E. D. Hutchinson, ed. pp. 3–38. London: Sage Publications Inc.

Inglemark, B. E., 1939   The Skeletons. In Arms and Armour from the Battle of Wisby, 1361. B. Thordeman, ed. pp. 149–209. Stockholm: Vitterhets Historie Och Antikvitets Akademien.

Insoll, T., ed., 2007   The Archaeology of Identities, A Reader. London: Routledge.

Jones, F. W., 1908   The Examination of Bodies of 100 Men Executed in Nubia in Roman Times. British Medical Journal March 28 (2465):736–737.

Jones, F. W., 1910   Fractured Bones and Dislocations. In The Archaeological Survey of Nubia. Report for 1907–1908. vol. 2: Report on the Human Remains. G. E. Smith and F. W. Jones, eds. pp. 293–342. Cairo: National Printing Department.

Judd, M. A., 2002a   Comparison of Long Bone Recording Methods. Journal of Archaeological Science 29:1255–1265.

Judd, M. A., 2002b   Ancient Injury Recidivism: An Example from the Kerma Period of Ancient Nubia. International Journal of Osteoarchaeology 12:89–106.

Judd, M. A., 2008   The Parry Problem. Journal of Archaeological Science 35:1658–1666.

Judd, M. A., and Roberts, C. A., 1998   Fracture Patterns at the Medieval Leper Hospital in Chichester. American Journal of Physical Anthropology 105:43–55.

Jurmain, R., 2005   Stories from the Skeleton: Behavioral Reconstruction in Human Osteology. London: Taylor and Francis.

Jurmain, R., Bartelink, E. J., Leventhal, A., Bellifemine, V., Nechayev, I., Atwood, M., and DiGiuseppe, D., 2009   Paleoepidemiological Patterns of Interpersonal Aggression in a Prehistoric Central California Population from CA-ALA-329. American Journal of Physical Anthropology 139:462–473.

Keenleyside, A., Bertulli, M., and Fricke, H. C., 1997   The Final Days of the Franklin Expedition: New Skeletal Evidence. Arctic 50:36–46.

Kilgore, L., Jurmain, R., and Van Gerven, D., 1997   Palaeoepidemiological Patterns of Trauma in a Medieval Nubian Skeletal Population. International Journal of Osteoarchaeology 7:103–114.

Kirmayer, L. J., Lemelson, R., and Barad, M., eds., 2008   Understanding Trauma. Integrating Biological, Clinical and Cultural Perspectives. New York: Cambridge University Press.

Knudson, K. J., and Torres-Rouff, C., 2009   Investigating Cultural Heterogeneity and Multiethnicity in San Pedro De Atacama, Northern Chile Through Biogeochemistry and Bioarchaeology. American Journal of Physical Anthropology 138:473–485.

Knüsel, C. J., 2002   More Circe than Cassandra: The Princess of Vix in Ritualized Social Context. European Journal of Archaeology 5:275–307.

Knüsel, C. J., 2005   The Physical Evidence for Warfare: A Subtle Stigmata? In Warfare, Violence and Slavery in Prehistory. M. Parker Pearson and I. J. N. Thorpe, eds. pp. 49–66. Oxford: British Archaeological Reports, International Series S1374.

Krukemeyer, M. G., Grellner, W., Gehrke, G., Koops, E., and Puschel, K., 2006   Survived Crossbow Injuries. American Journal of Forensic Medicine and Pathology 27:274–276.

Letourneux, C., and Petillon, J., 2008   Hunting Lesions Caused by Osseous Projectile Points: Experimental Results and Archaeological Implications. Journal of Archaeological Science 35:2849–2862.

Lewis, J. E., 2008   Identifying Sword Marks on Bone: Criteria for Distinguishing Between Cut Marks Made by Different Classes of Bladed Weapons. Journal of Archaeological Science 35:2001–08.

Lewis, M. E., 2007   The Bioarchaeology of Children. Cambridge:Cambridge University Press.

Lieger, O., Zix, J., Kruse, A., and Lizuka, T., 2009   Dental Injuries in Association With Facial Fractures. Journal of Oral and Maxillofacial Surgery 67:1680–1684.

Lindenbaum, S., 2009   Cannibalism, Kuru and Anthropology. Folia Neuropathologica 47:138–144.

Lotan, E., 2000   Feeding the Scavengers. Actualistic Taphonomy in the Jordan Valley, Israel. International Journal of Osteoarchaeology 10:407–425.

Lovejoy, C. O., and Heiple, K. G., 1981   The Analysis of Fractures in Skeletal Populations with an Example from the Libben Site, Ottowa, Ohio. American Journal of Physical Anthropology 55:529–541.

Lovell, N. C., 1997   Trauma Analysis in Paleopathology. Yearbook of Physical Anthropology 40:139–170.

Lovell, N. C., 2008   Analysis and Interpretation of Trauma. In Biological Anthropology of the Human Skeleton. M. A. Katzenberg, and S. A. Saunders, eds. pp. 341–386. Hoboken, NJ: Wiley-Liss.

Lucas-Championnière, J. M. M., 1878   La Trépanation Guidée par les Localisations Cérébrales. Paris: V. A. Delahaye et Co.

Lukacs, J. R., 2007   Dental Trauma and Antemortem Tooth Loss in Prehistoric Canary Islanders: Prevalence and Contributing Factors. International Journal of Osteoarchaeology 17:157–173.

Maat, G. J. R., and Huls, N., 2009   Histological Fracture Dating of Fresh and Dried Bone Tissue. In Forensic Aspects of Pediatric Fractures. R. A. C. Bilo, S. G. F. Robben, and R. R. Van Rijn, eds. pp. 194–201. London: Springer.

Maat, G. J. R., and Mastwijk, R. W., 2000   Avulsion Injuries of Vertebral Endplates. International Journal of Osteoarchaeology 10:142–152.

Malville, N. J., 2005   Mortuary Practices and the Ritual Use of Human Bone in Tibet. In Interacting With the Dead. Perspectives on Mortuary Archaeology for the New Millennium. G. F. M. Rakita, J. E. Buikstra, L. A. Beck, and S. R. Williams, eds. pp. 190–204. Florida: University Press of Florida.

Mann, R. W., and Murphy, S. P., 1990   Regional Atlas of Bone Disease. Springfield, Illinois: Charles C. Thomas.

Mays, S. A., 2000   Age-Dependent Cortical Bone Loss in Women From 18th and Early 19th Century London. American Journal of Physical Anthropology 112:349–361.

Mays, S. A., 2006   A Palaeopathological Study of Colles' Fracture. International Journal of Osteoarchaeology 16:415–428.

Mays, S. A., 2010   The Archaeology of Human Bones. London: Routledge.

McClennen, J. C., 2005   Domestic Violence between Same-Gender Partners. Journal of Interpersonal Violence 20:149–154.

McNeill, J. R., 2005   Putting the Dead to Work: An Examination of the Use of Human Bone in Prehistoric Guam. *In* Interacting with the Dead. Perspectives on Mortuary Archaeology for the New Millennium. G. F. M. Rakita, J. E. Buikstra, L. A. Beck, and S. R. Williams, eds. pp. 305–316. Florida: University Press of Florida.

Mechem, C. C., Shofer, F. S., Reinhard, S. S., Hornig, S., and Datner, E., 1999   History of Domestic Violence among Male Patients Presenting to an Urban Emergency Department. Academic Emergency Medicine 6:786–791.

Merbs, C. F., 1989   Trauma. *In* Reconstruction of Life from the Skeleton. M. Y. İşcan and K. A. R. Kennedy, eds. pp. 161–189. New York: Alan R Liss.

Mitchell, P. D., 2006   Trauma in the Crusader Period City of Caesarea: A Major Port in the Medieval Eastern Mediterranean. International Journal of Osteoarchaeology 16:493–505.

Mock, C. N., Adzotor, E., Denno, D., Conklin, E., and Rivara, F., 1995   Admissions for Injury at a Rural Hospital in Ghana: Implications for Prevention in the Developing World. American Journal of Public Health 85:927–931.

Molnar, P., 2006   Tracing Prehistoric Activities: Musculoskeletal Stress Marker Analysis of a Stone-Age Population on the Island of Gotland in the Baltic Sea. American Journal of Physical Anthropology 129:12–23.

Molnar, P., 2008   Dental Wear and Oral Pathology: Possible Evidence and Consequences of Habitual Use of Teeth in a Swedish Neolithic Sample. American Journal of Physical Anthropology 136:423–431.

Morton, S. G., 1839   Crania Americana; or, A Comparative View of the Skulls of Various Aboriginal Nations of North and South America. Philadelphia: Dobson.

Muelleman, R., Lenaghan, P., and Pakieser, R., 1996   Battered Women: Injury Locations and Types. Annals of Emergency Medicine 28:486–492.

Neri, R., and Lancellotti, L., 2004   Fractures of the Lower Limbs and their Secondary Skeletal Adaptations: A 20th Century Example of Pre-Modern Healing. International Journal of Osteoarchaeology 14:60–66.

Nerlich, A. G., Peschel, O., and Vigl, E. E., 2009   New Evidence for Özti's Final Trauma. Intensive Care Medicine 35:1138–1139.

Nordberg, E., 1994   Injuries in Africa: A Review. East African Medical Journal 71:339–345.

Novak, S. A., 2000   Battle-related Trauma. *In* Blood Red Roses. The Archaeology of a Mass Grave from the Battle of Towton AD 1461. V. Fiorato, A. Boylston, and C. Knüsel, eds. pp. 90–102. Oxford: Oxbow Books.

Oakley, A., 2007   Fracture. Adventures of a Broken Body. Bristol: The Policy Press.

Ortner, D. J., 2003   Identification of Pathological Conditions in Human Skeletal Remains. 2nd edition. San Diego: Academic Press.

Ortner, D. J., and Putschar, W. G. J., 1981   Identification of Pathological Conditions in Human Skeletal Remains. Washington, DC: Smithsonian.

Osifo, O. D., Iribhogbe, P., and Idiodi-Thomas, H., 2010   Falls From Heights: Epidemiology and Pattern of Injury at the Accident and Emergency Centre of the University of Benin Teaching Hospital. Injury 41:544–547.

Ousley, S., Billeck, W., and Hollinger, R. E., 2006   Federal Repatriation Legislation and the Role of Physical Anthropology in Repatriation. Yearbook of Physical Anthropology 48:2–32.

Oxford Medical Dictionary, 2000   Concise Oxford Medical Dictionary. 5th edition. E. A. Martin, ed. Oxford: Oxford University Press.

Park, V. M., Roberts, C. A., and Jakob, T., 2010   Palaeopathology in Britain: A Critical Analysis of Publications with the Aim of Exploring Recent Trends (1997–2006). International Journal of Osteoarchaeology 20:497–507.

Peden, M., McGee, K., and Sharma, G., 2002   The Injury Chart Book: A Graphical Overview of the Global Burden of Injuries. Geneva: World Health Organization.

Pernter P., Gostner, P., Vigl, E. E., and Rühli, F. J., 2007   Radiologic Proof for the Iceman's Cause Of Death (Ca. 5300 BP). Journal of Archaeological Science 34:1784–1786.

Pettigrew, T. J., 1834   History of Egyptian Mummies, and an Account of the Worship and Embalming of the Sacred Animal. London: Longman, Rees, Orme, Brown, Green, and Longman.

Prokopec, M., and Halman, L., 1999   Healed Fractures of the Long Bones in 15th to 18th Century City Dwellers. International Journal of Osteoarchaeology 9:349–356.

Redfern, R., 2009   Does Cranial Trauma Provide Evidence for Projectile Weaponry in Late Iron Age Dorset? Oxford Journal of Archaeology 28:399–424.

Redfern, R., 2010   A Regional Examination of Surgery and Fracture Treatment in Iron Age and Roman Britain. International Journal of Osteoarchaeology 20:443–471.

Reynolds, A., 2009   Anglo-Saxon Deviant Burial Customs. Oxford: Oxford University Press.

Reynolds, N., 1988   The Rape of the Anglo-Saxon Women. Antiquity 62:715–718.

Robb, J., 1997   Intentional Tooth Removal in Neolithic Italian Women. Antiquity 71:659–669.

Robb, J., 2002   Time and Biography: Osteobiography of the Italian Neolithic Lifespan. In Thinking Through the Body: Archaeologies of Corporeality. Y. Hamilakis, M. Pluciennik, and S. Tarlow, eds. pp. 153–172. London: Kluwer Academic/Plenum.

Roberts, C. A., 1991   Trauma and Treatment in the British Isles in the Historic Period: A Design for Multidisciplinary Research. In Human Paleopathology: Current Syntheses and Future Options. D. Ortner and A. Aufderheide, eds. pp. 225–240. Washington, DC: Smithsonian.

Roberts, C. A., 2006   Trauma in Biocultural Perspective: Past, Present and Future Work in Britain. In Human Osteology in Archaeology and Forensic Science. M. Cox, and S. Mays, eds. pp. 337–356. Cambridge: Cambridge University Press.

Roberts, C. A., 2010   Adaptation of Populations to Changing Environments: Bioarchaeological Perspectives on Health for the Past, Present and Future. Bulletins et Mémoires de la Société d'Anthropologie de Paris 22(1–2):38–46.

Roberts, C. A., and Manchester, K., 1995   The Archaeology of Disease. Ithaca: Cornell University Press.

Roberts, C. A., and Manchester, K., 2005   The Archaeology of Disease. Stroud: Sutton Publishing Ltd.

Sarauw, T., 2007   Male Symbols or Warrior Identities? The "Archery Burials" of the Danish Bell Beaker Culture. Journal of Anthropological Archaeology 26:65–87.

Saul, F. P., 1972   The Human Skeletal Remains from Altar de Sacrificios: An Osteobiographic Analysis. Papers of the Peabody Museum of Archaeology and Ethnology, Vol. 63, no. 2.

Saul, F. P., and Saul, J. M., 1989   Osteobiography: A Maya Example. In Reconstruction of Life from the Skeleton. M. Y. İşcan and K. A. R. Kennedy, eds. pp. 287–902. New York: Alan R Liss.

Sayer, D., 2010   Death and the Family. Developing Generational Chronologies. Journal of Social Archaeology 10:59–91.

Schell, L. M., 1997   Culture as a Stressor: A Revised Model of Biocultural Interaction. American Journal of Physical Anthropology 102:67–77.

Schenck, R. C., Jr., and Goodnight, J. M., 1996   Current Concept Review–Osteochondritis Dissecans. Journal of Bone and Joint Surgery (American) 78:439–456.

Schulting, R., 2006   Skeletal Evidence and Contexts of Violence in the European Mesolithic and Neolithic. In Social Archaeology of Funerary Remains. R. Gowland, and C. Knüsel, eds. pp. 224–237. Oxford: Oxbow Books.

Schwartz J. H., 2007   Skeleton Keys. New York.

Scott, G. R., and Winn, J. R., in press.   Dental Chipping: Contrasting Patterns of Microtrauma in Inuit and European Populations. International Journal of Osteoarchaeology. DOI: 10.1002/oa.1184.

Sen, A., 2007   Identity and Violence: The Illusion of Destiny. London: Penguin Books.

Shaw, C. N., and Stock, J. T., 2009   Habitual Throwing and Swimming Correspond with Upper Limb Diaphyseal Strength and Shape in Modern Human Athletes. American Journal of Physical Anthropology 140:160–172.

Shepherd, J., Shapland, M., Pearce, N. X., and Scully, C., 1990   Pattern, Severity and Aetiology of Injuries in Victims of Assault. Journal of the Royal Society of Medicine 83:75–78.

Siffert, R. S., and Katz, J. F., 1983   Growth Recovery Zones. Journal of Pediatric Orthopaedics 3:196–201.

Singer, P., 2006   Children at War: Los Angeles: University of California Press.

Šlaus, M., Novak, M., Vyroubal, V., and Bedić, Ž., 2010   The Harsh Life on the 15th Century Croatia–Ottoman Empire Military Border: Analyzing and Identifying the Reasons for the Massacre in Čepin. American Journal of Physical Anthropology 141:358–372.

Smith, G. E., and Jones, F. W., 1910   The Archaeological Survey of Nubia. Report for 1907–1908. Volume 2: Report on the Human Remains. Cairo: National Printing Department.

Smith, M. J., Brickley, M. B., and Leach, S. L., 2007   Experimental Evidence for Lithic Projectile Injuries: Improving Identification of an Under-Recognised Phenomenon. Journal of Archaeological Science 34:540–553.

Smith, M. O., 1996   Parry Fractures and Female-Directed Interpersonal Violence: Implications from the Late Archaic Period of West Tennessee. International Journal of Osteoarchaeology 6:84–91.

Sofaer, J., 2006   The Body as Material Culture. A Theoretical Osteoarchaeology. Cambridge: Cambridge University Press.

Sofaer Derevenski, J., ed., 2000   Children and Material Culture. London: Routledge.

Steckel, R., and Rose, J., eds., 2002   The Backbone of History: Health and Nutrition in the Western Hemisphere. Cambridge: Cambridge University Press.

Steckel, R., Sciulli, P., and Rose, J., 2002   A Health Index for Skeletal Remains. In The Backbone of History: Health and Nutrition in the Western Hemisphere. R. Steckel, and J. Rose, eds. pp. 61–93. Cambridge: Cambridge University Press.

Steinbock, R. T., 1976   Trauma. In Paleopathological Diagnosis and Interpretation. R. T. Steinbock, ed. pp. 17–59. Springfield, IL: Charles C Thomas.

Stirland, A. C., 2001   Raising the Dead: The Skeleton Crew of King Henry VIII's Great Ship The Mary Rose. Chichester: John Wiley & Sons.

Terrell, H. K., Hill, E. D., and Nagoshi, C. T., 2008   Gender Differences in Aggression: The Role of Status and Personality in Competitive Interactions. Sex Roles 5:914–826.

Thomas, S. P., 1995   Shifting Meanings of Time, Productivity and Social Worth in the Life Course in Meru, Kenya. Journal of Cross-Cultural Gerontology 10:233–256.

Tomoya, T., Tatsuhiko, H., Yoshiji, K., Koji, F., Takuya, M., Toshinori, S., and Yoshio, S., 2001   Myositis Ossificans of Hip: A Case Report. Clinical Orthopaedic Surgery 36:1091–1094.

Torres-Rouff, C., 2009   The Bodily Expression of Ethnic Identity: Head Shaping in the Chilean Atacama. In Bioarchaeology and Identity in the Americas. K. J. Knudson, and C. M. Stojanowski, eds. pp. 212–227. Gainesville: University of Florida Press.

Torres-Rouff, C., and Junqueira, M. A. C., 2006   Interpersonal Violence in Prehistoric San Pedro Do Atacama, Chile: Behavioral Implications of Environmental Stress. American Journal of Physical Anthropology 130:60–70.

Treherne, P., 1995   The Warrior's Beauty: The Masculine Body and Self Identity in Bronze-Age Europe. European Journal of Archaeology 3:105–144.

Turner, C. G., II, and Turner, J., 1998   Man Corn. Cannibalism and Violence in the Prehistoric American Southwest. Utah: University of Utah Press.

Ubelaker, D. H., 1989   Human Skeletal Remains: Excavation, Analysis and Interpretation. Chicago: Aldine Publishing Company.

Van der Merwe, A. E., Steyn, M., and L'Abbé, E. N., 2010   Trauma and Amputations in 19th Century Miners from Kimberley, South Africa. International Journal of Osteoarchaeology 20:291–306.

Verano, J. W., 2005   Human Sacrifice and Postmortem Modification at the Pyramid of the Moon, Moche Valley, Peru. *In* Interacting With The Dead. Perspectives On Mortuary Archaeology for the New Millennium. G. F. M. Rakita, J. E. Buikstra, L. A. Beck, and S. R. Williams, eds. pp. 277–289. Florida: University Press of Florida.

Walker, P. L., 1989   Cranial Injuries as Evidence of Violence in Prehistoric Southern California. American Journal of Physical Anthropology 80:313–323.

Walker, P. L., 1997   Wife Beating, Boxing, and Broken Noses: Skeletal Evidence for the Cultural Patterning of Violence. *In* Troubled Times: Violence and Warfare in the Past. D. L. Martin, and D. W. Frayer, eds. pp. 145–180. Amsterdam: Gordon and Breach.

Walker, P. L., Cook, D. C., and Lambert, P. M., 1997   Skeletal Evidence for Child Abuse: A Physical Anthropological Perspective. Journal of Forensic Sciences 42: 196–207.

Wendorf, F., 1968   Site 117: A Nubian Final Paleolithic Graveyard near Jebel Sahaba, Sudan. *In* The Prehistory of Ancient Nubia, Volume 2. F. Wendorf, ed. pp. 954–995. Dallas: Southern Methodist University Press.

Wheatley, B. P., 2008   Perimortem or Postmortem Bone Fractures? An Experimental Study of Fracture Patterns in Deer Femora. Journal of Forensic Sciences 53:69–72.

Wheeler, S. M., Williams, L., Beauchesne, P., and Molto, J. E., 2007   Fractured Childhood: A Case of Probable Child Abuse from Ancient Egypt. Society for Archaeological Sciences Bulletin 30:6–9.

White, T., 1992   Prehistoric Cannibalism at Mancos 5MTUMR-2346. Princeton: Princeton University Press.

World Health Organization, 2007   International Classification of Diseases and Related Health Problems. Geneva: World Health Organization. http://www.who.int/classifications/icd/en/.

CHAPTER 21 Developmental Disorders in the Skeleton

Ethne Barnes

INTRODUCTION

Developmental defects of the skeleton have been recognized and described, often treated as oddities in osteological case reports for over two centuries. By the beginning of the 20th century attention began to turn to the genetics behind these defects (Smith and Wood Jones 1910). Studies of different groups of ancient peoples suggested genetics played a role with some disorders, while extrinsic factors may have had an influence on others. Bess Miller (1928) surmised the importance of determining heritability of skeletal disorders based on the evidence then that revealed differing patterns of disorders apparently following different family lines. This was most obvious with minor developmental disorders.

Minor developmental disorders that do not threaten life are easily recognized in ancient human skeletal populations while more severe life threatening developmental disorders, better known as congenital defects, rarely survive in archaeological material. Many of the non-life-threatening disorders can cause some degree of functional disorder, yet other disorders are so minor as to cause no functional threat. Variation in development is an evolutionary necessity that allows for defective development whenever variation goes beyond an acceptable boundary of normal.

"Cookie cutter" development simply does not exist. Variations within a normal developmental theme occur in every part of the body, including the skeleton. Non-life-threatening variations are tolerated and promoted within genetic populations wherever they appear. This was what early researchers were discovering around the

A Companion to Paleopathology, First Edition. Edited by Anne L. Grauer.

beginning of the 20th century. They particularly recognized that human cranial morphology appeared to vary between different groups of people, and so a metric system with specific measurements and indices was devised to measure cranial variation for determining population and racial differences.

Craniometric trait analyses ruled genetic studies of ancient populations for over a century, despite early pleas to also use minor developmental anomalies found in the cranium (Dixon 1900; Russell 1900; Hepburn 1908; Sullivan 1922). These minor anomalies, such as variations in foramina and extra sutural ossicles (wormian bones), soon became known as nonmetric, discontinuous, quasicontinuous, discrete, or epigenetic traits, and most of them were recognized as part of an acceptable range of normal development. The major arguments against using nonmetric data were that such traits could be subject to extrinsic factors, while craniometric traits remained intrinsic.

During the 1970s the application of statistical techniques to the study of minor developmental variants in the cranium became a popular procedure in the field of population studies. Statistical analyses relied on scoring traits as present or absent (Saunders 1989). These types of studies were encouraged by Berry and Berry (1967) with their selective list of cranial nonmetric traits based on genetic studies carried out on mice, especially studies by Gruneberg (1952, 1954, 1955, 1963, 1964) and Searle (1954). Waddington (1956) described the process of progressive determination and differentiation of embryonic cells and tissues resulting from genetic instructions responsible for variations in development. Hauser and DeStefano (1989) emphasized this process as they described various minor cranial anomalies.

Less well known postcranial variations, although recognized early on (Oetteking 1922; Hrdlicka 1923; Trotter 1929, 1934; Stewart 1932) were not seriously considered for nonmetric population comparison studies until the late 1960s. Laughlin and Jorgensen (1956), Anderson (1968), and Brothwell and Powers (1968), encouraged analysis of postcranial developmental variations, especially in the vertebral column. Finnegan (1972, 1974, 1978) demonstrated that selected postcranial nonmetric traits could be used in statistical analyses for comparative population studies similar to cranial nonmetric studies. However, lacking an understanding of how and why these traits develop, the proponents of this approach encountered skepticism among many researchers (Buikstra and Ubelaker 1994).

The emphasis on select nonmetric traits, especially cranial traits, easily scored present or absent for statistical processing, overshadowed the huge quantity of genetic trait data available within the skeleton despite several reports of findings of a number of congenital disorders following familial lines (Allbrook 1955; Selby et al. 1955; Merbs and Wilson 1960; Ferembach 1963; Bennett 1973; Morse 1978). Variability exists in every part of the skeleton, offering ample opportunity for genetic studies that can be useful to paleopathology as well as to other anthropological studies. The nonmetric trait studies opened the door to understanding and evaluating the development of both nonpathological and pathological skeletal anomalies.

The problem researchers have faced is how to organize and classify the variable skeletal disorders for meaningful analytical discourse (Zimmerman and Kelley 1982; Manchester 1983; Turkel 1989). Definitions of developmental disorders in the skeleton require understanding of their underlying etiologies before any type of classification can occur. Different types of developmental disorders affect the skeleton based on etiology: acquired (extrinsic), and genetic mutations (intrinsic). Extrinsic factors—maternal infection, exposure to harmful chemicals or drugs, metabolic disorders, and nutritional deficiencies—can affect the developing fetus, such as with congenital syphilis. Extrinsic factors can also interfere with morphogenesis of the developing embryo but only if they cross the placental barrier at specific sensitive threshold events of development to modify genetic messages, as with the drug Thalidomide that disrupts limb bud development. Genetic disorders develop exclusively during embryogenesis, although some may not make an appearance until childhood or later. Many of the more serious defects are lethal to the developing embryo or fetus, and the vast majority of spontaneous abortions stem from malformed embryos.

THE ROLE OF GENES IN DEVELOPMENTAL DISORDERS

The term epigenetic can apply to either a mutant gene or extrinsic factor that exerts influence on a major threshold event of embryonic development governed by underlying genetics. The vast majority of epigenetic influences occur with inherited mutant genes, while some appear at random. Some mutant genes are evolutionary adaptive responses of a population to local environmental pressures. Congenital hemoglobin defects like the thalassemias and sickle cell are inherited hemoglobin mutations within populations in response to environmental pressure—endemic malaria—that can leave their marks on the skeleton. Extrinsic factors, such as abnormal maternal metabolism of folic acid combined with low maternal nutritional intake of the vitamin, can disrupt crucial threshold events of neural tube development in the embryo with susceptible underlying genetics, impacting on the developing axial skeleton. Timing is key to epigenetic disorders since they can only interfere with development at specific threshold events. These vulnerable threshold events involve times of rapid change during morphogenesis as specific, newly formed cells are proliferating, differentiating, or migrating.

Genetic disorders of the skeleton can be divided into tissue-specific defects and localized structural defects. Specific genes control the primordial cells leading to the development of bone tissue, while other master genes map out the developmental pathways of bone structures. Skeletal growth can also be affected by defective nonskeletal tissue in the developing embryo, as with the faulty release of growth hormone from the pituitary gland leading to such disorders as pituitary dwarfism and primordial dwarfism. Nonskeletal tissue or organ defects, can also interfere with adjacent structural development, as with neural tube defects that interrupt adjacent vertebral arch formation.

Tissue-specific skeletal disorders arise very early in embryogenesis with the altered formation of any one of the different types of precursor cells involved in creating the complex of bone tissue. The myriad of skeletal tissue disorders known as skeletal dysplasias develop according to the specific defective precursor bone or cartilage cells that accumulate within specific developing bone segments (Rubin 1964). Any of the varied precursor cells that go into forming bone tissue can become defective and produce a range of disorders within each type of defect, dependent on the timing of the alteration during a threshold event.

Bone tissue derives from embryonic mesenchymal cells. Some are activated to become primordial membranous tissue covering the cranium and face, and forming the clavicle (membranous bone formation), while other mesenchymal cells are programmed to form bone models of primitive hyaline cartilaginous tissue for the base of the skull and all other bones. Primordial periosteal membranous tissue spreads over the outer surface of these cartilaginous models, and its undersurface creates membranous bone tissue to interface with the cartilage bone models (endochondral bone formation). Disruptions of any segment of endochondral bone formation, especially within long bones, are the most commonly recognized forms of bone dysplasias (chondrodysplasias) with a wide assortment of different forms of dwarfism. Defective primordial cartilage tissue modeling in long bones often occurs at the growth plates, leading to disrupted long-bone growth and deformed limbs associated with achondroplasia dwarfism that has been identified in prehistoric human remains (Snow 1943; Hoffman 1976; Gladykowska-Rzeczycka 1980; Buikstra 1993). Other types of chondrodysplasias resulting in dwarfism have also been identified in ancient human remains (Frayer et al. 1988; Arcini and Frolund 1996). Defective precursor collagen cells within membranous bone formation, especially periosteal membranous bone covering normally developing cartilaginous models of long bones, lead to thin, fragile cortical bone development. This results in various forms of osteogenesis imperfecta, with severe forms leading to multiple deforming fractures, and a few cases have been identified in ancient human remains (Wells 1965; Gray 1969). Several types of bone dysplasias have been identified by the type of defective bone cells and the bone segments involved (Ortner and Putschar 1985; Aufderheide and Rodriguez-Martin 1998; Ortner 2003).

Dysplasias of skeletal tissue are rarely found in ancient human remains. Most of the skeletal disorders that we do see in human skeletal remains are caused by mutant genetic changes in directions for the development of structures within specific embryonic developmental fields (Gruneberg 1963, 1964). Close interaction of select developing tissues involved in the complex composition of a specific structure or set of closely related structures during morphogenesis, mark a developmental field (Barnes 1994a). Sometimes more than one developmental field is affected by linked genetic disturbances that present a set of developmental disorders identified as a syndrome. With an understanding of how bones of the skeleton evolve from various developmental fields within the growing embryo, we can determine cause and effect of structural developmental disorders, and classify them accordingly. This can be helpful with analyses of patterns of pathological developmental defects within a population.

Embryonic growth and development is a very complex process. Master genes trigger the pattern formation for structural development, creating a cascade of evolving events and signals along a specific pathway with a gradient of signals and interactions backed by a series of checks and balances. Along the way, some genetic factors activate or inhibit certain gene expressions, while others promote specific actions by regulating timing and location of the final genetic expression within a structure. Specific cells or tissues act as inducers to initiate change in other cells as they come into contact. Signals (cross talk) are exchanged between the different cell types to initiate the necessary cellular changes for morphogenesis. Abnormal mutant genetic signals, anywhere along the developmental pathway, can interfere with timing of events. Compensation for delayed or missing signals can take place with other isotropic members of the involved genetic factor, thus leading to variable expressions that may or may not interfere with normal development. Without compensation, missing steps in development will occur. The mapping of disturbances in various genetic pathways that lead to developmental disorders has been accomplished by microbiologists for over a decade, and their work provides us with credible evidence for the genetics behind developmental disorders of the skeleton (Larsen 2001; Sadler 2006).

SKELETAL MORPHOGENESIS

In itself, the embryonic development of body form with its organs is evolutionary, and the process carries the potential for variability at every stage of evolvement. Thus, variability is inevitable. Understanding the process of morphogenesis helps us to understand and classify variability within specific developmental fields as the patterns for the evolving primordial skeleton are determined in the first eight to ten weeks.

Morphogenesis begins with the formation of the primitive streak in the midline of the germ disc that divides the embryo into two halves along a cranial and caudal axis. Consequently much of the human skeleton derives from paired primordial tissues that develop concurrently: the two sides of the face and skull, both sides of the rib cage and the limbs. The vertebrae and sternum also evolve from paired primordial tissues. If the growth of one of a pair is delayed, asymmetrical growth takes place. One limb or portion of a limb can be shorter than the other limb, or asymmetry of the face with one side showing delay in development can occur. Failure of one side to develop produces a severe defect. Primordial segmentation is another aspect of morphogenesis in developing limbs, cranial and vertebral parts, sternum, and ribs. Delay, disorder, or absence of developmental segmentation results in abnormality.

The human skeleton is divided into two components. The axial skeleton, consisting of 80 bones, forms the skull, hyoid, vertebral column, ribs, and sternum. The appendicular skeleton, represented by 126 bones, forms the upper limbs (shoulder girdles, arms, wrists, and hands) and lower limbs (pelvic girdle, legs, and feet). The primordial axial skeleton is the first to make an appearance following the emergence of the future head from the slightly elevated cranial end of the primitive streak (the primitive node) with a depressed center (primitive pit). Several epiblast cells—the outer layer of blastoderm cells of the germ disc—migrate to the primitive streak and receive signals from the streak to proliferate, detach and move through the streak to differentiate into the germ layers endoderm and mesoderm, while remaining epiblast cells form ectoderm. Skeletal and cartilaginous tissue, along with connective and muscular tissue, arise from mesenchymal cells originating from mesoderm. By the third week, some mesenchymal cells migrate to the head end to form the prechordal plate that is responsible for induction of the forebrain, followed by formation of a dense midline tube just caudal to the prechordal plate. The primitive streak regresses as this tube grows in length and transforms into a solid rod known as the notochord. The notochord acts as scaffolding necessary for the development of the axial skeleton. Defective notochord development leads to disorganized development of the axial skeleton that usually creates a nonviable fetus.

The stage is now set for the appearance of the developmental fields that contribute to the formation of the axial skeleton. Mesenchymal cells migrating from the head end line up on both sides of the notochord to form a matching pair of cylindrical condensations, thus creating the paraxial mesoderm that forms the vertebral column and contributes to the skull base. Once the notochord serves its purpose, it regresses, but leaves behind rudimentary tissue to form the nucleus pulposus centered in the cartilaginous discs that separate the vertebral bodies.

Sequential cranial–caudal transformation within the paired columns of paraxial mesoderm form a series of whorl-like cell clusters known as somitomeres with the more cranial ones contributing to the cranial base. The somitomeres lined up on each side of the notochord form tissue block segments called somites: four occipital, eight cervical, twelve thoracic, five lumbar, five sacral, and eight to ten coccygeal pairs. The 1st occipital and last five to seven coccygeal segments disappear as the remaining vertebral somites form matching hemimetamere pairs in the vertebral column. Each somite separates into sclerotome, myotome and dermatome tissue. The vertebral bodies will evolve from the ventral portion of the paired sclerotomes as they surround the notochord, while the dorsal portions surround the developing neural tube alongside the notochord as it extends caudally to form the vertebral arches. Posterior and ventral sclerotome cells respond to different signals from their adjacent tissue structures that direct their development with changes in descending signaling for necessary shifts in vertebral form. The vertebral sclerotomes undergo resegmentation, splitting into cranial and caudal halves. The caudal halves fuse with the cranial halves of adjacent sclerotomes as sprouting nerves pass between them and the fibrous intervertebral discs develop. The thoracic costal processes are stimulated to produce primordial ribs.

Mesenchymal cells condense into paired tissue bands ahead of the developing ribs. The sternal bands fuse together in a cranial–caudal direction as they meet midline, beginning with the uppermost portions as they are contacted by the growing upper ribs. As the cranial ends fuse, they unite with a small condensation of mesenchymal cells, the precostal process, arising from the midline near the juncture of the sternal bands to form the primordial manubrium. The precostal process is joined by a pair of small mesenchymal condensations known as the suprasternal structures, to form the interfaces between the manubrium and the clavicles. The united sternal bands form the primordial mesosternum that segments into four sternebrae with the most caudal end forming the xiphoid process.

As the building blocks of the vertebral column develop, the cranial prechordal plate and uppermost portion of the notochord induce overlying epiblast cells to thicken and create the neural plate, the precursor to the central nervous system. The cranial ends broaden out into paired neural folds, creating the two sides of the developing brain, while the narrower caudal ends gradually fold towards each other to fuse into the developing neural tube that overlies the notochord. The narrowing neural tube will become the spinal cord. The neural tube gradually moves downward with the extension of the precursor structures of the vertebral column. The opening at the cranial end (cranial neuropore) remains open until signaled to close at a specific stage of development of the upper portion of the axial skeleton, followed in a few days with closure of the caudal opening (caudal neuropore), sealing the tubular structure with the developing brain. Failure or delay of closure of either neuropore can influence the developing axial skeleton.

The neural folds also play a role in the development of the blastemal desmocranium, the primordial cranial vault covering the brain. Cells along the lateral borders of the broad portions of the neural folds transform into neural crest cells, move away from the folds and are directed to provide the precursor bones consisting of membranous tissue for the cranium and face. Each area of the developing brain has a corresponding overlying sheet of membranous tissue of a specific cranial bone: frontal, parietals, and the interparietal squamosa of the occipital.

The upper mid portion of the face evolves from a bulging tissue concentration, the visocranium, just below and forward of the developing blastemal desmocranium, while the lower portion of the face develops from paired gill-like prominences, pharyngeal arches (branchial arches) with internal pouches separated by ectodermal grooves forming on the sides of the head end of the embryo. Five of these gill-like arches form, but only the first pair contribute to lower facial structures: maxilla, mandible, zygomatics, temporal squamosa, parts of the sphenoids, palatine bones, malleus and incus bones of the ears. The bulging visocranium swells to form the frontonasal prominence, and by the end of the fourth week, the face begins to take shape around a primitive mouth. The frontonasal prominence subdivides, forming a pair of nasal placodes that will become the nasal pits with raised rims dividing the two in the middle and forming lateral boundaries with the developing maxillae. The nasal bones, and bony septum with the vomar, develop from the upper medial aspect, while the inferior parts of the medial aspect expand below the nasal cavity to form the two halves of the intermaxillary process that becomes the premaxilla. Ectodermal grooves form at the lateral borders with the maxillae to eventually become the nasolacrimal grooves bounded by the lacrimal bones. The primordial maxillae move to meet with the frontonasal process, while the mandible halves develop around cartilaginous bars, known as Meckel's cartilage, that extend through each half, attracting neural crest cells to form the membranous bone precursors. The medial walls of the maxillae grow downward with the developing tongue to join the small primary palatal extensions of the premaxilla. The two halves then expand and come together to form the primordial palate, moving into place with the horizontal plates of the primordial palatine bones.

The groove for the 1st pharyngeal arch that forms the lower face produces the auditory meatus and auditory canal as the inner pouch develops into the inner ear's tympanic cavity. Membrane between the pouch and distal end of the groove forms the tympanic plate, appearing as membranous tissue enclosed by a bony tympanic ring at birth that ossifies by age five and provides the bony sheath for the styloid.

The remaining pharyngeal arches contribute to the laryngeal cartilages in the neck and the stylohyoid chains. The 2nd and 3rd pharyngeal arches produce a pair of primitive cartilages, Reichert's cartilages, which develop into the stylohyoid chains consisting of the styloids with connecting stylohyoid ligaments, each attached to the lesser cornua of the hyoid greater cornua that connect to the hyoid body. The remaining pharyngeal arches contribute to the laryngeal cartilages, while the associated pouches contribute to developing glands in the neck as the grooves disappear.

The base of the skull evolves from a cartilaginous plate that extends from the nasal region to the cranial end of the neural tube, hence cradling the developing brain. The prechordal plate arises from three basic pairs of primitive cartilage precursors. Ventral (trabecular) cartilage forms in the interorbital-nasal region with contributions to the nasal cartilages, ethmoid, and much of the sphenoids. Lateral or otic capsules develop around the otocysts and form the temporal petromastoids. Parachordal cartilages form just distal to the hypophyseal fossa and fuse together to form the basioccipital.

The occipital somites form nonsegmenting sclerotomes that fuse with the parachordals as the 1st cervical sclerotome splits into cranial and caudal halves, and the cranial half joins the occipital sclerotomes to form the lateral exoccipitals and supraoccipital. The supraoccipital fuses with the membranous interparietal squamosa at the mendosa line to form the upper portion of the occipital.

As the axial skeleton progresses in development, the appendicular skeleton begins its appearance under the direction of a set of master genes directing paired upper limb buds to form in the cervical region from lateral plate mesoderm. Soon afterwards, another set of master genes directs the formation of the paired lower limb buds in the lumbar region. Each limb bud consists of an outer ectodermal cap and inner mesodermal core. Apical ectodermal tissue of the growing limb bud forms a thickened ridge. This apical ectodermal ridge (AER) at the tip end of the bud borders with undifferentiated mesenchymal tissue called the progress zone. The AER is responsible for the outgrowth of the limb. The limb grows in a proximal–distal direction as genetic signals coming from the AER through the progressive zone reach cells furthest away from the distal region where they receive other genetic signals to progressively differentiate into the various segments of the limbs. Each segment is under its own genetic control, with each segment following a descending sequence of genetic instructions in a proximal–distal direction. The upper limb segments consist of 1) the shoulder girdle segment: scapula and clavicle; 2) arm segment: humerus; 3) forearm segment: ulna, radius and proximal carpals; and 4) the hand segment: distal carpals, metacarpals and phalanges. The lower limb segments include 1) the pelvic girdle segment: ilium, ischium, pubis; 2) thigh segment: femur, patella; 3) lower leg segment: tibia, fibula and proximal tarsals; and the 4) foot segment: distal tarsals, metatarsals and phalanges. The developing limb segments also follow genetic instructions for anterior–posterior axis and dorsal–ventral axis development. The terminal ends flatten out into disc shaped plates—the hand and foot plates. Fingers and toes arise from ridges within the plates that separate into distinct rays of tissue. Limb joints develop from arrested primordial cartilage condensations designated as interzones between limb segments.

The Morphogenetic Approach for Analysis of Skeletal Developmental Disorders

Developmental disorders of skeletal structures can be traced and defined according to disturbances in the underlying developmental fields (Barnes 1994a, 2008) (see Tables 21.1–21.5). Basic principles govern these disturbances. Knowing that most of the skeleton arises from paired synchronous developing tissues furthers our understanding of how bony structures can become defective. One side can be smaller or absent, thus creating abnormality such as asymmetry or clefting. Sometimes both sides of a developing structure are affected. Bone segments or parts of bone segments may fail to develop. Segmentation errors can create disorder especially in the vertebral column with a variety of segmentation errors. Failure of separation within a developing segment or within adjacent primordial bone elements, accompanied by the lack of joint development is not unusual. Abnormal splitting of a bone element into two parts with an interfacing abnormal joint can happen. Duplication of developing tissues can occur in the developing digits, and in vertebral segments, leading to extra digits

Table 21.1 Developmental field disorders of the skull

A. Blastemal desmocranium: Cranial vault—frontal, parietals, occipital interparietal squamosa
   Sutural agenesis—scaphocephaly, oxycephaly, trigonocephaly
   Persistent infantile feature—metopism
   Parietal thinning, bilateral or unilateral
   Enlarged parietal foramina
   Ectodermal inclusion cyst (dermoid/epidermoid)
   Hydrocephaly
   Microcephaly

B. Neural tube defect
   Anencephaly, meningocele, encephalocele

C. Visocranial frontonasal prominence: Upper face—nares, nasal bones, lacrimals, premaxilla
   Facial cleft
   Hypoplasia/aplasia
   Midline cleft lip (premaxilla cleft)

D. Pharyngeal arch I: Lower face—maxilla, mandible, palate, palatine bones, zygomatics
   Hypoplasia/aplasia
   Cleft lip (unilateral, bilateral)
   Cleft palate
   Cleft mandible
   Palate ectodermal inclusion cyst
   Mandibular inclusion cyst (stafne defect)
   Mandibular hypoplasia expressions—hemifacial microsomia
   Bifid mandibular condyle
   Mandibular coronoid hyperplasia

E. Chondocranium prechordal cartilages and cranial somites: Cranial base—basioccipital, lateral exoccipitals, supraoccipital
   Persistent fetal feature—mendosa suture (inca bone)
   Basioccipital hypoplasia/aplasia
   Basioccipital vertical cleft
   Basioccipital horizontal cleft (cranial border shifting)
   Occipital vertebral manifestations (cranial border shifting)
   Occipitalized atlas (caudal border shifting)
   Precondylar facet, corresponding facets axis dens and atlas rim (caudal border shifting)

F. Pharyngeal arch I groove: External auditory meatus
   Atresia (aplasia)/hypoplasia
   External auditory torus

G. Pharyngeal arch I closing membrane: Tympanic plate
   Persistent infantile trait—tympanic aperture
   Aplasia

H. Pharyngeal arch II–III: Stylohyoid chain
   Styloid hypoplasia/aplasia
   Ossified styloid ligament (complete/partial)
   Thyroglossal cyst

or vertebral segments. Disorders have variable expressions, ranging from mild to severe, incomplete to complete, or following a gradient from mild hypoplasia to aplasia. Very rarely, hyperplasia of a bony element can occur. Some disorders do not make an appearance until the immature skeleton reaches maturity, while many others are readily apparent in the infant or juvenile skeleton. Final bone expressions in the

Table 21.2 Developmental field disorders of the vertebral column

A. Paraxial mesoderm: Cervical, thoracic, lumbar, sacral, caudal vertebral segments
    Occipitocervical cranial border shifting—
      Failed union or agenesis of dens (odontoid) with axis
    Occipitocervical caudal border shifting—
      Occipitalized atlas
      Facets dens' tip and atlas rim with corresponding facet on occipital base
    Cervicothoracic cranial border shifting—
      7th Cervical rib expressions
    Cervicothoracic caudal border shifting—
      Rudimentary (hypoplasia) 1st thoracic and 1st rib
    Thoracolumbar cranial border shifting—
      Rudimentary (hypoplasia) or agenesis 12th rib facets and ribs
      Transitional facets 11th thoracic
    Thoracolumbar caudal border shifting—
      1st lumbar rib expressions
      Transitional facets 1st lumbar
    Lumbosacral cranial border shifting—
      Sacralized 5th lumbar expressions
    Lumbosacral caudal border shifting—
      Lumbarized 1st sacral expressions
    Sacrocaudal cranial border shifting—
      Caudalization 5th sacral expressions
    Sacrocaudal caudal border shifting—
      Sacralization 1st caudal expressions

Cleft/bifid neural arch without spina bifida
Atlas agenesis anterior arch
Hemivertebrae—solitary, contralateral, unilateral, bilateral
Ventral hypoplasia/aplasia
Lateral hypoplasia/aplasia—solitary, multiple with or without postlateral bar
Dorsal hypoplasia (wedge vertebra)
Block vertebra
Multiple cervical block vertebrae (Klippel–Feil syndrome) with or without other defects
Sagittal cleft (butterfly vertebra)
Extra vertebral segment (transitional vertebra)

B. Neural tube defect
    Meningocele/mengomyelocele spina bifida cystica/occulta

C. Sacral agenesis (embryonic caudal regression defect)
    Agenesis all or most sacral segments

mature skeleton are programmed during morphogenesis. Sometimes fetal or infantile traits persist into the mature skeleton, such as metopism, or the persistent fetal mendosa suture between the membranous interparietal squamosa and cartilaginous occipital base (commonly known as inca bone).

Disturbances in the development of soft tissue structures can affect developing bone structures, especially relative to defects affecting the neural tube or neural folds. Early-stage neurulation defects cause severe damage—cranial anencephaly and spinal meningomyelocele (a protrusion of meninges and spinal cord through a defect in the spinal column). Delayed closure of the anterior neuropore allows developing

Table 21.3   Developmental field disorders of the ribs and sternum

A. Paraxial mesoderm thoracic outgrowths: Ribs
   Sternal bifurcation/flaring
   Fusion
   Rudimentary intrathoracic rib
   Hyperplasia
   Hypoplasia/aplasia

B. Sternal bars, preocostal process, suprasternal ossicles: Sternal manubrium and sternebrae
   Cleft/bifurcated sternum (complete/incomplete)
   Suprasternal ossicle (separate/fused to manubrium)
   Misplaced manubriomesosternal joint
   Sternal aperture
   Hypoplasia/aplasia sternal segment
   Extra sternal segment

Table 21.4   Developmental field disorders of the upper limbs

A. Shoulder girdle segment: Scapula and clavicle
   Clavicle hypoplasia/aplasia
   Bifurcated clavicle
   Scapular glenoid neck hypoplasia
   Scapular os acromion

B. Arm segment: Humerus
   Hypoplasia/aplasia
   Hypoplasia proximal head

C. Forearm segment: Radius, ulna and proximal carpals—scaphoid, lunate, triquetral, pisiform
   Hypoplasia/aplasia—especially radius
   Radioulnar synostosis
   Fused carpals
   Bipartite carpals

D. Hand segment: Distal carpals—trapezium, trapezoid, capitate, hamate, metacarpals, phalanges
   Fused carpals
   Bipartite carpals
   Syndactyly
   Symphalangism
   Os metastyloideum 3rd metacarpal
   Hyperplasia/hypoplasia/aplasia hamulus of hamate
   Brachydactyly
   Polydactyly—postaxial/preaxial expressions
   Ectodactyly
   Triphalangeal thumb
   Cleft (lobster claw) hand

meningeal tissue covering the brain and or brain tissue to protrude in the form of a cyst through the developing cranium. The cyst is covered by ectodermal tissue and may contain both brain and meningeal tissue (encephalocele) or just meningeal tissue (meningocele) anywhere along the sagittal plane from nasion root to basioccipital.

Table 21.5   Developmental field disorders of the lower limbs

A. Pelvic girdle segment: Ilium, ischium, pubis
   Hypoplasia/partial aplasia
   Congenital hip dysplasia

B. Thigh segment: Femur and patella
   Femur proximal hypoplasia/partial aplasia
   Bipartite patella
   Patella hypoplasia/aplasia

C. Lower leg segment: Tibia, fibula, proximal tarsals—talus, calcaneus
   Hypoplasia/aplasia—especially fibula
   Tibiofibular synostosis
   Talus os trigonum
   Calcaneus os secondaris
   Fused talus-calcaneus
   Club foot (talipes equinovarus)

D. Foot segment: cuboid, navicular, cuneiforms, metatarsals, phalanges
   Fused tarsals
   Bipartite tarsals
   Syndactyly
   Symphalangism
   Fused 3rd cuneifrom–3rd metatarsal—osseous/non-osseous
   Os intermetatarsium
   Os navicular
   Brachydactyly
   Polydactyly—postaxial/preaxial expressions
   Ectodactyly
   Cleft (lobster claw) foot

The meningocele occurs most often at bregma, creating a saucer-like bony imprint with raised border and an irregular opening at its base. The encephalocele most often arises from the floor of the anterior and middle fossae of the nasal region, forcing widening of the nasal region, and developing nasal bones to protrude. They can also present in the region of the sella turcica or roof the eye orbit angle. Encephaloceles located at the base of the skull force bifurcation of the developing cranium.

Delayed closure of the posterior neuropore leads to displacement of a portion of the meninges covering the spinal cord within an ectodermal covered cyst outside the spinal canal. Spinal meningoceles most often occur in the lumbosacral region. The cyst impacts the developing adjacent neural arches by forcing them apart into a wide cleft that encircles the growth—known as spina bifida. Spina bifida cystica refers to major protruding cystic defects, while minor cystic defects that do not protrude outward are known as spina bifida occulta but still impacting adjacent neural arches. The resulting neural arch clefts differ from neural arch clefts or bifurcations formed by hypoplasia of one or both sides of developing neural arches that fail to meet and fuse together.

Expressions of sacral agenesis (caudal regression syndrome) are associated with failure in development of the caudal end of the embryo, crippling development of the sacrum. Complete sacral agenesis forces the ilea to meet, often fusing to form a bony ring. Most often partial agenesis is expressed with development of only one or two of the upper sacral segments that often fuse with the last lumbar. Sacral agenesis is extremely rare in paleopathology.

Overlying developing ectodermal tissue can sometimes fail to retreat, and thus becomes entrapped by developing paired bony structures that create an inclusion cyst. Some can be quite large, while others are so small they go undetected. Inclusion cysts can arise anywhere along the sagittal plane of the cranium, including the roof of the eye orbit angle, leaving a cystic impression with sharp border on affected bone, with or without a rounded connecting opening at the base. Dermoid types tend to occur within the diploë and sometimes between the periosteum and scalp. Epidermal types frequently affect the dura covering of the brain. Inclusion cysts can also form in the palate, especially within the incisive canal. Some occur at the posterior midline of the palate or at the junction of the premaxilla and maxilla, between the lateral incisor and first premolar. Inclusion cysts can also be found in the submandibular fossa of the inner aspect of the mandible (Stafne defect) from premature development of the sublingual gland impinging on adjacent bone.

Microcephalic crania accompany small brain formation, while hydrocephalic crania occur with abnormal accumulation of cerebrospinal fluid within the cranial vault that is often associated with neural tube defect.

The most prominent disorders of the cranial vault include sutural agenesis, parietal thinning, and enlarged parietal foramina. Failure of sutures to develop between adjacent cranial bones can lead to cranial deformity. Failed development of the sagittal suture creates scaphocephaly, while lack of a metopic suture creates trigono-cephaly. Combinations of failed sutures, such as agenesis of the coronal and lamb-doidal sutures, produces a tower skull—oxycephaly. Parietal thinning (unilateral or bilateral) occurs when the diploë fails to develop in the dorsal aspects of the parietals. Enlarged parietal foramina occur in a variety of expressions from large slits, a lozenge shape, or large ovoid openings involving the lack of ossification of the posterior membranous parietals near where Santorini's emissary vessels usually pass through the developing bones.

Synchronized development of the face can be disrupted with hypoplasia of one or both parts of paired elements that delay or disrupt union. Facial clefting is the most serious facial defect likely to be seen in human skeletal remains. Failure of a maxillary half to meet and fuse with the frontonasal prominence anywhere along the ectoder-mal lacrimal groove that separates them creates a cleft, from the lower border of the eye orbit to the nares. However this is rare, especially in ancient human skeletons. Cleft lip and cleft palate are more likely to be encountered. Failure of one or both maxillae to merge with the midface premaxilla produces unilateral or bilateral cleft lip. Failure of the two halves of primordial premaxilla to merge results in a midline cleft lip. Cleft lip can also include clefting of the palate. The palate can be cleft without cleft lip when the two palatal shelves fail to merge because of hypoplasia or aplasia, either unilaterally or bilaterally. Variability occurs with each type of clefting, depending on the severity of the defect involved. The mandible can also be cleft when the two halves fail to merge, but most often incomplete clefting is expressed over complete clefting. The mandible is susceptible to hypoplasia, usually involving only one side, and primarily affecting the ramus and dorsal portion of the mandibular body. This is often referred to medically as hemifacial microsomia. Expressions can vary from: type

I (mild) with small, narrow ramus and posterior body; type II, abnormally shaped narrow ramus and posterior body; to type III (severe) with aplasia of the ramus and a very narrow posterior body. Mandibular condyles can be bifid, and the coronoid process can be programmed to overdevelop during adolescence with coronoid hyperplasia, making it difficult to fully close the mouth.

Aplasia of the dorsal end of the ectodermal groove that creates the external auditory meatus, leaves the ear opening and inner ear canal absent (atresia) causing congenital deafness. The tympanic plate and styloid will be missing, and the petrous portion will be underdeveloped. Sometimes the tympanic plate fails to ossify, thus appearing absent. The more common nonpathological tympanic aperture, an expression of incomplete ossification, is seen in ancient populations, and can be referred to as a persistent infantile trait under morphogenetic programming.

The major defect of the stylohyoid chain is the complete or nearly complete ossification of the styloid ligament connecting the bony styloid to the hyoid. The hyoid body can be impacted by a thyroglossal cyst. The cyst develops from remnant primordial thyroid tissue as it moves along its midway migratory pathway from the dorsal area of the developing tongue to its position below the hyoid. It can develop anywhere along this pathway, including on the surface of the hyoid body where it leaves a rounded cystic impression (Moore 1985; Sadler 2006).

The base of the skull suffers the most serious disorder from shifting of the occipitocervical (O–C) border, especially when the border shifts downward, to incorporate the atlas into the skull base, either completely or partially. Another serious expression of O–C caudal border shifting occurs when a precondylar facet with matching facets on the tip of the dens of the axis and rim of the atlas reveal that the vertebrae did not descend away from the base of the skull as expected with normal development. The most serious of the cranial shifts of the O–C border appear as manifestations of an occipital vertebra that actually never separates into a separate entity, but instead forms raised bony areas around the foramen magnum. Transverse clefts can also form in the basioccipital as another expression of O–C cranial border shifting. Rarely, the parachordal cartilages fail to completely merge and form a vertical cleft. Other minor expressions of O–C border shifting do not pose a serious threat.

Developmental disorders occur more frequently in the vertebral column than in any other part of the skeleton, with a wide variety of both minor and serious expressions. Border shifting is the most common type of disorder affecting the vertebral column. We need to remember that it is the border, not the affected vertebra that shifts its identity, either up or down. This can involve just one border or more than one border. Each border follows its own genetic command, so that when more than one border is affected, there may be both cranial and caudal shifting in the same vertebral column or only one type of shifting. As mentioned with discussion of defects affecting the base of the skull, caudal shifting of the O–C border forces the atlas to unite with the skull base. In addition to expressions of the occipital vertebra, cranial O–C border shifting can also have a dramatic affect on the developing axis dens (odontoid). The dens arises separately from the axis body within the region of the primordial foramen magnum, and when the O–C border shifts cranially, it may fail to unite with the axis body. Sometimes just the apex is affected but usually the entire dens is involved. On rare occasions the dens or part of the dens may fail to develop. Either defect can cause neurological problems.

The most pronounced defect of the cervicothoracic border results from cranial border shifts that result in expressions of cervical rib, either jointed into a true rib or without a true joint. Either expression may or may not articulate with the 1st rib. Neurological disturbances often occur with cervical rib. Lesser expressions are not pathological. Cervical rib can be bilateral, but often appear unilateral. Caudal C–T border shifting is less dramatic, causing the 1st thoracic transverse process and 1st rib to appear rudimentary, much smaller than normally accepted. Lumbar rib expressions, either unilateral or bilateral, with or without true joints, result from caudal shifting at the thoracolumbar border. Large lumbar ribs can cause pathology. Cranial shifting at the T–L border may force the 12th rib to be rudimentary (hypoplasia), or absent (aplasia), reflected by the absence or small appearance of the rib facets on the 12th thoracic. Transitional facets, normally appear on the 12th thoracic, but can move upwards to the 11th thoracic with cranial border shifting, while caudal border shifting can force them downward to the 1st lumbar. Lumbosacral cranial border shifting forces the 5th lumbar downward to unite with the sacrum. Variable expressions include complete or incomplete union either unilaterally or bilaterally. The incompletely united 5th lumbar usually articulates with the sacrum, with corresponding facets, but it can also appear to have sacral-like alae without articulation. Caudal shifting at the L–S border forces the 1st sacral segment upward to completely or incompletely separate and join with the lumbars. Shifting of the L–S border often causes pathology, especially when asymmetrical. The sacrocaudal border is also affected by border shifts, with cranial border shifting forcing caudalization of the last sacral segment and caudal border shifting forcing the 1st caudal segment to unite with the sacrum.

Sometimes an extra vertebral segment forms to produce a transitional vertebra, usually at the thoracolumbar border or lumbosacral border that can take on characteristics of either adjacent vertebra. This can add confusion to deciphering cranial–caudal border shifting in the vertebral column.

As already noted, clefting or bifurcation occurs in neural arches without spina bifida when one or both precursors have hypoplasia (sometimes aplasia) that interferes with their osseous union. This occurs most often in the 1st and 2nd sacral segments, followed by a complete cleft sacrum. The 5th lumbar, and atlas posterior arch, are also common sites, but clefting can occur anywhere along the vertebral column. Functional stress on a cleft neural arch, especially in the lumbar area, often causes unilateral spondylolysis, while bilateral spondylolysis usually results from functional stress on a united but asymmetrically formed neural arch. The anterior arch of the atlas may fail to develop and appear cleft as the lateral processes attempt to meet midline in its absence.

Segmentation errors often cause expressions of block vertebra when adjacent developing vertebral segments fail to separate either completely or incompletely. Most often only one block vertebra appears, usually in the cervical spine, and this does not cause pathology. When several cervical vertebral segments form a block, this constitutes the pathology known medically as Klippel–Feil syndrome, and quite often it is associated with block vertebrae in other parts of the spine, and with other types of vertebral defects. Multiple defects, with different types of disorders occurring in the same vertebral column, can be challenging to decipher but the morphogenetic approach can unravel the various defective events.

Too often, ventral hypoplasia of a vertebral body gets confused with compression fracture, when actually the ventral portion of the vertebral body fails to develop in step with the dorsal part. Sometimes it does not develop at all, exhibiting ventral aplasia instead of hypoplasia. This defect often occurs in the lower thoracic spine, especially affecting the 12th thoracic, leading to kyphosis. Very rarely, the dorsal aspect of a vertebral body can be affected by hypoplasia, producing a dorsal wedge shaped vertebral body that is sometimes noted in the last lumbar vertebra, leading to lordosis. Lateral hypoplasia produces scoliosis when one side of the vertebral body is underdeveloped, or in some cases does not develop at all. When multiple vertebral bodies are affected by lateral aplasia and hypoplasia, a buttressing bony postlateral bar often forms on the affected side, forming a severe scoliosis. Developing ribs on the affected side are affected and often fuse together. Lateral hypoplasia generally affects the thoracic or lumbar spines, but it sometimes appears in the sacrum with one or more sacral segments forming an asymmetrically shaped sacrum.

Hemivertebra develop from somite hemimetamere segmentation error, appearing as a lateral wedge shaped vertebral half with a complete neural arch half. Thoracic hemivertebrae will also have its complimentary rib. Solitary hemivertebra usually are the product of an extra somite forming without a matching partner, while two or more hemivertebrae usually form when the counterparts of their hemimetamere precursors shift out of order before the designated time of fusion, leaving them unmatched. Multiple hemivertebrae can appear unilaterally or contralaterally, and they often are mixed in with other vertebral defects. Unilateral expressions of hemivertebrae cause severe scoliosis. Contralateral hemivertebra appear balanced, causing limited pathology. Sometimes bilateral hemivertebra occur opposite each other when the two halves fail to meet on time and unite. Hemivertebra often unite with an adjacent vertebra, or remain as a separate vertebral half segment.

Sagittal cleft vertebra is due to delayed regression of the notochord from a developing primordial vertebral body, leaving a cleft or circular pouch filled with notochordal tissue in the developing bone (Merbs 2004). This defect can mimic the appearance of bilateral hemivertebra when viewed radiographically (often referred to as butterfly vertebra for this reason). It usually appears in the thoracic or lumbar spines as a solitary sagittal cleft.

Ribs as outgrowths of primordial vertebral segments can be affected by vertebral disorders that unite them at the vertebral ends or with border shifting that creates hypoplasia or aplasia of adjacent ribs. Sternal ends can bifurcate or flare. Sometimes ribs can be quite wide from hyperplasia. The most unusual and rarest rib defect is development of the intrathoracic rib, a small rudimentary rib segmented from a parent rib.

The sternum displays many variable shapes and anomalies. The most striking, but very rare defect, occurs when the two mesosternal halves fail to unite either completely or partially to form a cleft or bifurcated sternum. Sometimes one or both of the suprasternal structures develop into separate ossicles that often fuse with the manubrium, either near their original destination, or upper dorsal or ventral side of the manubrium. Sometimes the manubriomesosternal joint shifts downward between the 1st and 2nd sternebrae. The familiar, nonpathological sternal aperture appears when the two halves of the mesosternum do not unite where the aperture is located. This anomaly has often been mistaken for a pathological sternal foramen. Hypoplasia or aplasia of sternebra segments of the mesosternum often occurs, producing a short sternum. Occasionally an extra segment forms giving the sternum an extra long appearance.

Developmental disorders of the upper and lower limbs can affect a whole limb, one or more limb segments, or just a few elements of a limb segment. Disturbance of the apical ectodermal cap can cause agenesis of an entire limb (amelia) or part of a limb (micromelia). The end plates can continue to develop abnormally as rudimentary hands or feet attached to the trunk by irregularly shaped bones (phocomelia). The forearm and lower leg segments are more vulnerable to developmental disturbance than are the upper segments. The distal thigh segment may not produce a patella (aplasia), or it may appear quite small (hypoplasia). Fusion disorders can occur in the forearm and lower leg segments, especially in the proximal ulna and radius (radioulnar synostosis), or in the proximal tibia and fibula (tibofibular synostosis). The radius and fibula may even be missing or suffer hypoplasia.

The shoulder girdle and pelvic girdle segments rarely suffer developmental disorders. The clavicle can be affected by hypoplasia or aplasia, and more rarely, bifurcation (Moore 1985). The most common disorder of the scapula is failure of the acromion to unite with it as the skeleton matures. Hypoplasia of the glenoid neck can affect the developing humeral head causing abnormal joint function. Developmental disruption of the fibrocartilaginous rim forming the acetabulum labrium that acts as a buttress for the femoral head within the acetabular joint space leads to dislocation of the developing femoral head and congenital hip dysplasia (Ferguson 1968). Defective formation of the femoral head and or neck can also lead to hip dysplasia (Mitchell and Redfern 2008).

The developing end segments for wrist and hand, and for ankle and foot, are the most affected by developmental disorders. The middle rays are affected more than the others: the middle or 3rd metacarpal with adjacent capitate, and the middle or 3rd metatarsal with adjacent cuneiform. The styloid process of the 3rd metacarpal can ossify as a separate bone, the os metastyloideum, that may fuse with the capitate or be absent (Case 1996). Absence of joint formation between the 3rd metatarsal and its adjacent cuneiform results in complete or partial osseous fusion. Most often a nonosseous fibrous connection forms with incomplete development of the joint (Regan et al. 1999). Defective development of the middle ray can result in a variety of cleft defects known as lobster claw defects in both the hand and foot. The middle ray in lobster claw defect is often absent while the lateral and medial digits fuse together (Smith 2002). Other precursor digits can also fail to develop (ectodactyly) or suffer hypoplasia to produce brachydactyly. Syndactyly in bony or soft tissue occurs when the developing digits fail to separate either completely or partially, causing altered growth of the associated metacarpals (Larsen 2001). Symphalangism refers to bony interphalangeal union of phalanges within the same digit, and it commonly appears in the 5th toe, followed by the 4th toe, and 5th finger of the hand (Case and Heilman 2005). Occasionally the 5th digit of the hand develops an extra phalange for a triphalangeal thumb (Smith 2002).

Extra digits, or polydactyly, can occur in both hands and feet, or just one or the other, and either unilaterally or bilaterally. Polydactyly may or may not cause pathology. There are many different expressions of digital duplication: a) completely separate digit; b) branching of the metacarpal or metatarsal with extra digit; c) double headed metacarpal or metatarsal with double set of digits; d) double headed phalanges with duplicate sets of distal phalanges. The most common site for development of polydactyly is the postaxial side of the developing hand or foot that involves the 5th digits. The pre-axial sides involving the 1st digits are rarely affected (Barnes 1994b; Case et al. 2006)

Carpals and tarsals can be affected by fusion or bifurcation, either completely or incompletely. The proximal rows of carpals and tarsals are most affected, particularly the schapoid, lunate, and triquetral wrist bones (Case 1996), and the talus and calcaneus in the ankle (Resnick 1989). Separation can also occur in marginal areas, with tuberosities, or with bony extensions such as with os metastyloideum in the hand. The talus os trigonum and calcaneous os secondarius (Anderson 1988) are examples of this in the foot, and bipartite patella with marginal separation of the superior lateral angle.

Hypoplasia, or aplasia, and sometimes hyperplasia, of a bony part can be expressed as with the hamulus of the hamate. Anomalous extra bones forming in the wrist or ankle are very rare. The os intermetatarseum may appear within the proximal borders of the 1st and 2nd metatarsals and 1st cuneiform, usually attaching to one of the three involved bones, to appear as a broad bony flap overlapping one or both of the other bones (Case et al. 1998). Club foot deformity has many expressions that appear to mirror mesenchymal disturbances in forming the muscles, ligaments, and bones of the ankle. Talipes equinovarus is the most common type with deformity of the talar head and neck affecting the developing calcaneus, forcing the foot to turn inward (Barnes 2008; Brothwell 1967; Morse 1978). A similar deformity can affect the hand—club hand or talipomanus, with the carpals of the wrist affected, developing in a twisted position with strong flexion and adduction of the hand.

SUMMARY

We have only recently moved beyond purely descriptive analyses of developmental disorders (congenital defects) as interesting case studies to discerning how and why they develop, and determining the frequency of their appearances in a given population. Developmental disorders follow two different types of classification: a) according to specific defective bone tissue involving developing bone segments; and b) according to localized bone structural defects within a developmental field. Bone-tissue defects in paleopathology are extremely rare while bone-structural defects are more often recognized. Thus, the emphasis of this chapter focuses on the latter. The interpretation and classification of a developmental disorder requires an understanding of skeletal morphogenesis, as briefly outlined. Only then can we place developmental disorders in a paleopathological perspective.

This discussion has focused on examples of the more serious structural developmental field disorders found in paleopathology, as well as common minor disorders frequently identified in paleopathological analyses. Quite often it is difficult to separate pathological and nonpathological developmental disturbances that follow similar, or the same developmental gradient. Many more developmental variants in each field have been identified (Barnes 1994; Case 1996), and most of them are not serious enough to interfere completely with normal functioning to be pathological, but they do provide additional genetic data that can be included in analyses of kinship relationships, population relationships, and marriage patterns. Bess Miller was right in 1928 with her assessment of the heritability of developmental variations, and modern microbiologists have proven her to have been correct. Nonmetric studies opened the door to the feasibility of using the vast number of developmental variations, both pathological and nonpathological, of the entire skeleton for genetic studies, particularly when invasive studies are not allowed.

Morphogenetic analysis of developmental disorders of the skeleton is a fairly recent introduction to paleopathology (Barnes 1994a, b), organizing localized structural defects according to specific disturbances in associated developmental fields (see Tables 21.1–21.5). The morphogenetic approach is open-ended, with room for new discoveries of developmental disorders and variations of recognized defects. Knowing how disorders arise within developmental fields provides the necessary steps for deciphering and classifying various structural disorders of the skeleton. Expressions for each type of developmental disorder or anomaly are important to note, because underlying genetic causes can vary, such as with the variable types of polydactyly. Percentages best represent frequencies of these expressions, whereas statistical analyses can distort findings. The appearance of just one or a few of the very rare expressions or types of developmental disorders within a population is just as important as the presence of more common expressions or types.

REFERENCES

Allbrook, D. B., 1955   The East African Vertebral Column: A Study in Racial Variability. American Journal of Physical Anthropology 13:489–513.

Anderson, J. E., 1968   Skeletal "Anomalies" as Genetic Indicators. In The Skeletal Biology of Earlier Human Populations. D. R. Brothwell, ed. pp. 135–147. Oxford: Pergamon Press.

Anderson, T., 1988   Calcaneus Secondarius: An Osteo-Archaeological Note. American Journal of Physical Anthropology 77:529–531.

Arcini, C., and Frolund, P., 1996   Two Dwarves from Sweden: a Unique Case. International Journal of   Osteoarchaeology 6:155–166.

Aufderheide, A. C., and Rodriguez-Martin, C., 1998   The Cambridge Encyclopedia of Human Paleopathology. Cambridge: Cambridge University Press.

Barnes, E., 1994a   Developmental Defects of the Axial Skeleton in Paleopathology. Niwot: University Press of Colorado.

Barnes, E., 1994b   Polydactyly in the Southwest. Kiva 59:419–431.

Barnes, E., 2008   Congenital Anomalies. In Advances in Human Paleopathology. R. Pinhasi, and S. Mays, eds. pp. 329–362. Chichester: John Wiley & Sons, Ltd.

Bennett, K. A., 1973   Lumbo-sacral Malformations and Spina Bifida Occulta in a Group of Pre-Historic Modoc Indians. American Journal of Physical Anthropology 36:435–440.

Berry, A. C., and Berry, R. J., 1967   Epigenetic Variation in the Human Cranium. Journal of Anatomy 101:361–379.

Brothwell, D. R., 1967   Major Congenital Anomalies of the Skeleton: Evidence from Earlier Populations. In Diseases in Antiquity. D. Brothwell and A. T. Sandison, eds. pp. 423–443. Springfield, IL: C. C. Thomas.

Brothwell, D.R., and Powers, R., 1968   Congenital Malformations of the Skeleton in Earlier Man. In The Skeletal Biology of Earlier Human Populations. D. R. Brothwell, ed. pp. 173–203. Oxford: Pergamon Press.

Buikstra, J. E., 1993   The Skeletal Remains of a Female Achondroplastic Dwarf from the Middle Woodland Period, West-Central Illinois. American Journal of Physical Anthropology Supplement 16:65.

Buikstra, J. E., and Ubelaker, D. H., 1994   Standards for Data Collection from Human Skeletal Remains. Fayetteville: Arkansas Archaeological Survey Research Series No. 44.

Case, D. T., 1996   Developmental Defects of the Hands and Feet in Paleopathology. Masters of Arts Thesis. Arizona State University.

Case, D. T., and Heilman, J., 2005   Pedal Symphalangism in Modern American and Japanese Skeletons. Journal of Homo 55:251–262.

Case, D. T., Hill, R. J., Merbs, C.F., and Fong, M., 2006   Polydactyly in the Prehistoric American Southwest. International Journal of Osteoarchaeology 16:221–235.

Case, D. T., Ossenberg, N. S., and Burnett, S. E., 1998   Os Intermetatarsium: A Heritable Accessory Bone of the Foot. American Journal of Physical Anthropology 107:199–209.

Dixon, A. F., 1900   On Certain Markings on the Frontal Part of the Human Cranium and Their Significance. Journal of the Royal Institute 30:96–97.

Ferembach, D., 1963   Frequency of Spina Bifida Occulta in Prehistoric Human Skeletons. Nature 199:100–101.

Ferguson, A. B., 1968   Orthopedic Surgery in Infancy and Childhood. 3rd edition. Baltimore: Williams & Wilkins.

Finnegan, M., 1972   Population Definition in the Northwest Coast by Analysis of Discrete Character Variation. Ph.D. Dissertation, Department of Anthropology, University of Colorado.

Finnegan, M., 1974   Cranial and Infracranial Non-Metric Traits: Those Traits Which are Most Important and How They may be Handled.   American Journal of Physical Anthropology 41:478–479.

Finnegan, M., 1978   Non-Metric Variation of the Infracranial Skeleton. Journal of Anatomy 125:23–37.

Frayer, D., Macchiarelli, R., and Mussi, M., 1988   A Case of Chondrodystrophic Dwarfism in the Italian Late Upper Paleolithic. American Journal of Physical Anthropology 75:549–565.

Gladyskowska-Rzeczycka, J., 1980   Remains of Achondroplastic Dwarf from Legnica of XI–XIIth Century. Ossa 7:71–74.

Gray, P. H. K., 1969   A Case of Osteogenesis Imperfecta, Associated with Dentiogenesis Imperfecta, Dating from Antiquity. Clinical Radiology 20:106–108.

Gruneberg, H., 1952   General Studies on the Skeleton of the Mouse. IV. Quasicontinuous Variation.   Journal of Genetics 51:95–114.

Gruneberg, H., 1954   Variation Within Inbred Strains of Mice. Nature 173:674–676.

Gruneberg, H., 1955   Genetical Studies on the Skeleton of the Mouse. XV. Relations Between Major and Minor Variants. Journal of Genetics 53:515–535.

Gruneberg, H., 1963   The Pathology of Development; a Study of Inherited Skeletal Disorders in Animals. New York: John Wiley & Sons.

Gruneberg, H., 1964   The Genesis of Skeletal Abnormalities. In Congenital Malformations. M. Fishbein, ed. pp. 219–223. New York: International Medical Congress.

Hauser, G., and DeStefano, G. F., 1989   Epigenetic Variants of the Human Skull. Stuttgart: E. Schweizerbartische Verlagsbuchhandlung (Nagele u. Obermiller).

Hepburn, D., 1908   Anomalies in the Supra-inial Portion of the Occipital Bone, Resulting from Irregularities of Its Ossification, with Consequent Variations of the Interparietal Bone. Journal of Anatomy 42:88–92.

Hoffman, J. M., 1976   An Achondroplastic Dwarf from the Augustine Site (CA-Sac-127). *In* Studies in California Paleopathology IV. J. M. Hoffman, and L. Brunker, eds. pp. 65–119. Berkeley: Contributions of the University of California Archaeological Research Facility.

Hrdlicka, A., 1923   Incidence of the Supracondyloid Process in Whites and Other Races. American Journal of Physical Anthropology 6:405–412.

Larsen, W. J., 2001   Human Embryology. 3rd edition. Philadelphia: Churchill Livingstone.

Laughlin, W. S., and Jorgensen, J. B., 1956   Isolate Variation in Greenlandic Eskimo. Acta Genetics 6:3–12.

Manchester, K., 1983   The Archaeology of Disease. Leeds: Arthur Wigley.

Merbs, C. F., 2004   Sagittal Clefting of the Body and Other Vertebral Developmental Errors in Canadian Inuit Skeletons. American Journal of Physical Anthropology 123:236–249.

Merbs, C. F., and Wilson, W. H., 1960   Anomalies and Pathologies of the Sadlermuit Eskimo. Bulletin of National Museum of Canada 180:154–179.

Miller, B. L., 1928   The Inheritance of Human Skeletal Anomalies. Journal of Heredity 19:28–46.

Mitchell, P. D., and Redfern, R.C., 2008   Diagnostic Criteria for Developmental Dislocation of The Hip in Human Skeletal Remains. International Journal of Osteoarchaeology 18:61–71.

Moore, K. L., 1985   Clinically Oriented Anatomy 2nd edition. Baltimore: Williams & Wilkins.

Morse, D., 1978   Ancient Disease in the Midwest. Springfield: Reports of Investigations No. 15, Illinois State Museum.

Oetteking, B., 1922   Anomalous Patellae. Anatomical Record 23:269–279.

Ortner, D. J., 2003   Identification of Pathological Conditions in Human Skeletal Remains. Amsterdam: Academic Press.

Ortner, D. J., and Putschar, W., 1985   Identification of Pathological Conditions in Human Skeletal Remains. Washington, DC: Smithsonian Institution Press.

Regan, M. H., Case, D. T., and Brundige, J. C., 1999   Articular Surface Defects in the Third Metatarsal and Third Cuneiform: Nonosseous Tarsal Coalition. American Journal of Physical Anthropology 109:53–65.

Resnick, D., 1989   Bone and Joint Imaging. Philadelphia: W.B. Saunders.

Rubin, P., 1964   Dynamic Classification of Bone Dysplasias. Chicago: Yearbook Medical Publications.

Russell, F., 1900   Studies in Cranial Variation. American Naturalist 34:737–745.

Sadler, T. W., 2006   Langman's Medical Embryology, 10th edition. Philadelphia: Lippincott Williams & Wilkins.

Saunders, S. R., 1989   Nonmetric Skeletal Variation. In Reconstruction of Life from the Skeleton. M. T. Y. Iscan, and K. A. R. Kennedy, eds. pp. 95–108. New York: Alan R. Liss, Inc.

Searle, A. G., 1954   Genetic Studies on the Skeleton of the Mouse. IX. Causes of Skeletal Variation Within Pure Lines. Journal of Genetics 52:68–102.

Selby, S., Garn, S. M., and Kanareff, V., 1955   The Incidence and Familial Nature of a Bony Bridge on the First Cervical Vertebra. American Journal of Physical Anthropology 13:129–141.

Smith, E. G., and Wood Jones, F., 1910   The Archaeological Survey of Nubia Report for 1907–1908, Volume II, Report On the Human Remains: Cairo.

Smith, P., 2002   The Hand Diagnosis and Indications, 4th edition. London: Churchill Livingstone.

Snow, C. E., 1943   Two Prehistoric Indian Dwarf Skeletons from Moundville. University of Alabama: Alabama Museum of Natural History Museum Paper 21.

Stewart, T. D., 1932   Vertebral Column of the Eskimo. American Journal of Physical Anthropology 17:123–136.

Sullivan, L. R., 1922   The Frequency and Distribution of Some Anatomical Variations in American Crania. Anthropological papers American Museum, New York 23:207–258

Trotter, M., 1929   The Vertebral Column in Whites and in American Negroes. American Journal of Physical Anthropology 13:95–108.

Trotter, M., 1934   Septal Aperture in the Humerus of American Whites and Negroes. American Journal of Physical Anthropology 19:213–228.

Turkel, S. J., 1989   Congenital Abnormalities in Skeletal Populations. In Reconstruction of Life from the Skeleton. M. Y. Iscan and K. A. R. Kennedy eds. pp. 109–127. New York: Alan R. Liss, Inc.

Waddington, C. H., 1956   The Principles of Embryology. London: Allen and Unwin.

Wells, C., 1965   Osteogenesis Imperfecta from an Anglo-Saxon Burial Ground at Burgh Castle, Suffolk.   Medical History 9:88–89.

Zimmerman, M. R., and Kelley, M. A., 1982   Atlas of Human Pathology. New York: Praeger.

CHAPTER **22** # Metabolic and Endocrine Diseases

Tomasz Kozłowski
and Henryk W. Witas

INTRODUCTION

In 1854, in the valley of the Düsel River, Germany, remains of a hominid, known today as *Homo neanderthalensis* or *Homo sapiens neanderthalensis* (Wolpoff 1999), were discovered. The skeleton triggered a lively debate. On one side of the controversy were researchers who concluded that the bones belonged to a human ancestor, not a modern human. On the other side were individuals such as Rudolf Virchow, a distinguished German pathologist, who held that the remains belonged to a contemporary individual who had suffered from rickets (caused by vitamin D deficiency) (Virchow 1872). Although anthropologists today agree that Virchow was incorrect in asserting that the skeleton was from an anatomically modern human, Ivanhoe (1970) suggested that the diagnosis of Vitamin D deficiency could be correct. Hence, one can claim that the discussion of metabolic diseases in ancient human remains dates back almost 150 years in the history of anthropology and pathology.

Metabolic and endocrine diseases pose serious medical issues today, with their etiology and pathogenesis at times remaining unknown. The term "metabolic bone disease" has been used since 1948, when Albright and Reifenstein introduced it to describe conditions affecting systemic bone formation and remodeling (Brickley and Ives 2008). In the field of pathology it is often restricted to disorders which cause defective bone tissue, and well as proliferative bone growth or loss (such as that seen in osteoporosis or osteomalacia). However, in paleopathology, this group of disorders is particularly problematic, as they often do not leave marks on the skeleton or manifest their presence with unique lesions.

A Companion to Paleopathology, First Edition. Edited by Anne L. Grauer.
© 2012 John Wiley & Sons, Ltd. Published 2016 by John Wiley & Sons, Ltd.

But before we launch into a discussion of metabolic and endocrine disorders in paleopathology, it is essential to become familiar with the biochemical and physiological bases of the diseases. Metabolism is a set of related biochemical reactions leading to the synthesis (anabolism) or decomposition (catabolism) of certain chemical compounds (Tortora and Derrickson 2009) through the use or emission of energy. These processes occur in all living cells (including bone cells) and, importantly, remain under the control of the endocrine system. Vitamins and ions of different chemical elements (e.g. calcium (Ca), phosphorus (P), magnesium (Mg), barium (B), zinc (Zn), and iron (Fe)) also play an active role. Thus, in clinical medicine metabolic diseases of bone, along with endocrine disorders are often treated together (e.g., Porat and Sherwood 1986).

Proper functioning of cells requires a relatively constant supply of building and nutritional substances. Many metabolic diseases, therefore, result from nutrient deficiencies which disrupt the cells' metabolic pathways. Similarly, disturbances in the function of endocrine glands, which might result in hormone deficiencies or the hyperactivity of glands, along with genetic mutations which alter the expression of individual genes or their groups, can significantly impact the activity of enzymes responsible for the transformation of chemical compounds within metabolic pathways (for example, resulting in celiac disease, diabetes, and cystic fibrosis). Bone, like any other tissue, must have its metabolic needs met in order to keep it "healthy." As a tissue, it is influenced by endogenous and exogenous factors which can alter the remodeling process—the lifelong balance between bone resorption and new bone formation. While endogenous and exogenous factors influence bone development and function throughout the life of the individual, the recognizable symptoms of metabolic and hormone disruption can vary depending upon the age of the individual. Metabolic or endocrine disruption occurring during childhood can affect the skeleton differently than if it occurred during adulthood. In both instances in the past, without adequate pharmacological support, these disorders could have led to the development of severe symptoms, and consequently, death.

Metabolic Diseases in Paleopathology

Currently, metabolic diseases most often investigated by paleopathologists focus on disorders induced by food deficiency, although other factors can be important in the etiology.

Angel (1978, 1982), for instance, proposed a way of estimating the nutritional status of an individual (thus the presence of metabolic stress caused by low food supply), based on metric parameters of bone. These parameters included the pelvic brim index (Angel 1978) and cranial base height (basion–porion height, ba–po) (Angel 1982). He posited that as a result of an inadequate diet during childhood and adolescence, platybasia (low values of the skull base height) and a decrease of the pelvic brim index (sagittal dimension) would ensue. Additionally, the presence of dental lesions, such as enamel hypoplasia, along with the reduction in overall body height were argued to be indicators of nutritional stress, and hence could indicate the presence of a metabolic disorder (Kelley and Angel 1987). Obviously, these characteristics measure the general level of nutritional status (metabolic stress) of an individual or within a population, at best. They do not indicate the particular nutritional components that might be missing from a diet, nor their pathological effects on the body. Thus, metric analyses appear to be most useful in providing a recognition of general life conditions of human populations in the past. They do not garner the enthusiasm of paleopathologists or paleoepidemiologists who wish to understand the nature or direct cause of metabolic diseases.

There are a number of specific disorders which have captured the attention of paleopathologists in recent years. These include diseases like scurvy, rickets, osteomalacia, anemia and osteoporosis. Arguably, they leave permanent, diagnosable traces in the human skeleton (Stuart-Macadam 1989; Aufderheide and Rodriguez-Martin 1998; Ortner 2003; Brickley and Ives 2008), and occur as a result of complex interactions between human physiology, culture, and diet. Finding traces of these conditions in skeletal remains is one of the most important means by with the reconstruction of health and living conditions of ancient populations is assessed. Moreover, the presence of these conditions provides insight into the synergistic relationship between metabolic disease and infection (Buckley 2000), as well as population genetics.

Scurvy

Scurvy, also known as Moller–Barlow's disease, is caused by a deficiency of ascorbic acid (Vitamin C) in the diet (Ortner 1984; Maat 2004; Brickley and Ives 2008). This vitamin is indispensable in a number of metabolic processes, such as the formation of collagen, which is the body's (and bone tissue's) main structural protein (Ortner and Ericksen 1997; Maat 2004; White and Folkens 2005; Mays 2008). The presence of Vitamin C in the human diet is essential. Humans and other primates, as well as, for example, guinea pigs, microbats from India (*Microchiroptera*),the Red-vented Bulbul (*Pycnonotus cafer*) and some species of trout and salmon are incapable of creating the enzyme necessary for the synthesis of Vitamin C within the organism (Stuart-Macadam 1989; Aufderheide and Rodriguez-Martin 1998). Thus, ascorbic acid must be constantly ingested and absorbed into the body. Its shortage leads to the reduction of osteoid formation and the weakening of connective tissue's structure (Stuart-Macadam 1989).

Controversies among paleopathologists often focus on the diagnostic criteria of the disease (see, for instance, Ortner 1984). Until the end of 20th century, the identification and presence of scurvy in human skeletal material was poorly documented (Ortner et al. 1999). Rarely were any cases of scurvy described in the anthropological literature (Schultz et al. 2007). Furthermore, paleopathologists focused their attention on the presence of the condition in adults, and had only a slight interest in the manifestation of the disease in juveniles (no doubt exacerbated by the poor preservation and sample bias of juvenile remains). Ironically, while scurvy can occur in people of all ages, it most often affects small children, particularly between the ages of 6 to 18 months (Buckley 2000, after Jelliffe).

The effects of scurvy on the human skeleton are subtle and often linked to soft tissue changes, especially in infants and children where bone growth is still occurring. For instance, the weakening and thinning of blood vessel walls associated with compromised connective tissue, leads to hemorrhaging in a number of areas of the body (Fain 2005; Mays 2008). The extravasated blood, irritating the periosteum, triggers inflammation and stimulates the formation of new, heterotrophic bone with porous structure (Ortner and Ericksen 1997). In paleopathology, lesions connected with scurvy are macroscopically found in the form of symmetrical, scattered traces of periosteal inflammation and porosity (Ortner and Ericksen 1997; Ortner et al. 1999; Aufderheide and Rodriguez-Martin 1998; Brickley and Ives 2008). The porous lesions (often less than 1 mm in diameter) are remnants of tubules of small blood vessels (Ortner 1984; Mays 2008), and appear most often on the greater wings of the sphenoid, the zygomatic, the posterior surface of maxilla, the coronoid process of mandible, and on orbital plates of the frontal (Ortner 1984; Ortner and Ericksen 1997; Ortner et al. 1999; Melikan and Waldron, 2003; Brickley and Ives 2008; Mays 2008). The localization appears to occur at sites where supporting muscles attach. Intracranial and subscleral hemorrhage can also occur in individuals with scurvy, creating hyperplasia of capillaries on the surfaces of the skull (Brickley and Ives 2006; Mays 2008). However, similar changes can result from dura mater irritation by other disease processes (e.g. specific and nonspecific infections, tumors, trauma), so one must be very cautious when attempting to distinguish scurvy from other pathological lesions (Lewis 2004). Schultz (2001) and Schultz et al. (2007) have argued that without the use of microscopy, the identification of scurvy as the cause of these types of general bone changes is very difficult (Schultz 2001; Schultz et al. 2007), as both a number of *in vivo* and postmortem taphonomic processes can create porous-looking bone.

The presence of scurvy has been linked to a number of postcranial changes in juveniles, including the presence of new bone formation at the joints, the development of Pelkan spurs at the metaphyses of long bones, abnormal porosity of the scapula, and the enlargement of the ribs at the costochondral end. In adults, the changes are difficult to distinguish from other conditions with different etiology, but might be recognized by the presence of new bone formation in the orbits, inflammation of the alveolar processes of the mandible and maxilla, and new bone formation at the ends of the long bones.

In practice, the examination for the presence of scurvy might benefit from gently touching the bone with the end of one's finger and from looking for changes in color. Unhealed porotic changes often have the feel of fine-grained sandpaper or pumice. The color of the bone, particularly at the ends of long bones, tends to be darker, sometimes grayish. Maat (2004), in his analysis of Dutch whalers found on Spitsbergen, concluded that almost 80 percent exhibited symptoms of active scurvy based on the presence of symmetrical black spots on the knee and ankle joints. The discoloration was argued to be residues of hematomas, characteristic of scurvy. It should be cautioned, however, that color changes can occur from organic and mineral dyes in the environment.

Although adult scurvy only leaves subtle and nonspecific traces, it appears that in some cases the diagnosis of the disease is possible (Van der Merwe et al. 2010). Between the 15th and 18th centuries, a period known for the launching of long sea voyages, scurvy was prevalent among sailors (Stuart-Macadam 1989; Aufderheide and Rodriguez-Martin 1998; Pimentel 2003; Maat 2004; Fain 2005). Their diet, rich in salted meat, but poor in fresh vegetables and fruits (superb sources of Vitamin C), caused the manifestation of the first symptoms (fatigue and lethargy) of scurvy only 3–4 months into their voyages. By 6–8 months after leaving port, their gums began to bleed (Henschen 1966; Stuart-Macadam 1989; Fain 2005). In some areas and in some human groups scurvy could be endemic. For example in northern Europe, there was a general lack of fresh fruits and vegetables, particularly in late winter and early spring (Melikan and Waldron 2003; Maat 2004). Eskimos also suffered from a shortage of Vitamin C (Ortner 1984), as did African Americans from rural areas of the southern USA who adhered to their traditional diets (Kelley and Angel 1987). Pathological changes in 19th-century miners from Kimberley, South Africa, which were confirmed by histological analysis, were interpreted by Van der Merwe et al. (2010) as indications of overcoming the effects of scurvy.

Interestingly, data from the Global History of Health Project, which provides a meta-analysis of 10,724 skeletons from Europe, display a mean frequency of scurvy of 1.37 percent (Brickley et al. 2009). However, it should be emphasized that these results probably do not give a clear and or reliable epidemiological picture of the occurrence of scurvy in the past, as the diagnostic criteria are relatively new and the skeletal changes are often subtle. Clear and properly defined criteria of paleopathological diagnosis have not been fully developed until now.

Rickets

Rickets is a disease primarily affecting infants and children. It is characterized by disruption of the mineralization process of growing cartilage and bone tissue. It is often caused by insufficient Vitamin D (calciferol), which is involved in the regulation of calcium and phosphorus homeostasis in the organism (Tortora and Derrickson 2009). Vitamin D is naturally synthesized by the body with exposure to ultraviolet light from the sun. The reduction of exposure to sunlight (ultraviolet light) due to cultural practices and/or the environment is considered the main etiological factor of rickets (Mays et al. 2007). The lack of sufficient Vitamin D causes the osteoid produced by osteoblasts and cartilage formed by chondroblasts to be insufficiently mineralized. For juveniles, this can result in a delay in the linear growth of bones. The reduced mineralization can also lead to bending and deformation of the bones, especially when mechanical forces acting on the bones are maintained (this can occur to long bones and pelvis due to continued bipedal posture, or to the bones of the upper extremity due to crawling, or to cranial bones of infants who spend considerable amount of time on their backs) (Stuart-Macadam 1989; Ortner and Mays 1998; Brickley et al. 2005; Brickley and Ives 2008). A delay or disorder in the sequence of deciduous dentition, along with poor mineralization and/or the presence of enamel hypoplasia, can also be associated with this vitamin deficiency. On ribs, between the costal cartilage and rib, bone can become thick, known as rachitic rosary. Bone fractures may also appear on the ribs. Moreover, the exacerbation of thoracic kyphosis may occur, sometimes accompanied by scoliosis—bending of the spine in the frontal plane (Stuart-Macadam 1989; Aufderheide and Rodriguez-Martin 1998; Ortner and Mays 1998). Harris lines might develop (Ameen et al. 2005), indicating delayed or disrupted long bone growth. Other generalized bone changes can be found in the skull, as the frontal, parietal, occipital, and facial bones (especially the jaws) can begin to display porous lesions and the deposition of woven bone. Differential diagnosis for rickets can be especially difficult (Brickley and Ives 2008), with the situation becoming even more complicated when there are simultaneously overlapping conditions caused by the deficiency of other nutrients and vitamins (such as Vitamin C, iron, Vitamin B_{12} and B_6). This is not uncommon in human populations, as lack of a single nutrient is unlikely in the face of undernutrition and/or starvation. In such cases individual symptoms can overlap and mask the typical picture of rickets, hindering the assessment of the presence of this disease based on morphological criteria.

Rickets in adults is referred to as osteomalacia (Aufderheide and Rodriguez-Martin 1998; Schamall et al. 2003; Brickley and Ives 2008). The rachitic lesions in adults are often more subtle than in children. Under the influence of osteomalacia, bones become thinner, more delicate, and structurally weaker. This can lead to the compression of vertebral bodies and deepening of thoracic kyphosis (Aufderheide and Rodriguez-Martin 1998; Brickley et al. 2005). However, it is difficult to consider these macroscopic skeletal changes as absolutely pathognomonic to rickets, as differential diagnosis practically always demands radiological analysis. Microscopic studies (Schamall et al. 2003) can also reveal changes in density and volume of bones, particularly of spongy (trabecular) tissue, and can reveal increased bone resorption; changes that are associated with rickets and osteomalacia. However, some methods used in clinical medicine to estimate the bone mineral density (BMD) have proven to be of little value (e.g. qCT, DEXA) (Schamall et al. 2003), as distinguishing osteomalacia from the presence of osteoporosis (which is associated with age, not a vitamin deficiency), can be difficult, if not impossible, at times.

Anemia

Anemia is a general condition without a specific etiology. Rather, its presence is detected by reduced red blood cells, or the reduced function of red blood cells (RBC). It is triggered by a suite of conditions and/or situations. As cells of the body rely upon oxygen "delivered" by RBC, reduction in the number of cells, or compromised ability to bond and carry oxygen, can profoundly affect the body. Iron is one of the most important chemical elements in this process. It influences the regulation of oxygen transmission to cells, immunocompetency, neurotransmission, and collagen synthesis (Stuart-Macadam 1989). Humans acquire iron primarily through ingested food, with egg yolks, animal organs, legumes and seafood providing substantial amounts. With insufficient amounts of iron in an organism, caused by genetic mutation which affects red blood cell production, or caused by shortage of this element in the diet, or disorders such as parasitic infestation which compromise its absorption from the gastrointestinal tract, or infection which triggers the body to reduce serum iron levels, or intensive hemorrhage, erythrocytes produced in marrow become hypochromatic (achromatic) and smaller (microcytosis) (Stuart-Macadam 1989). Thus the symptoms of anemia, regardless of its etiology, in the skeletal material do not produce highly specific changes. Rather, they are most often characterized by the proliferation of spongy bone caused by the hyperplasia of bone marrow which is localized on the superior aspect of the orbits (known as cribra orbitalia), or on parietal bones and occipital squama (known as porotic hyperostosis).

Numerous authors discuss cribra orbitalia and porotic hyperostosis as manifestations of acquired iron deficiency anemia, especially occurring during childhood (Hengen 1971; Stuart-Macadam 1985, 1992; Walker et al. 2009). But it has also been linked to the presence of infectious diseases and parasitic infestation

(Stuart-Macadam 1987, 1989, 1992), as well as genetically inherited anemias (e.g. thalassemia) (Lagia et al. 2006). In the latter case, changes are particularly intensive and may affect the whole skeleton. Walker et al. (2009), however, have more recently argued that the main cause of porotic changes is not due to iron-deficiency anemia. Rather, they argue that more probable causes of the osseous changes may be hemolytic anemias, such as thalassemia, sickle-cell anemia and megaloblastic anemias, whose main cause is Vitamin B_{12} (cobalamin) and Vitamin B_9 (folic acid) shortage.

The presence of cribra orbitalia and porotic hyperostosis have frequently been treated as indicators of the overall health and living standards, as well as its hygiene and sanitary conditions, in prehistoric and early-historic populations. The high frequency of cribra orbitalia is interpreted by some researchers as an adaptation to generally harsh environmental conditions and exposure to infections (Piontek and Kozłowski 2002). For instance, based on an analysis of a large sample of medieval skeletons from Central and Eastern Europe (Poland) (N = 1753), the frequency of cribra orbitalia in adults varies between 22.7 percent and 30.2 percent (Piontek and Kozłowski 2002). The mean frequency of cribra orbitalia in the six compared populations is 26.9 percent. However, a significantly higher frequency has been observed on children's skulls. In an early-medieval cemetery from Gruczno (12th–14th-century Poland) it reaches 47.8 percent (N = 92) (Piontek and Kozłowski 2002). In the light of these findings, it appears that skeletal changes associated with anemia are widespread, and can be used to infer an important component of the environment influencing the human organism in the past.

Other vitamin and element deficiencies

In the light of research on metabolic diseases related to chemical elements and vitamin deficiencies, analyses of dental and osseous tissues seem exceptionally promising. These studies are based on determining the content and proportions of elements such as calcium (Ca), phosphorus (P), barium (B), zinc (Zn), strontium (Sr), cadmium (Cd), copper (Cu), iron (Fe), lead (Pb) and stable isotopes of carbon (C) and nitrogen (N) (Szostek 2009). Although results of these analyses are used mainly in diet reconstruction, they are particularly useful in paleopathological research, allowing hypotheses concerning the etiology of pathological changes to bone to be tested. Some examples are comparative studies which examine orbital hypertrophy (cribra orbitalia) and multielement profiles characteristic of individuals with and without the skeletal changes (Gleń-Haduch et al. 1997; Szostek 2006). These analyses have shown that hyperplasia and hypertrophy of the orbital diploë are accompanied by dysfunction of elemental homeostasis. They also indicated that examining only iron or any other single element did not reliably explain the presence of cribra orbitalia (Szostek 2006).

The formation and mineralization of osseous tissue, forming upon previously synthesized proteins, is a complicated biochemical process. The biosynthesis of collagen and hydroxyapatite requires the involvement of many enzymes, critical levels of essential elements, and hormonal balance providing metabolic homeostasis. The chemical analysis of bones of a prehistoric Indian child revealed changes in the proline and lysine hydroxylation process (Von Endt and Ortner 1982). This led the authors to conclude that in children showing hyperplasic changes of bone, the disorders may involve protein synthesis and are a result of shortage or lack of some coenzymes necessary in this process. It appears that porotic lesions may indicate dysfunction of metabolic homeostasis, possibly caused by many factors, the determination of which is problematic.

Osteoporosis

Osteoporosis is a metabolic disorder of bones. It poses serious health issues today, and has been called one of the most important skeletal stress markers of the past (Goodman et al. 1988). Osteoporosis refers to a quantifiably low bone mass. It occurs due to an imbalance between the process of bone formation and resorption (Stini 1990), with a net loss in bone density.

The etiology of osteoporosis is complex. The condition is divided into two groups: primary and secondary. Primary osteoporosis includes idiopathic and juvenile osteoporosis (Dent–Friedman disease), and both Types I and II osteoporosis. Type I osteoporosis commonly occurs in postmenopausal women and is linked to hormonal changes. Type II osteoporosis appears to occur as part of the natural ageing process, and appears in both men and women of advanced age (Riggs 1991). Secondary osteoporosis occurs as a result of other conditions such as neurogenic paralysis of muscles (e.g. as a result of polio) (Kozłowski and Piontek 2000), prolonged immobilization, which causes muscle atrophy due to inactivity, diabetes, hyperthyroidism and kidney disorders (Aufderheide and Rodriguez-Martin 1998; Ortner 2003; Brickley and Ives 2008). There are important physiological factors associated with osteoporosis, including sex, family history, ethnic descent, diet (supply of Ca and Na, Vitamin D and particularly C, and amino acids) and physical activity (Zaki et al. 2009).

Since osteoporosis decreases bone's mechanical strength, it substantially increases the probability of bone fractures. Particularly common are vertebral body compression fractures, extensory fracture of the radial bone distal epiphysis (Colles' fracture) and femoral neck fracture. In archaeological material the first two are often encountered, while femoral neck fractures are less common (Brickley 2002)

Gout

Among the metabolic disorders rarely described by paleopathologists is gout (uric arthritis). In the past it was probably one of the most important causes of arthritic joint destruction (Klepinger 1980). Today, as well, the disease is of great clinical concern. It is caused by excessive accumulation of uric acid (Ortner 2003; Swinson et al. 2010). The illness occurs most often in males over the age of 40 Gout is often accompanied by hyperlipidemia (Horton 1986). As a result of the precipitation of crystals of uric acid (hydrated sodium uricate) in joint and extra-joint tissues, inflammation and damage to cartilage and bone ensues. This is known as gouty arthritis. The metatarsal–phalangeal joint of the first toe is one of the most common anatomical locations of this disease.

Paget's Disease

Another condition that is important but rarely described by paleopathologists is Paget's disease (Stirland 1991), also known as osteitis deformans or osteodystrophia deformans. Its precise etiopathogenesis remains enigmatic, but the cause may be linked to genetic predisposition, environmental conditions, or possibly a virus (Krämer 1996; Roches et al. 2002; Ortner 2003; Brickley and Ives 2008; Wade et al. 2009). Particularly interesting is the possibility that the disease has a zoonotic origin (Brickley and Ives 2008), with the vector responsible for the transmission to humans being dogs, cats, birds and cattle. Paget's disease can be defined as a progressive pathological reconstruction of bones (Krämer 1996), where normal bone is destroyed and replaced by mechanically defective osseous tissue. The areas of the skeleton most often affected are the pelvic girdle, femora, and the axial skeleton, including the cranium, lumbar, and thoracic vertebrae. Specific bone alterations include the thickening and increasing sclerotic character of cranial flat bones, accompanied by lytic lesions (Aufderheide and Rodriguez-Martin 1998; Ortner 2003). These changes are macroscopically visible, but differential diagnosis is enhanced by microscopy (Roches et al. 2002), since Paget's disease may resemble the alterations arising from treponematosis (Pinto and Stout 2009), or with rickets or osteomalacia when femoral and tibial bending is present (Aufderheide and Rodriguez-Martin 1998). Paget's disease is most commonly found in men between 50 and 60 years of age. This may explain its exceptional rarity in ancient skeletal material (Stirland 1991), as reaching this age was uncommon.

ENDOCRINE DISEASES IN PALEOPATHOLOGY

Endocrine diseases are only occasionally described in the paleopathological literature. In today's clinical practice, however, they are frequently encountered (Dacre and Kopelman 2002). Hormone disorders leaving traces within the skeleton include hypofunction and hyperfunction of the pituitary gland, hypofunction of the thyroid gland, hypofunction and hyperfunction of the parathyroids, endocrine disorders of adrenals, and dysfunction of the gonads, especially the ovaries (Aufderheide and Rodriguez-Martin 1998; Ortner 2003). The endocrine system, together with the nervous system, coordinates the functions of all the other systems of the body (Tortora and Derrickson 2009), and plays the most important role in balancing the inner homeostasis of the organism. The endocrine system controls and regulates reproduction, growth and development, as well as the balance of energy (Jordan and Kohler 1986). Thus, disorders of the endocrine system are complexly reflected in clinical symptoms. Unfortunately, paleopathologists are unlikely to detect most of them (e.g. rapid increase of body mass, weight loss, increased thirst, diarrhea, frequent urination, perspiration, headache etc.).

In bioarchaeology and paleopathology, the study of endocrine disorders is relatively new and focuses on determining diagnostic criteria and providing careful descriptions. Contemporary skeletons of people suffering from hormonal disorders are also analyzed in an effort to recognize important phenotypic features of endocrine changes which can be applied to archaeological materials.

A few endocrine diseases have been found in human skeletons from the past. One of them is hyperpituitarism. The pituitary gland produces the growth hormone somatotropin. Depending on the phase of the individual's development when the disorder manifests itself, excessive somatotropin secretion leads to gigantism (in children) or acromegaly (in adults). Acromegaly can be recognized by growth of facial bones, particularly the mandible accompanied by substantial prognathism, growth of the osseocartilaginous scaffolding of the nose, excessive development, protrusion and massiveness of supraciliary ridge, and growth of morphological structures of the occipital bone. In postcranial bones, deposition of new bone tissue and thickening may affect all the bones (Aufderheide and Rodriguez-Martin 1998).

Gigantism, on the other hand, is characterized by enormous long bone growth, well beyond that of the average population, leading to obvious great stature, but with the maintenance of normal body proportions (Mulhern 2005). Secondary degenerative alterations in the joints also occur. It is, however, relatively rare. Sometimes, however, features of acromegaly expressed to a varying degree can be observed on the skull (Gładykowska-Rzeczycka et al. 1998, 2001; Mulhern 2005).

Insufficient level of the growth hormone in children who are still growing creates variable growth deficiency and reduction in body size. This condition creates proportional reduction of the human body, known as pituitary dwarfism. In paleopathology such cases are rare. However, Roberts (1987) reports on a Roman-period skeleton from Gloucester which may display this condition, but the differential diagnosis in this specimen is difficult in light of the presence of new bone formation on the ectocranial and endocranial surfaces. In fact, the etiology of hypopituitarism is can be complex and may not always be unambiguously determined.

A change in the level of secretion of thyroid and parathyroid hormones can also affect the skeleton. These hormones are responsible for the calcium–phosphate balance within the body (Tortora and Derrickson 2009), similar to the role of Vitamin D discussed earlier in this chapter. Reduction of the synthesis of the thyroid hormone thyroxin leads to cretinism-dwarfism, the most frequently observed skeletal symptom of this disorder (Ortner and Hotz 2005). Shortage of the hormone stops the activity of epiphyseal cartilage plates, thus inhibiting long bone growth. People suffering from this condition are short, have infantile body proportions, a short (brachycephalic) cranium and prognathic face (Aufderheide and Rodriguez-Martin 1989). Obliteration of cranial sutures is delayed. Most of the epiphyses show multifocal irregular ossification centers (Ortner 2003). Skeletal anomalies are usually symmetric (Ortner and Hotz 2005), although in some patients certain groups of bones may show significant anomalies, while others remain intact. This may point to the presence of different susceptibilities of human skeletal elements to the same pathogenic factor: a low level of thyroid hormones.

Endemic cretinism, often caused by thyroid failure triggered by iodine deficiency, has been found in human populations from mountain regions (e.g. Alpine regions of Europe). However, similar effects may be due to genetic defects, infections, and thyroid tumors. Other causes of thyroid failure are very rare. Determining the etiology of thyroid hormone deficiency, based only on skeletal changes, is difficult at best. In paleopathology, described cases are often alleged and extraordinarily rare.

Another symptom of metabolic-endocrine disorder is hyperostosis frontalis interna (HFI), causing hyperplastic changes most often on the endocranial surface of the frontal bone. The lesions take the form of nodular, irregular overgrowth of the inner surface of the frontal bone, with the simultaneous thickening. These alterations may be present also on other bones of the skull (Belcastro et al. 2006). Today, the condition mostly affects postmenopausal women (Rühli and Henneberg 2004), leading some researchers to informally regard HFI as a useful feature for sex identification.

In comparison to contemporary populations, HFI is less commonly encountered in archaeological material (Rühli and Henneberg 2002). Several dozen documented cases come from a very broad time-range, from the Pleistocene, where the occurrence of HFI is suggested in *Homo erectus* and Neanderthals (Anton 1997), up to the pre-industrial era (Belcastro et al. 2006). In contrast to the present day association between HFI and females, it has been argued that HFI in prehistoric and early-historic populations was more frequent in males (Watrous et al. 1993), and that HFI in women living today might be a manifestation of microevolutionary changes in hormonal balance (Rühli and Henneberg 2002, 2004). Rühli and Henneberg (2002) hypothesize that a decrease in selective pressure favored an increased metabolic rate caused by higher levels of leptin, which produced an increase in bone growth. These authors argue for the development of a precise definition of HFI based on macroscopic, microscopic and radiological criteria as a foundation for future paleopathological research. They also encourage the analysis of as many skeletal populations as possible, the inclusion of other osteological features which may be related to hormone balance, and to undertake studies of the relationship between the level of leptin, metabolism, and the growth of osseous tissue (e.g. are leptin receptors present on the frontal bone?). Undoubtedly, research on the occurrence of HFI in the past seems to be of particular interest not only for paleopathologists, but also scientists investigating the evolution of our species, clinical aspects of metabolism, and even the causes of obesity in contemporary populations, since obesity and HFI are often found to be linked in clinical studies

Another endocrine disorder which induces adrenal function changes and can affect the skeleton is hyperadrenalism leading to Cushing syndrome. The skeletal symptoms, are not very specific and include osteoporosis, spontaneous bone fractures and necrosis of the femoral head (Krämer 1996).

METABOLIC AND ENDOCRINE DISEASES AND ANCIENT DNA STUDIES

Although we are beginning to understand much about the human past, it is obvious that our knowledge is far from complete. Much remains to be unraveled since the history of diseases appears to be an integral part of human evolution and adaptation. Some authors suggest that some diseases (not only infectious) were heavily responsible for shaping today's modern genome. But a myriad of questions remain. For instance, why are humans disconcertingly susceptible to diseases? Why does it appear that humans have not adapted well to certain pathogenic environments, and how might a particular evolutionary change occur over time and space? Until recently, osseous manifestations of change have been providing the only useful insight for paleopathologists. Human skeletal and mummifed remains have enabled disease identification to be undertaken using traditional macroscopic techniques along with CT-scans, X-rays, etc. (see Wanek et al. Chapter 18 in this volume, for more information on imaging techniques). Seeking new methods, anthropologists have found them in the labs of molecular biologists and geneticists. Below, we discuss metabolic and endocrine diseases which leave no traces in the human skeleton, but whose presence and evolution might become better understood through ancient DNA (aDNA) studies.

Cystic fibrosis and the profile of the most frequent mutation in the past

Cystic fibrosis (CF; MIM no. 219700), caused by the mutation of the *CFTR* gene (MIM no. 602 421), is regarded as one of the most common lethal autosomal recessive disorders in European Caucasian populations, affecting 1 in every 2,500 individuals (CFGAC 2005). The overall frequency of mutated allele carriers in modern European populations is about 1 in 25 individuals. Despite great progress in our basic knowledge about CF and its treatment, the disease is still associated with substantial morbidity and high fatality rates. Clinical symptoms of CF include the production of abnormally thick and viscous mucus and elevated sweat electrolytes secreted by the exocrine epithelia glands. The disease often manifests itself as progressive respiratory and gastrointestinal problems due to chronic obstructive pulmonary infections and pancreatic insufficiency, which are the result of limited secretion of digestive enzymes (Welsh and Ramsey 1998).

Today, as many as 1600 possible sequence alterations in the *CFTR* gene are recognized. Among them is that of *CFTRΔF508*, which accounts for as much as 70 percent of all mutations in USA and Western Europe. Patient survival is increasing as new clinical treatments are introduced. However, patient survival depends largely on the type of the *CFTR* gene mutation carried. Today, the lives of patients with *CFTRΔF508* is greatly prolonged (from 5–15 to even 30–40 years (CFF 2008)), certainly when compared to times past.

The antiquity of the *CFTRΔF508* mutation, however, is highly controversial, with estimates ranging from 150–300 generations, i.e. 3–6 kyr (Serre et al. 1990) to 2,600 generations, representing approximately 52 kyr (Morral et al. 1994), including the interval between 11 and 34 kyr, as calculated by Wiuf (2001).

One of the hypotheses, proposed by Modiano and co-workers (Modiano et al. 2007), attempting to elucidate the high frequency of *CFTRΔF508* in modern populations (0.02), suggests that the selection of the already-present allele took place due to the advantage it conveyed to heterozygous carriers in ever-increasing cattle milk production economies. As humans, along with all mammals, were naturally lactose-intolerant after weaning, the growing amount of dairy milk was likely associated with rather severe bouts of diarrhea. So, the dairy milk could have played a selective role, triggering the increased frequency of this allele due to natural selection. Thus, it seems that the *ΔF508* allele, beneficial for survival and reproduction, was likely introduced to the population of Neolithic farmers much earlier, even before milk became a component of the adult diet during the domestication of goats, sheep, or cattle. This seems to coincide with the finding that the genotype for lactase persistence was absent among early European farmers 7–8 kyr, as determined by aDNA analysis (Burger et al. 2007).

While aDNA has not been successfully isolated from human remains dating to the Neolithic, a few attempts have been made to find the *CFTR ΔF508* mutation in more recent remains. Witas et al. (2006) report that they did not find the mutation in specimens from Stary Brzesc Kujawski and Dziekanowice, medieval sites in Central Poland.

However, more recent but unpublished analyses of material from the western frontier of Poland (from the medieval site of Cedynia) has successfully found the mutation. This analysis revealed three *CFTR ΔF508* homozygous juveniles among the 23 studied in the population, suggesting a four-times-higher frequency than in modern populations. The high frequency of *CFTR ΔF508* frequency is probably the effect of gene drift—produced randomly due to the small size and geographical isolation of the population. Cedynia is believed to have had approximately 320 individuals, who were using the cemetery for over nine generations (Malinowska-Łazarczyk 1982).

Did the mutation originate from the Fertile Crescent? It could have spread out to Europe with the migratory waves of early farmers (known as the demic diffusion model) (EWGCFG 1990) since its frequency coincides with the southeast–northwest cline of their movement. Nevertheless, the reverse situation may also be possible. A Paleolithic origin of the mutation has been suggested as an alternative hypothesis (Casals et al. 1992). Thus the frequency of the mutation could have been diluted due to the introduction of other CF mutations in some populations by Neolithic migrations. Further analysis of aDNA isolated from remains of the carriers of the Funnel Beaker Culture, the first settlers to appear on the North European plain, should bring us closer to an answer regarding the origin, epidemiology, and the role of this mutated allele in human adaptation.

Diabetes mellitus and genetic predisposition to the disease in the past

Diabetes mellitus appears to have plagued humans for more than 3,500 years, with the first direct reference to the condition being found in the Ebers Papyrus, an Egyptian medical treatise dated to circa 1500 B.C. Writings originating from the earliest civilizations (Asia Minor, Egypt, China or India) frequently describe manifestations of the disease as loss of weight, thirst, and excessive urination, which triggered the misclassification of the disease and its purported cause to be dysfunction of the kidneys until the late 19th century. For a long time, clinical signs of diabetes were an invariable death sentence. The development of the fields of physiology and chemistry, and the change from observational to investigative data, led to defining diabetes as an endocrine disease.

There has been continuous increase in the frequency of Type 1 diabetes (T1D, IDDM1, MIM no. 220 100) worldwide (Onkamo et al. 1999), particularly in central–eastern Europe (Green et al. 2001), which mirrors the increasing prevalence of other autoimmune diseases (Ivarsson et al. 2000; Haynes et al. 2004) in modern populations. The reasons for this increase are not known. However, it appears that a combination of genetic and exogenous factors is essential for its development. A polygenic predisposition to T1D is linked to a number of genes (more than 15), which are sensitive to environmental agents. These genes include variation within the *HLA* (human leukocyte antigens), *CTLA4* (cytotoxic T lymphocyte associated antigen-4, MIM no. 601388) and *PTPN22* (protein tyrosine phosphatase nonreceptor type 22, MIM no. 600716) genes (Pearce and Merriman 2006). The 20th- century increase in the frequency of these predisposing alleles, is in part due to the introduction of insulin. Access to this hormone has prolonged the lifespan of sufferers, allowing them to pass the alleles to their offspring, thus influencing the gene pool.

Natural selection might also have played a role in the presence of diabetes in human populations. For instance, the allele *HLA DQB1* (MIM no. 604 305), encoding aspartic acid at a key location of the protein molecule, is also involved in protection against Type I diabetes. However, its presence has also been associated with a weakened response of the immune system in the presence of *Mycobacterium tuberculosis* (Delgado et al. 2006). Thus, it seems plausible that successive epidemics of tuberculosis in Europe from the 16th to the 19th centuries constituted a selecting force which decreased the frequency of this beneficial allele in protecting against the development of Type I diabetes. On the other hand, other environmental factors might influence morbidity (either positively or negatively) despite the genotype of an individual. Among a number of suggested environmental factors triggering altered autoimmune responses are pathogens, drugs, pollutants, diet, recently introduced chemicals, as well as stress coming from increasing population density and lifestyle challenges (e.g. Roivainen 2006; Dahlquist 2006).

Thus, predicting the sensitivity or resistance of an individual or population based on analyses on the genetic level is difficult, at best, especially in light of our complex modern environments. Analysis of aDNA isolated from medieval archaeological sites around Poland showed significantly higher frequency of alleles *HLA DQB57* and *CTLA-4* (which are believe to be predisposing to Type 1 diabetes) than in a modern control group (Witas et al. 2007; Witas et al. 2010). In another study, Hermann et al. (2003) note a decreased frequency of the protective alleles *HLA DRB1-DQA1-DQB1* in Finns diagnosed with Type I diabetes prior to 1965 and at the beginning of the 21st century. However, since their study involves more recent time periods, it also suggests that complicated mechanisms of pathogenesis, which include the presence of both predisposing and protective alleles, along with environmental variables, must be considered when exploring the evolution and presence of Type I diabetes in past and present populations (Pitkäniemi et al. 2004).

The discussions above, which explore two different endocrine disorders with hereditary components, clearly indicate the usefulness of aDNA studies. These studies not only have the potential to assist in our understanding of predisposing alleles on population and individual levels, but also to follow diseases occurrence, epidemiology, and routes of transmission and spread. Clearly, they also have the potential to help us unravel evolutionary forces, selection mechanisms, and adaptation processes of the past.

CONCLUSION

A characteristic feature of metabolic and endocrine diseases is the development of skeletal alterations induced by systemic factors. These large-scale factors influence our interpretation of the pathological changes, as they require particular attention be paid to the qualitative character, distribution within the skeleton, age, sex, ethnic descent and biocultural environment within which the sufferer operated. The preservation of skeletal material is also of direct relevance. While in the case of bone fracture only a single bone might be sufficient for a conclusion to be formed, in the case of metabolic and endocrine diseases, whole bodies are often needed in order to understand the scope and potential cause of the disorder. Metabolic, and possibly endocrine disorders ones were likely major causes of morbidity in the past, as they are today. However, their history has not been sufficiently investigated and documented in the paleopathological literature. This concerns even the most intensively studied metabolic diseases such as Vitamin C and D deficiency (Ortner and Mays 1998). It appears that in case of metabolic and endocrine diseases, the field of paleopathology has far to go, since we are still at the stage of collecting and describing skeletal changes rather than undertaking attempts to synthesize data using paleoepidemiological premises. The titles of paleopathologic papers often contain the word "probable" or "possible".

Many of the metabolic and hormone disorders known in contemporary medicine remain out of reach to paleopathologists due to the almost exclusive involvement of soft tissue. However, new methods and analytic techniques (e.g. aDNA studies, chemical analyses, modern imaging methods, microscopic techniques), along with new emphases on interdisciplinary work, will undoubtedly allow us to broaden our knowledge base and to add new modern diagnoses to ancient specimens. Attempts to uncover the history of metabolic and endocrine diseases, and to explore their relationship to the quality and evolution of human life in the past, are particularly interesting directions warranting our attention. In light of gaps in our knowledge, it is also an extremely promising area for further scientific exploration by paleopathologists.

REFERENCES

Ameen, S., Staub, L., Ulrich, S., Vock, P., Balmer, F., and Anderson, S. E., 2005   Harris Lines of the Tibia Across Centuries: A Comparison of Two Populations, Medieval and Contemporary in Central Europe. Skeletal Radiology 34:279–284.

Angel, J. A., 1978   Pelvic Inlet Form: A Neglected Index of Nutritional Status (Abstract). American Journal of Physical Anthropology 48:378.

Angel, J. A., 1982   A New Measure of Growth Efficiency: Skull Base Height. American Journal of Physical Anthropology 58:297–305.

Anton, S. C., 1997   Endocranial Hyperostosis in Sangiran 2, Gibraltar 1, and Shanidar 5. American Journal of Physical Anthropology 102:111–122.

Aufderheide, A. C., and Rodriguez-Martin, C., 1998   The Cambridge Encyclopedia of Human Paleopathology, Cambridge: Cambridge University Press.

Belcastro, G. M., Facchini, F., and Rastelli, E., 2006   Hyperostosis Frontalis Interna and Sex Identification of Two Skeletons from Early Middle Ages Necropolis of Vicenne–Campochiaro (Molise, Italy). International Journal of Osteoarchaeology 16:506–516.

Brickley, M., 2002   An Investigation of Historical and Archaeological Evidence for Age-related Bone Loss and Osteoporosis. International Journal of Osteoarchaeology. 12: 364–371.

Brickley, M., and Ives, R., 2006   Skeletal manifestation of Infantile Scurvy. American Journal of Physical Anthropology 129:163–172.

Brickley, M., and Ives, R., 2008   The Bioarchaeology of Metabolic Bone Disease. Oxford: Elsevier.

Brickley, M., Kozłowski, T., et al., 2009   Socio-Culturally Mediated Disease: Rickets and Scurvy. American Journal of Physical Anthropology 138(S48): 97–98.

Brickley, M., Mays, S., and Ives, R., 2005   Skeletal Manifestations of Vitamin D Deficiency Osteomalacia in Documented Historical Collections. International Journal of Osteoarcheology 15:389–403.

Buckley, H. R., 2000   Subadult Health and Disease in Prehistoric Tonga, Polynesia. American Journal of Physical Anthropology 113:481–505.

Burger, J., Kirchner, M., Bramanti, B., Haak, W., and Thomas, M. G., 2007   Absence of the Lactase-Persistence-Associated Allele in Early Neolithic Europeans. Proceedings of the National Academy of Sciences, USA 104(10):3736–3741.

Casals, T., Vazquez, C., Lizaro, C., Girbau, E., Gimenez, F. J., and Estivill, X., 1992   Cystic Fibrosis in the Basque Country: High Frequency of Mutation Delta F508 in Patients of Basque Origin. American Journal of Human Genetics 50:404–410.

CFF (Cystic Fibrosis Foundation), 2008   What is the life expectancy for people who have CF (in the United States)? Electronic document. http://www.cff.org. Accessed June 3, 2011.

CFGAC (Cystic Fibrosis Genetic Analysis Consortium), 2005   Cystic Fibrosis Mutation Database. Electronic document. http://www.genet.sickkids.on.ca/cftr/. Accessed June 3, 2011.

Dacre, J., and Kopelman, P., 2002   A Handbook of Clinical Skills. London: Manson Publishing.

Dahlquist, G., 2006   Can We Slow the Rising Incidence of Childhood-Onset Autoimmune Diabetes? The Overload Hypothesis. Diabetologia 49(1):20–24.

Delgado, J. C., Baena, A., Thim, S., and Goldfeld, A. E., 2006   Aspartic Acid Homozygosity at Codon 57 of HLA-DQ Beta Is Associated with Susceptibility to Pulmonary Tuberculosis in Cambodia. Journal Immunology 176:1090–1097.

EWGCFG (European Working Group on CF Genetics), 1990   Gradient of Distribution In Europe of The Major CF Mutation, and of Its Associated Haplotype. Human Genetics 85:436–441.

Fain, O., 2005   Musculoskeletal Manifestations of Scurvy. Joint Bone Spine 72:124–128.

Gładykowska-Rzeczycka, J., Smoczyński, M., Dubowik, M., and Mechlińska, J., 2001   Rare Developmental Disorder of the Sound-Conducting System of the Skeleton from an Early Medieval Cemetery in Poland: Endoscopy and CT findings. The Mankind Quarterly, XLII 1:3–16.

Gładykowska-Rzeczycka, J., Wrzesińska, A., and Sokół, A., 1998   A Giant from Ostrów Lednicki (12th–13th c.) dist. Lednogóra, Poland. The Mankind Quarterly 39 (2):147–172.

Gleń-Haduch, E., Szostek K., and Głąb H., 1997   Cribra Orbitalia and Trace Element Content in Human Teeth from Neolithic and Early Bronze Age graves in Southern Poland, American Journal of Physical Anthropology, 103:201–207.

Goodman, A. H., Brooke, T. R., Swelung, A. C., and Armelagos, G. J., 1988   Biocultural Perspectives on Stress in Prehistoric, Historical and Contemporary Population Research. Yearbook of Physical Anthropology 31:169–202.

Green, A., and Patterson, C. C., EURODIAB TIGER Study Group, 2001   Europe and Diabetes. Trends in the Incidence of Childhood-Onset Diabetes in Europe 1989–1998. Diabetologia 44 (3):B3–B8.

Haynes, A., Bower, C., Bulsara, M. K., Jones, T. W., and Davis, E. A., 2004   Continued Increase in the Incidence of Childhood Type 1 Diabetes in a Population-Based Australian Sample (1985–2002). Diabetologia 47(5):866–870.

Hengen O. P., 1971   Cribra Orbitalia: Pathogenesis and Probable Etiology, Homo – Journal of Comparative Human Biology, 22:11–16.

Henschen, F., 1966   The History of Diseases. London: Longmans, Green and Co LTD.

Hermann, R., Knip, M., Veijola, R., Simell, O., Laine, A. P., et al., FinnDiane Study Group 2003   Temporal Changes in the Frequencies of HLA Genotypes in Patients with Type 1 Diabetes–Indication of an Increased Environmental Pressure? Diabetologia 46(3):420–425.

Horton, E. S., 1986   Obesity. In Clinical Endocrinology. P. O. Kohler, ed. pp. 451–464. New York: John Wiley and Sons.

Ivanhoe, F., 1970   Was Virchow Right about Neandertal? Nature 227:577–579.

Ivarsson, A., Persson, L. A., Nyström, L., Ascher, H., Cavell, B., et al., 2000   Epidemic of Coeliac Disease in Swedish Children. Acta Paediatrica 89(2):165–171.

Jordan, M. R., and Kohler, P. O., 1986   General Principles of Endocrinology. In Clinical Endocrinology. P. O. Kohler, ed. pp. 1–9. New York: John Wiley & Sons.

Kelley, J. O., and Angel, J. L., 1987   Life Stress of Slavery. American Journal of Physical Anthropology 74:199–211.

Klepinger, L. L., 1980   The Evolution of Human Disease: New Findings and Problems. Journal of Biosocial Science 12:481–486.

Kozłowski, T., and Piontek, J., 2000   A Case of Atrophy of Bones of the Right Lower Limb of a Skeleton from a Medieval (12th–14th Centuries) Burial Ground in Gruczno, Poland. Journal of Paleopathology 12 (1):5–16.

Krämer, J., 1996   Orthopadie. Berlin: Springer-Verlag.

Lagia A., Elipoulos C., and Manolis S., 2006   Thalassemia: Macroscopic and Radiological Study of a Case, International Journal of Osteoarcheology, 17:269–285.

Lewis, M. E., 2004   Endocranial Lesions in Non-adult Skeletons: Understanding Their Aetiology. International Journal of Osteoarchaeology 14:82–97.

Maat, G. J. R., 2004   Scurvy in Adults and Youngsters: the Dutch Experience. A Review of the History and Pathology of a Disregarded Disease. International Journal of Osteoarchaeology 14:77–81.

Malinowska-Łazarczyk, H., 1982   Report of Archaeology Department of Upper Silesian Museum. Silesia Antiqua 26:120.

Mays, S., 2008   A Likely Case of Scurvy from Early Bronze Age Britain. International Journal of Osteoarchaeology 18:178–187.

Mays, S., Brickley, M., and Ives, R., 2007   Skeletal Evidence for Hyperparathyroidism in a 19th-Century Child with Rickets. International Journal of Osteoarchaeology 17:73–81.

Melikan, M., and Waldron, T., 2003   An Examination of Skulls from Two British Sites for Possible Evidence of Scurvy. International Journal of Osteoarcheology 13:207–212.

Modiano, G., Ciminelli, B. M., and Pignatti, P. F., 2007   Cystic Fibrosis and Lactase Persistence: A Possible Correlation. European Journal of Human Genetics 15:255–259.

Morral, N., Bertranpetit, J., Estivill, X., Nunes, V., Casals, T., Giménez, J., et al., 1994   The Origin of the Major Cystic Fibrosis Mutation (Delta F508) in European Populations. Nature Genetics 7(2):169–75.

Mulhern, M. D., 2005   A Probable Case of Gigantism in a Fifth Dynasty Skeleton from the Western Cemetery at Giza, Egypt. International Journal of Osteoarchaeology 15:261–275.

Onkamo, P., Vaananen, S., Karvonen, M., and Tuomilehto, J., 1999   Worldwide Increase in Incidence of Type I Diabetes – The Analysis of the Data on Published Incidence Trends. Diabetologia 42:1395–1403.

Ortner, D. J., 1984   Bone Lesions in a Probable Case of Scurvy from Metlavik, Alaska. Museum Applied Science Center for Archaeology 3:79–81.

Ortner, D. J., 2003   Identification of Pathological Conditions in Human Skeletal Remains. San Diego: Academic Press.

Ortner, D. J., and Ericksen, M. F., 1997   Bone Changes in the Human Skull Probably Resulting from Scurvy in Infancy and Childhood. International Journal of Osteoarchaeology 7:212–220.

Ortner, D. J., and Hotz, G., 2005   Skeletal Manifestations of Hypothyroidism from Switzerland. American Journal of Physical Anthropology 127:1–6.

Ortner, D. J., Kimmerle, E. H., and Diez, M., 1999   Probable Evidence of Scurvy in Subadults From Archeological Sites in Peru. American Journal of Physical Anthropology 108:321–331.

Ortner, D. J., and Mays, S., 1998   Dry-Bone Manifestations of Rickets in Infancy and Early Childhood. International Journal of Osteoarchaeology 8:45–55.

Pearce, S. H., and Merriman, T. R., 2006   Genetic Progress Towards the Molecular Basis of Autoimmunity. Trends in Molecular Medicine, 12:90–98.

Pimentel, L., 2003   Scurvy: Historical Review and Current Diagnostic Approach. American Journal of Emergency Medicine 21(4):328–332.

Pinto, D. C., and Stout, S. D., 2009   Paget's Disease in Pre-Contact Florida? Revisiting the Briarwoods Site in Gulf Coast Florida. International Journal of Osteoarchaeology DO1:10. 1002/oa. 1043.

Piontek, J., and Kozłowski, T., 2002   Frequency of Cribra Orbitalia in the Subadult Medieval Population from Gruczno, Poland. International Journal of Osteoarchaeology 12:202–208.

Pitkäniemi, J., Onkamo, P., Tuomilehto, J., and Arjas, E., 2004   Increasing Incidence of Type 1 Diabetes – Role for Genes? BMC Genetics 5:5.

Porat, A., and Sherwood, L. M., 1986   Disorders of Mineral Homeostasis and Bone. In Clinical Endocrinology. P. K. Kohler, ed. pp. 377–426. New York: John Wiley and Sons.

Riggs, B. L., 1991   Overview of Osteoporosis. Western Journal of Medicine 154:63–77.

Roberts, C., 1987   Possible Pituitary Dwarfism from the Roman Period. British Medical Journal 295:1659–1660.

Roches, E., Blondiaux, J., Cotton, A., Chastanet, P., and Flipo, R.-M., 2002   Microscopic Evidence for Paget's Disease in Two Osteoarchaeological Samples from Early Northern France. International Journal of Osteoarchaeology 12:229–234.

Roivainen, M., 2006   Enteroviruses: New Findings on the Role of Enteroviruses in Type 1 Diabetes. The International Journal of Biochemistry and Cell Biology 38(5–6):721–725.

Rühli, J. F., and Henneberg, M., 2002   Are Hyperostosis Frontalis Interna and Leptin Linked? A Hypothetical Approach about Hormonal Influence on Human Microevolution. Medical Hypotheses 58(5):378–381.

Rühli, J. P., and Henneberg M., 2004   Hyperostosis frontalis interna: archaeological evidence of possible microevolution of human steroids? Homo – Journal of Comparative Human Biology, 55:91–99.

Schamall, D., Teschler-Nicola, M., Kainberger, F., Tangl, S., Brandstatter, F., Patzak, B., Muhsil, J., and Plenk H., Jr., 2003   Changes in Trabecular Bone Structure in Rickets and Osteomalacia: The Potential of a Medico-Historical Collection. International Journal of Osteoarchaeology 13:283–288.

Schultz, M., 2001   Paleohistopathology of Bone. A New Approach to the Study of Ancient Disease. Yearbook of Physical Anthropology 44:106–147.

Schultz, M., Timme, U., and Schmidt-Schultz, H. T., 2007   Infancy and Childhood in the Pre-Columbian North American Southwest – First Results of the Palaeopathological Investigation of the Skeletons from the Grasshopper Pueblo, Arizona. International Journal of Osteoarchaeology 17:369–379.

Serre, J. L., Simon-Bouy, B., Mornet, E., Jaume-Roig, B., Balassopoulou, A., Schwartz, M., Taillandier, A., Boué, J., and Boué, A., 1990   Studies of RFLP Closely Linked to the Cystic Fibrosis Locus Throughout Europe Lead to New Considerations in Populations Genetics. Human Genetics 84(5):449–454.

Stini, W. A., 1990   Bone Mineral Loss in Middle and Old Age: Mechanisms and Treatments. Collegium Anthropologicum 14:263–272.

Stirland, A., 1991   Paget's Disease (Osteitis Deformans): A Classic Case? International Journal of Osteoarchaeology 1(3–4):173–177.

Stuart-Macadam, P., 1985   Porotic Hyperostosis: Representative of a Childhood Condition, American Journal of Physical Anthropology 66:391–398.

Stuart-Macadam, P., 1987   Porotic Hyperostosis: New Evidence to Support the Anemia Theory, American Journal of Physical Anthropology 74:521–526.

Stuart-Macadam, P. L., 1989   Nutritional Deficiency Diseases: A Survey of Scurvy, Rickets, and Iron-Deficiency Anemia. In Reconstruction of Life from the Skeleton. M. Y. İşcan, and K. A. R. Kennedy, eds. pp. 201–222. New York: Alan R. Liss.

Stuart-Macadam P., 1992   Porotic Hyperostosis: a New Perspective, American Journal of Physical Anthropology, 87:39–47.

Swinson, D., Snaith, J., Buckberry, J., and Brickley, M., 2010   High-Performance Liquid Chromatography (HPLC) in the Investigation of Gout in Palaeopathology. International Journal of Osteoarchaeology 20:135–143.

Szostek, K., 2006   Rekonstrukcja Ogólnego Stanu Biologicznego Historycznych i Przedhistory-Cznych Grup Ludzkich na Podstawie Analiz Makro i Mikroelementów w Materiale Odontologicznym. Kraków, PiT.

Szostek, K., 2009   Chemical Signals and Reconstruction of Life Strategies from Ancient Human Bones and Teeth – Problems and Perspectives, Anthropological Review, 72:3–30.

Tortora, J. G., and Derrickson, H. B., 2009   Principles of Anatomy and Physiology, 12th Edition. New York: John Wiley and Sons.

Van der Merwe, A. E., Steyn, M., and Maat, G. J. R., 2010   Adult Scurvy in Skeletal Remains of late 19th Century Mineworkers in Kimberley, South Africa. International Journal of Osteoarchaeology 20:307–316.

Virchow, R., 1872   Untersuchung des Neanderthal-Schadels. Verhandlungen der Berliner Gasellschaft fur Anthropologie, Ethnologie und Urgeschichte, Berlin: Wiegandt und Hempel.

Von Endt, W. D., and Ortner J. D., 1982   Amino Acid Analysis of Bone From a Possible Case of Prehistoric Iron Deficiency Anemia From the American Southwest, American Journal of Physical Anthropology, 59:377–385.

Wade, A. D., Holdsworth, D. W., and Garvin, G. J., 2009   CT and Micro-CT Analysis of a Case of Paget's Disease (Osteitis Deformans) in the Grant Skeletal Collection. International Journal of Osteoarchaeology, n/a. doi:10.1002/oa.1111.

Walker, P. L., Bathurst, R. R., Richman, R., and Gjerdrum, T., 2009   The Cause of Porotic Hyperostosis and Cribra Orbitalia: A Reappraisal of the Iron-Deficiency-Anemia Hypothesis. American Journal of Physical Anthropology 139:109–125.

Watrous, A. C., Anton, S. C., and Plourde, A. M., 1993   Hyperostosis Frontalis Interna in Ancient Egyptians. American Journal of Physical Anthropology 16:205.

Welsh, M. J., and Ramsey, B. W., 1998   Research on Cystic Fibrosis: A Journey from Hart House. American Journal of Respiratory and Critical Care Medicine 157:S148–S154.

White, T. D., and Folkens, P. A., 2005   The Human Bone Manual. Burlington: Elsevier Academic Press.

Witas, H. W., Jędrychowska-Dańska, K., and Zawicki, P., 2007   Extremely High Frequency of Autoimmune-Predisposing Alleles in Medieval Specimens. Journal of Zhejiang University-Science B 8(7):512–514.

Witas, H. W., Jatczak, I., Jędrychowska-Dańska, K., Żądzińska, E., Wrzesińska, A., Wrzesiński, J., and Nadolski, J., 2006   Sequence of DF508 CFTR Allele Identified as Present is Lacking in Medieval Specimens from Central Poland. Preliminary Results. Anthropologischer Anzeiger 1:1–9.

Witas, H. W., Jędrychowska-Dańska, K., and Zawicki, P., 2010   Changes in Frequency of IDDM-Associated HLA DQB, CTLA4 and INS Alleles. International Journal of Immunogenetics. 37(3):155–158.

Wiuf, C., 2001   Do Delta F508 Heterozygotes Have a Selective Advantage? Genetics Research 78:41–47.

Wolpoff, M., 1999   Paleoanthropology. Boston:McGraw-Hill.

Zaki, M. E., Hussein, F. H., and El-Shafy El-Banna, R., 2009   Osteoporosis Among Ancient Egyptians. International Journal of Osteoarchaeology 19:78–89.

23 Tumors: Problems of Differential Diagnosis in Paleopathology

Don Brothwell

Although finds of tumors and tumor-like processes are uncommon in archaeology, it is nevertheless a major disease category which deserves some attention in paleopathology. The clinical and radiological literature is extensive, and I might selectively refer to Resnick (1995) and Stoker (1986) as worthwhile reference works. In the field of anthropology, the dry-bone pathology, as well as radiology, has been reviewed in a growing body of literature (for instance, Brothwell 1967, 2008; Roberts and Manchester 1995; Strouhal 1998; Ortner 2003). But there are major diagnostic problems and as yet too few cases reported to allow even the crudest epidemiological assessment. Unlike modern cancer research, we are concerned not only with major geographic, environmental and cultural differences between peoples (Shivas 1967), but also a long time-dimension. I therefore believe that we have a positive contribution to make to this field. Although "Paleo-oncology" (Retsas 1986) embraces ancient literature as well as ancient tissue, I restrict myself here to bone and tooth pathology.

TUMORS AND THE HISTORY OF PALEOPATHOLOGY

The search for ancient tumors was well established during the nineteenth century, both in the Americas and in the European area. Joseph Jones (1833–96), Frederic Putnam (1839–1915), Aleš Hrdlička (1869–1943), Marc Ruffer (1859–1917), Roy Moodie (1880–1934), and Herbert Williams (1866–1938), were just some of the

A Companion to Paleopathology, First Edition. Edited by Anne L. Grauer.
© 2012 John Wiley & Sons, Ltd. Published 2016 by John Wiley & Sons, Ltd.

early paleopathologists who searched for different categories of disease, including tumors (see the review by Jarcho 1966). But for all the searching, only a modest number of tumor cases came to light, which led Henry Sigerist (1951) to conclude that tumor evidence "is very small" (Sigerist 1951:58). Walter Putschar (1966) was also unimpressed with the evidence, and concludes that "Bone tumors are rare in prehistoric material, with exception of the "button osteomas" of the vault of the skull" (Putschar 1966:62).

Tumor evidence in fossil hominins occurs, but all cases are debatable. The irregular mass of bone on the shaft of the first Homo erectus femur from Java is most probably of traumatic, not neoplastic origin. However, the additional bone in the chin region of the Pleistocene Kanam mandible has been described by Tobias (1960) as a "sub-periosteal ossifying sarcoma." In the case of the 120,000-year-old parietal from Lazaret cave in Nice, a comparative radiological study concluded that changes to the inner and outer tables of the skull bone are at least suggestive of a meningioma (Duplay et al. 1970). Because most skeletal material is dated only to the past 10,000 years, it is to be expected that more cases are described from historic periods, where considerably more material has become available for study.

In terms of ancient writings, both the Ebers and Hearst papyri from Egypt, refer to swellings which could have been tumors. The ancient physician was trained to observe the nature of the swelling, its shape, temperature and so on, but non-neoplastic conditions would equally have been included. Early Greek literature, including Hippocratic writings (ca.410–360 B.C.), was also aware of swellings and possible tumors. Nevertheless, David and Zimmerman (2010) conclude that neither the ancient literature, nor the mummy and skeletal evidence, enables a conclusion other than that the disease is rare in antiquity.

Up to a point I would agree with these conclusions, but there remains a need for caution. Many malignancies affect only the soft tissues. Where metastatic deposits cause bone modification, changes can be mistaken for inflammatory reactions, or postmortem changes. In the case of calcification from urinary bladder tumors (Ferris and O'Connor 1965), the calcifications within the pelvic basin would probably be interpreted by field archaeologists as heterogeneous soil matrix.

Regrettably, it is not easy to support the case for more commonly occurring tumors in the past, from the data on neoplasm prevalence today. Some of the commonly occurring tumors today vary considerably from region to region, and skeletal evidence may hardly occur where percentages are high. Also, there have been significant changes in life expectancy over the past few centuries. In the case of prostate cancer, for instance, it is most destructive in men over fifty or sixty, but few men in the past lived that long. Environmental changes have also occurred, from accuracy of diagnosis affecting the last few decades, to greater hygiene, consumption of less pickled foods and salt, greater alcohol intake, less smoking and smoked foods, and, probably, with greater urban development there has been wider distribution of carcinogenic viruses.

The situation is clearly complex, and for that reason I would certainly hesitate to say that all tumors have been rare in the past. Lower life-expectancy has been an important influence on the prevalence of malignancies, but it remains to be seen what neoplasms occurred in the under-50s of at least some ancient populations. There was not great importance given to finding an ancient tumor in human remains fifty years ago, but

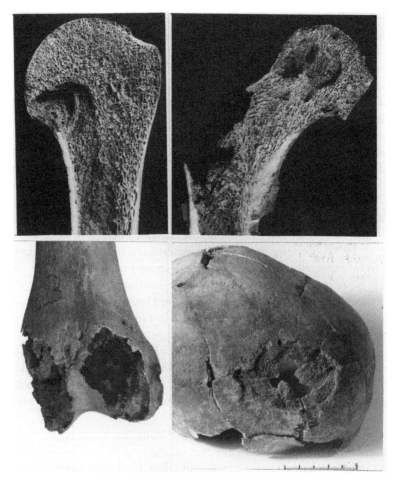

Figure 23.1 Examples of pathology and pseudopathology: a and b, lytic lesions in a Hungarian skeleton; c, insect damage to a Socotran femur; d, postmortem skull damage.

interests are changing, especially now that the biology of cancer and benign tumors is becoming better understood. We can now see more clearly how they are associated with such etiological factors as genetic background, diet and viruses.

DANGERS OF MISIDENTIFICATION

Both in bone destruction and additional bone growth, it is possible to mistake non-neoplastic processes for actual tumors. Perhaps the worst common problem is distinguishing postmortem erosion and damage from the osteolytic removal of bone by a malignant tumor. The three specimens in Figure 23.1 illustrate this point. The cavitation in the humerus and femur (Figures 23.1a and 23.1b) is from Grave 38, Kérpuszta, in Hungary. It could be interpreted as postmortem erosion, but the multiple lytic lesions in the skull strongly suggest a malignant tumor. The distal femur from the island of Socotra displays destruction without any marginal reaction (Figure 23.1c). Insect remains were found in the cavities, and it can be concluded that this is evidence of postmortem damage by insects. In the case of the skull from Sussex (Figure 23.1d), the irregular crater exposes a large area of diploic tissue displays no antemortem bone changes. Additional breaks in the skull support the view that this is simply postmortem damage.

The three cranial abnormalities seen in Figure 23.2 provide further challenges in terms of diagnosis. The changes in two areas of skull 9513 (Berkeley collection), a prehistoric Native American, are not typical examples of treponemal cavitation (Figure 23.2a). There are no clear surface reactions and bone destruction expands within the diploic tissue. The external margins are also very irregular. Metastatic tumors appear to be more likely than treponemal lesions in this case. An inner view of the maxillary sinus of Jewbury 4420 (Figure 23.2b) shows a series of smooth rounded mounds of bone. Are these multiple benign osteomas? On the outer surface of the cranial vault, this could be a likely diagnosis, but in the sinus area, it is far more likely that the additional bone is indicative of a chronic, late stage sinusitis.

The final case (Figure 23.2c) shows the endocranial surface of a Jewbury frontal bone (4149), with deep pitting on the upper left side. This would not be a normal position for a meningioma, and the pits are deep but rounded, not ragged and lytic in appearance. In fact, these are very probably the result of bone remodeling in relation to ageing in a cluster of normal pachionian bodies.

Distribution of Tumors Within the Skeleton

At times there is value in considering the position of tumors within the body, in relation to diagnosis. Stoker (1986) has provided an excellent series of diagrams showing the percentage incidence of various tumors at various positions within the skeleton. The patterns are based on relatively large samples, but may not represent all world populations to the same degree. It is always important to remember that without the specialist knowledge of an oncologist, it is best not to be dogmatic in diagnosis, although there is nothing wrong in attempting a tentative differential diagnosis. Tumors are capable of mimicking other types of pathology, which is another reason to be cautious. For instance, an X-ray of a hemangioma may look remarkably like an osteosarcoma, with a similar "sunburst" appearance of new bone. Also, while the majority of types of tumor will occur either throughout the age range or mainly in specific age ranges, there are always exceptions. Sex ratios vary between tumor types, and occasionally change slightly during the adult lifespan (Table 23.1).

If we consider the skeletal distribution and incidences of some of the more common tumors, as diagrammatized by Stoker (1986), it is seen that osteoid osteomas are mainly in the legs (although the small "button" osteoma is normally seen on the skull). Solitary bone cysts are mainly situated at the proximal humerus and femur, whereas the aneurysmal bone cyst is especially in the lower spine and lower leg. The giant cell tumor has a preference for the knee, and to a lesser extent the pelvis and arms. In contrast again, the nonossifying fibroma is mainly found below the mid-thigh.

Regarding the more common malignant neoplasms, chondrosarcomas particularly involve the proximal femur, pelvis and ribs, in contrast to osteosarcomas which especially appear at the knee, and to a lesser extent at the proximal femur and humerus.

Figure 23.2 Examples of pathology and normal variation: a, external surface of a prehistoric Native American skull showing osteolytic lesions; b, maxillary sinus with remodeled bone indicative of sinusitis; c, endocranial view of a frontal bone displaying deep pachionian impressions.

Again, there is a contrast with the fibrosarcoma, which strongly involves the pelvis and long bones of the legs, and to a lesser degree the scapula and humerus. A surprisingly different pattern is seen in the primary malignant lymphoma (reticulum cell sarcoma), which avoids articular ends of long bones and strongly involves femur, tibia and humerus shafts, and to a lesser degree ribs, scapula and skull vault.

Table 23.1 Some major classes of tumor which can affect the skeleton. Adapted from Stoker (1986)

	C = Common R = Rare	Usual age	Sex ratio
Benign			
Chondroma	C	10–80	Equal
Osteochondroma	C	All ages	Male : Female 1.4 : 1
Chondroblastoma	R	Usually under 20	2 : 1
Chondromyxoid fibroma	R	Usually 10–30	1 : 1
Osteoma	C	All ages?	Equal
Osteoid osteoma	C	10–30	2.5 : 1
Osteoblastoma	R	10–30	2 : 1
Solitary bone cyst	C/R	0–20	2 : 1
Aneurysmal bone cyst	C	5–20	1 : 1
Giant cell tumor	C	15–45	1 : 1.5?
Non-ossifying fibroma	C	0–20	? 1.5 : 1
Eosinophilic granuloma	R	Under 10	2 : 1
Malignant			
Chondrosarcoma	C	Adults	2 : 1
Osteosarcoma	C	10–30	2 : 1
Fibrosarcoma	C	20+	1 : 1
Ewing's sarcoma	R	5–20	2 : 1
Long bone adamantinoma	R	10–50	1.5 : 1
Chordoma	R	All ages	1 : 1
Primary malignant lymphoma	R	10–80	3 : 2
Leukemia	C	Under 5, over 45	1.2 : 1
Hodgkin's disease	C	All ages	1.1 : 1
Myelomatosis	C	50+	2 : 1

PATHOLOGICAL CHANGES IN BONE TUMORS AND TUMOR-LIKE LESIONS

Whenever possible, diagnosis should be based on a study of the actual external appearance of the bone pathology, together with X-rays and possibly CT scans. In the case of benign tumors, most are solitary in nature. They are normally slow in growth, and expand only at the initial site of development. In contrast, the malignant form often spreads to other areas, and can invade other tissues, eventually causing the death of the individual. Secondary tumors are of two malignant forms; either metastatic spread from the primary neoplasm (in soft tissue), or by transforming from a benign tumor into a destructive malignant type. Benign tumors can have rounded margins, but malignant types are likely to show ragged (or "moth-eaten") margins.

While access to X-ray facilities of any kind was not easy a few decades ago, even digital radiography is becoming more possible now in human skeletal studies (Chhem and Brothwell, 2008; and see Chapter 18 by Wanek et al. in this volume, for more information on imaging techniques). Because much detail of tumors is hidden in the inner structures of bones, it is essential that some form of radiographic evaluation is undertaken. The translucent areas which can for instance be seen in a hemangioma, or a nonosteogenic fibroma, or an aneurysmal bone cyst, can only be seen in X-rays. The translucency may also be associated with bone swelling, as in multiple enchondromas, and all of these internal changes also need to be differentially considered in relation to non-neoplastic processes such as tuberculous abscess and hydatid cyst. Extra ossification in tumors may present as a rounded bone surface mass, as in an osteoid osteoma, or as a "sun-ray" or spicular development seen in the osteosarcoma. Both of these forms could be mistaken for an ossifying haematoma or severe periostitis. More than in any other major disease category, it is therefore especially important to arrive at a differential diagnosis after a broad based review of alternatives (Brothwell 2010).

Benign Tumor Forms

By far the commonest benign tumor is the osteoma, usually a simple rounded mound of normal cortical bone, seen on the skull. Of a similar kind is the osteoid osteoma, usually found on the long bones of the legs, and consisting of dense cortical bone within which is a small epicentre of osteoid. The osteoblastoma is similar in the bone changes, but also involves the spine and is comparatively rare.

Cysts can be mentioned here, but are of variable etiology and are not true tumors. They appear in X-ray as simple or multilocular rounded cavities. The aneurysmal bone cyst can also involve long bones, but also the feet and vertebral column. Giant cell tumors also display a somewhat cystic form in X-ray, but are not as well defined. Benign tumors of a cartilage origin are of a number of kinds. Chondromas are usually singular, but occasionally multiple, and are usually restricted to bones of the hands or feet. The tumor expands (usually in a globular form) beyond bone margins, with thinning of the cortex. In contrast, the osteochondrama can expand into a substantial mass, taking on a "bath sponge" or "cauliflower-like" appearance.

The nonossifying fibroma is usually within the long bones of the legs, and begins within the cortex and expands into what appears in X-ray to be a cyst. It is in contrast with the much rarer desmoplastic fibroma which has an expanded trabeculated appearance. Although angiomas are regarded as the result of congenital vascular malformations, it is worth emphasizing that hemangiomas can produce a fine spicular "sunburst" appearance very like that of an osteosarcoma. Mycotic bone infection can also mimic this pathology.

Malignant Lesions

It should perhaps be mentioned at the outset of reviewing this destructive group of tumors, that it is important to remember that solitary secondary tumors involving bone are relatively common (circa 10 percent), mainly as a result of prostate and breast cancer. It is therefore advisable when attempting a differential diagnosis of possible malignant bone tumors, to keep metastatic lesions in mind as an alternative possibility. In the case of multiple bone metastases, they can occur in over 70 percent of individuals with some forms of carcinoma (Stoker 1986). Metastases are mainly found in the skull, pelvis, vertebrae, ribs and proximal femur and humerus. They can mimic other forms of tumor at times. They can also be osteolytic, osteoblastic, or of mixed form.

The chondrosarcoma is mainly a tumor of more mature adult years. The pathology could be confused with a multilocular bone cyst or an enchondroma, so care is needed in even a tentative diagnosis. The osteosarcoma involves a younger age-group, with the knee being especially involved. In the later stage of its development, the tumor displays massive cortical destruction with the elaboration of coarse irregular "sunburst" spiculation. A marginal reaction to the tumor mass may result in a Codman's triangle, an elevation of the periosteum, marginal to the tumor, which initiates further bone growth.

The fibrosarcoma is a tumor of adults, especially involving the upper legs and pelvis. It is highly osteolytic in form, but the margins of the bone destruction can appear somewhat "moth-eaten." The bone pathology can be rather similar to a number of other tumors, including the giant cell tumor and a more lytic osteosarcoma. The primary malignant lymphoma (reticulum cell sarcoma of bone), Ewing's sarcoma, chordoma and long bone adamantinoma are rare but might nevertheless be found in archaeological human remains. However, because of the complex classification of tumors, and the at-times similarities of bone changes, producing diagnostic problems, it would seem best to concentrate on the commoner tumors as given in Table 23.1. If, after study, the dry bone pathology and X-rays do not suggest a satisfactory differential diagnosis, then I suggest that further study of medical texts or specialist consultation would be desirable.

The final short series of tumors are separated as they are linked to the proliferation of bone marrow and allied conditions. The etiology of leukemia has been much discussed, but is included as a neoplasm here. The incidence has probably fluctuated through time. Bone changes are commonest in children, where they can be diffuse and with osteolytic and osteoblastic changes, sometimes with marked erosive cortical changes at the ends of the metaphyses. Hodgkin's disease (lymphadenoma) eventually produces bone changes in 75 percent of cases. Multiple lesions are common, and especially involve the ribs, spine and pelvis. Lytic lesions are probably the commonest, but osteoblastic and periosteal reactions can occur. A differential diagnosis would include myelomatosis, leukemia and lymphosarcoma, and it should be said that this is not an easy condition to distinguish.

Myelomatosis (multiple myeloma) has been described more than most tumors in archaeological material, but caution is needed as it can be mistaken for secondary metastases. Much of the skeleton can be involved, including the skull, and presents itself as multiple somewhat rounded "punched-out" lesions. Eosinophilic granuloma, a type of histiocytosis, produces bone changes commonly, especially in the skull. A distinctive erosion of the cortex can occur, but except in the skull, a firm diagnosis may not be possible.

Tumor Examples by Case Studies

It would seem worthwhile to discuss here a number of possible tumor cases from archaeological sites in order to demonstrate the problems of differential diagnosis when considering neoplasms. In the case of this kind of ancient material, recourse can be made to the considerable range of radiological texts which consider tumors. It is also advisable to consult more recent reviews which consider archaeological material and include Ortner (2003) and Brothwell (2008). The specimen descriptions which follow are not in any classificatory order, but might be taken as a random sample which could turn up in a large archaeological sample of human remains.

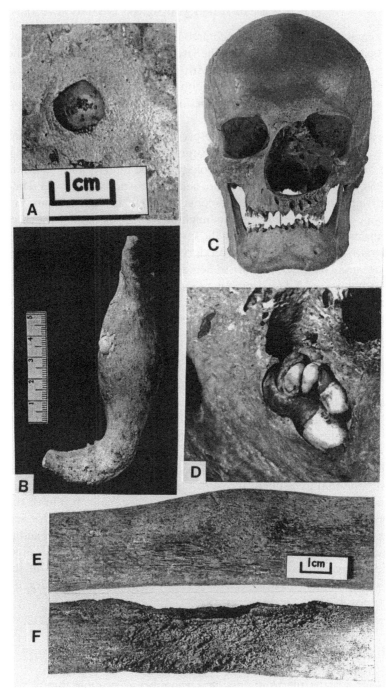

Figure 23.3 Variation in archaeological tumors: (a) small osteoma on the external surface of a medieval skull; (b) much expanded mandibular body of a prehistoric native American, with a displaced molar. Probably a dentigorous cyst; (c) a Neolithic skull from Denmark, displaying bone remodeling in the face as a result of a soft tissue tumor; (d) an odontome in the mandible of a prehistoric Socotran; (e) a medieval tibia from York displaying changes which are probably indicative of an osteoid osteoma; (f) midshaft of a post-medieval femur from York, displaying deep cortical destruction, indicative of a malignancy.

Case 1   This was seen on Skull 2357, from the medieval cemetery of Jewbury, in York (Figure 23.3a). The small rounded mound was smooth and composed of cortical bone. It might be considered to be the most identifiable of the benign tumors, but it could nevertheless be confused with a mature ossified haematoma, perhaps resulting from a blow to the head, followed by restricted subperiosteal hemorrhage and ossificiation.

Case 2   The view seen in Figure 23.3b is the base of a mandibular body which shows very considerable expansion along most of the tooth row. There is also a molar tooth protruding on the lower lingual aspect. This prehistoric Californian Amerindian (3604, Univ. Cal. Berkeley collection) appears to display a dentigorous cyst.

Case 3   This is a possible Neolithic tumor from Slagslunde in Denmark (Bennike 1985). The expanding soft tissue neoplasm was probably benign, but has resulted in the rounded remodeling with some bone destruction to the left of the nasal aperture, maxillary sinus and orbital margin (Figure 23.3c). A differential diagnosis would need to include a cyst, granuloma or fibroma.

Case 4   In the mandible of an ancient Socotran is an irregular composite tumorous mass of dental tissue (Figure 23.3d). Paul Broca created the term "odontome" in 1867, and the term is considered to mean an abnormal benign neoplastic growth derived from dental formative organs. The case emphasizes the diverse nature of pathology which falls within the range of tumors and tumor-like processes.

Case 5   A tibia from the medieval skeleton 2380, excavated at Jewbury, York (Figure 23.3e). The pathology is in the form of rounded additional cortical bone extending approximately 65 mm along the shaft. Thickness of extra bone does not exceed 7.0 mm. Striations marginal to the extra bone raises the question of an alternative traumatic or inflammatory response to the more probable diagnosis of an osteoid osteoma. However a pathognomonic feature of these osteomas is a small internal lytic focus, but before reburial X-ray of such detail was not possible in this case. As Ortner (2003) points out, a small intracortical abscess can mimic this lytic feature.

Case 6   The shaft of a post-medieval femur from All Saints Church, York, displays deep irregular cortical destruction (Figure 23.3f). Marginal to this aggressive lytic removal of cortical bone, there is shallow irregular subperiosteal new bone (Codman's triangle). The lack of extensive inflammatory-type changes along the shaft, and deep lytic destruction of cortical bone, strongly argues for a malignant tumor. Ewing's sarcoma is one possibility, but is rare, and thus less likely to be the cause. An alternative involving single bones, primary malignant lymphoma, can be similar in expression but also rare. A more lytic form of osteosarcoma is another possibility, but clearly a more certain differential diagnosis is not easy in cases such as this.

Case 7   This young adult female from Peru is diagnostically interesting in that the bone changes are probably indicative of metastatic tumors, but most metastases seen in archaeological material are of an osteolytic kind. In this case, there is a rash of osteoblastic activity to be seen (Figure 23.4a), probably the result of a metastatic carcinoma, the primary site being the intestinal tract, ovary or lung.

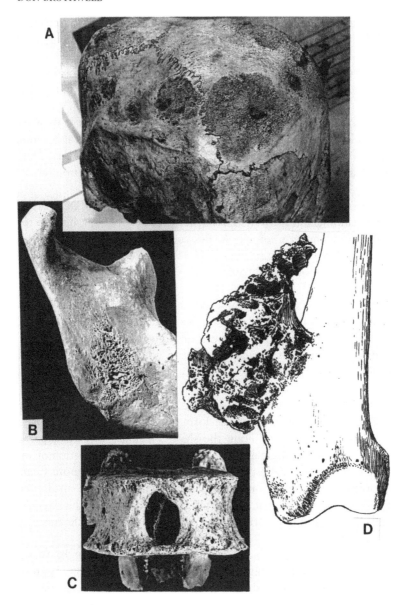

Figure 23.4 Further examples of bone changes in tumor development: (a) zones of osteoblastic surface changes in an ancient Peruvian female, probably indicative of a metastatic carcinoma; (b) mandible from Saxon Winchester, with a lytic lesion and reactive marginal bone (Codman's triangle), probably indicating a metastatic deposit; (c) a Nubian lumbar vertebra with a rounded inner chamber and external opening, possibly indicating a hemangioma; (d) the femur of a Vth Dynasty Egyptian, displaying changes typical of an osteochondroma.

Case 8   In the immature Saxon skeleton from Grave 30 at Wincluster, there are lytic lesions on the skull, pelvis, vertebrae and other areas (Figure 23.4b). The lesions appear to be mainly destructive of bone, except for slight marginal new bone from the periosteal lifting by the soft tumor mass. However, the pelvis appears to display a mixed blastic and lytic response. In a previous evaluation (Brothwell 1967), the pathology was presumed to be indicative of metastatic deposits. But are there any reasonable alternatives? The destructive nature of the bone pathology rules out a benign neoplasm. The shape of the lesions rules out multiple myelomas. A number of other diagnostic possibilities do not usually involve the skull; and others produce very different bone pathology. I am therefore left again to suggest that metastatic deposits derived from, for instance, a malignant lymphoma, neuroblastoma or eosinophilic granuloma, seem to be the most likely cause.

Case 9 A Middle Kingdom Nubian site discussed by Smith and Dawson (1924), included in its pathology an upper lumbar vertebra with a large cavity exposed to the anterior bone surface (Figure 23.4c). While it was initially considered to be indicative of tuberculosis, there is no clear inflammatory change and it seems far more likely that the pathology was indicative of a hemangioma or aneurysmal bone cyst.

Case 10 Although this is a well known example from Vth Dynasty Egypt (Figure 23.4d), it is nevertheless useful in discussing the differential diagnosis of tumors. In fact, this massive bone development at the distal end of a femur had originally been considered to be an osteosarcoma. But in this latter case, the bone appearance is usually of a spicular mass with a "sunburst" appearance. In this Egyptian specimen, the pathology is of a more globular "bath sponge" form, typical of an osteochondroma.

CONCLUSIONS

Vast sums of research money worldwide go into the study of "cancer." It is now seen by the public as the most alarming group of diseases, although progress has been made in the treatment of some tumors, with survival considerably improved. Structurally, the cells of benign tumors are well differentiated and resemble the original tissue. In contrast, in a malignant tumor, there is imperfect differentiation of tissue and much variation. The benign tumor is well defined and grows by expansion but is clearly confined. The malignant form extends by invasion and infiltrates other cellular structures. Benign tumors are usually slow growing, but there is often rapid growth in malignancies. The benign form may cease growth, continue slowly, or even regress. Malignant tumors rarely show spontaneous regression, but normally continue to grow and, if left untreated, will cause death. So benign tumors are normally harmless unless they affect normal physiology, lead to serious hemorrhage or impede the functioning of other body structures (e.g. the brain). Benign tumors do not give rise to secondary tumors (i.e. they do not metastasize), whereas malignant forms almost always do eventually. All kinds of tissue, connective, muscular, nervous, epithelial, hemopoietic, and others, may become neoplasic, with both benign and malignant forms possible. The classification of tumors is thus linked to the tissue form. For instance, fibroma (benign) and fibrosarcoma (malignant); osteoma (benign) and osteosarcoma (malignant). It is the destructive nature of malignant tumors which holds most research interest, both for oncology today, but also for the field of paleopathology. It is to be hoped that eventually even studies on ancient tumors may help to provide a long perspective to the antiquity of these conditions in humans, and perhaps will even assist in the proper evaluation of the various etiological factors.

Although tumors and tumor-like processes are some of the least common pathologies to be encountered archaeologically, they are nevertheless an important group to identify. In my student years, it was "popular" mythology that malignant growths were linked to recent advanced societies. But the archaeological evidence demonstrates clearly that a wide range of tumors extend back into prehistory and may be associated with simpler as well as more complex societies. In some instances, the nature of the bone pathology clearly indicates the form of tumor involved, but in other examples, only a very tentative diagnosis or a number of alternative possibilities have to be given.

Patterns of tumor expression within the skeleton can vary. But this is based on modern data and we don't know how applicable such schemes are to the past. At a statistical level, an osteosarcoma or chondrosarcoma is most likely to involve the femur. A chondromyxoid fibroma favors the lower leg, while an adamantinoma nearly completely aims for the tibia. Osteoblastomas favor the vertebral column. But again this is on modern clinical evidence, and often with a bias towards European groups and environments. Archaeological data therefore has the potential to establish if tumor patterns and prevalences have changed over time, and might therefore significantly contribute to medical history.

In a recent review of cancer research, Bertolaso (2009) evaluates different academic factions which hold an interest in this disease group. She writes: "....the history of cancer research does not indicate there has been a game between rivals, but a balance between intellectual fields when investigated according to shared aims and purposes" (Bertolaso 2009:93). I would like to think that eventually tumor research in paleopathology will be accepted as one of these significant fields of research.

Acknowledgments

I am very grateful to the photographic departments of the Smithsonian Institution (Washington), Natural History Museum (London), and University of York, for their photographic assistance. Also, I wish to record the kindness and help of the late Dr. J. Nemeskéri in providing the Hungarian case. Finally, an apology to any colleagues I have inadvertently failed to acknowledge. Figures 23.3 and 23.4 are produced by permission of Churchill Livingstone and Dr. Dennis Stoker.

REFERENCES

Bennike, P., 1985   Palaeopathology of Danish Skeletons. Copenhagen: Akademisk Forlag.
Bertolaso, M., 2009   Towards an Integrated View of the Neoplastic Phenomena in Cancer Research. History and Philosophy of the Life Sciences 31:79–98.
Brothwell, D. R., 1967   The Evidence for Neoplasms. In Diseases in Antiquity. D. R. Brothwell, and A. T. Sandison, eds. pp 320–345. Springfield, IL: C. C. Thomas.
Brothwell, D. R., 2008   Tumours and tumour-like processes. In Advances in Human Palaeopathology. R. Pinhasi, and S. A. Mays, eds. pp 253–281. Chichester: John Wiley.
Brothwell, D. R., 2010   On Problems of Differential Diagnosis in Palaeopathology, as Illustrated by a Case from Prehistoric Indiana. International Journal of Osteoarchaeology, 20:621–622.
Chhem, R. K., and Brothwell, D. R., 2008   PaleoRadiology, Imaging Mummies and Fossils. Berlin: Springer.

David, A. R., and Zimmerman, M. R., 2010   Cancer: an Old Disease, a New Disease or Something in Between? Nature Reviews/Cancer 10:728–733.

Duplay, J., de Lumley, M. A. and Julliard, G., 1970   Un Pariétal Anténéandertalien: Problème Diagnostique. Neuro-chirurgie 16:5–13.

Ferris, E. J. and O'Connor, S. J., 1965   Calcification in Urinary Bladder Tumors. American Journal of Roentgenology 95:447–449.

Jarcho, S., 1966   The Development and Present Condition of Human Palaeopathology in the United States. *In* Human Palaeopathology. S. Jarcho, ed. pp. 3–42. New Haven: Yale University Press.

Ortner, D. J., 2003   Identification of Pathological Conditions in Human Skeletal Remains. London: Academic Press.

Putschar, W. G. J., 1966   Problems in the Pathology and Palaeopathology of Bone. *In* Human Palaeopathology. S. Jarcho, ed. pp. 57–83. New Haven: Yale University Press.

Resnick, D., ed., 1995   Diagnosis of Bone and Joint Disorders. Philadelphia: W. B. Saunders.

Retsas, S., ed., 1986   Palaeo-Oncology: The Antiquity of Cancer. Farrand Press, London.

Roberts, C., and Manchester, K., 1995   The Archaeology of Disease. Ithaca, NY: Cornell University Press.

Shivas, A. A., ed., 1967   Racial and Geographical Factors in Tumour Incidence. Edinburgh: Edinburgh University Press.

Sigerist, H. E., 1951   A History of Medicine. Vol 1. Oxford: Oxford University Press.

Smith, G. E., and Dawson, W. R., 1924   Egyptian Mummies. London: Allen and Unwin.

Stoker, D. J., 1986   Bone Tumours (pp. 1273–1326) and Myeloproliferative and Similar Disorders (pp. 1327–1348). *In* Diagnostic Radiology. Vol 2. R. G. Grainger, and D. J. Allison, eds. Edinburgh: Churchill Livingstone.

Strouhal, E., 1998   Survey and Analysis of Malignant Tumours of Past Populations in England and Scotland. Journal of Palaeopathology 10:101–109.

Tobias, P. V., 1960   The Kanam jaw. Nature 185:946–947.

24 Re-Emerging Infections: Developments in Bioarchaeological Contributions to Understanding Tuberculosis Today

Charlotte Roberts

It must be a matter of concern that 20 years after it was realized that tuberculosis was out of control across much of the developing world, the tide of tuberculosis shows no sign of being controlled.

(Davies 2008:xv)

INTRODUCTION

Tuberculosis (TB) is an infectious disease that affects humans, and is caused by bacteria of the *Mycobacterium* complex, most often *M. tuberculosis* and *M. bovis*. The bacillus was first described in 1882 and established as the cause of the disease by Robert Koch. The *Mycobacterium tuberculosis* complex consists of the following organisms: *M. tuberculosis* (humans), *M. bovis* (wild and domesticated animals), *M. canettii* (humans), *M. microti* (voles, hyrax, llama), *M. caprae* (goats), *M. pinnipedii* (seals and sea lions), and *M. africanum* (humans in Africa: Type I in the west and Type II in the east). Humans can contract *M. bovis* and *M. caprae* as secondary hosts

A Companion to Paleopathology, First Edition. Edited by Anne L. Grauer.
© 2012 John Wiley & Sons, Ltd. Published 2016 by John Wiley & Sons, Ltd.

(e.g. see Rodríguez et al 2009), but *M. microti* and *M. pinnipedii* only rarely (Grange 2008, and also see Panteix et al. 2010).

It is suggested that TB is caused by *M. bovis* today in probably around 0.5 percent of reported TB cases (Pfeiffer 2008). Whether this can be applied to the past needs exploring. Prior to 2002 and Brosch et al.'s seminal paper, it was believed that *M. tuberculosis* developed from *M. bovis* as a result of domestication of animals by humans several thousand years ago. Since 2002, however, and analysis of the TB genome, it is believed to be more likely that *M. tuberculosis* evolved from *M. bovis* well before the domestication of animals. This naturally has implications for our understanding of the co-evolution of TB, humans and their animals and the impact of domestication and agriculture on health (see below). TB is classed as a re-emerging infection with a long global history and it is of particular concern to clinicians in both developed and developing countries. It is also of a constant and increasing interest to bioarchaeologists who recognize the infection in archaeological human remains, and to medical historians who focus on it in written and artistic representations.

TB is contracted via inhalation of bacteria laden exhaled droplets from an infected person (usually *Mycobacterium tuberculosis*) or ingestion of infected products of an animal (usually *Mycobacterium bovis*). Once in the lungs or intestinal tract, the bacteria then spread via the circulatory and lymphatic systems to other parts of the body, including the skeleton. Primary TB usually occurs in childhood and in people who have never been exposed to TB before; the primary lesion (lungs or intestines) tends to heal, or the person dies in the acute stages. However, later in life reactivation of the primary lesion can occur, or re-infection may develop (post-primary/secondary TB); this is what is seen in bioarchaeology. Paleopathologists tend to mainly macroscopically analyze skeletal remains as their primary data source for disease, but TB has also been diagnosed in mummies (e.g. Salo et al. 1994), and using more advanced methods such as those from histology, imaging and biomolecular science (aDNA especially, but also mycolic acid analysis—see below, and Chapter 8 by Spigelman et al. in this volume).

Bone changes in TB occur in the post-primary stage of infection, usually in adulthood, and bones and joints may be affected 3–5 years following infection (Walgren 1948 in Ormerod 2008). In bioarchaeology, TB is recognized primarily through destructive lesions of the vertebral bodies (Pott's disease), usually in the lower thoracic and lumbar vertebrae; there is little new bone formation. Resnick and Niwayama (1995a) suggest that 25–40 percent of people with skeletal TB will have spinal damage. Some scholars have suggested that the early stages of vertebral change may also be recognized (Baker 1999) although the apparent destruction described on the anterior vertebral bodes may be the remnant holes resulting from developmental fusion of the two halves of the vertebral body. There are naturally a number of differential diagnoses of destructive lesions of the vertebral bodies that are also considered (e.g. fracture, osteoporosis, osteomyelitis, neoplastic disease, brucellosis, and see Roberts and Buikstra 2003:Table 3.3).

Non-specific skeletal changes that could be associated with TB have also been identified in skeletal remains and include: periosteal pitting and new bone formation on visceral rib surfaces (e.g. Roberts et al. 1994; see also clinical radiological data suggesting new bone formation enlarging ribs: Eyler et al. 1996), new bone formation on long bones (hypertrophic pulmonary osteoarthropathy—Resnick and Niwayama 1995b,c), new bone formation on the endocranial surface of the skull (suggested by some to be tuberculous meningitis—but see Lewis 2004 on the possible range of etiologies for endocranial bone formation, and Roberts et al. 2009), calcified pleura (e.g. Donoghue et al. 1998), bone changes suggestive of lupus vulgaris or TB of the skin (Padmavathy et al. 2003; Umredkar et al. 2010), septic arthritis of hip and knee joints (Resnick and Niwayama 1995a), and tuberculous dactylitis (spina ventosa) in the short bones of the hands and feet (Resnick and Niwayama 1995a). However, all these skeletal changes may be caused by other pathological conditions. To try and prove an association between TB and some of these nonspecific indicators, research has attempted to amplify tuberculous aDNA from a bone sample of individuals with these nonspecific changes (see below). While some data have suggested a positive identification of TB through aDNA, this unfortunately does not prove a direct link between TB and the bone lesions. It should, however, be noted that potentially any bone in the body can be affected with TB, that only about 3–5 percent of those with untreated TB may develop bone changes (Resnick and Niwayama 1995a:2462), and that *M. bovis* infection is more likely to cause changes to the skeleton (Stead 2000).

Thus, evidence for TB in paleopathology is the tip of the iceberg of the TB experience of our ancestors; even if pathogen DNA analysis (see below) could be successfully applied to all skeletal remains to identify all those who suffered, this would still not identify all affected people because data are soley reliant on excavated human remains, which are biased in so many ways. This is notwithstanding all the caveats of inferring health from the skeleton outlined by Wood et al. in their seminal paper of 1992. These include individuals dying before bone changes occurred, or in the acute stages of the disease, only having soft tissue involvement, lack of knowledge of individual frailty, or indeed the strength of immune systems in the past, and the impact of these factors on contraction of disease and development of bone changes. It is also important to have well-preserved human remains so that distribution patterns of bone forming and bone destroying lesions can be assessed and differential diagnoses inferred. However, there are also some parts of the world where funerary rituals may not preserve human remains well, for example where the dead were cremated.

This chapter aims to take a clinical base for understanding tuberculosis in the past but recognizes that the characteristics of the disease, and the experiences of people with TB today, may be very different to how it was recognized in the past and how our ancestors viewed this infection. Being healthy today enables society to function effectively, but being unhealthy, including suffering from tuberculosis, has (and will have had) physical, mental and socioeconomic impacts (e.g. see Rajeswari et al.'s 2005 study in India). Whether it is possible to measure all those impacts on individual and community function in the past is debatable. However, a paleopathological approach to TB is essential to attempt to achieve these outcomes.

TB frequency and risk factors today

While TB is an infection that was thought to be controlled by the late 1980s (Smith 1988), it began to rise in frequency again in the early 1990s, especially in relation to people with HIV (human immunodeficiency virus) whose immune systems are compromised (Harries and Zachariah 2008; Pozniak 2008). In 1993 the World Health Organization (WHO) declared TB to be a global emergency. Even though there is a difference between infection and manifest disease (WHO 1994), it is now suggested

Table 24.1   Risk factors for tuberculosis: potential for recognition in the past (✓ or —)

Risk factor	Past
Older and younger people (depending on transmission rates)	✓ (age indicators)
Males > females	✓ (morphological features in adults)
Ethnicity	✓ (morphological and metrical analysis)
Social status	✓ (grave goods, documentary data)
Low body mass index	✓ (morphological and metrical analysis)
Poverty	✓ (enamel hypoplasia, Harris lines, dietary deficiency disease, documentary data, and stature)
Animal interaction	✓ (archaeological animal remains, structures, documentary data)
Ingestion of infected animal products and infected animals' remains	✓ but challenging — *M. bovis* TB (pathogen aDNA analysis)
Overcrowding	✓ but challenging (archaeological and documentary data)
Urban environment	✓ (diseases of the urban environment, archaeological and documentary data)
Poor hygiene	✓ but challenging (diseases associated with poor hygiene, documentary data)
Poor diet	✓ (enamel hypoplasia, Harris lines, dietary deficiency disease, documentary data, and stature)
Iron deficiency	✓ but challenging (indirectly through C and N stable isotope analysis, archaeological and documentary data)
Occupation	✓ but challenging (occupational related disease, documentary data)
Travel/migration	✓ (stable isotope analysis (strontium and oxygen), dental and skeletal variation, aDNA)
Vitamin D deficiency	✓ (rickets and osteomalacia)
Lactose tolerance	✓ Lactase gene identification
Poor air quality, including use of biomass fuels	✓ sinusitis, rib periostitis
Climate/weather/season	✓ but challenging (climate records)
Concepts of disease	✓ but challenging (documentary data)
Poor access to health care	✓ but challenging (documentary data)
Non-compliance (treatment)	—
HIV/AIDS	—
Multidrug resistance	—
Immunosuppressive therapy	—
Excessive alcohol consumption	—
Low level of education	—
Unemployment	—

that nearly one-third of the world population is infected with *Mycobacterium tuberculosis*. In 2008 WHO estimated that there were 9.4 million incident cases of TB, 11.1 million prevalent cases, 1.3 million deaths in HIV-negative people, and 0.52 million deaths in HIV-positive people (WHO 2009a). In a later report (WHO 2009b) they updated the global situation to reveal that the regions of Central and Eastern Europe, Latin

America, the Western Pacific, and high-income countries, had managed to halve the 1990 mortality rate from TB in advance of the target year of 2015. In addition, prevalence and mortality rates in all other regions were declining except in African countries. In high-burden countries it appears that the Asian countries of India, China and Indonesia account for almost 50 percent of the world TB burden (Banavaliker 2008). However, in a recent report the number of new cases for 2010 is said to be now approaching 10 million (Dye and Williams 2010), although WHO has made a recent statement on the action and resources needed to control TB (WHO 2011a).

There are many risk factors that predispose people to TB today, some of which were present in the past. Table 24.1 shows those risk factors and illustrates that there is indeed much overlap between past and present, although for some factors it is not possible to identify whether they were important in the past (e.g. the presence of HIV co-infection, which is classed as the most important risk factor today—Dye 2008:36; Crofts et al. 2010). It should also be emphasized that for any one person affected by TB there can be a combination of factors that have predisposed that person, and this will vary between developed and developing countries.

THE EVIDENCE FOR TUBERCULOSIS IN THE PAST

Documentary and iconographic evidence

The earliest documentary data for TB come from Egypt and India dated to the 2nd millennium B.C. and from China dated to the 3rd millennium B.C. (Keers 1981; Evans 1998); what appears to be the signs and symptoms of TB are described. Artistic representations may be in the form of paintings, sculpture and reliefs in the form of people with obvious kyphotic spines (e.g. 4th–3rd millennium B.C. Egypt: Schrumpf-Pierron 1933; Morse et al. 1964) and pale thin people (e.g. in the 19th century). However, both documentary descriptions and artistic representations of people with supposed TB can be hard to interpret because the signs and symptoms could represent other disease processes (e.g. coughing, shortness of breath, pallor or wasting of the body). It should be noted that the earliest evidence for TB by necessity rest with human remains from archaeological sites.

Paleopathological evidence

As a result of the infrequency with which TB affects the skeleton, in paleopathology, TB is usually reported most commonly in adult individuals (Lewis 2007). Nonetheless, together these "case reports" contribute much to our understanding of the origin and history of TB from a global perspective. This aspect of the publication of data on palaeopathology in general has recently been discussed by Mays (in press). He found that case studies are common but cited less often than other types of publication in palaeopathology. He concluded that although fewer case reports may be desirable, well-executed case reports will continue to play a role, especially in synthetic works (e.g. as seen in Roberts and Buikstra 2003; Roberts and Cox 2003).

TB in Eurasia   Most evidence in Eurasia comes from Europe, perhaps reflecting higher levels of bioarchaeological "activity," although there are recent new data in

Japan (Suzuki and Inoue 2007), Korea (Suzuki et al. 2008: 1st century B.C.), and Thailand (Tayles and Buckley 2004: Iron Age– first two centuries A.D.). China has its first evidence in the 2nd century B.C., and in Japan it is dated to between 454 B.C. and the 1st century A.D. (Suzuki and Inoue 2007). In the rest of Eurasia, the earliest evidence comes from Italy (Canci et al. 1996, Formicola et al. 1987) dated to as early as the 6th millennium B.C., with later Bronze Age (1700–1500 B.C.) and Roman data (e.g. Canci et al. 2005: 1st century A.D.). The 6th millennium early Neolithic evidence from Italy (Arma dell'Aquila, Liguria 5800 ± 90 B.C.) is the earliest skeletal evidence in the world (Canci et al. 1996).

Spain and Poland also have Neolithic evidence for TB (Santoja 1975 and Gladykowska-Rzeczycka 1999, respectively) and, in Jordan, Ortner and Frohlich (2008) document two skeletons with TB dated to 3150–2200 B.C. Iron Age (400–230 B.C.) data come from England (Mays and Taylor 2003) and Denmark (Bennike 1985: 500–1 B.C.). However, most evidence for TB comes later in time, from the Roman period in Austria, Lithuania, England, and France (2nd–5th centuries), and later into the medieval period in many areas. Hungary, particularly, has produced much evidence for tuberculosis spanning the Avar Age (7th–8th centuries A.D.) through to the 19th century A.D. (e.g. see Pálfi and Marcsik 1999), as has Lithuania (e.g. see Jankauskas 1999).

TB in Africa and Oceania   In Africa (the Nile Valley) there is evidence for TB dating back to 4500 B.C. onwards (Derry 1938; Morse et al. 1964; Morse 1967; Buikstra et al. 1993; Strouhal 1999; Dabernat and Crubézy 2010) but no evidence in subSaharan Africa. In Oceania there is some evidence in Hawaii of pre-European contact TB (e.g. Pietreuwsky and Douglas 1994; Trembly 1997), but in the Americas, like Eurasia, there are considerable data on tuberculosis from human remains, again mostly from skeletons.

TB in The Americas   The earliest convincing evidence until recently appeared to come from South America around A.D. 700 (Allison et al. 1973, 1981), with southern Peru and northern Chile providing the most data (Stone et al. 2009). Unresolved issues include the apparent absence of TB in Mesoamerica, and why the earliest evidence of TB occurs in South America when data suggests a north-to-south migration. However, recently published data (Dabbs 2009) has documented TB in skeletons from Point Hope in Alaska dated to 100 B.C.– A.D. 500 Until this evidence appeared, it was believed that in North America evidence for TB postdated A.D. 900; however, most of the current data cluster in eastern North America and the Southwest (see summary and references in Stone et al. 2009 and Figure 4.1 in Roberts and Buikstra 2003). The recent evidence from Alaska is suggested to indicate transmission from Russia or Europe when, in Alaska, there was much change that put the population at risk from contracting new diseases (Dabbs 2009). Mesoamerica has virtually no evidence and a number of reasons for this are suggested such as non-survival of the evidence because of poor preservation or burial in unidentified areas (Stone et al. 2009).

Thus, there is much evidence for TB in specific parts of the world but it is absent from parts of North and South America, most of Africa, the north Atlantic islands, much of eastern Europe and most of Asia. The absence of evidence can be explained by a number of factors which include: a real absence of evidence for TB, lack of survival of evidence due to poor preservation, burial practice that is invisible or such that the skeletal evidence for TB is not present (e.g. cremated individuals), burial in rural cemeteries that are not a target for excavation, lack of training/ infrastructure/ finance for bioarchaeology, and few people working in the field in a particular place. It is expected that more evidence will be discovered as palaeopathology develops as a discipline in many regions of the world (see Buikstra and Roberts 2012).

FUTURE DIRECTIONS IN THE PALEOPATHOLOGY OF TUBERCULOSIS

As discussed above, the key focus in ancient TB research has been identifying where, when and why TB first made its appearance on the world's continents and how, as an infection, it developed a hold on the human population. Many of these studies, especially in more recent years, have attempted to think of the wider sociocultural, political and economic context for TB, and have grounded the data in questions of settlement pattern, diet, occupation, climate, mobility, poverty and the many other factors potentially responsible for its presence.

Tuberculosis has been studied in archaeological human remains for almost 100 years and yet there are avenues of research that remain to be explored. Bioarchaeologists have focused on individual skeletons showing signs of TB (e.g. Stirland and Waldron 1990, Suzuki et al. 2008), on frequencies at a population level (e.g. see Buikstra and Williams 1991, and studies in Pálfi et al. 1999), and in animals (Mays 2005). There have been studies developing diagnostic criteria (Roberts et al. 1994), and more recently extensive use of ancient DNA (aDNA) analysis for diagnosis (e.g. Murphy et al. 2009). However, there are parts of the world where evidence is limited for various reasons, not least because of the absence of paleopathological or bioarchaeo-logical work due to a lack of trained people or the resources to analyze skeletal remains. Nevertheless, with the development in analytical techniques, bioarchaeologists are starting to answer questions about past TB that could not have been answered before. This has been prompted by research on living people, and also the sequencing of the TB genome (see below). There are a number of key thematic areas that are now being (or could be) explored.

Vitamin D and tuberculosis
A lack of vitamin D has been documented to be associated with respiratory infections, including TB. For example, Talat et al. (2010) found 79 percent of 129 people with TB in Pakistan had low vitamin D levels, and calculated that this created a fivefold increased risk for progression to TB. Bartley (2010) documents early-20th-century treatment of TB of the skin using ultraviolet light therapy, and the need to have the right levels of the vitamin to prevent respiratory infections, and there have even been suggestions that vitamin D should be used to treat TB (Morcos et al. 1998, Shapira et al. 2010). There is, in addition, an association between levels of vitamin D in the body and latitude. This is because the production of vitamin D_3 in the skin is influenced by the action of ultraviolet light on the skin (50–90 per cent of total D requirement), along with

ingestion of vitamin D in the diet (Lips 2010). However, the amount of clothing covering the body, amount of exposure to the sun, and skin pigmentation are highly relevant to how much of the vitamin is produced.

Currently high rates of vitamin D deficiency are seen in India, China, Africa and the Middle East. Interestingly, higher D levels are seen in the northern latitudes of Europe than in southern areas; this is suggested to be due to people having a lighter skin in the north and making an effort to seek out ultraviolet light (Lips 2010). Pregnant women, children, older people, and migrants appear to be the most affected today by D-deficiency. Additionally, a recent finding is that there appears to be a correlation between the *ApoE4* allele in people in China, they being less likely to develop vitamin D deficiency, and more likely to live in northern latitudes (Hu et al. 2010). There has been virtually no link between latitude and vitamin D deficiency explored in paleopathology, although preliminary data from the Global History of Health Project (http:www.global.sbs.ohio-state.edu) suggest that more children with rickets derived from cemetery sites in the northern latitudes. There is certainly potential for exploring the relationship between TB, latitude, the *ApoE4* allele, and vitamin D deficiency in paleopathology to see whether the pattern seen today is mirrored in the past.

Stigma and TB

TB remains a disease associated with stigma in some parts of the world. This association hinders access to effective treatment because there is a reluctance to access diagnosis and treatment (Courtwright and Turner 2010). While in paleopathology and bioarchaeology there have been many skeletal remains attributed a TB diagnosis, there has been very little work done on analyzing relationships between those with TB and possible stigmatization. For example analyzing the funerary context of those with a diagnosis of TB may highlight marginalization and how communities dealt with those who died with TB.

Nutrition and TB

It is well known that the strength of the immune system and health are directly affected by the quality of the diet (Schwander and Ellner 2008; WHO 2011b), and for TB this is no exception. A person with a depressed immune system, such as somebody suffering HIV, is more likely to contract other infections such as TB. It is also well recognized that mycobacteria need a supply of iron in the host to grow and initiate disease (Ratledge 2004).

In bioarchaeology it is only possible to assess a person's nutritional status to a certain extent. This may be done by observing dietary excess (e.g. gout, possibly diffuse idiopathic skeletal hyperostosis), deficiency (e.g. dental enamel hypoplasia, scurvy), or gaining a general indication of the type of diet a person was eating (stable carbon and nitrogen isotope analysis—e.g. Katzenberg 2008). Linking these dietary indicators with disease, for example TB, is fairly commonly done, but the problem of directly associating the quality of the diet at the point at which a person contracted TB is challenging. This is because identifying when a person was infected cannot be known through examination of skeletal indicators of TB. However, novel ways of exploring this interaction have been attempted (Wilbur et al. 2008). While recognizing that iron and also protein are important for immune system function and the outcome of infection with TB, they highlighted that nutritional status may influence the dissemination of TB bacilli to the skeleton and ultimately the development of bone lesions used for diagnosis in paleopathology. They hypothesize that with adequate iron stores (lack of iron protects against infection—see Stuart-Macadam 1992), along with protein malnutrition, an infected person would suffer disseminated disease around the body, including into the skeleton. If this hypothesis is correct then with protein and iron deficiency there would be rapid death after TB infection, but no bone changes, and with adequate protein and iron a person would be susceptible to disseminated TB. Thus, the presence or absence of bone changes due to TB provides a window on a person's nutritional state at the time of infection.

Another diet-linked factor can be explained by the fact that one of the two ways that humans contract TB is via meat and milk of infected animals. Lactase is essential in the body to enable a person to digest lactose, the milk sugar (Wiley 2008). In many parts of the world today many people are lactose-intolerant because they do not produce the enzyme lactase. However, lactase persistence (LP) is most common in people from Europe (ca. 95 percent estimated prevalence—Wiley 2008), but is also seen in people who practice pastoralism in Africa (c.80–85 percent compared to ca.10–20 percent in nonpastoralists—Wiley 2008), the Middle East and South Asian areas (Itan et al. 2009). LP is also associated with a change in the *MCM6* gene (*13,910 T* allele) and the allele frequency varies. One expects to find LP in people who consume milk, and there are suggestions that the age of the allele fits well with the origin of dairying. As an extension of these facts, if people are consuming milk then they might also expect to contract TB from infected herds, and thus an association between TB lesions in a bioarchaeological context and lactase persistence would be expected. Likewise, people without LP would be expected not to have any TB lesions because they would not have tolerated dairy foods. This potentially now could be tested in the bioarchaeological record. Vitamin D deficiency might also be expected at high latitudes when milk would be consumed for both calcium and vitamin D, when lack of UV light is present.

Mobility and TB

A link between poorer health and migrants has also been identified (WHO 2010, 2011c). Specifically, there is a higher risk of TB in people who travel (Albert and Davies 2008), and those who migrate to other countries (Menzies et al. 2010). For example, some studies of people travelling by air have documented a link between people infected with TB and transmission to uninfected passengers (Driver et al. 1994), but other studies have highlighted the problem of quantifying the risks of air travel (Abubakar 2010; Marienau et al. 2010). However, airlines have become more aware of the risks that passengers take when travelling, and have outlined how those risks may be mitigated (Dowdall et al. 2010).

People have travelled around and between regions for thousands of years and for many reasons, and they will have taken their infections with them, and have been exposed to new infections when they arrived in new places. In recent years bioarchaeologists have started to track those movements mainly via stable isotope analyses of strontium and oxygen (e.g. Katzenberg 2008), but it is only in the past few years that a link has been explored between those isotope levels and infectious disease (e.g. see Roberts et al. 2013 and Richards and Montgomery 2012). Therefore, there appears to be great potential, albeit with challenges in interpretation, for studies of mobility in the past and its impact on tuberculosis transmission and spread.

Occupation and TB

There is undoubtedly a correlation between certain occupations and poor health, including TB (see WHO 2011d), and specific occupations have been explored to establish links with a higher risk of TB (e.g. Rees et al. 2010). There appear to be three types of occupation that produce a risk of acquiring TB (Bowden and McDiarmid 1994): unskilled laboring (e.g. food-handlers), work that increases susceptibility to the TB bacilli (e.g. making pottery), and occupations that increased exposure to TB (e.g. working in hospitals, prisons, and farms). Likewise, bioarchaeologists have documented skeletal changes on bones that have been interpreted as the result of "activity" or even certain occupations (see Jurmain 1999 for a balanced view, along with Chapter 29 by Jurmain et al, and Chapter 28 by Waldron, in this volume). While there may be more convincing indications of what work a particular person did in the past from their remains (e.g. see Brown and Molnar 1990 on dental changes), there has been debate about the validity of making activity-related interpretations from human skeletal changes (e.g. see Jurmain and Roberts 2008). However, due to the link between occupation and TB today, it is perhaps worth exploring this association in the past.

Housing/settlement/use of space and TB

The ways that people use space and the impact that might have on their health is well studied and understood for the living (e.g. see Wanyeki et al. 2006) but for the past we have much to learn. For example, were the urban living environments of the late medieval period in Europe most conducive for TB to spread via droplet transmission, or did contemporary rural living equally predispose people to transmission? Wanyeki et al. (2006) in a study of people with TB and their residences found that people with TB were more likely to live in dwellings more than five stories high in poorer areas, with more people per room than those without TB; they were also more likely to be migrants. Pokhrel et al. (2010) also report an association between women with TB and the use of biomass fuel inside houses in Nepal. García-Sancho et al. (2009) further document that women with TB exposed to biomass fuel smoke for over 20 years were three times more likely to have TB than those who were not. However, the links are not always that clear (e.g. see Slama et al. 2010).

There is increasing research, too, focusing on spatial clustering of people and populations with TB using spatial scanning and geographical information systems (GIS; see Tiwari et al. 2006), using collected data to help develop effective care of those affected (e.g. in the Gambia—Touray et al. 2010).Bioarchaeologists could do more to explore the detailed living environments of our ancestors to assess the risks posed by their dwellings and settlements. For example, this would include the materials used for building, ventilation, insulation, light, the design of houses, the arrangement and distribution of activities within houses, and differences between the experiences of the sexes, ages and statuses, in addition to the sharing of housing with animals. Linked to housing and settlement is a focus on the impact of climate, weather, and the seasons, on the frequency of TB (e.g. see Epstein 1999; Luquero et al. 2008; Naranbat et al. 2009; Jay and Marmot 2009). It is suggested that there is a strong link between health and climate, and that colder weather, mist and fog can predispose in general to respiratory conditions (Howe 1997). Furthermore, people with vitamin D deficiency, possibly living in more northern latitudes with shorter winter days and less UV light, and spending more time indoors in poorly ventilated housing, are at greater risk of developing lung disease, including TB. As there are much data on climate history now available, it is perhaps time for bioarchaeologists to integrate those data better with the evidence for TB to explore whether climate was at least partly responsible for TB in the past.

Tuberculosis as a Zoonotic Disease

The World Health Organization documents extensively the risks of disease from other animals (WHO 2011e) and it is becoming clear that many bacteria affecting animals co-evolved with humans. Research exploring TB in other animals today is prolific, but work on the remains of past animals is rare (see Chapter 11 by Upex and Dobney in this volume; and Thomas 2012). It is appreciated that a range of animals today with varying susceptibility, wild and domesticated, can become infected with bovine TB (O'Reilly and Daborn 1995) and potentially can transmit that infection to humans via secondary products such as milk, blood, meat, dung and hides. For example, in England there is concern about the rise in bovine TB in cattle, described to be due to infection from badgers, and the potential risk to humans (DEFRA 2010). However, it is suggested that a larger herd size can actually mean increased frequency of TB in that herd (Brooks-Pollock and Keeling 2009). Thus, badgers may not be totally responsible for TB contraction and transmission in cattle.

First described in A.D. 40 by Columnella (Pfeiffer 2008), *M. bovis* has been a plague of mammals other than humans for over 2,000 years. In archaeological contexts, Mays (2005) and Lignereux and Peters (1999) have described the potential bone changes that may be apparent for animals affected by TB. However, zooarchaeologists do not routinely observe animal bones for evidence of TB, and their observations are confounded by disarticulated fragmentary bones (i.e. no complete skeletons to observe bone change distribution), the fact that animals were probably also slaughtered if sick in the past, and the lack of established diagnostic criteria for recognizing TB in animal remains (O'Connor 2000). There are of course exceptions (see for example, Bathurst and Barta 2004, Von Hunnius 2009). Nevertheless, more efforts should be made to explore the zoonoses, such as TB, to establish whether it was a disease that posed a significant risk to humans in the past via gastrointestinal infection.

Biomolecular Analysis

Thanks to the development of analytical techniques in other disciplines such as biomolecular science, and their application to answering bioarchaeological questions, we are now in a position to advance our knowledge about the origin and evolution of this infectious disease that remains so prevalent.

Understanding of the TB genome's structure, biology and evolution, along with studies of molecular epidemiology, have provided a better knowledge of variation within the *M. tuberculosis* complex (Cole et al. 1998). Developments in the biomolecular science of TB since 1998 have contributed to diagnosis and treatment, and understanding susceptibility and resistance to TB in the living. Furthermore, research over the last 12 years has initiated questions about how long TB has been a problem for the human population, and the relationship between domestication and the *M. bovis* and *M. caprae* strains (Stone et al. 2009). For example, Gutierrez et al. (2005) proposed that the progenitor species for tuberculosis originated in Africa 2.5–3 mya, but how fast it evolved is not yet known. Furthermore, Gordon et al. (1999 in Stone et al. 2009) suggested that *M. tuberculosis* did not evolve from *M. bovis*, that *M. tuberculosis* and *M. africanum* were seen to be closely related, *M. microti* and *M. bovis* more distantly, and that *M. microti* was closely related to *M. bovis*. However, in 2002, Brosch et al. examined the assumption that *M. bovis* came first in mammals and then adapted to the human host. They analyzed 20 insertion-deletions in 100 diverse modern strains of the *M. tuberculosis* complex and found that it was likely that *M. canettii* and *M. tuberculosis* were older than the *M. microti*, *M. africanum*, and *M. bovis* lineages. This overturned our assumptions that the human form of TB developed from the bovine at the time of domestication of animals.

Research has also focused on exploring sequences of structural genes in strains of organisms in the *M. tuberculosis* complex (see Stone et al. 2009 for a summary). Thus, a lot of research on modern DNA of TB has led to a better appreciation of the diversity of the *M. tuberculosis* complex, and how the different strains within the bacteria of the complex are related. Modern DNA analysis of TB strains is also helping establish patterns of migration of those affected, susceptibility and resistance of people, patterns of transmission, and strains that are resistant to drugs (e.g. Daley 2008; Magana-Arachchi et al. 2010). For example, Barnes et al. (2010) have recently suggested that people who have been urbanized for a long time develop increased resistance to diseases such as TB due to natural selection. This is attributed to the presence of an allele of the *SLC11A1* gene found to be associated with natural resistance to TB in living people.

There is no doubt that out knowledge about TB today is being enhanced by developments in biomolecular science, and this is also the case for exploring the past. The aDNA of TB has been proved to be preserved in bones, teeth and other body tissues of excavated human remains, and its analysis is currently revolutionizing our understanding of the archaeology of TB. While the focus here is mainly on aDNA, very useful research, albeit limited relative to aDNA analysis, has also documented mycolic acids specific to the TB bacteria in human remains (e.g. Gernay et al. 1999, 2001; Redman et al. 2009; Mark et al. 2010; Minnikin et al. 2010). However, ancient pathogen DNA analysis is not without its problems (e.g see Roberts and Ingham 2008; Wilbur et al. 2009; also see Hershkovitz et al. 2008; Donoghue et al. 2009; and Taylor et al. 2010). For example, Roberts and Ingham (2008) document some key areas to improve such as adhering to proper authentication procedures, controlling for contamination, and independently replicating results. Nevertheless, developments in TB aDNA research may be limited by the public availability of data on modern strains of TB for comparative purposes: very little is known about genetic variation in environmental mycobacteria in soil and water which are likely contaminants of ancient skeletal samples, and the phylogeny of the *M. tuberculosis* complex is incompletely understood (Stone et al. 2009).

Most research on ancient tuberculous DNA has attempted to identify the *IS6110* fragment, which is a specific repeat element in the *M. tuberculosis* complex. Work has often used this repeat element to confirm a diagnosis of pathological changes in human tissues suggestive or diagnostic of TB. The first undisputed example of this type of work came from the study of a 1,000-year-old mummy from Peru by Salo et al. (1994, but also see Arriaza et al. 1995; Taylor et al. 1996; Braun et al. 1998; Mays and Taylor 2002, 2003; Klaus et al. 2010; and Bathurst and Barta 2004 on dog remains). There have also been other mycobacterial loci studied to attempt to distinguish to which phylogenetic group a strain of *M. tuberculosis* belongs, or to distinguish between *M. tuberculosis* and *M. bovis* (Stone et al. 2009). These are single-copy gene sequences (fragments that come from several different genes), and polymorphic direct repeat (DR) regions, where direct repeats in strains and the presence and the absence of nonrepetitive spacers between those regions can suggest regions of difference and patterns of variation (Stone et al. 2009). For example, Mays et al. (2001) examined single-copy gene sequences and found *M. tuberculosis* rather than *M. bovis* was present in individuals buried at the rural late-post-medieval site of Wharram Percy in England. Zink et al. (2004) also interestingly found both *M. tuberculosis* and *M. africanum* in individuals buried in Egypt between 2050 and 1650 B.C., but that *M. africanum* infected the later burials. More recently, semi-nomadic pastoralist burials were analyzed from Aymyrlyg, a Siberian site dated to the Iron Age (4th century B.C. to 4th century A.D.). Taylor et al. (2007) and Murphy et al. (2009) distinguished between *M. tuberculosis* and M. *bovis* and reported the first evidence of ancient *M. bovis* affecting a human. Of particular interest at this site was the new bone formation on the ilia of a juvenile individual, potentially representing gastrointestinal infection. In terms of DR region analysis, Taylor et al. (1999) found the TB affecting medieval people in England had a closer relationship to *M. tuberculosis* than *M. bovis*.

The examination of the co-evolution of humans and their animals with TB is important; the development of drug-resistant strains and very virulent strains of the bacteria today need to be explored and explained, and ancient pathogen DNA analysis is a way of achieving this goal. Of particular importance is the study of how the TB organism has evolved and changed and, although studies so far have been limited in palaeopathology, there is huge potential and ongoing projects that may help the control of TB in the present and future (e.g. see Durham University 2011). Furthermore, there are still genetic relationships between the Old and New Worlds that need to be explored, an area that this project on TB is addressing. This type of study might guide us as to how to deal with the present problem from a bioarchaeological perspective, for example whether strains of TB have changed through time globally, whether animals are key to transmission, and what predisposing factors we should look out for in the future.

Some examples from the literature show this potential. Taylor et al. (1999) examined three medieval bone samples from England and found the strain infecting these people in London was similar to the modern strain of *M. tuberculosis*. Likewise, Fletcher et al. 2003 analyzed individuals from an 18th–19th century post-medieval church crypt in Hungary; fragments of the *gyrA* and *katG* genes were identified, strains belonging to modern TB groups 2 and 3 (Sreevatsan et al. 1997), with group 3 representing a recently evolved strain. Again, Zink et al. (2007), in their study of human remains from Germany and Egypt, determined that some of the strains of TB represented were modern. Finally, Taylor et al. (2005), in an analysis of an Iron Age individual from England (2200 B.P.), found *M. tuberculosis* was the strain that affected this person and that the strain was related to modern strains in lineages 1, 2, and 3 (Baker et al. 2004).

The study of TB strains can of course suggest the origin and movement of people with the infection, and much research is being carried out on living people with TB. For example, Gallego et al. (2010) found that the *M. tuberculosis* genotypes isolated in over 900 New South Wales individuals in Australia reflected their migration patterns, and from over 4,000 individuals assessed from the Midlands in England, 45 percent had Euro-American strains and 34 percent had East African-Indian strains (Evans 2010). Other research is focusing on genetic susceptibility of people to TB by analyzing samples in living populations for specific susceptibility genes (e.g. *NRAMP1, VDR*; Möller et al. 2010; Yim and Selvaraj 2010). Both these avenues of research have potential in paleopathology and bioarchaeology for examining the spread of TB around the world and assessing how susceptible people were throughout time.

Beyond identifying whether human remains have preserved aDNA of TB for a diagnosis to be made or determining what organism caused the infection, there have been attempts to identify TB in human remains where there are no pathognomonic signs. For example, Faerman et al. (1997) analyzed samples from tuberculous and nontuberculous 15th–17th-century Lithuanian skeletons, and found a larger number of individuals with evidence of TB overall than the bone changes suggested. This is important because, as we have seen, only 3–5 percent of people with TB will develop the bone signatures (Resnick and Niwayama 1995a), although many more people in a population exposed to TB will harbor the disease. Gernaey et al. (1999), using mycolic acid analysis, also found that the resulting TB frequency was similar to the frequency of TB recorded in historical documentation associated with a post-medieval site in Newcastle, England. This was compared to the skeletal evidence, which was much lower.

Furthermore, individuals with what are acknowledged as nonspecific bone changes of TB (see Roberts and Buikstra 2003) have been examined, with conclusions suggesting that the lesions can be related to TB. For example, Haas et al. (2000) examined 7th–8th century A.D. skeletons from Hungary with and without pathognomonic changes of TB. They found that some of the individuals with positive results had vertebral changes suggested to be early signs of the infection. Another study, however, did not find a correlation between a positive result and lesions on the ribs, purported by some to be the result of TB (Mays et al. 2001), and the nonspecific change of hypertrophic pulmonary osteoarthropathy was seen to be associated with a positive aDNA result in one of two individuals with the bone changes (Mays and Taylor 2002). Both the last two papers focused on the 10th–16th century site of Wharram Percy, England. Finally, analysis of calcified pleura from a 1,400-year-old male individual from Israel, another possible nonspecific change for TB, gave a positive aDNA result (Donoghue et al. 1998). However, the point to bear in mind here, as indicated earlier, with all these studies is that positive pathogen aDNA analysis for TB cannot be taken as proof that TB caused the lesions observed. Firstly, many causes can be attributed to each nonspecific change and, secondly, even if the person had suffered TB they may not have shown any pathognomonic bone changes before they died. A recent development in biomolecular analysis has focused on mycobacterial proteins in Hungarian skeletal remains dated from A.D. 700–1600 (Boros-Major et al. 2011). Here *M. tuberculosis* proteins were identified, and mass spectrometric analysis of these proteins is proposed as a more reliable method to diagnose TB. More research will surely support or refute this statement

While aDNA analysis of TB in past humans is providing new information about this devastating infection, we must not move towards a situation in paleopathology where only evidence for tuberculosis aDNA in an individual is accepted for a definitive diagnosis in human remains. In spite of the advances in knowledge described above, we should also remember that strict protocols must be adhered to for macroscopic diagnosis before destructive sampling occurs, and for the actual sampling and analysis in biomolecular research.

CONCLUSIONS

Controlling tuberculosis remains a real challenge today. Its status as a disease of poverty suggests that dealing with poverty around the world would solve a lot of the problem (Wilkinson and Pickett 2009), along with providing better access to treatment and DOTS (directly observed therapy short course—directly observing patients taking the drugs), dealing with other factors leading to it, and multi drug resistance, and developing new drugs and vaccines. In paleopathology we can do more to understand how we have come to see TB remaining a problem. We can provide the time-depth that modern studies cannot, and by using biomolecular analysis it is possible to see how the TB organism has evolved and changed. There is still much to do, but one of the emphases in palaeopathology should be increasingly on using bioarchaeological data for TB to understanding the infection today, and for planning for a TB future. As Mario Raviglione, the Director of the WHO's "Stop TB" campaign, said in 2007 (Glusker 2007): "The positive is that the incidence ... is declining ... but the bad news is that the decline is slow, too slow."

Providing more population-based studies of TB is still important in paleopathology, and trying to fill gaps in our knowledge of TB in countries where there is little evidence is also a priority. We have proved TB's immense antiquity, and biomolecular analyses are providing useful data for specific populations. There is a need to have more comparative studies of these data around the world now; this will give a much better picture of the evolution and transmission of this infection with the movement of people. Linking these data with socio-cultural variables will help us better understand the key predisposing factors in regions and time periods. Context is key to interpretation, but so is the involvement in bioarchaeological research of experts from multiple relevant disciplines such as evolutionary medicine (Nesse and Williams 1994; Gluckman et al. 2009; and Zuckerman et al., Chapter 3 in this volume), biomolecular science (Brown 2004), earth sciences (Faure and Mensing 2005), clinical epidemiology (Sackett 1991, Boldsen and Milner, Chapter 7 in this volume), medical history (Porter 1999), climatology (Epstein 1999), medical anthropology (McElroy and Townsend 2009), and medical geography (Brown et al. 2010; Rupke 2000; Howe 1997).

Yes, paleopathologists do need to work hard to produce valid data from our ancestors remains, but there is a lot to offer the present and future, hopefully before too many human remains are repatriated and reburied (Roberts and Mays 2011, and see Lambert, Chapter 2, this volume). This of course would be a great loss to our discipline and to our descendants; perhaps now is also the time to create a biobank for samples from threatened human remains for future work when methods of analysis have developed even further (UK Biobank 2011). Biobanks are cryogenic storage facilities for biological samples for future research. They are increasing in number for accumulating samples from the living, and there may be a place now for bioarchaeology to think about such a development.

REFERENCES

Abubakar, I., 2010   Tuberculosis and Air Travel: A Systematic Review and Analysis of Policy. Lancet Infectious Disease 10:176–183.

Albert, P., and Davies, P. D. O., 2008   Tuberculosis and Migration. *In* Clinical Tuberculosis. 4th edition. P. D. O. Davies, P. F. Barnes, and S. B. Gordon, eds. pp. 367–381. London: Hodder Arnold.

Allison, M. J., Gerszten, E., Munizaga, J., Santoro, C., and Mendoza, D., 1981   Tuberculosis in Pre-Columbian Andean Populations. *In* Prehistoric Tuberculosis in the Americas. J. E. Buikstra, ed. pp. 49–51. Evanston, IL: Northwestern University.

Allison, M. J., Mendoza, D., and Pezzia, A., 1973   Documentation of a Case of Tuberculosis in Pre-Columbian America. American Review of Respiratory Disease 107:985–991.

Arriaza, B. T., Salo, W., Aufderheide, A. C., and Holcomb, T. A., 1995   Pre-Columbian Tuberculosis in Northern Chile: molecular and skeletal evidence. American Journal of Physical Anthropology 98:37–45.

Baker, B. J., 1999   Early Manifestations of Tuberculosis in the Skeleton. *In* Tuberculosis. Past and Present. G. Pálfi, O. Dutour, J. Deák, and I. Hutás, eds. pp. 301–307. Budapest: TB Foundation and Szeged: Golden Book Publishers.

Baker, L., Brown,T., Maiden, M. C., and Drobniewski, F., 2004   Silent Nucleotide Polymorphisms and a Phylogeny for *Mycobacterium tuberculosis*. Emerging Infectious Diseases 10:320–1577.

Banavaliker, J. N., 2008   Control of Tuberculosis in High-Prevalence Countries. *In* Clinical Tuberculosis.4thedition.P.D.O.Davies,P.F.Barnes,andS.B.Gordon,eds.pp.457–480.London: Hodder Arnold.

Barnes, I., Duda, A., Pybus, O., and Thomas, M. G., 2010   Ancient Urbanization Predicts Genetic Resistance to Tuberculosis. Evolution (International Journal of Organic Evolution) DOI: 10. 1111/j. 1558–5646. 2010. 01132. x.

Bartley, J., 2010   Vitamin D, Innate Immunity and Upper Respiratory Tract Infection. Journal of Laryngology and Otolaryngology 124:465–469.

Bathurst, R., and Barta, J. L., 2004   Molecular Evidence of Tuberculosis-Induced Hypertrophic Osteopathy in a 16th-Century Iroquian Dog. Journal of Archaeological Science 31:917–925.

Bennike, P., 1985   Palaeopathology of Danish Skeletons. A Comparative Study of Demography, Disease and Injury. Copenhagen: Akademisk Forlag.

Boros–Major, A., Bona, A., Lovasz, G., Molnar, E., Marcsik, A., Palfi,G., and Mark, L., 2011   New perspectives in Biomolecular Paleopathology of Ancient Tuberculosis: A Proteomic Approach. Journal of Archaeological Science 38(1): 197–201.

Bowden, K. M., and McDiarmid, M. A., 1994   Occupationally Acquired Tuberculosis. What's Known. Journal of Occupational Medicine 36:320–325.

Braun, M., Cook, D. C., and Pfeiffer, S., 1998   DNA From *Mycobacterium tuberculosis* Complex Identified in North American, Pre-Columbian Skeletal Remains. Journal of Archaeological Science 25:271–277.

Brooks–Pollock, E., and Keeling, M., 2009   Herd Size and Bovine Tuberculosis Persistence in Cattle Farms in Great Britain. Preventive Veterinary Medicine 92:360–365.

Brosch, R., Gordon, S. V., Marmiesse, M., Brodin, P., Buchrieseer, C., Eiglmeier, K., Garnier, T., Gutierrez, C., Hewinson, G., Kreemer, K., Parsons, L. M., Pym, A. S., Samper,S., Van

Soolingen, D., and Cole, S. T., 2002   A New Evolutionary Sequence for the *Mycobacterium tuberculosis* Complex. Proceedings of the National Academy of Sciences 99:3684–3689.

Brown, T., 2004   Genetics: A Molecular Approach. Oxon: Garland Science/BIOS Scientific.

Brown, T., McLafferty, S., and Moon, G., eds., 2010   A Companion to Health and Medical Geography. Chichester: Wiley Blackwell.

Brown, T., and Molnar, S., 1990   Interproximal Grooving and Task Activity in Australia. American Journal of Physical Anthropology 81:545–553.

Buikstra, J. E., Baker, B. J., and Cook, D. C., 1993   What Diseases Plagued Ancient Egyptians? A Century of Controversy Considered. *In* Biological Anthropology and the Study of Ancient Egypt. W. V. Davies, and R. Walker, eds. pp. 24–53. London: British Museum Press.

Buikstra, J. E., and Roberts, C. A., eds., 2012   A Global History of Paleopathology: Pioneers and Prospects. New York: Oxford University Press.

Buikstra, J. E., and Williams, S., 1991   Tuberculosis in the Americas: Current Perspectives. *In* Human Paleopathology: Current Syntheses and Future Options. D. Ortner, and A. C. Aufderheide, eds. pp. 161–172. Washington, DC: Smithsonian Institution Press.

Canci, A., Minozzi,S., and Borgognini Tarli, S., 1996   New Evidence of Tuberculous Spondylitis From Neolithic Liguria (Italy). International Journal of Osteoarchaeology 6:497–501.

Canci, A., Nencioni, L., Minozzi, S., Catalano, P., Caramella, D., and Fornaciari, G., 2005   A Case of Healing Spinal Infection From Classical Rome. International Journal of Osteoarchaeology 15:77–83.

Cole, S. T., Brosch, R., Parkhill, J., Garnier,T., Churcher, C., Harris, D., Gordon, S. V., Eiglmeier, K., Gas, S., Bary, E., Tekaia, F., Badcock, K., Basham, D., Brown, D., Chillingworth, T., Connor, R., Davies, R., Devlin, T., Feltwell,T., Gentles, S., Hamlin, N., Holroyd, S., Hornsby, T., Jagels, K., Krogh, A., McLean, J., Moule, L., Murphy, K., Oliver, J., Osborne, J., Quail, M. A., Rajandream, M–A, Rogers,J., Rutter, S., Seeger, J., Skelton, R., Squares, S., Sulston, J. E., Taylor, K., Whitehead, S., and Garrell, B. G., 1998   Deciphering the Biology of *Mycobacterium tuberculosis* from the Complete Genomic Sequence. Nature 393:537–544.

Courtwright, A., and Turner, A. N., 2010   Tuberculosis and Stigmatization: Pathways and Interventions. Public Health Reports 125 (Suppl 4):34–42.

Crofts, J. P., Andrews, N. J., Barker, R. D., Delpech, V., and Akubaker, I., 2010   Risk Factors For Recurrent Tuberculosis in England and Wales, 1998–2005. Thorax 65:310–314.

Dabbs, G. R., 2009   Resuscitating the Epidemiological Model of Differential Diagnosis: Tuberculosis at Prehistoric Point Hope, Alaska. Paleopathology Association Newsletter 148: 11–24.

Dabernat, H., and Crubézy, E., 2010   Multiple Bone Tuberculosis in a Child from Predynastic Upper Egypt (3200 B. C.). International Journal of Osteoarchaeology 20(6): 719–730.

Daley, C. L., 2008   Genotyping and Its Implications for Transmission Dynamics and Tuberculosis Control. *In* Clinical Tuberculosis. 4th edition. P. D. O. Davies, P. F. Barnes, and S. B. Gordon, eds. pp. 45–63 London: Hodder Arnold.

Davies, P. D. O., 2008   Preface. *In* Clinical Tuberculosis. 4th edition. P. D. O. Davies, P. F. Barnes, and S. B. Gordon, eds. pp. xv–xvi. London:Hodder Arnold.

DEFRA (Department for Environment Food and Rural Affairs), 2010   Research Report on Bovine TB in Badgers. Electronic document http://www.defra.gov.uk/news/2010/11/08/bovine-tb-reports (Accessed June 15, 2011).

Derry, D. E., 1938   Pott's Disease in Ancient Egypt. Medical Press and Circular 197:196–199.

Donoghue, H.D., Hershkovitz, I., Minnikin, D. E., Besra, G. S., Lee, O. Y–C., Galili, E., Greenblatt, C. L., Lemma, E., Spigelman, M., and Bar–Gal, G. K., 2009   Biomolecular Archaeology of Ancient Tuberculosis: Response to "Deficiencies and Challenges in the Study of Ancient Tuberculosis DNA" by Wilbur et al. (2009). Journal of Archaeological Science 36:2797–2804.

Donoghue, H. D., Spigelman, M., Zias, J., Gernaey–Child, A. M., and Minnikin, D. E., 1998   *Mycobacterium tuberculosis* Complex DNA in Calcified Pleura From Remains 1400 Years Old. Letters in Applied Microbiology 27:265–269.

Dowdall, N. P., Evans, A. D., and Thibeault, C., 2010   Air Travel and TB: An Airline Perspective. Travel Medicine and Infectious Disease 8:96–103.

Driver, C. R., Valway, S. E., Morgan, W. M., Onorato, I. M., and Castro, K. G., 1994   Transmission of *Mycobacterium tuberculosis* Associated With Air Travel. Journal of the American Medical Association 272:1031–1035.

Durham University, 2011   Department of Archaeology: Research: Research Projects. Electronic document http://www.dur.ac.uk/archaeology/research/projects/?mode=project&id=353 (Accessed June 15, 2011).

Dye, C., 2008   Epidemiology. *In* Clinical Tuberculosis. 4th edition. P. D. O. Davies, P. F. Barnes, and S. B. Gordon, eds. pp. 21–41. London: Hodder Arnold.

Dye, C., and Williams, B. G., 2010   The Population Dynamics and Control of Tuberculosis. Science 328:856–861.

Epstein, P., 1999   Climate and Health. Science 285:347–348

Evans, C. C., 1998   Historical Background. *In* Clinical tuberculosis. 2nd edition. P. D. O. Davies, ed., pp. 1–19. London: Chapman and Hall Medical.

Evans, J. T., 2010   Global Origin of *Mycobacterium tuberculosis* in the Midlands, UK. Emerging Infectious Diseases 16:542–545.

Eyler, W. R., Monsein, L. H., Beute, G. H., Tilley, B., Schultz, L. R., and Schmitt, W. G. H., 1996   Rib Enlargement in Patients With Chronic Pleural Disease. American Journal of Radiology 167:921–926.

Faerman, M., Jankauskas, R., Gorski, A., Bercovier, H., and Greenblatt, C. L., 1997   Prevalence of Human Tuberculosis in a Medieval Population of Lithuania Studied by Ancient DNA Analysis. Ancient Biomolecules 1:205–214.

Faure, G., and Mensing, T. M., 2005   Isotopes. Principles and Applications. 3rd edition. Hoboken, New Jersey: Wiley.

Fletcher, H. A., Donoghue, H. D., Holton, H., Pap, I., and Spigelman, M., 2003   Widespread Occurrence of *Mycbacterium tuberculosis* DNA from 18th–19th-century Hungarians. American Journal of Physical Anthropology 120:144–152.

Formicola, V., Milanesi, Q., and Scarsini, C., 1987   Evidence of Spinal Tuberculosis at the Beginning of the Fourth Millennium B. C. From Arene Candide Cave (Liguria, Italy). American Journal of Physical Anthropology 72:1–6.

Gallego, B., Sintchenko, V., Jelfs, P., Coiera, E., and Gilbert, G. L., 2010   Three–Year Longitudinal Study of Genotypes of *Mycobacterium tuberculosis* in a Low-Prevalence Population. Pathology 42:267–272.

García-Sanchez, M. C., García–García, L., Báez-Saldaña, R., Ponce-De-León, A., Sifuentes-Osomio, J., Bobadilla-Del-valle, M., Ferreyra-Reyes, L., Cano-Arellano, B., Canizales-Qunitero, S., Palacios–Merino, Ldel C., Juárez-Sandino, L., Ferreira-Guerrero, E., Cruz-Hervert, L. P., Small, P. M., and Pérez-Padilla, J. R., 2009   Indoor Pollution as an Occupational Risk Factor For Tuberculosis Among Women: A Population-Based, Gender-Oriented, Case-Control Study in Southern Mexico. Revista de Investigación Clínica 61:392–398.

Gernaey, A. M., Minnikin, D. E., Copley, M. S., Ahmed, A. M. S., Robertson, D. J., Nolan, J., and Chamberlain, A. T., 1999   Correlation of the Occurrence of Mycolic Acids With Tuberculosis in an Archaeological Population. *In* Tuberculosis: Past and Present. G. Pálfi, O. Dutour, J. Deák, and I. Hutás, eds. pp. 275–282. Budapest: TB Foundation and Szeged: Golden Book Publishers.

Gernaey, A. M., Minnikin, D. E., Copley, M. S., Dixon, R. A., Middleton, J. C., and Roberts, C. A., 2001   Mycolic Acids and Ancient DNA Confirm an Osteological Diagnosis of Tuberculosis. Tuberculosis 81:259–265.

Gladykowska–Rzeczycka, J. J., 1999   Tuberculosis in the Past and Present in Poland. *In* Tuberculosis: Past and Present. G. Pálfi, O. Dutour, J. Deák, and I. Hutás, eds. pp. 561–573. Budapest: TB Foundation and Szeged: Golden Book Publishers.

Gluckman, P., Beedle, A., and Hanson, M., 2009   Principles of Evolutionary Medicine. Oxford: Oxford University Press.

Glusker, A., 2007   Global Tuberculosis Levels Plateau While Extensively Drug Resistant Strains Increase. British Medical Journal 334:659.

Gordon, S. V., Brosch, R., Billault, A., Garnier, T., Eiglmeier, K., and Cole, S. T., 1999   Identification of Variable Regions in the Genomes of Tubercle Bacilli Using Bacterial Artificial Chromosome Arrays. Molecular Microbiology 32:643–655.

Grange, J. M., 2008   *Mycobacterium tuberculosis*: The Organism. *In* Clinical Tuberculosis. 4th edition. P. D. O. Davies, P. F. Barnes, and S. B. Gordon, eds. pp. 65–78. London: Hodder Arnold.

Gutierrez, M. C., Brisse,S., Brosch, R., Fabre, M., Omais, B., Marmiesse,M., Supply, P., and Vincent, V., 2005   Ancient Origin and Gene Mosaicism of the Progenitor of *Mycobacterium tuberculosis*. PLOS Pathogens 1:e5.

Haas, C. J., Zink, A., Molnar, E., Szeimes, U., Reischl, U., Marcsik, A., Ardagna, Y., Dutour, O., Pálfi,G., and Nerlich, A., 2000   Molecular Evidence for Different Stages of Tuberculosis in Ancient Bone Samples From Hungary. American Journal of Physical Anthropology 113:293–304.

Harries, A. D., and Zachariah, R. 2008   The Association Between HIV and Tuberculosis in the Developing World, With a Special Focus on Sub–Saharan Africa. *In* Clinical Tuberculosis. 4th edition. P. D. O. Davies, P. F. Barnes, and S. B. Gordon, eds. pp. 315–342. London: Hodder Arnold.

Hershkovitz, I., Donoghue, H. D., Minnikin, D. E., Besra, G. S., Lee,O.-C., Gernaey, A,M., Galili, E., Eshed, V., Greenblatt, C. L., Lemma, E., Kahila Bar-Gal. G., and Spigelman, M., 2008   Detection and Molecular Characterization of 9000-Year-Old *Mycobacterium tuberculosis* From a Neolithic Settlement in the Eastern Mediterranean. PLoS ONE 3(10):e3426.

Howe, G. M., 1997   People, Environment, Disease and Death. Cardiff: University of Wales Press.

Hu, P., Qin, Y. H., Jing, C. X., Lu, L., Hu, B., and Du, P. F., 2010   Does the Geographical Gradient of ApoE4 Allele Exist in China? A Systematic Comparison Among Multiple Chinese Populations. Molecular Biology Reports PMID:20354905.

Itan, Y., Powell, A., Beaumont, M. A., Burger, J., and Thomas, M. G., 2009   The Origins of Lactase Persistence in Europe. PLoS Computational Biology 5(8):e1000491. doi:10. 1371/journal. pcbi. 1000491.

Jankauskas, R., 1999   Tuberculosis in Lithuania: Paleopathological and Historical Correlations. *In* Tuberculosis: Past and Present. G. Pálfi, O. Dutour, J. Deák, and I. Hutás, eds. pp. 551–558. Budapest: TB Foundation and Szeged: Golden Book Publishers.

Jay, M., and Marmot, M. G., 2009   Health and Climate Change. British Medical Journal 339:b3669.

Jurmain, R. D., 1999   Stories From the Skeleton. Behavioural Reconstruction in Human Osteology. New York: Gordon and Breach Publishers.

Jurmain, R. D., and Roberts, C. A., 2008   Juggling the Evidence: The Purported "Acrobat" from Tell Brak. Antiquity 82.

Katzenberg, M. A., 2008   Stable Isotope Analysis: A Tool for Studying Past Diet. *In* Biological Anthropology of The Human Skeleton. M. A. Katzenberg, and S. R. Saunders, eds. pp. 413–441. New York: Wiley.

Keers, R. Y., 1981   Laënnec: A Medical History. Thorax 36:91–94.

Klaus, H. D., Wilbur, A. K., Temple, D. H., Buikstra, J. E., Stone, A. C., Fernandez, M., Wester, C., and Tam, M. E., 2010   Tuberculosis on the North Coast of Peru: Skeletal and Molecular Paleopathology of Late Pre-Hispanic and Postcontact Mycobacterial Disease. Journal of Archaeological Science 37:2587–2597.

Lewis, M. E., 2004   Endocranial Lesions in Non–Adult Skeletons: Understanding Their Aetiology. International Journal of Osteoarchaeology 14:82–97.

Lewis, M. E., 2007   The Bioarchaeology Children. Perspectives From Biological Anthropology and Forensic Anthropology. Cambridge: Cambridge University Press.

Lignereux, Y., and Peters, J. 1999   Elements for the Retrospective Diagnosis of Tuberculosis on Animal Bones from Archaeological Sites. *In* Tuberculosis: Past and Present. G. Pálfi, O. Dutour, J. Deák, and I. Hutás, eds. pp. 339–348. Budapest: TB Foundation and Szeged: Golden Book Publishers.

Lips, P., 2010   Worldwide Status of Vitamin D Nutrition. Journal of Steroid Biochemistry and Molecular Biology 121:297–300.

Luquero, F. J., Sanchez-Padilla, E., Simon-Soria, F., Eiros, J. M., and Glub, J. E., 2008   Trend and Seasonality of Tuberculosis in Spain, 1996–2001. International Journal of Tuberculosis and Lung Disease 12:221–224.

Magana-Arachchi, D. N., Perera, A. J., Senaratne, V., and Chandrasekharan, N. V., 2010   Patterns of Drug Resistance and RFLP Analysis of *Mycobacterium tuberculosis* Strains Isolated from Recurrent Tuberculosis Patients in Sri Lanka. Southeast Asian Journal of Tropical Medicine and Public Health 41:583–589.

Marienau, K. J., Burgess, G. W., Cramer, E., Averhoff, F. M., Buff, A. M., Russell, M., Kim, C., Neatherlin, J. C., and Lipman, H., 2010   Tuberculosis Investigations Associated With Air Travel: U. S. Centers for Disease Control and Prevention, January 2007–June 2008. Travel Medicine and Infectious Disease 8:104–112.

Mark, L., Patonai, Z., Vaczy, A., Lorand, T., and Marcsik, A., 2010   High-Throughput Mass Spectrometric Analysis of 1400-Year-Old Mycolic Acids as Biomarkers for Ancient Tuberculosis Infection. Journal of Archaeological Science 37:302–305.

Mays, S. A., 2005   Tuberculosis As a Zoonotic Disease in Antiquity. *In* Diet and Health in Past Animal Populations. Current Research and Future Directions. J. Davies, M. Fabiš, I. Mainland, M. Richards, and R. Thomas, eds. pp. 125–134.Oxford: Oxbow Books.

Mays, S. A., 2012   The Impact of Case Reports Relative to Other Types of Publication in Palaeopathology. International Journal of Osteoarchaeology 22:81–85.

Mays, S. A., and Taylor, G. M., 2002   Osteological and Biomolecular Study of Two Possible Cases of Hypertrophic Osteoarthropathy from Mediaeval England. Journal of Archaeological Science 29:1267–1276.

Mays, S. A., and Taylor, G. M., 2003   A First Prehistoric Case of Tuberculosis from Britain. International Journal of Osteoarchaeology 13:189–196.

Mays, S. A., Taylor, G. M., Legge, A. J., Young,D. B., and Turner-Walker, G., 2001   Paleopathological and Biomolecular Study of Tuberculosis in a Medieval Skeletal Collection from England. American Journal of Physical Anthropology 114:298–311.

McElroy, A., and Townsend, P. K., 2009   Medical Anthropology in Ecological Perspective. Boulder, CO: Westview Press.

Menzies, H. J., Winston,C. A., Holtz, T. H., Cain, K. P., and MacKenzie, W. R., 2010   Epidemiology of Tuberculosis Among US and Foreign-Born Children and Adolescentsin the United States, 1994–2007. American Journal of Public Health 100:1724–1729.

Minnikin, D. E., Lee, OY–C., Pitts, M., Baird, M. S., and Besra, G. S., 2010   Essentials in the Use of Mycolic Acid Biomarkers For Tuberculosis Detection: Response to "High-Throughput Mass Spectrometric Analysis of 1400-Year-Old Mycolic Acids as Biomarkers For Ancient Tuberculosis Infection" by Mark et al., 2010. Journal of Archaeological Science 37(10): 2407–2412.

Möller, M., de Wit, E., and Hoal, E. G., 2010   Past, Present and Future Directions in Human Genetic Susceptibility to Tuberculosis. FEMS Immunology and Medical Microbiology 58:3–26.

Morcos, M. M., Gabr, A. A., Samuel, S., Kamel, M., el Baz, M., el Beshry, M., and Michail, R. R., 1998   Vitamin D Administration to Tuberculous Children and Its Value. Bollettino Chimico Farmaceutico 137:157–164.

Morse, D., 1967   Tuberculosis. *In* Diseases in Antiquity. D. R. Brothwell, and A. T. Sandison, eds. pp. 249–271 Illinois: C. C. Thomas.

Morse, D., Brothwell, D. R., and Ucko, P. J., 1964   Tuberculosis in Ancient Egypt. American Review of Respiratory Diseases 90:526–541.

Murphy, E. M., Chistov, Y. K., Hopkins, R., Rutland, P., and Taylor, G. M., 2009   Tuberculosis Among Iron Age Individuals from Tyva, South Siberia: Palaeopathological and Biomolecular findings. Journal of Archaeological Science 36:2029–2038.

Naranbat, N., Nymadawa, P., Schopfer, K., and Rieder, H. L., 2009   Seasonality of Tuberculosis in an Eastern–Asian Country With an Extreme Continental Climate. The European Respiratory Journal 34:921–925.

Nesse, R. M., and Williams, G. C., 1994   Why We Get Sick. The New Science of Darwinian Medicine. London: Penguin Books.

O'Connor, T. P., 2000   The Archaeology of Animal Bones. Stroud, Gloucestershire: Sutton Publishing.

O'Reilly, L. M., and Daborn, C. J., 1995   The Epidemiology of *Mycobacterium bovis* Infections in Animals and Man. Tubercle and Lung Disease 76(Supplement 1):1–46.

Ormerod, P., 2008   Non-Respiratory Tuberculosis. *In* Clinical Tuberculosis. 4th edition. P. D. O. Davies, P. F. Barnes, and S. B. Gordon, eds. pp. 163–188. London: Hodder Arnold.

Ortner, D. J., and Frohlich, B., 2008   The Early Bronze Age I Tombs and Burials of Bab edh-Dhra. Lanham, MD: Rowman and Littlefield.

Padmavathy, L., Rao, L., and Veliath, A., 2003   Utility of Polymerase Chain Reaction as a Diagnostic Tool in Cutaneous Tuberculosis. Indian Journal of Dermatology, Venereology, and Leprology 69:214–216.

Pálfi, G., Dutour, O., Deák, J., and Hutás, I., eds. 1999   Tuberculosis. Past and Present. Budapest: TB Foundation and Szeged: Golden Book Publishers.

Pálfi, G., and Marcsik, A., 1999   Paleoepidemiological Data of Tuberculosis in Hungary. *In* Tuberculosis. Past and Present. G. Pálfi, O. Dutour, J. Deák, and I. Hutás, eds. pp. 533–539. Budapest: TB Foundation and Szeged: Golden Book Publishers.

Panteix, G., Gutierrez, M. C., Boschiroli, M. L., Rouviere, M., Plaidy, A., Pressac, D., Porcheret, H., Chyderiotis, G., Ponsada, M., Van Oortegem, K., Salloum, S., Cabuzel, S., Bañuls, A. L., Van de Perre, P., and Godreuil, S., 2010   Pulmonary Tuberculosis Due to *Mycobacterium microti*: A Study of Six Recent Cases in France. Journal of Medical Microbiology 59:984–989.

Pfeiffer, D. U., 2008   Animal tuberculosis. *In* Clinical Tuberculosis. 4th edition. P. D. O. Davies, P. F. Barnes, and S. B. Gordon, eds. pp. 519–528. London; Hodder Arnold.

Pietreuwsky, M., and Douglas, M. T., 1994   An Osteological Assessment of Health and Disease in Precontact and Historic (1778) Hawai'i. *In* In The Wake of Contact: Biological Responses to Conquest. C. S. Larsen, and G. R. Milner, eds. pp. 179–196. New York: Wiley–Liss.

Pokhrel, A. K., Bates, M. N., Verma, S. C., Joshi, H. S., and Sreeramareddy, C. T., and Smith, K. R., 2010   Tuberculosis and Indoor Biomass and Kerosene Use in Nepal: A Case-Control Study. Environmental Health Perspectives 118:558–564.

Porter, R., 1999   The Greatest Benefit to Mankind: A Medical History of Humanity From Antiquity to the Present. London: Fontana Press.

Pozniak, A., 2008   HIV and TB in Industrialized Countries. *In* Clinical Tuberculosis. 4th edition. P. D. O. Davies, P. F. Barnes, and S. B. Gordon, eds. pp. 343–365. London: Hodder Arnold.

Rajeswari, R., Muniyandi, M., Balasubramanian, R., and Narayanan, P. R., 2005   Perceptions of Tuberculosis Patients about Their Physical, Mental and Social Well-Being: A Field Report From South India. Social Science and Medicine 60:1845–1853.

Ratledge, C., 2004   Iron, Mycobacteria and Tuberculosis. Tuberculosis 84:110–130.

Redman, J. E., Shaw, M. J., Mallet, A. I., Santos, A. L., Roberts, C. A., Gernaey, A. M., and Minnikin, D. E., 2009   Mycocerosic Acid Biomarkers for the Diagnosis of Tuberculosis in the Coimbra Skeletal Collection. Tuberculosis 89(4): 267–277.

Rees, D., Murray, J., Nelson, G., and Sonnenberg, P., 2010   Oscillating Migration and the Epidemics of Silicosis, Tuberculosis and HIV Infection in South African Gold Miners. American journal of Industrial Medicine 53:398–404.

Resnick, D., and Niwayama, G., 1995a   Osteomyelitis, Septic Arthritis, and Soft Tissue Infection: Organisms. *In* Diagnosis of Bone and Joint Disorders. 3rd edition. D. Resnick, ed. pp. 2467–2474. Edinburgh: W. B. Saunders.

Resnick, D., and Niwayama, G., 1995b   Osteomyelitis, Septic Arthritis, and Soft Tissue Infection: Axial Skeleton. *In* Diagnosis of Bone and Joint Disorders. 3rd edition. D. Resnick, ed. pp. 2419–2447. Edinburgh: W. B. Saunders.

Resnick, D., and Niwayama, G., 1995c   Enostosis, Hyperostosis and Periostitis. *In* Diagnosis of Bone and Joint Disorders. 3rd edition. D. Resnick, ed. pp. 4396–4466. Edinburgh: W. B. Saunders.

Richards, M. P., and Montgomery, J., 2012   Isotope Analysis and Palaeopathology: A Short Review and Future Developments. *In* A Global History of Paleopathology: Pioneers and Prospects. J. E. Buikstra, and C. A. Roberts, eds. pp. 718–731. New York: Oxford University Press.

Roberts, C. A., and Buikstra, J. E., 2003   The Bioarchaeology of Tuberculosis. A Global View on a Reemerging Disease. Gainesville: University Press of Florida.

Roberts, C. A., and Cox, M., 2003   Health and Disease in Britain. From Prehistory to the Present Day. Stroud: Sutton Publishing.

Roberts, C. A., and Ingham, S., 2008   Using Ancient DNA Analysis in Palaeopathology: A Critical Analysis of Published Papers With Recommendations for Future Work. International Journal of Osteoarchaeology 18(6): 600–613.

Roberts, C. A., Lucy, D., and Manchester, K., 1994   Inflammatory Rib Lesions of Ribs: An Analysis of the Terry Collection. American Journal of Physical Anthropology 95: 169–182.

Roberts, C. A., and Mays, S. A., 2011   Study and Restudy of Curated Skeletal Collections in Bioarchaeology: A Perspective on the UK and the Implications for Future Curation of Human Remains. International Journal of Osteoarchaeology 21:626–630.

Roberts, C. A., Millard, A. R., Nowell, G. M., Grocke, D., Macpherson, C., Pearson, G., and Evans, D., 2013   The Origin and Mobility of People With Venereal Syphilis Buried in Hull, England in the Late Medieval Period. American Journal of Physical Anthropology 150:273–285.

Roberts, C. A., Pfister, L.-A., and Mays, S. A., 2009   Letter to the Editor: Was Tuberculosis Present in *Homo erectus* in Turkey? American Journal of Physical Anthropology 139: 442–444.

Rodríguez, E., Sánchez, L. P., Pérez, S., Herrera, L., Jiménez, M. S., Samper, S., Iglesias, M. J., 2009   Human Tuberculsois Due to *Mycobacterium bovis* and *M. caprae* in Spain, 2004–2007. International Journal of Tuberculosis and Lung Disease 13:1536–1541.

Rupke, N. A., ed., 2000   Medical Geography in Historical Perspective. London: Wellcome Institute for the History of Medicine.

Sackett, D. L., 1991   Clinical Epidemiology: A Basic Science for Clinical Medicine. Boston: Little Brown.

Salo, W. L., Aufderheide, A. C., Buikstra, J., and Holcomb, T. A., 1994   Identification of *Mycobacterium tuberculosis* DNA in a Pre-Columbian Mummy. Proceedings of the National Academy of Sciences USA 91:2091–2094.

Santoja, M., 1975   Estudio Antrológico. *In* Excavaciones de la Cueva de la Vaquera, Torreiglesias, Segovia (Edad del Bronce). pp. 74–87. Segovia.

Schrumpf-Pierron, B., 1933   Le Mal de Pott en Egypte 4,000 Ans Avant Notre Ère. Aesculape (Paris):295–299.

Schwander, S. K., and Ellner, J. J., 2008   Human Immune Response to *M. tuberculosis*. *In* Clinical Tuberculosis. 4th edition. P. D. O. Davies, P. F. Barnes, and S. B. Gordon, eds. pp. 121–141. London; Hodder Arnold.

Shapira, Y., Agmon-Levin, N., and Schoenfeld, Y., 2010   *Mycobacterium tuberculosis*, Autoimmunity, and Vitamin D. Clinical Reviews In Allergy & Immunology 38:169–177.

Slama, K., Chiang, C. Y., Hinderaker, S. G., Bruce, N., Vedal, S., and Enarson, D. A., 2010   Indoor Solid Fuel Combustion and Tuberculosis: Is There An Association? International Journal of Tuberculosis and Lung Disease 14:6–14.

Smith, E. R., 1988   The Retreat of Tuberculosis 1850–1950. London: Croom Helm.

Sreevatasan, S., Pan, X., Stockbauer, K. E., Connell, N. D., Kreiswirth, B. N., Whittam, T. S., and Musser, J. M., 1997   Restricted Structural Gene Polymorphism in the *Mycobacterium tuberculosis* Complex Indicates Evolutionarily Recent Global Dissemination. Proceedings of the National Academy of Sciences USA 94:9869–9874.

Stead, W. W., 2000   What's In a Name? Confusion of *Mycobacterium tuberculosis* and *Mycobacterium bovis* in Ancient DNA Analysis. Paleopathology Association Newsletter 110:13–16.

Stirland, A., and Waldron, T., 1990   The Earliest Cases of Tuberculosis in Britain. Journal of Archaeological Science 17:221–230.

Stone, A. C., Wilbur, A. K., Buikstra, J. E., and Roberts, C. A., 2009   Tuberculosis and Leprosy in Perspective. Yearbook of Physical Anthropology 52: 66–94.

Strouhal, E., 1999   Ancient Egypt and Tuberculosis. *In* Tuberculosis. Past and Present. G. Pálfi, O. Dutour, J. Deák, and I. Hutás, eds. pp. 453–460. Budapest: TB Foundation and Szeged: Golden Book Publishers.

Stuart-Macadam, P., 1992   Porotic Hyperostosis: A New Perspective. American Journal of Physical Anthropology 87:39–47.

Suzuki, T., Fujita, H., and Choi, J. G., 2008   Brief Communication: New Evidence of Tuberculosis From Prehistoric Korea – Population Movement and Early Evidence of Tuberculosis in Far East Asia. American Journal of Physical Anthropology 136:357–360.

Suzuki, T., and Inoue, T., 2007   Earliest Evidence of Spinal Tuberculosis from the Neolithic Yayoi Period in Japan. International Journal of Osteoarchaeology 17:392–402.

Talat, N., Perry, S., Parsonnet, J., Dawood, G., and Hussain, R., 2010   Vitamin D Deficiency and Tuberculosis Progression. Emerging Infectious Diseases 16:853–855.

Tayles, N., and Buckley, H. R., 2004   Leprosy and Tuberculosis in Iron Age Southeast Asia? American Journal of Physical Anthropology 125:239–256.

Taylor, G. M., Crossey, M., Saldanha, J., and Waldron, T., 1996   DNA from *Mycobacterium tuberculosis* Identified in Mediaeval Human Skeletal Remains Using Polymerase Chain Reaction. Journal of Archaeological Science 23:789–798.

Taylor, G. M., Goyal, M., Legge, A. J., Shaw, R. J., and Young, D., 1999   Genotypic Analysis of *Mycobacterium tuberculosis* from Medieval Human Remains. Microbiology 145:899–904.

Taylor, G. M., Mays, S. A., Huggett, J. F., 2010   Ancient DNA (aDNA) Studies of Man and Microbes: General Similarities, Specific Differences. International Journal of Osteoarchaeology 20(6):747–751.

Taylor, G. M., Murphy, E., Hopkins, R., Rutland, P., and Chistov, Y., 2007   First Report of *Mycobacterium bovis* DNA in Human Remains From the Iron Age. Microbiology 153:1243–1249.

Taylor, G. M., Young, D. B., and Mays, S. A., 2005   Genotypic Analysis of the Earliest Known Prehistoric Case of Tuberculosis in Britain. Journal of Clinical Microbiology 43:2236–2240.

Thomas, R., 2012   Non-Human Paleopathology. *In* A Global History of Paleopathology: Pioneers and Prospects. J. E. Buikstra, and C. A. Roberts, eds. pp. 652–664. New York: Oxford University Press.

Tiwari, N., Adhikari, C. M. S., Tewari, A., and Kandpal, V., 2006   Investigation of Geo-Spatial Hotspots for the Occurrence of Tuberculosis in lmora District, India, Using GIS and Spatial Scan Statistic. International Journal of Health Geographics 5:33.

Touray, K., Adetifa, I. M., Jallow, A., Rigby, J., Jeffries, D., Cheung, Y. B., Donkor, S., Adegbola, R. A., and Hill, P. C., 2010   Spatial Clustering of Tuberculosis in an Urban West African Setting: Is There Evidence of Clustering? Tropical Medicine & International Health 15:664–672.

Trembly, D., 1997   A Germ's Journey to Isolated Islands. International Journal of Osteoarchaeology 7:621–624.

UK Biobank, 2011   UKBiobank—What Is It? Electronic document http://www.ukbiobank.ac.uk/about/what.php (Accessed June 15, 2011).

Umredkar, A., Mohindra, S., Chhabra, R., and Gupta, R., 2010   Vertebral Body Hyperostosis as a Presentation of Pott's Disease: A Report of Two Cases and Literature Review. Neurology India 58:125–127.

Von Hunnius, T., 2009   Using Microscopy to Improve a Diagnosis: An Isolated Case of Tuberculosis-Induced Hypertrophic Osteopathy in Archaeological Dog Remains. International Journal of Osteoarchaeology 19:397–405.

Walgren, A., 1948   The Timetable of Tuberculosis. Tubercle 29:245–251.

Wanyeki, I., Olson, S., Brassard, P., Menzies, D., Ross, N., Behr, M., and Schwartzman, K., 2006   Dwellings, Crowding, and Tuberculosis in Montreal. Social Science and Medicine 63:501–511.

WHO (World Health Organization), 1994   Framework for Effective TB Control. Tuberculosis Programme. Geneva, Switzerland: World Health Organization.

WHO (World Health Organization), 2009a   WHO Report 2009: Global TB Control – Epidemiology, Strategy, Financing. Geneva, Switzerland: World Health Organization.

WHO (World Health Organization), 2009b   Global TB Control: A Short Update to the 2009 Report. Geneva, Switzerland: World Health Organization.

WHO (World Health Organization), 2010   International Travel and Health. Geneva, Switzerland: World Health Organization.

WHO (World Health Organization), 2011a   The Global Plan to Stop TB. Electronic document http://www.stoptb.org/global/plan/ (Accessed June 15, 2011).

WHO (World Health Organization), 2011b   Health Topics: Nutrition. Electronic document http://www.who.int/topics/nutrition/en/ (Accessed June 15, 2011).

WHO (World Health Organization), 2011c   Heath Topics: Travel and Health. Electronic document http://www.who.int/topics/travel/en/ (Accessed June 15, 2011).

WHO (World Health Organization), 2011d   Health Topics: Occupational Health. Electronic document http://www.who.int/topics/occupational_health/en/ (Accessed June 15, 2011).

WHO (World Health Organization), 2011e   Health Topics: Zoonoses. Electronic document. URL: http://www.who.int/topics/zoonoses/en/ (Accessed June 15, 2011).

Wilbur, A. K., Bouwman, A. S., Stone, A. C., Roberts, C. A., Pfister, L., Buikstra, J. E., and Brown, T. A., 2009   Deficiencies and Challenges in the Study of Ancient Tuberculosis DNA. Journal of Archaeological Science 36: 1990–1997.

Wilbur, A. K., Farnbach, A. W., Knudson, K. J., and Buikstra, J. E., 2008   Diet, Tuberculosis and the Palaeopathological Record. Current Anthropology 49:963–991.

Wiley, A., 2008   Cow's Milk Consumption and Health. An Evolutionary Perspective. In Evolutionary Medicine and Health. Trevathan, W. R., Smith, E. O., and McKenna, J. J., eds. pp. 117–133. Oxford: Oxford University Press.

Wilkinson, R., and Pickett. K., 2009   The Spirit Level. Why Equality is Better for Everyone. London: Penguin Books.

Wood, J. W., Milner, G. R., Harpending, H. C., and Weiss, K. M., 1992   The Osteological Paradox: Problems of Inferring Prehistoric Health from Skeletal Samples. Current Anthropology 33:343–370.

Yim, J. J., and Selvaraj, P., 2010   Genetic Susceptibility in Tuberculosis. Respirology 15:241–256.

Zink, A. R., Molnár, E., Motamedi, N., Pálfi, G., Marcsik,A., and Nerlich, A. G., 2007   Molecular History of Tuberculosis From Ancient Mummies and Skeletons. International Journal of Osteoarchaeology 17:380–391.

Zink, A. R., Sola, C., Reischl, U., Grabner, W., Rastogi, N., Wolf, H., and Nerlich, A. G., 2004   Molecular Characterization of *Mycobacterium tuberculosis* Complex in Ancient Egyptian Mummies. International Journal of Osteoarchaeology 14:404–413.

CHAPTER **25** # Leprosy (Hansen's disease)

*Niels Lynnerup
and Jesper Boldsen*

INTRODUCTION

Hansen's disease, or leprosy, is caused by the bacillus *Mycobacteria leprae*. Surprisingly, in spite of its antiquity, it is little understood today. At present, the World Health Organization estimates that up to two million people are affected by the disease (WHO 2008). Leprosy is a severely debilitating and crippling disease, made worse by the overt symptoms which often lead to the social and physical ostracizing of its sufferers. Today, it is largely confined to developing countries, but its presence appears to have been common in Europe and the Levant in the Middle Ages. Thus, leprosy is a particularly interesting disease for paleopathologists to study.

Leprosy is a chronic disease, with patients often developing bony changes. These changes can serve as strong indicators of the presence of the disease, especially in the past, leading at times to relatively confident paleopathological diagnoses. Paleopathological diagnosis has been further assisted by the fact that throughout the European medieval period (and even up until the 20th century in some areas of the world), the forced segregation and confinement of patients to leprosaria, or "leper colonies", often meant that the individuals suffering from leprosy were buried in segregated cemeteries (see Cule 2002; Trabjerg 1993). The excavation of these cemeteries has yielded skeletal remains which uniquely represent individuals who all, arguably, suffered from the same disease, and allows us to better ascertain the role of this debilitating disease in history, and its significance in past societies.

A Companion to Paleopathology, First Edition. Edited by Anne L. Grauer.
© 2012 John Wiley & Sons, Ltd. Published 2016 by John Wiley & Sons, Ltd.

HISTORY

A burial in India from the Chalcolithic period, dating to 2000–2500 B.C., has been presented as one of the earliest skeletally documented finds of leprosy (Robbins et al. 2009). A genetic diagnosis of another suspected early case of leprosy was made on a burial from 50 B.C. to A.D 16 in Israel (Matheson et al. 2009). While the early presence of the disease in humans has been argued based on biblical references, it has also been disputed. It appears that the term used in the Bible may reflect several diseases, not just leprosy (Aufderheide and Rodriguez-Martin 1998). The term leprosy itself probably derives from Greek, meaning scaly or peeling skin (Aufderheide and Rodriguez-Martin 1998). Recent DNA research on the bacillus genome indicates that leprosy arose millions of years ago in humans in East Africa (Monot et al. 2009). It has been speculated that the disease became endemic in the Levant between 300 and 100 B.C. (Stanford and Standford 2002), and then spread to Europe and Asia. Although the spread is likely to be related to human migration and trade, it does not appear that its spread to Europe was caused by the crusades. Leprosy was already established in Europe before the crusades began (Mitchell 2002), but does not appear to have been well-established in Pre-Columbian New World populations (Ortner 2003). Most paleopathological cases of leprosy, by far, are derived from the excavations of European medieval leprosy asylums, or leprosaria. The first leprosaria were established in France in the seventh century (Likovsky et al. 2006), and archaeological excavations have yielded many skeletons associated with leprosaria in the UK (Roberts 2002), Denmark (Møller-Christensen 1953, 1958), Italy (Belcastro et al. 2005), Hungary (Palfi 1991) and the Czech Republic (Dokladal 2002).

Interestingly, leprosy virtually disappeared in Europe by the 17th century, although Scotland, Iceland, and Norway experienced many new cases into the 19th century (Richards 1960; Sharp 2007). It is puzzling why leprosy was endemically present in these northern regions long after it virtually disappeared from the rest of Europe. One piece of the puzzle may be the fact that the leprosy bacillus has been shown to thrive in peat (Kazda et al. 1990). In fact, leprosy constituted such a major public health danger in rural Norway that the authorities built leprosy hospitals and appointed a Medical Supervisor up to the beginning of the twentieth century (Irgens 2006). This position was filled by Dr. Gerhard Henrik Armauer Hansen from 1875 to 1912.

THE CAUSE OF LEPROSY

Leprosy is caused by the bacillus *Mycobacterium leprae*, first identified by G. H. A. Hansen (hence the term Hansen's disease) in 1873 (Hansen 1875). Perusing patient histories, and the unique Norwegian leprosy patient registrar (Irgens 2006), he ascertained that the disease must be caused by an infectious agent; finally succeeding in identifying the agent and showing it to be a bacterium within the genus *Mycobacterium* (Irgens 2006). Considerable debate about the discovery ensued between Hansen and a German doctor by the name of Neisser, who is accredited with the isolation of the gonoccocal bacterium. But Hansen eventually became the acknowledged discoverer of the pathogen's identity (Irgens 2006).

Hansen's work continued beyond identifying the cause of the disease. In subsequent years he concentrated on the mechanism of contagion, and attempted to inoculate two patients already suffering from tuberculoid leprosy (see below) with tissue from a patient suffering from lepromatous leprosy (and where Hansen had detected the bacteria) in an effort to understand the disease's transmission. Hansen had not sought the formal consent of the patients, a measure contrary to ethical standards practiced now and then. It is possible that he was driven to take these measures since researchers were unable to culture *M. leprae* in any type of media or in animals. Indeed, Hansen's predecessor had inoculated himself, members of his staff, and patients suffering of other diseases; always with negative results (Irgens 2006). Hansen was indicted in 1880, and later convicted for having broken medical rules of ethics. He was removed from his position at the hospital, but not from his position as Chief Medical Officer for Leprosy (Irgens 2006).

The extreme difficulty in culturing and inoculating the bacillus is one of the reasons why leprosy has remained an enigmatic disease. In modern medical research the nine-banded armadillo became a focus of leprosy research, as it is one of the few known animals to serve as host to the pathogen (Kirchheimer and Storrs 1971). Interestingly, armadillos have low body temperature—an important variable for *M. leprae* (see below) (Storrs 1972). Still, due to the relative slow growth rate of the animal, many experiments have been carried out using the foot pads (again, the lowered temperature) of laboratory mice (Shepard 1960).

Leprosy's exact path of contagion is still uncertain. What is known is that affected patients may expel the bacterium in great quantities along with upper airway mucosal secretions (e.g., when sneezing). This would indicate that infection can occur through aerosol aspiration, or by direct contact, for instance, if an affected person sneezes into her hands and then touches another person's hands, who then touches the nose, mouth or eyes, thus depositing mucus with the bacteria. It has also been speculated that bacteria may enter the body through cuts in the skin (Miller 1994). Among close family contacts of untreated lepromatous patients, the risk of disease is increased approximately eightfold, and the attack rate can be as high as 10 percent (Miller 1994), although adult-to-child transmission rates may be five times greater (Aufderheide and Rodriguez-Martin 1998).

As a member of the genus *Mycobacterium*, *M. leprae* is part of a large family of bacilli which are ubiquitous in the environment. Some cause major animal and human diseases, such as *M. tuberculosis*. *M. tuberculosis* alone is responsible for 8–10 million new cases of tuberculosis a year (Stanford and Stanford 2002). Genomic analyses of *M. leprae* has shown that it is derived from a single clone evolving asexually; i.e. the bacillus expanded into a new pathogenic niche, humans, but at the expense of participating in interspecific and intraspecific genetic exchange (Cole et al. 2001). Thus, all strains of *M. leprae* are very similar, their few differences being caused by mutations. This so-called single-clonal niche invasion probably resulted in extensive gene decay. It is estimated that *M. leprae's* genome is 25 percent smaller than that of *M. tuberculosis*. This has led to a loss of metabolic capacity, which may be the reason for the extreme difficulty in cultivating the bacteria outside human hosts (Cole et al. 2001). Paleopathologically, it is interesting to note that single-clone niche invasion is thought to be the mechanism whereby *M. tuberculosis*, *Yersinia pestis* (linked with plague) and *Treponema pallidum* (causing endemic and venereal syphilis) also became human diseases (Cole et al. 2001).

THE CLINICAL PICTURE OF LEPROSY

Just as there is uncertainty concerning contagion, not much is known about the events following entry of *M. leprae* into the host. It seems that only 10–20 percent of those infected develop signs of indeterminate leprosy (see below), and only 50 percent of these will progress to full-blown clinical leprosy (Miller 1994). The incubation period is equally perplexing, as it appears to range from 6 months in some patients to several decades in others. An incubation period, however, of 3–5 years appears to be the most common (Miller 1994). In the clinical record, it appears that the detection rate of leprosy is higher in males than among females, and that the male/female ratio approaches 1.0 the higher the prevalence (Irgens and Skjaerven 1985).

Regardless of the incubation period or the sex of the individual, the progression of the disease can lead to a wide spectrum of clinical manifestations. These have been classified as follows:

- **Early or indeterminate leprosy**
  Early signs of leprosy are usually limited to the skin. Patients develop hypopigmented or hyperpigmented plaques, which may further become hyposensitive. These maculae or plaques may disappear inside a year (Miller 1994).
- **Tuberculoid leprosy (TT)**
  Tuberculoid leprosy is characterized by a sharply demarcated hypopigmented and hypesthetic macule, which proceeds to become elevated, and spreads. Sweat glands, skin follicles and the peripheral nerves disappear completely in the lesions. Infection may then spread to the larger peripheral nerves, especially the ulnar, the fibular and the greater auricular nerves. This neurologic involvement leads to muscle atrophy and hyposensibility (Miller 1994).
- **Lepromatous leprosy (LL)**
  Lepromatous leprosy is characterized by dermal involvement, especially of the face, but also of the upper limbs. The skin of the face may become thickened and furrowed (at times referred to as "leonine face"). Involvement is also seen in the upper airway mucosa, especially in the nose, but also in the larynx. Septal perforation and collapse of the nose bridge, along with resorption of the nasal spine and palate (referred to as facies leprosa), may develop. At later stages, distal nerve involvement, as in tuberculoid leprosy, may develop (Miller 1994).
- **Borderline leprosy**
  The classification of borderline leprosy occurs when the case appears to fall in between tuberculoid leprosy (TT) and lepromatous leprosy (LL). Hence, TT and LL are seen in this classification system as the "outer" poles of a clinical spectrum, with borderline leprosy in the middle. However, researchers and clinicians may choose to further subclassify the manifestations of the disease, with terms such as borderline TT and borderline LL.

Another means of classification, created by the World Health Organization, has been built upon differentiating the bacillary load of the patient. This classification system became possible when, based on Hansens's work, the bacilli itself could be demonstrated in the tissues, not just by the body's reaction to its presence (Hansen and Looft 1895). The terms multibacillary (or bacilliferous) and nonbacillary (or paucibacillirous) refer to a high or low bacterial load, respectively. Individuals classified as multibacillary are usually associated with symptoms commensurate with lepromatous leprosy or borderline lepromatous leprosy, while individuals classified as paucibacillary usually display lesions associated with tuberculoid leprosy or borderline tuberculoid leprosy. Cellular immunity response and level of circulating antibodies can also assist in the classification of the different manifestations of the disease.

Why patients develop one or the other of these manifestations is still not clear, although it is known that genetic factors (specifically the so-called HLA-associated genes) play a role (Miller 1994; Aufderheide and Rodriguez-Martin 1998). The role of the immune system also appears to be an important variable in the manifestation of the disease, with individuals with compromised immune systems more likely to develop lepromatous leprosy than immunologically strong individuals. It is also known that borderline cases may shift across the clinical spectrum (towards the lepromatous form if untreated and towards the tuberculoid form during treatment). However, shift from one end of the spectrum to the other (e.g., lepromatous to tuberculoid) has never been documented (Miller 1994).

Other clinical varieties of leprosy, which are quite rare, such as a diffuse leprosy known as Lucio and Lapati, comprise Lucio's phenomenon, a necrotizing vasculitis, seen among patients with the multibacillary forms (Stanford and Stanford 2002). Similarly, Wade's histoid leprosy, another rare form of multibacillary leprosy, was first described in patients treated with a short course of sulfones (Wade 1960, 1963).

PATHOLOGY

The primary sites of infection are the skin, peripheral nerves, anterior aspects of the eyes, the upper airways, the testes, and the distal aspects of the upper and lower extremities. A commonality for all these anatomical sites is the slightly lowered temperature (Miller 1994). While involvement of the skin, eyes, airways, and testes deeply impacts patients suffering from leprosy, the recognition of these conditions by the paleopathologist analyzing human skeletal remains is unlikely. Alternatively, nerve involvement, with its ultimate impact on the use of extremities, has the potential to be discerned. Hence, there are three anatomical sites of special importance for the paleopathologist:

- **Craniofacial involvement**
  LL often starts as chronic rhinitis and sinusitis. The mucosa becomes swollen, ulcerated, and in turn leads to septal perforation, and periostitis of the maxillary palate. Resorption of the anterior nasal spine and alveolar process, leading to the loss of the maxilla incisors (Nah et al. 1985; Ortner 2002, 2003) ensues. These manifestations are known as facies leprosa. Along with the bony changes, patients also experience severe facial skin involvement, characterized by localized swollen and furrowed areas, sometimes affecting the sclera of the eyes (Aufderheide and Rodriguez-Martin 1998). Møller-Christensen was able to directly relate skeletal changes seen on medieval skeletons with those of the pathologic changes seen among living leprosy patients in Africa, India and southeast Asia (Møller-Christensen 1978).

- **Ulnar nerve involvement**
  The ulnar nerve rests very superficially as it passes along the lateral humeral epicondyle, and this, due to the predilection for slightly lower temperatures, may be why affliction of this nerve is often seen. The bacteria accumulate inside the nerve sheath, leading to disruption of motor and sensory nerves, and subsequently cause numbness and/or complete lack of sensation (Miller 1994; Aufderheide and Rodriguez-Martin 1998).

- **Fibular (peroneal) nerve involvement**
  Similar to the ulnar nerve, the fibular nerve has a very superficial location as it passes around the head of the fibula below the knee. And, like the ulnar nerve, bacteria may accumulate inside these nerve sheaths.

Bone lesions due to peripheral nerve involvement occur earlier and more intensively for TT than for LL (Aufderheide and Rodriguez-Martin 1998). However, in a paleopathological setting, it would be difficult to differentiate between TT and LL based solely on the bony lesions: both TT and LL may display varying degrees of all the above-mentioned bone changes (Aufderheide and Rodriguez-Martin 1998). Illarramendi et al. (2002) found that in a modern patients from Brazil, 7 of 95 (7.5 percent) had bone loss of the extremities, and of these seven cases, four were diagnosed as LL, and none as TT. On the other hand, it seems that TT is the form most prevalent when leprosy is first introduced to a new population, i.e. as a true epidemic. Upon later becoming an endemic disease most individuals afflicted develop LL (Mitchell 2002).

Examinations of male leprous patients have shown that the patients have lower concentrations of free testosterone and estradiol, and that these concentrations are further correlated with low BMD (bone mineral density) and the length of time the patients has been affected by leprosy (Ishikawa 1997, 1999). This is perhaps due to the fact that a very high proportion of male leprosy patients have involvement of the testes (Saporta and Yuksel 1994).

It may well be speculated that in historic times the lesions of early and indeterminate leprosy, usually skin lesions, may have been strongly stigmatizing, as they might be seen as the first signs of leprosy. However, there may have been a great number of misdiagnoses at this stage, since other skin diseases display similar discrete characteristics. Differential diagnoses today include lupus erythematosus and lupus vulgaris, sarcoidosis, yaws, dermal leishmaniasis and more banal skin diseases (Miller 1994). While it is possible that some individuals displaying skin lesions of a different cause might have been thought to have leprosy, an argument can be made that the skin lesions of leprosy are rather pathognomonic, e.g. hypesthesia is always present, and peripheral nerve involvement is common, and hence diagnosis during the medieval period was possible. The more serious symptoms associated with lepromatous leprosy would certainly have been unambiguous (facies leprosa, peripheral paresis and atrophy, subsequent loss of digits due to secondary infection, and blindness).

THE PALEOPATHOLOGY OF LEPROSY

Given the above special circumstances of leprosy, especially the segregation of sufferers and their subsequent burial in leprosaria cemeteries, paleopathologists have been able to identify a number of specific skeletal features indicative of the disease.

Early modern studies of skeletal material from leprosaria date back to the 19th century (Ehlers 1898). One of the pioneers of paleopathological research into leprosy was the Danish physician Vilhelm Møller-Christensen (1961, 1967, 1978). Møller-Christensen carefully described facies leprosa, which is still accepted as a pathognomonic symptom for the disease (Andersen and Manchester 1992). He demonstrated, for example, that 75 percent of the adult skulls from the material from the well-known leprosarium in Næstved displayed facies leprosa (Møller-Christensen 1961).

Skeletally, the pre-eminent signs of leprosy in the skull include periostitis of the palatine process of the maxilla, resorption and remodeling of the margins of the nasal aperture, loss of the anterior nasal spine, and recession of the alveolar process of the maxilla (Møller-Christensen et al. 1952; Møller-Christensen 1978; Andersen and Manchester 1988; Andersen and Manchester 1992; Ortner 2003). The involvement of the anterior aspect of the maxilla and alveolar process may lead to the loss of the upper incisors (Møller-Christensen et al. 1952). If the onset of leprosy occurs during dental development, the dental roots may not develop to their full size (Ortner, 2003).

Extremities of the body are also directly affected by leprosy, and therefore potentially recognizable in human skeletal remains. Changes to the hands and feet, in particular, can be strongly indicative of the disease (Ortner 2003). The accumulation of bacteria in the nerve sheaths of the peripheral ulnar and fibular nerves leads to a loss of motor function, with the hand assuming the so-called "claw-hand" configuration (hyperextension in the metacarpo-phalangeal joints and hyperflexion in the inter-phalangeal joints), and a similar paresis of the foot phalanges. The contraction of hands and fingers, feet and toes, can be seen directly on the phalangeal bones (Ortner 2003). Accumulation of bacteria in the nerve sheaths also leads to profound loss of sensation. Without the ability to feel, sufferers often fall victim to pressure trauma, as they are unable to gauge the pressure that they are placing on tissues during simple tasks such as holding a glass of water. Injuries, large and small, can go undetected. This leads to high rates of secondary infection at the sites of injury, further leading to tissue necrosis and ultimately to the loss of limbs (Aufderheide and Rodriguez-Martin 1998; Ortner 2003). However, it is likely an oversimplification to attribute the characteristic atrophy and loss of phalanges only to trauma and secondary infections. It has been suggested that more complex neurovascular alterations and episodes of heightened inflammatory response take place in the presence of *M. leprae* as well (Stanford and Standford 2002), leading to the development of grooves and exostoses on the phalanges (Andersen and Manchester 1987, 1988). Still, the end result is a loss of bone at the distal phalanges, with a rather diagnostic morphology (occasionally referred to as "pencil ends") of the remaining proximal phalanges.

Along with the neurological changes brought about by the accumulation of the bacilli in the nerve sheaths, other changes can occur. The presence of the bacilli may lead to periosteal reaction of the tibia and fibula. Ortner (2003) states that periosteal reaction is more pronounced near the distal ends of the tibia and fibula (nearer to the ankle joint) than it is near the proximal ends (where the afflicted nerves are located), which suggests that the lesions might be the result of the body's more general response to this chronic and systemic condition (Andersen et al. 1992, 1994). Hence, it is important to caution that periosteal reaction is characteristic of, but not specific to, leprosy (Andersen 1969; and see Chapter 27 by Weston in this volume). In an analysis of skeletal material from leprosaria in Denmark, 20 percent of the adult sample displayed periosteal reaction, and 100 percent of the skeletons with signs clearly indicative of leprosy displayed periosteal reactions (Brander and Lynnerup 2002). These statistics correspond well with the results presented by Lewis et al. (1995) on the leprosarium in Chichester, England. Likewise, studies conducted by Boldsen on Danish medieval cemeteries, both at leprosaria and common urban and rural churchyards, suggest that periosteal reaction on the fibula and fifth metatarsal may be especially indicative of leprosy (Boldsen 2005a).

Finally, due to the involvement of the testis in males and the subsequent lowering of bone mineral content, fractures may be seen more often in sufferers of leprosy. Judd and Roberts (1998) reported fractures in 2.6 percent of the material at the Chichester leprosarium.

THE PALEOEPIDEMIOLOGY OF LEPROSY

Why leprosy, of all human diseases, became so physically and socially feared is not quite clear. What is clear is that throughout history, even up to the 20th century, great measures have been taken to segregate the sufferers. Some countries went so far as to create laws to enforce segregation of the afflicted into specific facilities or into designated isolated geographic areas (Ehlers 1898; Cule 2002). Leprosaria, the name often given to these facilities, were frequently placed outside, but near, towns and subsisted by residents' farming produce and on alms from the town and/or church (Trabjerg 1993). Upon entering, patients might experience a "burial ceremony," whereby they were effectively declared "dead to the world." Removal from one's family and home, whether as a young child or an adult, must have been an excruciating experience in spite of the charitable shelter and support that leprosaria hoped to provide for the patients.

Whatever the reasons and logic behind the segregation, the results allow for unique paleopathological and paleoepidemiological analyses to be made. With cemeteries linked to medieval leprosaria, it is possible to statistically evaluate the occurrence of the disease in the past, as well as the diagnostic probabilities. In spite of these possibilities, there are limitations to studying these cemeteries. First, clinical research has shown that not all leprosy patients will show definite bony changes. Hence, residents of the leprosaria may, or may not, have had bone involvement associated with the disease. As leprosy directly causes neurological damage, and may indirectly contributes to bone changes, especially to fingers and toes, loss of function of the fingers, even without loss of the digits, is common. So, too, is blindness (Miller 1994). Both blindness and nerve damage has the capacity to severely handicap a patient without affecting bone, and might in the past have led to the sequestering of sufferers into the facility. Second, not all patients brought to a medieval leprosarium may in fact have had leprosy. As mentioned previously, a number of other diseases and conditions can mimic symptoms associated with leprosy. In spite of these limitations, a number of insights into the history of leprosy have been made through the careful analysis of medieval skeletal remains.

A common dictum among clinicians is that leprosy is a disease one dies *with* rather than *of*. Modern clinical data, however, indicates that the presence of leprosy is associated with reduced longevity: patients with leprosy die 25 years earlier on average than the population from which they come (Miller 1994; and see Chapter 7 by Boldsen and Milner in this volume, for further discussion of epidemiological approaches and interpretations of the effects of leprosy in medieval populations). Hence, even if the general lifespan in prehistoric and historic populations was shorter than it is today, this would mean for many medieval populations that the disease was ultimately deadly and that it would have impacted the overall mean lifespan of the population. It has been shown that the age at onset of leprosy increases when the incidence of the disease decreases (Irgens et al. 1985): that is, when leprosy occurs infrequently in a population, the age at which a person is likely to contract it increases. Alternatively, a high prevalence of leprosy (and, in fact many other infectious diseases) is associated with low (i.e., younger) age at onset. These epidemiological and demographic correlations correspond with the material analyzed from the Danish leprosarium at Næstved. Here it has been shown that most of the patients had a childhood onset (Ell 1988), which was calculated using the degree of loss of the maxillary bone and comparing these data with the rate of maxillary bone loss in modern leprosy patients from Malaysia (Nah 1985). The high frequency of cribra orbitalia (55 percent of the skulls) found in Næstved (Møller-Christensen 1961) could also indicate a childhood onset in medieval Scandinavia. Cribra orbitalia has been associated with the presence of childhood anemia (Mittler and van Gerven 1994; Stuart-Macadam 1985), and a study of untreated leprosy patients indicated that 19 percent were diagnosed with anemia, while 85.7 percent of the individuals displaying lepromatous leprosy were anemic (Lapinsky 1992).

Further understanding leprosy in the past has been based on comprehensive studies of over 3,000 skeletons from eleven medieval Danish cemeteries, and 99 skeletons from the North Scandinavian medieval site of Westerhus. It was found, for instance, that in urban communities sufferers of leprosy were frequently institutionalized in leprosaria (one was established, for instance, in Odense, Denmark around 1275). In rural communities this did not happen. Rather, it appears that upon death there was an internal segregation of sufferers within rural cemeteries from the rest of the population (Boldsen 2005b). In the early Middle Ages (A.D. 1150–1350) the point prevalence at death (a measure of the proportion of people in a population who have a disease or condition at a particular time) among adults with leprosy in rural villages was higher (25–40 percent) than it was in urban (10–20 percent) communities. Villages situated close to larger towns displayed lower frequencies of leprosy than villages which were situated further away from larger centers (Boldsen 2001). The somewhat counterintuitive pattern discerned from the data is that leprosy declined first in larger towns and cities in the late Middle Ages, and then later in rural communities. Indeed, Muir (1968) described leprosy as a disease of villages (Muir 1968). In Odense, Denmark, and in Malmö, Sweden, both large urban centers, it appears that leprosy was effectively eliminated by 1350, whereas leprosy appears to still be present at Øm Kloster, Denmark (a small community), around 1550. Geographic latitude also appears to play a role in the presence of leprosy, as it is found to be less common in North Scandinavia than in South Scandinavia (Boldsen 2009).

Why would leprosy first decline in areas with dense populations? This is certainly not an expected pattern of many infectious diseases, as large host populations often tend to successfully harbor pathogens more effectively than small populations. In the case of medieval leprosy in Scandinavia, however, it is suggested that the rapid and early decline of leprosy in larger towns and cities was caused by breaking the "chains of infection". This was accomplished by segregating and institutionalizing the most recognizably afflicted sufferers of the disease (Boldsen 2009). In rural communities, however, where it appears that individuals were less likely to be segregated during life even if they displayed signs of leprosy, the pathogen continued to find suitable hosts until, perhaps, a decline of leprosy was brought about as a consequence of the natural cross-immunity between *M. leprae* and a related mycobacterium: the cattle-infecting *Mycobacterium bovis* (Manchester 1991). It has been posited that individuals contracting and waging an immunological battle against *M. bovis* or *M. tuberculosis*, were provided with a natural immunity to *M. leprae*. If inhabitants of rural communities in Scandinavia were more likely and increasingly contracted *M. bovis*, due to their close association with cattle, they might be less likely to contract or suffer from *M. leprae*. Importantly, however, cross-immunity does not appear to work in both directions; meaning that individuals suffering from *M. leprae* appear to have no natural immunity to TB. In fact, infection by *M. tuberculosis* is not rare in leprosy patients, and may indeed be the infection which kills the patient. Mortality due to tuberculosis was found to account for 21 percent (Glaziou et al. 1993) and 26 percent (Nigam et al. 1979; Kumar et al. 1982) of deaths in more recent populations of leprosy suffers. This finding is not new, as it was noted by Hansen and Looft in 1895, that TB was the most frequent complication for leprosy patients. More recently it has been found that infection with *M. leprae* and *tuberculosis* may even increase or decrease in tandem (Wilbur et al. 2002). Based on these observations, and aDNA data from archaeological bone samples from Egypt, Israel, Hungary and Sweden, and spanning Roman to medieval times, Donoghue et al. (2005) propose that the weakened state of the leprosy patients made them especially vulnerable to infection with tuberculosis, and hence led to a speedier death. This, in turn, leads to a marked decline in the number of people suffering from leprosy.

CONCLUSION

A full understanding of the rise and decline of leprosy through time and geographic location remains on our horizon. While there appears to be an important connection between leprosy and tuberculosis, other factors play a major role. Frustratingly little is known about the exact mechanisms of contagion, infection and pathogenesis of leprosy, in spite of the attention it has been given over the past decades. Furthermore, leprosy is still a disease to be reckoned with today regardless of the fact that it is treatable. Perhaps today, being regarded as a disease of developing countries and of times past, it has been removed from the forefront of modern medical research. As such, the paleopathological analysis of leprosy has been relegated to medicohistorical and sociohistorical importance. This is a dangerous misperception of the importance of research into leprosy. The many skeletal collections gathered from leprosaria, with their often unambiguous pathological changes, may perhaps hold the key, by way of modern DNA-analyses and proteinomics, to uncover the mechanisms by which this bacillus emerged as a powerful human pathogen. Understanding the evolution of host–pathogen relationships of leprosy might very well provide clues as to how other bacterial pathogens, responsible for some of the most deadly diseases in human history and today, emerged.

REFERENCES

Andersen, J. G., 1969   Studies in the Medieval Diagnosis of Leprosy in Denmark. Copenhagen: Costers Bogtrykkeri.

Andersen, J. G., and Manchester, K., 1987   Grooving of the Proximal Phalanx in Leprosy: a Palaeopathological and Radiological Study. Journal of Archaeological Science 14:77–82.

Andersen, J. G., and Manchester, K., 1988   Dorsal Tarsal Exostoses in Leprosy: A Palaeopathological and Radiological Study. Journal of Archaeological Science 15:51–56.

Andersen, J. G., and Manchester, K., 1992   The Rhinomaxillary Syndrome in Leprosy. A Clinical, Radiological and Palaeopathological Study. International Journal of Osteoarchaeology 2:121–129.

Andersen, J. G., Manchester, K., and Ali, S., 1992   Diaphyseal Remodeling in Leprosy: A Radiological and Palaeopathological Study. International Journal of Osteoarchaeology 2:211–219.

Andersen, J. G., Manchester, K., and Roberts, C., 1994   Septic Changes in Leprosy: A Clinical, Radiological and Palaeopathological Review. International Journal of Osteoarchaeology 4:21–30.

Aufderheide, A. C., and Rodriguez-Martin, C., 1998   The Cambridge Encyclopedia of Human Paleopathology. pp. 141–154. Cambridge: Cambridge University Press.

Belcastro, M. G., Mariotti, V., Facchini, F., and Dutour, O., 2005   Leprosy in a Skeleton from the 7th-Century Necropolis of Vicenne–Campochiaro (Molise, Italy). International Journal of Osteoarchaeology 15:16–34.

Boldsen, J. L., 2001   Epidemiological Approach to the Paleopathological Diagnosis of Leprosy. American Journal of Physical Anthropology 115(4):380–387.

Boldsen, J. L., 2005a   Leprosy and Mortality in the Medieval Danish Village of Tirup. American Journal of Physical Anthropology 126(2):159–68.

Boldsen, J. L., 2005b   Testing Conditional Independence in Diagnostic Palaeoepidemiology. American Journal of Physical Anthropology 128(3):586–592.

Boldsen, J. L., 2009   Leprosy in Medieval Denmark: Osteological and Epidemiological Analyses. Anthropologischer Anzeiger 67(4):407–425.

Brander, T., and Lynnerup, N., 2002   A Possible Leprosy Hospital in Stubbekøbing, Denmark. In The Past and Present of Leprosy: Archaeological, Historical, Palaeopathological and Clinical Approaches. C. A. Roberts, M. E. Lewis, and K. Manchester, eds. pp 145–148. British Archaeological Reports International Series 1054. Oxford: Archaeopress.

Cole, S. T., Eiglmeier, K., Parkhill, J., James, K. D., Thomson, N. R., Wheeler, P. R., Honoré, N., Garnier, T., Churcher, C., Harris, D., Mungall, K., Basham, D., Brown, D., Chillingworth, T., Connor, R., Davies, R. M., Devlin, K., Duthoy, S., Feltwell, T., Fraser, A., Hamlin, N., Holroyd, S., Hornsby, T., Jagels, K., Lacroix, C., Maclean, J., Moule, S., Murphy, L., Oliver, K., Quail, M. A., Rajandream, M. A., Rutherford, K. M., Rutter, S., Seeger, K., Simon, S., Simmonds, M., Skelton, J., Squares, R., Squares, S., Stevens, K., Taylor, K., Whitehead, S., Woodward, J. R., and Barrell, B. G., 2001   Massive Gene Decay in the Leprosy Bacillus. Nature 409(6823):1007–1011.

Cule, J., 2002   The Stigma of Leprosy: Its Historical Origins and Consequences with Particular Reference to the Laws of Wales. In The Past and Present of Leprosy. Archaeological, Historical, Palaeopathological and Clinical Approaches. C. A. Roberts, M. E. Lewis, and K. Manchester, eds. pp. 149–154. British Archaeological Reports International Series 1054. Oxford: Archaeopress.

Dokladal, M., 2002   The History of Leprosy in the Territory of the Czech Republic. In The Past and Present of Leprosy: Archaeological, Historical, Palaeopathological and Clinical

Approaches. C. A. Roberts, M. E. Lewis, and K. Manchester, eds. pp. 155–156. British Archaeological Reports International Series 1054. Oxford: Archaeopress.

Donoghue, H. D., Marcsik, A., Matheson, C., Vernon, K., Nuorala, E., Molto, J. E., Greenblatt, C. L., and Spigelman, M., 2005   Co-Infection of *Mycobacterium tuberculosis* and *Mycobacterium leprae* in Human Archaeological Samples: A Possible Explanation for the Historical Decline of Leprosy. Proceedings of the Royal Society B: Biological Sciences 272(1561):389–94.

Ehlers, E., 1898   Danske St. Jørgensgaarde i Middelalderen. Bibliotek Laeger 90:243–288, 331–371, 639–644.

Ell, S. R., 1988   Reconstructing the Epidemiology of Medieval Leprosy: Preliminary Efforts with Regard to Scandinavia. Perspectives in Biology and Medicine 31(4):496–506.

Glaziou, P., Cartel, J. L., Moulia-Pelat, J. P., Ngoc, L. N., and Chanteau, S., 1993   Tuberculosis in Leprosy Patients Detected Between 1902 and 1991 in French Polynesia. International Journal of Leprosy and Other Mycobacterial Diseases 61:199–204.

Hansen, G. H. A., 1875   On the Aetiology of Leprosy. British and Foreign Medico-Chirurgical Review, 55:459–489 (from Irgens, 2006).

Hansen, G. H. A., and Looft, C., 1895   Leprosy and Its Clinical and Pathological Aspects. Bristol: John Wright (cit. from Irgens, 2006).

Illarramendi, X., Nery, J. A.C., Vieira, L. M. M., and Sarno, E. N., 2002   Acral Bone Resorption in Multibacillary Patients. Aretrospective Clinical Study. *In* The Past and Present of Leprosy: Archaeological, Historical, Palaeopathological and Clinical Approaches. C. A. Roberts, M. E. Lewis, and K. Manchester, eds. pp. 43–50. BAR International Series 1054. Oxford: Archaeopress.

Irgens, L.M., Skjaerven, R. 1985   Secular Trends in Age at Onset, Sex Ratio, and Type Index in Leprosy Observed During Declining Incidence Rates. American Journal of Epidemiology 122(4):695–705.

Irgens, L. M., 2006   The Discovery of *Mycobacterium leprae*. *In* Leprosy. L. M. Irgens, Y. Nedrebo, S. Sandmo, and A. Skivenes, eds. pp. 33–39. Bergen, Norway: Selia Forlag.

Ishikawa, S., 1997   Osteoporosis Due to Testicular Atrophy in Male Leprosy Patients. Acta Medica Okayama 51(5):279–283.

Ishikawa, S., 1999   Osteoporosis in Male and Female Leprosy Patients. Calcified Tissue International 64:144–147.

Judd, M. A., and Roberts, C. A., 1998   Fracture Patterns at the Medieval Leper Hospital in Chichester. American Journal of Physical Anthropology 105:43–55.

Kadza, J., Irgens, L. M., and Kolk, A. H., 1990   Acid-Fast Bacilli Found in Sphagnum Vegetation of Coastal Norway Containing *Mycobacterium leprae*-Specific Phenolic Glycolipid–I. International Journal of Leprosy and Other Mycobacterial Diseases. 58(2):353–357.

Kirchheimer, W. F., and Storrs, E. E., 1971   Attempts to Establish the Armadillo as a Model for the Study of Leprosy. Report of Lepromatoid Leprosy in an Experimentally Infected Armadillo. International Journal of Leprosy 39:693–702.

Kumar, B., Kaur, S., Kataria, S., and Roy, S. N., 1982   Concomitant Occurrence of Leprosy and Tuberculosis – A Clinical, Bacteriological and Radiological Evaluation. Indian Journal of Leprosy 54:671–676.

Lapinsky, S. E., 1992   Anemia, Iron-Related Measurements and Erythropoietin Levels in Untreated Patients with Active Leprosy. Journal of Internal Medicine 232:273–278.

Lewis, M. E., Roberts, C. A., and Manchester, K., 1995   Inflammatory Bone Changes in Leprous Skeletons from the Medieval Hospital of St. James and St. Mary Magdalene, Chichester, England. International Journal of Leprosy 63(1):77–85.

Likovsky, J., Urbanova, M., Hajek, M., Cerny, V., and Cech, P., 2006   Two Cases of Leprosy from Zatec (Bohemia), Dated to the Turn of the 12th Century and Confirmed by DNA Analysis for *Mycobacterium leprae*. Journal of Archaeological Science 33:1276–1283.

Manchester, K. M., 1991   Tuberculosis and Leprosy: Evidence for Interaction of Disease. *In*. Human Paleopathology: Current Syntheses and Future Options. D. J. Ortner, and A. C. Aufderheide, eds. pp. 23–35. Washington, DC: Smithsonian Institution Press.

Matheson, C. D., Vernon, K. K., Lahti, A., Fratpietro, R., Spigelman, M., et al., 2009   Molecular Exploration of the First-Century Tomb of the Shroud in Akeldama, Jerusalem. PLoS ONE 4(12):e8319. doi:10.1371/journal.pone.0008319

Miller, R. A., 1994   Leprosy (Hansen's Disease). In Harrison's Principles of Internal Medicine. B. Isselbacher, M. Wilson, and K. Fauci, eds. pp. 718–722. New York, McGraw Hill.

Mitchell, P. D., 2002   The Myth of the Spread of Leprosy with the Crusades. In The Past and Present of Leprosy: Archaeological, Historical, Palaeopathological and Clinical Approaches. C. A. Roberts, M. E. Lewis, and K. Manchester, eds. pp. 171–178. British Archaeological Reports International Series 1054. Oxford: Archeopress.

Mittler, D. M., and Van Gerven, D. P., 1994   Developmental, Diachronic, and Demographic Analysis of Cribra Orbitalia in the Medieval Christian Populations of Kulubnarti. American Journal of Physical Anthropology 93(3):287–97.

Møller-Christensen, V., 1953   Location and Excavation of the First Danish Leper Graveyard from the Middle Ages. Bulletin of the History of Medicine 17:112–123.

Møller-Christensen, V., 1958   Bogen om Æbelholt Kloster. København: Dansk Videnskabs forlag.

Møller-Christensen, V., 1961   Bone Changes in Leprosy. Copenhagen: Munksgaard.

Møller-Christensen, V., 1967   Evidence of Leprosy in Earlier Peoples. In Diseases in Antiquity. D. R. Brothwell, and A. T. Sandison, eds. pp. 295–307. Springfield, IL: C. C. Thomas.

Møller-Christensen, V., 1978   Leprosy Changes in the Skull. Odense: Odense University Press.

Møller-Christensen, V., Bakke, S. N., Melsom, R. S., and Waaler, E., 1952   Changes in the Anterior Nasal Spine and the Alveolar Process of the Maxillary Bone in Leprosy. International Journal of Leprosy 20:335–340.

Monot, M., Honoré, N., Garnier, T., Zidane, N., Sherafi, D., Paniz-Mondolfi, A., Matsuoka, M., Taylor, G. M., Donoghue, H. D., Bouwman, A., et al., 2009   Comparative Genomic and Phylogeographic Analysis of Mycobacterium leprae. Nature Genetics 41:1282–1289.

Muir, E., 1968   Relationship of Leprosy to Tuberculosis. Leprosy Review 28:11–19.

Nah, S. H., Marks, S. C., and Subramaniam, K., 1985   Relationship Between the Loss of Maxillary Anterior Alveolar Bone, and the Duration of Untreated Lepromatous Leprosy in Malaysia. Leprosy Review 56:51–55.

Nigam, P., Dubey, A. L., Dayal, S. G., Goyal, B. M., Saxena, H. N., and Samuel, K. C., 1979   The Association of Leprosy and Pulmonary Tuberculosis. Indian Journal of Leprosy 51:65–73.

Ortner, D. J., 2002   Observations on the Pathogenesis of Skeletal Disease in Leprosy. In The Past and Present of Leprosy: Archaeological, Historical, Palaeopathological and Clinical Approaches C. A. Roberts, M. E. Lewis, and K. Manchester, eds. pp. 73–80. British Archaeological Reports International Series 1054. Oxford: Archaeopress.

Ortner, D. J., 2003   Identification of Pathological Conditions in Human Skeletal Remains. New York: Academic Press.

Palfi, G., 1991   The First Osteoarchaeological Evidence of Leprosy in Hungary. International Journal of Osteoarchaeology 1:99–102.

Richards, P., 1960   Leprosy in Scandinavia: A Discussion of Its Origins, Its Survival, and Its Effect on Scandinavian Life over the Course of Nine Centuries. Centaurus 7:101–33.

Robbins, G., Mushrif-Tripathy, V., Misra, V. N., Mohanty, R. K., Shinde, V. S., Gray, K. M., and Schug, M. D., 2009   Ancient Skeletal Evidence for Leprosy in India (2000 B.C.). Public Library of Science ONE 4(5):8.

Roberts, C. A., 2002   The Antiquity of Leprosy in Britain: The Skeletal Evidence. In The Past and Present of Leprosy: Archaeological, Historical, Palaeopathological and Clinical Approaches. C. A. Roberts, M. E. Lewis, and K. Manchester, eds. pp. 213–222. British Archaeological Reports International Series 1054. Oxford: Archaeopress.

Saporta L., and Yuksel, A., 1994   Androgenic Status in Patients with Lepromatous Leprosy. British Journal of Urology 74(2):221–4.

Sharp, D., 2007   Lerposy Lessosn from Old Bones. The Lancet 369:807–8.

Shepard, C. C., 1960   The Experimental Disease that Follows the Injection of Human Leprosy Bacilli into Foot Pads of Mice. Journal of Experimental Medicine 11(2):445–454.

Stanford, J. L., and Standford, C. A., 2002   Leprosy: A Correctable Model of Immunological Perturbation. *In* The Past and Present of Leprosy: Archaeological, Historical, Palaeopathological and Clinical Approaches. C.A. Roberts, M. E. Lewis, and K. Manchester, eds. pp. 25–38. British Archaeological Reports International Series 1054. Oxford: Archaeopress.

Storrs, E. E., 1971   The Nine-Banded Armadillo: A Model for Leprosy and Other Biomedical Research. International Journal of Leprosy and Other Mycobacterial Diseases 39(3):703–14.

Stuart-Macadam, P., 1985   Porotic Hyperostosis: Representative of a Childhood Condition. American Journal of Physical Anthropology 66:391–398.

Trabjerg, L., 1993   Middelalderens Hospitaler i Danmark. Afd. for Middelalder-arkæologi og Middelalder-arkæologisk nyhedsbrev, Den juridiske Faggruppes Trykkeri, Århus Universitet, Denmark.

Wade, H. W., 1960   The Histoid Leproma. International Journal of Leprosy 28:469.

Wade, H. W., 1963   Histoid Variety of Lepromatous Leprosy. International Journal of Leprosy 31:129–142.

WHO (World Health Organization), 2008   World Leprosy Situation. Weekly Epidemiological Record 83:293–300.

Wilbur, A. K., Buikstra, J. E., and Stojanowski, C., 2002   Mycobacterial Diseases in North America: An Epidemiological Test of Chaussinand's Cross-Immunity Hypothesis. *In* The Past and Present of Leprosy. Archaeological, Historical, Palaeopathological and Clinical Approaches. C. A. Roberts, M. E. Lewis, and K. Manchester, eds. pp. 247–258. British Archaeological Reports International Series 1054. Oxford: Archaeopress.

CHAPTER 26 Treponematosis: Past, Present, and Future

*Della Collins Cook
and Mary Lucas Powell*

INTRODUCTION

From the early mid-20th century onwards, four distinct treponemal syndromes—pinta, yaws, bejel and syphilis, known collectively as the treponematoses—have been distinguished in the clinical and epidemiological literature (Hackett 1963). The identification of these four closely related (but not identical) disease entities emerged in the early 1940s, at a time when the prospects for antibiotic control of the treponematoses and other infectious diseases seemed bright. The four-syndrome model was first formulated by Cecil James Hackett (1905–1995), a physician who worked with aboriginal groups in his native Australia before World War II and took part in the WHO eradication campaign against yaws in Uganda, Indonesia, and elsewhere. After he retired from WHO in 1965, he became an affiliate of the Wellcome Museum in London and devoted the remainder of his career to the study of bone lesions of treponemal disease (e.g., Hackett 1976), making numerous important contributions to paleopathology and medical history (Fairley 1994). Hackett's model of the biological relationship of these four syndromes became the standard consensus view of the treponematoses in the 20th century in both the clinical literature and in paleopathology (e.g., Skerman et al. 1980; Brothwell 1981; Baker and Armelagos 1988; Kiple 1993a; Dutour et al. 1994; Aufderheide and Rodríguez-Martín 1998; Ortner 2003). However, new discoveries pertaining to the genetics and epidemiology of the treponematoses around the beginning of the 21st century prompt reconsideration of this classic model.

A Companion to Paleopathology, First Edition. Edited by Anne L. Grauer.
© 2012 John Wiley & Sons, Ltd. Published 2016 by John Wiley & Sons, Ltd.

Each one of the four syndromes is caused by some form of the pathogenic bacterium, *Treponema pallidum*, an organism with a single long flagellum that is highly mobile under a dark-field microscope. These different forms are currently classified as subspecies (Skerman et al. 1980; Centurion-Lara et al. 2000). All four syndromes manifest as chronic infections that begin with a small unobtrusive skin lesion; if untreated, each may progress through increasingly harmful secondary and tertiary stages (see Table 26.1)(Kiple 1993b; and see Powell and Cook (2005) for more detailed descriptions of the syndromes). Variable degrees of cross-immunity have been claimed among the four modern syndromes but not conclusively documented (Schell and Musher 1983). All four are easily cured by early treatment with long-acting penicillin or with tetracycline or chloramphenicol in patients allergic to penicillin. Fortunately for modern sufferers, *T. pallidum* displays a remarkable immunologic stability, in sharp contrast to the drug resistance skillfully developed by *Mycobacterium tuberculosis* (see Roberts, Chapter 24 in this volume, for more detailed information about tuberculosis), *Staphylococcus aureus*, and other pathogens. Syphilis has shown almost no altered response to penicillin since its first therapeutic use in 1942 (Centurion-Lara et al. 1998; Meheus and Tikhomirov 1999). However, the recent emergence of strains of syphilis with antibiotic resistance (Stamm 2010) suggests that this fortunate circumstance may be ending.

Human hosts have no natural immunity to pathogenic treponemes (Kiple 1993b), in contrast to the many genetic adaptations that protect against other diseases, such as malaria. However, the development of clinical symptoms after exposure depends upon multiple factors, including postcontact hygiene and size of the inoculum. Recent epidemiological studies suggest that there may be 2½ million people infected with one of the endemic treponematoses worldwide (Antal et al. 2002), and as many as 260,000 persons may suffer disability as a result of these infections (Meheus and Tikhomirov 1999). In comparison, there were about 12 million new cases of venereal syphilis worldwide in 1999 (Stamm 1999).

Pinta

Pinta has historically been the most geographically limited of the treponematoses, and an energetic public health effort employing antibiotic therapy in the mid-20th century may well have rendered it extinct. It was formerly restricted to relatively isolated communities in lowland Central and South America. Regional names include "mal de pinto" (spotted sickness), "carate," "tiña," "empeines," and "puru-puru": all terms describing the characteristic dark and light skin lesions.

Two geographically regional manifestations of pinta are particularly well-documented. "Puru-puru" has been documented among certain ethnic groups living in a very moist riverine environment in the Rio Negro region of the Amazon from as early as 1775 to the recent past. This skin disease was so geographically localized that it served as an ethnic marker and was intentionally transmitted through tattooing (Biocca 1945; Guimares and Rodrigues 1948). We cannot help but regret the cultural losses surrounding its extinction: "The Paumari evoke the curing of the pinta skin infection by the missionaries as a historical-temporal landmark that enabled the inauguration of a 'new' era marked by compliance with Christian morality, cultivated by the Evangelicals, and the consequent rupture with the 'old culture' (Bonilla

Table 26.1 Pathological changes caused by treponemal syndromes

Treponemal syndrome	Bone lesions	Skin/mucosal tissue lesions	Other organ systems affected	Ref #
Pinta				
Secondary	None	Papules, pintids, generalized rash	Regional lymphadenopathy	12, 13
Tertiary	None	Depigmented patches; pintids (rare)	Regional lymphadenopathy	12, 13
Yaws				
Secondary	Distal limb segments; hands and feet; maxilla	Papules, papilomas, granulomas; macular lesions (depigmented or hyperpigmented); palmar or plantar lesions; condylomatous lesions in moist regions of the body	Regional lymphadenopathy	1
Tertiary	Tibia (46%), fibula (20%), femur (13%), ulna (10%), humerus (9%), radius (7%), spine (5%), clavicle (4%), hands and feet (4%), skull vault (3%), ribs (3%), pelvis (2%); joints (mainly in children)	Nodular, lupoid skin lesions with granulation and depigmentation, persistent gummatous ulcers may invite bacterial and/or fungal superinfection; destructive nasal and palate lesions (nasopharyngitis mutilans) may cause *gangosa*	Regional lymphadenopathy	2, 3
Endemic Syphilis				
Secondary	Tibia most common, often with ostalgia; dactylitis (rare)	Oropharyngeal mucosal lesions, oral papules, condylomatous lesions in moist regions of the body	Regional lymphadenopathy	4,5
Tertiary	Tibia, fibula (67%); ulna, radius (18%); hands, feet (4%); joints (6%); knee most frequent; frontal (4%), palate & nose (rare)	Gummateous lesions of the skin and nasopharynx	None	6,7
Venereal Syphilis				
Secondary	Periostitis and osteitis, often with ostalgia	Macular, pustular, and papular rashes on the skin and genitals, condyloma lata, mucous patches, erosions, and ulcers of the oropharyngeal cavity, alopecia	Fever of unknown origin, lymphadenopathy, anorexia	11

Tertiary	Distal limb segments commonly affected; skull and vertebral lesions less so; gummas and Charcot joints (rare)	Gummas (granulomatous lesions with coagulated necrotic centers) of the skin	Gummas in the liver or spleen; cardiovascular system (aortitis and aneurysms); CNS (hemiplegia, hemiparesis, aphasia, seizures, syphilitic dementia, tabes dorsalis and other locomotor and sensory symptoms; congenital transmission	11
Congenital Syphilis				
Early (0–4 years)	Bones and joints (17%): generalized periostitis, osteitis, and osteochondritis	Diffuse macropapular rash	Chronic inflammation of orofacial region (rhinitis); anemia, hepatosplenomegaly, jaundice, neurological problems, and lymphadenopathy	8, 14
Late (+4 years)	Bones and joints (15–28%). In adults, tibia bowing (4%), cranial bossing (87%), medial clavicle shaft expanded (39%), saddle nose (73%), Hutchinson's incisors, Moon's molars			8,9, 10

References
1. Maegraith 1965:218–219
2. Hunter et al. 1966: 134
3. Goldmann and Smith 1943
4. Murray et al 1956: 1006; Hudson 1958: 77; Csonka 1953: 98
5. Hudson 1958: 78
6. Murray et al 1956: 1001, 1006
7. Hudson 1958: 108–109
8. Moore 1941: 503–504
9. Fiumara and Lessell 1970
10. Vogt 1931
11. Howles 1943: 373; Aegerter and Kirkpatrick 1968: 304
12. Brothwell 1993a
13. Chulay 1990
14. Tramont 1990

2009:135). High levels of immune response to tests for treponematosis in persons without lesions elsewhere in the Amazon suggest that there may yet be strains of pinta that are even less pathogenic than "puru-puru" (Lee et al. 1978). In contrast to this strikingly autochthonous account, pinta in lowland Peru was an occupational disease of fishermen, workers in sugar estates and a few other rural occupations, and is reported to have low levels of cardiovascular and arthritic complications. Pinta was described in the earliest colonial records in many ethnic groups, not all of them indigenous to Peru (Weiss 1947). In Peru as well as in Mexico, pinta has been diagnosed by physicians from symptoms depicted in Pre-Columbian art objects (Weiss 1947;Verut 1973).

Three features of pinta are important in the context of paleopathology. First, it is, or was, a skin disease lacking the osseous, cardiovascular, and neurological complications of the other treponematoses. Second, it proved even more difficult to transmit to lab animals than the other pathogenic treponemes, so that cultures of pinta are not available for modern phylogenetic studies. Third, cross-immunity among the treponematoses may have made pinta a factor in constraining susceptibility to the other treponematoses in the small, isolated populations that served as hosts. We deeply regret that a case described in recent travel medicine was not cultured, as the genetic data would been valuable for comparisons with examples of the other modern syndromes (Woltsche-Kahr et al. 1999).

Yaws

Yaws is a treponemal disease common to tropical regions world-wide, known by many different local names including parangi, pian, bouba, and framboesia. Despite extensive eradication campaigns undertaken by the World Health Organization in the 1950s, it remains endemic in several isolated localities (Fegan et al. 2010; Satter and Tokarz 2010). The prevalence of bone lesions in late secondary or tertiary yaws is estimated at 1–5 percent (Table 26.1) (Aufderheide and Rodríguez-Martín 1998; Ortner 2003).

The best descriptions of yaws in bone are provided by Hackett (1951) from his clinical work in Uganda. Yaws produces severely destructive ulcerative lesions (nasopharyngitis mutilans) which may penetrate the nasal cavity and the palate and alter the resonance of the nasopharyngeal region, giving a harsh quality to the voice; this particular form of pathology is called "gangosa," from the Spanish word "gangoso" meaning "a twangy, nasal voice." Along with physical changes to the sufferer, facial deformities may alter the social identity of the victim, as is witnessed by their depiction in African masks in the early 20th century (Simmons 1957). The clinical literature includes assessments of the social costs of yaws deformity (Pandey and Roy 2002), and it has recently been reported that gorillas similarly handicapped by facial lesions of yaws are reproductively compromised (Levréro et al. 2007). Concepts of disability and social stigma have received little attention in the paleopathology of yaws, but a recent paper by Buckley and Tayles (2003) explores this aspect of ancient lives in the Pacific.

Diagnosing yaws in the archaeological record is not easy. The eccentric (and tibiacentric) methods of diagnosis advocated by Rothschild and Rothschild (1995) (see Chapter 12 by Powell and Cook, this volume), have deflected diagnostic attention from the distinctive facial lesions of yaws. For instance, diagnosing yaws in an australopithecine solely on the basis of periosteal thickening (Rothschild et al. 1995) is problematic, as many other systemic conditions are plausible in this case. Similarly, in a recent publication by Phillips and Sivilich (2006), the authors discuss palate perforation with extensive scarring of the nasal aperture in a Late Woodland male from Indiana. The cause, they assert, is cleft palate, regardless of the fact that this type of destructive nasal lesion is common in yaws. In this specimen, dental occlusion is clearly normal, an impossibility in untreated cleft palate, leading us to agree with Brothwell (2010) that this case is better diagnosed as endemic treponematosis.

Endemic (nonvenereal) syphilis

Endemic syphilis was widespread throughout Europe (e.g., Anderson et al. 1986), the Middle East (Hudson 1958), and dry temperate regions in Africa (Grin 1956; Willcox 1955) and Australia (Domett et al. 2006) until advances in medical treatment and hygiene in the 19th century began its slow eradication. After World War II, the World Health Organization undertook a vigorous campaign to eliminate it from southwestern Europe (Guthe et al. 1972), but it is still seen in some dry areas of the Middle East and in central and southern Africa today (Tabbara 1990; Meheus and Tikhomirov 1999).

Endemic syphilis has had many local names, including "sibbens" in Scotland, "button scurvy" in Ireland, "radesyge" in Norway, "saltfluss" in Sweden, "spirocolon" in Greece and Russia, "skerljevo" and "frenga" in the Balkans "ukwekwe" in South Africa, "dichuchwa" in Botswana, "njovera" in Zimbabwe, and "irkintja" in Australia. The Greek example appears to have been limited to a brief outbreak during the war of independence (1820–25), and "skrljevo" is a local place name, more familiar today as Sarajevo, attached to a similar early-19th-century outbreak (Slavec 1996). Sources behind this litany of names are cited in Willcox's careful reviews (Willcox 1955, 1960).

Sibbens in northwest Europe is a particularly interesting example. Historical accounts of this disease are particularly rich (Pollock 1953, among others), and show that it seems to have followed Cromwell's army across Britain and persisted until the recent past (Morton 1967). Canadian physicians familiar with sibbens recognized the late 18th Century epidemic of *mal de la Baie St. Paul* in Quebec province as a local form of endemic syphilis, "disseminated extragenitally and innocently" within families by asexual contacts, and P. W. Matthews, medical officer to the Hudson's Bay Company suggested that it may have come to northeastern Canada along with the large numbers of Company employees recruited from northern Scotland (Horne (2005). This behavior suggests a burnt-out or endemicized transformation of venereal syphilis more than it does an ancestral endemic species.

Victims of endemic syphilis may be infected through skin-to-skin contact or by mechanical transmission of pathogens by insects (Chulay 1990). After the initial pathology subsides, a latent period of months or years may be followed by tertiary-stage gummatous lesions of the skin, nasopharynx, and bones (Table 26.1) (Csonka and Pace 1985; Erdelyi and Molla 1984). The best account of bejel in dry bones remains Marcus Goldstein's description of a small collection of recent Bedouin remains from the Negev in Israel (Goldstein et al. 1976). Goldstein had the advantage of prior experience with diagnosing treponemal skeletal lesions in Pre-Columbian archaeological series from Texas (Goldstein 1957). In contrast, Neolithic remains from Israel generally lack evidence for bejel (Arensberg et al. 1986).

Ellis Herndon Hudson (1890–1992), a physician and the son of American missionaries, wrote extensively on bejel, based on his long service at a Presbyterian mission hospital that he founded at Deir-es-Zor, Syria. In his clinical practice he noted two distinct patterns: urban residents were typically afflicted with venereal syphilis while the nomadic rural Bedouins displayed an endemic form acquired through asexual contacts (e.g., sharing contaminated dishes or eating utensils, drinking vessels, pipes, toothpicks, cigarettes, or towels harboring treponemal pathogens shed from lesions in the oral mucosa and nasopharyngeal tissues shared drinking vessels, etc.). The Bedouins' voluminous robes covered almost all of their bodies, leaving few opportunities for infection through skin-to-skin contact, as in yaws. Deir-es-Zor, located on the Euphrates River, is less arid in winter than the stereotype of very dry environments associated with bejel (e.g., the Negev (Goldstein et al. 1976)) and is thus more similar to areas in temperate Europe where endemic treponematoses had flourished in previous centuries. Hudson's work illustrates many destructive lesions of the nasopharynx as well as tertiary long bone lesions, and most "textbook" illustrations of bejel come from his work. His numerous authoritative publications (Hudson 1932–33, 1958) firmly established the Arabic name for this disease, "bejel," in the medical and anthropological literature.

Venereal syphilis

Tramont (1990) describes venereal syphilis as "a complex systemic illness with protean clinical manifestations caused by the spirochete *Treponema pallidum* subsp. *pallidum*. It holds a special place in the history of medicine as the "great imitator" or the "great imposter." It is most often transmitted by sexual contact, and unlike most other infectious diseases, it is rarely diagnosed by isolation and characterization of the causative organism." (Tramont 1990:1794). It is, by far, the most dangerous of the four modern treponemal syndromes because unlike pinta, yaws, and endemic (nonvenereal) syphilis, its ravages are not restricted to the skin, mucosal tissues, and the skeletal system but may invade any organ system and even cross the placenta to infect a developing fetus (Table 26.1).

Congenital infection contributes substantially to high fetal and perinatal mortality through miscarriages, premature births, and severe illness in the first weeks of life. By the mid-19th century, European physicians had observed empirically that pregnant women infected with venereal syphilis shortly before or after conception were far more likely to pass the infection to their fetuses than were women whose infections were several years old; as a result, married women whose husbands had been diagnosed with syphilis were routinely advised to avoid pregnancy for at least five years after the diagnosis to minimize the risk of congenital infection (Quétel 1990). In his excellent review of the clinical aspects and epidemiology of yaws and endemic syphilis, Grin (1956) argued that the absence of documented congenital transmission in the endemic treponemal syndromes reflects the fact that they are typically contracted a decade or more before puberty; as a consequence, the pathogen load is very low or absent in previously infected young women at the time of their first pregnancies (see also

Willcox 1955). His model suggests that there should be rare cases of congenital transmission in previously unexposed pregnant women. However, recent molecular comparisons of the pathogens associated with yaws and venereal syphilis indicate that both genetic differences in the organisms themselves and genetic differences among lab animal strains influence the age structure of susceptibility to infection, suggesting that the biological situation is more complicated than Grin imagined (Wicher et al. 2000) and that transplacental infection is an intrinsic characteristic of *T. pallidum* spp.

As many as one-third of untreated patients with venereal syphilis may develop devastating tertiary-stage symptoms such as Charcot joints, gummas (granulomatous lesions with coagulated necrotic centers) of the skin, skeletal system, liver, or spleen, or lesions in the cardiovascular system (e.g., aortic aneurysm, etc.) or central nervous system (e.g., dementia, paresis, etc.). Although tertiary syphilis has been claimed as the culprit behind the "madness" of numerous historical figures (Gilman 1985, 1988), only two of the modern strains of venereal syphilis have been associated clinically with symptoms of neurosyphilis and persistence of the organism in the cerebrospinous fluid (Stamm 2010), a fascinating enigma.

THEORIES OF EVOLUTION OF THE TREPONEMATOSES

Four theories—the Columbian, the Pre-Columbian, the Unitarian, and the Evolutionary—have been proposed to account for this cluster of closely related disease syndromes, and in particular, the appearance of venereal syphilis. The medical historian Claude Quétel provides an extensive discussion of these theories in his *History of Syphilis* (1990), as do Powell and Cook (2005).

The *Columbian Hypothesis* posited that venereal syphilis originated in the New World and was introduced to the Old World through Columbus's voyages of discovery to the Americas at the end of the 15th century. This theory was advanced by Oviedo, De Isla, and las Casas and other 16th-century physicians and historians who described the explosive outbreak of a supposedly new virulent sexually transmitted disease in southern Europe in 1495 (Quétel 1990). However, not all 16th-century physicians were persuaded that this "new" disease was either really new or that it came from the New World: their opinions formed the basis of the countervailing *Pre-Columbian Hypothesis*, which argues that one or more forms of treponemal disease, including venereal syphilis, existed throughout the Old World prior to 1492 but was diagnostically confused with other diseases, particularly with one called "venereal leprosy."

In the mid-20th century, this second hypothesis was favored by C. J. Hackett and R. R. Wilcox (mentioned above) and T. Aidan Cockburn (1961, 1963), physicians whose extensive clinical and epidemiological experience and interest in the evolution of infectious diseases persuaded them that various forms of treponemal disease had afflicted humans worldwide for at least several millennia.

A third hypothesis, the *Unitarian Theory*, was formulated by E. H. Hudson (mentioned above for his research on endemic syphilis); it posits that the four modern syndromes are not in fact distinct diseases but instead are variable environmentally determined expressions of "a single and extremely flexible disease whose permutations are directly related to man's physical and cultural status …" (Hudson 1965:888). However, paleopathological and biomolecular research undertaken since Hudson's time on past and present varieties of treponemal disease has revealed subtleties of clinical and genetic variation that do not match his model, as, for example, Arturo Centurion-Lara's discoveries (1998, 2000) that the various strains of *T. pallidum* differ in their genetic profiles, which may influence specific tissue pathogenicity and virulence.

The most recent hypothesis, called the *Evolutionary Theory* was formulated by C. J. Hackett (1963), who proposed two alternative scenarios. In one, pinta represents a modern geographically restricted survival of the earliest world-wide form of treponematosis, predating the human occupation of the New World, the biological precursor of yaws in the tropics and endemic syphilis in temperate regions. Venereal syphilis, the most recent of the syndromes to appear, represents the pathogen's successful adaptation to a decrease in endemic transmission due to specific cultural changes: improved hygiene and the adoption of clothing, which protected against promiscuous skin-to-skin contacts, and increased "sexual laxity." Hackett's alternative hypothesis also placed pinta as the earliest human treponemal disease, but viewed yaws as a somewhat later parallel mutation from the original ancestral pathogenic organism, which in turn gave rise to endemic and venereal syphilis. Hackett's figure illustrating his hypotheses (1963:25) features two little dragons, each anchoring a vertical pole that bears a series of banners representing the different phylogenetic relationships of the four modern treponematoses. Brothwell (1981) expanded upon Hackett's model, proposing instead that the four modern syndromes emerged at different times in the New World, Asia, Africa, and Europe; his model is visualized as a sort of hydra with seven heads. In his most recent assessment of this complicated phylogenetic conundrum, Brothwell commented "If I were a betting man, armed with our knowledge of ancient treponemal pathology I would suggest that the genetic picture for the pathogenetic treponemes (to be confirmed or refuted by future DNA studies) would turn out to be far more complex and variable than previously imagined." (Brothwell 2005:490). We agree with Brothwell that this beast has, and has had, *many* heads, and we cannot know just how many without much more extensive sampling of the surviving strains. The best-equipped hunters for this many-headed dragon are our colleagues in tropical medicine.

THE NEW HISTORY AND THE NEW BIOLOGY

The two oldest hypotheses, the Columbian and the Pre-Columbian, are explicitly "syphiliocentric" and "Eurocentric," focused specifically on the circumstances surrounding the first appearance of venereal syphilis in Europe. Some historians of medicine have recently questioned whether some or all of these scenarios may not be largely social constructs. For example, Sander Gilman has seen the question of the origins of the treponematoses as a reflection of a 19th-century European obsession with race, purity, pollution, degeneracy, and disease (1985), and he has pointed out profound changes in the gender symbolism of the iconography of syphilis from the 15th to the 19th centuries (1988). Other historians have viewed the epidemiological literature on the treponematoses as heavily influenced by cultural stereotypes that contrast rural innocence (represented by the non-venereally transmitted treponemal syndromes) with urban depravity (represented by sexually transmitted venereal syphilis) (Engelstein 1986; Vaughn 1992; Solomon 1993). Many of the older accounts of congenital syphilis do partake of the notion of a "hereditary taint" that has more to do with "degenerationism," a miasma of negative ideas about the innate immorality of the "lower classes" (or races, or castes) than with any biologically based concept of infection (Lomax 1979). Local and national narratives of race, class, and power also inform scientific understandings of disease, as Jochelson has shown in southern Africa (2001), Vaughn in East and Central Africa (1992), Engelstein in the Russian dependencies in Central Asia (1986), Solomon for Russian colonial medicine in the Buriat region (1993) and Chaplin for colonial North America (1997). Each of these studies portrays endemic syphilis, often misidentified as the venereal form (as described in Vaughn (1992)) as a disease that separates an educated European elite from native peoples who were disparaged as degenerate, filthy, promiscuous and negligent in childcare. In both medicine and paleopathology, a narrowly positivistic approach to evidence for disease is the norm; nevertheless, we should keep in mind that historical accounts of the various treponematoses are inevitably products of their social context, and so are paleopathologists' accounts of ancient peoples and their diseases.

The genus *Treponema* is very large; it includes a number of free-living saprophytic species and a few zoonotic species, as well as several species that are oral commensals in humans. These organisms are distantly related to *T. pallidum* and *T. carateum*, and they are not included in the concept of treponematosis we are using here. Until very recently, chemical and microbiological analyses of treponemes associated with the four modern syndromes failed to reveal consistent morphological and serological differences indicative of species or subspecies variations (Turner and Hollander 1957; Schell and Musher 1983). Part of this failure may reflect errors during the many decades of curation of the laboratory strains of these difficult-to-maintain organisms (Wicher et al. 2000). However, comparisons of DNA sequences within and between different pathogenic *Treponema* organisms (e.g., Hardham et al. 1997, Centurion-Lara et al. 1998, 2000), distantly related "nonpathogenic" and "cultivable" treponemes, and more distantly related spirochetes such as *Leptospira* and *Borrelia* (Schmid 1989) have now begun to reveal microvariations that may affect pathogenicity. If so, the different pathological profiles of the modern syndromes may indeed be keyed to specific genetic mutations, e.g., the ability of *Treponema pallidum* subsp. *pallidum* to survive at higher host-tissue temperatures than *Treponema carateum* may explain why venereal syphilis, unlike pinta, is able to affect internal organ systems and is therefore far more dangerous to its hosts.

The new genome-centered bacterial systematics is characterizing new species of *Treponema* at a very rapid rate, as a check of the National Library of Medicine's Taxonomy Browser will attest. New organisms from human dental plaque and periodontal lesions, from coronet band lesions in cows, and from sheep's and cow's rumens, have joined *Treponema paraluiscuniculi*, a venereal disease of rabbits, and the simian treponemes as members of the genus. It is noteworthy that the range of mammals from which treponemes have been isolated is quite narrow, and no species with carnivores as hosts appear in the literature.

As yet, there appears to be no comprehensive study of molecular phylogeny revealing the relationships among these organisms. The complete sequencing of the genome of *Treponema pallidum* subspecies *pallidum* (Fraser et al. 1998) has already shed light on the biological affinities between *T. pallidum* subsp. *pallidum* and the related spirochete *Borrelia burgdorferi*, the agent of Lyme disease, and should greatly facilitate direct comparisons with *T. carateum, T. pallidum* subspecies *endemicum*, and *T. pallidum* subspecies *pertenue*. Efforts to isolate and characterize organisms responsible for pinta, bejel, and the apparently nonpathogenic infections, as well as additional strains from yaws patients, are needed before many of the remaining questions about the human treponematoses can be answered. It is tempting to interpret a recent demonstration of remarkably high strain variability in *T. pallidum* in South Africa as evidence for survival of endemic nonvenereal syphilis (Pillay et al. 2002).

The Paleopathology of Treponematosis: Where Have We Been, Where Are We Now, and Where Shall We Go in the Future?

Much of the clinical literature on treponemal disease that is most useful to paleopathologists comes from the era *before* the introduction of effective antibiotics in the developed world, and the World Health Organization's eradication campaigns in the 1950s–1970s also provided invaluable epidemiological data for modeling premodern treponematoses. If what we are doing in paleopathology is science, we must set out objective criteria for identifying diseases in ancient human remains and make our identification procedures explicit.

The paleopathological identification of treponemal disease faces two problems: first, the diagnostic distinctions between treponematosis and *other* diseases (e.g., osteomyelitis) which affect the skeletal system in similar ways, and second, the similarity of the osseous lesions produced by the three modern treponemal syndromes that affect bone. Both problems have plagued researchers attempting to trace the antiquity of specific syndromes in specific geographical settings. Hackett (1976), Ortner (2003), and Aufderheide and Rodríguez-Martín (1998) provide valuable guides for differential diagnosis of treponematosis vs. similar skeletal pathology, and Powell (1992) has contrasted significant differences in skeletal pathology and biological costs of endemic treponematosis and tuberculosis in Late Prehistoric populations in the southeastern USA. Steinbock (1976: Figures 41, 58) illustrated the contrasting bone lesion patterns for venereal syphilis and yaws in a particularly useful form. Frequency data for the distinctive dental lesions in congenital syphilis are reviewed by Jacobi and colleagues (1992); while reported frequencies and diagnostic criteria vary considerably from study to study, Moon molars and Hutchinson incisors are sufficiently common to be useful evidence for congenital syphilis in paleopathology (Hillson et al. 1998), and should be present in large skeletal series if congenital transmission was common. Neurotrophic joint involvement in late congenital syphilis and tertiary acquired syphilis distinguishes these conditions from the other treponematoses, but can occur in other conditions that result in sensory impairment, including diabetes, syringomyelia, leprosy, and pernicious anemia (Aegerter and Kirkpatrick 1968:752–753; Sommer and Lee 2001). Syphilitic aortitis is likewise restricted to late venereal syphilis among the treponematoses, but there are many other causes of enlargement of the aorta (Kelley 1979).

In 1976, in his landmark study of the problem of differentiating among the modern treponematoses in dry bones, Hackett described the most common destructive and proliferative lesions appearing in cranial and post-cranial elements, and identified a series of cranial lesions (the "caries sicca" sequence) characteristic of progressive involvement of the calvarium. The final, most distinctive stages of this sequence had been nominated by the eminent 19th-century pathologist Rudolf Virchow (1858, 1896) as pathognomonic of venereal syphilis. Hackett accepted Virchow's judgment: "If a master criterion of syphilis were requested, the contiguous series of the caries sicca sequence in the calvariae would provide the answer." (Hackett 1976:112). Despite Hackett's efforts, the term caries sicca has disappeared from the medical literature since the mid-20th century, and persists only in the literature of paleopathology.

Hackett's perspective on paleopathological diagnosis may be termed "processual," because he emphasized that the varying morphology of cranial lesions represents *progressive* stages of the infection; thus, the forms of the lesions are transformed one into another through time in the natural course of the disease. He cautioned emphatically against attempts to differentiate syphilis, yaws, and treponarid (endemic syphilis) based upon isolated dry bone specimens, noting in his Summary that "The bone lesions of these three closely related diseases cannot at present be separated" although they may be useful for distinguishing treponematosis as a category of disease from other diseases such as tuberculosis, pyogenic osteomyelitis, chronic ulcers, or Paget's disease (Hackett 1976:113). The naivety of the claim made in some paleopathological reports that, in effect, "periostitis = treponematosis" (e.g., Rothschild and Rothschild 1995, and *contra*, Weston 2008) is immediately apparent when viewed in the light of Hackett's detailed description of the progressive osteopathology of this complex disease.

Questions remain as to the role played by mercury therapy, widely used throughout Europe since medieval times to treat skin lesions of venereal syphilis and other diseases (Quétel 1990) in producing the florid cranial lesions of caries sicca (Ortner 2003). In addition, the overlap of caries sicca lesions and healed scalping lesions (Ortner 2003) poses a particular diagnostic problem for paleopathologists working in North America.

In paleopathology a case study by Anderson and colleagues from 19th-century Norway (Anderson et al. 1986) is a particularly good example of how fitting historical accounts of sibbens to bone lesions can enrich our understanding of the treponematoses. Exciting recent research in Britain (von Hunnius et al. 2006; Cole and Waldron 2011) and France (Molnar et al. 1998) has focused on establishing Pre-Columbian syphilis in Britain, but these analyses might benefit from attention to sibbens, and to whether it is an ancient endemic treponematosis or more recent introduction. The heated argument over the dating of the British specimens would be moot if earlier contact, for example via the cod trade, is considered (see Fagan 2006).

Endemic syphilis as a unifying concept has a problematic relationship to the various theories of the evolutionary history of the treponematoses, and paleopathology provides the only means of testing them. European physicians of the 16th century recognized that venereal syphilis could be spread by asexual contact, e.g., an infant born with congenital syphilis could infect its wet nurse through minor skin trauma to her nipple, and that same wet nurse could subsequently infect her own healthy infant from her breast lesion (Quétel 1990); some four centuries later, Eisenberg and colleagues (1949) and Luger (1972) documented examples of asexual transmission of syphilis in mid-20th-century Chicago and Vienna. If endemic syphilis is simply a burnt-out form of venereal syphilis transmitted through casual contact under conditions of poor hygiene, endemic syphilis should be genetically similar throughout its extensive geographical range. However, if the local variants are ancient and ancestral, then they should be differentiated in their symptoms, reflecting thousands of years of adaptation to local populations, and there should be many species or subspecies, many of them extinct. We find it parsimonious to interpret the interesting case of putative yaws recently reported from Guyana (Harper 2008, Mulligan 2008) as an endemicized or burnt-out syphilis, rather than an endemic treponematosis. An ambitious Bayesian interpretation (de Melo et al. 2010) of the genetic data, still wedded to the current taxonomy of the treponematoses, reaches a very interesting conclusion regarding the date of divergence of the strains (16,500 to 5,000 years before present). How these dates fit the puzzles of history and epidemiology is still problematic, but broadening the concept of the treponematoses and seeking more samples for genetic analysis certainly seem desirable.

Molecular Paleopathology

A new methodology now increasingly applied in the study of ancient diseases is the search for biomolecular residues of host–pathogen interactions in the past. Tuberculosis is an ideal subject for such investigations, because human hosts who exhibit recognizable skeletal pathology harbor a relatively high mycobacterial pathogen load within the diagnostic bone lesions. Unfortunately, treponematosis presents a diametrically opposite clinical profile as regards pathogen loads: the diagnostic bone lesions typically accompany the tertiary stage of disease, at a time when the actual treponemal pathogen load in the host's body is relatively low. TB organisms protect themselves with a resistant, waxy coat; treponemes are notoriously fragile. Recovery of ancient treponemal DNA from ancient human skeletal material is, therefore, far more problematical. There is just one successful demonstration of isolation of ancient DNA in a treponematosis case, and that case is only 200 years old (Kolman et al. 1999; von Hunnius et al. 2007).

A more promising source of molecular evidence for prehistoric treponematosis is the identification of distinctive immunoglobulins from ancient bone. Two such studies (Fornaciari et al. 1989; Ortner et al. (1992) have linked treponemal soft tissue and skeletal pathology with specific immunological products in human hosts. We regard the result of an earlier attempt by Rothschild and Turnbull (1987) to use off-the-shelf immunological tests on bone lesions in a fossil bear as a "false positive" outcome because it lacked the positive and negative controls needed when extending a clinical test to fossil bones from a mammalian order for which (a) the test was not designed and in which (b) there are no known treponematoses (see also Neiberger 1984, 1988; Edwards et al. 2003; Valdez et al. 2000)! The recent announcement of yaws in gorillas (Levréro et al. 2007) merits similar scrutiny; not all destructive lesions of the nasal cavity are caused by treponemes, even if treponemes are present, and the specificity of clinical tests for syphilis in gorillas that perhaps carry several other pathogens may be questioned. Ideally, these lesions ought to be cultured to demonstrate treponemes, and the resulting cultures should be evaluated phylogenetically in rigorous protocols like those of Centurion-Lara and colleagues (1998, 2000).

Diagnosis and interpretation: not too little, not too much

In an earlier chapter in this volume, the editor invited us to speculate on the question, "How Does The History of Paleopathology Predict Its Future?" In that chapter, we addressed two very different (but, in our eyes, equally problematic) recently published approaches to the paleopathological interpretation of skeletal lesions representative of treponemal disease. The first approach is, to our minds, unrealistically narrow (Rothschild and Rothschild 1995), focused as it is on a single bone, the tibia, and employing a set of diagnostic criteria called SPIRAL that are drawn from problematic samples.

The second approach we find, conversely, unrealistically broad (Goodman and Martin 2002, in Steckel and Rose 2002:33; Armelagos and van Gerven 2003), as it views a broad range of proliferative bone lesions as "nonspecific stress markers" that signal the presence of the otherwise undifferentiated pathological category, "infection." These authors view attempts at differential diagnosis as futile because "Bones and teeth do not often respond with the kind of specificity necessary for a clinical diagnostic approach to all diseases." (Armelagos and van Gerven 2003:59). However, treponematosis is one specific infectious disease that *does* produce distinctively diagnostic skeletal pathology, and moreover, its various syndromes have imposed differential burdens of morbidity and mortality upon human populations for at least several millennia. While we readily acknowledge that diagnosis in paleopathology is an often equivocal process fraught with doubts and difficulties, we find the blanket refusal to attempt appropriate differential diagnoses reminiscent of adages about babies and bathwater.

WHAT'S NEXT?

There are some exciting recent developments in the paleopathology of the endemic treponematoses. Suzuki (1991) has shown that lesions of treponematosis in Japan are Post-Columbian and support the scenario of the epidemic spread of venereal syphilis from west to east. As yet we are aware of no paleopathologist who has followed-up on Frazier and Li's (1949) demonstration that treponematosis was endemic in rural China and venereal in urban areas in the early 20th century. In Central and South America, evidence for endemic treponematosis is substantially less common than in the North American record, and diagnosis is complicated by the similar skeletal lesions of leishmaniasis. Nevertheless, there are several recent studies, interestingly from relatively arid environments, that indicate that some populations experienced yaws-like manifestations (Standen and Arriaza 2000; Pechenkina et al. 2007; Pineda et al. 2009).

Looking forward, we predict that new techniques in microbiology and "molecular paleopathology" (under development even as we write these words) will yield additional phylogenetic information about the evolutionary history of the genus *Treponema* and its world-wide range. Archaeological excavations will no doubt turn up more examples of treponemal skeletal pathology in hitherto undocumented contexts. And, with a nod to the past, we encourage our colleagues interested in gaining a true picture of the clinical and epidemiological diversity of this protean disease entity called treponematosis to read carefully the many fascinating accounts written over the past 500 years by physicians, by historians, and by lay victims of the pathogenic members of its vast tribe. We close this chapter with the admonition of the eminent archaeologist William Y. Adams to his colleagues who seek to develop ever more precise classificatory typologies for ceramics and other categories of material remains: "Don't mistake the pigeonholes for the pigeons!" By no means do we denigrate the heuristic value of diagnostic pigeonholes, but our present dovecote (the four modern syndromes) is surely too small to contain all of the forms of treponemal infections that have ever existed. We look forward with pleasure to following future pathways of research into the natural history of the treponemes, bending and twisting in their timeless dark-field dance, and their co-evolution with that other fascinating organism, *Homo sapiens.*

REFERENCES

Aegerter, E., and Kirkpatrick, J. A., 1968   Orthopedic Diseases. Philadelphia: W.B. Saunders.

Anderson, T., Arcini, C., Anda, S., Tangerud, A., and Robertson, G., 1986   Suspected Endemic Syphilis (Treponarid) in Sixteenth-Century Norway. Medical History 30:41–350.

Antal, G. M., Lukehart, S.A., and Meheus, A. Z., 2002   The Endemic Treponematoses. Microbes and Infection 4(1):83–94.

Arensberg, B., Garfinkel, Y., and Herzkovitz, I., 1986   Neolithic Skeletal Remains at Yiftahel, Area C (Israel). Paleorient 12(1):73–81.

Armelagos, G. J., and van Gerven, D. P., 2003   A Century of Skeletal Biology and Paleopathology: Contrasts, Contradictions, and Conflicts. American Anthropologist 105(1):53–64.

Aufderheide, A. C., and Rodríguez-Martín, C., 1998   The Cambridge Encyclopedia of Human Paleopathology. Cambridge: Cambridge University Press.

Baker, B. J., and Armelagos, G. J., 1988   The Origin and Antiquity of Syphilis: Paleopathological Diagnosis and Interpretation. Current Anthropology 29(5):2–79.

Biocca, E., 1945   Estudos Etno-biologicos Sobre os Indios da Região do Alto Rio Negro-Amazonas. Arquivos de Biologia, São Paulo 29:7–12.

Bonilla, O., 2009   The Skin of History: Paumari Perspectives on Conversion and Transformation. *In* Native Christians: Modes and Effects of Christianity among Indigenous Peoples of the Americas. A. Vilaça, and R. M. Wright, eds. pp. 127–145. Aldershot: Ashgate.

Brothwell, D. R., 1981   Microevolutionary Change in the Human Pathogenic Treponemes: An Alternative Hypothesis. International Journal of Systematic Bacteriology 31(1):82–87.

Brothwell, D. R., 2005   North American Treponematosis Against the Bigger World Picture. *In* The Myth of Syphilis: The Natural History of Treponematosis in North America. M. L. Powell, and D. C. Cook, eds. pp. 480–496. Gainesville: University Press of Florida.

Brothwell, D. R., 2010   Correspondence: On Problems of Differential Diagnosis in Palaeopathology, as Illustrated by a Case from Prehistoric Indiana. International Journal of Osteoarchaeology 20(5):621–622.

Buckley, H. R., and Tayles, N. G., 2003   The Functional Cost of Tertiary Yaws (*Treponema pertenue*) in a Prehistoric Pacific Island Skeletal Sample. Journal of Archaeological Science 30 (10):1301–14.

Centurion-Lara, A., Castro, C., Castillo, R., Shaffer, J. M., Van Voorhis, W. C., and Lukehart, S. A., 1998   The Flanking Region Sequences of the 15-kDA Lipoprotein Gene Differentiate Pathogenic Treponemes. Journal of Infectious Diseases 177:1036–1040.

Centurion-Lara, A., Godornes, C., and Castro, C., Van Voorhis, W. C., Lukehart, S.A., 2000   The tprK Gene Is Heterogenous Among *Treponema pallidum* Strains and Has Multiple Alleles. Infection and Immunity 68:824–31.

Chaplin, J. E., 1997   Natural Philosophy and an Early Racial Idiom in North America: Comparing English and Indian Bodies. William and Mary Quarterly 54(1):229–252.

Chulay, J. D., 1990   *Treponema* Species (Yaws, Pinta, Bejel). *In* Principles and Practice of Infectious Diseases. G. L. Mandell, R. G. Douglas, and J. E. Bennett, eds. pp. 1808–1812. New York: Churchill and Livingstone.

Cockburn, T. A., 1961   The Origin of the Treponematoses. Bulletin of the World Health Organization 24:221–228.

Cockburn, T. A., 1963   The Evolution and Eradication of Infectious Diseases. Baltimore: Johns Hopkins University Press.

Cole, G., and Waldron, T., 2011   Apple Down 152: A Putative Case of Syphilis in Sixth Century AD Anglo-Saxon England. American Journal of Physical Anthropology. 144(1):72–79.

Cook, D. C., and Powell, M. L., 2005   Piecing the Puzzle Together: North American Treponematosis in Overview. *In* The Myth of Syphilis: The Natural History of Treponematosis in North America. M. L. Powell and D. C. Cook, eds. pp.442–479. Gainesville: University Press of Florida.

Csonka, G., and Pace, J., 1985   Endemic Nonvenereal Treponematosis in Saudi Arabia. Reviews of Infectious Disease 7:S260–S265.

De Melo, F.L., Moreira de Mello, J.C., Fraga, A.M., Nunes, K., and Eggars, S., 2010   Syphilis at the Crossroad of Phylogenetics And Paleopathology. PLoS Neglected Tropical Diseases, Jan 5, 4(1): e575, 1–11.

Domett, K. M., Wallis, L. A., Kynuna, D., Kynuna, A., and Smith, H., 2006   Late Holocene Human Remains from Northwest Queensland, Australia: Archaeology and Palaeopathology. Archaeology of Oceania 41:25–36.

Dutour, O., Pálfi, G., Berato, J., and Brun, J.-P., eds., 1994   L'Origine de la Syphilis en Europe – avant ou après 1493? Paris: Editions Errance.

Edwards, A. M., Dymock, D., and Jenkinson, H. F., 2003   A Review: From Tooth to Hoof: Treponemes in Tissue-Destructive Disease. Journal of Applied Microbiology 92:767–780.

Eisenberg, H., Plotke, F., and Baker, A. H., 1949   Asexual Syphilis in Children. The Journal of Venereal Disease Information January:7–11.

Engelstein, L., 1986   Syphilis, Historical and Actual: Cultural Geography of a Disease. Review of Infectious Diseases 8:1036–48.

Erdelyi, R. I., and Molla, A. A., 1984   Burned-Out Endemic Syphilis (Bejel): Facial Deformities and Defects in Saudi Arabia. Plastic and Reconstructive Surgery 74(5):589–602.

Fagan, B., 2006   Fish on Friday: Feasting, Fasting and the Discovery of the New World. New York: Basic Books.

Fairley, D. H., 1994   Cecil John Hackett and the Treponematoses. *In* L'Origine de la Syphilis en Europe – avant ou après 1493? O. Dutour, G. Pálfi, J. Berato, and J-P. Brun, eds. pp.20–22. Paris: Editions Errance.

Fegan, D., Glennon, M. J., Thami, Y., and Pakoa, G., 2010   Resurgence of Yaws in Tanna Vanuatu: Time for a New Approach? Tropical Doctor 40:68–9.

Fornaciari, G., Castagna, M., A. Tognetti, Tornaboni, D., and Bruno, J., 1989   Syphilis in a Renaissance Italian Mummy. The Lancet 2(8663):614.

Fraser, C. N., Norris, S.J., Weinstock, G. M., White, O., Sutton, G. G., Dodson, R., Gwinn, M., Hickey, E. K., Clayton, R., Ketchum, K. A., Sodergren, E., Hardham, J. M., McLeod, M. P., Salzberg, S., Peterson, J., Khalak, H., Richardson, D., Howell, J., Chidambaram, M., Utterback, T., McDonald, L., Artiach, P., Bowman, C., Cotton, M. D., Fujii, C., Garland, S., Hatch, B., Horst, K., Roberts, K., Sandusky, M., Weidman, J., Smith, H. O., and Ventner, C. J., 1998   Complete Genome Sequence of *Treponema pallidum*, the Syphilis Spirochete. Science 281:375–388.

Frazier, C. N., and Hung-Chiung, L., 1948   Racial Variations in Immunity to Syphilis. Chicago: University of Chicago Press.

Gilman, S. L., 1985   Difference and Pathology: Stereotypes of Sexuality, Race and Madness. Ithaca: Cornell University Press.

Gilman, S. L. 1988   Disease and Representation: Images of Illness from Madness to AIDS. Ithaca: Cornell University Press.

Goldstein, M. S., 1957   The Skeletal Pathology of Early Indians in Texas. American Journal of Physical Anthropology 15:299–312.

Goldstein, M. S., Arensburg, B., and Nathan, H., 1976   Pathology of Bedouin Skeletal Remains from Two Sites in Israel. American Journal of Physical Anthropology 45(3, 2):621–639.

Goodman, A. H., and Martin, D. L., 2002   Reconstructing Health Profiles from Skeletal Remains. In The Backbone of History: Health and Nutrition in the Western Hemisphere. R. H. Steckel, and J. C. Rose, eds. pp. 11–60. Cambridge: Cambridge University Press.

Grin, E. I., 1956   Endemic Syphilis and Yaws. Bulletin of the World Health Organization 15:959–973.

Guimares, F. N., and Rodrigues, B. A., 1948   O Puru-Puru da Amazonia (Pinta, Carate, Mal del Pinto etc.). Memorias do Instituto Oswaldo Cruz 46:135–197.

Guthe, T., Ridet, J., Vorst, F., D'Costa, J., and Grab, B., 1972   Methods for the Surveillance of Endemic Treponematoses and Sero-Immunological Investigations of "Disappearing" Disease. Bulletin of the World Health Organization 46:1–14.

Hackett, C. J., 1951   Bone Lesions of Yaws in Uganda. Oxford: Blackwell Scientific Publications.

Hackett, C. J., 1963   On the Origin of the Human Treponematoses (Pinta, Yaws, Endemic Syphilis and Venereal Syphilis). Bulletin of the World Health Organization 29:7–41.

Hackett, C. J., 1976   Diagnostic Criteria of Syphilis, Yaws and Treponarid (Treponematoses) and Some Other Diseases in Dry Bones (for Use in Osteo-Archaeology). In Sitzungsberichte der Heidelberger Akademie der Wissenschaften Mathematisch-naturwissenschaftliche Klasse, Abhandlung 4. Berlin: Springer-Verlag.

Hardham, J. M., Frye, J. G., Young, N. R., and Stamm, L. V., 1997   Identification and Sequences of the Treponema pallidum flhA, flhF, and orf304 Genes. DNA Sequence 7:107–116.

Harper, K. N., Ocampo, P. S., Steiner, B. M., George, R. W., Silverman, M. S., Bolotin, S., Pillay, A., Saunders, N. J., and Armelagos, G. J., 2008   On the Origin of the Treponematoses: A Phylogenetic Approach. PLoS Neglected Tropical Diseases 2(1):3148.

Hillson, S., Grigson, C., and Bond, S., 1998   Dental Defects of Congenital Syphilis. American Journal of Physical Anthropology 107:25–40.

Horne, P. D., 2005   Endemic Syphilis in Colonial Canada. In The Myth of Syphilis: The Natural History of Treponematosis in North America. M. L. Powell, and D. C. Cook, eds. pp. 345–349. Gainesville: University Press of Florida.

Hudson, E. H., 1932–33   Syphilis in the Euphrates Arab. American Journal of Syphilis 16:447–469, 17:10–14.

Hudson, E. H., 1958   Non-Venereal Syphilis, a Sociological and Medical Study of Bejel. Edinburgh: E. S. Livingstone, Limited.

Hudson, E. H., 1965   Treponematosis and Man's Social Evolution. American Anthropologist 67:885–901.

Jacobi, K., Cook, D. C., Corruccini, R. S., and Handler, J. S., 1992   Congenital Syphilis in the Past: Slaves at Newton Plantation, Barbados, West Indies. American Journal of Physical Anthropology 89:145–158.

Jochelson, K., 2001   The Colour of Disease: Syphilis and Racism in South Africa, 1880–1950. Oxford: Palgrave.

Kelley, M. A., 1979   Skeletal Changes Produced by Aortic Aneurysms. American Journal of Physical Anthropology 51(1):35–38.

Kiple, K. F., 1993a   The Treponematoses. In The Cambridge World History of Human Disease. K. F. Kiple, ed. pp. 1053–1055. Cambridge: Cambridge University Press.

Kiple, K. F., 1993b   Syphilis. In The Cambridge World History of Human Disease. K. F. Kiple, ed. pp. 1025–33. Cambridge: Cambridge University Press.

Kolman, C. J., Centurion-Lara, A., Lukehart, S. A., Owsley, D. W., and Tuross, N., 1999   Identification of *Treponema pallidum* Subspecies *pallidum* in a 200-Year-Old Skeletal Specimen. Journal of Infectious Diseases 180(6):2060–2063.

Lee, R. V., Black, F. L., Hierholzer, W. J., and West, B. L., 1978   A Novel Pattern of Treponemal Antibody Distribution in Isolated South American Indian Populations. American Journal of Epidemiology 107:46–53.

Levréro, F., Gatti, S., Gautier-Hion, A., and Ménard, N., 2007   Yaws Disease in a Wild Gorilla Population and Its Impact on the Reproductive Status of Males. American Journal of Physical Anthropology 132(4):568–75.

Lomax, E., 1979   Infantile Syphilis as an Example of the Nineteenth Century Belief in the Inheritance of Acquired Characteristics. Journal of the History of Medicine 34:23–39.

Luger, A., 1972   Non-Venereally Transmitted 'Endemic' Syphilis in Vienna. British Journal of Venereal Diseases 48:356–360.

Meheus, A., and Tikhomirov, E., 1999   Endemic Treponematoses. *In* Sexually Transmitted Diseases. K. K. Holmes, P.-A. Mårdh, P. F. Sparling, S. M. Lemon, W. E. Stamm, P. Piot, and J. N. Wasserheit, eds. pp. 511–513. New York: McGraw-Hill.

Molnár, E., Dutour, O., and Pálfi, G., 1998   Diagnostic Paléopathologique des Tréponématoses: á Propos D'Un Cas Bien Conserve. Bulletins et Mémoires de la Société d'anthropologie de Paris 10(1–2):17–28.

Morton, R. S., 1967   The Sibbens of Scotland. Medical History 11:374–380.

Mulligan, C. J., Norris, S. J., and Lukehart, S. A., 2008   Molecular Studies in *Treponema pallidum* Evolution: Toward Clarity? PLoS Neglected Tropical Diseases 2(1): e184.

Neiberger, E. J., 1984   Lesions in a Prehistoric Bear: Differential Diagnosis. Paleopathology Newsletter 48:8–11.

Neiberger, E. J., 1988   Syphilis in a Pleistocene Bear? Nature 333:603.

Ortner, D.J., 2003   Identification of Pathological Conditions in Human Skeletal Remains. 2nd edition. Amsterdam: Academic Press.

Ortner, D. J., Tuross, N., and Stix, A. I., 1992   New Approaches to the Study of Disease in Archaeological New World Populations. Human Biology 63(3):337–360.

Pandey, D. G., and Roy, J., 2002   Some Aspects of Socio-Cultural Associated with the Persistence of Yaws in the Abujhmaria. South Asian Anthropologist 2(2):91–96.

Pechenkina, E., Vradenberg, J. A., Benfer, R. A., and Farnum, J. F., 2007   Skeletal Biology of the North Peruvian Coast: Consequences of Changing Population Density and Progressive Dependence on Maize Agriculture. *In* Ancient Health: Skeletal Indicators of Agricultural and Economic Intensification. M. N. Cohen, and G. M. M. Crane-Kramer, eds. pp. 92–112. Gainesville: University Press of Florida.

Phillips, S. M., and Sivilich, M., 2006   Cleft Palate: A Case Study of Disability and Survival in Prehistoric North America. International Journal of Osteoarchaeology 16:528–535.

Pillay, A., Lui, H., Ebrahim, S., Chen, C. Y., Lai, W., Fehler, G., Ballard, R. C., Steiner, B., Sturm, A. W., Morse, S. A., 2002   Molecular Typing of *Treponema pallidum* in South Africa: Cross-Sectional Studies. Journal of Clinical Microbiology 40(1):256–258.

Pineda, C., Mansilla-Lory, J., Martínez-Lavín, M., Leboreiro, I., Pijoan, C., 2009   Rheumatic Diseases in the Ancient Americas: The Skeletal Manifestations of Treponematoses. Journal of Clinical Rheumatology 15(6):280–283.

Pollock, J. S. M., 1953   Sibbens or Sivvens – The Scottish Yaws. Transactions of the Royal Society of Tropical Medicine and Hygiene 47:431–436.

Powell, M. L., 1992   Endemic Treponematosis and Tuberculosis in the Prehistoric Southeastern United States: The Biological Costs of Chronic Endemic Disease. *In* Human Paleopathology: Current Syntheses and Future Options. D. J. Ortner, and A. C. Aufderheide, eds. pp. 173–180. Washington: Smithsonian Institution Press.

Powell, M. L., and Cook, D. C., eds., 2005   The Myth of Syphilis: The Natural History of Treponematosis in North America. Gainesville: University Press of Florida.

Quétel, C., 1990   History of Syphilis. J. Braddock, and B. Pike, trans. Baltimore: Johns Hopkins University Press.

Rothschild, B. M, Hershkovitz, I., and Rothschild, C., 1995   Origin of Yaws in the Pleistocene. Nature 378(6555):343–4.

Rothschild, B. M., and Rothschild, C., 1995   Treponemal Disease Revisited: Skeletal Discriminators for Yaws, Bejel, and Venereal Syphilis. Clinical Infectious Diseases 20:1402–8.

Rothschild, B. M., and Turnbull, W., 1987   Treponemal Infection in a Pleistocene Bear. Nature 329:61–62.

Satter, E. K., and Tokarz, V. A., 2010   Secondary Yaws: An Endemic Treponemal Infection. Pediatric Dermatology 27(4):364–7.

Schell, R. F., and Musher, D. M., 1983   Pathogenesis and Immunology of Treponemal Infection. (Immunology Series 20.) New York: Marcel Dekker.

Schmid, G. P., 1989   Epidemiology and Clinical Similarities of Human Spirochetal Diseases. Reviews of Infectious Diseases 11(6): S1460–S1469.

Simmons, D. C., 1957   The Depiction of Gangosa on Efik-Ibibio Masks. MAN 57(18):17–20.

Skerman, V. B. D., McGowan, V., and Sneath, P. H. A., 1980   Approved Lists of Bacterial Names. International Journal of Systematic Bacteriology 30:225–420.

Slavec, Z. Z., 1996   Morbus Skerljevo – An Unknown Disease Among Slovenians in the First Half of the 19th Century. Wiener klinische Wochenschrift 108(23):764–70.

Solomon, S. G., 1993   The Soviet–German Syphilis Expedition to Buriat Mongolia, 1928: Scientific Research on National Minorities. Slavic Review 52:204–232.

Sommer, T. C., and Lee, T. H., 2001   Charcot Foot: The Diagnostic Dilemma. American Family Physician 64:1591–8.

Stamm, L. V., 1999   Biology of Treponema pallidum. In Sexually Transmitted Diseases. K. K. Holmes, P.-A. Mardh, P. F. Sparling, S. M. Lemon, W. E. Stamm, P. Piot, and N. Wasserheit, eds. pp. 467–472. New York: McGraw-Hill.

Stamm, L. V., 2010   Global Challenge of Antibiotic-Resistant Treponema pallidum. Antimicrobial Agents and Chemotherapy 54(2):583–9.

Standen, V. G., and Arriaza, B. T., 2000   La Treponematosis (Yaws) en las Poblaciones Prehispánicas del Desierto de Atacama (Norte de Chile). Chungará 32(2):185–92.

Steckel, R.H., and J.C. Rose (eds), 2002   The Backbone of History: Health and Nutrition in the Western Hemisphere. Cambridge: Cambridge University Press.

Steinbock, R. T., 1976   Paleopathological Diagnosis and Interpretation. Bone Diseases in Ancient Human Populations. Springfield: C. C. Thomas.

Suzuki, T., 1991   Paleopathological Study of Infectious Diseases in Japan. In Human Paleopathology, Current Syntheses and Future Options. D. J. Ortner, and A. C. Aufderheide, eds. pp. 128–139. Washington DC: Smithsonian Institution Press.

Tabbara, K. E., 1990   Endemic Syphilis (Bejel). International Ophthalmology 14:379–381.

Tramont, E. C., 1990   Treponema Pallidum (Syphilis). In Principles and Practice of Infectious Diseases, 3rd edition. G. L. Mandell, R. G. Douglas, J. E. Bennett, eds. pp. 1794–1808. New York: Churchill and Livingstone.

Turner, T. B., and Hollander, D. H., 1957   Biology of the Treponematoses. Geneva: World Health Organization Monograph 35.

Valdez, M., Haines, R., Riviere, K. H., Riviere, G. R., and Thomas, D. D., 2000   Isolation of Oral Spirochetes from Dogs and Cats and Provisional Identification Using Polymerase Chain Reaction (PCR) Analysis Specific for Human Plaque Treponema. Journal of Veterinary Dentistry 17(1):23–26.

Vaughn, M., 1992   Syphilis in Colonial East and Central Africa: The Social Construction of an Epidemic. In Epidemics and Ideas, Essays on the Historical Perception of Pestilence. T. Ranger, and P. Slack, eds. pp. 269–302. Cambridge: Cambridge University Press.

Verut, D. D., 1973   Precolumbian Dermatology and Cosmetology in Mexico. Schering Pub.

Virchow, R., 1858   Über die Natur der constitutionell-syphilitischen Affectionen. Virchows Archiv für Pathologische Anatomie und Physiologie und für Klinische Medizin 15: 227–336.

Virchow, R., 1896   Beiträge zur Geschichte der Lues. Dermatologische Zeitschrift 3(317):1–9.

Von Hunnius, T. E., Roberts, C. A., Boyleston, A., and Saunders, S. R., 2006   Histological Identification of Syphilis in Pre-Columbian England. American Journal of Physical Anthropology 129:559–566.

Von Hunnius, T. E., Yang, D., Eng, B., Waye, J. S., Saunders, S. R., 2007   Digging Deeper into the Limits of Ancient DNA Research on Syphilis. Journal of Archaeological Science 34:2091–2100.

Weiss, P., 1947   Contribucion al Estudo del Mal de Pinto, Pinta, Cara, Oviera, o Enfermidad de Leon Blanco en el Peru. Revista de Medicina Experimental 6:1–75.

Weston, D. A., 2008   Investigating the Specificity of Periosteal Lesions in Pathology Museum Specimens. American Journal of Physical Anthropology 137(1):48–59.

Wicher, K., Wicher,V., Abbruscato, F., and Baughn, R. E., 2000   *Treponema pallidum* subsp. *pertenue* Displays Pathogenic Properties Different from Those of *T. pallidum subsp. pallidum*. Infection and Immunity 68:3219–3225.

Willcox, R. R., 1955   The Non-Venereal Treponematoses. Journal of Obstretics and Gynaecology in the British Empire 62(6):853–862.

Willcox, R. R., 1960   Evolutionary Cycle of the Treponematoses. British Journal of Venereal Diseases 36:78–91.

Woltsche-Kahr, I., Schmidt, B., Aberer, W., Aberer, E., 1999   Pinta in Austria (or Cuba?) Import of an Extinct disease? Archives of Dermatology 135:685–689.

CHAPTER 27

Nonspecific Infection in Paleopathology: Interpreting Periosteal Reactions

Darlene A. Weston

Introduction

"Periostitis," or periosteal new bone formation, is one of the most commonly reported pathological lesions in archaeological human skeletal remains. It can affect any bone in the skeleton, but is most often seen on the long bones, particularly the tibiae. In response to pathological stimuli, the osteogenic periosteum first creates woven bone, which remodels over time into lamellar bone. Despite their high reported frequency, periosteal reactions have not often been the primary focus of research. In the clinical literature they are often mentioned in passing, simply as a sign of a disease or pathological condition. In the bioarchaeological literature, periosteal reactions are commonly interpreted as a sign of "nonspecific infection," despite a wealth of clinical literature pointing to multiple etiologies (e.g. Resnick 1995). Additionally, periostitis is integral to the "stress-indicator hypothesis" (Goodman et al. 1984b; Goodman et al. 1988), as a sign of infectious disease. The interpretation of periosteal reactions within the stress-indicator framework has had far reaching implications, resulting in generalizations made about pathogen load models in archaeological populations (e.g. Lallo et al. 1978; Lallo and Rose 1979).

Periosteal new bone production is often termed "periostitis" in the anthropological and clinical literature, and can be defined as, "inflammation of the periosteum"

A Companion to Paleopathology, First Edition. Edited by Anne L. Grauer.
© 2012 John Wiley & Sons, Ltd. Published 2016 by John Wiley & Sons, Ltd.

("periostitis," *Merriam-Webster's Medical Dictionary*). The term "periostitis" implicitly refers to the soft-tissue membrane, not the bone itself, and to a very specific pathological response, inflammation. Accordingly, it is probably not the best word to use to describe the phenomenon of the pathological deposition of new bone. As an alternative, Bush (1989) advocated the use of "periostosis," meaning new bone produced by the periosteum. It is a more precise definition that does not imply an etiological process, but unfortunately, the term has not been widely adopted. Ragsdale (1993:465) also dislikes the term "periostitis," stating its usage obscures the true mechanisms involved, and that it "connotes a histopathologic change (inflammation) not usually present." Ragsdale prefers the term "periosteal new bone production," and it is this term, or alternatively "periosteal reaction," which for the above-mentioned reasons, will be used for the remainder of this chapter.

The aim of this chapter is to outline the interpretation of periosteal reactions in human skeletal remains, showing how their interpretation as being pathognomonic of nonspecific infection is potentially dangerous, creating false paradigms for human health and disease in the past. In order to better understand how periosteal reactions have been interpreted, some background regarding the nature of periosteal new bone formation will be provided, specifically mechanisms for its formation, how it is viewed from a clinical viewpoint, and the common paleopathological perspective, including the interpretation of periosteal new bone production within the stress-indicator hypothesis.

INFLAMMATION VS. INFECTION

One way to consider disease in the skeleton is to think of it as the effect of an injury and a reaction to that injury at a cellular level. Factors initiating tissue damage can be divided into two main categories, endogenous and exogenous (Štvrtinová et al. 1995). Endogenous factors can include immunopathological reactions, neurological and genetic disorders, while exogenous factors can be subdivided into five causative agents of cell injury: mechanical (e.g. trauma), physical (e.g. low or high temperature, radiation), chemical (e.g. caustic agents, poisons, venoms), nutritive (e.g. deficiency of oxygen, vitamins) or biological (e.g. viruses, microorganisms, parasites) (Štvrtinová et al. 1995). The human body has evolved several ways in which it can protect itself from injury, the so-called "four vital reactions": hemostasis (to minimize the risk of hemorrhage), regeneration (to replace damaged cells), the immune response, and most importantly, in the case of periosteal new bone production, inflammation (Mitchinson et al. 1996).

Studies of archaeological skeletons have frequently attributed the majority of periosteal lesions to infection, perhaps due to confusion between the definitions of inflammation and infection (i.e. Lallo 1973; Mensforth et al. 1978; Littleton 1998). Inflammation can be defined as a vascular response to tissue damage from a number of causes, while infection only occurs when the body has to deal with pathogenic organisms, such as bacteria, viruses, parasites, etc. As a response to infection, the body will embark on an inflammatory response, which aims to neutralize the pathogenic organism and repair or heal the resultant damage. Inflammatory responses are very frequently caused by infection, but they are far from the only cause (Bush 1989; Ortner 2003), and occur no matter what the initial cause of injury is.

The inflammatory response (see Mitchinson et al. 1996 for a detailed account) is localized at the site of injury and is acute in nature, lasting several days at most and always following the same path: vasodilatation (widening of the blood vessels), fluid exudation, and phagocyte recruitment (though note that the inflammatory process can become chronic, and more complicated, if the causative agent persists more than 2 weeks). Vasodilatation results in hyperoxia (an increase in blood oxygen), which in turn stimulates osteoclast function (bone-destroying cells). Edema, excess accumulation of fluid, results from fluid exudation and in turn causes hypoxia (a decrease in blood oxygen). Hypoxia is one of the main agents for the stimulation of osteoblast function (bone forming cells). It is this sequence of oxygen states and its effect on osteoclasts and osteoblasts that is integral to the bony changes resulting from inflammation of the skeleton (Roberts and Manchester 1995).

Healing begins if the inflammatory process is successful. Dead cells and other tissues are the inevitable result of the inflammatory process. The healing process replaces this material through scavenging, regeneration and repair (Mitchinson et al. 1996). Macrophages and osteoclasts scavenge the dead tissue, followed by tissue repair via the formation of granulation tissue, a loose connective tissue filled with capillary buds, osteoblasts and macrophages. The disorganized, injured tissue is given structure by the granulation tissue, which eventually becomes less cellular and vascular. The inflammatory cells disappear, the osteoblasts turn into osteocytes, and the tissue fluids become reabsorbed. The granulation tissue emerges as periosteal new bone, starting as woven bone and in time remodeling into lamellar bone.

Mechanisms of Pathological Periosteal New Bone Production

Although inflammation is one of the mechanisms for the production of pathological periosteal new bone, generally new bone formation throughout the skeleton is the result of the activation of resting osteoblasts or the proliferation and differentiation of preosteoblast cells, regardless of cause. Once these cells are activated, hormones regulate the rate and duration of bone matrix production by each osteoblast. Hormonal regulation influences bone formation at the level of cell replication and osteoblast function (Canalis et al. 1993; Raisz and Kream 1983). However, as well as stimulating osteoblast function, hormones can inhibit it. Osteoclasts are also hormonally controlled in a similar way. Calcium-regulating hormones (e.g. parathyroid hormone), systemic hormones (e.g. glucocorticoids), growth factors (e.g. growth hormone), local factors (e.g. prostaglandins), and ions (e.g. calcium) can all play a role.

Although hormones play a crucial role in stimulating periosteal new bone production, other mechanisms have parts to play as well. In fact, almost anything that tears, breaks, stretches, or even merely touches the periosteum can stimulate it to initiate new bone formation (Richardson 1994). Additionally, the periosteum does not always need to be the source of osteogenic cells. Histological analysis of healing fractures has shown that surrounding soft tissues are major contributors to periosteal new bone production as the external fibrous layer of the periosteum can be replenished from surrounding fascia, fat, and muscle tissues (Ragsdale et al. 1981).

The physical elevation of the periosteum from the bone is often cited as a prerequisite for periosteal new bone production. It may be present, but it is not always required for new bone production to occur, meaning other factors must be operative as well (Ragsdale 1993). These factors may include compensation for weakness due to osteolysis, mechanical adaptation, changes in blood circulation, efforts at tumor containment, and even osteogenic signals emanating from the tumors themselves (Ragsdale 1993; Ragsdale et al. 1981).

Periosteal new bone that is formed secondary to the elevation of the fibrous outer layer of the periosteum may be a result of the compression and stretching of blood vessels by blood, granulation tissue, neoplasm, pus or trauma (Bush 1989). This may result in hemorrhage beneath the periosteal membrane, with a subsequent reduced blood supply to the bone. If this situation is of sufficient duration, the periosteal bone tissue will die. Amelioration will restore osteoblastic activity, producing new subperiosteal bone that is deposited on the normal cortical bone surface (Bush 1989; Jaffe 1972). A row of refilled Howship's lacunae forming a reversal line can usually be seen on the original cortical surface beneath the newly added bone, testifying to an initial resorptive phase probably due to active hyperemia (Ragsdale 1993).

CLINICAL INTERPRETATIONS OF PERIOSTEAL REACTIONS

The clinical literature pertaining to periosteal new bone production provides little or no description of the phenomenon itself. For the most part, it is merely stated that "periostitis" is present as part of the manifestations of the clinical disease. In part, this lack of information is due to the reliance on imaging techniques to report on and diagnose pathological conditions. In its earliest stages, periosteal new bone does not always appear on radiographs or in other images. It is also likely to be overlooked during autopsy (Kelley 1989).

Medical writings on inflammation of the periosteum can be traced back to at least the beginning of the 19th century, with works by Crampton (1817) and Parsons (1849). By the time Senn wrote "Clinical Lecture on Periostitis" in the *Philadelphia Medical Times* in 1886, medical knowledge had progressed sufficiently that a detailed and knowledgeable account of periosteal new bone production was offered. Decades later, radiographic properties of periosteal reactions drew clinical attention, with papers by Edeiken et al. (1966), Edeiken (1981), Ragsdale et al. (1981), Ragsdale (1993), and Resnick (1995) providing descriptions and illustrations of the condition.

For instance, Edeiken et al. (1966) and Edeiken (1981) separated periosteal reactions into a number of descriptive categories based around two types, the solid and the interrupted form (see Table 27.1). Of particular help to paleopathology, Ragsdale et al. (1981) and Ragsdale (1993) categorize periosteal reactions as either productive or negative (see Table 27.2 and Figure 27.1). They provide the reader with a link between the morphology and constitution of periosteal lesions and the processes creating them. Recognizing the different types leads to insight into etiology. Similarly, Resnick (1995) offers seven types of periosteal reaction (see Figure 27.2): 1. a single layer of periosteal new bone, which may be seen in benign or malignant tumors, infection, and secondary hypertrophic osteoarthropathy; 2. multiple layers of periosteal new bone or "onion-skinning," which can be seen in infection, malignant tumors, hypertrophic osteoarthropathy and other conditions; 3. a thick linear osseous deposit, which can be separate from or mixed with the underlying cortex, as seen in hypertrophic osteoarthropathy and

Table 27.1   Types of periosteal new bone production (after Edeiken et al. 1966, Edeiken 1981)

1	*Solid*	The hallmark of a benign disease process.
a	Dense undulating	Often occurring with long-standing varicosities or arterial disease.
b	Thin undulating	Located primarily on the concave aspect of long bones.
c	Dense elliptical	Present in long-standing osteoid osteomas.
d	Cloaked	Found in long-standing benign conditions.
e	Codman's triangle	Resulting from anything lifting the periosteum whether benign or malignant.
2	*Interrupted*	Caused by periosteal elevation in active conditions.
a	Lamellated	Occurs due to alternating periods of rapid and slow growth.
b	Perpendicular	The result of new bone growing at right angles to the shaft of the host bone.
c	Amorphous	Due to malignant tumours.

Table 27.2   Types of periosteal new bone production (after Ragsdale et al. 1981; Ragsdale 1993)

Negative	Continuous	Productive Interrupted	Complex
Concentric atrophy	Ridged shell	Buttress	Divergent spiculated
Saucerized with endostosis	Lobulated shell	Codman angle	Combined
Saucerised with sharp margin	Smooth shell	Lamellated	
Cortical erosion with ill-defined margin	Solid Single lamella	Spiculated	
Cortical destruction with medullary invasion	Lamellated		
	Parallel spiculated		

venous stasis; 4. an irregular osseous excrescence with a spiculated contour that merges with the underlying cortex, as seen in thyroid acropachy or primary hypertrophic osteoarthropathy; 5. a thin, linear osseous deposit that extends in a direction perpendicular to the underlying cortex, as seen in Ewing's sarcoma; 6. a sunburst pattern, in which linear deposits fan out from a single focus, as seen in osteosarcoma; and 7. a Codman's triangle, consisting of a triangular elevation of the periosteum with one or more layers of new bone, as often seen in various malignant lesions.

Periosteal new bone production is also germane to pediatric radiology, as incidental observation of periosteal new bone on long bones is not uncommon (Shopfner 1966). In a study of 335 full-term and 75 premature infants, 115 (35 percent) and 75 (34 percent) respectively, had bilateral periosteal new bone present on the femora, humeri, tibiae, ulnae, and radii, and fibulae (in order of frequency). With the exception of one premature infant, no periosteal new bone production was seen in infants less than 1 month old. In infants less than 3 months of age, the periosteal new bone appeared as a delicate, smooth layer on the long bone diaphyses and metaphyses. It was often quite thick, but did not form multiple layers. Shopfner concluded that the periosteal new bone formation was a manifestation of normal bone growth and surmised that the reason it was not roentgenologically visible in all infants was due to a normal

Periosteal reactions

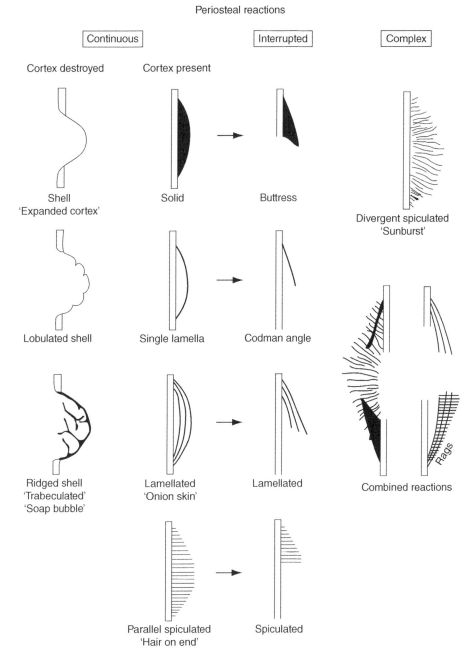

Figure 27.1   Types of periosteal new bone production (after Ragsdale et al. 1981).

differential in the rate of bone formation, with only those showing the most rapid growth being visible on X-ray. Shopfner postulated that increased activity of the periosteum may arise when it is not tightly bound to the underlying cortex, possibly as a result of sparse or shortened Sharpey's fibers. This increased periosteal activity may occur following asynchrony in growth rates of the physis cartilage and the

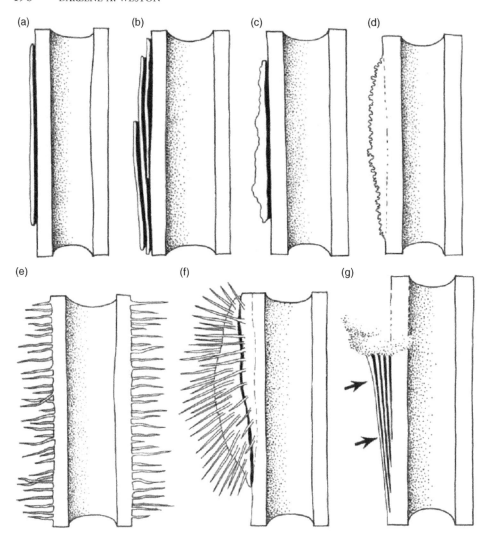

Figure 27.2 Various types of periosteal reaction (after Resnick 1995).

periosteum, resulting in periosteal tension and periosteal new bone production (Jones, personal communication in Scheuer and Black 2000).

Shopfner (1966) cautioned that proliferative periosteal reactions seen on the long bones of infants between the ages of 1 and 6 months were not rare or abnormal, and stated that normal periosteal bone growth may have been wrongly described as components of some diseases. Several researchers in paleopathology have heeded this message, with Mann and Murphy (1990) cautioning that the rapidly growing bones of infants, children, and adolescents can mimic periosteal new bone production. Lewis (2000, 2004) similarly states that it is almost impossible to differentiate between normal and abnormal bone in infant skeletal remains, particularly emphasizing changes on the endocranium, while Ribot and Roberts (1996) advise against attempting to identify periosteal reactions in juveniles fewer than 2 years old.

The warnings issued by Shopfner (1966) and his like-minded successors have important implications for the interpretation of pathology in juvenile skeletal remains. For example, in their analysis of the age distribution of periosteal reactions at the Libben site, Mensforth et al. (1978) state that the highest frequency of unremodeled lesions (those consisting of woven bone) occurred in infant skeletons less than 1 year of age. The frequency of unremodeled lesions dropped steadily for infants between the ages of 1 and 3 years, rarely occurring in juveniles older than 3 years of age. The authors concluded that the distribution of unremodeled periosteal reactions indicated that chronic malnutrition was absent among the Libben population, and that the nutritional stress seen among the infants' skeletons was a synergistic response to nutrient depletion and the increased demands which accompany rapid growth, microbial infection and malabsorption due to weaning diarrhea. Another way to interpret Mensforth et al.'s (1978) assertions is to note that the high frequency of proliferative periosteal reactions in infants under 1 year mirrors Shopfner's (1966) findings, leading to the possibility that the lesions in the Libben population are, in fact, not pathological.

PALEOPATHOLOGICAL INTERPRETATION OF PERIOSTEAL REACTIONS

Since clinicians rarely examine periosteal reactions on dry bone, and often overlook them during autopsy, it leaves paleopathologists with few clinical models to apply to ancient skeletal material. Consequently, paleopathologists have to generate their own models to characterize and interpret periosteal reactions. It appears from the published paleopathological literature that the tibiae are most frequently affected by periosteal reactions. A random sample of published and unpublished archaeological skeletal reports from the U.K. yielded prevalence rates for tibial periostitis ranging from 5.9 to 64 percent (Weston 2008: Table 2). However, the number of skeletons exhibiting tibial periosteal reactions is often difficult to assess, as there are no universal standards for recording and reporting the lesion.

Although periosteal new bone production is frequently recorded on human remains, there have been relatively few studies focusing on the properties of periosteal reactions themselves. Two recent studies (Weston 2008; 2009) investigating the specificity of periosteal reactions in pathology museum specimens demonstrated that the morphological and quantitative properties of periosteal reactions, at a macroscopic, radiographic, and histological level cannot be correlated to specific disease processes, as the periosteum responds in a similar manner regardless of the etiology. Differences in periosteal new bone formation were deemed to likely be a response to the chronicity of the pathological episode.

Anthroposcopic characterization of periosteal reactions

The terminology used to describe and classify periosteal reactions within paleopathology has been inconsistent. However, there are a few researchers who have devised strategies for scoring periosteal reactions in skeletal remains. For instance, Hackett (1976) developed a well-known classification system involving plaques, striations, nodes, and expansions (particularly in reference to the treponematoses). A number of other

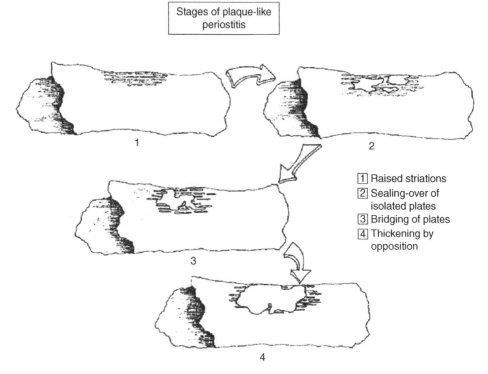

Figure 27.3   An illustration of Strothers' and Metress' (1975) four stages of plaque-like periostitis.

researchers have proposed classification systems, such as Strothers and Metress (1975), who divided periosteal reactions into two types and four subtypes as illustrated in (see Table 27.3 and Figure 27.3). Lallo (1973) devised a scoring system to represent the stages of severity of infectious pathology, devising nine stages of severity based on the "location of the infection and various characteristic manifestations on the periosteal surface resulting from infection" (see Table 27.4) (Lallo 1973:207). In Cook's (1976) system, periosteal lesions were divided into "fiber-bone periostitis" and "sclerotic periostitis" (see Table 27.5). Cook characterizes "fiber-bone periostitis" as unconsolidated, matte-surfaced new bone with a woven appearance. This woven appearance indicates that the periosteal reaction was active at or near the time of death. "Sclerotic periostitis" is deemed to be dense "periostitis" with a surface texture like normal bone, resulting from the remodeling and replacement of woven bone with normal or sclerotic tissue. This form of reaction indicates an inactive or slowly progressing lesion. Finally, Buikstra and Ubelaker (1994) provide a protocol for the recording of abnormal bone formation, requiring recording the bone, side, section, and aspect of the bone affected, along with the type of bony lesion (lamellar, spicular, or shell-type reactions).

All of the recording methods for periosteal new bone formation described above have advantages and deficiencies. Most importantly, however, none are consistently used by paleopathologists. This renders the assessment of authors' conclusions impossible at times, and certainly curtails the possibility of comparing one population

Table 27.3 Types of periosteal new bone production (Strothers and Metress 1975)

1	Lesion totally exterior to cortex and appears as a layered plaque which is bark-like in appearance.	i	Fine, almost imperceptible, fiber-like striations which always run parallel to the longitudinal axis of the shaft.
		ii	The formation of continuous but isolated bony plates, separated by intervening striations.
		iii	Bridging over of a number of isolated bony plates to form one continuous plaque-like formation.
		iv	Thickening of the bony plaque through growth by apposition.
2	Lesion of very fine porosity which covered the exterior surface of a bone but does not involve the underlying cortex.		

Table 27.4 Stages of severity of tibial infection (Lallo 1973)

Stage 1	Smooth and undamaged periosteal surface	Represents appearance of tibia not yet afflicted by infection
Stage 2	Longitudinal striations begin to appear in at least three quarters, two of which must be continuous up and down	Initial stages of disease wherein periosteal damage and destruction is only beginning
Stage 3	From one to three noncontinuous quarters with mild pitting and swelling	
Stage 4	At least four noncontinuous quarters with moderate pitting and swelling	
Stage 5	At least one local large swelling with moderate pitting in three quarters	From Stage 5 onward periosteal damage and destruction becomes more acute
Stage 6	Three or more noncontinuous quarters exhibit swelling; or two continuous quarters (up and down) exhibit swelling; or one zone may exhibit swelling and scaling	
Stage 7	Heavy pitting in at least one zone (zones may consist of noncontinuous quarters)	
Stage 8	Heavy pitting in at least two continuous quarters	Stages 8 and 9 represent extremely heavy periosteal destruction
Stage 9	Heavy pitting of periosteum over the entire area of three continuous quarters and part of the fourth	

to another. Perhaps the best guide to recording periosteal reactions would be to follow Ortner's (no date: 1) advice: "To be effective, terms used to describe abnormalities encountered in human remains must have a clear definition that is easily understood and applied by most potential users. A classificatory system must have categories that: (1) distinguish relevant, biologically significant features, (2) include

Table 27.5 System for scoring periosteal new bone production and periosteal vascularisation (Cook 1976)

Fiber-bone periostitis and sclerotic periostitis
1 Normal.
2 Isolated elevated plaque or plaques covering less than one third of the functional/vascular surface.
3 As above covering one to two thirds of the functional/vascular area.
4 Uniform elevation of two thirds or more of the area with little increase in diameter.
5 As above with elevation of more than 2–3 mm.

Periosteal vascularization
1 Normal
2 Multiple small striae
3 Multiple small foramina
4 Multiple large striae
5 Multiple large foramina
6 Mixed abnormal vascularization

all conditions encountered, and (3) are, largely, mutually exclusive." In order to describe and classify periosteal new bone production in a satisfactory manner, I suggest that all the following questions be answered:

1. Which bones in the skeleton are affected and which are not?
2. Which part of the bone is affected—epiphysis, metaphysis, distal third of the diaphysis, etc.? Anatomically precise terms must be used.
3. Which aspect of the bone is affected—medial, lateral, anterior, posterior, etc.? Again, anatomical precision is paramount.
4. Is the lesion focal or diffuse?
5. What percentage of the bone is affected?
6. Does the lesion consist of woven bone, lamellar bone, or a mixture of the two?
7. What type of vascularization does the lesion have (large or small striae, large or small foramina, a mixture)?

If the skeleton under investigation is incomplete, at best a general pathological category can be assigned to the periosteal lesions, and at worst the lesions can simply be noted without imposing unjustified interpretation.

Periosteal reactions and nonspecific infection

In the paleopathological literature, periosteal new bone production has occasionally been mentioned as a component of specific infectious diseases such as tuberculosis (Kelly and Micozzi 1984; Pfeiffer 1991; Kelly et al. 1994; Roberts et al. 1994), the treponematoses (Hackett 1976, 1978, 1983; Elting and Starna 1984; Reichs 1989; Palfi et al. 1992; Hutchinson 1993; Rothschild and Heathcote 1993; Lewis 1994; Schermer et al. 1994; Mansilla and Pijoan 1995) and leprosy (Lewis et al. 1995; Roberts and Manchester 1995). Sporadically, periosteal reactions have been mentioned in association with syndromes such as hypertrophic osteoarthropathy (Fennell and

Trinkaus 1997), with metabolic diseases like scurvy (Ortner et al. 2001) or as a component of trauma (Detweiler-Blakely 1988). However, periosteal new bone formation is most frequently mentioned as being a result of "nonspecific" infection (Powell 1988, 1991, 1992; Rathbun 1987; Pfeiffer 1991; Suzuki 1991; Pfeiffer and Fairgrieve 1994; Pietrusewsky and Douglas 1994; Williams 1994; Roberts and Manchester 1995; Larsen 1997; Larsen et al. 2007).

When periosteal reactions are described as being caused by "nonspecific" infection, the authors are implying that an infection was present but its cause was unknown. Thus "nonspecific infection" is actually a moniker for lack of etiological knowledge. Many researchers have come out in opposition of the practice, with Birkett (1983:100) describing the term, "nonspecific infection" as "a misnomer." Similarly, Powell (1988:148) stated that, "...reporting the prevalence of infectious response or "periostitis" in general terms deliberately ignore[s] an aspect of population health of major significance: the influence of specific infectious diseases upon levels of morbidity and mortality."

Rothschild and Rothschild (1996:559) support Powell's position, stating that, "The hypothesis of non-specific periosteal reaction no longer appears relevant in population studies." Furthermore, they state that the concept of the nonspecific periosteal reaction was derived from the study of bones from archaeological sites and that it "...does not appear to have a recognized counterpart in clinical medicine [e.g. Resnick 1995]" (p. 559). They believe that confusion over Hackett's (1976) use of the term "nonspecific," used to describe bones whose appearance was not absolutely diagnostic of disease, may have popularized its [mis]use.

Larsen's (1997:83) equating "periostitis" with "nonspecific infection" serves as a case in point. At first the author asserts that, "Periostitis represents a basic inflammatory response that may result from bacterial infection, but traumatic injury is also implicated in its etiology." Later in the text, he eschews other noninfectious causes of periosteal new bone production, and mentions only infection as the cause. He then links periosteal new bone production with "nonspecific infection" stating that, "Although periosteal reactions [...] are nonspecific, their documentation has proven highly useful for assessing levels and patterns of community health. Nonspecific infections provide a rather incomplete and undiagnostic picture of a population's disease experience, but documentation of prevalence and changing patterns reflect the health costs of specific lifeways, such as sedentary agriculture" (Larsen 1997:84).

Similar assertions have been made by Goodman et al. (1984a:291), who categorically equated "periostitis" with infection and by Roberts (2000:146), who discusses "periostitis" under the heading of infectious disease. Roberts (2000:148), however, briefly acknowledges in passing that periosteal reactions can have multiple etiologies.

The assumption that the quantification of the amount of periosteal new bone formation present in a skeletal population is representative of the general health of that population is a dangerous one to make. This point is stated by Wood et al. (1992) and has been supported by some researchers (Hutchinson 1992; Ubelaker 1992; see also Chapter 7, by Boldsen and Milner, in this volume) and criticized by others (Goodman 1993; Cohen 1994). In fact, Wood et al. assert that the equation of the frequency of any pathological change with population health is problematic. Aside from overestimating the prevalence of pathological conditions in a population, they argue that lesion frequencies can underestimate population prevalence due to the low sensitivity of skeletal lesions in reflecting disease states. They continue by arguing that individuals who exhibit healed periosteal lesions are of lower frailty and thus have a lower risk of death. These individuals might have been robust enough to manifest a healed periosteal lesion, while a frailer individual would have died before a bony response occurred or while the lesion was still active. Thus, they argue, "the presence of a healed lesion may sometimes indicate a state of comparatively good overall health" (Wood et al. 2002:353).

Aside from infectious diseases, it is clear that various forms of trauma, including fractures, dislocations, displacements, hematomas, and artificially produced deformities can be etiological factors in periosteal new bone formation (Ortner 2003). Trauma should be the first factor eliminated when attempts are made to narrow down the etiological origins of periosteal reactions, as the generally focal, unilateral nature of traumatic periosteal reactions makes them relatively straightforward to identify. Hence, the importance of considering the complete skeleton when studying periosteal new bone formation cannot be overstated, as the examination of only a few bony elements may present a false impression of the nature of periosteal new bone formation throughout the skeleton. Studies that derive prevalence rates for "nonspecific infection" based on the presence of periosteal reactions on individual bones (i.e. see studies in Cohen and Armelagos 1984; Cohen and Crane-Kramer 2007) may undoubtedly lead to an overestimation of the prevalence of infectious disease in a population, as individual bones with periosteal lesions could have been subject to trauma, localized ulceration, hypertrophic osteoarthropathy, or other noninfectious conditions, for which differential diagnoses could be ascertained with a complete skeleton.

Periosteal reactions and the stress indicator hypothesis

The stress indicator hypothesis, as used in human skeletal studies, is based on the "Selyean" concept of stress and the General Adaptation Syndrome (GAS) (Selye 1957; Buikstra and Cook 1980; Armelagos and Goodman 1991; Bush 1991; Powell 1988). Many of the difficulties in stress-based research stem from the problems surrounding the definition of the terms "stress," "stressor," and "stressful." Most of this confusion originates from Selye's (1957) own definitions, which include "an abstraction" (p. 43), "not a nonspecific reaction" (p. 54), and "not a specific reaction" (p. 54). Selye (1976:53) finally decided on the definition of stress as being the "nonspecific response of the body to any demand." Evidenced by the fact that Selye himself had enormous difficulty in defining the term, stress has been called "…an exercise in futility" (Levine et al. 1989), with Bush (1991) suggesting that biological anthropologists would benefit from striking the word "stress" from their vocabularies altogether.

Nevertheless, Selye saw stress as an individual's response to the demands of his or her environment, with the primary concern being the physiological mechanisms involved in the response. The physiological manifestations of the GAS are primarily hormonal responses. According to Selye (1957), when specific organs of the body are induced to function intensely or when tissues are damaged, the positive evidence of this occurrence is the increased secretion of adrenocorticotrophic hormone (ACTH), the adrenal-stimulating pituitary hormone.

Subsequent research has refined the physiological stress response and focused attention on the pituitary-adrenal cortical system and the sympathetic-adrenal medullary system. The adrenal medulla is closely connected to the sympathetic nervous system and produces adrenaline and noradrenaline. Adrenaline increases metabolism, causes the liver to release glycogen as glucose, and dilates the small arteries of the heart, muscles, and bronchial tubes. The amount of adrenaline secreted varies and is elevated by fear, anger or excitement (Cox 1978). Subsequent studies have shown that enjoyable experiences are almost as effective as unpleasant ones in intensifying catecholamine excretion, but this is rarely referred to in the clinical literature and has not been noted in paleopathological studies (Bush 1991).

In a clinical environment, the activity of the neuroendocrine system can be directly measured by assessing the level of hormone excretion in a urine sample. Obviously this is not possible for archaeological populations; so biological anthropologists must rely on the indirect evidence of neuroendocrine activity that can be detected in bones and teeth. As neuroendocrine activity causes the release of growth-inhibiting catabolic hormones, the effects of this activity are what are sought in the human skeleton (Bush 1991). It must be remembered that there is a broad range of hormones and endocrine systems that also respond concurrently to stress: the pituitary–thyroid, pituitary–gonadal, growth hormone and insulin systems are also involved (Mason 1975; and for research on the physiological stress phenomenon see Chrousos and Gold 1992; McEwen and Stellar 1993).

Unfortunately, as remarked by several researchers (Armelagos and Goodman 1991; Bush 1991; Mason 1971), the GAS does not adequately describe the body's response to stress. In concentrating his attention on the body's physiological response to stressor agents, Selye, for the most part, ignored the role of psychological processes. Most in the field of stress research now agree that much of the physiological response is not directly determined by the actual presence of the stressor agent, but by its psychological impact on the person (Cox 1978). Though Selye (1976) did state that psychological stressors, and particularly emotional arousal, were of primary importance, he argued that they did not always form a common pathway to the stress response. Hence, by the mid-1970s, the enthusiasm for Selye's nonspecific (physiological) stress response had waned among stress researchers. Alternative stress models began to be developed by clinicians (e.g. Cox and Mckay 1976 in Cox 1978; Lazarus 1976; Howarth 1978)—but just as the medical and psychological stress researchers were constructing these new stress models and abandoning Selye's model and his General Adaptation Syndrome, the bioarchaeologists were embracing it.

Researchers in the field of human skeletal studies devised a method to evaluate the health of past populations through the interpretation of dental and skeletal features ("stress indicator" or "stress markers") caused by what they believe was the body's adaptive response to stress. Buikstra and Cook (1980) state that paleopathological research incorporating stress indicators was stimulated by Scrimshaw et al.'s (1968) work, which emphasized the synergism between disease and dietary stress and their impact on human biological adaptation. In their review of the state of the art of paleopathology, Buikstra and Cook (1980) summarize a number of "nonspecific" stress indicators: Harris lines, dental hypoplasia, bone mineral density, growth attainment, sexual dimorphism and developmental asymmetry. The list of stress markers was expanded by Huss-Ashmore et al. (1982), though much of the theory for this paper was based on Goodman's ecological model of general stress (Armelagos et al. 1980). Goodman et al. (1984b) took the idea of the "stress indicator" further, becoming the primary reference to which the vast majority of subsequent papers on the topic of skeletal stress refer.

Within the framework of the stress indicator model, periosteal new bone production is seen as primarily representative of nonspecific infectious disease, with the infectious reactions being acute and localized or chronic and systemic, depending on the virulence of the infection and the resistance of the host (Goodman et al. 1984b). Today, these assertions are seen to be problematic. First, as previously stated, recording the prevalence of infectious response or periosteal new bone in general terms, deliberately ignores an aspect of population health of major significance: the influence of specific infectious diseases upon levels of morbidity and mortality (Powell 1988), not to mention the fact that all other etiologies of periosteal new bone formation are ignored. Second, the human body's physiological response to stress is not conducive to the formation of periosteal new bone. While not all is known about the precise mechanisms which trigger periosteal new bone production (Caron et al. 1987), as previously mentioned, it has been determined that hormones and ions play an essential role. Hormones and ions assert their control at the cellular level, determining the actions of osteoblasts and preosteoblasts. They influence the rate and duration of bone matrix formation, but they can also cause the inhibition of bone matrix formation (Canalis et al. 1993; Raisz and Kream 1983).

The interpretation of proliferative periosteal reactions as a stress indicator is inherently flawed due to the way in which glucocorticoids participate in the human body's physiological response to stress. Stress results in the stimulation of glucocorticoid secretion, which in turn inhibits bone formation, and therefore inhibits periosteal new bone production. Periosteal reactions should not be considered stress indicators, because the human body cannot produce new bone when it is stressed.

The adoption of Goodman et al.'s (1984b; 1988) skeletal stress indicator hypothesis as a framework for the investigation of human skeletal remains has been widespread (Hodges 1987; Powell 1988, 1991; De la Rua et al. 1995; Ribot and Roberts 1996; Byers 1997; Hitzemann et al. 1997; Pietrusewsky et al. 1997; Keenleyside 1998; Suzuki 1998; Lewis 1999; Bennike et al. 2005). But despite its popularity, many researchers are not completely satisfied with the model and have commented on its shortcomings. Jankauskas and Cesnys (1992:360) question the validity of the skeletal stress indicators, stating, "...there can be no simple way of interpreting these markers." Referring to skeletal stress indicators in general, Pecotte (1982) believes that it is naïve to assume that frequencies of stress indicators in a skeletal mortality sample are directly indicative of stress levels suffered by the general population, and that considerations as to the nature of the skeletal sample must first be given before any attempt is made to interpret the data.

Bush (1991:16) is strong in her disapproval of Selyean-based stress models, stating that the Selyean stress response is "not an adequate view of stress upon which to build biocultural interpretations." She underlines the importance of recognizing that psychological factors may play an important role in the formation of stress markers, particularly the psychological impact of weaning. Also important to consider is the individual's perception of stress. Bush (1991) emphasizes the fact that a universally stressful situation does not exist—each situation is dependent on the individual's response to the stressor.

Powell (1988) speaks of the limitations of stress markers as diagnostic tools in the investigation of population adaptation. She cites their nonspecificity, their potential loss through subsequent remodeling of bone or loss of dental tissues, their inability to act as measures of the extent and intensity of the stress episode, and the lack of any demonstrated predictable association between a particular stress episode and the formation of a particular stress marker as important shortcomings. Based on these arguments, perhaps stress indicators should be used as only very general indicators of population health and the quest for the identification of specific stress episodes should be abandoned.

Conclusions

Despite repeated admonitions, paleopathologists have largely ignored the wealth of clinical literature which mentions periosteal new bone formation as part of numerous disease complexes. Consequently, most paleopathological interpretations of periosteal reactions have been overly simplistic. Instead of trying to understand the role in which periosteal new bone production plays within different pathological contexts, many have concluded that it manifests as a result of "nonspecific infection." Most research does not describe the characteristics of the periosteal reactions encountered—it merely reports them as present. If qualifiers are placed on the reactions, they are simply described as being "mild," "moderate," or "severe," with no descriptions of these categories. Researchers may pay lip-service to the multifactorial etiology of periosteal reactions, but generally proceed as normal, continuing to perpetuate the assumption that "periostitis" can be simply equated with "nonspecific infection."

It was this assumption that led to periosteal reactions being included as one of the "indicators of skeletal stress," despite the fact that everything about the physiological stress response, particularly the effect of glucocorticoids, leads to the arrest of bone formation. The physiological stress response is designed to arrest growth in the body so that that energy can be allocated to more important coping mechanisms. Thus skeletal stress indicators such as stature, growth curves, and enamel hypoplasia are far better stress indicators than periosteal reactions. Paleopathology needs to standardize the reporting of periosteal reaction, move away from the simplistic interpretations, and recognize that these lesions are simply more than indicators of "nonspecific infection."

REFERENCES

Armelagos, G. J., and Goodman, A. H., 1991   The Concept of Stress and Its Relevance to Studies of Adaptation in Prehistoric Populations. Collegium Antropologicum 15:45–58.
Armelagos, G. J., Goodman, A. H., Bickerton, S., 1980   Determining Nutritional and Infectious Disease Stress in Prehistoric Populations. American Journal of Physical Anthropology 52:201 (abstract).
Bennike, P., Lewis, M. E., Schutkowski, H., Valentin, F., 2005   A Comparison of Childhood Morbidity in Two Contrasting Medieval Cemeteries from Denmark. American Journal of Physical Anthropology 128:734–746.
Birkett, D. A., 1983   Non-Specific Infections. In Disease in Ancient Man. G. D. Hart, ed. pp. 99–105. Toronto: Clarke Irwin.
Buikstra, J. E., and Cook, D. C., 1980   Paleopathology: An American Account. Annual Review of Anthropology 9:433–470.
Buikstra, J. E., and Ubelaker, D. H., eds. 1994   Standards for Data Collection from Human Skeletal Remains. Fayetteville: Arkansas Archeological Survey.

Bush, H. M., 1989   The Recognition of Physiological Stress in Human Skeletal Material: A Critique of Method and Theory with Specific Reference to the Vertebral Column. Ph.D. Dissertation, University of Sheffield.

Bush, H. M., 1991   Concepts of Health and Stress. *In* Health in Past Societies: Biocultural Interpretations of Human Skeletal Remains in Archaeological Contexts. H. Bush, and M. Zvelebil, eds. pp. 11–21. Oxford: Tempus Reparatum.

Byers, S. N., 1997   The Relationship Between Stress Markers and Adult Skeletal Size. American Journal of Physical Anthropology S24:85–6 (abstract).

Canalis, E., McCarthy, T. L., and Centrella, M., 1993   Factors that Regulate Bone Formation. *In* Physiology and Pharmacology of Bone. G. R. Mundy, and T. J. Martin, eds. pp. 249–266. Berlin: Springer-Verlag.

Caron, J. P., Barber, S. M., Doige, C. E., and Pharr, J. W., 1987   The Radiographic and Histologic Appearance of Controlled Surgical Manipulation of the Equine Periosteum. Veterinary Surgery 16:13–20.

Chrousos, G. P., and Gold, P. W., 1992   The Concept of Stress and Stress System Disorders. Journal of the American Medical Association 267:1244–1252.

Cohen, M. N., 1994   The Osteological Paradox Reconsidered. Current Anthropology 35:629–631.

Cohen, M. N., and Armelagos, G. J., eds. 1984   Paleopathology at the Origins of Agriculture. Orlando, FL: Academic Press.

Cohen, M. N., and Crane-Kramer, G. M. M., eds. 2007   Ancient Health: Skeletal Indicators of Agricultural and Economic Intensification. Gainesville: University Press of Florida.

Cook, D. C., 1976   Pathologic States and Disease Process in Illinois Woodland Populations: An Epidemiologic Approach. Ph.D. Dissertation, University of Chicago.

Cox, T., 1978   Stress. London: The Macmillan Press Ltd.

Crampton, P., 1817   On Periostitis or Inflammation of the Periosteum. Dublin: Hodges and McArthur.

De la Rua, C., Izagirre, N., and Manzano, C., 1995   Environmental Stress in a Medieval Population of the Basque Country. Homo 45:268–289.

Detweiler-Blakely, B., 1988   Stress and the Battle Casualties. *In* The King Site: Continuity and Contact in Sixteenth Century Georgia. R. L. Blakely, ed. pp. 87–98. Athens: University of Georgia Press.

Edeiken, J., 1981   Roentgen Diagnosis of Diseases of Bone. 3rd edition. Baltimore: Williams & Wilkins.

Edeiken, J., Hodes, P. J., and Caplan, L. H., 1966   New Bone Production and Periosteal Reaction. American Journal of Roentgenology, Radium Therapy and Nuclear Medicine 97:708–718.

Elting, J. J., and Starna, W. A., 1984   A Possible Case of Pre-Columbian Treponematosis from New York State. American Journal of Physical Anthropology 65:267–273.

Fennell, K. J., and Trinkaus, E., 1997   Bilateral Femoral and Tibial Periostitis in the La Ferrassie 1 Neanderthal. Journal of Archaeological Science 24:985–995.

Goodman, A. H., 1993   On the Interpretation of Health from Skeletal Remains. Current Anthropology 34:281–288.

Goodman, A. H., Lallo, J., Armelagos, G. J., and Rose, J. C., 1984a   Health Changes at Dickson Mounds, Illinois (A.D. 950–1300). *In* Paleopathology at the Origins of Agriculture. M. N. Cohen, and G. J. Armelagos, eds. pp. 277–291. New York: Academic Press.

Goodman, A. H., Martin, D. L., Armelagos, G. J., and Clark, G., 1984b   Indications of Stress from Bone and Teeth. *In* Paleopathology at the Origins of Agriculture. M. N. Cohen, and G. J. Armelagos, eds. pp. 13–49. New York: Academic Press.

Goodman, A. H., Thomas, R. B., Swedlund, A. C., and Armelagos, G. J., 1988   Biocultural Perspectives on Stress in Prehistoric, Historical, and Contemporary Population Research. Yearbook of Physical Anthropology 31:169–202.

Hackett, C. J., 1976   Diagnostic Criteria of Syphilis, Yaws and Treponarid (Treponematoses) and of Some Other Diseases in Dry Bones (for Use in Osteo-Archaeology). Berlin: Springer-Verlag.

Hackett, C. J., 1978   Treponematoses (Yaws and Treponarid) in Exhumed Australian Aboriginal Bones. Records of South Australia Museum 17:387–405.

Hackett, C. J., 1983   Problems in the Palaeopathology of the Human Treponematoses. *In* Disease in Ancient Man. G. D. Hart, ed. pp. 106–128. Toronto: Clarke Irwin.

Hitzemann, N. J., Langford, D. M., Lesley, B. P., and Specht, W. J., 1997   Bioarchaeological Analysis of a Late Woodland, Illinois Site: A Synthesis of Multiple Stress Indicators. American Journal of Physical Anthropology S24:130 (abstract).

Hodges, D. C., 1987   Health and Agricultural Intensification in the Prehistoric Valley of Oaxaca, Mexico. American Journal of Physical Anthropology 73:323–332.

Howarth, C. I., 1978   Environmental Stress. *In* The Uses of Psychology. C. I. Howarth and W. C. Gillham, eds. London: Allen and Unwin.

Huss-Ashmore, R., Goodman, A. H., and Armelagos, G. J., 1982   Nutritional Inference from Paleopathology. *In* Advances in Archaeological Method and Theory, vol. 5. M. B. Schiffer, ed. pp. 395–474. New York: Academic Press.

Hutchinson, D. L., 1992   Comment on: "The Osteological Paradox," by J. W. Wood, et al. Current Anthropology 33:360.

Hutchinson, D. L., 1993   Treponematosis in Regional and Chronological Perspective from Central Gulf Coast Florida. American Journal of Physical Anthropology 92:249–61.

Jaffe, H. L., 1972   Metabolic, Degenerative and Inflammatory Diseases of Bones and Joints. Philadelphia: Lea & Febiger.

Jankauskas, R., and Cesnys, G., 1992   Comment on: "The Osteological Paradox," by J. W. Wood, et al. Current Anthropology 33:360.

Keenleyside, A., 1998   Skeletal Evidence of Health and Disease in Pre-Contact Alaskan Eskimos and Aleuts. American Journal of Physical Anthropology 107:51–70.

Kelley, M. A., 1989   Infectious Disease. *In* Reconstruction of Life from the Skeleton. M. Y. Işcan, ed. pp. 191–199. New York: Alan R Liss, Inc.

Kelley, M. A., and Micozzi, M. S., 1984   Rib Lesions in Chronic Pulmonary Tuberculosis. American Journal of Physical Anthropology 65:381–386.

Kelley, M. A., Murphy, S. P., Levesque, D.R., and Sledzik, P. S., 1994   Respiratory Disease Among Protohistoric and Early Historic Plains Indians. *In* Skeletal Biology in the Great Plains: Migration, Warfare, Health and Subsistence. D. W. Owsley, and R. L. Jantz, eds. pp. 123–130. Washington: Smithsonian Institution Press.

Lallo, J. W., 1973   The Skeletal Biology of Three Prehistoric American Indian Societies from Dickson Mounds. Ph.D. Dissertation, University of Massachusetts, Amherst.

Lallo, J. W., Armelagos, G. J., and Rose, J. C., 1978   Paleoepidemiology of Infectious Disease in the Dickson Mounds Population. Medical College of Virginia Quarterly 14:17–23.

Lallo, J. W., and Rose, J. C., 1979   Patterns of Stress, Disease and Mortality in Two Prehistoric Populations from North America. Journal of Human Evolution 8:323–335.

Larsen, C. S., 1997   Bioarchaeology. Cambridge: Cambridge University Press.

Larsen, C. S., Hutchinson, D. L., Stojanowski, C. M., Williamson, M. A., Griffin, M. C., Simpson, S. W., Ruff, C. B., Schoeninger, M. J., Norr, L., Teaford, M. F., Driscoll, E. M., Schmidt, C. W., and Tung, T. A., 2007   Health and Lifestyle in Georgia and Florida: Agricultural Origins and Intensification in Regional Perspective. *In* Ancient Health: Skeletal Indicators of Agricultural and Economic Intensification. M. N. Cohen, and G. M. M. Crane-Kramer, eds. pp. 20–34. Gainesville: University Press of Florida.

Lazarus, R. S., 1976   Patterns of Adjustment. New York: McGraw-Hill.

Levine, S., Coe, C., and Wiener, S. G., 1989   Psychoneuroendocrinology of Stress: A Psychobiological Perspective. *In* Psychoendrocrinology. F. R. Brush, and S. Levine, eds. pp. 341–377. New York: Academic Press.

Lewis, B., 1994   Treponematosis and Lyme Borreliosis Connections: Explanation for Tchefuncte Disease Syndromes? American Journal of Physical Anthropology 93:455–475.

Lewis, M. E., 1999   Measures of Environmental Stress in Non-Adult Skeletons from Pre- and Post-Industrial Communities in England. American Journal of Physical Anthropology S28:183 (abstract).

Lewis, M. E., 2000   Non-Adult Palaeopathology: Current Status and Future Potential. *In* Human Osteology in Archaeology and Forensic Science. M. Cox, and S. Mays, eds. pp. 39–57. London: Greenwich Medical Media Ltd.

Lewis, M. E., 2004   Endocranial Lesions in Non-Adult Skeletons: Understanding Their Aetiology. International Journal of Osteoarchaeology 14:82–97.

Lewis, M. E., Roberts, C. A., and Manchester, K., 1995   Inflammatory Bone Changes in Leprous Skeletons from the Medieval Hospital of St. James and St. Mary Magdalene, Chichester, England. International Journal of Leprosy 63:77–85.

Littleton, J., 1998   Skeletons and Social Composition Bahrain 300 B.C.–A.D. 250. Oxford: Archaeopress.

Mann, R. W., and Murphy, S. P., 1990   Regional Atlas of Bone Disease: A Guide to Pathologic and Normal Variation in the Human Skeleton. Springfield: C. C. Thomas.

Mansilla, J., and Pijoan, C. M., 1995   Brief Communication: A Case of Congenital Syphilis during the Colonial Period in Mexico City. American Journal of Physical Anthropology 97:187–195.

Mason, J. W., 1971   A Re-Evaluation of the Concept of "Non-Specificity" in Stress Theory. Journal of Psychosomatic Research 8:323–334.

Mason, J. W., 1975   A Historical View of the Stress Field.   Journal of Human Stress 1(1): 6–12, 1(2):22–36.

McEwen, B. S., and Stellar, E., 1993   Stress and the Individual: Mechanisms Leading to Disease.   Archives of Internal Medicine 153:2093–2101.

Mensforth, R. P., Lovejoy, C. O., Lallo, J. W., and Armelagos, G. J., 1978   The Role of Constitutional Factors, Diet, and Infectious Disease in the Etiology of Porotic Hyperostosis and Periosteal Reactions in Prehistoric Infants and Children. Medical Anthropology 2:1–59.

Merriam Webster's Medical Dictionary, 1995   Periostitis. Springfield: Merriam-Webster Inc.

Milner, G. R., 1991   Health and Cultural Change in the Late Prehistoric American Bottom, Illinois. *In* What Mean These Bones? Studies in Southeastern Bioarchaeology. M. L. Powell, P. S. Bridges, and A. M. Wagner Mires, eds. pp. 52–69. Tuscaloosa: University of Alabama Press.

Mitchinson, M. J., Arno, J., Edwards, P. A. W., LePage, R. W. F., and Minson, A. C., 1996   Essentials of Pathology. Oxford: Blackwell Science Ltd.

Ortner, D. J., no date   Classification of Bone Lesions. Draft document. Washington, DC: Smithsonian Institution.

Ortner, D. J., 2003   Identification of Pathological Conditions in Human Skeletal Remains, 2nd edition. London: Academic Press.

Ortner, D. J., Butler, W., Cafarella, J., and Milligan, L., 2001   Evidence of Probable Scurvy in Subadults from Archaeological Sites in North America. American Journal of Physical Anthropology 114:343–351.

Palfi, G., Dutour, O., Borreani, M., Brun, J.-P., and Berato, J., 1992   Pre-Columbian Congenital Syphilis from the Late Antiquity in France. International Journal of Osteoarchaeology 2:245–64.

Parsons, U., 1849   Prize Dissertations on 1. Inflammation of the Periosteum 2. Eneuresis Irritata 3. Cutaneous Diseases 4. Cancer of the Breast 5. Malaria, 2nd edition. Providence: BT Albro.

Pecotte, J. K., 1982   Nutritional Stress and Health in Ancient Egypt: Methodology. Antropologia Contemporanea 5:147–154.

Pfeiffer, S., 1991   Rib Lesions and New World Tuberculosis. International Journal of Osteoarchaeology 1:191–198.

Pfeiffer, S., and Fairgrieve, S. I., 1994   Evidence from Ossuaries: The Effect of Contact on the Health of Iroquoians. *In* In the Wake of Contact: Biological Responses to Conquest. C. S. Larsen, and G. R. Milner, eds. pp. 47–61. New York: Wiley-Liss, Inc.

Pietrusewsky, M., and Douglas, M. T., 1994   An Osteological Assessment of Health and Disease in Precontact and Historic (1778) Hawai'i. *In* In the Wake of Contact: Biological Responses to Conquest. C. S. Larsen, and G. R. Milner, eds. pp. 179–196. New York: Wiley-Liss, Inc.

Pietrusewsky, M., Douglas, M. T., and IkeharaQuebral, R. M., 1997 An Assessment of Health and Disease in the Prehistoric Inhabitants of the Mariana Islands. American Journal of Physical Anthropology 104:315–342.

Powell, M. L., 1988 Status and Health in Prehistory: A Case Study of the Moundville Chiefdom. Washington, DC: Smithsonian Institution Press.

Powell, M. L., 1991 Ranked Status and Health in the Mississippian Chiefdom at Moundville. *In* What Mean These Bones? Studies in Southeastern Bioarchaeology. M. L. Powell, P. S. Bridges, and A. M. Wagner Mires, eds. pp. 22–51. Tuscaloosa: University of Alabama Press.

Powell, M. L., 1992 In the Best of Health? Disease and Trauma among the Mississippian Elite. *In* Lords of the Southeast: Social Inequality and the Native Elites of Southeastern North America. A. W. Barker, and T. R. Pauketat, eds. pp. 81–97.Washington: Archaeology Papers of the American Anthropology Association, No. 3.

Ragsdale, B. D., 1993 Morphologic Analysis of Skeletal Lesions: Correlation of Imaging Studies and Pathologic Findings. Advances in Pathology and Laboratory Medicine 6:445–490.

Ragsdale, B. D., Madewell, J. E., and Sweet, D. E., 1981 Radiologic and Pathologic Analysis of Solitary Bone Lesions, Part II: Periosteal Reactions. Radiologic Clinics of North America 19:749–783.

Raisz, L. G., and Kream, B. E., 1983 Regulation of Bone Formation (First of Two Parts). New England Journal of Medicine 309:29–35.

Rathbun, T. A., 1987 Health and Disease at a South Carolina Plantation: 1840–1870. American Journal of Physical Anthropology 74:239–53.

Reichs, K. J., 1989 Treponematosis: a Possible Case from the Late Prehistoric of North Carolina. American Journal Physical Anthropology 79:289–303.

Resnick, D., 1995 Diagnosis of Bone and Joint Disorders, 3rd edition. Philadelphia: W. B. Saunders Co.

Ribot, I., and Roberts, C., 1996 A Study of Non-Specific Stress Indicators and Skeletal Growth in Two Mediaeval Subadult Populations. Journal of Archaeological Science 23:67–79.

Richardson, M., 1994 Approaches to Differential Diagnosis in Musculoskeletal Imaging: Periosteal Reaction, University of Washington. Electronic document. http://www.rad.washington.edu/academics/academic-sections/msk/teaching-materials/online-musculoskeletal-radiology-book/periosteal-reaction (Accessed June 12, 2010).

Roberts, C. A., 2000 Infectious Disease in Biocultural Perspective: Past, Present and Future Work in Britain. *In* Human Osteology in Archaeology and Forensic Science. M. Cox, and S. Mays, eds. pp. 145–162. London: Greenwich Medical Media Ltd.

Roberts, C. A., Lucy, D., and Manchester, K., 1994 Inflammatory Lesions of Ribs: An Analysis of the Terry Collection. American Journal of Physical Anthropology 95:169–82.

Roberts, C. A., and Manchester, K., 1995 The Archaeology of Disease. 2nd edition. Stroud: Sutton Publishing.

Rothschild, B. M., and Heathcote, G. M., 1993 Characterization of the Skeletal Manifestations of the Treponemal Disease Yaws as a Population Phenomenon. Clinical Infectious Diseases 17:198–203.

Rothschild, B. M., and Rothschild, C., 1996 Treponemal Disease in the New World. Current Anthropology 37:556–561.

Schermer, S. J., Fisher, A. K., and Hodges, D. C., 1994 Endemic Treponematosis in Prehistoric Western Iowa. *In* Skeletal Biology in the Great Plains: Migration, Warfare, Health and Subsistence. D. W. Owsley, R. L. Jantz, eds. pp. 109–121. Washington, DC: Smithsonian Institution Press.

Scheuer, L., and Black, S. M., 2000 Developmental Juvenile Osteology. London: Academic Press.

Scrimshaw, N. S., Taylor, C. E., and Gordon, J. E., 1968 Interactions of Nutrition and Infection. Geneva: World Health Organization.

Selye, H., 1957   The Stress of Life. London: Longmans, Green and Co.

Selye, H., 1976   Forty Years of Stress Research: Principal Remaining Problems and Misconceptions. The Canadian Medical Association Journal 115:53–56.

Senn, N., 1886   Clinical Lecture on Periostitis. Philadelphia Medical Times 16 (22):769–774.

Shopfner, C. E., 1966   Periosteal Bone Growth in Normal Infants: A Preliminary Report. American Journal of Roentgenology 97:154–63.

Strothers, D. M., and Metress, J. F., 1975   A System for the Description and Analysis of Pathological Changes in Prehistoric Skeletons. Ossa 2:3–9.

Štvrtinová, V., Jakubovský, J., Hulín, I., 1995   Inflammation and Fever. In Pathophysiology. Principles of diseases. Bratislava: SAP (ISBN 80-967366-1-2).

Suzuki, T., 1991   Paleopathological Study on Infectious Diseases in Japan. In Human Palaeopathology: Current Syntheses and Future Options. D. J. Ortner, and A. C. Aufderheide, eds. pp. 128–139. Washington, DC: Smithsonian Institution Press.

Suzuki, T., 1998   Indicators of Stress in Prehistoric Jomon Skeletal Remains in Japan. Anthropological Science 106:127–137.

Ubelaker, D. H., 1992   Comment on: "The Osteological Paradox," by J. W. Wood, et al. Current Anthropology 33:363–364.

Weston, D. A., 2008   Investigating the Specificity of Periosteal Reactions in Pathology Museum Specimens. American Journal of Physical Anthropology 137:48–59.

Weston, D. A., 2009   Paleohistopathological Analysis of Pathology Museum Specimens: Can Periosteal Reaction Microstructure Explain Lesion Etiology? American Journal of Physical Anthropology 140:186–193.

Williams, J. A., 1994   Disease Profiles of Archaic and Woodland Populations in the Northern Plains. In Skeletal Biology in the Great Plains: Migration, Warfare, Health and Subsistence. D. W. Owsley, and R. L. Jantz, eds. pp. 91–107. Washington, DC: Smithsonian Institution Press.

Wood, J. W., Milner, G. R., Harpending, H. C., Weiss, K. M., 1992   The Osteological Paradox: Problems of Inferring Prehistoric Health from Skeletal Samples. Current Anthropology 33:343–358.

CHAPTER 28

Joint Disease

Tony Waldron

INTRODUCTION

Not surprisingly, joint disease is by far the most common form of pathology found in human remains and, with dental disease, accounts for about two-thirds of all the lesions found in an assemblage. There are many descriptions of joint disease in the paleopathological literature and I do not intend to add to them in great detail here, but, instead, highlight some of the more interesting and problematic features of this group of diseases.

It is usual to differentiate the joint diseases into those that are predominantly proliferative and those in which erosions form a significant part of the pathology. The archetypical proliferative joint disease is osteoarthritis, while rheumatoid arthritis can be considered as the 'type specimen' of the erosive disorders. The erosive arthropathies can be further subdivided into the seropositive and the seronegative forms, a distinction that was originally made depending on whether the patient's blood was positive or negative for the so-called rheumatoid factor. The seronegative group of diseases are characterized further by involvement of the sacroiliac joint and by some degree of spinal fusion; on this account they are increasingly referred to as the spondyloarthropathies (or spondylarthropathies). A further distinction between the proliferative and erosive arthropathies is that inflammatory change is a major feature of the latter but is a relatively minor component of the former; there is, however, a form of osteoarthritis in which inflammatory change is more prominent than usual, and this is known (unsurprisingly) as erosive osteoarthritis and in a sense, forms a bridge between the two groups (see Figure 28.1).

A Companion to Paleopathology, First Edition. Edited by Anne L. Grauer.
© 2012 John Wiley & Sons, Ltd. Published 2016 by John Wiley & Sons, Ltd.

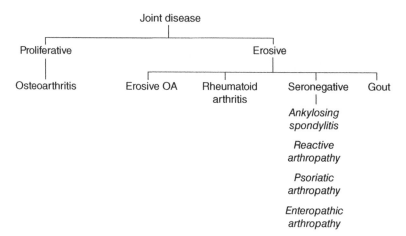

Figure 28.1   Classification of joint disease.

OSTEOARTHRITIS

Osteoarthritis (which some authorities prefer to call osteoarthrosis to avoid the implications of inflammation that the suffix –*itis* implies) is fundamentally a disease of the articular cartilage, and the morphological changes that take place secondarily to the breakdown of the cartilage are essentially reparative in nature. So, osteoarthritis (OA) should *not* be considered as a degenerative condition, even though the majority of sources still refer to it this context. In a very few cases, the changes in the joint are reversible (Bland 1983) although repair seems to be limited to small joints and is not to be expected in the great majority of cases. With the exception of the erosive form, inflammatory change is not prominent although inflammatory cytokines can be recovered from fluid aspirated from osteoarthritic joints.

Morphology and distribution

Osteoarthritis only affects the synovial joints and almost certainly affected animals with such joints from the very earliest times; dinosaurs and other fossil bones have been found with OA and it has undoubtedly afflicted humans and their ancestors from the time that any lived sufficiently long to develop it. So far as one can tell, the major pathological characteristics of the disease have not changed over time; thus, affected joints will show a suite of morphological features that include marginal osteophytes, new bone on the joint surface, pitting on the joint surface, alteration of the joint contour, and eburnation. Eburnation is a particularly important sign since it implies that the articular cartilage has completely worn away over at least part of the joint surface and that under the eburnated area bare bone was articulating with bare bone to produce the polished surface that sometimes resembles the surface of a billiard ball. Nor do the more important epidemiological characteristics seem to have changed greatly over time (Waldron 1995); thus it can readily be shown that the prevalence of the disease increases with age, and is slightly more common in females (especially "elderly" females) than in males.

The morphological changes in OA may vary considerably from case to case; they may be hypertrophic of atrophic depending on the amount of proliferation of new bone; and they may be either unifocal or multifocal. Any case can be considered to occupy position somewhere in a space defined by these two axes and a third which will indicate the degree of inflammatory change present. There do not seem to have been any differences in the proportion of cases that tend towards one or other of the proliferation endpoints, but there is some evidence that as it affects the hands, the disease has tended to become more towards the multifocal end of the scale in recent times. As mentioned below, there seems to be some evidence that inflammatory change is more frequent in modern populations than it was in the past.

Although any synovial joint is potentially at risk of developing OA (so long as it moves), the risk does not seem to be shared equally among the different joints of the body. Thus, in the hand, it is the joints on the radial side of the hand that are most commonly involved; while in the elbow, OA rarely affects the ulnohumeral joint, being almost always confined to the radio-humeral joint, while in the knee, the patellofemoral compartment is more frequently affected that either of the tibiofemoral compartments. In addition, there are some large joints that seem seldom to be affected, including the shoulder (glenohumeral) joint, and the ankle joint. That the ankle joint is almost invariably spared is especially remarkable considering that the stress on it greater than in any other joint; it remains the case, however, that OA of the ankle joint almost never occurs unless it has suffered previous trauma or the architectonics of the joint have been altered by some other disease; I have seen a case where OA of the ankle supervened due to bending of both tibias as the result of Paget's disease, for example. Although various explanations have been put forward to account for the seemingly erratic distribution of OA around the body, none seems entirely satisfactory although Hutton's (1987) hypothesis to account for the pattern in the hand is perhaps the most original if not entirely plausible.

Prevalence

One feature of the epidemiology of OA that would be interesting to examine is whether or not the prevalence has altered over time and how prevalence in the past compares with that of the present day. Regrettably, the means have not yet been found to achieve this goal. One stumbling block is that in many instances it is not at all clear how those who examine and report on human remains derive the frequency of OA in their assemblage; this is due not only to differences in diagnosing the disease but also in considerably uncertainty as to which denominators have been used in the calculation of the prevalence. The diagnosis of OA must obviously depend on recognizing the morphological changes that are present within the joint, almost always determined by direct observation, radiology not having a great deal to offer in this respect. Authors will vary, however, in which of the changes they consider need to be present for a diagnosis to be made; some appear content to make the diagnosis if marginal osteophytes are present, others may require additional signs, while we take a more stringent view that the disease should be diagnosed only when eburnation is present. And it is by no means unusual to find that authors will be rather coy about their diagnostic criteria, assuming no doubt that since everyone knows what OA is, there is no need to elaborate further. But unless this *is* known, there can be no certainty that comparisons are valid; in short, there is a lack of a commonly (much less a universally) accepted operational definition, and one is urgently needed to propel the subject forward. On the matter of denominators, there is an equal lack of clarity. If the frequency of OA of a particular joint is referred to, it is often difficult to determine what denominator has been used; is it the total number of skeletons; the total number of adults; the total number of joints; or what? Although this would seem to be a simple matter to decide, the decision is complicated by the inevitable absence of some joints due to poor preservation and, again, this is a matter on which general agreement is needed to avoid misleading conclusions being drawn from interstudy comparisons. I have made some suggestions elsewhere as to how both these problems can be dealt with, but I am not holding my breath. It is most usual to report the prevalence of OA in an assemblage, joint by joint; arriving at a summary statistic by which to express the prevalence of OA in terms of the complete assemblage is a much more complicated exercise, again because variable preservation of the joints wreaks havoc with denominators; one solution has been proposed, but I have yet to see any evidence that anyone has seen fit to try to work it.

The difficulties in determining the prevalence of OA in bone reports or papers that were published long ago in the past—say prior to the Second World War—are so great as to be virtually insurmountable. This is not only because in these early account, OA might masquerade under a variety of aliases, since authors in those days were more likely to try to write elegant prose and avoid repetition where possible, but they might be entirely misleading and refer to OA by a name which to us has a completely different connotation. For example, in the report on the human remains from Nubia, Wood Jones (1908) stated that rheumatoid arthritis was the most common disease found, a statement that would be remarkable, except that when one goes into it a bit further, they are obviously referring to osteoarthritis, which makes a good deal more sense; even then, they do not provide any data by which it might be possible to determine *how* common it was. Secondly—and this will not be surprising in view of what has been written above—when the frequency *is* given, it is hard to see what denominator has been used and very often, the term incidence is used instead of prevalence. Indeed, I wish that I could confine my criticism on this last point to the past, but even now, authors will refer to the incidence of OA (or indeed other diseases) in an assemblage for which they cannot possibly know the data to make the necessary calculation. This is due to the fact that the incidence of a disease is determined as the number of new cases arising in a population at risk over the study period. It is self-evident that *new* cases can never be defined in a skeletal assemblage, and hence incidence can never be determined. Such elementary errors are not likely to increase the confidence that one might have in these authors' epidemiological expertise.

Comparing the prevalence of OA in past populations (or at least such of their members as forms an assemblage of remains) with that of the contemporary population also raises some intriguing problems, not least how valid a comparison between a living and a dead population can be, a difficulty that arises because of the manner in which data are collected by modern epidemiologists. In the first place, it is important to determine whether the modern study has measured incidence or prevalence; secondly, the study base that was used; and thirdly, how the disease was diagnosed. The first problem should be immediately obvious from the title or abstract of the study, but even modern epidemiologists have been known to get it wrong. Secondly, the study base will almost certainly *not* be directly comparable to a study of human remains even ignoring their vital status. Thus the modern study might be conducted on subjects recruited from a general practice list, a hospital clinic, the general population, or—occasionally and nowadays increasingly rarely— from autopsies. The diagnosis may be determined by clinical (pain, swelling of the joints, crepitus) or radiological (marginal osteophytes and joint space narrowing) criteria, or a mixture of both. In very few cases will be resultant data be comparable with those that a paleopathologist might collect. Perhaps the most directly comparable modern data are those which have been obtained by radiographic examination of random samples of the general population, such as those carried out by Lawrence and his colleagues in the north of England in the 1960s and 1970s (Lawrence 1977). Even here, some caution has to be used as the diagnostic criteria may not coincide exactly with those used by the paleopathologist (see Chapter 7 by Boldsen and Milner in this volume, for further discussion of the use of an epidemiological approach in paleopathology).

Even if direct quantitative comparisons cannot be made between ancient and modern data, however, it may be possible to make qualitative comparisons and these may provide some useful information. The easiest way to make qualitative comparisons is by rank-ordering the frequency with which different joints are affected but even here, there are some pitfalls. The modern clinical literature suggests that in the modern population, OA is most common in the hand, hip and knee and this is certainly not the case with archaeological assemblages. The reason for this apparent discrepancy is not that there has been a profound change in the distribution of OA but that clinicians only see patients with painful joints. A more useful basis of comparison are the radiological surveys referred to above, and conducted on a population basis in the U.K., the Netherlands, and the United States. From studies such as these, OA of the various joints can be ranked in order of frequency and then compared with a similar ranking derived from skeletal assemblages. In Table 28.1, data have been extracted from these studies and compared with archaeological data from my own studies. There are some differences in the results of the radiological surveys although OA of the hand and foot are common, and OA of the knee more common than OA of the knee, except in the case of the U.K. males, in whom OA of the knee was the least common of all for reasons that are by no means clear. The archaeological data show that OA of the acromioclavicular joint (ACJ), spine and hand occupy the first three ranks in both the medieval and post-medieval periods. Osteoarthritis of the foot seems much less common in the medieval than in the post-medieval period, and there is a reversal of the order for OA of the hip and the knee between the two periods. There is much more work to be done, particularly to refine the archaeological data and, preferably, to be able to combine data from studies carried out on assemblages from similar time periods and geographical locations (using comparable methodology) so as to increase the number of cases, since, although the disease is commonly encountered in assemblages, cases of hip and knee OA, for example, are never very numerous and where small numbers are involved, small fluctuations may considerably affect ranking.

Table 28.1 Rank order of osteoarthritis at different sites; modern (radiological) and archaeological data

Rank order of sites affected	United Kingdom[a]		The Netherlands[b]		United States[c]		Archaeological[d]	
	Male	Female	Male	Female	Male	Female	Medieval	Post-medieval
1	Hand	Hand	Hand	Hand	Knee	Knee	ACJ	ACJ
2	Spine	Knee	Foot	Foot	Hand	Hand	Spine	Spine
3	Hip	Spine	Knee	Knee	Foot	Foot	Hand	Hand
4	Foot	Foot	Spine	Spine	Hip	Hip	Hip	Foot
5	Wrist	Hip	Wrist	Wrist			Knee	SCJ
6	Knee	Wrist	Hip	Hand			SCJ	Knee
7							Wrist	TMJ
8							Foot	Hip
9							Elbow	Elbow
10							TMJ	Wrist

Notes: [a]Lawrence, 1977 [b]van Saase et al, 1989 [c]Lawrence et al, 1998 [d]Author's data; both sexes combined
ACJ = acromioclavicular joint; SCJ = sterno-clavicular joint; TMJ = temporomandibular joint.

Relation between morphology and clinical effects

There is a tendency to assume that florid changes in a joint can be taken to mean that the joint was either painful during life, or that mobility was affected, or both. In fact, this is seldom the case. My late colleague Juliet Rogers was fond of showing an X-ray of a pelvis at meetings. One of the hips on the film showed extensive changes with marginal osteophyte, joint space narrowing and marked sclerosis (indicating eburnation), the other appeared virtually normal. The catch, of course, was that the patient was complaining of pain in the hip that showed minimal radiological change and that was the one that was replaced. Juliet was also able to collect a series of arthritic hips that had been surgically removed and reduce them to archaeological status by removing all the soft tissue and drying them. The result was that more than one that showed so little change in the dried state that they would probably have been passed as normal by a paleopathologist, suggesting that cases of OA that may have been clinically significant during life are being missed.

Several studies have shown that there is a poor correlation between radiological appearances and symptoms, the knee being the site where the correlation is greatest, although even here, the correlation is relatively weak (Watt 2000). We do know that OA is some sites is more likely to be clinically significant than in others; OA of the thumb base is often painful and so is OA of the medial compartment of the knee, and of the hip. The hip is an interesting case in that it is often the site at which change is minimal, presenting the so-called atrophic form of the disease. Occasionally there may be such a massive amount of marginal osteophyte that it actually impairs the movement of the joint—I have seen this in the hip joint, for example; under these conditions suggesting that this would have limited movement during life does not seem rash or unlikely. In other cases, there may be cortical thinning of the humerus in the presence

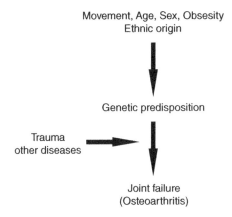

Figure 28.2   Factors contributing to joint failure (osteoarthritis).

of OA of the shoulder, for example; if this was interpreted as being due to disuse atrophy consequent upon restriction of movement at the shoulder joint, there is a high probability of this being correct. In most other circumstances, however, it would be sensible to err on the side of caution and leave speculation to those who care less about their reputation

Movement ... and occupation

To all intents and purposes, movement is a *sine qua non* for the development of OA although there are a number of other important precipitants. These include age, sex, obesity, ethnic origin and a genetic predisposition; it has also been suggested that vascular (Conaghan et al. 2005) and neurological factors may be involved (Tait and Bird 1994) but this has by no means been certainly established. In so-called primary OA—when there is no known proximate cause—any number of the precipitants interact in those with a genetic predisposition to cause joint failure which manifests itself in the spectrum of clinical and pathological signs that we call osteoarthritis. In the case of trauma, and in the presence of some other conditions that affect the skeleton, including Paget's disease of bone, OA is the direct consequence of the pre-existing condition and on this account is usually referred to as secondary OA (see Figure 28.2).

It is probably the central importance of movement in the development of OA that prompted the notion that it might be directly related to occupations in which repetitive forceful movements were a prominent feature. This then led to the notion that it might be possible to deduce the occupation of particular individuals from the disposition of OA around their skeletons (see Chapter 29 by Jurmain et al. in this book for further discussion of this argument and analysis of studies seeking to reconstruct activity in past populations). There have been a great many studies of OA in modern occupational groups, usually with results that are entirely unsurprising; that is to say, miners tend to have a greater than normal prevalence of OA of the spine, carpet layers of the knee, ballet dancers of the foot, and so on. There are, in addition, some results that are not so obvious, the most convincing being that farmers have a much greater frequency of OA of the hip than the general population (Schouten et al. 2002). This association has not been adequately explained and seems to apply to whatever group is being studied—arable, dairy, or mixed farmers—but the result has been confirmed in many studies and the relative risk is of the order of nine. There have also been studies which have not shown any increased risk in occupational groups, and it has to be remembered that by no means all those who engage in hard physical work get OA and, conversely, that those who lead entirely sedentary lives may do so.

The most important features of all this endeavor, however, are that there is absolutely *no* form of OA that is unique to one occupational group, and secondly, that even in those occupations in which there is a greatly increased risk of developed OA of a particular site, there are many more individuals outside that occupation with the condition than inside it. Thus, although there seems little doubt that farmers are greatly at risk of developing OA of the hip, the majority of those with the condition are *not* farmers. These considerations have not been seen as a bar by those who continue to infer the occupation of a skeleton with OA but it is very easy to demonstrate how futile an exercise this is. I can do this with a simple example. Suppose that you are invited to attend a rheumatology clinic with the usual preponderance of patients with OA. You are not allowed to talk to, or examine the patients but you are presented with a sheaf of X-rays which shows which of their joints are affected. Let us further suppose that there are fifty such patients and that on an accompanying sheet of paper you are invited to suggest the occupation of each on the basis of your study of their X-rays. How many would you expect to get correct? Very few—if any—I would suggest; and the reason is that here, as in paleopathological (or paleoepidemiological) studies, one must of necessity argue from what might be called "wrong end" epidemiology; that is, arguing from effect to cause, rather than from cause to effect which is the normal route. But this is exactly what those who affect to be able to tell occupation from the skeleton are trying to do, and unless they are very lucky, they are certain to be wrong, and even if they are not, there is no means of verifying their conclusion because there is no evidence which can be used to validate the conclusion. I have had it explained to me that it is alright to make the inference because in the past, there were relatively few occupations to chose from so that, presumably, this would increase the chances of getting the answer right. This too, is a gravely mistaken view, as a glance at a list of medieval occupations should make abundantly clear (see, for example, the list at Vincent 1999–2005).

It *may* be possible to infer something about the etiology of OA on a population basis, however, and some years ago, my colleague Ann Stirland and I were able to do this in the case of the crew of the *Mary Rose*, Henry VIII's flag ship that sank with great loss of life in 1545 (Stirland and Waldron 1997). The frequency of disease in the spines of the crew members was comparable in many respects to that found in an older group of male skeletons from a medieval site in Norwich (UK). It was concluded that age-related changes in the *Mary Rose* spines had been accelerated and, having matched for sex, and rejected that there could be any differences caused by genetic predisposition or ethnic origin, it seemed most reasonable to conclude that the effects were the result of working life on board ship. This seems a valid inference to make and one that could be made when comparing the distribution of OA or other forms of pathology between two groups. It should be noted even here, however, that it would not be possible to attribute any particular occupation to individual skeletons and the inference must be confined to the group as a whole.

The Erosive Arthropathies

In this group of joint diseases there is a major inflammatory component, unlike most cases of OA. The erosive arthropathies are much less common than OA and on that account alone, of more interest to both clinician and bone specialist alike. It is unlikely that the prevalence of any of the more common forms exceeds 2 percent of the population and for some (psoriatic arthropathy and reactive arthropathy, for example) it is substantially less. Extrapolating these data to skeletal assemblages, means that they are not likely to be encountered very frequently. Indeed, many assemblages will contain too few adults for these conditions to be likely to appear at all, as can be demonstrated with a simple calculation. Suppose, for example, that the prevalence of ankylosing spondylitis is 2 percent in the adult population, and that we need to estimate this to within $\pm$ 1 percent (1–3 percent), then the number required for the estimate (Wade 2001) is:

$$4 \times \frac{prevalence\,(100 - prevalence)}{1^2}$$
$$= 4 \times (2 \times 98)$$
$$= 784 \text{ adults.}$$

Given that in most assemblages the subadults account for about one-third of the total, to be sure of getting 784 adults would require a total assemblage in excess of a thousand; and assuming that 10–20 percent of the adults would be too poorly preserved to permit an adequate examination—well, the numbers keep increasing and there are relatively few assemblages which will provide sufficient numbers to be reasonably certain of seeing even a single case. There would be a greater chance were one able to combine the results of several assemblages, but this requires that the different examiners are using the same criteria to diagnose their cases, and this is all too often, *not* so.

Erosions

The defining characteristic of the erosive arthropathies is the presence of erosions, which are areas of bone loss, often at the joint margins or, in some cases, on the joint surface, the result of osteoclastic activity stimulated by inflammatory cytokines. A true erosion has some or all of the following characteristics:

- cortical destruction;
- undercut edges;
- exposed trabeculae;
- sharp or scalloped ridges, and
- a scooped floor.

The most important criterion is the presence of cortical destruction, and any hole in a bone in which the cortex is intact, *cannot* be a true erosion. Vascular foramina which are sometimes found around joints should not be confused because the cortex is intact; postmortem damage, however, is sometimes more difficult to differentiate but the fact that the edges of the putative lesion will have a much lighter color than the surrounding bone should eliminate it from consideration.

Erosive osteoarthritis

Erosive OA provides the bridge between the proliferative and erosive forms of joint disease. It is confined to the hands and very much more common in females than in males, and nowadays affects a substantial proportion of those with OA of the hands (Pattrick et al. 1989). The erosions appear first in the centre of the small joints of the hand, especially the interphalangeal joints and ankylosis may supervene, which it never does in the uncomplicated form of OA. There are some characteristic radiological appearances in the hand, including the so-called gull-wing and saw-tooth signs. The condition has been known in the modern population only since the 1960s and there are not many descriptions in the paleopathological literature. In an assemblage dating between 950 and 1850 A.D. recovered from St. Peter's Church in Barton-on-Humber, a small town in the northeast of England it had a prevalence of 2.7 percent (95 percent confidence interval 0.9–7.5 percent), which is at least of the same order of magnitude as in the contemporary population, implying that it is possible that it has tended to be overlooked. It is not a particularly difficult disease to diagnose, requiring only the presence of eburnation in any distal interphalangeal joints (Waldron 2007).

Rheumatoid arthritis

Rheumatoid arthritis (RA) is the archetypal erosive arthropathy and, indeed, for many years other erosive arthropathies were considered simply to be variants of RA. The first recognizable clinical description of RA was given by Augustin Jacob Landré-Beauvais (1772–1840) in his MD thesis of 1800, a volume of some mere ten pages, a fact that shocks modern aspirants to the degree. Landré-Beauvais described a group of nine female patients with chronic arthritis whom he had attended at the Salpêtrière hospital in Paris (Landré-Beauvais, 2001). He was certain that he had discovered a new form of arthritis and he proposed that it should be called goutte asthénique primitive (primary asthenic gout, gout being then the rubric under which all forms of joint disease were included). As with all so-called *new* diseases, the first description heralded the appearance of many others and RA seemed to become much more common during the 19th and 20th centuries. At its peak, it affected at least 1 percent of the adult population, females much more frequently than males. In recent years both the prevalence and incidence of the disease have decreased noticeably in the countries of the developing world but have increased in subSaharan Africa, although the disease there runs a milder clinical course (Kalla and Tikly 2003). Whatever the cause for these changes—and they are not at all clear—the disease may well be returning to the condition that it experienced prior to Landré-Beauvais's description. Had it been at all common prior to 1800 or been of the severity seen before disease-modifying medication was available, then it seems highly unlikely that the changes of severe RA in the hands could have been overlooked by the early physicians, who were generally good observers of clinical signs and symptoms.

Rheumatoid arthritis *has* been found in assemblages prior to 1800 but the prevalence seems to be very low and it probably was of a less severe form than has been the case until relatively recently. The condition is a disease of the synovial membrane, characterized by symmetrical marginal erosions which affect the small joints of the hands or the feet. There is almost no proliferation of new bone, the sacroiliac joints are not affected and there is no spinal fusion. Larger joints, such as the elbow or knee may be affected, and so may the odontoid peg. The last is a potentially dire complication since it has been known to lead to subluxation of the atlantoaxial joint, compression of the spinal cord, and sometimes, death. Ankylosis of joints is uncommon and it has been suggested that when it does occur, it may be indicative of effective treatment (Barbieri et al. 2010), presumably not a consideration when considering RA in the past.

The majority of patients with RA have an antibody in their blood known as rheumatoid factor (RF) which reacts with the Fc fragment of IgG. Following the discovery of RF it was found that the blood of patients with the so-called variants of RA did not contain this antibody, and such patients came to be referred to as seronegative, and their joint diseases as the seronegative arthropathies (see below). Rheumatologists were not quick to accept the existence of RF as a real phenomenon and it was not until the early 1960s that notice was first made of it in the rheumatology textbooks; it was not until the fourth edition of one popular British textbook (published in 1969) that the author was able to write that the nature of RF was "so well established that it is no longer necessary to discuss the several arguments against it" (Copeman 1969). The clinical literature prior to the 1960s will thus contain no reference to rheumatoid factor or to the seronegative arthropathies. This again illustrates the problematic nature of trying to determine either the frequency or the characteristics of disease in the past from the written record.

The seronegative spondyloarthropathies

By contrast with RA, the primary pathology in the spondyloarthropathies (or spondylarthropathies, the terms are equivalent) is enthesitis, that is to say, inflammation of the point at which the tendon of a muscle inserts into bone. In some, there is also a tendency towards bone proliferation. (Incidentally, the term spondyloarthropathy was not introduced until 1974 by Moll and his colleagues (Moll et al. 1974) so there will be no reference to this group of diseases in the prior literature.)

The contrast between RA and the seronegative group can be seen in the action of the cytokines that affect osteoclast and osteoblast formation. In RA, cytokines that stimulate bone resorption, such as RANKL, are predominant, whereas those that stimulate bone production such as Wnt and BMP are present in low concentration. In a seronegative arthropathy such as ankylosing spondylitis, the converse is the case; Wnt and BMP are present in high concentration, while that of RANKL is low (Schett 2007).

The group of spondyloarthropathies includes ankylosing spondylitis (which is the most common), reactive arthropathy (formerly, and sometimes still, referred to as Reiter's disease), psoriatic arthropathy and enteropathic arthropathy. There are also other forms of spondyloarthropathy that do not conform to the criteria established to distinguish the separate entities and these—with the true cunning of the medical profession—are known as the undifferentiated spondyloarthropathies. The prevalence of the group as a whole is not precisely known, but is probably about twice that of ankylosing spondylitis (which is about 2 percent of the general adult population).

Although it is conventional to split the spondyloarthropathies (SpA) into separate diseases, in many respects they represent a continuum which is linked by the presence of the HLA-B27 antigen, by sacroiliitis, asymmetric peripheral joint involvement, spinal fusion, and, of course, enthesitis. They are split on the basis of their etiology or association with other diseases—infection with *Campylobacter*, *Chlamydia*, *Clostridium*, *Salmonella*, *Shigella* or *Yersinia* species in the case of reaction arthropathy; ulcerative colitis or Crohn's disease in the case of enteropathic arthropathy and, of course, psoriasis for psoriatic arthropathy. The prevalence of the spondyloarthropathies is a reflection of the prevalence of HLA-B27 in the general population, but whereas it is the case that majority of patients with SpA are positive for HLA-B27, by no means all those with the antigen develop SpA, for reasons which remain to be clarified (Reveille 2004).

Ankylosing spondylitis

Ankylosing spondylitis is the most common of the SpA, and probably the easiest to diagnose. In it, there is bilateral involvement of the sacroiliac joints which commonly fuse and the spine is fused by the production of syndesmophytes, which are ossifications of the annulus fibrosus of the intervertebral disc. Spinal fusion always begins in the lowest part of the spine and may progress inexorably upwards without any normal vertebra intervening, that is to say, there are no so-called skip lesions. If the thoracic vertebrae are fused, then it is often found that the costovertebral and costotransverse joints are also fused. In the fully developed state, the end-result is truly impressive, with the pelvis, spine and ribs fused into a solid block. It is a disease of considerable antiquity but, at the risk of becoming repetitive, I will warn again against taking reports of the disease in the older literature at face value. To the early physicians, the term was used for any condition in which there was spinal fusion, and it is almost certain that some cases referred to as ankylosing spondylitis in the older literature were not what we would recognize by the term nowadays, but were other conditions such as diffuse idiopathic skeletal hyperostosis (DISH), or indeed, others of the group of SpA. An illustration of the condition in such instances is invariably more helpful than the text.

Reactive arthropathy

The first description of reactive arthropathy is usually attributed to Hans Reiter, who in 1916 described the triad of arthritis, urethritis and uveitis in a German officer who was suffering from diarrhea and a non-gonococcal urethral infection. He was not the first to describe this syndrome, this having been achieved by the English surgeon Sir Benjamin Brodie in 1818. Nevertheless, the combination of symptoms was for many years associated with Reiter's name, at least in the German- and English-speaking world. The French generally preferred to associate the condition with Feissinger and Leroy who described in independently of Reiter in the same year. The disease was for a long time notoriously associated with gonococcal urethritis even though the symptoms appeared in patients with a variety of other infections, especially those affecting the gut, and the use of the term Reiter's syndrome (or disease) was gradually phased out, reactive arthropathy being preferred and now generally used. The disuse of Reiter's name was also encouraged by the fact that he was an active Nazi and had taken part in what were referred to as "medical experiments" in the concentration camp at Buchenwald during the Second World War (Iglesias-Gammara et al. 2005).

Reactive arthropathy is an HLA-B27 related autoimmune disease, with many of the typical characteristics of the SpAs; the sacroiliac joints are affected, usually asymmetrically, and there is spinal fusion, although the fused segments are interspersed with normal (unfused) segments, so that there are said to be skip lesions present in the spine; enthesitis is prominent especially around the calcaneum. Proliferation of new bone is more common in reactive arthropathy than in some of the other SpAs and may be seen *inter alia* on the metatarsal shafts. The changes in ReA tend to affect the joints in the lower extremities more than in the upper and this is one of the features that may help to distinguish it from other forms of SpA although the distribution of the lesions is by no means universally consistent or unique to the different varieties.

There is a particularly severe form of ReA in which there is subluxation of the metatarsophalangeal joints known as Lanois's deformity and which may affect up to one in twelve of those with the condition (Martel 1979). This form of the disease may readily be confused with either RA or PsA, adding further to the diagnostic difficulties.

Psoriatic arthropathy

Psoriatic arthropathy (PsA) affects a subset of those who suffer from the skin disease psoriasis, which is present in about 1–3 percent of the general population. The complication of joint disease seems to occur predominantly in those whose nails are affected. There is no consensus on the proportion of those with psoriasis who develop joint disease but it is likely to be no more than a third. The arthropathy has the usual link with HLA B-27, the sacroiliac joints and the spine are involved, the latter showing the presence of skip lesions as in ReA. The principal distinguishing feature between ReA and PsA is that in the latter the joints of the upper limb tend to be preferentially affected.

The existence of PsA as an entity in its own right was debated for a number of years, some authors considering that since both psoriasis and seronegative joint disease were relatively common, the occurrence of both in the same individual was merely a coincidence; the authoritative review by Fitzgerald and Dougados (2006) has settled the matter, however, and there seems no longer any reason for PsA not to be considered an entity in its own right.

Psoriatic arthropathy comes in a number of guises, depending on the number of joints affected, and their distribution. Five subsets have been described (Table 28.2) and this makes it a particularly difficult disease to diagnose in the skeleton. The most common forms are those that present either as an asymmetric oligoarthritis, or as a symmetric arthritis somewhat similar to RA. There is a third subset of patients in whom the distal interphalangeal joints are predominantly involved, a fourth in which spondylitis is the major features, and finally, a very destructive form in the hands which results in considerable deformity, which is referred to as arthritis mutilans and which may closely resemble the end point of RA.

Table 28.2   Subsets of psoriatic arthropathy in order of occurrence

Order of occurrence	Subset
1	Asymmetric oligoarthropathy.
2	Symmetric arthropathy similar to rheumatoid arthritis.
3	Distal interphalangeal joints predominantly affected.
4	Spondylitis predominant feature.
5	Arthritis mutilans.

The best known changes in PsA occur in the distal interphalangeal joints of the hand. Marginal erosions are the first to appear but may spread to involve the central portions of the joint and resorption of the distal tufts of phalanges is characteristic; progressive bone resorption may result in shortening of the phalanges, and it is possible that these changes may have been mistaken for leprosy in the past. The proximal joint surface of the distal phalange may be widened and in association with osteolysis of the proximal phalange may give rise to the so-called cup and pencil sign on a radiograph. Proliferation of new bone is common, especially on the metacarpal or metatarsal shafts and so is bony ankylosis. The sacroiliac joints are generally affected bilaterally but almost any form of involvement—unilateral, bilateral, symmetric, asymmetric—can be found. One feature of PsA is that the cervical spine is frequently involved; erosions, fusion, subluxation of the atlantoaxial joint, erosion of the odontoid peg, and ankylosis are all likely to be found.

Psoriatic arthropathy is probably the most difficult of the distinct SpAs to be diagnosed in the skeleton, particularly given the fact that it may present in one of a number of subsets, and it is likely that the form that predominantly affects the distal interphalangeal joints will be the one that is most readily recognized. The affected bones are very delicate and may easily be damaged by excavation or postexcavation activities and this will generally do little to help with the diagnosis.

Enteropathic arthropathy

The association between chronic gut disease and joint disease was first recognized by W H White in 1895 and it now occurs most commonly in patients with ulcerative colitis or Crohn's disease. The joint changes occur some months after the onset of the gut disease and most closely resemble those seen in AS although the association with HLA B-27 is not as strong as in the idiopathic form of AS. There is little reason to suppose that the paleopathologist could distinguish between the two conditions.

Undifferentiated spondyloarthropathy

Patients are said to have an undifferentiated SpA if they fail to fulfill the criteria for the diagnosis of the four main categories referred to above, and it is probably as common as AS. It is likely that the majority of skeletons with sacroiliac involvement and spinal fusion will fall into the undifferentiated category, particularly if the skeleton is poorly preserved. It has to borne in mind that the full-blown signs and symptoms of SpA (or indeed any other disease) do no arise suddenly but will develop over a number of weeks or months. Clinicians have the advantage that they can follow up their patients and watch as their disease develops before arriving at a definitive diagnosis (although even then, they may fail to do so). Paleopathologists on the other hand, see their "patients" on one occasion only and that a particularly unfortunate one, the time of their death. There may be no telling at what stage they were in their disease when they died; was the disease beginning, developing, or fully developed? Finding skeletons with suggestive signs of an erosive arthropathy is not that uncommon, but relatively seldom can they be unequivocally attributed to one of the well recognized groups and the bone specialist should perhaps not strive too mightily to do so.

Juvenile arthritis

Juvenile arthritis has a prevalence of about 1 per thousand in the modern population, with two peaks of onset, one between the ages of 1 and 2, and the second between 9 and 15. In some cases rheumatoid factor is present, in which case the condition is sometimes referred to as Still's disease, while others more closely resemble a SpA; there is also a form associated with psoriasis. Although there is no reason to suppose that the condition did not occur in the past—a painting by Botticelli is said to show a young man with juvenile arthritis, for example (Alarcón-Segovia et al. 1983)—it would be an extremely difficult condition to diagnose in the juvenile skeleton (Rothschild et al. 1997) especially since the epiphyses will usually be unfused and often not recovered during excavation even if they have survived burial.

Gout

Gout is one of the oldest diseases to be recorded in the medical literature, several references to it appearing in the Hippocratic corpus. The 17th-century English physician Thomas Sydenham was himself a sufferer and he provided one of the best descriptions of the disease from self-observation. The disease seems to have become particularly common during the 18th century and the cartoonists of the day reveled in depicting members of the upper classes with their foot encased in bandages on a stool, suffering the penalties of their wealth and indulgence. It is one of those diseases that tends to excite mirth rather than pity, although the joke is seldom shared by the sufferer.

Gout is the result of a disorder of uric acid metabolism in which either the production of uric acid is increased or the excretion through the urine is decreased, the result being a rise in the concentration of uric acid in the blood. The high prevalence of gout during the 18th century is thought to have been caused in part at least, by the adulteration of food and drink with lead, lead having the effect of inhibiting the excretion of uric acid through the kidney (Ball 1971).

Uric acid is a by-product of purine metabolism and is formed from xanthine and hypoxanthine under the influence of the enzyme, xanthine oxidase. In most animals, uric acid is further metabolized to allantoin, but humans lack urate oxidase, the enzyme necessary for this conversion. Although the development of gout is dependent on the concentration of urate in the blood, only about one-fifth of those with high uric acid levels actually develop gout, so that other factors are obviously important in the etiology.

Gouty arthropathy results from the precipitation of uric acid crystals in structures either within or around a joint, stimulating inflammatory changes in the affected soft tissues. Acute gout most commonly affects the first metatarsophalangeal joint and, although exquisitely painful, is usually self-limiting. There follows an interval of months or years during which no further attacks occur but then about half the patients enter a chronic phase characterized by the production of tophi, which are agglomerates of inflammatory tissue and uric acid crystals, and which may settle out in any soft tissue including the synovial membrane, the articular cartilage or para-articular structures. Tophi deposited within or around joints cause erosions which may occupy any position with relation to the joint, within it or around it. The erosions are generally well defined, round or oval in shape and usually oriented in the long axis of the affected bone. They often have an overhanging edge which, when seen on X-ray, is referred to as a Martel hook and which represents bone forming over the tophus. The erosions may be multiple but asymmetric and commonly affect the foot, ankle, knee, hand and wrist.

The prevalence of gout varies worldwide but is probably 3–5 per thousand. It is more common in males than in females and the prevalence increases considerably with age; there is good evidence that it is increasing in prevalence, at least in the western world, probably in association with co-morbidities such as obesity and diabetes, and a purine-rich diet (Weaver 2008). Relatively few cases seem to have been found in skeletons, although the diagnosis should not prove unduly taxing; the finding of para-articular erosions which cross the joint margin is a particularly helpful sign. Radiography may be helpful in coming to a diagnosis; the presence of a Martel hook may be demonstrated and the erosion may have sclerotic margins. If the paleopathologist is having an especially lucky day, there may even be uric acid within the lesion; uric acid is sparingly soluble in water so theoretically it could survive burial but it may escape recovery and the chances of finding it would be greater in mummified rather than buried material.

Conclusions

"When a man is tired of London he is tired of life", Dr. Johnson famously remarked, and one might paraphrase him by saying that when a man (or woman, naturally) is tired of joint disease he (or she) is tired of studying bones. There is still a good deal to discover about this set of diseases in the past, particularly their epidemiological features, providing that some way can be found to standardize both diagnosis and the calculation of prevalence. It would be very desirable to be able to determine the changes in the frequency of the joint diseases as they relate to time and place but it is only by being able to combine data from different studies that sufficiently large numbers will be provided to enable confidence intervals to be reduced and thus allow the detection of what are likely to be relatively small differences in prevalence. It is easy to demonstrate how large the study base has to be in order to be certain of making accurate estimates of prevalence and that many studies lack the power to be able to make such estimates reliably. This is, of course, especially true of those conditions in which the prevalence is low, but even with osteoarthritis, by the time a seemingly large assemblage has been reduced to its various age and sex categories (and by phase, if the site permits), the numbers rapidly become very small. For example, at the site of Barton-on-Humber to which reference was given earlier, of the 2,750 inhumations, no more than about 400 could be attributed to any of the five discrete phases, and of those, only about half were adults, thus greatly reducing the chances of finding age-related pathology. Perhaps one outcome of the present volume might be to establish a group who would be willing to agree on methodology and create a database from which valid conclusions as to the characteristics of joint disease in the past could be drawn.

REFERENCES

Alarcón-Segovia, D., Laffón, A., and Alcocer-Varela, J., 1983   Probable Depiction of Juvenile Arthritis by Sandro Botticelli. Arthritis and Rheumatism 26:1266–1269.

Ball, G. V., 1971   Two Epidemics of Gout. Bulletin of the History of Medicine 45:401–408.

Barbieri, F., Parodi, M., Zampogna, G., Paparo, F., and Cimmino, M. A., 2010   Bone Ankylosis of the Wrist as a Possible Indicator of Treatment Efficacy in Rheumatoid Arthritis Rheumatology 49(7):1414–1416.

Bland, J. H., 1983   The Reversibility of Osteoarthritis: A Review. American Journal of Medicine 14:16–26.

Conaghan, P. G., Vanharanta, H., and Dieppe, P., 2005   Is Progressive Osteoarthritis an Atheromatous Vascular Disease? Annals of the Rheumatic Diseases 64:1539–1541.

Copeman, W. S. C., 1969   Textbook of the Rheumatic Diseases, 4th edition. Edinburgh: Livingstone.

Fitzgerald, O., and Dougados, M., 2006   Psoriatic Arthritis: One or More Diseases? Best Practice and Research Clinical Rheumatology 20:435–450.

Hutton, C. W., 1987   Generalised Osteoarthritis: An Evolutionary Problem. Lancet 1: 1463–1465.

Iglesias-Gammara, A., Pestrepo, J. F., Valle, R., and Matteson, E. L., 2005   A Brief History of Stoll–Brodie–Fiessinger–Leroy Syndrome (Reiter's syndrome) and Reactive Arthritis With a Translation of Reiter's Original 1916 Article in English. Current Rheumatology Reviews 1:71–79.

Kalla, A. A., and Tikly, M., 2003   Rheumatoid Arthritis in the Developing World. Best Practice and Research Clinical Rheumatology 17:863–875.

Landré-Beauvais, A. J., 2001   The First Description of Rheumatoid Arthritis. Unabridged text of the doctoral dissertation presented in 1800. Joint Bone Spine 68:130–143.

Lawrence, J. S., 1977   Rheumatism in Populations. London: Heinemann.

Lawrence, R. C., Helmick, C. G., Arnett, F. C., Deyo, R. A., Felson, D. T., Giannini, E. H., Heyse, S. P., Hirsch, R., Hochberg, M. C., Hunder, G. G., Liang, M. H., Pillemer, S. R., Steen, V. D., and Wolfe, F., 1998   Estimates of the Prevalence of Arthritis and Selected Musculoskeletal Disorders in the United States. Arthritis and Rheumatism 41:778–799.

Martel, W., 1979   Radiological Manifestations of Reiter's Syndrome. Annals of the Rheumatic Diseases 38:12–23.

Moll, J. M., Haslock, I., McCrae, J. F., and Wright, V., 1974   Associations Between Ankylosing Spondylitis, Psoriatic Arthritis, Reiter's Disease, the Intestinal Arthropathies, and Behçet's Syndrome. Medicine (Baltimore) 53:343–364.

Pattrick, M., Aldridge, S., Hamilton, E., Manhire, A., and Doherty, M., 1989   A Controlled Study of Hand Function in Nodal and Erosive Osteoarthritis. Annals of the Rheumatic Diseases 48:978–982.

Reveille, J. D., 2004   The Genetic Basis of Spondyloarthritis. Current Rheumatology Reports 6:117–125.

Rothschild, B. M., Herskovitz, I., Bedford, L., Latimer, B., Dutour, O., Rothschild, C., and Jellema, L. M., 1997   Identification of Childhood Arthritis in Archaeological Material:

Juvenile Rheumatoid Arthritis Versus Juvenile Spondyloarthropathy. American Journal of Physical Anthropology 102:249–264.

Schett, G., 2007   Joint Remodelling in Inflammatory Disease. Annals of the Rheumatic Diseases 66:42–44.

Schouten, J. S., de Bie, R. A., and Swaen, G., 2002   An Update on the Relationship Between Occupational Factors and Osteoarthritis of the Hip and Knee. Current Opinion in Rheumatology 14:89–92.

Stirland, A., and Waldron, T., 1997   Evidence for Activity-Related Markers in the Vertebrae of the Crew of the *Mary Rose*. Journal of Archaeological Science 24:329–335.

Tait, T. J., and Bird, H. A., 1994   Asymmetrical Osteoarthritis in a Patient With Ehlers-Danlos Syndrome and Poliomyelitis. Clinical and Experimental Rheumatology 12:425–427.

Van Saase, J. L. C. M., van Romunde, L. K. J., Cats, A., Vandenbroucke, J. P., and Valkenburg, H. A., 1989   Epidemiology of Osteoarthritis: Zoetermeer Survey. Comparison of Radiological Osteoarthritis in a Dutch Population with that in 10 other Populations. Annals of the Rheumatic Diseases 48:271–280.

Vincent, S., 1999–2005   What Did People Do in a Medieval City? Electronic document. http://www.svincent.com/MagicJar/Economics/MedievalOccupations (Accessed June 15 2011)

Wade, A., 2001   Study Size. Sexually Transmitted Infections 77:332–334.

Waldron, T., 1995   Changes in the Distribution of Osteoarthritis Over Historical Time. International Journal of Osteoarchaeology 5:383–389.

Waldron, T., 2007   St. Peter's Barton-upon-Humber, Lincolnshire. A Parish Church and Its Community. Volume 2. The Human Remains. Oxford: Oxbow Books.

Watt, I., 2000   Bone Disorders: A Radiological Approach. Baillière's Clinical Rheumatology 14:173–199.

Weaver, A., 2008   Epidemiology of Gout. Cleveland Clinic Journal of Medicine 75:S9–S12.

White, W. H., 1895   Colitis. Lancet 1:537–541.

Wood Jones, F., 1908   Pathological Report. In: The Archaeological Survey of Nubia. Bulletin No. 2. Cairo: Ministry of Finance.

CHAPTER 29

Bioarchaeology's
Holy Grail: The
Reconstruction
of Activity

*Robert Jurmain, Francisca
Alves Cardoso, Charlotte
Henderson, and
Sébastien Villotte*

INTRODUCTION

Human skeletal research aimed at reconstructing past activities has had a long and checkered history. In recent years, in fact, it has been embraced as a kind of "Holy Grail" by an entire subfield of human osteology. The quest, however, has proven more hazardous than most of us would have imagined. Many types of bone changes have been proposed as "markers" of activity, occupation, or mechanical stress. Three major research areas have been thought to show the most promise and, consequently, have been most actively pursued: osteoarthritis, musculoskeletal stress markers, and cross-sectional bone geometry.

As might be expected, most contributors to this volume focus on pathological conditions as identified in human remains. As has been widely recognized, some of the bone changes seen in osteoarthritis that have routinely been used to reconstruct activity are also sometimes pathological. In addition, much of the complex physiological change that occurs at muscle attachment sites (entheses) also involves inherently pathological processes—a basic consideration that has not yet been generally acknowledged by bioarchaeologists. In recent years many paleopathologists have actively pursued this type of research under the broad rubric of what Jane Buikstra

A Companion to Paleopathology, First Edition. Edited by Anne L. Grauer.
© 2012 John Wiley & Sons, Ltd. Published 2016 by John Wiley & Sons, Ltd.

(1977) called *bioarchaeology*. In fact, many enthusiastic young researchers have seized upon this subfield and redefined it to characterize and justify a "new" research orientation with a primary goal to reconstruct ancient lifestyles from skeletal evidence. It is essential to point out, however, that the main thrust of bioarchaeology emphasizes a broader and much more scientifically based approach (see Chapter 4 by Buzon in this volume).

Although central to this type of research, activity is rarely rigorously defined in the anthropological literature. Firstly, is it habitual or exceptional? Although most studies tacitly assume that the relevant bone changes result from habitual activity, the influence of exceptional, acute behavioral episodes, as reflected in traumatic lesions, has occasionally been discussed under the broad topic of activity reconstruction (e.g., Jurmain 1999; Walker et al. 2009; see Chapter 20 by Judd and Redfern in this volume). In addition, activity needs to be more fully and more clearly characterized in terms of duration (total exposure time), frequency (number of repetitions per unit time) and mechanical overloading (Luttmann et al. 2003). Other factors such as intensity, age of onset, and postural demands need to be considered. Ideally, this broader understanding should be concordant with clinical definitions.

It is the apparent simple cause-and-effect relationship between occupation and impaired health, which has lured bioarchaeologists into looking for similar trends in past populations. However, modern clinicians are less content to accept a cause and effect relationship between occupation and many of the types of bone changes used by bioarchaeologists; the same or similar conditions can be caused by other factors. A classic example of such a musculoskeletal disorder is carpal tunnel syndrome. This condition has been widely linked to typing, but other causes exist, including fractures, cysts, hypothyroidism and fluid retention. Similar multifactorial etiologies exist for virtually all of the changes used to study activity by bioarchaeologists, but such inconvenient complexity has rarely been seriously acknowledged. By recognizing complexity, steps can be taken to minimize the impact of these other factors and improve the scientific integrity of bioarchaeology.

In addition to osteoarthritis (OA), musculoskeletal stress markers (more appropriately termed, "entheseal changes"), and cross-sectional geometry, many other morphological features of bone have been suggested as useful indicators of past activities (Capasso et al. 1999 and references therein). Among these is a wide range of osseous changes, including: Schmorl's nodes, external auditory exostoses, humeral shaft alterations, fractures, spondylolysis, myositis ossificans, and certain nonmetric traits. Dental wear patterns have also been used to reconstruct activities in which the teeth are used as a tool or are used to act as a "third hand."

It would be constructive at this point to ask some basic questions regarding activity reconstruction studies in human osteology. Why have they been so attractive to researchers, indeed, almost irresistible? What has led to so many simplistic assumptions and hasty interpretations? And, why have the mistakes made in earlier work been so often repeated in later research? All bioarchaeological researchers have been tempted at some time by what Tony Waldron (1994) has called the "alluring prospect." Additionally, widely held societal and media expectations further push otherwise rational scientists further into the abyss. The rewards, both professional and emotional, are so enticing that unavoidable constraints arising from the complexities of bone biology, limited samples, or fragmentary preservation are often ignored. Waldron also commented incisively regarding this tendency, "There is a perfectly understandable drive to make the most of what little evidence survives in the skeleton and this sometimes has the effect of overwhelming critical faculties" (Waldron 1994:98, and see Chapter 28 by Waldron in this volume, for discussion specifically on joint disease).

Besides Waldron, many other paleopathologists have voiced similar concerns over the direction of activity-oriented research. Disturbingly though, in the last decade, the overall quality of such work has deteriorated. We do not, however, wish to be harsh and belittle colleagues. Indeed, all four co-authors of this chapter did their dissertation research on what we initially considered to be useful activity-related bone changes. As a result of our experiences, each of us came to better understand the complexities of such research. The learning curve was not identical, and from the thrust of our first publications and presentations, the senior author was the slowest to appreciate the most serious limitations relating to this type of scientific investigation. We also want to emphasize that the most recent trend in much of this research is one that introduces more rigor. Accordingly, the future is brighter than the disappointing results of recent years might lead one to conclude.

History of Activity Studies

Osteoarthritis (OA)

Prior to 1970 most studies of OA evaluated overall prevalence (e.g., Hooton 1930; Anderson 1963), with most early reports focused on the spine (e.g., Stewart 1947; Chapman 1972). Behavioral interpretations, however, were only rarely attempted. In particular, the influence of Angel (1966, 1971) and Wells (1962, 1972) stimulated much more interest, beginning with Ortner's study of patterns of elbow OA (1968), leading to broad population analyses by Robert Jurmain (1977, 1980). Probably the most influential work during this period was that by Merbs (1983), where his explicit approach emphasized activity reconstruction, most specifically using patterns of peripheral OA.

These studies quickly generated considerable enthusiasm to use OA as a marker of activity/occupational stress, as exemplified by numerous researchers (e.g., Walker and Hollimon 1989; Bridges 1991). Some studies more superficially concluded that OA provided clear information regarding activity in past cultures (e.g., Tainter 1980; Robb 1994), including several reports on historical populations proposing links with specific activities (e.g., Larsen et al. 1995; Hershkovitz et al. 1996). More specific hypotheses suggested a relationship to the effects of changes in subsistence economy (Cohen and Armelagos 1984). However, during this same period, several other researchers voiced concern and urged caution (Bennike 1985; Jurmain 1999). As summarized comprehensively by Bridges (1992), none of these early attempts produced systematic results linking broad subsistence behaviors to patterns of OA, nor have more recent studies done so (Lieverse et al. 2007). Nevertheless, from its beginnings, the Global History of Health Project assumed OA was a clear indicator of activity/stress, as made clear by Goodman and Martin when they concluded, "While many factors may contribute to the breakdown of skeletal tissue, the primary cause of osteoarthritis is related to biomechanical wear and tear and functional stress" (2002:41).

The second stage of this admirably large and comprehensive research venture is focused on European populations and employs more sensitive age controls, and considers the multifactorial origin of OA. Nonetheless, an unwavering bias favoring behavioral interpretations has continued, as clearly stated by Larsen and colleagues, "Degenerative joint disease (osteoarthritis) is central to behavioral reconstruction in past societies, especially as it relates to mechanical function and lifestyle" (Larsen et al. 2009:172). As seen in the above quotes, many researchers combine the bone changes observed in different types of joints under the overall term of *osteoarthritis*. However, intervertebral body joints are not synovial (diarthroses), but are cartilaginous (amphiarthroses); most clinical researchers view the differing physiology as significant. Technically, *OA* should be used specifically for synovial joints, while the typical bone changes in vertebral body joints are better termed *vertebral osteophytosis* (VOP). For convenience, we will include historical perspectives for both types of joint change under the umbrella of *osteoarthritis*, but will specifically use *VOP* to refer to the etiopathogenesis and interpretations of alterations in vertebral body joints. Waldron (Chapter 28 in this volume) also makes this distinction, and his review focuses exclusively on OA of synovial joints.

The clinical literature on OA is enormous, and a comprehensive review is beyond the scope of this chapter. One topic particularly relevant to understanding behavioral influences on skeletal patterning of OA relates to epidemiological studies of at-risk groups participating in particular occupational or sports activities. Much of this literature has been summarized in prior reviews (Jurmain 1999; Weiss and Jurmain 2007). Included in these summaries, a total of 41 studies evaluated the relationship of OA to general levels of activity, and another 70 studies considered the prevalence of OA in at-risk groups engaged in more specific activities. Results show only barely half (54 percent) of the studies found a significantly higher prevalence in groups that exhibited higher levels of general activity. The results from the more specific studies showed a somewhat more consistent pattern, with 69 percent showing a positive correlation. However, it must be recalled that these studies chose to investigate particular sports/occupational groups because they participated in intense and repetitive actives that would be expected to show a definitive pattern. Further updates for the last four years (not included in earlier reviews) do not alter the overall findings regarding general activities, but do show an overall lower percentage of those studies focused on more specific activities (reduced to only 60 percent in a sample of 82 total studies). Looking at the overall trends from this body of evidence, there is no support to enable the simplistic assumption that OA derives directly from habitual activity. Despite the fact that much of this evidence has been known for several years, many osteologists (particularly those identifying themselves as bioarchaeologists) have continued to make interpretations as though these clinical findings did not exist or were not worthy of comprehensive consideration.

Entheseal changes (EC)

The theory that occupations or activities affected enthesis morphology was propounded in early anatomical literature (Lane 1888; Schneider 1956). Excepting an early French report (Testut 1889) and work by Jones (1910), the majority of archaeological reports, seeking to link entheseal changes with behavior, are more recent (Wells 1963,

1964, 1965; Angel 1964, 1966; Hawkes and Wells 1975). It was not until the 1980s that the concept that entheses could be used to reconstruct behavioral patterns over the course of a lifetime took off (e.g., Dutour 1986). Since this time, a considerable body of literature has been published using these changes to reconstruct activity patterns and study the division of labor in osteological samples (e.g., Kennedy 1983; Dutour 1986; Hawkey and Merbs 1995; Nagy 2000; al-Oumaoui et al. 2004; Molnar 2006, 2010; Mariotti et al. 2007; Cardoso 2008; Pany 2008; Villotte 2009; Villotte et al. 2010a)

Entheseal changes, formerly called musculoskeletal stress markers (MSM) and, in the clinical literature enthesopathies, enthesiopathies, and enthesophytes, are changes to the muscle attachment sites (entheses). The term *musculoskeletal stress marker* originated from the influential publication of Hawkey and Merbs (1995) and gained wider acceptance through its use in the majority of papers published in the special issue of the International Journal of Osteoarchaeology in 1998. Prior to this publication, other terms were used in the bioarchaeological literature, all of which explicitly stated the relationship between these changes and activity-related stress, e.g., *activity-induced pathology* (Merbs 1983), *evidence for occupation* (Kelley and Angel 1987), *skeletal makers of occupational stress* (Kennedy 1989) and *activity-induced stress markers* (Hawkey and Street 1992). The recent decision to change the terminology from *musculoskeletal stress markers* to *entheseal changes* was taken to avoid explicitly specifying the etiology of these changes (Jurmain and Villotte 2010). It is now widely accepted that they can be multifactorial in origin: acute trauma and disease (e.g., seronegative spondyloarthropathies) are the most common other causes and cannot be distinguished from repetitive activity by appearance alone. In many respects, the naming process itself reflects the history of this subfield: a move from an anecdotal to a scientific approach.

Entheseal changes have been used to study intrapopulation variation, e.g., sexual (e.g., Hawkey and Merbs 1995; Robb 1998; Molnar 2006; Lieverse et al. 2009) and hierarchical division of labor (e.g., Robb 1998; Molnar 2010). Interpopulation studies have predominantly focused on differentiating subsistence strategies (e.g., Dutour 1986; Hawkey and Merbs 1995; Villotte et al. 2010a).

Entheseal changes can take two forms: new bone formation or bone destruction. These changes can vary in shape, size, and distribution. They also vary between the two classifications of enthesis: fibrocartilaginous and fibrous. This categorization, although only very recently accepted in the bioarchaeological literature (Henderson 2003, 2009a; Villotte 2006, 2009; Havelková and Villotte 2007; Alves Cardoso and Henderson 2010; Villotte et al. 2010b), refers to the anatomical structure of these sites originally described in the early 20th century (Dolgo-Saburoff 1929), but recently reclassified by interface structure (Benjamin et al. 2002). Recognizing that different interface structures exist between the soft and hard components of the enthesis has been an important step for bioarchaeologists, which will be discussed in detail below.

Recording methods and methodological considerations will be described in further detail in a later section. However, it is important to note that methodology has changed through time concomitant with the increase in interest in these changes. Initially nonsystematic methods predominated, i.e. only entheses with changes were recorded (e.g., Hawkes and Wells 1975). Indeed, similar methods are still found in more recent publications (e.g., Oates et al. 2008), particularly in nonspecialist literature. In general, the nonsystematic method is used to tell or to support a story. Systematic methods became much more popular with the widespread implementation of the method devised by Hawkey (1988). Nonvisual recording methods, although discussed more than a decade ago (Stirland 1998; Wilczak 1998), are only now becoming more widespread (Zumwalt 2005, 2006; Henderson and Gallant 2007; Henderson, 2009a; Pany et al. 2009; Wilczak 2009).

Cross-sectional bone geometry (CSBG)

The most recent general approach used to evaluate potential evidence of activity relates to architectural properties of bone, most especially changes in cross-sectional geometry of long bone diaphyses. From nearly the beginning of biomedical interest in using this approach (Jones et al. 1977), anthropologists became directly involved (e.g., Lovejoy et al. 1976; Ruff and Jones 1981). Unlike the other most commonly used skeletal indicators of activity, CSBG has been used frequently to evaluate bone morphology, especially as relating to locomotion in fossil hominins (e.g., Lovejoy and Trinkaus 1980; Galik et al. 2004; Ruff 2009). As a result of early and significant contributions by paleoanthropologists, CSBG research reflected more rigorous aspects of skeletal biology than was generally the case for either OA or EC analyses. The more clearly empirical approach of CSBG research was also likely enhanced by the use of much better defined (interval) data.

The first published examples of CSBG and activity in contemporary samples (Jones et al. 1977; Ruff et al. 1994) gave considerable impetus to initial confidence in CSBG as a good indicator of activity in numerous skeletal samples from a variety of archaeological populations (e.g., Bridges 1989; Larsen and Ruff 1994; Rhodes and Knüsel 2005). Some of the more recent articles made conclusions relating to general levels of activity (e.g., Maggiano et al. 2008), while others were more enthusiastic, suggesting more defined activities (e.g., Sládek et al. 2007; Weiss 2009).

Several other studies evaluated the variation in cortical bone in athletes (e.g., Emslander et al. 1998; Heinonen et al. 2002). Recently, the systematic research of Shaw and Stock (2009a, b) on university athletes, as compared to matched controls, has provided some of the most encouraging results. Their results on both upper- and lower-limb patterning are highly suggestive and provide paleoanthropologists and other skeletal biologists with a solid basis to interpret CSBG as related to at least some aspects of activity. However, as much emphasized by Pearson and Lieberman (2004), bone geometry is far more prone to significant biomechanical influences during adolescence than is the case for adults. Thus, it is no surprise that the most definitive results from experimental animals as well as those seen in human athletes came from samples of young individuals.

For example, the well-known sample of elite tennis players first evaluated in the 1970s (Jones et al. 1977) was re-evaluated by Ruff and colleagues (1994). In this more complete analysis, they showed the mean age at which the individuals began playing was 9.6 years. Likewise, in the studies of university athletes done by Shaw and Stock (2009a, b), the mean age at which intense competitive training began was between 9.5 and 13.7 years. A general consensus has emerged in recent years that, "Much of what we see in an adult's cortical bone morphology is a result of the history of skeletal loading during adolescence" (Pearson and Lieberman 2004:89).

Methodological Considerations

The fundamental considerations and constraints relating to all of the major presumed indicators of activity are their multifactorial etiology and their widespread expression in all human populations. That is, all populations will have some individuals with OA and EC; moreover, it is obvious every individual has bone geometry. Accordingly, the real questions are: 1) What is the morphological pattern of expression of these conditions within and between human populations? and 2) Can these patterns, as seen in skeletal samples, reliably be linked to activity?

Many studies of activity have assumed the etiology of these conditions is simple with one primary influence (i.e., activity), a convenient approach, especially if one chooses to ignore the complexities of bone biology and the voluminous biomedical literature that documents these realities. It is fair to point out that not all studies have been seriously flawed, nor has research on the three major indicators of activity displayed the same degree of problems. EC research has been the most plagued by poorly conceived and poorly executed studies. The number of OA studies focusing on activity has generally declined in recent years, and the recognition of a more multifactorial etiology has increased. Nevertheless, in the last few years a resurgence of more biased and inherently simplistic research has begun to re-emerge. Lastly, most investigations of CSBG have not been so narrowly focused and intrinsically biased as has been the case for EC and OA. As we noted, one reason relates to the more empirical approach used in CSBG analyses. Another likely factor is CSBG research is more difficult, more time-consuming, and more expensive (since it requires, at minimum, some radiography). As a result, those researchers wishing the quickest and easiest path to activity reconstruction have been more drawn to EC and/or OA than they have been to CSBG. Nonetheless, even some CSBG studies have not been as rigorous or as cautious as has been the norm.

Because all these skeletal expressions are found universally in human groups, a population/epidemiological approach is required. Consequently, the necessity for large samples is inescapable, and these samples must be carefully stratified and statistically analyzed to control simultaneously for numerous confounders (including age, sex, presence of previous pathological conditions, and, if possible, genetic influences). Without such viable and carefully controlled samples, many investigations, along with their manufactured hypotheses, are doomed from the outset.

An obvious crucial aspect in human skeletal biology concerns how the processes of bone remodeling are influenced by specific factors. Indeed, perhaps the most intriguing feature likely to contribute to a more complete understanding of these processes focuses on their potentially interrelated etiologies. For example, the development of marginal osteophytes in OA and hypertrophic bone formation at entheses (enthesophytes), could be governed by similar (or even the same) genetic mechanisms. Rogers and colleagues (1997) suggested such an etiological explanation for the development of both osteophytes and enthesophytes, especially in certain individuals they termed *bone formers*. However, the lack of rigorous age controls in this study makes this conclusion unconvincing.

Further evidence suggests rates of bone remodeling differ dramatically, depending on the age of the individual. In OA, osteophytes, in particular, develop after age 35 and accelerate dramatically after age 50 (Cooper et al. 1994). Two phases appear discernible for age influences on EC. The first is observed in early adulthood (30–35 years) and concerns the tendon. Gradually, the tendinous resistance decreases, the tendon is less well vascularized and hydrated, and amounts of collagen and glycoproteins decrease (Rodineau 1991; Bard 2003). These processes combine to produce mechanical injury. The second occurs from the sixth decade onwards, as entheses (at least fibrocartilaginous ones) undergo tissue disruption (Durigon and Paolaggi 1991), and major skeletal changes appear (Villotte 2009; Alves Cardoso and Henderson 2010; Villotte et al. 2010b). Age influences on CSBG appear to differ dramatically from those involving OA and EC and are far more influential during early life, particularly affecting bone remodeling during childhood and adolescence (Pearson and Lieberman 2004). As often mentioned, but not necessarily taken into account, geometrical properties are highly correlated with age, as was demonstrated by measuring the percentage of cortical area in a large sample of identified skeletons (Perréard Lopreno 2007). This finding challenges the accuracy of interpopulation comparisons without first controlling for age distribution and consequently for the methods of age-at-death assessment. Any consideration of how activity might influence ultimate bone morphology must thus consider the likely influence of age, sex, genetic variation, pathology, and the age at which the presumed activity began (Shaw and Stock 2009b).

Systematically recognizing confounding variables will greatly aid in more rigorous research on activity reconstruction. The main problem needing to be addressed could be termed a lack of *specificity*, as defined in archaeological samples by Dutour (1992) and occasionally used in this way by some epidemiologists (e.g., Cooper et al. 1994). The question of the specificity of a marker is closely linked to its inherently multifactorial etiology. For a given subject, the degenerative process associated with senescence, a strenuous daily physical activity, a particular body conformation, and genetic predispositions (the supposed *bone formers*) may promote the development of enthesophytes and osteophytes. In the same vein, a given marker (e.g., an enthesopathy at the lateral humeral epicondyle) could be observed in three subjects of the same age, the first being a professional tennis player, the second a barber, and the last a white collar worker with DISH.

The second major pitfall in making accurate interpretations derives from the fact that the sensitivity of a marker is never absolute, indeed, far from it (Dutour 1992). It takes time for bone changes to develop, and this will vary among individuals. For example, if enthesopathies of the shoulders and the elbows are observed in some current archers, they are expressed extremely rarely, as compared to the number of participants (ref. in Villotte et al. 2010a).

Clearly, the major issue for many of the supposed "markers" is the near-total lack of contemporary reference samples to permit accurate estimation of their specificity relative to particular activities. In fact, the activities for which we have anything approaching reasonable evidence are extremely rare (Dutour 1992; Villotte 2008) and potentially include running, throwing by hand, horseback riding, archery, kayaking, canoeing, immersion in a cold water environment, and a rather special case of squatting. Based on current evidence, not all the authors of this chapter are entirely comfortable including all (or even most) of these possibilities.

Osteoarthritis

Useful research on the skeletal patterning of OA requires a rigorous methodology from the outset. As noted above, paleoepidemiological research demands large and well-controlled samples. In any OA osteological research design, the first consideration is what types of bone changes should be evaluated. Various anatomical components of joints have been suggested, but these can be broadly summarized as those that involve 1) marginal hypertrophic changes (osteophytes) and 2) articular surface changes (pitting/porosity and eburnation). It is crucial to recognize that marginal and surface changes can occur separately or in combination. From well-supported clinical analogues, skeletal biologists concluded some time ago that osteophytes alone are not good indicators of OA (Rogers et al. 1987). Expression of osteophytes in vertebral bodies (vertebral osteophytosis, VOP) is an especially poor marker of activity, since it develops typically from universal biomechanical forces associated with bipedal locomotion (Merbs 1983; Jurmain 1999, 2000) rather than reflecting any influence falling under the rubric of "habitual activity." The questionable utility of VOP in activity reconstruction is further reinforced by its extremely high correlation with age as well as the near total lack of supporting clinical evidence. For all these reasons, most (but not all) osteologists have abandoned VOP in activity reconstruction of skeletal samples. Moreover, conclusions drawn from peripheral joint involvement are also suspect, if data are not differentiated between marginal and surface changes. Indeed, the most widely used scoring system recommends separate evaluation and recording of osteophytes, surface pitting, and eburnation (Buikstra and Ubelaker 1994). In fairness, while many osteologists have followed this or very similar scoring methods, these more precise data are usually not included or evaluated in formal reports (e.g., Merbs 1983). Scoring of OA changes has not been particularly controversial, since ordinal scaling systems have been almost universally used. Moreover, as noted, the "Standards" publication edited by Buikstra and Ubelaker (1994) has promoted further methodological consistency. Nevertheless, as noted by Waldron in this volume, serious issues regarding consistency remain.

The underlying difficulties lie primarily in which joints are evaluated, whether marginal and articular surface changes are evaluated separately, and whether level (severity) of involvement is reported and considered in interpretations. This latter point merits further explication. Commonly, the ordinal scaling of OA severity involves a three- or four-stage system. Basically, these levels follow typical clinical practice and can be summarized as slight, moderate, and severe. The difficulty arises if hasty interpretations are made without differentiating among these various severity levels. A potentially highly inaccurate conclusion could be reached, when most (if not all) OA involvement is merely slight. In standard epidemiology, usually this level of severity is recorded but ignored in analyses (where only moderate and severe expression are considered). At minimum, skeletal biologists should confirm if moderate and, particularly, severe degenerative involvement follow the same patterns as shown when slight expression is considered separately. The likelihood of misinterpretation is compounded when conclusions are primarily based on slight involvement reflected largely in osteophyte development. It is both remarkable and sobering that many imaginative behavioral scenarios have been based on such flimsy evidence.

The final methodological issue is perhaps the most crucial paleoepidemiological one, involving, as it does, adequate controls of potential confounders. And, far and away the most crucial of these is rigorous and effective control for the influence of age. Admirably, in recent years, many studies have addressed this issue and attempted to more systematically control for age. Nevertheless, problems continue.

Entheseal changes

Entheses have two distinct morphologies: fibrous and fibrocartilaginous. A spectrum between the two exists, and different portions of the same enthesis can have different characteristics (Benjamin et al. 1986). Generally, fibrocartilaginous entheses occur close to joints at the epiphysis of the bones, but also on short bones and some parts of vertebrae. In contrast, fibrous entheses occur mainly on diaphyses and the cranium (Benjamin and McGonagle 2001). It has been proposed that the difference in enthesis type is directly related to differences in mechanical requirements (Benjamin et al. 2002).

Fibrocartilaginous entheses can be divided into four histological zones (Cooper and Misol 1970; Benjamin et al. 1986): 1) tendon or ligament, 2) uncalcified fibrocartilage, 3) calcified fibrocartilage, and 4) subchondral bone. Zones 2 and 3 are avascular and separated by a regular calcification front called the *tidemark*. However, this applies only to the innermost part of the enthesis, devoid of periosteum, but at the periphery the collagen fibers merge with the periosteum (Benjamin et al. 1986) and vessels may be present (Benjamin and McGonagle 2001; Dörfl 1969). The tidemark is the point at which soft tissues are removed during maceration (Benjamin et al. 1986), and the zone of calcified fibrocartilage has been found to be preserved in some archaeological skeletal remains (Henderson and Gallant 2005). Thus, it is the surface of the calcified zone that is observed on dry bones, and, as the tidemark is relatively straight and avascular, a healthy enthesis appears on the skeleton as well circumscribed, smooth, without vascular foramina (Benjamin et al. 2002). EC may occur in the outer and/or inner part of a fibrocartilaginous enthesis. Enthesophytes, the most recognized of these changes, generally form in the most fibrous, outer portion of the enthesis.

Mechanically induced changes are relatively well known for fibrocartilaginous entheses (e.g., Dupont et al. 1983; Potter et al. 1995; Selvanetti et al. 1997). Clinical literature has also reported the presence of EC for fibrocartilaginous entheses in many diseases (listed in Henderson 2008 and Villotte 2009). Fibrous entheses attach soft tissues to bone directly or via a mediating layer of periosteum (Benjamin et al. 2002). This layer often disappears with age and leaves the soft tissue attaching directly to bone, without a mediating layer (Benjamin et al. 2002). Few clinical studies have been undertaken to determine what causes EC at fibrous entheses. Recently, it has been hypothesized that the physiological transition from a periosteal to a bony attachment in early adulthood may explain the high frequency of skeletal changes (irregularity) seen in young/middle-aged adults (Villotte 2009). Later in adulthood, a second process related to cellular degeneration could explain major EC, such as bony ridges. Genetic factors and body mass may also play an important role in EC at fibrous entheses (Weiss 2003; Villotte et al. 2010b), and diseases have also been found to be related to their appearance (Henderson 2009b). It is noteworthy that fibrous entheses seem to be less vulnerable than fibrocartilaginous attachments to overuse injuries (Benjamin et al. 2002), and that, in fact, few medical descriptions of overuse injuries for these attachments could be found (Villotte 2008, 2009; Henderson 2009a).

Given these considerations, recognizing that different interface structures exist between the soft and hard components of the enthesis is crucially important for bioarchaeologists. Not only do these different structures affect the response of the enthesis to loading, but also their normal appearance on bone. Thus, the same methodology cannot be applied for both types of entheses and outcomes of EC cannot be pooled together. Little is currently known about fibrous entheses, and further study is required before they can be used to reconstruct past activity.

Pathological aspects of entheseal changes   Detailed investigations concerning the precise pathological nature of EC have been rare in clinical medicine and almost completely ignored by bioarchaeologists. In the approaches proposed by Crubézy (1988) and Dutour (1986, 1992), both medically trained, EC were seen as pathological alterations of the insertion sites (i.e. enthesopathies). Conversely, according to Hawkey and Merbs (1995) and Robb (1998), only the most exuberant changes are actually pathological. This discrepancy is, in our opinion, linked to the important differences between the two types of entheses. According to clinical findings, for some fibrocartilaginous entheses, EC are rarely seen for subjects younger than 50 years of age. Moreover, for such entheses there is a relative correspondence between experimental, clinical and osteological data. For instance, in the experiments conducted by Nakama and colleagues (2005, 2007), microtears were more common at the outer part of the enthesis. Traumatic and microtraumatic injuries in humans, as well as skeletal changes, are also more frequent at this location. These correspondences allow us to consider these entheseal changes as pathological. However, this conclusion perhaps does not apply to fibrous entheses and to several other fibrocartilaginous entheses (e.g., attachment of the ligamenta flava, for which skeletal changes are very common).

Limits and achievements of recording methods   As previously discussed, the most commonly utilized recording method is a ranking method developed by Hawkey. Several authors have criticized this method previously (Robb 1998; Mariotti et al. 2004, 2007; Cardoso 2008; Henderson 2009a; Villotte 2009). According to Hawkey and Merbs (1995), changes to the enthesis can be related to the process that produced them, and mean scores for fibrous and fibrocartilaginous attachments can be compared. The differences in anatomy described above make it obvious that this is a gross oversimplification. In fact, most of the previous published methods (e.g., Crubézy 1988; Robb 1998; al-Oumaoui et al. 2004; Mariotti et al. 2004, 2007) suffer from the same ignorance of anatomical structures and, until very recently, the more specific classification into fibrous and fibrocartilaginous entheses was not utilized or even recognized by bioarchaeologists. As a result of these deficiencies, in the last few years new binary visual recording methods for fibrocartilaginous entheses have been proposed (Henderson 2009a; Alves Cardoso and Henderson 2010; Villotte et al. 2010b).

Nonvisual recording methods incorporate the various methods aimed at quantifying various aspects of entheses, e.g., size and shape (Stirland 1998; Wilczak 1998, 2009; Zumwalt 2005, 2006; Henderson and Gallant 2007; Henderson 2009a, 2010a, b; Pany et al. 2009). The underlying concept in quantitative studies emphasizes that the dimensions of an enthesis are fixed during development; therefore an enthesis too small for the muscle mass will be stressed, leading to EC. Surface shape and cortical bone thickness, on the other hand, may be adaptable. The recent reduction in costs of computing power and three-dimensional laser scanners has facilitated this research, and it is probable this is the direction that future research will take. The primary limitation of these methods is their time-consuming nature, the data processing involved, and computing power and storage required.

Reference samples and the influence of physical activity   Several studies of EC using identified skeletal samples have been undertaken in the past. Cunha and Umbelino (1995), used the Crubézy's method on a part (n = 151) of the Coimbra collection for which five categories of occupation have been defined. This study failed to show significant differences between categories. Mariotti and collaborators, applying their method on about 100 identified skeletons from the Sassari and Bologna collections, were also unable to demonstrate any conclusive relationship between specific occupations and a particular distribution of EC (Mariotti et al. 2004, 2007). Niinimäki (2011) applied Hawkey's method on Finnish identified skeletons (n = 108). She found that only age and muscle size were relevant factors explaining combined MSM score. However, this score is usually greater for individuals from the heavy labor group in early life, compared to the less active group.

These negative results could indicate that the influence of occupational stresses on the occurrence of EC is far from obvious. However, inappropriate methods, anatomical issues and statistical tools, small sample size, or poor definition of the occupational groups may have obscured meaningful results. Villotte and colleagues (2010b) applied a new visual method of studying changes for fibrocartilaginous entheses to 367 males from the Spitalfields, Coimbra, Sassari, and Bologna collections. Men with occupations involving heavy manual tasks were found to present significantly more EC than non-manual and light-manual workers. This difference, however, was not found for fibrous entheses (Villotte 2009).

Different results were obtained by Cardoso and Henderson (2010), as they tested the relationship between activities, the ageing process and EC in 111 males from the Coimbra and Lisbon collections. They tested not only manual versus nonmanual workers, but also specific activities, and found that age-related degeneration, rather than degeneration caused by activities, may have been the primary cause of EC formation.

Cross sectional bone geometry

Serious methodological problems have not been as apparent or as contentious in CSBG investigations as they have in OA and EC studies. The greater degree of methodological consistency found in CSBG research was likely influenced by it deriving from a relatively small group of initial researchers who quite precisely defined their methodologies. Nevertheless, some potential difficulties have arisen involving the various techniques used to assess CSBG in archaeological samples. Direct sections, standard (biplanar) radiography, and CT have all been used. Directly sectioning bone is a destructive technique, and one not likely to be popular with collection curators. Plane radiography, even with standardized multiple views, is not as accurate as is CT. Moreover, results obtained from precise CT in living populations (Shaw and Stock 2009a, b) can be more easily compared to archaeological samples when a similar methodology is used.

Understandably, the added expense and more limited availability of CT have made it less practical in many studies of archaeological materials. The greater inaccuracy of standard (biplanar) radiography is generally not seen as a major difficulty, at least when the goal is to evaluate bilateral asymmetry (Trinkaus et al. 1994). However, biplanar radiography alone can produce results that are not as accurate and require correction. An alternative method, using latex casts of the external bone diaphysis, is more accurate and has been proposed as an adjunct to reinforce radiographic data (O'Neill and Ruff 2004).

Some difficulties could arise when the specimen is not well preserved. For instance, exfoliation of the original surface or post-depositional accretions could slightly alter cross-sectional dimensions. The precise orientation of a section when specimens are missing one or both ends could also be a problem. Moreover, the position of cross-sections is commonly defined relatively to the bone length, and as, long bones are often incomplete, the position of the cross-section frequently has to be estimated. Recently Sládek and colleagues (2010) evaluated the effect of inaccurately located femoral and tibial midshafts and found that important variations in the assessment of tibial second moments of area arise when midshaft locations are not precise. They concluded that "If tibial midshaft cannot be estimated within a 14–20 mm interval, then it should be considered whether to include the individual in the analysis" (Sládek et al. 2010:331). Moreover, the variables measured are extremely sensitive to the acquisition parameters (window width and level). As area (mm^2) and moments of area (mm^4) are used, small changes in these parameters induce high percentages of measurement errors, as was demonstrated for a wide sample of identified skeletons (Perréard Lopreno 2007).

A more serious issue is the tendency to succumb to the temptation to "tell stories" about earlier populations. Paleoanthropologists have largely resisted this trap, although investigators of more recent human samples have not been as cautious. Attributing some general level of activity as a primary factor leading to identifiable patterns of CSBG is one thing. Proposing specific activities in the absence of specific evidence is quite another (e.g., Fresia et al. 1990; Larsen and Ruff 1994; Weiss 2009). The problem, as for all the so-called markers of activity, concerns the sensitivity and specificity of such bone changes. No doubt, CSBG does have sufficient sensitivity in certain situations to at least allow inference of general levels of activity. The degree of specificity that CSBG permits remains undetermined. Experience with OA and, more recently, with EC would caution against over-optimism.

A particularly intriguing hypothesis, first made by Ruff (1987), proposed a widespread change in femoral shaft shape linked to major shifts in subsistence (i.e., from hunter-gathering to agriculture). Whether such changes in bone geometry result from altered activities is not firmly established (since diet and other factors could also contribute). However, recent results showing that cross-country runners develop a distinctive tibial shaft shape (Shaw and Stock 2009b) lend weight to this hypothesis.

Prospects For Future Research

In the last four decades research on activity markers has swung widely from enthusiastic expectations to disappointment. Progress has been slow, with prior mistakes often repeated in later research. However, with the application of more rigorous research approaches in the last decade, at least some studies have made improvements concerning all the major markers. From this foundation we can foresee yet further advancements. Firstly, the use of well-documented skeletal collections has been crucial in establishing more rigor in analyses of all the major indicators of activity. Moreover, particularly for OA (and to a lesser degree for EC), the clinical literature from sports and occupational medicine provides a rich resource deriving from dozens of controlled studies. A detailed understanding of this literature (and its varied results) is essential for bioarchaeologists, as is continued mastery of it as new work is published.

For EC, a number of comprehensive research programs have been developed which aim to interpret bone changes in well-documented collections that provide data on age at death, sex, as well as occupation. For instance, Henderson (2010a, b) has been using identified collections in London to test the relationship between EC appearance, size and shape, the aging process and activity-patterns. Along these same lines, a "methods group" was explicitly created after a workshop focusing on EC took place in Coimbra, Portugal in 2009.

While early support for activity-related interpretation of CSBG was based on limited contemporary samples, more recently well-controlled studies that rely on detailed collection of data and advanced CT technology (see, for example, Shaw and Stock (2009a, b) set outstanding examples for future researchers. Similarly, experimental research on nonhuman animals is another avenue that has been successfully pursued in an effort to understand and interpret the presence of OA (e.g., Newton et al. 1997), EC (e.g., Nakama et al. 2005, 2007; Zumwalt 2006), and CSBG (e.g., Pearson and Lieberman 2004) in archaeological populations. Zumwalt's (2006) recent experimental work on EC in sheep that were exercised on a treadmill is noteworthy. Ten animals exercised for a total of 900 hours were compared with matched sedentary controls. Despite a clear increase in muscle size, "The results of the study demonstrate no effect of the exercise treatment used in this experiment on any measure of enthesis morphology (p. 444)." Other recent research at the microscopic level (Nakama et al. 2005, 2007) indicates that cyclical loading, as induced by electrically stimulated muscle contractions in rabbits, produced significantly more tendon microtears.

These and other experimental results are limited in a variety of ways. For example, the degree of exercise to which Zumwalt subjected her experimental sheep was considerable for domestic animals, and to have attempted to do more (both in duration and intensity) would have been impractical in the former case and likely impossible within animal welfare guidelines in the latter regard. The degree of stress experienced by the sheep may well bear little correlation with contemporary human athletes or, presumably, with some members of ancient societies who subjected their bodies to extreme overuse. Likewise, Nakama and colleagues' results show development of pathological changes, but, again, the applicability to EC in humans experiencing habitual activity remains unclear. Nevertheless, the onset of tissue failure may be informative regarding pathogenesis in human EC, and also serves to remind us that EC is often the result of an intrinsically pathological process.

Recent advances in medical imaging, e.g., magnetic resonance imaging with ultrashort echo time (Benjamin et al. 2008) suggest that there may be reliable methods to analyze entheses in living individuals. Therefore, future research should aim to use these methods on contemporary athletes or people undertaking traditional crafts to study the effect of mechanical stress on EC, in contrast to the confounding influences of ageing and disease, as was done for CSBG (Shaw and Stock, 2009a).

Another productive line of research involves direct collaboration between bioarchaeologists and clinical colleagues. As noted, for CSBG such collaboration has long characterized basic research. Notable collaborative research has also been accomplished for OA, especially the research of Rogers, Waldron, and colleagues (e.g., Waldron and Cox 1989; Rogers et al. 1997). A more recent excellent model of such research has been completed by Robson-Brown and colleagues (Brown et al. 2008). For the most part, however, in recent years the less prominent participation of rheumatologists and orthopedists in the Paleopathology Association has accompanied a concomitant reduction in collaborative research on OA between bioarchaeologists and their medical colleagues.

The direction of contemporary skeletal biological research clearly is being transformed by advances in molecular biology. For OA especially, new findings show a very clear and major influence of specific genes on the ultimate expression of degenerative bone changes (Riancho et al. 2010). Preliminary experimental work has also recently been undertaken to delineate genetic influences on EC (Chen et al. 2007). Bioarchaeologists would be well served by mastering a fundamental understanding of "bone basics," or at least the basics relating to molecular/regulatory mechanisms influencing the phenotypic expressions observed on bone (See Chapter 5 in this volume, by Gosman for further discussion on this topic). Indeed, productive collaborative efforts with molecular biologists could yield exciting and far-reaching discoveries.

The future of bioarchaeological research is one filled with new opportunities. The path, however, will require new approaches and broader expertise. The old assumptions leading to an "activity-only" mindset are no longer tenable. Further research on OA, EC, and CSBG will likely yield other types of pertinent information about how the human skeleton develops and is altered during the life course. Skeletal biologists already know that ascertaining activity from archaeological remains is highly complicated. Indeed, determining clear and consistent conclusions may well prove to be generally impossible. Nevertheless, as the field moves forward, bioarchaeological researchers may find they have much more to contribute to overall skeletal biology than they would by persisting with a fruitless quest for the unattainable.

REFERENCES

Al-Oumaoui, I., Jiménez-Brobeil, S., and du Souich, P., 2004   Markers of Activity Patterns in Some Populations of the Iberian Peninsula. International Journal of Osteoarchaeology 14:343–359.

Alves Cardoso, F., and Henderson, C. Y., 2010   Enthesopathy Formation in the Humerus: Data from Known Age–at–Death and Known Occupation Skeletal Collections. American Journal of Physical Anthropology 141:550–560.

Anderson, J. E., 1963   The People of Fairty. An Osteological Analysis of an Iroquois Ossuary. National Museum of Canada, Bull. No. 193, Contributions to Anthropology 1961-1962:28–129.

Angel, J. L., 1964   The Reaction Area of the Femoral Neck. Clinical Orthopaedics and Related Research 32:130–142.

Angel, J. L., 1966   Early Skeletons from Tranquility, California. Smithsonian Contributions to Anthropolology 2:1–19.

Angel, J. L., 1971   The People of Lerna. Washington, D.C.: Smithsonian Institution Press.

Bard, H., 2003   Physiopathologie, Réparation, Classification des Tendinopathies Mécaniques. In Tendons et Enthèses. H. Bard, A. Cotten, J. Rodineau, G. Saillant, and J.-J. Railhac, eds. pp. 165–178. Montpellier: Sauramps Médical.

Benjamin, M., Copp, L., and Evans, E. J., 1986   The Histology of Tendon Attachments to Bone in Man. Journal of Anatomy 149:89–100.

Benjamin, M., Milz, S., and Bydder, G. M., 2008   Magnetic Resonance Imaging of Entheses. Part 2. Clinical Radiology 63:704–711.

Benjamin, M., Kumai, T., Milz, S., Boszczyk, B. M., Boszczyk, A. A., and Ralphs, J. R., 2002   The Skeletal Attachment of Tendons–Tendon "Entheses". Comparative Biochemistry and Physiology, Part A: Molecular & Integrative Physiology 133:931–945.

Benjamin, M., and McGonagle, D., 2001   The Anatomical Basis for Disease Localisation in Seronegative Spondyloarthropathy at Entheses and Related Sites. Journal of Anatomy 199:503–526.

Benjamin, M., and McGonagle, D., 2009   Entheses: Tendon and Ligament Attachment Sites. Scandinavian Journal of Medicine and Science in Sports 19:520–527.

Bennike, P., 1985   Paleopathology of Danish Skeletons. Copenhagen: Akademisk Forlag.

Bridges, P. S., 1989   Changes in Activity with the Shift to Agriculture in the Southeast United States. Current Anthropology 30:385–394.

Bridges, P. S., 1991   Degenerative Joint Disease in Hunter-Gatherers and Agriculturists From the Southeastern United States. American Journal of Physical Anthropology 85:379–391.

Bridges, P. S., 1992   Prehistoric Arthritis in the Americas. Annual Reviews of Anthropology 21:67–91.

Brown, K. R., Pollintine, P., and Adams, M. A., 2008   Biomechanical Implications of the Degenerative Joint Disease in the Apophyseal Joints of the Human Thoracic and Lumbar Vertebrae. American Journal of Physical Anthropology 136:318–326.

Buikstra, J. E., 1977   Biocultural Dimensions of Archeological Study. In Biocultural Adaptation in Prehistoric America. R. L. Blakely, ed. Pp. 67–84. Athens: University of Georgia Press.

Buikstra, J. E., and Ubelaker, D. H., 1994   Standards for Data Collection from Human Skeletal Remains, 44. Fayetteville: Arkansas Archeological Survey Research Series.

Capasso, L., Kennedy, K. A. R., and Wilczack, C. A., 1999   Atlas of Occupational Markers on Human Remains. Teramo: Edigrafital S.p.A.

Cardoso, F. A., 2008   A Portrait of Gender in Two 19th and 20th Century Portuguese Populations: a Palaeopathological Perspective. Ph.D. dissertation, Durham University.

Chapman, F. H., 1972   Vertebral Osteophytosis in Prehistoric Populations of Central and Southern Mexico. American Journal of Physical Anthropology 36:31–38.

Chen, X., Macica, C., Nasari, A., Judex, S., and Bloadus, A. E., 2007   Mechanical Regulation of PTHrP Expression in Enthesis. Bone 41:752–759.

Cohen, M. N., and Armelagos, G. J., eds., 1984   Paleopathology at the Origins of Agriculture. Orlando, FL: Academic Press.

Cooper, C., McAlindon, T., Coggon, D., Egger, P., and Dieppe, P., 1994   Occupational Activity and Osteoarthritis of the Knee. Annals of Rheumatic Diseases 53:90–93.

Cooper, R. R, and Misol, S., 1970   Tendon and Ligament Insertion. A Light and Electron Microscopic Study. The Journal of Bone and Joint Surgery [Am.] 52:1–20.

Crubézy, E., 1988   Interactions entre Facteurs Bio-Culturels, Pathologie et Caractères Discrets. Exemple d'une Population Médiévale: Canac (Aveyron). Ph.D. dissertation, Université de Montpellier.

Cunha, E., and Umbelino, C., 1995   What Can Bones Tell About Labour and Occupation: The Analysis of Skeletal Markers of Occupational Stress in the Identified Skeletal Collection of the Anthropological Museum of the University of Coimbra (Preliminary Results). Antropologia Portuguesa 13:49–68.

Dolgo-Saburoff, B., 1929   Über Ursprung und Insertion der Skelettmuskeln. Anatomischer Anzeiger 68:80–87.

Dörfl, J., 1969   Vessels in the Region of Tendinous Insertions. I. Chondroapophyseal Insertion. Folia Morphologica 17:74–78.

Dupont, M., Pasteels, J. L., Duchateau, M., and Szpalski, M., 1983   Tendinites Corporéales et Ostéotendinites, Essai de Définition des Lésions et de leur Traitement Chirurgical. Acta Orthopædica Belgica 49:30–41.

Durigon, M., and Paolaggi, J.-B., 1991   Enthèse au Cours de la Vie. In Pathologie des Insertions et Enthésopathies. L. Simon, C. Hérisson, and J. Rodineau, eds. Pp. 12–17. Paris: Masson.

Dutour, O., 1986   Enthesopathies (lesions of muscular insertions) as Indicators of the Activities of Neolithic Saharan Populations. American Journal of Physical Anthropology 71:221–224.

Dutour, O., 1992   Activités Physiques et Squelette Humain: le Difficile Passage de l'Actuel au Fossile. Bulletins et Mémoires de la Société d'Anthropologie de Paris n.s., 4:233–241.

Emslander, H., Siukai, M., Muhs, J., Chao, E., Wahner, H., Bryant, S., Riggs, L., and Eastell, R., 1998   Bone Mass and Muscle Strength in Female College Athletes (Runners and Swimmers). Mayo Clinic Proceedings 73:1151–1160.

Fresia, A. E., Ruff, C. B., and Larsen, C. S., 1990   Temporal Decline in Bilateral Asymmetry of the Upper Limb on the Georgia Coast. In The Archaeology of Mission Santa Catalina de Guale: 2 Biocultural Interactions of a Population in Transition. C.S. Larsen, ed. pp.121–132. Anthropological Papers 68. New York: American Museum of Natural History.

Galik, K., Senut, B., Pickford, M., Gommery, D., Treil, J., Kuperavage, A. J., and Eckhardt, R. B., 2004   External and Internal Morphology of the BAR 1002'00 Orrorin tugenensis Femur. Science 305:1450–1453.

Galtés, I., Jordana, X., Manyosa, J., and Malgosa, A., 2009   Functional Implications of Radial Diaphyseal Curvature. American Journal of Physical Anthropology 138:286–292.

Goodman, A. H., and Martin, D. B., 2002   Reconstructing Health Profiles from Skeletal Remains. In The Backbone of History. Health and Nutrition in the Western Hemisphere. R. H. Steckel, and J. C. Rose eds. pp. 11–60. Cambridge: Cambridge University Press.

Havelková, P., and Villotte, S., 2007   Enthesopathies: Test of Reproducibility of the New Scoring System Based on Current Medical Data. Slovenská Antropológia 10:51–57.

Hawkes S. C., and Wells, C., 1975   Crime and Punishment in an Anglo–Saxon Cemetery? Antiquity 49:118–122.

Hawkey, D. E., and Merbs, C. F., 1995   Activity–Induced Musculoskeletal Stress Markers (MSM) and Subsistence Strategy Changes among Ancient Hudson Bay Eskimos. International Journal of Osteoarchaeology 5:324–338.

Hawkey, D. E, and Street, S., 1992   Activity-Induced Stress Markers in Prehistoric Human Remains from the Eastern Aleutian Islands. American Journal of Physical Anthropology [Suppl]14:89.

Henderson, C. Y., 2003   Rethinking Musculoskeletal Stress Markers. Poster presented at the British Association for Biological Anthropology and Osteoarchaeology Annual Conference, Southampton, September 5–17.

Henderson, C. Y., 2008   When Hard Work is Disease: The Interpretation of Enthesopathies. In Proceedings of the Eighth Annual Conference of the British Association for Biological Anthropology and Osteoarchaeology. M. Brickley, and M. Smith, eds. pp. 17–25. Oxford: British Archaeological Reports International Series 1743.

Henderson, C. Y., 2009a   Musculo-Skeletal Stress Markers in Bioarchaeology: Indicators of Activity Levels or Human Variation? A Re-Analysis and Interpretation. Ph.D. dissertation, Durham University.

Henderson, C. Y., 2009b   Disease-Related Entheseal Remodelling (Enthesopathies) at Fibrous Entheses. Poster presented at the British Association for Biological Anthropology and Osteoarchaeology Annual Conference, Bradford, September 18–20.

Henderson, C. Y., 2010a   Handedness and Enthesis Size: a Relationship? Poster presented at the 79th Annual Meeting of the American Association of Physical Anthropologists, Albuquerque, April 4–17.

Henderson, C. Y., 2010b   Enthesis Roughness in Documented Skeletal Remains. Paper presented at the 37th Annual Meeting of the Paleopathology Association, Albuquerque, April 12–14.

Henderson, C. Y., and Gallant, A. J., 2005   A Simple Method of Characterising the Surface of Entheses. Poster presented at the 32nd Paleopathology Association Meeting, Milwaukee, April 5–6.

Henderson, C. Y., and Gallant, A. J., 2007   Quantitative Recording of Entheses. Paleopathology Newsletter 137:7–12.

Heinonen, A., Sievanen, H., Kannus, P., Oja, O., and Vuori, I., 2002   Site-Specific Skeletal Response to Long–Term Weight Training Seems to be Attributable to Principal Loading Modality: A pQCT Study of Female Weightlifters. Calcified Tissue International 70: 469–474.

Hershkovitz, I., Bedford, L., Jellema, L. M., and Latimer, B., 1996   Injuries to the Skeleton Due to Prolonged Activity in Hand-to-Hand Combat. International Journal of Osteoarchaeology 6:167–178.

Hooton, E. A., 1930   The Indians of Pecos Pueblo. New Haven: Yale University Press.

Jones, H. H., Priest, J. D., Hayes, W. C., Tichenor, C. C., and Nagel, D. A., 1977   Humeral Hypertrophy in Response to Exercise. Journal of Bone and Joint Surgery 59A:204–208.

Jurmain, R. D., 1977   Stress and the Etiology of Osteoarthritis. American Journal of Physical Anthropology 46:353–366.

Jurmain, R. D., 1980   The Pattern of Involvement of Appendicular Degenerative Joint Disease. American Journal of Physical Anthropology 53:143–150.

Jurmain, R. D., 1999   Stories from the Skeleton. Behavioral Reconstruction in Human Osteology. Amsterdam: Gordon and Breach.

Jurmain, R. D., 2000   Degenerative Joint Disease in African Great Apes: An Evolutionary Perspective. Journal of Human Evolution 39:185–203.

Jurmain, R. D., and Villotte, S. 2010   Terminology—Entheses in Medical Literature: a Brief Review. In Workshop in Musculoskeletal Stress Markers (MSM): Limitations and Achievements in the Reconstruction of Past Activity Patterns. http://www.uc.pt/en/cia/msm/msm_after. Accessed June 13, 2011.

Kelley, J. O., and Angel, J. L., 1987   Life Stresses of Slavery. American Journal of Physical Anthropology 74:199–211.

Kennedy, K. A. R., 1983   Morphological Variations in Ulnar Supinator Crests and Fossae as Identifying Markers of Occupational Stress. Journal of Forensic Sciences 28:871–876.

Kennedy, K. A. R., 1989   Skeletal Markers of Occupational Stress. *In* Reconstruction of Life from the Skeleton. M. Y. Iscan, and K. A. R. Kennedy, eds. pp. 129–160. Wiley–Liss: New York.

Lane, W. A., 1888   The Anatomy and Physiology of the Shoemaker. Journal of Anatomy and Physiology 22:593–628.

Larsen, C. S., Craig, J., Sering, L. E., Schoeninger, M. J., Russell, K. F., Hutchinson, D.L., and Williamson, M. A., 1995   Cross Homestead: Life and Death on the Midwestern Frontier. *In* Bodies of Evidence: Reconstructing History through Skeletal Analysis. A. Grauer, ed. pp. 139–159. New York: Wiley–Liss.

Larsen, C. S., and Ruff, C. B., 1994   The Stresses of Conquest in Spanish Florida: Structural Adaptation and Change Before and After Contact. *In* In the Wake of Contact. Biological Responses to Conquest. C. S. Larsen, and G. R. Milner, eds. pp 21–34. New York: Wiley–Liss.

Larsen, C. S., Walker, P. L., Steckel, R. H. et al., 2009   History of Degenerative Joint Disease in Europe: Inferences About Lifestyle and Activity. Poster presented at the Annual Meeting of the American Association of Physical Anthropologists, Chicago, April 2. [Abstract: American Journal of Physical Anthropology 138(S48):172]

Lieverse, A. R., Weber, A. W., Bazaliiskii, V. I., Goriunova, O. I., and Savel'ev, N. A., 2007   Osteoarthritis in Siberia's Cis-Baikal: Skeletal Indicators of Hunter-Gatherer Adaptation and Cultural Change. American Journal of Physical Anthropology 132:1–16.

Lieverse, A.R., Bazaliiskii, V. I., Goriunova, O. I., and Weber, A. W., 2009   Upper Limb Musculoskeletal Stress Markers among Middle Holocene Foragers of Siberia's Cis-Baikal Region. American Journal of Physical Anthropology 138:458–472.

Lovejoy, C. O., Burstein, A. H., and Heiple, K. G., 1976   The Biomechanical Analysis of Bone Strength: A Method and its Application to Platycnemia. American Journal of Physical Anthropology 44:489–506.

Lovejoy, C. O., and Trinkaus, E., 1980   Strength and Robusticity of the Neandertal Tibia. American Journal of Physical Anthropology 53:465–470.

Luttmann, A., Jäger, M., Griefahn, B., Caffier, G., Liebers, F., and Steinberg, U., 2003   Protecting Workers' Health Series No. 5: Preventing Musculoskeletal Disorders in the   Workplace. World Health Organization http://www.who.int/occupational_health/publications/en/oehmsd3.pdf, accessed June 13, 2011.

Maggiano, I. S., Schultz, M., Kierdorf, H., Sierra Sosa, T., Maggiano, C. M., and Tiesler Blos, V., 2008   Cross-Sectional Analysis of Long Bones, Occupational Activities and Long-Distance Trade of the Classic Maya from Xcambó – Archaeological and Osteological Evidence. American Journal of Physical Anthropology 136:470–477.

Mariotti, V., Facchini, F., and Belcastro, M. G., 2004   Enthesopathies – Proposal of a Standardized Scoring Method and Applications. Collegium Antropologicum 28:145–159.

Mariotti, V., Facchini, F., and Belcastro, M. G., 2007   The study of Entheses: Proposal of a Standardised Scoring Method for Twenty–Three Entheses of the Postcranial Skeleton. Collegium Antropologicum 31:191–313.

Merbs, C. F., 1983   Patterns of Activity-Induced Pathology in a Canadian Inuit Population. Archaeological Survey of Canada, Paper 119. Ottawa: National Museums of Canada.

Molnar, P., 2006   Tracing Prehistoric Activities: Musculoskeletal Stress Marker Analysis of a Stone–Age Population on the Island of Gotland in the Baltic Sea. American Journal of Physical Anthropology 129:12–23.

Molnar, P., 2010   Patterns of Physical Activity and Material Culture on Gotland, Sweden, During the Middle Neolithic. International Journal of Osteoarchaeology 20:1–14.

Nagy, B., 2000   The Life Left in Bones: Evidence of Habitual Activity Patterns in Two Prehistoric Kentucky Populations. Ph.D. dissertation, Arizona State University.

Nakama, L. H., King, K. B., Abrahamsson, S., and Rempel, D. M., 2005   Evidence of Tendon Microtears Due to Cyclical Loading in an In Vivo Tendinopathy Model. Journal of Orthopaedic Research 23:1199–1205.

Nakama, L. H., King, K. B., Abrahamsson, S., and Rempel, D. M., 2007 The Effect of Repetition Rate on the Formation of Microtears in Tendon in an In Vivo Cyclical Loading Animal Model. Journal of Orthopaedic Research 25:1176–1184.

Newton, P. M., Mow, V. C., Gardner, J. T., Buckwalter, R. A., and Albright, J. P., 1997 The Effect of Lifelong Exercise on Canine Articular Cartilage. American Journal of Sports Medicine 25:282–287.

Niinimäki, S., 2011 What Do Muscle Marker Ruggedness Scores Actually Tell Us? International Journal of Osteoarchaeology 21(3):292–299.

Oates, J., Molleson, T., and Sołtysiak, A., 2008 Equids and an Acrobat: Closure Rituals at Tell Brak. Antiquity 82:390–400.

O'Neill, M.C., and Ruff, C. B., 2004 Estimating Human Long Bone Cross-sectional Geometric Properties: A Comparison of Noninvasive Methods. Journal of Human Evolution 47:221–235.

Ortner, D. J., 1968 Description and Classification of Degenerative Bone Changes in the Distal Joint Surfaces of the Humerus. American Journal of Physical Anthropology 28: 139–156.

Pany, D., 2008 "Working in a Saltmine…" – Erste Ergebnisse der Anthropologischen Auswertung von Muskelmarken an den menschlichen Skeletten aus dem Gräberfeld Hallstatt. In Interpretierte Eisenzeiten. Fallstudien, Methoden, Theorie. Tagungsbeiträge der 1. Linzer Gespräche zur interpretativen Eisenzeitarchäologie. K. Raimund, and J. Leskovar, eds. pp. 101–111.Studien zur Kulturgeschichte von Oberösterreich, Folge 18.Linz. http:// www.schlossmuseum.at/eisenzeiten/eisenzeiten%20I%20pdfs/Pany.pdf, accessed June 13, 2011.

Pany D., Viola, T., and Teschler-Nicola, M., 2009 The Scientific Value of Using a 3D Surface Scanner to Quantify Entheses. Paper presented at the Workshop in Musculoskeletal Stress Markers (MSM): Limitations and Achievements in the Reconstruction of Past Activity Patterns. University of Coimbra, July 2–3.

Pearson, O. M, and Lieberman, D. L., 2004 The Aging of Wolff's "Law": Ontogeny and Response to Mechanical Loading. Yearbook of Physical Anthropology 47:63–99.

Perréard Lopreno, G., 2007. Adaptation Structurelle des Os du Membre Supérieur et de la Clavicule à l'Activité: Analyse de l'Asymétrie des Propriétés Géométriques de Sections Transverses et de Mesures Linéaires dans une Population Identifiée (Collection SIMON).

Potter H. G., Hannafin, J. A., Morwessel, R. M., DiCarlo, E. F., O'Brien, S. J., and Altchek, D. W., 1995 Lateral Epicondylitis: Correlation of MR Imaging, Surgical and Histopathologic Findings. Radiology 196:43–46.

Rhodes, J. A., and Knüsel, C. J., 2005 Activity-Related Skeletal Change in Medieval Humeri: Cross-Sectional and Architectural Alterations. American Journal of Physical Anthropology 128:536–546.

Riancho, J. A., García-Ibarbia, C., Gravani, A., Raine, E. V., Rodríguez-Fontenla, C., Soto-Hermida, A., Rego-Perez, I., Dodd, A. W., Gómez-Reino, J. J., Zarrabeitia, M. T., Garcés, C. M., Carr, A., Blanco, F., González, A., and Loughlin, J., 2010 Common Variations In Estrogen-Related Genes Are Associated with Severe Large Joint Osteoarthritis: A Multicenter Genetic and Functional Study. Osteoarthritis and Cartilage. 18:927–933.

Robb, J., 1994 Skeletal Signs of Activity in the Italian Metal Ages: Methodological and Interpretive Notes. Human Evolution 9:215–229.

Robb, J., 1998 The Interpretation of Skeletal Muscle Sites: a Statistical Approach. International Journal of Osteoarchaeology 8:363–377.

Rodineau, J., 1991 Pathogénie des Enthésopathies du Membre Supérieur. In Pathologie des Insertions et Enthésopathies. L. Simon, C. Hérisson, and J. Rodineau, eds. pp. 166–171. Paris: Masson.

Rogers, J., Shepstone, L., and Dieppe, P., 1997 Bone Formers: Osteophyte and Enthesophyte Formation are Positively Associated. Annals of the Rheumatic Diseases 56:85–90.

Ruff, C. B., 1987   Sexual Dimorphism in Human Lower Limb Bone Structure: Relationship to Subsistence Strategy and Sexual Division of Labor. Journal of Human Evolution 16:391–416.

Ruff, C. B., 2009   Relative Limb Strength and Locomotion in *Homo habilis*. American Journal of Physical Anthropology 138:90–100.

Ruff, C. B., and Jones, H. H., 1981   Bilateral Asymmetry in Cortical Bone of the Humerus and Tibia – Sex and Age Factors. Human Biology 53:69–86.

Ruff, C. B., Walker, A., and Trinkaus, E., 1994   Postcranial Robusticity in Homo. III. Ontogeny. American Journal of Physical Anthropology 93:35–54.

Schneider, H., 1956   Zur Struktur der Sehnenansatzzonen. Zeitschrift für Anatomie und Entwicklungsgeschichte 119: 431–456.

Selvanetti, A., Cipolla, M., and Puddu, G., 1997   Overuse Tendon Injuries: Basic Science and Classification. Operative Techniques in Sports Medicine 5:110–117.

Shaw, C. N., and Stock, J. T., 2009a   Habitual Throwing and Swimming Correspond with Upper Limb Diaphyseal Strength and Shape in Modern Athletes. American Journal of Physical Anthropology 140:160–172.

Shaw, C. N., and Stock, J. T., 2009b   Intensity, Repetitiveness, and Directionality of Habitual Adolescent Mobility Patterns Influence the Tibial Diaphysis Morphology of Athletes. American Journal of Physical Anthropology 140:149–159.

Sládek, V., Berner, M., Galeta, P., Friedl, L., and Kudrnová, S., 2010   The Effect of Midshaft Location on the Error Ranges of Femoral and Tibial Cross-Sectional Parameters. American Journal of Physical Anthropology 141:325–332.

Sládek, V., Berner, M., Sosna, D., and Sailer, R., 2007   Human Manipulative Behavior in the Central European Late Eneolithic and Early Bronze Age: Humeral Bilateral Asymmetry. American Journal of Physical Anthropology 133:669-681.

Sosna, D., and Sailer, R., 2007   Human Manipulative Behavior in the Central European Late Eneolithic and Early Bronze Age: Humeral Bilateral Asymmetry. American Journal of Physical Anthropology 133:669–681.

Stirland, A. J., 1998   Musculoskeletal Evidence for Activity: Problems of Evaluation. International Journal of Osteoarchaeology 8:354–362.

Tainter, J. A., 1980   Behavior and Status in a Middle Woodland Mortuary Population from the Illinois Valley. American Antiquity 45:308–313.

Testut, L., 1889   Recherches Anthropologiques sur le Squelette Quaternaire de Chancelade. (Dordogne). Lyon: Pitrat aîné.

Trinkaus, E., Churchill, S. E., and Ruff, C. B., 1994   Postcranial Robusticity in Homo. II. Humeral Bilateral Asymmetry and Bone Plasticity. American Journal of Physical Anthropology 93:1–34.

Villotte, S., 2006   Connaissances Médicales Actuelles, Cotation des Enthésopathies: Nouvelle Méthode. Bulletins et Mémoires de la Société d'Anthropologie de Paris n.s., 18:65–85.

Villotte, S., 2008   Les Marqueurs Ostéoarticulaires d'Activité. *In* Ostéo-Archéologie et Techniques Médico-Légales: Tendances et Perspectives. P. Charlier, ed. pp. 383–389. Paris: Editions De Boccard.

Villotte, S., 2009   Enthésopathies et Activités des Hommes Préhistoriques – Recherche Méthodologique et Application aux Fossiles Européens du Paléolithique Supérieur et du Mésolithique. BAR International Series 1992. Oxford: Archaeopress.

Villotte, S., Churchill, S. E., Dutour, O., and Henry-Gambier, D., 2010a   Subsistence Activities and the Sexual Division of Labor in the European Upper Paleolithic and Mesolithic: Evidence from Upper Limb Enthesopathies. Journal of Human Evolution 59:35–43.

Villotte, S., Castex, D., Couallier, V., Dutour, O., Knüsel, C.J., and Henry-Gambier, D., 2010b   Enthesopathies as Occupational Stress Markers: Evidence from the Upper Limb. American Journal of Physical Anthropology 142:224–234.

Walker, P., and Hollimon, S. E., 1989   Changes in Osteoarthritis Associated with the Development of a Maritime Economy among Southern California Indians. International Journal of Anthropology 4:171–183.

Walker, P. L., Steckel, R. H., Larsen, C. S., et al., 2009   Historical Patterns of Injury and Violence in Europe. Poster presented at the Annual Meeting of the American Association of Physical Anthropologists, Chicago, April 2.

Waldron, H. A., and Cox, M., 1989   Occupational Arthropathy: Evidence from the Past. British Journal of Industrial Medicine 46:420–422.

Waldron, T., 1994   Counting the Dead. The Epidemiology of Skeletal Populations. New York: John Wiley.

Weiss, E., 2003   Understanding Muscle Markers: Aggregation and Construct Validity. American Journal of Physical Anthropology 121:230–240.

Weiss, E., 2009   Sex Differences in Humeral Bilateral Asymmetry in Two Hunter-Gatherer Populations: California Amerinds and British Columbian Amerinds. American Journal of Physical Anthropology 140:19–24.

Weiss, E., and Jurmain, R. D., 2007   Osteoarthritis Revisited: A Contemporary Review of Aetiology. International Journal of Osteoarchaeology 17:437–450.

Wells, C., 1962   Joint Pathology in Ancient Anglo-Saxons. Journal of Bone and Joint Surgery 44B:948–949.

Wells, C., 1963   Hip Disease in Ancient Man. Report of Three Cases. Journal of Bone and Joint Surgery. 45B:790–791

Wells, C., 1964   Bones, Bodies, and Disease. Evidence of Disease and Abnormality in Early Man. New York: Praeger.

Wells, C., 1965   Disease of the Knee in Anglo–Saxons Medical and Biological Illustration 15:100–107.

Wells, C., 1972   Ancient Arthritis. M&B Pharmaceutical Bulletin (December):1–4.

Wilczak, C. A., 1998   Consideration of Sexual Dimorphism, Age, and Asymmetry in Quantitative Measurements of Muscle Insertion Sites. International Journal of Osteoarchaeology 8:311–325.

Wilczak, C. A., 2009   New Directions in the Analysis of Musculoskeletal Stress Markers. Paper presented at the Workshop in Musculoskeletal Stress Markers (MSM): Limitations and Achievements in the Reconstruction of Past Activity Patterns. University of Coimbra, July 2–3.

Wood Jones, F., 1910   Fractured Bones and Dislocations. In The Archaeological Survey of Nubia, Volume 2. G. Elliot–Smith, and F. Wood Jones eds. pp. 293–342. Cairo: National Printing Department.

Zumwalt, A., 2005   A New Method for Quantifying the Complexity of Muscle Attachment Sites. The Anatomical Record Part B: The New Anatomist 286B:21–28.

Zumwalt, A., 2006   The Effect of Endurance Exercise on the Morphology of Muscle Attachment Sites. The Journal of Experimental Biology 209:444–454.

CHAPTER 30

Oral Health in Past Populations: Context, Concepts and Controversies

John R. Lukacs

INTRODUCTION

This contribution to the literature of paleopathology has three goals: to provide context and perspective on the field of dental paleopathology, giving consideration to its links with allied disciplines, to review select examples of recent advances and enduring problems in the field, and to identify promising and prospective areas for future research. The introduction contextualizes the study of dental diseases and developmental anomalies in past populations by defining and delimiting this topically broad discipline and by providing historical perspectives on specific issues. Significant trends and accomplishments are noted and critically evaluated. The review of recent research achievements and enduring problems confronting the field focuses on four topics: a) dental diseases and anomalies in pre-Holocene hominins, b) diet, subsistence and dental disease, c) advances in caries research and sex differences in etiology, and d) developmental and genetic dental anomalies. Examples from my own research and from the current literature in dental paleopathology are included. The chapter identifies problematic issues in the field and suggests prospective areas for further research.

This broad, thematic, issues-oriented approach to the study of dental disease in past populations was adopted because existing literature already contains abundant

A Companion to Paleopathology, First Edition. Edited by Anne L. Grauer.
© 2012 John Wiley & Sons, Ltd. Published 2016 by John Wiley & Sons, Ltd.

resources devoted to the recognition, diagnosis, scoring, interpretation and reporting of pathological dental lesions. Some of these resources provide overviews treating multiple lesion types (Lukacs 1989; Langsjoen 1998; Hillson 2000), while others are dedicated to specific lesions: abscesses and granulomas (Dias and Tayles 1997; Ogden 2008), calculus (Lieverse 1999), caries (Lukacs 1995; Erdal and Duyar 1999; Hillson 2001, 2008; Duyar and Erdal 2003), and enamel hypoplasia (Goodman and Rose 1991; Goodman and Song 1999), to cite some examples. Another motivation for pursuing an expansive range of topics is to counter the perception that dental paleopathology is restricted to, or exclusively focused on, issues regarding the relationship between dental disease and diet, nutrition, or subsistence. By defining the field broadly and illustrating the diversity of techniques and applications embraced by dental paleopathology, new ideas and prospects for creative and innovative research will be fostered.

DEFINING AND DELIMITING THE FIELD OF STUDY

Dental paleopathology strives to identify and interpret diseases and anomalies of the teeth and jaws of past populations. The analysis of dental diseases and anomalies is conducted within two broad arenas or research traditions. One tradition is focused on the relationship between dental diseases and cultural factors such as diet, nutrition and subsistence. Another is centered on anomalies of dental development that are influenced to a greater extent by genetic factors. Each of these research traditions has a long and well established history and each has benefitted from research accomplishments of both anthropologists and clinical investigators.

Within the diet and dental disease research paradigm a diverse array of specific goals motivate research. These include: a) determining subsistence and diet of skeletal series with few associated cultural remains (Turner 1979; Lukacs 1989), b) understanding how differences in food preparation can influence the frequency of pathological lesions (Powell 1985), c) reconstructing trends in oral pathology across major changes in subsistence, such as the shift from foraging to agriculture (Cohen and Armelagos 1984; Cohen and Crane-Kramer 2007), d) determining the impact of colonization on diet and oral health of indigenous populations (Larsen and Milner 1994; Klaus and Tam 2010), and perhaps most commonly, e) providing an integrated biocultural perspective on subsistence, diet and nutrition of past populations. The last objective is an integral component of any multifaceted bioarchaeological research project. The basic data that come from such studies ultimately permits regional and global synthetic analyses of oral health from a comparative, paleoepidemiological perspective.

A second somewhat less common, yet valuable, area of enquiry in dental paleopathology involves genetic and developmental anomalies of the teeth and jaws. The anthropological analysis of developmental and genetic anomalies of teeth entails a wide range of goals and objectives. These include dental morphogenesis and pathogenesis, evolutionary trends and variation in tooth size and number of teeth, how population size and isolation influences breeding systems and the frequency of dental anomalies, physiological stress and developmental dental aberrations, and forensic identification. A comprehensive review of hereditary dental anomalies included anomalies of tooth form and size, disorders of tooth number, anomalies of tooth position (malalignment) and arch relationship (malocclusion), hereditary disturbance of tooth structure (amelogenesis and dentinogenesis imperfecta), disturbances in tooth eruption, and congenital defects and syndromes involving dental anomalies (Alt and Türp 1998). Recent developments in each of these two major research paradigms will be summarized and critically discussed once the classification of oral diseases, the interdisciplinary nature of research and some important historical issues are considered.

A CLASSIFICATION OF DENTAL DISEASES

The broad subdivision of dental diseases and anomalies into two main research paradigms, dental diseases associated with cultural factors and dental diseases linked to genetic and developmental anomalies, is not always clearly discernable in classifications of dental pathology. Nevertheless, classifications of dental diseases are useful because they define the nature and breadth of pathological conditions and processes studied, and also reveal how limited is the list of diseases and anomalies that can be investigated in ancient skeletal samples. For example, Chapter XI of the WHO's International Classification of Diseases (2007) is devoted to diseases, disorders and anomalies of the digestive system and includes a "block" (K00-K14) entitled "Diseases of oral cavity, salivary glands and jaws." An abridged version of the WHO–ICD list of diseases of the teeth, jaws and salivary glands is provided in Table 30.1 to illustrate the range of pathological lesions subsumed under the rubric of dental disease.

While this list is comprehensive, it does not clearly differentiate diseases and anomalies into the two research paradigms described above. The diet and dental disease paradigm includes WHO–ICD disease categories K02 through K06, while the genetic and developmental diseases paradigm would consist of categories K00-01, K07, K09 and K10. Some WHO–ICD categories consist exclusively of soft tissue pathological conditions, such as stomatitis, and diseases of the lip, oral mucosa, and tongue (K 12–14), and therefore cannot be studied in skeletal samples. Other categories (K08, for example) include a mix of systemic and genetic conditions (exfoliation) with lesions such as tooth loss that may be secondary to nonsystemic factors including injury or diseases with a dietary association (periodontal disease, scurvy).

Clearly, dental paleopathology as a field of research is interdisciplinary in nature and derives significant methodological and theoretical input from anthropology, dentistry and evolutionary biology (see Figure 30.1). Depending on the nature of the research problem, dental paleopathology benefits from methodological and theoretical developments in many allied fields. For example, ethnographic documentation of diet and food preparation methods may provide indispensable contextual data for better understanding differences in the frequency of pathological dental lesions. Use of teeth as tools in a wide range of nondietary occupational tasks, such as the use and making of tools or processing of reeds or sinew, is enhanced and amplified by ethnographic research (Larsen 1985; Brown and Molnar 1990). An ethnographic component in the assessment of diet, dental health and culture change has been adopted by Walker and colleagues in the study of African hunter-gatherers (Aka, Mbuti) and their farming neighbors, the Bantu (Walker and Hewlett 1990). Subsequently, investigations among South American native hunter-horticulturalists,

Table 30.1 World Health Organization's International Classification of Disease

Block K: Diseases of oral cavity, salivary glands and jaws

K00   *Disorders of tooth development and eruption*
         Anomalies of number, size and form; mottled teeth; enamel hypoplasia; peg-shaped teeth; enamel pearls; hereditary disturbances of tooth structure.

K01   *Embedded and impacted teeth.*
         Failure to erupt with (impacted) or without (embedded) obstruction.

K02   *Dental caries*
         Caries of enamel, dentine, cementum.

K03   *Other diseases of hard tissues of teeth*
         Excessive attrition, abrasion, erosion; hypercementosis; calculus; accretions.

K04   *Diseases of pulp and periapical tissues*
         Pulpitis, abscesses, radicular cysts.

K05   *Gingivitis and periodontal diseases*
         Acute and chronic gingivitis, periodontitis.

K06   *Other disorders of gingiva and edentulous alveolar ridge*
         Gingival enlargement or recession, alveolar ridge lesions associated with trauma.

K07   *Dentofacial anomalies [including malocclusion]*
         Anomalies of arch position, tooth position, temporomandibular joint disorders.

K08   *Other disorders of teeth and supporting structures*
         Exfoliation from systemic cause, tooth loss from injury, extraction, periodontitis.

K09   *Cysts of oral region, not elsewhere classified*
         Developmental, odontogenic and non-odontogenic cysts.

K10   *Other diseases of jaws*
         Maxillary and mandibular tori, Stafne's cyst, exostoses, osteitis, periostitis.

K11   *Diseases of salivary glands*
         Atrophy / hypertrophy of salivary gland, gland stones and calculus, xerostomia.

K12   *Stomatitis and related lesions*

K13   *Other diseases of lip and oral mucosa*

K14   *Diseases of tongue*
         http://apps.who.int/classifications/apps/icd/icd10online/

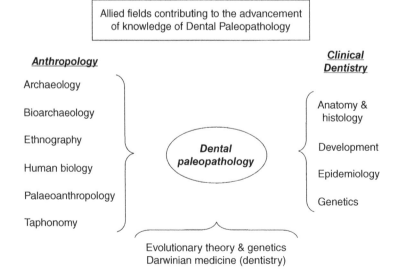

Figure 30.1 Allied fields contributing to the advancement of knowledge of dental paleopathology.

made extensive use of ethnographic data on gender differences in behavior and oral health, diet and food preparation (Walker et al. 1998). In this study investigators stressed the need for more extensive "ethno-bioarchaeological" research, "By this we mean studies in which ethnographic and physical anthropological data are collected as part of collaborative efforts to answer ethnographic and bioarchaeological questions." (Walker et al. 1998: 381).

THE HISTORY OF DENTAL PALEOPATHOLOGY

A comprehensive history of research in dental paleopathology has not been written. However, the historical development of interactive research in North American bioarchaeology and dental anthropology was reviewed by Rose and Burke (2006), who include dental paleopathology as a central theme (Rose and Burke 2006:323–346). Entitled "The dentist and the archaeologist", this comprehensive historical review includes pathological dental lesions, together with developments in the anthropological study of tooth size, tooth wear (macro- and micro-), microstructure (histology), and morphology, with special attention to ancient North American samples and investigators. In their analysis, the history of dental anthropology is organized into four chronological periods and three stages of research development. Following Willey and Sabloff (1993) the chronological periods include: a) Classificatory / descriptive (1840–1914), b) Classificatory / historical (1914–1940), c) Contextual / functional (1940–1960) and d) Modern (1960–present). Conceptually, the progressive development of research in dental anthropology is viewed as conforming to a general developmental sequence of stages: from plausibility to methodology to application. The stage of plausibility relies on the application of uniformitarian principles to establish that an idea, or cause-and-effect relationship, observed in living groups is also discernable in prehistoric skeletal samples. For example, variation in degree of occlusal wear or nature of striae in modern groups is linked to food consistency or to food-preparation methods. Plausibility establishes that degree of wear in skeletal samples reveals information about food texture and preparation practices in the past. The methodology stage develops and refines standardization of data collection, analytic techniques and interpretation, thereby improving reliability of observations and fine tuning causal associations between diet and dental disease or between frequency of dental anomalies and evolutionary principles such as mating patterns. Application is the third stage in this developmental sequence. Fundamental relationships between cause and effect established during the plausibility and methodology stages can now be applied with reliability and confidence. Application stage goals include either systematically documenting pathological dental lesions or anomalies in archaeologically derived human skeletal series, or addressing specific problem-oriented research questions employing extensive comparative samples.

Rose and Burke's (2006) historical study of bioarchaeology and dental anthropology provides useful insights regarding the development of dental paleopathology; they include a) significant advances made in different subfields by pioneering researchers (Mummery, Ruffer), b) early investigator's preempting of dental wear analysis for the purpose of age estimation, thus precluding this variable from a more prominent role in dietary reconstruction, and c) emphasizing Brothwell's role in synthesizing and disseminating research in dental anthropology, thus acting as a catalyst for further research in dental paleopathology.

The origin of the term "dental anthropology" has been attributed to a symposium of the Society for the Study of Human Biology at the Natural History Museum (London; Hillson 2007: xxv). The published proceedings volume of this symposium, entitled *Dental Anthropology*, was edited by D. R. Brothwell, who has played a central role in the development of the field of dental paleopathology in the Modern Era. The catalytic impact of this volume is clearly demonstrated by Rose and Burke (2006:338–39, 342). While the prominent role of *Dental Anthropology* is well established, the relative neglect of a volume published ten years earlier by two dentists is paradoxical. *The Human Masticatory Apparatus: An Introduction to Dental Anthropology* employs principles of evolutionary theory and culture change to understand the current condition of modern human teeth and jaws (Klatsky and Fischer 1953). The growth and development, and the structure and function of the masticatory system are addressed before the authors consider dietary variation and its impact on in the masticatory system. The prime focus of the volume is dental pathology— explaining the prevalence of modern human dental diseases within an evolutionary paradigm. In sequence they consider the causes, incidence and prevention of dental caries, periodontal disease and malocclusion. A causal association between "the rise of civilization" and increasing frequency of dental diseases is a central theme of the volume. Though developments in dental paleopathology are not featured in Scott and Turner's (2008) history of dental anthropology, the impact of Brothwell's (1963a) *Dental Anthropology* is regarded as a stimulus for subsequent research, while Klatsky and Fischer's (1953) volume, subtitled *An Introduction to Dental Anthropology* is not mentioned.

What accounts for the significant difference in impact these volumes had on the field of dental paleopathology? The possible answers are diverse yet speculative in nature. Basic differences in format, publisher and distribution may be involved. *The Masticatory Apparatus* was co-authored by two dentists with evolutionary anthropological interests. It was published by a small press (Dental Items of Interest, Brooklyn, NY), with limited distribution. By contrast, *Dental Anthropology*, an edited volume with 15 chapters and 18 authors, was published by Macmillan, with a more extensive distribution. When topical coverage is considered, just three chapters of *Dental Anthropology* focus on aspects of dental pathology (Brothwell et al. on congenital absence, Clement on microstructure and biochemistry, and Brothwell on macroscopic dental pathology), while *The Masticatory Apparatus* gives extensive attention to caries, periodontal disease, and malocclusion in evolutionary perspective. Possibly the more tightly focused, problem-oriented chapters of *Dental Anthropology* appealed more to anthropologists than the clinical– evolutionary paradigm of *The Masticatory Apparatus*, which championed the view that modern levels of dental morbidity result from the rise of civilization. Perhaps Klatsky and Fisher's "Introduction to Dental Anthropology" was unappreciated by its readership—clinicians and practicing dentists, or its distribution was limited and did not reach anthropologists conducting research in dental pathology. The minimal impact of this volume on the field of dental paleopathology is in some respects surprising and enigmatic.

The Scope of Dental Paleopathology

Dental paleopathology is a diverse field of research. Hence, different topics within the domain of dental paleopathology appeal to different investigators. As a guide to the literature and for easy reference, frequently cited literature in bioarchaeology and paleopathology has been selected to provide a review of subject matter and topical coverage embraced by the field of dental paleopathology. The goal here is to illustrate the range of subjects, to identify popular topics of research, and to provide a guide to the literature for interested students and researchers. For instance, definitions and current understanding of the etiology of commonly studied pathological dental lesions are provided in Table 30.2, while Figure 30.2 illustrates representative examples of pathological dental lesions commonly recorded in the analysis of dental disease in past populations. Similarly, in Table 30.3, lists dental diseases related to diet, subsistence and cultural modification of teeth along with key sources that address each affliction. While this survey is neither comprehensive, nor complete, it intentionally includes a wide range of publication outlets, from monographs with a geographic or ethnic focus (Bennike 1985; Webb 1995) to book chapters (Lukacs 1989; Hillson 2000;); from textbooks in bioarchaeology (Larsen 1997), paleopathology (Roberts and Manchester 2005), and osteology (Mays 1998), to encyclopedia entries (Langsjoen 1998). Historical sources (Klatsky and Fisher 1953; Brothwell 1963b; Wells 1975) and recent publications (Ogden 2008) are included as well. Differences in topical coverage are remarkable. For example, Mays' (1998) introductory text in human osteology gives attention to two lesion types: caries and enamel hypoplasia; while Roberts and Manchester's (2005) introduction to paleopathology provides extensive coverage of diet-related dental diseases. The paleopathology of Aboriginal Australians specifically avoids dental disease with one exception—dental enamel hypoplasia (Webb 1995). A valuable, yet seemingly underutilized, resource is Langsjoen's (1998) review of dental pathology in the *Cambridge Encyclopedia of Human Paleopathology* (Aufderheide and Rodriguez-Martin 1998), which covers a broad range of dental diseases and includes attention to diet-related diseases, genetic and syndromic dental disorders, and cultural modification of teeth. By contrast, Ogden's (2008) chapter on dental pathology centers on of a specific rare type of enamel hypoplasia (cuspal) and on diagnostic and interpretive problems associated with periodontal disease, abscesses, and granulomas.

Genetic and developmental dental anomalies—(K00-01, K10) are not well covered in most anthropological treatments of dental paleopathology, yet they have considerable significance for understanding evolutionary and population dynamics of earlier human groups. Though not truly "pathological" in nature, evidence of the cultural treatment or modification of teeth is of interest to anthropologists for the insights they yield regarding dental aesthetics in past societies and for the patterning of such practices by gender and status. In addition to the sources in Table 31.3, extensive treatment of topics such as ablation, implants, therapy and occupational wear are discussed elsewhere (Milner and Larsen 1991; Alt and Pichler 1998). Occlusal variation, under the pejorative label "malocclusion," received greater attention among earlier researchers, especially clinically oriented investigators (Brace 1977). An interesting difference in interpreting the etiology of occlusal variation is discernable. Dentists,

Table 30.2 Pathological lesions commonly observed in teeth and jaws of archaeologically derived human skeletons

Lesion Type	Definition / Etiology	Source
Abscess	A smooth-walled sinus cavity resulting from chronic infection of the pulp which channels through an exit (fenestra) in cortical bone (periapical / alveolar).	Dias and Tayles 1997, Ogden 2008
Alveolar recession	Reduction in height of alveolar crest due to chronic periodontitis (inflammation of periodontal tissues).	Odgen 2008, Hillson 2000
Antemortem tooth loss	Loss of teeth during life, evidenced by progressive resorption of the alveolus. May result from periodontal disease, penetrating caries, severe occlusal wear or trauma.	Lukacs 1989
Calculus	Mineralized plaque that accumulates at the basal surface of a living plaque deposit and is attached to the surface of the tooth. Adherence of plaque influenced by biochemical components of saliva.	Hillson 1996, 2000, Lieverse 1999
Caries	A disease process involving progressive, focal demineralization of dental hard tissues by organic acids derived from bacterial fermentation of dietary carbohydrates, especially refined sugars.	Featherstone 1987, Larsen 1997
Crowding	Displacement of teeth from their standard anatomical position due to inadequate developmental space. Results from a discrepancy between tooth size and jaw size.	Lukacs 1989
Dislocation	Tilting of the tooth (crown lingual / roots buccal) due to progressive loss of support from continuous eruption as a result of severe occlusal wear and alveolar recession. Roots may ultimately remain functional.	Clarke and Hirsch 1991a, b
Enamel hypoplasia	A deficiency in enamel thickness visible as transverse grooves or pits on the outer enamel surface. Results from disrupted ameloblast function and reduced secretion of enamel matrix during amelogenesis.	Goodman and Rose 1991
Granuloma	A small spherical soft-tissue lesion surrounding the root apex (periapical) in which breakdown products of necrotic pulp accumulate creating a space in alveolar bone. (See abscess above.)	Dias and Tayles 1997, Odgen 2008
Hypercementosis	Deposition of excessive amounts of cementum. Results from tooth mobility and compensates for reduction in crown height due to severe occlusal wear (continuous physiological eruption).	Brothwell 1963b, Langsjoen 1998, Corruccini 1991
Malocclusion	Deviation from a rarely attained ideal arrangement of teeth within and between jaws, not always resulting in functional problems. More appropriately referred to as occlusal variation.	Odgen 2008
Periodontal disease (periodontitis)	Inflammation and destruction of periodontal tissues that anchor the tooth to the jaw, results from the accumulation of bacterial plaque at the gingiva (gum margins). Causes reduction in height of alveolar crest.	Langsjoen 1998
Pulp exposure	Perforation of the pulp chamber allowing bacterial invasion and contact with connective tissue, nerves and blood vessels. May result from penetrating caries, severe occlusal wear, complicated dental trauma, and other factors.	Langsjoen 1998
Temporomandibular joint (TMJ) disease	An osteoarthritic response to chronic stress of the TMJ, resulting in porosity, marginal lipping or eburnation and altered shape of the articular surface of the glenoid fossa and / or mandibular condyle.	Langsjoen 1998

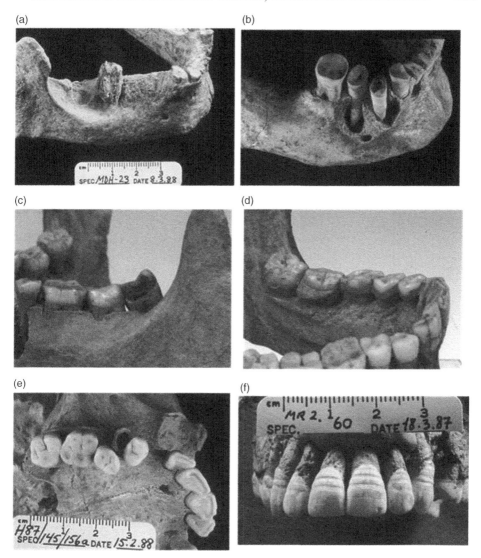

Figure 30.2 Examples of pathological conditions found in human dental remains: a) dislocation of RM_1 and antemortem loss of teeth due to severe wear (MDH 23); b) antemortem loss (RM_{2-3}), pulp exposure and periapical abscesses resulting from severe occlusal wear (P_{3-4}; HAR 148a); c) large carious lesion (LM_3) exposing pulp chamber (SKH 5); d) medium-size calculus deposits, lingual aspect, left P_4 through M_2 (SKH 12) (note calculus removed from LM_3, only traces present); e) transposed upper left canine. Canine crown carious, large LI^2–P^3 diastema (HAR 156a); and f) linear enamel hypoplaisa (LEH) four defects visible in LI^1. Severe episodes matched on I^1, I^2 and C indicating systemic stress (MR2 60) (all photos J. R .Lukacs).

who often treat members of a family from one generation to the next, tend to favor genetic factors. By contrast, anthropologists, who more often combine epidemiological and genetic perspectives in cultural context, find strong environmental factors at work. The epidemiological transition in occlusion can be seen in contemporary groups: with rural communities more frequently exhibiting better occlusion while

Table 30.3 Tabular guide to topical coverage of oral pathology and related conditions

Source	Abscesses	Alveolar recession / periodontal disease	AMTL	Calculus	Caries	Excessive wear / tooth dislocation	Enamel hypoplasia	Hypercementosis	Pulp exposure	'Malocclusion' crowding	Temporomandibular joint disease	Cultural dental modification
Bennike (1985)	–	✓	✓	–	✓	✓	–	–	–	–	–	–
Brothwell (1963)	–	–	✓	–	✓	–	✓	✓	✓	–	–	–
Hillson (2000)	–	✓	–	✓	✓	✓	✓	–	–	✓	–	–
Klatsky and Fisher (1953)	–	✓	–	✓	✓	✓	–	–	–	–	✓	✓
Langsjoen (1998)	–	✓	✓	✓	–	✓	✓	✓	✓	✓	✓	✓
Larsen (1995)	✓	✓	✓	✓	✓	✓	✓	✓	✓	✓	–	–
Lukacs (1998)	–	–	–	–	✓	–	✓	–	–	–	–	–
Mays (1998)	–	✓	✓	✓	✓	–	✓	–	–	–	–	–
Odgen (2008)	–	✓	–	–	–	–	✓	–	–	–	–	–
Roberts and Manchester (1995)	–	–	✓	✓	✓	✓	✓	–	–	–	✓	✓
Webb (1995)	–	–	–	–	–	–	✓	–	–	–	–	–
Wells (1975)	–	✓	✓	✓	✓	–	–	–	–	–	✓	–

urban samples display higher frequencies of malocclusion. A clinal patterning of occlusal variation across the urban–rural continuum is evident in diverse geographical and cultural settings including India (Kaul and Corruccini 1984) and Kentucky (Corruccini and Whitley 1981), and provides valuable context for understanding the temporal transition in occlusion from past to present (Corruccini 1991). The anthropological analysis of occlusal variation in past populations requires excellent preservation enabling articulation of undeformed upper and lower jaws. While the study of occlusal variation in skeletal samples continues and new methods are being used, the complex biomechanics of occlusion and the impact of severe tooth wear combine to complicate the analysis (Alt and Rossbach 2009).

DENTAL DISEASE IN PAST HUMAN POPULATIONS

Dental diseases and anomalies in pre-Holocene early hominins

This section provides a "deep-time" perspective on diet-based dental diseases (WHO–ICD categories K02 through K06), and on developmental dental anomalies (categories K00-01, K07, K09 and K10). While the primary focus of this chapter is on dental diseases and anomalies of post-Pleistocene, archaeologically derived modern humans, evidence of the longstanding nature of our present-day dental afflictions are becoming better known from the pre-Holocene fossil record. This brief review provides a glimpse of the range of dietary and developmental dental afflictions of early hominins. Representative noteworthy examples include: bilateral temporomandibular joint disease of the mandibular condyles of *Australopithecus afarensis* (MAK-VP-1/12) from Maka, Ethiopia at 3.4 mya (White et al. 2000). South African early hominins have yielded evidence of a supernumerary upper lateral incisor (SK 83, UI2) and interproximal dental caries, unassociated with enamel hypoplasia, in robust australopithecines at 1.7–1.5 mya (Grine et al. 1990; Ripamonti et al. 1999). Approximately 3.0 percent of a *Paranthropus robustus* dental sample (over 100 teeth) from Swartkrans has caries lesions, mostly in association with enamel hypoplasia (Grine et al. 1990). While periodontal disease has been reported in *Australopithecus africanus*, dental caries have not been documented in this taxon (Ripamonti and Petit 1991; Ripamonti et al. 1997). Dental anomalies and pathological lesions such as axial rotation (90 degrees) of the maxillary left second premolar (D2700), lingual tilting and torsomolar rotation of the mandibular right third molar (D211), abscesses and periodontal disease (D2600), and edentulism (D3444 / D3900) have been described in the extensive fossil sample of *Homo erectus* (cf. *ergaster*) from Dmanisi at 1.7 mya (Lordkipanidze et al. 2006; Rightmire et al. 2006). Bilateral, ninety degree rotation of maxillary P⁴s has been documented in the enigmatic hominin taxon *Homo floresiensis* (specimen LB1) (Brown et al. 2004:1058), as well as in living and archaeological samples of modern humans (see Figure 30.3).

While caries are generally rare in pre-Holocene fossils, they have been reported in Neanderthal and early modern human specimens (Brothwell 1963b; Hillson 2008). One unrepresentative yet provocative specimen, the Kabwe skull (Broken Hill 1), exhibits rampant dental caries (with 10 of 16 maxillary teeth affected). In addition, this unique individual exhibits multiple periapical abscesses, exposure of the pulp chamber and secondary (reparative) dentin (Koritzer and St. Hoyme 1979; Sperber

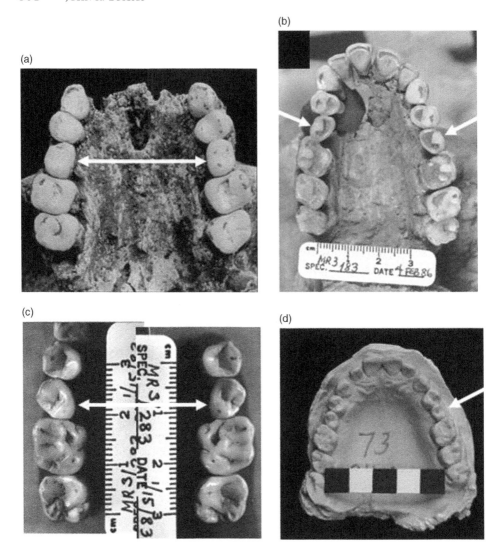

Figure 30.3 Examples of rotation of maxillary P⁴: a) maxilla, *Homo floresiensis* (LB 1), showing bilateral rotation of P⁴s (Brown et al. 2004; photo courtesy of Peter Brown); b) maxillary dentition, adult female (MR3 183) Neolithic Mehrgarh, with bilaterally rotated P⁴s (photo J. R. Lukacs); c) maxillary dentition (MR3 283), Neolithic Mehrgarh, with bilateral 45 degree rotation of P⁴ inferred from interproximal wear facets (photo J. R. Lukacs); and d), maxillary dentition of an adult Chenchu female (CHU 73) with bilateral rotation and reduced crown size of P⁴ associated with reduced crown size of RI² (photo G. C. Nelson).

1986). The poor oral health of Kabwe has been attributed by some to lead poisoning (Bartsiokas and Day 1993) and by others to lack of dental hygiene—specifically not knowing the use of tooth picks (Puech et al. 1980). Developmental anomalies including tooth rotation, supernumerary and congenital absence of teeth, occlusal variation and anomalies of eruption have been documented for Middle Paleolithic early modern humans from Dolní Veˇstonice and Pavlov (Hillson 2006). Finally, the analysis and interpretation of developmental enamel defects known as linear enamel hypoplasia (LEH) in hominin fossils has burgeoned, yielding new insights on the timing and relative frequency of physiological stress and growth disruption in early hominins (*Australopithecus* and *Paranthropus*)(Guatelli-Steinberg 2003, 2004) and in Neandertals (Guatelli-Steinberg et al. 2004). Significantly, evidence of defective enamel formation comes from some of the earliest hominins and from an individual subadult specimen. For example, enamel hypoplasia suggestive of physiological stress is present in the late Miocene hominin *Sahelanthropus* from Chad at 7.0–6.0 mya (Brunet et al. 2002) and amelogenesis imperfecta, a hereditary disorder of enamel formation has been documented in a *Homo erectus* child (Garba IV) recovered from Melka Kunture, Ethiopia and dated to 1.5 mya (Zilberman et al. 2004). These examples clearly demonstrate that pathological dental lesions can be preserved in fossil hominins for millions of years, can be reliably diagnosed (differentially distinguished from postburial diagenetic effects), and can inform us of the antiquity and in some cases relative frequency of dental diseases and anomalies in pre-Holocene times.

Holocene changes in subsistence, diet and dental disease: the impact of agriculture

Changes in prevalence of a complex group of diet-related dental diseases (WHO–ICD disease categories K02 through K06) is the main theme of this section. The nature of research conducted by pioneers in the field of dental paleopathology reveals that many questions motivating 21st-century research have long and well-established histories. An early phase in this history involved the recognition that pathological lesions and developmental anomalies in ancient specimens could be equated with diseases and abnormalities in living populations. The study of dynastic Egyptians by medically trained British anthropologists is noteworthy in this regard, with initial studies by Marc Armand Ruffer (1913, 1920; Sandison 1967) and Grafton Elliot Smith and Fredric Wood Jones (Smith and Jones 1910). Both lesion categories, dental diseases relating to diet and developmental genetic anomalies were included in Ruffer's analyses. The study of dynastic Egyptian dental disease continues into the modern era, with the work of Greene (1967, 1972), Hillson (1979), and Forshaw (2009).

While research on diet, subsistence and dental disease has many potential foci, the most intensively investigated question is, "How does a shift in subsistence and related changes in diet impact dental health?" Significant changes in dental health comprise one of several complex and interrelated biological consequences of agriculture and sedentism (Larsen 1995). Key distinctions in dental pathology profile have been documented for hunting and foraging groups and farmers. The foundation for such studies included early "ethno-bioarchaeology" research on the relationship between diet, subsistence and dental disease among native North Americans by (Leigh 1925), which helped provide context for the analysis of diet among prehistoric groups (Leigh 1929). The associations between diet and dental pathology, and to a lesser extent dental wear, were incorporated in the analysis of living groups (Aleut) (Moorrees 1957) and of prehistoric skeletal samples (Pecos Pueblo) (Hooton 1930). The idea that an epidemiological transition in dental health accompanied the subsistence shift from hunting and foraging to agriculture received extensive support in the influential volume *Paleopathology and the Origins of Agriculture* (Cohen and Armelagos 1984). Two theoretical propositions were addressed in this volume: a) that agriculture was not adopted by choice, but rather resulted from need and stress, and b) that the

Table 30.4 Changing global coverage of health and subsistence transition

Continent	Region	Cohen and Armelagos (1984)		Cohen and Crane-Kramer (2007)	
		Number	%	Number	%
America	North	9	47.4	5	23.8
	South and central	4	21.2	3	14.3
	Total	13	68.4	8	38.1
Africa	Sub-Saharan	1	5.3	1	4.8
Asia	Central	0	0.0	1	4.8
	Eastern	0	0.0	1	4.8
	Southern	3	15.8	7	33.3
	Total	3	15.8	9	42.9
Europe	All regions	2	10.5	3	14.3

trajectory of human health following the adoption of farming is better characterized by decline rather than progressive improvement. The idea that the origin and intensification of agriculture was accompanied by an array of biological costs, that could be measured in terms of longevity, stature, physiological stress, relative work load, and oral health was demonstrated in skeletal and dental samples from diverse regions of the globe. Of the 19 studies in this volume, nearly half (47.4 percent) focused on sites and samples from North America (see Table 30.4). Problems of small sample-size and regional representation, uneven coverage of chronological periods and subsistence categories, and diversity of research methods muted the impact of this volume for some. However the consensus regarding dental health was clear—that the transition to agriculture was associated with a decline in oral health. The decline in oral health with farming was patently evident in the increasing frequency of dental caries, but other dental diseases and dental wear were also involved. These findings may derive in part from the North American focus of the volume and the detrimental consequences of maize-dependent diets on general health and nutrition and on oral health in particular.

In April 2004, on the 20th anniversary of the publication of *Paleopathology at the Origins of Agriculture*, an international group of bioarchaeologists convened in Clearwater, Florida at a Wenner-Gren-sponsored symposium to update and reassess skeletal evidence of agricultural and economic intensification (Cohen and Crane-Kramer 2007). Organizers intentionally broadened the geographic diversity of sites and samples with a significant increase in attention to data from Asia appearing in the published volume, *Ancient Health*. While only three chapters (15.8 percent) of the 1984 volume dealt with Asian samples (India, Levant, and southwest Asia) the 2007 update had nine chapters (42.9 percent) on Asia, the majority of these (seven chapters) centered on southern Asian samples (Bahrain, India, Levant, Malaysia, Thailand (2), United Arab Emirates) (see Table 30.4). The relationship between dental health, diet and subsistence received attention from most investigators (17/21 chapters; 81 percent). Perhaps not surprisingly the clear correlation between farming and oral health (especially caries) resulting from the 1984 volume was less clear in the sequel, *Ancient*

Health, due to an increase in the distinctiveness of regional variations including differences in ecology, culture (food preparation) and staple cultigen. Figure 22.4 in *Ancient Health* (Cohen and Crane-Kramer 2007:329) compares temporal trends in frequency of dental caries in all regions discussed. While caries were not reported for three areas (Bahrain, Mongolia, and South Africa), the majority—twelve regions— show a gradual or steep upward trend in caries frequencies, and two (England and UAE) show short-term variability with fluctuating increases and decreases in caries frequency. Thailand is represented by two studies. Despite archaeological evidence for increasing rice production from the Bronze Age (Ban Lum Khao) to the Iron Age (Noen U-Loke), tooth count and individual caries rates are very similar (Domett and Tayles 2007). Results from two additional sites in northeast Thailand reveal contrasting trends, with caries frequency decreasing at Ban Chiang and increasing at Non Nok Tha (Douglas and Pietrusewsky 2007). Earlier indications suggested that the close association between oral health and agriculture found among native North Americans is not directly applicable to the southeast Asia context, where rice is the primary cultigen (Tayles et al. 2000). Several contributors to the *Bioarchaeology of Southeast Asia* (Oxenham and Tayles 2006), found little support for the predicted decline in oral health and rise in caries frequency with the origin and intensification of rice agriculture. The increased emphasis on Asian sites and samples in *Ancient Health* confirms that caries should not be interpreted unequivocally as an index indicator of farming in all regions or for all cultigens. A summary of changing perspectives on rice cultivation and caries frequencies in the decade since the initial study by Tayles and colleagues (2000), asks "Can dental caries be interpreted as evidence of farming?: The Asian experience." (Tayles et al. 2009). The important object lesson here is that the current understanding of disease–subsistence and diet relationships are dependent on the geographic, cultural, ecological and temporal context of samples comprising the database. Furthermore, generalizations are likely to be inaccurate or wrong without critical consideration of all variables, and the advancement of research will require periodic reassessment of consensus opinion, especially when analyzing new samples from poorly documented regions.

Two concerns regarding research on diet and dental pathology that should be important themes for future investigation are: a) closer examination of hypothesized differences in dental pathology profile between coastal and inland groups, and b) more attention to changes in frequency of enamel hypoplasia with agriculture and related mortality effects. Several contributions to *Ancient Health* reported on dental health of coastal samples, from Bahrain, the southeastern U.S., Peru, and the United Arab Emirates, others examined inland and coastal samples from the same region Florida, Georgia, and North Carolina with some consistent and some divergent results depending on lesion considered, sample size and local ecological variables.

Building on prior research by Turner (1979) on Jomon health and subsistence, and using the concept of dental pathology profile (Lukacs 1989), a new and potentially valuable study constructed a theoretical dental pathology profile for marine-dependent groups (Selwood 2010). After first identifying groups with marine-dependent diets using stable isotopes, a suite of pathological dental lesions associated with marine dependent diets was established and labeled the marine-dependent dental pathology profile (MDDPP). This pathology profile included a combination of low-frequency (abscesses, antemortem tooth loss and caries) and high-frequency (calculus, periodontal disease, and severe wear) conditions. Despite omnipresent obstacles such as disparities in sample size, some small or fragmentary samples, and limitations relating to differences in method of reporting frequencies of pathological lesions, the idea of refining and narrowing the applicability of dental pathology profile to more specific dietary regimens is promising. Though the use of stable isotopes in concert with data on dental pathology to more precisely reconstruct diet in past populations has an established history (Lubell and Jackes 1994; Keenleyside 2008), the field would benefit from more extensive and creative application of this combination of methods.

Enamel hypoplasia exhibits a less consistent trend than caries with the onset and intensification of agriculture, as evidenced by the papers in *Ancient Health* and summarized by Cohen and Crane-Kramer (2007:326, Fig. 22.1). Nevertheless, linear enamel hypoplasias are especially useful in reconstructing the chronological timing of stress events during growth and development, because the age at which a defect formed can be estimated from its position on the labial surface of the tooth crown. The most reliable method of estimating the time of defect formation, and therefore the age at which a stress event occurred, remains controversial (Reid and Dean 2000; Martin et al. 2008; Ritzman et al. 2008). Nevertheless, bioarchaeological analysis of hypoplastic enamel is said to be an "untapped" source of valuable information that now provides tentative support for a concept known as the Barker Hypothesis, also more fully described as the Developmental Origins of Health and Disease Hypothesis (DOHaD) (Armelagos et al. 2009). Currently only a handful of studies support the contention that physiological disruptions in growth and development that occur *in utero* or in early childhood (indicated by enamel hypoplasia) are associated with increased morbidity later in life and with early mortality in adolescence or early adulthood. These linkages between developmental stress, as inferred from the presence and frequency of enamel hypoplasia, should be analyzed in relationship to early mortality in skeletal samples of foragers and farmers.

Caries etiology: new insights, complex etiology

This section is dedicated to new developments in understanding the etiology of a single, yet important dental disease: caries (WHO–ICD category K02). The relationship between diet, subsistence and dental disease has progressed over time through the research stages of plausibility, methodology and application (see above). The nature and extent of this relationship continues to be a major research initiative, in part because dental caries is a prevalent and costly public health concern (Vieira et al. 2008). Dental caries is a disease process involving progressive, focal demineralization of dental hard tissues by organic acids derived from bacterial fermentation of dietary carbohydrates, especially refined sugars (Featherstone 1987, 2000). Research developments in allied fields are continually and rapidly adding to knowledge regarding the multifactorial and complex etiology of dental caries. It is essential that anthropologists actively involved in reconstructing diet from dental pathology keep abreast of developments in clinical and genetic research that continue to broaden the range of etiological factors and influence the expression and frequency of dental caries. Advances on several fronts are helping clarify the mechanisms underlying cariogenesis and improving our understanding of sex differences in oral health. These include: a) variation in genes influencing enamel formation, b) genetic determinants of oral ecology, and c) genomic diversity of oral bacteria.

Genome-wide scans detecting associations between gene variants (quantitative trait loci) and every conceivable aspect of human variation have recently flourished and the number of publications is expected to peak in 2010 (Miller 2009). Not surprisingly, dental caries is among the diseases included in genome-wide association studies (GWAS). However, contrary to Miller's pessimistic disappointment in the accomplishments of GWAS research, the findings and implications for caries research hold promise. The first such study of 46 families from the Philippines found caries suggestive loci for genes influencing saliva flow and diet preferences (Vieira et al. 2008). The study documents a significant sex difference in mean number of decayed, missing and filled teeth (DMFT) between fathers (10.96) and mothers (14.45) in the Philippine families sampled. A protective locus for caries was identified on the X chromosome (Xq27.1) and has implications for the global and widely reported sex differences in caries experience (Lukacs and Largaespada 2006; Lukacs and Thompson 2008). Subsequent studies of variation in selected candidate genes involved in amelogenesis have focused on ameloblastin, amelogenin, enamelin, tuftelin-1, and tuftelin-interacting protein 11, in a Guatemalan-Mayan sample (Deeley et al. 2008) and in a sample of Turkish children (Patir et al. 2008). Collectively, these studies suggest that variation in gene loci controlling enamel formation, especially ameloblastin, amelogenin, and tuftelin, contribute to observed differences in caries susceptibility. The authors hypothesize that variation in these gene loci contributes to microstructural alterations in enamel that may result in higher mineral loss under acidic conditions and may facilitate bacterial attachment to biofilms (Patir et al. 2008). This observation is especially interesting in view of Rose and Burke's (2006) repeated references to early studies in the Classificatory / Descriptive (1840–1914) and Classificatory / Historical (1914–1940) periods concluding that increasing rates of dental decay (dental caries) in "civilized groups" might result from imperfect enamel or a decline in the quality of enamel. In the Contextual / Functional period (1940–1960), "The notion that a decline in the quality of enamel was the cause of the modern epidemic of dental disease was finally laid to rest." (Rose and Burke 2006: 337). This transformation of an idea, from early broad explanatory statements regarding the diachronic decline in quality of enamel and rise of caries frequency, to a modern population and molecular genetic understanding is enlightening and informative. Current research suggests that enamel microstructure is influenced by genetic factors and is variable within and between populations and sexes resulting in differential susceptibility to caries. This important advance in knowledge of caries etiology has direct implications for researchers seeking associations between dental caries frequencies and diet in past populations.

Recent advances in genomic research have dramatically improved our view of caries microbiology by revealing the genome sequences of 15 different oral bacteria (Russell 2008). Understanding the genomics of plaque bacteria permits valuable insights into their evolution, physiology and regulatory mechanisms, some of which may be linked to virulence. Preliminary analyses of global diversity in the human salivary biome (Nasidze et al. 2009), coupled with advances in the ecology and succession of plaque microbial biofilms help elucidate proximate mechanisms involved in individual and sex differentials in caries susceptibility (Kolenbrander and Palmer 2004; Marsh 2004).

Prior research in the field of bioarchaeology has documented a significant sex difference in oral health in past human populations, and across a diverse array of cultures and subsistence systems (Larsen 1983, 1998; Walker and Erlandson 1986; Lukacs 1996; Lukacs and Thompson 2008). Clinical research reveals a similar and consistent sex differential in oral health, especially in caries experience in diverse global samples of living populations (Haugejorden 1996; Lukacs 2008). The role that female sex hormones, pregnancy, and women's reproductive history play in contributing to the sex difference in oral health is supported by clinical research, but often unappreciated by anthropologists (Lukacs and Largaespada 2006). An analysis of sex differences in dental caries prevalence in Hungary and India, and among the hunting and foraging Xavante adds to the database in support of the patterning of dental caries by sex (Lukacs in press). A growing appreciation for the sex difference in oral health and the importance of women's reproductive history (total fertility) in contributing to differences in caries and tooth loss is encouraging (Arantes et al. 2009; Fields et al. 2009; Watson et al. 2010). The multifactorial and complex etiology of dental caries complicates analyses of the association between caries experience and pregnancy (parity). Consequently, contradictory and conflicting reports reveal a nearly equal balance of early studies in support and against a causal linkage (see early sources in Mangi 1954). However, recently published results of age-controlled samples of pregnant and nonpregnant women in Chang Mai, Thailand found significantly higher caries frequency among pregnant women (Rakchanok et al. 2010; and sources cited therein). Indirect evidence in support of a relationship between caries experience and parity comes from the widespread pattern in which the sex difference in caries and tooth loss increases with age, especially during women's reproductive years. Seeking evidence of a causal relationship between parity and poorer oral health in women will require greater care in research design that incorporates and controls for the diverse array of confounding variables, such as related oral lesions (periodontal disease and tooth-loss), the details of women's reproductive histories, and contributing social and economic variables (Russell et al. 2010). A recent meta-analysis of gender differences in caries prevalence and experience in South Asia revealed significantly higher caries in girls associated with the onset of puberty (Lukacs 2011). Preliminary results of this meta-analysis of the sex disparity in dental caries in India exhibit the complex nature of confounding social and religious variables. These include son preference (daughter neglect) in health care and feeding regimens (Miller 1981), the popularity of fasting as religious practice among Hindu women (Pearson 1996), and the belief that dietary restriction during pregnancy is desirable because it will result in a less difficult birth (Vallianatos 2006). These beliefs and behaviors take on added significance when coupled with clinical evidence that fasting, undernutrition and malnutrition result in changes in oral ecology promoting caries and in increased caries rates (Alvarez 1995; Johansson et al. 1984; Psoter et al. 2005, 2008). While social and religious factors such as these contribute to observed sex differences in caries experience, they are secondary to the synergistic impact of sex differences in diet and the influence of hormones and other systemic changes associated with pregnancy.

In sum, the results of genome-wide association studies hold promise for illuminating the causal mechanisms and pathways through which sex differences may in part originate. A significant outcome of the genetic research is that the sex difference in caries experience may be caused by: a) variation in the quality of tooth enamel (genes controlling enamel formation), b) variation in oral ecology (saliva flow and composition), c) variation in dietary preferences (olfaction and gustatory senses), and d) variation in the pathogenic microorganisms of the oral cavity. Clinical and epidemiological research into the sex differential in oral health needs to accommodate and control for as many contributing variables as possible. It is encouraging to see that fertility and women's reproductive biology are now being considered in the analysis of oral health of contemporary hunter gatherers (Arantes et al. 2009) and when sex differences are encountered in skeletal samples both female reproductive biology and sex differences in dietary behavior are considered (Klaus and Tam 2010; Temple and Larsen 2007). Future research into the sex bias in oral health holds promise for refining our understanding of the relative importance of key variables that contribute to it (genes, physiology, pregnancy, and culture).

Developmental dental anomalies: anthropological insights

This section focuses on the anthropological importance of documenting developmental dental anomalies (WHO–ICD categories K00-01, K07, K09 and K10) in past populations and emphasizes the value of clinical research to anthropological inference. The earliest descriptive accounts of dental pathology among Egyptians included attention to specimens with supernumerary teeth (fourth molars) and deficiencies in tooth number (third molar agenesis) (Ruffer 1920). Unerupted maxillary canine teeth (also known as suppressed or embedded) were noted in five male (2.98 percent) and two female (2.47 percent) adult skulls from an early analysis of the osteology of the Guanches, ancient inhabitants of Tenerife in the Canary Islands (Hooton 1925). Brief attention to dental occlusion, tooth rotation, impaction, and congenital absence was included in Nelson's (1938) analysis of the Pecos Pueblo dentition. These early descriptions of developmental dental anomalies do not discuss causal factors or hypothesize etiology. An early survey of congenital absence of teeth cites examples from fossil hominins, as well as prehistoric and living humans (Brothwell et al. 1963). Importantly, Brothwell and colleagues suggest potential factors contributing to the observed patterns of congenital absence, including evolutionary and genetic mechanisms such as natural selection, genetic drift, and inbreeding in isolated populations.

Relative to studies of diet and dental disease, the documentation and analysis of developmental anomalies may be less common simply because: a) the traits under investigation (hypodontia, supernumerary teeth, tooth displacement and rotation) occur in low frequency, and therefore large sample sizes are required for analysis, and b) dentitions recovered from archaeological contexts are often affected by postburial diagenesis and are either fragmentary, incomplete or distorted. Nevertheless, study of dental anomalies in prehistoric samples provides valuable insight into diverse aspects of early population dynamics and evolution. One benefit of documenting dental anomalies in past populations is better understanding the nature and origin of these afflictions. This approach relies on evidence that many dental anomalies of shape, form, number and position have a significant genetic component. Representative examples of the kinds of pathological dental anomalies that have been documented in the course of bioarchaeological research and their relevance to paleopathology are briefly summarized.

The frequency and etiology of a developmental anomaly of tooth position, known as canine–premolar transposition (Mx.C.P1) has been described among living samples in India (Chattopadhyay and Srinivas 1996) and North America (Peck et al. 1993; Peck and Peck 1995). Variation in trait frequency ranges from a low of 0.12–0.13 percent among Asian Indian and Saudi Arabian samples (respectively), to a high of

8.49 percent among native Americans from Santa Cruz Island, California. The high trait frequency among Santa Cruz Islanders has been interpreted to result from an inbreeding effect due to population isolation (Nelson 1992). Individual cases of canine–premolar transposition occur in Bronze Age and Iron Age samples from South Asia (Lukacs 1998) and the trait is present in low frequency (1.8 percent; 9 of 500) in the Pecos Pueblo sample from New Mexico (Burnett and Weets 2001).

Developmental dental anomalies commonly observed in prehistoric and living samples include reduction in size, displacement (or axial rotation) and congenital absence of teeth. Two classes of teeth have received special attention with regard to prevalence of rotation, displacement, or agenesis: maxillary lateral incisors and third molars (Garn et al. 1963). These teeth frequently exhibit significant variation in size and demonstrate a predisposition to dental reduction referred to as "peg-shaped" incisors or "barrel-shaped" molars (Montagu 1940; Le Bot and Salmon 1977; Le Bot et al. 1980). Permanent dental agenesis in modern humans is most frequent in third molars, followed by lateral incisors and second premolars (Polder et al. 2004). The high frequency of missing lateral incisors in an Iron Age sample from Noen U-Loke (Thailand) was cautiously attributed to agenesis, rather than pathological loss or ritual ablation (Nelsen et al. 2001). The authors infer that endogamy, possibly due to isolation, may have contributed to the high prevalence (79 percent of adults; 30 of 38) of individuals with at least one missing incisor in either the maxilla or mandible in this sample.

In a clinical sample of 1620 subjects, Baccetti (1998) found a significant association between dental agenesis (aplasia) and tooth-rotation anomalies. Maxillary lateral incisor agenesis was significantly associated with premolar rotation, and the reverse relationship, P^4 agenesis in association with I^2 rotation, was also significant (Baccetti 1998). This association of dental anomalies has been documented in a prehistoric specimen from Tuscany. Two independent studies describe the association of developmental anomalies in an Etruscan adolescent from the 6th century B.C. (Baccetti and Moggi-Cecchi 1995; Kocsis et al. 1995). Agenesis of second maxillary premolar teeth in this individual is associated with maxillary lateral incisors that are reduced in size and have a 'peg-shaped' form.

In addition to its presence in fossil hominin taxa *Homo erectus* (Dmanisi D2700; Rightmire et al. 2006) and *H. floresiensis* (LB 1), axial rotation of maxillary P^4s has been observed among living and Neolithic South Asians (see Figure 30.3), prehistoric Guanches of the Canary Islands (Lukacs et al. no date), and Rampasasa pygmies from Flores (Jacob 2006:13424). Clinical evidence supports a genetic influence on the expression of UP4 agenesis and rotation and an association between it and the absence or diminution of UI2 and UM3. Appreciation of an interpretative model that regards intertrait associations as complex, interdependent, and genetic in origin (Baccetti 1998; Kotsomitis et al. 1996), can add significantly to the understanding of past populations structure. However, paleopathologists must exercise vigilance in observing and recording dental agenesis, axial rotation and tooth size reduction at multiple loci in the dental arcade (I, P, M). These attributes—axial rotation, reduced or absent maxillary lateral incisors and third molars—are more frequent in females and exhibit increased frequencies in genetic isolates where inbreeding and/or founder effect can increase expression of rare genetically influenced conditions. Consequently, systematic observation of these developmental genetic anomalies should be included in dental pathology research protocols, for the insights they provide regarding prehistoric population dynamics.

CONCLUSION

The field of dental paleopathology consists of two primary research foci. One documents the prevalence and distribution of dental diseases in past populations to better understand the history of oral health in relation to cultural changes in behavior, especially diet and subsistence. The other investigates the frequency of developmental dental anomalies in fossil hominins and prehistoric skeletal samples for insights into evolutionary and genetic mechanisms influencing dental variation.

The field of dental paleopathology is broadly interdisciplinary and relies on developments in evolutionary theory and genetics, as well as advances in epidemiology and the clinical study of dental diseases and anomalies. In historical perspective, the field continues to be concerned with some of the same issues and research questions that intrigued early researchers interested in oral health and dental anomalies. This survey of dental paleopathology focused on four specific topics:

1) Examples of a wide range of dental diseases and anomalies in pre-Holocene hominins illustrates the great antiquity of afflictions that also occur in modern humans (caries, supernumerary teeth, enamel hypoplasia and periodontal disease, for example). While some examples represent single unique occurrences, others such as the analysis of linear enamel hypoplasia is based on adequate samples that allow insight on the relative level of developmental stress in early hominins (*Australopithecus* vs. *Paranthropus*) and in later premodern hominins (Neandertals vs. modern humans).
2) An assessment of the current status of research on diet, subsistence and dental disease confirmed the general pattern of declining oral health on the adoption and intensification of agriculture. However, rice-dependent societies of Southeast Asia appear to be an exception to this pattern and demonstrate that generalizations regarding subsistence and oral health from one geographical region (native North America) may not be applied universally.
3) Recent advances in understanding of caries etiology have yielded new insights from research in genetics (genome wide scans), clinical dentistry and epidemiology. Variation in enamel protein gene loci may result in altered enamel microstructure, influencing plaque adherence and rate of demineralization, thereby contributing to variation in cariogenesis. Pregnancy and malnutrition have a multifactorial, yet often unappreciated, impact on caries development. For example, both pregnancy and malnutrition have been documented to influence saliva flow rate and composition, key components in the etiology of dental caries that may contribute to the observed differential prevalence of caries by sex and by nutritional status.
4) Developmental and genetic dental anomalies have had an important place in dental paleopathology from the earliest pioneering research on ancient Egyptians. Today better understanding of the genetics underlying dental agenesis, tooth size reduction, and axial rotation of teeth allows for more precise inferences regarding the role of evolutionary forces on the expression and frequency developmental dental anomalies.

If this survey generates new questions and stimulates creative and innovative research resulting in further advancement of the field of dental paleopathology, it will have been a successful endeavor.

Acknowledgments

I would like to thank colleagues and institutions who have provided access to prehistoric South Asian skeletal series over the years, especially the Archaeology Department, Deccan College (Pune, India), the Department of Ancient History, Culture and Archaeology, University of Allahabad (Allahabad, India), and the Department of Archaeology, Government of Pakistan. Access to the Guanche specimens of Tenerife was courtesy of the Museo Archaeologico de Tenerife. Colleagues facilitating or collaborating in dental anthropology and paleopathology research include the Gyani Lal Badam, late George Franklin Dales, M.A. Halim, Brian E. Hemphill, Jean-Francois Jarrige, Kenneth A.R. Kennedy, Mark Kenoyer, Nancy C. Lovell, Richard Meadow, Greg C. Nelson, J. N. Pal, Conrado Rodriguez-Martín, Michael Schultz, and Subhash R. Walimbe.

Financial support for field research on the bioarchaeology and dental anthropology of South Asia comes from American Institute of Indian Studies, Alexander von Humboldt Foundation, National Geographic Association, National Science Foundation–International Program, Smithsonian Institution (foreign currency program), and the Wenner-Gren Foundation. Special thanks are due: to Anne Grauer for inviting me to contribute to this volume, to Greg Nelson for commenting on an earlier draft of this manuscript, and to Sarah E. Miller for assistance with scanning images and producing illustrations.

REFERENCES

Alt, K. W., and Pichler, S. L., 1998   Artificial Modifications of Human Teeth. *In* Dental Anthropology: Fundamentals, Limits, and Prospects. K. W. Alt, F. W. Rösing, and M. Teschler-Nicola, eds. pp. 387– 415. Wien: Springer.

Alt, K. W., and Rossbach, A., 2009   Nothing in Nature is as Consistent as Change. *In* Comparative Dental Morphology. T. Koppe, G. Meyer, and K. W. Alt, eds. pp. 190–196. Basel: Karger.

Alt, K. W., and Türp, J. C., 1998   Hereditary Dental Anomalies. *In* Dental Anthropology: Fundamentals, Limits, and Prospects. K. W. Alt, F. W. Rösing, and M. Teschler-Nicola, eds. pp. 95–128. Wein: Springer.

Alvarez, J. O., 1995   Nutrition, Tooth Development, and Dental Caries. American Journal of Clinical Nutrition 61:410S–416S.

Arantes, R., Santos, R. V., Frazao, P., and Coimbra, C. E. A., 2009   Caries, Gender, and Socio-Economic Change in the Xavante Indians from Central Brazil. Annals of Human Biology 36(2):162–175.

Armelagos, G. J., Goodman, A. H., Harper, K. N., and Blakey, M. L., 2009   Enamel Hypoplasia, and Early Mortality: Bioarchaeological Support for the Barker Hypothesis. Evolutionary Anthropology 18:261–271.

Aufderheide, A. C., and Rodriguez-Martin, C., 1998   The Cambridge Encyclopedia of Human Paleopathology. Cambridge (UK): Cambridge University Press.

Baccetti, T., 1998   Tooth Rotation Associated with Aplasia of Nonadjacent Teeth. Angle Orthodontist. 68:471–474.

Baccetti, T., and Moggi-Cecchi, J., 1995   Associated Dental Anomalies in an Etruscan Adolescent. Angle Orthodontist. 65:75–80.

Bartsiokas, A., and Day, M. H., 1993   Lead Poisoning, and Dental Caries in the Broken Hill Hominid. Journal of Human Evolution 24(3):243–249.

Bennike, P., 1985   Paleopathology of Danish Skeletons: A Comparative Study of Demography, Disease, and Injury. Denmark: Akademisk Forlag.

Brace, C. L., 1977   Occlusion to the Anthropological Eye. *In* The Biology of Occlusal Development. J. A. McNamara, ed. pp. 179–209. Craniofacial Growth Series, Monograph No. 7. Ann Arbor: Center for Human Growth, and Development.

Brothwell, D. R., ed, 1963a   Dental Anthropology. New York: Macmillan (Pergamon Press).

Brothwell, D. R., 1963b   The Macroscopic Dental Pathology of Some Earlier Human Populations. *In* Dental Anthropology. D. R. Brothwell, ed. pp. 271–288. New York: Macmillan (Pergamon Press).

Brothwell, D. R., Carbonell, V. M., and Goose, D. H., 1963   Congenital Absence of Teeth in Human Populations. *In* Dental Anthropology. D. R. Brothwell, ed. pp.179–190. New York: Macmillan (Pergamon Press).

Brown, P., Sutikna, T., Morwood, M. J., Soejono, R. P., Jatmiko, E. W., and Due, R. A., 2004   A New Small-Bodied Hominin from the Late Pleistocene of Flores, Indonesia. Nature 431(7012):1055–1061.

Brown, T., and Molnar, S., 1990   Interproximal Grooving, and Task Activity in Australia. American Journal of Physical Anthropology 81:545–554.

Brunet, M., Fronty, P., Sapanet, M., De Bonis, L., and Viriot, L., 2002   Enamel Hypoplasia in a Pliocene Hominid From Chad. Connective Tissue Research 43:94–97.

Buikstra J. E., and Ubelaker, D. H., 1994   Standards for Data Collection from Human Skeletal Remains. Fayetteville: Arkansas Archaeological Survey.

Burnett, S. E., and Weets, J. D., 2001   Maxillary Canine–First Premolar Transposition in Two Native American Skeletal Samples from New Mexico. American Journal of Physical Anthropology 116(1):45–50.

Chattopadhyay, A., and Srinivas, K., 1996   Transposition of Teeth, and Genetic Etiology. Angle Orthodontist 66:147–152.

Clarke N., and Hirsch R. S., 1991a   Tooth Dislocation: The Relationship With Tooth Wear, and Dental Abscesses. American Journal of Physical Anthropology 85(3):293–298.

Clarke N., and Hirsch R. S., 1991b   Physiological, Pulpal, and Periodontal Factors Influencing Alveolar Bone. In: M. A. Kelley, and C. S. Larsen, eds. Advances in Dental Anthropology. pp. 241–266. New York: Wiley–Liss.

Cohen, M. N., and Armelagos, G. J. eds., 1984   Paleopathology at the Origins of Agriculture. Orlando, FL: Academic Press.

Cohen, M. N., and Crane-Kramer, G M. M., eds., 2007   Ancient Health: Skeletal Indicators of Agricultural, and Economic Intensification. Gainsville: University Press of Florida.

Corruccini, R. S., 1991   Anthropological Aspects of Orofacial, and Occlusal Variations, and Anomalies. *In* Advances in Dental Anthropology. M. A. Kelley, and C. S. Larsen, eds. pp. 295–323. New York: Wiley–Liss.

Corruccini, R. S., and Whitley, L. D., 1981   Occlusal Variation in a Rural Kentucky Community. American Journal of Orthodontics 79:250–262.

Deeley, K., Letra, A., Rose, E. K., Brandon, C. A., Resick, J. M., Marazita, M. L., and Vieira, A. R., 2008   Possible Association of Amelogenin to High Caries Experience in a Guatemalan–Mayan Population. Caries Research 42:8–13.

Dias, G., and Tayles, N., 1997   "Abscess Cavity" – A Misnomer. International Journal of Osteoarchaeology 7:548–554.

Domett, K., and Tayles, N., 2007   Population Health from the Bronze to the Iron Age in the Mun Valley, Northeastern Thailand. *In* Ancient Health: Skeletal Indicators of Agricultural, and Economic Intensification. M. N. Cohen, and G. M. M. Crane-Kramer, eds. pp. 286–299. Gainsville: University Press of Florida.

Douglas, M. T., and Pietrusewsky, M., 2007   Biological Consequences of Sedentism: Agricultural Intensification in Northeast Thailand. Ancient Health: Skeletal Indicators of

Agricultural, and Economic Intensification. M. N. Cohen, and G. M. M. Crane-Kramer, eds. pp. 300–319. Gainsville: University Press of Florida.

Duyar, I., and Erdal, Y. S., 2003   A New Approach for Calibrating Dental Caries Frequency of Skeletal Remains. Homo 54(1):57–70.

Erdal, Y. S., and Duyar, I., 1999   A New Correction Procedure for Calibrating Dental Caries Frequency. American Journal of Physical Anthropology 108(2):237–240.

Featherstone, J. D. B., 1987   The Mechanism of Dental Decay. Nutrition Today (May/June):10–16.

Featherstone, J. D. B., 2000   The Science, and Practice of Caries Prevention. Journal of the American Dental Association 131:887–899.

Fields, M., Herschaft, E. E., Martin, D. L., and Watson, J. T., 2009   Sex, and the Agricultural Transition: Dental Health of Early Farming Females. Journal of Dentistry, and Oral Hygiene 1(4):42–51.

Forshaw, R. J., 2009   Dental Health, and Disease in Ancient Egypt. British Dental Journal 206(8):421–424.

Garn, S. M., Lewis, A. B., and Kerewsky, R. S., 1963   Third Molar Polymorphism, and its Significance in Dental Genetics. Journal of Dental Research 42:1344–1363.

Goodman, A. H., and Rose, J. C., 1991   Dental Enamel Hypoplasias as Indicators of Nutritional Status In Advances in Dental Anthropology. M.A. Kelley, and C.S. Larsen, eds. 279–293. New York: Alan R Liss.

Goodman, A. H., and Song, R.-J., 1999   Sources of Variation in Estimated Ages at Formation of Linear Enamel Hypoplasias. In Human Growth in the Past: Studies from Bones, and Teeth. R. D. Hoppa, and C. M. Fitzgerald, eds. pp. 210–240. Cambridge: Cambridge University Press.

Greene, D. L., 1967   Dentition of Meroitic, X-group, and Christian Populations from Wadi Halfa, Sudan. University of Utah, Anthropological Papers. No. 85. pp. i–xi, 1–71.

Greene, D. L., 1972   Dental Anthropology of Early Egypt, and Nubia. Journal of Human Evolution 1:315–324.

Grine, F. E., Gwinnett, A. J., and Oaks, J. H., 1990   Early Hominid Dental Pathology: Interproximal Caries in 1.5 Million-Year-Old *Paranthropus robustus* From Swartkrans. Archives of Oral Biology 35(5):381–386.

Guatelli-Steinberg, D., 2003   Macroscopic, and Microscopic Analyses of Linear Enamel Hypoplasia in Plio-Pleistocene South African Hominins With Respect to Aspects of Enamel Development, and Morphology. American Journal of Physical Anthropology 120(4):309–322.

Guatelli-Steinberg, D., 2004   Analysis, and Significance of Linear Enamel Hypoplasia in Plio-Pleistocene Hominins. American Journal of Physical Anthropology 123(3):199–215.

Guatelli-Steinberg, D., Larsen, C. S., and Hutchinson, D. L., 2004   Prevalence, and the Duration of Linear Enamel Hypoplasia: A Comparative Study of Neandertals, and Inuit Foragers. Journal of Human Evolution 47(1–2):65–84.

Haugejorden, O., 1996   Using the DMF Gender Difference to Assess the "Major" Role of Fluoride Toothpastes in the Caries Decline in Industrialized Countries: A Meta-Analysis. Community Dentistry, and Oral Epidemiology 24:369–375.

Hillson, S., 1979   Diet, and Dental Disease. World Archaeology 11(2):147–162.

Hillson, S., 1996   Dental Anthropology. Cambridge: Cambridge University Press.

Hillson, S., 2000   Dental Pathology. In Biological Anthropology of the Human Skeleton. M. A. Katzenberg, and S. R. Saunders, eds. pp. 249–286. New York: Wiley–Liss.

Hillson, S., 2001   Recording Dental Caries in Archaeological Human Remains. International Journal of Osteoarchaeology 11(4):249–289.

Hillson, S., 2006   Dental Morphology, Proportions, and Attrition. In Early Modern Human Evolution in Central Europe. E. Trinkaus, and J. Svoboda, eds. pp. 179–223. New York: Oxford University Press.

Hillson, S., 2007   Introduction. *In* Dental Perspectives on Human Evolution: State-of-the-Art Research in Dental Paleoanthropology. S. E. Bailey, and J.-J. Hublin, eds. pp. xxiii–xxviii. Dordecht: Springer.

Hillson, S., 2008   The Current State of Dental Decay. *In* Technique, and Application in Dental Anthropology. J. D. Irish, and G. C. Nelson, eds. pp. 111–135. Cambridge: Cambridge University Press.

Hooton, E. A., 1925   The Ancient Inhabitants of the Canary Islands. Cambridge, MA: Peabody Museum of Harvard University.

Hooton, E. A., 1930   The Indians of Pecos Pueblo: A Study of their Skeletal Remains. Papers of the Southwestern Expedition, 4. New Haven, CT: Yale University Press.

Jacob, T., Indriati, E., Soejono, R. P., Hsü, K., Frayer, D. W., Eckhardt, R. B., Kuperavage, A. J., Thorne, A., and Henneberg, M., 2006   Pygmoid Australomelanesian *Homo sapiens* Skeletal Remains From Liang Bua, Flores: Population Affinities, and Pathological Abnormalities. Proceedings of the National Academy of Sciences of the United States of America. 103(36):13421–13426.

Johansson, I., Ericson, T., and Steen, L., 1984   Studies of the Effect of Diet on Saliva Secretion, and Caries Development – The Effect of Fasting on Saliva Composition of Female Subjects. Journal of Nutrition 114:2010–2020.

Kaul, S., and Corruccini, R. S., 1984   The Epidemiological Transition in Dental Occlusion in North Indian Population. *In* The People of South Asia. J. R. Lukacs, ed. pp. 201–216. New York: Plenum Press.

Keenleyside, A., 2008   Dental Pathology, and Diet at Apollonia, A Greek Colony on the Black Sea. International Journal of Osteoarchaeology 18(3):262–279.

Klatsky, M., and Fischer, R. L., 1953   The Human Masticatory Apparatus: An Introduction to Dental Anthropology. Brooklyn: Dental Items of Interest.

Klaus, H. D., and Tam, M. E., 2010   Oral Health, and the Post-Contact Adaptive Transition: A Contextual Reconstruction of Diet in Mórrope, Peru. American Journal of Physical Anthropology 141(4):594–609.

Kocsis, A., Oláh, S., and Cencetti, S., 1995   Abnormal Dental Characteristics on an Etruscan Specimen (6th Century B.C.): Case report. *In* Aspects of Dental Biology: Palaeontology, Anthropology, and Evolution. J. Moggi-Cecchi, ed., pp. 373–378. Cortona: International Institute for the Study of Man.

Kolenbrander, P. E., and Palmer, R. J., Jr., 2004   Human Oral Bacterial Biofilms. *In* Microbial Biofilms. M. Ghannoum, and G.A. O'Toole, eds. pp. 85–117. Washington, DC: ASM Press.

Koritzer, R. T., and St. Hoyme, L. E., 1979   Extensive Caries in Early Man Circa 110,000 Years Before Present. Journal of the American Dental Association 99:642–643.

Kotsomitis, N., Dunne, M. P., and Freer, T. J., 1996   A Genetic Aetiology for Some Common Dental Anomalies: A Pilot Twin Study. Australian Orthodontics Journal 14:172–178.

Langsjoen, O., 1998   Diseases of the Dentition. *In* The Cambridge Encyclopedia of Human Paleopathology. A. C. Aufderheide, and C, Rodríguez-Martín, eds. pp. 393–412. Cambridge: Cambridge University Press.

Larsen, C. S., 1983   Behavioral Implications of Temporal Change in Cariogenesis. Journal of Archaeological Science 10:1–8.

Larsen, C. S., 1985   Dental Modification and Tool Use in the Western Great Basin. American Journal of Physical Anthropology 67(4):393–402.

Larsen, C. S., 1995   Biological Changes in Human Populations with Agriculture. Annual Review of Anthropology 24185–213.

Larsen, C. S., 1997   Bioarchaeology: Interpreting Behavior from the Human Skeleton. Cambridge: Cambridge University Press.

Larsen, C. S., 1998   Gender, Health, and Activity in Foragers, and Farmers in the American Southeast: Implications for Social Organization in the Georgia Bight. *In* Sex, and Gender in

Paleopathological Perspective. A. L. Grauer, and P. Stuart-Macadam, eds. pp. 165–187. Cambridge: Cambridge University Press.

Larsen, C. S., and Milner, G. R., 1994   In the Wake of Contact: Biological Responses to Conquest. New York: Wiley–Liss.

Le Bot, P., and Salmon, D., 1977   Congenital Defects of the Upper Lateral Incisor: Condition, and Measurements of the Upper Teeth. American Journal of Physical Anthropology 46: 231–244.

Le Bot, P., Gueguen, A., and Salmon, D., 1980   Congenital Defects of the Upper Lateral Incisors, and the Morphology of Other Teeth in Man. American Journal of Physical Anthropology 53:479–486.

Leigh, R. W., 1925   Dental Pathology of Indian Tribes of Varied Environmental, and Food Conditions. American Journal of Physical Anthropology 8(2):179–199.

Leigh, R. W., 1929   Dental Morphology, and Pathology of Prehistoric Guam. Memoirs of the Bernice P. Bishop Museum XI (3)3–19.

Lieverse, A. R., 1999   Diet, and the Aetiology of Dental Calculus. International Journal of Osteoarchaeology 9(4):218–232.

Lordkipanidze, D., Vekua, A., Ferring, R., Rightmire, G. P., Zollikofer, C. P. E., De Leon, M. S. P., Agusti, J., Kiladze, G., Mouskhelishvili, A., Nioradze, M., and Tappen, M., 2006   A Fourth Hominin Skull From Dmanisi, Georgia. Anatomical Record, Part a – Discoveries in Molecular Cellular, and Evolutionary Biology 288A(11):1146–1157.

Lubell, D., and Jackes, M., 1994   The Mesolithic–Neolithic Transition in Portugal: Isotopic, and Dental Evidence of Diet. Journal of Archaeological Science 21:201–216.

Lukacs, J. R., 1989   Dental Paleopathology: Methods for Reconstructing Dietary Patterns. *In* Reconstruction of Life from the Skeleton. M. Y. Iscan, and K. A. R. Kennedy, eds. pp. 261–286. New York: Alan R. Liss.

Lukacs, J. R., 1995   The 'Caries Correction Factor': A New Method of Calibrating Dental Caries Rates to Compensate for Antemortem Loss of Teeth. International Journal of Osteoarchaeology 5:151–156.

Lukacs, J. R., 1996   Sex Differences in Dental Caries Rates with the Origin of Agriculture in South Asia. Current Anthropology 37:147–153.

Lukacs, J. R., 1998   Canine Transposition in Prehistoric Pakistan: Bronze Age, and Iron Age case Reports. Angle Orthodontist 68(5):475–479.

Lukacs, J. R., 2008   Fertility and Agriculture Accentuate Sex Differences in Dental Caries Rates. Current Anthropology 49:901–914.

Lukacs, J. R. 2011   Gender Differences in Oral Health in South Asia: Metadata Imply Multifactorial Biological and Cultural Causes. American Journal of Human Biology 23: 398–411.

Lukacs, J. R., in press.   Sex Differences in Dental Caries Experience: Clinical Evidence, and Complex Etiology. Clinical Oral Investigations. online first (DOI: 10.1007/s00784-010-0445-3) http://www.springerlink.com/content/k3t0643r47556414/.

Lukacs, J. R., and Largaespada, L., 2006   Explaining Sex Differences in Dental Caries Rates: Saliva, Hormones, and "Life History" Etiologies. American Journal of Human Biology 18:540–555.

Lukacs, J. R., and Thompson, L. M., 2007   Dental Caries Prevalence by Sex in Prehistory: Magnitude, and Meaning. *In* Technique, and Application in Dental Anthropology. J. D. Irish, and G. C. Nelson, eds. pp. 136–177. Cambridge: Cambridge University Press.

Lukacs, J. R., Nelson, G. C., and Walker, C., no date.   Developmental Anomalies in Modern Humans, Fossil Hominins, and *Homo floresiensis*. unpublished MS. pp. 1–30.

Mangi, S.L., 1954   The Effect of Pregnancy on the Incidence of Dental Caries in Indian women. Journal of the All-India Dental Association 26:1–4.

Marsh, P. D., 2004   Dental Plaque as a Microbial Biofilm. Caries Research 38:204–211.

Martin, S. A., Guatelli-Steinberg, D., Sciulli, P. W. and Walker, P. L., 2008   Brief Communication: Comparison of Methods for Estimating Chronological Age at Linear

Enamel Formation on Anterior Dentition. American Journal of Physical Anthropology 135(3):362–365.

Mays, S. A., 1998   The Archaeology of Human Bones. London: Routledge.

Miller, B. D., 1981   The Endangered Sex: Neglect of Female Children in Rural North India. Ithaca: Cornell University Press.

Miller, G., 2009   The Looming Crisis in Human Genetics. The Economist: The World in 2010. pp. 151–152.

Milner, G. R., and Larsen, C. S., 1991   Teeth as Artifacts of Human Behavior: Intentional Mutilation, and Accidental Modification. In Advances in Dental Anthropology. M. A. Kelley, and C. S. Larsen, eds. pp. 357–378. New York: Wiley–Liss.

Montagu, M. F. A., 1940   The Significance of the Variability of the Upper Lateral Incisor in Man. Human Biology 12:323–358.

Moorrees, C. F. A., 1957   The Aleut Dentition: A Correlative Study of Dental Characteristics in an Eskimoid People. Cambridge (MA): Harvard University Press.

Nasidze, I., Li, J., Quinque, D., Tang, K., and Stoneking, M., 2009   Global Diversity in the Human Salivary Microbiome. Genome Research 19(4):636–643.

Nelsen, K., Tayles, N., and Domett, K. M., 2001   Missing Lateral Incisors in Iron Age South-East Asians as Possible Indicators of Dental Agenesis. Archives of Oral Biology 46(10):963–971.

Nelson, C., 1938   The Teeth of the Indians of Pecos Pueblo. American Journal of Physical Anthropology 23261–293.

Nelson, G. C., 1992   Maxillary Canine / Third Premolar Transposition in a Prehistoric Population from Santa Cruz Island, California. American Journal of Physical Anthropology 88(2)135–144.

Ogden, A., 2008   Advances in the Paleopathology of Teeth, and Jaws. In Advances in Human Palaeopathology. Ron Pinhasi, and Simon Mays, eds. pp. 283–307. Chichester: John Wiley.

Oxenham, M., and Tayles, N., 2006   Bioarchaeology of Southeast Asia. Cambridge: Cambridge University Press.

Patir, A., Seymen, F., Yildirim, M., Deeley, K., Cooper, M. E., Marazita, M. L., and Vieira, A. R., 2008   Enamel Formation Genes are Associated with High Caries Experience in Turkish Children. Caries Research 42(5):394–400.

Pearson, A. M., 1996   "Because It Gives Me Peace of Mind": Ritual Fasts in the Religious Lives of Hindu Women. Albany: State University of New York (SUNY).

Peck, S., and Peck, L., 1995   Classification of Maxillary Tooth Transpositions. American Journal of Orthodontics Dentofacial Orthopedics 107:505–517.

Peck, S., Peck, L., and Yves, A., 1993   Maxillary Canine–First Premolar Transposition, Associated Dental Anomalies, and Genetic Basis. Angle Orthodontist 63:99–101.

Polder, B. J., Van't Hof, M. A., Van der Linden, F. P. G. M., and Kuijpers-Jagtman, A. M., 2004   A Meta-Analysis of the Prevalence of Dental Agenesis of Permanent Teeth. Community Dentistry, and Oral Epidemiology 32: 217–226.

Powell, M. L., 1985   The Analysis of Dental Wear, and Caries for Dietary Reconstruction. In The Analysis of Prehistoric Diets. R. I. Gilbert Jr., and J. H. Mielke, eds. pp. 307–338. Orlando, FL: Academic Press.

Psoter, W. J., Reid, B. C., and Katz, R. V., 2005   Malnutrition, and Dental Caries: A Review of the Literature. Caries Research 39:441–447.

Psoter, W. J., Spielman, A. L., Gebrian, B., St. Jean, R., and Katz, R. V., 2008   Effect of Childhood Malnutrition on Salivary Flow, and pH. Archives of Oral Biology 53:231–237.

Puech, P. F., Albertini, H., and Mills, N. T. W., 1980   Dental Destruction in Broken-Hill Man. Journal of Human Evolution 9(1):33–39.

Rakchanok, N., Amporn, D., Yoshida, Y., Harun-or-Rashid, M. D., and Sakamoto, J., 2010   Dental Caries, and Gingivitis Among Pregnant, and Non-Pregnant Women in Chiang Mai, Thailand. Nagoya Journal of Medical Science 72(1–2):43–50.

Reid, D. J., and Dean, M. C., 2000   The Timing of Linear Hypoplasias on Human Anterior Teeth. American Journal of Physical Anthropology 113(1):135–139.

Rightmire, G. P., Lordkipanidze, D., and Vekua, A., 2006   Anatomical Descriptions, Comparative Studies, and Evolutionary Significance of the Hominin Skulls from Dmanisi, Republic of Georgia. Journal of Human Evolution 50(2): 115–141.

Ripamonti, U., and Petit, J. C., 1991   Periodontal-Diseases Amongst Australopithecine Fossil Remains. Anthropologie 95(2–3):391–400.

Ripamonti, U., Kirkbride, A. N., Yates, S. C., and Thackeray, J. F., 1997   Further Evidence of Periodontal Bone Pathology in a Juvenile Specimen of Australopithecus Africanus From Sterkfontein, South Africa. South African Journal of Science 93(4):177–178.

Ripamonti, U., Petit, J. C., and Thackeray, J. F., 1999   A Supernumerary Tooth in a 1.7 Million-Year-Old *Australopithecus robustus* from Swartkrans, South Africa. European Journal of Oral Sciences 107(5):317–321.

Ritzman, T. B., Baker, B. J., and Schwartz, J. T., 2008   A Fine Line: A Comparison of Methods for Estimating Ages of Linear Enamel Hypoplasia Formation. American Journal of Physical Anthropology 135(3):348–361.

Roberts, C. A., and Manchester, K., 2005   The Archaeology of Disease.3rd edition. Ithaca NY: Cornell University Press.

Rose, J. C., and Burke, D. L., 2006   The Dentist, and the Archaeologist: The Role of Dental Anthropology in North American Bioarchaeology. *In* Bioarchaeology: The Contextual Analysis of Human Remains. J. E. Buikstra, and L. A. Beck, eds. pp. 323–346. Amsterdam: Elsevier / Academic.

Ruffer, M. A., 1913   On Pathological Lesions Found in Coptic Egyptian Bodies. Journal of Pathology, and Bacteriology 18149–162.

Ruffer, M. A., 1920   Study of Abnormalities, and Pathology of Ancient Egyptian Teeth. American Journal of Physical Anthropology 3:335–382.

Russell, R. R. B., 2008   How Has Genomics Altered Our View of Caries Microbiology? Caries Research 42:319–327.

Russell, S. L., Ickovics, J. R., and Yaffee, R. A., 2008   Exploring Potential Pathways Between Parity, and Tooth Loss Among American Women. American Journal of Public Health 98:1263–1270.

Russell, S. L., Ickovics, J. R., and Yaffee, R. A., 2010   Parity and Untreated Dental Caries in US Women. Journal of Dental Research 89:1091–1096.

Sandison, A. T., 1967   Sir Marc Armand Ruffer (1859–1917) Pioneer of Palaeopathology. Medical History 11(2):150–156.

Selwood, K., 2010   Is There a Definable Signature in Dental Pathology for Marine Dependent Diets?: A View from the Pacific. Master of Arts Thesis; Auckland: University of Auckland; Department of Anthropology.

Smith, G. E., and Jones, F. W., 1910   Report on the Human Remains. Archaeological Survey of Nubia Report 1907–08, Cairo.

Sperber, G. H., 1986   Paleodontopathology, and Paleodontotherapy. Journal of the Canadian Dental Association 10:835–838.

Tayles, N., Domett, K. M., and Halcrow, S., 2009   Can Dental Caries be Interpreted as Evidence of Farming? The Asian experience. *In* Comparative Dental Morphology. T. Koppe, G. Meyer, and K. W. Alt, eds. pp. 162–166. Basel: Karger.

Tayles, N., Domett, K. M., and Nelsen, K., 2000   Agriculture, and Dental Caries? The Case of Rice in Prehistoric Southeast Asia. World Archaeology 32(1):68.

Temple, D. H., and Larsen, C. S., 2007   Dental Caries Prevalence as Evidence for Agriculture, and Subsistence Variation During the Yayoi Period in Prehistoric Japan: Biocultural Interpretations of an Economy in Transition. American Journal of Physical Anthropology 134(4):501–512.

Turner, C. G. II, 1979   Dental Anthropological Indications of Agriculture Among the Jomon People of Central Japan. American Journal of Physical Anthropology 51(4):619–636.

Vallianatos, H., 2006   Poor, and Pregnant in New Delhi, India. Edmonton: Qual Institute Press.

Vieira, A. R., Marazita, M. L., and Goldstein-McHenry, T., 2008   Genome-wide Scan Finds Suggestive Caries Loci. Journal of Dental Research 87(10):915–918.

Walker, P. L., and Erlandson, J., 1986   Dental Evidence for Prehistoric Dietary Change on the Northern Channel Islands, California. American Antiquity 51:375–383.

Walker, P. L., and Hewlett, B., 1990   Dental Health, Diet, and Social Status Among Central African Foragers, and Farmers. American Anthropologist 92:383–398.

Walker, P. L., Sugiyama, L., and Chacon, R., 1998   Diet, Dental Health, and Culture Change Among Recently Contacted South American Indian Hunter-Horticulturalists. *In* Dental Development, Morphology, and Pathology: A Tribute to Albert A. Dahlberg. J. R. Lukacs, ed. pp. 355–386. Eugene: University of Oregon Anthropology Papers, No. 54.

Watson, J. T., Fields, M., and Martin, D.L., 2010   Introduction of Agriculture, and Its Effects on Women's Oral Health. American Journal of Human Biology 22:92–102.

Webb, S., 1995   Palaeopathology of Aboriginal Australians. Cambridge: Cambridge University Press.

Wells, C., 1975   Prehistoric, and Historical Changes in Nutritional Diseases, and Associated Conditions. Progress in Food, and Nutrition Science 1(11/12):729–779.

White, T. D., Suwa, G., and Simpson, S. A. B., 2000   Jaws, and Teeth of *Australopithecus afarensis* from Maka, Middle Awash, Ethiopia. American Journal of Physical Anthropology 11:145–68.

Willey, G. R., and Sabloff, J. A., 1993   A History of American Archaeology. 3rd edition. New York: W.H. Freeman.

WHO (World Health Organization), 2007   International Statistical Classification of Diseases, and Related Health Problems. 10th revision. http://apps.who.int/classifications/apps/ icd/icd10online/ Accessed June 13, 2011.

Zilberman, U., Smith, P., Piperno, M., and Condemi, S., 2004   Evidence of *amelogenesis imperfecta* in an Early African *Homo erectus*. Journal of Human Evolution 46:647–653.

Index

Page numbers in bold, e.g. **388**, indicate figures or tables

Printed and bound by CPI Group (UK) Ltd, Croydon, CR0 4YY

16/04/2025

14658833-0001